WileyPLUS

T0174522

WileyPLUS is a research-based online environment for effective teaching and learning.

WileyPLUS builds students' confidence because it takes the guesswork out of studying by providing students with a clear roadmap:

- what to do
- how to do it
- if they did it right

It offers interactive resources along with a complete digital textbook that help students learn more. With *WileyPLUS*, students take more initiative so you'll have greater impact on their achievement in the classroom and beyond.

For more information, visit www.wileyplus.com

Now available for

Bb
Blackboard

WileyPLUS

ALL THE HELP, **RESOURCES**, AND PERSONAL **SUPPORT** YOU AND YOUR STUDENTS NEED!

www.wileyplus.com/resources

1st DAY OF CLASS
...AND BEYOND!

2-Minute Tutorials and all of the resources you and your students need to get started

WileyPLUS

Student Partner Program

Student support from an experienced student user

Wiley Faculty Network

Collaborate with your colleagues, find a mentor, attend virtual and live events, and view resources
www.WhereFacultyConnect.com

WileyPLUS

Quick Start

Pre-loaded, ready-to-use assignments and presentations created by subject matter experts

Technical Support 24/7 FAQs, online chat, and phone support
www.wileyplus.com/support

© Courtney Keating/iStockphoto

Your *WileyPLUS* Account Manager providing personal training and support

DIFFERENTIAL EQUATIONS
An Introduction to Modern Methods and Applications

THIRD EDITION

James R. Brannan
Clemson University

William E. Boyce
Rensselaer Polytechnic Institute

with contributions by
Mark A. McKibben
West Chester University

WILEY

PUBLISHER	Laurie Rosatone
ACQUISITIONS EDITOR	David Dietz
FREELANCE DEVELOPMENT EDITOR	Anne Scanlan-Rohrer
EDITORIAL ASSISTANT	Michael O'Neal
MARKETING	Jesse Adler
CREATIVE DIRECTOR	Harry Nolan
SENIOR DESIGNER/COVER DESIGN	Madelyn Lesure
COVER PHOTO	© PhilipYb / Shutterstock
SENIOR PRODUCTION EDITOR	Laura Abrams
SENIOR PHOTO EDITOR	James Russiello

This book was set in 10/12 STIX by Aptara®, Inc.

This book is printed on acid free paper. ∞

Founded in 1807, John Wiley & Sons, Inc. has been a valued source of knowledge and understanding for more than 200 years, helping people around the world meet their needs and fulfill their aspirations. Our company is built on a foundation of principles that include responsibility to the communities we serve and where we live and work. In 2008, we launched a Corporate Citizenship Initiative, a global effort to address the environmental, social, economic, and ethical challenges we face in our business. Among the issues we are addressing are carbon impact, paper specifications and procurement, ethical conduct within our business and among our vendors, and community and charitable support. For more information, please visit our website: www.wiley.com/go/citizenship.

Copyright © 2015, 2011, 2007 John Wiley & Sons, Inc. All rights reserved.

No part of this publication may be reproduced, stored in a retrieval system or transmitted in any form or by any means, electronic, mechanical, photocopying, recording, scanning or otherwise, except as permitted under Section 107 or 108 of the 1976 United States Copyright Act, without either the prior written permission of the Publisher, or authorization through payment of the appropriate per-copy fee to the Copyright Clearance Center, Inc., 222 Rosewood Drive, Danvers, MA 01923, website www.copyright.com. Requests to the Publisher for permission should be addressed to the Permissions Department, John Wiley & Sons, Inc., 111 River Street, Hoboken, NJ 07030-5774, (201)748-6011, fax (201)748-6008, website www.wiley.com/go/permissions.

Evaluation copies are provided to qualified academics and professionals for review purposes only, for use in their courses during the next academic year. These copies are licensed and may not be sold or transferred to a third party. Upon completion of the review period, please return the evaluation copy to Wiley. Return instructions and a free of charge return shipping label are available at www.wiley.com/go/returnlabel. Outside of the United States, please contact your local representative.

ISBN 978-1-119-65763-7
ISBN 978-1-118-98122-1 (BRV)

The inside back cover will contain printing identification and country of origin if omitted from this page. In addition, if the ISBN on the back cover differs from the ISBN on this page, the one on the back cover is correct.

Printed in Great Britain by TJ International Ltd, Padstow, Cornwall

10 9 8 7 6 5 4 3 2 1

PREFACE

This is a textbook for a first course in differential equations. The book is intended for science and engineering majors who have completed the calculus sequence, but not necessarily a first course in linear algebra. It emphasizes a systems approach to the subject and integrates the use of modern computing technology in the context of contemporary applications from engineering and science.

Our goal in writing this text is to provide these students with both an introduction to, and a survey of, modern methods, applications, and theory of differential equations that is likely to serve them well in their chosen field of study. The subject matter is presented in a manner consistent with the way practitioners use differential equations in their work; technology is used freely, with more emphasis on methods, modeling, graphical representation, qualitative concepts, and geometric intuition than on theory.

Notable Changes in the Third Edition

This edition is a substantial revision of the second edition. The most significant changes are:

▶ Enhanced Page Layout We have placed important results, theorems, definitions, and tables in highlighted boxes and have put subheadings just before the most important topics in each section. This should enhance readability for both students and instructors and help students to review material for exams.

▶ Increased Emphasis on Qualitative Methods Qualitative methods are introduced early. Throughout the text, new examples and problems have been added that require the student to use qualitative methods to analyze solution behavior and dependence of solutions on parameters.

▶ New Chapter on Numerical Methods Discussions on numerical methods, dispersed over three chapters in the second edition, have been revised and reassembled as a unit in Chapter 8. However, the first three sections of Chapter 8 can be studied by students after they have studied Chapter 1 and the first two sections of Chapter 2.

▶ Chapter 1: Introduction This chapter has been reduced to three sections. In Section 1.1 we follow up on introductory models and concepts with a discussion of the art and craft of mathematical modeling. Section 1.2 has been replaced by an early introduction to qualitative methods, in particular, phase lines and direction fields. Linearization and stability properties of equilibrium solutions are also discussed. In Section 1.3 we cover definitions, classification, and terminology to help give the student an organizational overview of the subject of differential equations.

▶ Chapter 2: First Order Differential Equations New mathematical modeling problems have been added to Section 2.3, and a new Section 2.7 on subsitution methods has been added. Sections on numerical methods have been moved to Chapter 8.

▶ Chapter 3: Systems of Two First Order Equations The discussion of Wronskians and fundamental sets of solutions has been supplemented with the definition of, and relationship to, linearly independent solutions of two-dimensional linear systems.

▶ Chapter 4: Second Order Linear Equations Section 4.6 on forced vibrations, frequency response, and resonance has been rewritten to improve its readability for students and instructors.

▶ Chapter 10: Orthogonal Functions, Fourier Series and Boundary-Value Problems
This chapter gives a unified treatment of classical and generalized Fourier series in the
framework of orthogonal families in the space PC[a, b].

▶ Chapter 11: Elementary Partial Differential Equations   Material and projects on the
heat equation, wave equation, and Laplace's equation that appeared in Chapters 9 and
10 of the second edition, have been moved to Chapter 11 in the third edition.

▶ Miscellaneous Changes and Additions   Changes have been made in current problems,
and new problems have been added to many of the section problem sets. For ease in
assigning homework, boldface headings have been added to partition the problems into
groups corresponding to major topics discussed in the section.

Major Features

▶ Flexible Organization.   Chapters are arranged, and sections and projects are structured,
to facilitate choosing from a variety of possible course configurations depending on
desired course goals, topics, and depth of coverage.

▶ Numerous and Varied Problems.   Throughout the text, section exercises of varying levels of difficulty give students hands-on experience in modeling, analysis, and computer
experimentation.

▶ Emphasis on Systems.   Systems of first order equations, a central and unifying theme
of the text, are introduced early, in Chapter 3, and are used frequently thereafter.

▶ Linear Algebra and Matrix Methods.   Two-dimensional linear algebra sufficient for the
study of two first order equations, taken up in Chapter 3, is presented in Section 3.1.
Linear algebra and matrix methods required for the study of linear systems of dimension
n (Chapter 6) are treated in Appendix A.

▶ Optional Computing Exercises.   In most cases, problems requesting computer-
generated solutions and graphics are optional.

▶ Visual Elements.   The text contains a large number of illustrations and graphs. In addition, many of the problems ask the student to compute and plot solutions of differential
equations.

▶ Contemporary Project Applications.   Optional projects at the end of all but one of
Chapters 2 through 11 integrate subject matter in the context of exciting, often contemporary, applications in science and engineering.

▶ Laplace Transforms.   A detailed chapter on Laplace transforms discusses systems, discontinuous and impulsive input functions, transfer functions, feedback control systems,
poles, and stability.

▶ Control Theory.   Ideas and methods from the important application area of control theory are introduced in some examples, some projects, and in the last section on Laplace
transforms. All this material is optional.

▶ Recurring Themes and Applications.   Important themes, methods, and applications,
such as dynamical system formulation, phase portraits, linearization, stability of equilibrium solutions, vibrating systems, and frequency response, are revisited and reexamined
in a variety of mathematical models under different mathematical settings.

▶ Chapter Summaries.   A summary at the end of each chapter provides students and
instructors with a bird's-eye view of the most important ideas in the chapter.

▶ Answers to Problems.   Answers to selected odd-numbered problems are provided at
the end of the book; many of them are accompanied by a figure.

Problems that require the use of a computer are marked with ◈. While we feel that students
will benefit from using the computer on those problems where numerical approximations

or computer-generated graphics are requested, in most problems it is clear that use of a computer, or even a graphing calculator, is optional. Furthermore there are a large number of problems that do not require the use of a computer. Thus the book can easily be used in a course without using any technology.

Relation of This Text to Boyce and DiPrima

Brannan and Boyce is an offshoot of the well-known textbook by Boyce and DiPrima. Readers familiar with Boyce and DiPrima will doubtless recognize in the present book some of the hallmark features that distinguish that textbook.

To help avoid confusion among potential users of either text, the primary differences are described below:

▶ Brannan and Boyce is more sharply focused on the needs of students of engineering and science, whereas Boyce and DiPrima targets a somewhat more general audience, including engineers and scientists.

▶ Brannan and Boyce is intended to be more consistent with the way contemporary scientists and engineers actually use differential equations in the workplace.

▶ Brannan and Boyce emphasizes systems of first order equations, introducing them earlier, and also examining them in more detail than Boyce and DiPrima. Brannan and Boyce has an extensive appendix on matrix algebra to support the treatment of systems in *n* dimensions.

▶ Brannan and Boyce integrates the use of computers more thoroughly than Boyce and DiPrima, and assumes that most students will use computers to generate approximate solutions and graphs throughout the book.

▶ Brannan and Boyce emphasizes contemporary applications to a greater extent than Boyce and DiPrima, primarily through end-of-chapter projects.

▶ Brannan and Boyce makes somewhat more use of graphs, with more emphasis on phase plane displays, and uses engineering language (e.g., state variables, transfer functions, gain functions, and poles) to a greater extent than Boyce and DiPrima.

Options for Course Structure

Chapter dependencies are shown in the following block diagram:

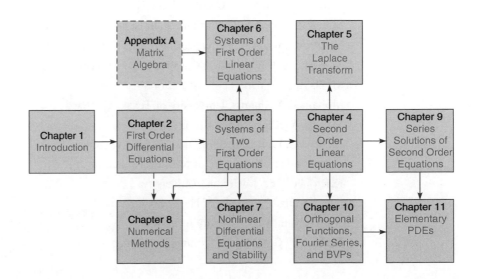

The book has much built-in flexibility and allows instructors to choose from many options. Depending on the course goals of the instructor and background of the students, selected sections may be covered lightly or even omitted.

▶ Chapters 5, 6, and 7 are independent of each other, and Chapters 6 and 7 are also independent of Chapter 4. It is possible to spend much class time on one of these chapters, or class time can be spread over two or more of them.

▶ The amount of time devoted to projects is entirely up to the instructor.

▶ For an honors class, a class consisting of students who have already had a course in linear algebra, or a course in which linear algebra is to be emphasized, Chapter 6 may be taken up immediately following Chapter 2. In this case, material from Appendix A, as well as sections, examples, and problems from Chapters 3 and 4, may be selected as needed or desired. This offers the possibility of spending more class time on Chapters 5, 7, and/or selected projects.

Acknowledgments

It is a pleasure to offer our grateful appreciation to the many people who have generously assisted in the preparation of this book.

To the individuals listed below who reviewed parts or all of the third edition manuscript at various stages of its development:

Miklós Bóna, University of Florida

Mark W. Brittenham, University of Nebraska

Yanzhao Cao, Auburn University

Doug Cenzer, University of Florida

Leonard Chastkofsky, University of Georgia

Jon M. Collis, Colorado School of Mines

Domenico D'Alessandro, Iowa State University

Shaozhong Deng, University of North Carolina at Charlotte

Patricia J. Diute, Rochester Institute of Technology

Behzad Djafari Rouhani, University of Texas at El Paso

Alina N. Duca, North Carolina State University

Marek Z. Elżanowski, Portland State University

Vincent Graziano, Case Western Reserve University

Mansoor A. Haider, North Carolina State University

M. D. Hendon, University of Georgia

Mohamed A Khamsi, University of Texas at El Paso

Marcus A. Khuri, Stony Brook University

Richard C. Le Borne, Tennessee Technological University

Glenn Ledder, University of Nebraska-Lincoln

Kristopher Lee, Iowa State University

Jens Lorenz, University of New Mexico

Aldo J. Manfroi, University of Illinois

Marcus McGuff, Austin Community College

William F. Moss, Clemson University

Mike Nicholas, Colorado School of Mines

Mohamed Ait Nouh, University of Texas at El Paso

Francis J. Poulin, University of Waterloo

Mary Jarratt Smith, Boise State University

Stephen J. Summers, University of Florida

Yi Sun, University of South Carolina Kyle Thompson, North Carolina State University

Stella Thistlethwaite, University of Tennessee, Knoxville

Vincent Vatter, University of Florida

Pu Patrick Wang, University of Alabama

Dongming Wei, Nazarbayev University

Larissa Williamson, University of Florida

Hafizah Yahya, University of Alberta

Konstantin Zuev, University of Southern California.

To Mark McKibben, West Chester University; Doug Meade, University of South Carolina; Bill Siegmann, Rensselaer Polytechnic Institute, for their contributions to the revision.

To Jennifer Blue, SUNY Empire State College, for accuracy checking page proofs.

To the editorial and production staff of John Wiley and Sons, Inc., identified on page iv, who were responsible for turning our manuscript into a finished book. In the process, they maintained the highest standards of professionalism.

We also wish to acknowledge the less tangible contributions of our friend and colleague, the late Richard DiPrima. Parts of this book draw extensively on the book on differential equations by Boyce and DiPrima. Since Dick DiPrima was an equal partner in creating the early editions of that book, his influence lives on more than thirty years after his untimely death.

Finally, and most important of all, we thank our wives Cheryl and Elsa for their understanding, encouragement, and patience throughout the writing and production of this book. Without their support it would have been much more difficult, if not impossible, for us to complete this project.

James R. Brannan
Clemson, South Carolina

William E. Boyce
Latham, New York

Supplemental Resources for Instructors and Students

An Instructor's Solutions Manual, includes solutions for all problems in the text.

A Student Solutions Manual, ISBN 9781118981252, includes solutions for selected problems in the text.

A Companion website, www.wiley.com/college/brannan, provides a wealth of resources for students and instructors, including:

▶ PowerPoint slides of important ideas and graphics for study and note taking.

▶ Online Only Projects—these projects are like the end-of-chapter projects in the text. They present contemporary problems that are not usually included among traditional differential equations topics. Many of the projects involve applications derived from a variety of disciplines and integrate or extend theories and methods presented in core material.

▶ Mathematica, Maple, and MATLAB data files are provided for selected end-of-section or end-of-chapter problems in the text allowing for further exploration of important ideas in the course utilizing these computer algebra and numerical analysis packages. Students will benefit from using the computer on problems where numerical approximations or computer generated graphics are requested.

▶ Review of Integration—An online review of integration techniques is provided for students who need a refresher.

WileyPLUS: Expect More from Your Classroom Technology

This text is supported by *WileyPLUS*—a powerful and highly integrated suite of teaching and learning resources designed to bridge the gap between what happens in the classroom and what happens at home. *WileyPLUS* includes a complete online version of the text, algorithmically generated exercises, all of the text supplements, plus course and homework management tools, in one easy-to-use website.

Organized around the everyday activities you perform in class, *WileyPLUS* helps you:

▶ Prepare and Present: *WileyPLUS* lets you create class presentations quickly and easily using a wealth of Wiley-provided resources, including an online version of the textbook, PowerPoint slides, and more. You can adapt this content to meet the needs of your course.

▶ Create Assignments: *WileyPLUS* enables you to automate the process of assigning and grading homework or quizzes.

▶ Track Student Progress: An instructor's gradebook allows you to analyze individual and overall class results to determine students' progress and level of understanding.

▶ Promote Strong Problem-Solving Skills: *WileyPLUS* can link homework problems to the relevant section of the online text, providing students with context-sensitive help. *WileyPLUS* also features mastery problems that promote conceptual understanding of key topics and video walkthroughs of example problems.

▶ Provide Numerous Practice Opportunities: Algorithmically generated problems provide unlimited self-practice opportunities for students, as well as problems for homework and testing.

▶ Support Varied Learning Styles: *WileyPLUS* includes the entire text in digital format, enhanced with varied problem types to support the array of different student learning styles in today's classrooms.

▶ Administer Your Course: You can easily integrate *WileyPLUS* with another course management system, gradebooks, or other resources you are using in your class, enabling you to build your course, your way.

CONTENTS

© Ingram Publishing / Newscom

Introduction

In this introductory chapter we formulate several problems that illustrate basic ideas that reoccur frequently in this book.

In Section 1.1 we discuss two mathematical models, one from physics and one from population biology. Each mathematical model is a differential equation—an equation involving the rate of change of a variable with respect to time. Using these models as examples, we introduce some basic terminology, explore the notion of a solution of a differential equation, and end with an overview of the art and craft of mathematical modeling.

It is not always possible to find analytic, closed-form solutions of a differential equation. In Section 1.2 we look at two graphical methods for studying the qualitative behavior of solutions: phase lines and direction fields. Although we will learn how to sketch direction fields by hand, we will use the computer to draw them.

Sections 1.1 and 1.2 give us a glimpse of two of the three major methods of studying differential equations, the **analytical** method and the **geometric** method, respectively. We defer study of the third major method—**numerical**—to Chapter 8. However, you may study the first three sections of Chapter 8 immediately after Chapter 1.

In Section 1.3 we present some important definitions and commonly used terminology in conjunction with different ways of classifying differential equations. Classification schemes provide organizational structure for the book and help give you perspective on the subject of differential equations.

1.1 Mathematical Models and Solutions

Many of the principles, or laws, underlying the behavior of the natural world are statements, or relations, involving rates in which one variable, say, y, changes with respect to another variable, t, for example. Most often, these relations take the form of equations containing y and certain of the derivatives $y', y'', \ldots, y^{(n)}$ of y with respect to t. The resulting equations are then referred to as **differential equations**. Some examples of differential equations that will be studied in detail later on in the text, are:

$$y' = r\left(1 - \frac{y}{K}\right)y, \qquad \text{an equation for population dynamics,}$$

$$my'' + \gamma y' + ky = 0, \qquad \text{the equation for a damped spring-mass system, and}$$

$$\theta'' + \frac{g}{l}\sin(\theta) = 0, \qquad \text{the pendulum equation.}$$

The subject of differential equations was motivated by problems in mechanics, elasticity, astronomy, and geometry during the latter part of the 17th century. Inventions (or discoveries) in theory, methods, and notation evolved concurrently with innovations in calculus. Since their early historical origins, the number and variety of problems to which differential equations are applied have grown substantially. Today, scientists and engineers use differential equations to study problems in all fields of science and engineering, as well as in several of the business and social sciences. Some representative problems from these fields are shown below.

Applications of Differential Equations

- airplane and ship design
- earthquake detection and prediction
- controlling the flight of ships and rockets
- modeling the dynamic behavior of nerve cells
- describing the behavior of economic systems
- forecasting and managing the harvesting of fish populations
- designing optimal vaccination policies to prevent the spread of disease
- heat transfer
- wave propagation
- weather forecasting
- designing medical imaging technologies
- determining the price of financial derivatives

The common thread that links these applications is that they all deal with systems that evolve in time. Differential equations is the mathematical apparatus that we use to study such systems.

We often refer to a differential equation that describes some physical process as a **mathematical model** of the process; many such models are discussed throughout this book. In this section we construct a model from physics and a model from population biology. Each model results in an equation that can be solved by using an integration technique from calculus. These examples suggest that even simple differential equations can provide useful models of important physical systems.

Heat Transfer: Newton's Law of Cooling

EXAMPLE 1

If a material object is hotter or colder than the surrounding environment, its temperature will approach the temperature of the environment. If the object is warmer than the environment, its temperature will decrease. If the object is cooler than the environment, its temperature will increase. Sir Isaac Newton postulated that the rate of change of the temperature of the object is negatively proportional to the difference between its temperature and the temperature of the surroundings (the **ambient temperature**). This principle is referred to as **Newton's law of cooling**.

Suppose we let $u(t)$ denote the temperature of the object at time t, and let T be the ambient temperature (see Figure 1.1.1). Then du/dt is the rate at which the temperature of the object changes. From Newton, we know that du/dt is proportional to $-(u - T)$. Introducing a positive constant of proportionality k called the **transmission coefficient**, we then get the differential equation

$$\frac{du}{dt} = -k(u - T), \qquad \text{or} \qquad u' = -k(u - T). \tag{1}$$

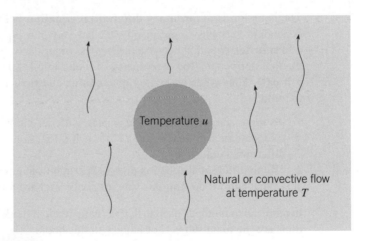

Temperature u

Natural or convective flow at temperature T

FIGURE 1.1.1 Newton's Law of Cooling: The time rate of change of u, du/dt, is negatively proportional to $u - T$: $du/dt \propto -(u - T)$.

Note that the minus sign on the right side of Eq. (1) causes du/dt to be negative if $u(t) > T$, while du/dt is positive if $u(t) < T$. The transmission coefficient measures the rate of heat exchange between the object and its surroundings. If k is large, the rate of heat exchange is rapid. If k is small, the rate of heat exchange is slow. This would be the case, for example, if the object was surrounded by thick insulating material.

The temperatures u and T are measured in either degrees Fahrenheit (°F) or degrees Celsius (°C). Time is usually measured in units that are convenient for expressing time intervals over which significant changes in u occur, such as minutes, hours, or days. Since the left side of Eq. (1) has units of temperature per unit time, k must have the units of $(\text{time})^{-1}$.

Newton's law of cooling is applicable to situations in which the temperature of the object is approximately uniform at all times. This is the case for small objects that conduct heat easily, or containers filled with a fluid that is well mixed. Thus, we expect the model to be reasonably accurate in predicting the temperature of a small copper sphere, a well-stirred cup of coffee, or a house in which the air is continuously circulated, but the model would not be very accurate for predicting the temperature of a roast in an oven.

Terminology

Let us assume that the ambient temperature T in Eq. (1) is a constant, say, $T = T_0$, so that Eq. (1) becomes

$$u' = -k(u - T_0). \tag{2}$$

In Section 1.2 we consider an example in which T depends on t. Common mathematical terminology for the quantities that appear in this equation are:

time	t	is an **independent variable**,
temperature	u	is a **dependent variable** because it depends on t,
	k and T_0	are **parameters** in the model.

The equation is an **ordinary differential equation** because it has one, and only one, independent variable. Consequently, the derivative in Eq. (2) is an ordinary derivative. It is a **first order** equation because the highest order derivative that appears in the equation is the first derivative. The dependency of u on t implies that u is, in fact, a function of t, say, $u = \phi(t)$. Thus when we write Eq. (2), three questions may, after a bit of reflection, come to mind:

1. "Is there actually a function $u = \phi(t)$, with derivative $u' = d\phi/dt$, that makes Eq. (2) a true statement for each time t?" If such a function exists, it is called a solution of the differential equation.
2. "If the differential equation does have a solution, how can we find it?"
3. "What can we do with this solution, once we have found it?"

In addition to methods used to derive mathematical models, answers to these types of questions are the main subjects of inquiry in this book.

Solutions and Integral Curves

By a **solution** of Eq. (2), we mean a differentiable function $u = \phi(t)$ that satisfies the equation. One solution of Eq. (2) is $u = T_0$, since Eq. (2) reduces to the identity $0 = 0$ when T_0 is substituted for u in the equation. In other words, "It works when we put it into the equation." The constant solution $u = T_0$ is referred to as an **equilibrium solution** of Eq. (2). Although simple, equilibrium solutions usually play an important role in understanding the behavior of other solutions. In Section 1.2 we will consider them in a more general setting.

If we assume that $u \neq T_0$, we can discover other solutions of Eq. (2) by first rewriting it in the form

$$\frac{du/dt}{u - T_0} = -k. \tag{3}$$

By the chain rule the left side of Eq. (3) is the derivative of $\ln|u - T_0|$ with respect to t, so we have

$$\frac{d}{dt} \ln|u - T_0| = -k. \tag{4}$$

Then, by integrating both sides of Eq. (4), we obtain

$$\ln|u - T_0| = -kt + C, \tag{5}$$

where C is an arbitrary constant of integration. Therefore, by taking the exponential of both sides of Eq. (5), we find that

$$|u - T_0| = e^{-kt+C} = e^C e^{-kt}, \tag{6}$$

or

$$u - T_0 = \pm e^C e^{-kt}. \tag{7}$$

Thus

$$u = T_0 + ce^{-kt} \tag{8}$$

is a solution of Eq. (2), where $c = \pm e^C$ is also an arbitrary (nonzero) constant. Note that if we allow c to take the value zero, then the constant solution $u = T_0$ is also contained in the expression (8). The expression (8) contains all possible solutions of Eq. (2) and is called the **general solution** of the equation.

Given a differential equation, the usual problem is to find solutions of the equation. However, it is also important to be able to determine whether a particular function is a solution of the equation. Thus, if we were simply asked to verify that u in Eq. (8) is a solution of Eq. (2), then we would need to substitute $T_0 + ce^{-kt}$ for u in Eq. (2) and show that the equation reduces to an identity, as we now demonstrate.

EXAMPLE 2

Verify by substitution that $u = T_0 + ce^{-kt}$, where c is an arbitrary real number, is a solution of Eq. (2),

$$u' = -k(u - T_0), \tag{9}$$

on the interval $-\infty < t < \infty$.

Substituting $\phi(t) = T_0 + ce^{-kt}$ for u in the left side of the equation gives $\phi'(t) = -kce^{-kt}$ while substituting $\phi(t)$ for u into the right side yields $-k(T_0 + ce^{-kt} - T_0) = -kce^{-kt}$. Thus, upon substitution, Eq. (2) reduces to the identity

$$\underbrace{-kce^{-kt}}_{\phi'(t)} = \underbrace{-kce^{-kt}}_{-k(\phi(t)-T_0)}, \qquad -\infty < t < \infty,$$

for each real number c and each value of the parameter k.

▶ **Integral Curves.** The geometrical representation of the general solution (8) is an infinite family of curves in the tu-plane called **integral curves**. Each integral curve is associated with a particular value of c; it is the graph of the solution corresponding to that value of c.

Although we can sketch, by hand, qualitatively correct integral curves described by Eq. (8), we will assign numerical values to k and T_0, and then use a computer to plot the graph of Eq. (8) for some different values of c. Setting $k = 1.5$ day^{-1} and $T_0 = 60°$F in Eq. (2) and Eq. (8) gives us

$$\frac{du}{dt} = -1.5(u - 60), \tag{10}$$

with the corresponding general solution

$$u = 60 + ce^{-1.5t}. \tag{11}$$

In Figure 1.1.2 we show several integral curves of Eq. (10) obtained by plotting the graph of the function in Eq. (11) for different values of c. Note that all solutions approach the equilibrium solution $u = 60$ as $t \to \infty$.

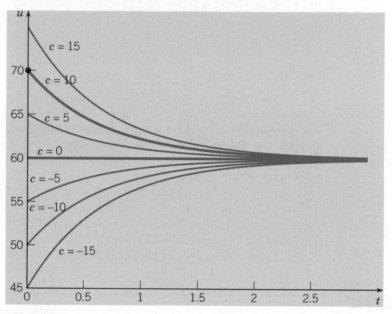

FIGURE 1.1.2 Integral curves of $u' = -1.5(u - 60)$. The curve corresponding to $c = 10$ in Eq. (11) is the graph of $u = 60 + 10e^{-1.5t}$, the solution satisfying the initial condition $u(0) = 70$. The curve corresponding to $c = 0$ in Eq. (11) is the graph of the equilibrium solution $u = 60$, which satisfies the initial condition $u(0) = 60$.

Initial Value Problems

Frequently, we want to focus our attention on a single member of the infinite family of solutions by specifying the value of the arbitrary constant. Most often, we do this by specifying a point that must lie on the graph of the solution. For example, to determine the constant c in Eq. (11), we could require that the temperature have a given value at a certain time, such as the value 70 at time $t = 0$. In other words, the graph of the solution must pass through the point $(0, 70)$. Symbolically, we can express this condition as

$$u(0) = 70. \tag{12}$$

Then, substituting $t = 0$ and $u = 70$ into Eq. (11), we obtain

$$70 = 60 + c.$$

Hence $c = 10$, and by inserting this value in Eq. (11), we obtain the desired solution, namely,

$$u = 60 + 10e^{-1.5t}. \tag{13}$$

The graph of the solution (13) is the thick curve, labeled by $c = 10$, in Figure 1.1.2. The additional condition (12) that we used to determine c is an example of an **initial condition**.

The differential equation (10) together with the initial condition (12) form an **initial value problem**.

Note that the solution of Eq. (10) subject to the initial condition $u(0) = 60$ is the equilibrium solution $u = 60$, the thick curve labeled by $c = 0$ in Figure 1.1.2.

Population Biology

Next we consider a problem in population biology. To help control the field mouse population in his orchards, in an economical and ecofriendly way, a fruit farmer installs nesting boxes for barn owls, predators for whom mice are a natural food supply. In the absence of predators we assume that the rate of change of the mouse population is proportional to the current population; for example, if the population doubles, then the number of births per unit time also doubles. This assumption is not a well-established physical law (such as the laws of thermodynamics, which underlie Newton's law of cooling in Example 1), but it is a common initial hypothesis[1] in a study of population growth. If we denote time by t and the mouse population by $p(t)$, then the assumption about population growth can be expressed by the equation

$$\frac{dp}{dt} = rp, \tag{14}$$

where the proportionality factor r is called the **rate constant** or **growth rate**.

As a simple model for the effect of the owl population on the mouse population, let us assume that the owls consume the mice at a constant predation rate a. By modifying Eq. (14) to take this into account, we obtain the equation

$$\frac{dp}{dt} = rp - a, \tag{15}$$

where both r and a are positive. Thus the rate of change of the mouse population, dp/dt, is the net effect of the growth term rp and the predation term $-a$. Depending on the values of p, r, and a, the value of dp/dt may be of either sign.

EXAMPLE 3

Suppose that the growth rate for the field mice is 0.5/month and that the owls kill 15 mice per day. Determine appropriate values for the parameters in Eq. (15), find the general solution of the resulting equation, and graph several solutions, including any equilibrium solutions.

We naturally assume that p is the number of individuals in the mouse population at time t. We can choose our units for time to be whatever seems most convenient; the two obvious possibilities are days or months. If we choose to measure time in months, then the growth term is $0.5p$ and the predation term is $-(15 \text{ mice/day}) \cdot (30 \text{ days/month}) = -450$ mice/month, assuming an average month of 30 days. Thus Eq. (15) becomes

$$\frac{dp}{dt} = 0.5p - 450, \tag{16}$$

where each term has the units of mice/month.

By following the same steps that led to the general solution of Eq. (2), we find that the general solution of Eq. (16) is

$$p = 900 + ce^{t/2}, \tag{17}$$

where c is again a constant of integration.

---

[1] A somewhat better model of population growth is discussed in Section 2.5.

Integral curves for Eq. (16) are shown in Figure 1.1.3. For sufficiently large values of p it can be seen from the figure, or directly from Eq. (16) itself, that dp/dt is positive, so that solutions increase. On the other hand, for small values of p the opposite is the case. Again, the critical value of p that separates solutions that increase from those that decrease is the value of p for which dp/dt is zero. By setting dp/dt equal to zero in Eq. (16) and then solving for p, we find the equilibrium solution $p = 900$ for which the growth term and the predation term in Eq. (16) are exactly balanced. This corresponds to the choice $c = 0$ in the general solution (17).

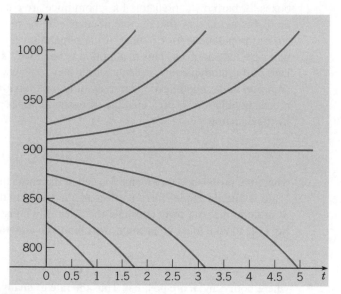

FIGURE 1.1.3 Integral curves, including the equilibrium solution $p = 900$, for $p' = 0.5p - 450$.

Solutions of the more general equation (15), in which the growth rate and the predation rate are unspecified, behave very much like those of Eq. (16). The equilibrium solution of Eq. (15) is $p = a/r$. Solutions above the equilibrium solution increase, while those below it decrease.

Constructing Mathematical Models

Mathematical modeling is the craft, and art, of using mathematics to describe and understand real-world phenomena. A viable mathematical model can be used to test ideas, make predictions, and aid in design and control problems that are associated with the phenomena. For instance, in Example 1, we constructed the differential equation

$$\frac{du}{dt} = -k(u - T) \tag{18}$$

to model heat exchange between an object and its surroundings. Recall that $u(t)$ is the time-dependent variable representing the temperature of the object and T is the temperature of the surroundings. If the value of u is known at time $t = 0$, and the values of the parameters T and k are known, solutions of this differential equation tell us what the temperature of the object will be for times $t > 0$.

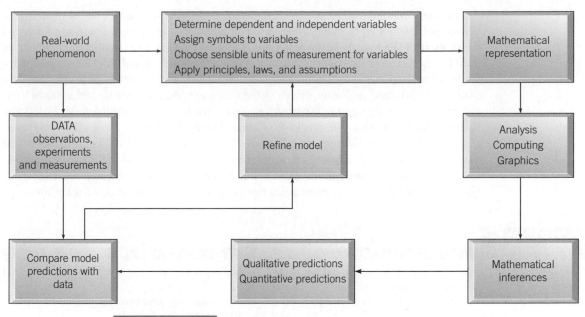

FIGURE 1.1.4 A diagram of the modeling process.

The steps used to arrive at Eq. (18) are typical of the steps used to construct any mathematical model. It is, therefore, worthwhile to illustrate the general process by a system flow diagram, as in Figure 1.1.4.

In the Problems for this section, and for many other sections of this textbook, we ask you to construct differential equation models of various real-world phenomena. In constructing mathematical models, you will find that each problem is different. Although the modeling process, in broad outline, is well represented by the above diagram, it is not a skill that can be reduced to the observance of a set of prescribed rules. Successful modeling usually requires that the modeler be intimate with the field in which the problem originates. However experience has shown that the very act of attempting to construct a mathematical model forces the modeler to ask the most cogent questions about the phenomenon being investigated:

1. What is the purpose of the model?
2. What aspects of the phenomenon are most important for the intended uses of the model?
3. What can we measure or observe?
4. What are the relevant variables; what is their relationship to the measurements?
5. Are there well-established principles (such as physical laws, or economic laws) to guide us in formulating the model?
6. In terms of the variables, how do we mathematically represent the interaction of various components of the phenomenon?
7. What simplifying assumptions can we make?
8. Do conclusions and predictions of the model agree with experiment and observations?
9. What additional experiments are suggested by the model?
10. What are limitations of the model?

For many applied mathematicians, engineers, and scientists, mathematical modeling is akin to poetry—an art form and creative act employing language that adheres to form and conventions. Likewise, there are rules (e.g., physical laws) that the mathematical modeler must follow, yet he or she has access to a myriad of mathematical tools (the language) for describing the phenomenon under investigation. History abounds with the names of scientists, mathematicians, and engineers, driven by the desire to understand nature and advance technology, who have engaged in the practice of mathematical modeling: Newton, Euler, von Kármán, Verhulst, Maxwell, Rayleigh, Navier, Stokes, Heaviside, Einstein, Schrödinger, and so on. Their contributions have literally changed the world. Nowadays, mathematical modeling is carried out in universities, government agencies and laboratories, business and industrial concerns, policy think tanks, and institutes dedicated to research and education. For many practitioners of mathematical modeling, it is, in a sense, their *raison d'être*.

PROBLEMS

1. **Newton's Law of Cooling.** A cup of hot coffee has a temperature of 200°F when freshly poured, and is left in a room at 70°F. One minute later the coffee has cooled to 190°F.

(a) Assume that Newton's law of cooling applies. Write down an initial value problem that models the temperature of the coffee.
(b) Determine when the coffee reaches a temperature of 170°F.

2. Blood plasma is stored at 40°F. Before it can be used, it must be at 90°F. When the plasma is placed in an oven at 120°F, it takes 45 minutes (min) for the plasma to warm to 90°F. Assume Newton's law of cooling applies. How long will it take the plasma to warm to 90°F if the oven temperature is set at 100°F?

3. At 11:09 p.m. a forensics expert arrives at a crime scene where a dead body has just been found. Immediately, she takes the temperature of the body and finds it to be 80°F. She also notes that the programmable thermostat shows that the room has been kept at a constant 68°F for the past 3 days. After evidence from the crime scene is collected, the temperature of the body is taken once more and found to be 78.5°F. This last temperature reading was taken exactly one hour after the first one. The next day the investigating detective asks the forensic expert, "What time did our victim die?" Assuming that the victim's body temperature was normal (98.6°F) prior to death, what does she tell the detective?

4. **Population Problems.** Consider a population p of field mice that grows at a rate proportional to the current population, so that $dp/dt = rp$.

(a) Find the rate constant r if the population doubles in 30 days.
(b) Find r if the population doubles in N days.

5. The field mouse population in Example 3 satisfies the differential equation

$$dp/dt = 0.5p - 450.$$

(a) Find the time at which the population becomes extinct if $p(0) = 850$.
(b) Find the time of extinction if $p(0) = p_0$, where $0 < p_0 < 900$.
(c) Find the initial population p_0 if the population is to become extinct in 1 year.

6. **Radioactive Decay.** Experiments show that a radioisotope decays at a rate negatively proportional to the amount of the isotope present.

(a) Use the following variables and parameters to write down and solve an initial value problem for the process of radioactive decay: $t =$ time; $a(t) =$ amount of the radioisotope present at time t; $a_0 =$ initial amount of radioisotope; $r =$ decay rate, where $r > 0$.
(b) The **half-life**, $T_{1/2}$, of a radioisotope is the amount of time it takes for a quantity of the radioactive material to decay to one-half of its original amount. Find an expression for $T_{1/2}$ in terms of the decay rate r.

7. A radioactive material, such as the isotope thorium-234, disintegrates at a rate proportional to the amount currently present. If $Q(t)$ is the amount present at time t, then $dQ/dt = -rQ$, where $r > 0$ is the decay rate.

(a) If 100 milligrams (mg) of thorium-234 decays to 82.04 mg in 1 week, determine the decay rate r.
(b) Find an expression for the amount of thorium-234 present at any time t.
(c) Find the time required for the thorium-234 to decay to one-half its original amount.

8. **Classical Mechanics.** The differential equation for the velocity v of an object of mass m, restricted to vertical motion and subject only to the forces of gravity and air resistance, is

$$m\frac{dv}{dt} = -mg - \gamma v. \qquad \text{(i)}$$

In Eq. (i) we assume that the drag force, $-\gamma v$ where $\gamma > 0$ is a drag coefficient, is proportional to the velocity.

Acceleration due to gravity is denoted by g. Assume that the upward direction is positive.

(a) Show that the solution of Eq. (i) subject to the initial condition $v(0) = v_0$ is

$$v = \left(v_0 + \frac{mg}{\gamma}\right) e^{-\gamma t/m} - \frac{mg}{\gamma}.$$

(b) Sketch some integral curves, including the equilibrium solution, for Eq. (i). Explain the physical significance of the equilibrium solution.

(c) If a ball is initially thrown in the upward direction so that $v_0 > 0$, show that it reaches its maximum height when

$$t = t_{max} = \frac{m}{\gamma} \ln\left(1 + \frac{\gamma v_0}{mg}\right).$$

(d) The terminal velocity of a baseball dropped from a high tower is measured to be 33 m/s. If the mass of the baseball is 145 grams (g) and $g = 9.8$ m/s^2, what is the value of γ?

(e) Using the values for m, g, and γ in part (d), what would be the maximum height attained for a baseball thrown upward with an initial velocity $v_0 = 30$ m/s from a height of 2 m above the ground?

9. For small, slowly falling objects, the assumption made in Eq. (i) of Problem 8 that the drag force is proportional to the velocity is a good one. For larger, more rapidly falling objects, it is more accurate to assume that the drag force is proportional to the square of the velocity.[2]

(a) Write a differential equation for the velocity of a falling object of mass m if the drag force is proportional to the square of the velocity. Assume that the upward direction is positive.

(b) Determine the limiting velocity after a long time.

(c) If $m = 0.025$ kilograms (kg), find the drag coefficient so that the limiting velocity is -35 m/s.

Mixing Problems. Many physical systems can be cast in the form of a mixing tank problem. Consider a tank containing a solution—a mixture of solute and solvent–such as salt dissolved in water. Assume that the solution at concentration $c_i(t)$ flows into the tank at a volume flow rate $r_i(t)$ and is simultaneously pumped out at the volume flow rate $r_o(t)$. If the solution in the tank is well mixed, then the concentration of the outflow is $Q(t)/V(t)$, where $Q(t)$ is the amount of solute at time t and $V(t)$ is the volume of solution in the tank. The differential equation that models the changing amount of solute in the tank is based on the principle of conservation of mass,

$$\underbrace{\frac{dQ}{dt}}_{\text{rate of change of } Q(t)} = \underbrace{c_i(t)r_i(t)}_{\text{rate in}} - \underbrace{\{Q(t)/V(t)\}\, r_o(t)}_{\text{rate out}}, \quad \text{(i)}$$

where $V(t)$ also satisfies a mass conservation equation,

$$\frac{dV}{dt} = r_i(t) - r_o(t). \quad \text{(ii)}$$

If the tank initially contains an amount of solute Q_0 in a volume of solution, V_0, then initial conditions for Eqs. (i) and (ii) are $Q(0) = Q_0$ and $V(0) = V_0$, respectively.

10. A tank initially contains 200 liters (L) of pure water. A solution containing 1 g/L enters the tank at a rate of 4 L/min, and the well-stirred solution leaves the tank at a rate of 5 L/min. Write initial value problems for the amount of salt in the tank and the amount of brine in the tank, at any time t.

11. A tank contains 100 gallons (gal) of water and 50 ounces (oz) of salt. Water containing a salt concentration of $\frac{1}{4}\left(1 + \frac{1}{2}\sin t\right)$ oz/gal flows into the tank at a rate of 2 gal/min, and the mixture flows out at the same rate. Write an initial value problem for the amount of salt in the tank at any time t.

12. A pond initially contains 1,000,000 gal of water and an unknown amount of an undesirable chemical. Water containing 0.01 g of this chemical per gallon flows into the pond at a rate of 300 gal/h. The mixture flows out at the same rate, so the amount of water in the pond remains constant. Assume that the chemical is uniformly distributed throughout the pond.

(a) Write a differential equation for the amount of chemical in the pond at any time.

(b) How much of the chemical will be in the pond after a very long time? Does this limiting amount depend on the amount that was present initially?

13. **Pharmacokinetics.** A simple model for the concentration $C(t)$ of a drug administered to a patient is based on the assumption that the rate of decrease of $C(t)$ is negatively proportional to the amount present in the system,

$$\frac{dC}{dt} = -kC,$$

where k is a rate constant that depends on the drug and its value can be found experimentally.

(a) Suppose that a dose administered at time $t = 0$ is rapidly distributed throughout the body, resulting in an initial concentration C_0 of the drug in the patient. Find $C(t)$, assuming the initial condition $C(0) = C_0$.

(b) Consider the case where doses of C_0 of the drug are given at equal time intervals T, that is, doses of C_0 are administered at times $t = 0, T, 2T, \dots$. Denote by C_n the concentration immediately after the nth dose. Find an expression for the concentration C_2 immediately after the second dose.

(c) Find an expression for the concentration C_n immediately after the nth dose. What is $\lim_{n \to \infty} C_n$?

---

[2] See Lyle N. Long and Howard Weiss, "The Velocity Dependence of Aerodynamic Drag: A Primer for Mathematicians," *American Mathematical Monthly 106*, no. 2 (1999), pp. 127–135.

14. A certain drug is being administered intravenously to a hospital patient. Fluid containing 5 mg/cm³ of the drug enters the patient's bloodstream at a rate of 100 cm³/h. The drug is absorbed by body tissues or otherwise leaves the bloodstream at a rate proportional to the amount present, with a rate constant of 0.4 (h)⁻¹.

(a) Assuming that the drug is always uniformly distributed throughout the bloodstream, write a differential equation for the amount of the drug that is present in the bloodstream at any time.

(b) How much of the drug is present in the bloodstream after a long time?

Continuously Compounded Interest. The amount of money $P(t)$ in an interest bearing account in which the principal is compounded continuously at a rate r per annum and in which money is continuously added, or subtracted, at a rate of k dollars per annum satisfies the differential equation

$$\frac{dP}{dt} = rP + k. \qquad\qquad \text{(i)}$$

The case $k < 0$ corresponds to paying off a loan, while $k > 0$ corresponds to accumulating wealth by the process of regular contributions to an interest bearing savings account.

15. Show that the solution to Eq. (i), subject to the initial condition $P(0) = P_0$, is

$$P = \left(P_0 + \frac{k}{r}\right)e^{rt} - \frac{k}{r}. \qquad\qquad \text{(ii)}$$

Use Eq. (ii) in Problem 15 to solve Problems 16 and 17.

16. According to the International Institute of Social History (Amsterdam), the amount of money used to purchase Manhattan Island in 1626 is valued at \$1,050 in terms of today's dollars. If that amount were instead invested in an account that pays 4% per annum with continuous compounding, what would be the value of the investment in 2020? Compare with the case that interest is paid at 6% per annum.

17. How long will it take to pay off a student loan of \$20,000 if the interest paid on the principal is 5% and the student pays \$200 per month. What is the total amount of money repaid by the student?

18. Derive Eq. (ii) in Problem 15 from the discrete approximation to the change in the principal that occurs during the time interval $[t, t + \Delta t]$,

$$P(t + \Delta t) \cong P(t) + (r\Delta t)P(t) + k\Delta t,$$

assuming that $P(t)$ is continuously differentiable on $t \geq 0$. [*Hint:* Substitute $P(t + \Delta t) = P(t) + P'(t)\Delta t + (1/2)P''(\hat{t})$ $(\Delta t)^2)$, where $t < \hat{t} < t + \Delta t$, simplify, divide by Δt, and let $\Delta t \to 0$.]

Miscellaneous Modeling Problems

19. A spherical raindrop evaporates at a rate proportional to its surface area. Write a differential equation for the volume of the raindrop as a function of time.

20. Archimedes's *principle of buoyancy* states that an object submerged in a fluid is buoyed up by a force equal to the weight of the fluid displaced. An experimental, spherically shaped sonobuoy of radius 1/2 m with a mass m kg is dropped into the ocean with a velocity of 10 m/s when it hits the water. The sonobuoy experiences a drag force due to the water equal to one-half its velocity. Write down a differential equation describing the motion of the sonobuoy. Find values of m for which the sonobuoy will sink and calculate the corresponding terminal sink velocity of the sonobuoy. The density of seawater is $\rho_0 = 1.025$ kg/L.

1.2 Qualitative Methods: Phase Lines and Direction Fields

In Section 1.1 we were able to find solutions of the differential equations

$$\frac{du}{dt} = -k(u - T_0) \qquad \text{and} \qquad \frac{dp}{dt} = rp - k \qquad\qquad \text{(1)}$$

by using a simple integration technique. Do not assume that this is always possible. Finding closed-form analytic solutions of differential equations can be difficult or impossible. Fortunately, it is possible to obtain information about the qualitative behavior of solutions by using elementary ideas from calculus and graphical methods; we consider two such methods in this section—phase line diagrams and direction fields.

   Qualitative behavior refers to general properties of the differential equation and its solutions such as existence of equilibrium points, behavior of solutions near equilibrium points, and long-time behavior of solutions.[1] Qualitative analysis is important to the mathematical modeler because it can provide insight into even a very complicated model without having to find an exact solution or an approximation to an exact solution. It can show, often with only a small amount of effort, whether the equations are a plausible model of the phenomenon being studied. If not, what changes need to be made in the equations?

Autonomous Equations: Equilibrium Solutions and the Phase Line

A first order **autonomous** differential equation is an equation of the form

$$\frac{dy}{dt} = f(y). \tag{2}$$

The distinguishing feature of an autonomous equation is that the independent variable, in this case t, does not appear on the right side of the equation. For instance, the two equations appearing in (1) are autonomous. Other examples of autonomous equations are

$$p' = rp(1 - p/K), \qquad x' = \sin x, \qquad \text{and} \qquad y' = \sqrt{k^2/y - 1},$$

where r, K, and k are constants. However, the equations

$$u' + ku = kT_0 + kA \sin \omega t, \qquad x' = \sin(tx), \qquad \text{and} \qquad y' = -y + t$$

are not autonomous because the independent variable t does appear on the right side of each equation.

Equilibrium Solutions.   The first step in a qualitative analysis of Eq. (2) is to find constant solutions of the equation. If $y = \phi(t) = c$ is a constant solution of Eq. (2), then $dy/dt = 0$. Therefore any constant solution must satisfy the algebraic equation

$$f(y) = 0. \tag{3}$$

These solutions are called **equilibrium solutions** of Eq. (2) because they correspond to no change or variation in the value of y as t increases or decreases. Equilibrium solutions are also referred to as **critical points**, **fixed points**, or **stationary points** of Eq. (2).

Equilibrium solutions, although simple, are usually important for understanding the behavior of other solutions of the differential equation. To obtain information about other solutions, we draw the graph of $f(y)$ versus y. Figure 1.2.1 shows a generic plot of $f(y)$, where the equilibrium points are $y = a$, b, and c. It is convenient to think of the variable y as the position of a particle whose motion along the horizontal axis is governed by Eq. (2). The corresponding velocity of the particle, dy/dt, is prescribed by Eq. (2).

At points where the velocity of the particle $dy/dt = f(y) > 0$, so that y is an increasing function of t, the particle moves to the right. This is indicated in Figure 1.2.1 by placing on the y-axis arrows that point to the right in the intervals $y < a$ and $b < y < c$ where $f(y) > 0$. At points where the velocity of the particle $dy/dt = f(y) < 0$, so that y is a decreasing function of t, the particle moves to the left. This is indicated in Figure 1.2.1 by placing on the y-axis arrows that point to the left in the intervals $a < y < b$ and $y > c$, where $f(y) < 0$.

---

[1]In addition, the qualitative properties of differential equations include results about existence and uniqueness of solutions, intervals of existence, and dependence of solutions on parameters and initial conditions. These issues will be addressed in Sections 2.4 and 2.5.

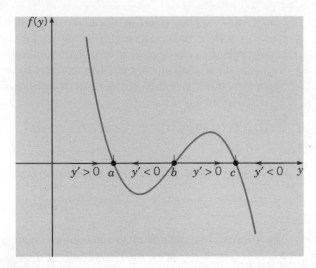

FIGURE 1.2.1 A generic graph of the right side of Eq. (2). The arrows on the y-axis indicate the direction in which y is changing [given by the sign of $y' = f(y)$] for each possible value of y. At the equilibrium points $y = a$, b, and c, $dy/dt = 0$.

The particle is stationary at the equilibrium points $y = a, b$, and c since $dy/dt = 0$ at each of those points.

The horizontal line in Figure 1.2.1 is referred to as the **phase line**, or the **one-dimensional phase portrait** of Eq. (2). The information contained in the phase line can be used to sketch the qualitatively correct integral curves of Eq. (2) by drawing it vertically just to the left of the ty-plane, as shown in Figure 1.2.2. We first draw the equilibrium solutions $y = a, b$ and c; then we draw a representative sampling of other curves that are increasing when $y < a$ and $b < y < c$ and decreasing when $a < y < b$ and $y > c$, as shown in Figure 1.2.2b.

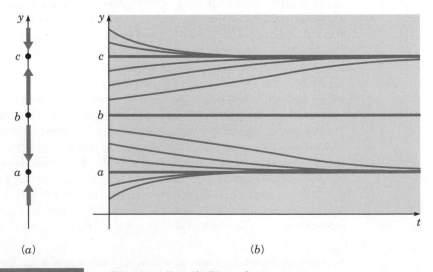

    (a)   (b)

FIGURE 1.2.2 (a) The phase line. (b) Plots of y versus t.

Stability of Equilibrium Points. In the drawings of the phase line notice that arrows drawn on either side of the equilibrium point $y = a$ point toward $y = a$. Consequently, solution curves in Figure 1.2.2*b* that start sufficiently close to $y = a$ approach $y = a$ as $t \to \infty$. Similarly, arrows drawn on either side of $y = c$ in Figures 1.2.1 and 1.2.2*a* point toward $y = c$. It follows that solution curves that start sufficiently close to $y = c$ approach $y = c$ as $t \to \infty$, as shown in Figure 1.2.2*b*. The equilibrium points $y = a$ and $y = c$ are said to be **asymptotically stable**. On the other hand, arrows in the phase line that lie on either side of the equilibrium point $y = b$ point away from $y = b$. Correspondingly, solution curves that start near $y = b$ move away from $y = b$ as t increases. The equilibrium point $y = b$ is said to be **unstable**.

To facilitate our understanding of asymptotically stable and unstable equilibrium points, it is again useful to think of y as the position of a particle whose dynamics are governed by Eq. (2). A particle, perturbed slightly via some disturbance, from an asymptotically stable equilibrium point, will move back toward that point. However, a particle situated at an unstable equilibrium point, subjected to any disturbance, will move away from that point. *All real-world systems are subject to disturbances, most of which are unaccounted for in a mathematical model. Therefore, systems residing at unstable equilibrium points are not likely to be observed in the real world.*

EXAMPLE 1

Draw phase line diagrams for Eq. (2) of Section 1.1,

$$\frac{du}{dt} = -k(u - T_0), \qquad \text{where } k > 0, \tag{4}$$

and use it to discuss the behavior of all solutions as $t \to \infty$. Compare behaviors for two different values of k, $0 < k_1 < k_2$.

As shown in Figure 1.2.3*a*, the graph of $f(u) = -k_1(u - T_0)$ versus u is a straight line with slope $-k_1 < 0$ that intersects the phase line at $u = T_0$, the only equilibrium solution of Eq. (4). Since $u' > 0$ if $u < T_0$ and $u' < 0$ if $u > T_0$, all arrows on the phase line point toward $u = T_0$, which is therefore asymptotically stable. Consequently, any solution $u = \phi(t)$ of Eq. (4) satisfies

$$\lim_{t \to \infty} \phi(t) = T_0.$$

Equation (4) and Figure 1.2.3*a* also show that the absolute value of the instantaneous rate of heat exchange (as measured by $|u'|$) is an increasing function of the difference between the temperature of the object and the temperature of the surroundings,

$$|u'| = k_1|u - T_0|.$$

Thus the slope of any solution curve will be steeper at points far away from T_0 compared to points that are close to T_0. Furthermore the slope will approach zero as $|u - T_0| \to 0$. Solution curves consistent with these observations are shown in Figure 1.2.3*b*.

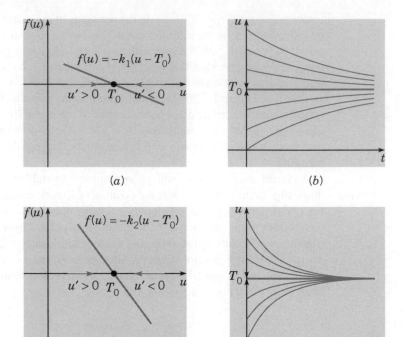

FIGURE 1.2.3 (a) and (c) Phase lines for $du/dt = -k(u - T_0)$, $k = k_1$ and k_2, where $k_1 < k_2$. The heavy blue arrows on the u-axis indicate the direction in which u is changing [given by the sign of $u'(t)$] for each possible value of u. For a given temperature difference $u - T_0$, the instantaneous rate of heat exchange depends on the slope $-k$ of the line. The parameter k is called the transmission coefficient. (b) and (d) Corresponding solutions of $du/dt = -k(u - T_0)$, where the phase line information in (a) and (c) is overlaid on the vertical axes. The rate of approach to equilibrium is governed by k. If k is small, the rate of heat exchange is slow. If k is large, the rate of heat exchange is rapid.

EXAMPLE 2

Draw a phase line diagram for the mouse population growth model, Eq. (15) of Section 1.1,

$$\frac{dp}{dt} = rp - a, \qquad \text{where } r, a > 0, \tag{5}$$

and use it to describe the behavior of all solutions as $t \to \infty$. Discuss implications of the model for the fruit farmer.

The only equilibrium solution of Eq. (5) is $p = a/r$. A plot of $f(p) = rp - a$ versus p in Figure 1.2.4a illustrates that $p' < 0$ when $p < a/r$, and $p' > 0$ when $p > a/r$. Thus the arrows on the p-axis point away from the equilibrium solution, which is unstable. Corresponding solution curves are shown in Figure 1.2.4b; note that the phase line diagram is overlaid on the p-axis.

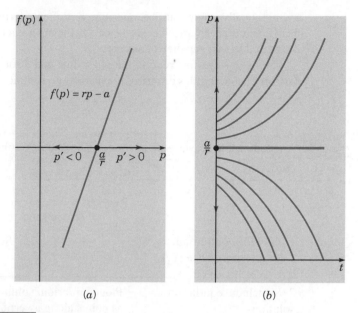

FIGURE 1.2.4   (a) The phase line for Eq. (5), $dp/dt = rp - a$, where $r, a > 0$. The slope r of the line corresponds to the growth rate of the mouse population. The direction of the arrows on the p-axis shows that the equilibrium solution $p = a/r$ is unstable. (b) Integral curves for Eq. (5).

Since the equilibrium solution is unstable, as time passes, an observer may see a mouse population either much larger or much smaller than the equilibrium population, but the equilibrium solution itself will not, in practice, be observed. Without the possible benefits of a more accurate and complex population model,[2] one inference that the fruit farmer might draw is that if he wants to control the mouse population, then he must install enough nesting boxes for the owls, thereby increasing the harvest rate a, to ensure that the mouse population $p(t)$ is always less than a/r. Thus a/r is a threshold value that should never be exceeded by $p(t)$ if the control strategy is to succeed.

This model also suggests a number of questions that the fruit farmer may wish to pursue, perhaps with assistance from a biologist who is knowledgeable about life cycles and habitats of field mice and owls:

▶ What is the growth rate of a field mouse population when there is an abundant food supply?

▶ How many mice per day does a barn owl consume?

▶ How do we estimate the size of the mouse population?

▶ Should we model the owl population?

▶ What will be a sustainable owl population if the mouse population drops to an economically acceptable level.

In each of the above examples, equilibrium solutions are important for understanding how other solutions of the given differential equation behave. An equilibrium solution may be thought of as a solution that serves as a reference to other, often nearby, solutions. An

---

[2]More elaborate population models appear in Sections 2.5 and 7.4.

asymptotically stable equilibrium solution is often referred to as an **attractor** or **sink**, since nearby solutions approach it as $t \to \infty$. On the other hand, an unstable equilibrium solution is referred to as a **repeller** or **source**.

The main steps for creating the phase line and a rough sketch of solution curves for a first-order autonomous differential equation are summarized in Table 1.2.1.

TABLE 1.2.1 Procedure for drawing phase lines and sketching solution curves for an autonomous equation.

Step	Illustration	
	Phase Line	Solution Curves
1. Find the equilibrium solutions of $dy/dt = f(y)$.	Solve $f(y) = 0$.	
2. Sketch the equilibrium solutions. These partition the phase line and ty-plane into disjoint regions.	Plot equilibrium solutions as points along a vertical line in increasing order as you move upward along the line. For instance, if $y_1 < y_2$ are equilibrium solutions, the phase line looks like	Plot equilibrium solutions as dashed horizontal lines in the ty-plane. For instance, if $0 < y_1 < y_2$ are equilibrium solutions, the ty-plane looks like
3. In each region, assess the sign of $f(y)$.		
(a) If $f(y) > 0$, then the solution curves passing through points in that region are increasing for all t, and either: (i) $\lim_{t \to \infty} y(t) = \infty$ if there is no larger equilibrium solution.	Affix arrowheads appropriately in each region.	Sketch a representative solution curve in each region.
(ii) $\lim_{t \to \infty} y(t) = y_2$ if y_2 is the next larger equilibrium solution.		

(continued)

	Procedure for drawing phase lines and sketching solution curves for an autonomous
TABLE 1.2.1 	equation. *(continued)*

	Illustration	
Step	Phase Line	Solution Curves
(b) If $f(y) < 0$, then the solution curves passing through points in that region are decreasing for all t, and either:	Affix arrowheads appropriately in each region.	Sketch a representative solution curve in each region.
(i) $\lim\limits_{t \to \infty} y(t) = -\infty$ if there is no smaller equilibrium solution.		
(ii) $\lim\limits_{t \to \infty} y(t) = y_1$ if y_1 is the next smaller equilibrium solution.		

Classification of Equilibrium Solutions

There are four possible arrow patterns that can encase a given equilibrium point of Eq. (2). The behavior of the solution curves "nearby" is different for each arrow pattern, resulting in different classifications of the corresponding equilibrium points. Suppose y_1 is an equilibrium point of Eq. (2). We illustrate the four possibilities in Table 1.2.2.

Remark. We use the same classification for the equilibrium *solution curve* $y = \phi(t) = y_1$ as we do for the equilibrium *point* y_1, with the only change being that the word "point" is replaced by "solution" in each case in Table 1.2.2.

	Classification of equilibrium points of (2).
TABLE 1.2.2	

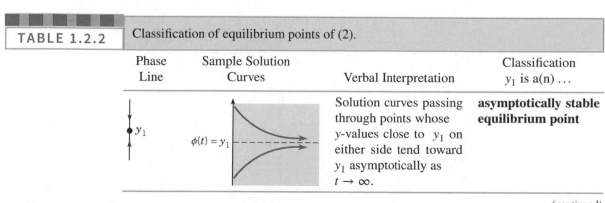

Phase Line	Sample Solution Curves	Verbal Interpretation	Classification y_1 is a(n) ...
y_1	$\phi(t) = y_1$	Solution curves passing through points whose y-values close to y_1 on either side tend toward y_1 asymptotically as $t \to \infty$.	**asymptotically stable equilibrium point**

(continued)

TABLE 1.2.2	Classification of equilibrium points of (2). *(continued)*

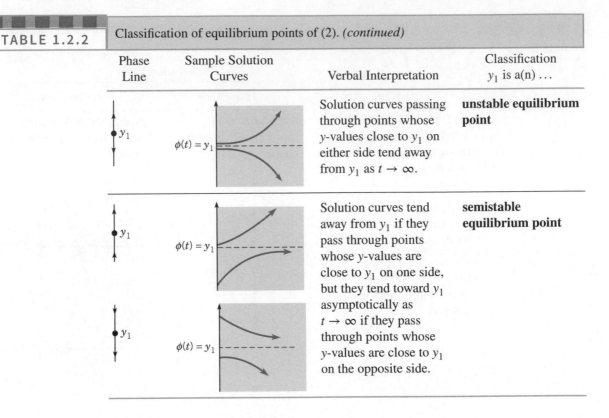

Phase Line	Sample Solution Curves	Verbal Interpretation	Classification y_1 is a(n) …
$\phi(t) = y_1$		Solution curves passing through points whose y-values close to y_1 on either side tend away from y_1 as $t \to \infty$.	**unstable equilibrium point**
$\phi(t) = y_1$		Solution curves tend away from y_1 if they pass through points whose y-values are close to y_1 on one side, but they tend toward y_1 asymptotically as $t \to \infty$ if they pass through points whose y-values are close to y_1 on the opposite side.	**semistable equilibrium point**
$\phi(t) = y_1$			

Linearization About an Equilibrium Point

Since the classification of an equilibrium point y_1 depends only on the behavior *near* y_1, we can extract its classification from certain features of the graph of f near y_1. Assume that f is differentiable in a vicinity of y_1 and suppose that $f'(y_1) < 0$. Then the graph of f in this vicinity resembles its tangent line, which has a negative slope. So, the graph of f is decreasing in this vicinity and hence the continuity of f implies that

▶ $f(y)$ is positive when $y < y_1$ and y is close by y_1.
▶ $f(y)$ is negative when $y > y_1$ and y is close by y_1.

Thus, the phase line must look like the one for an asymptotically stable equilibrium point. Similar reasoning shows that if $f'(y_1) > 0$, then y_1 must be an unstable equilibrium point. If $f'(y_1) = 0$, classifying y_1 from this information alone is impossible because the graph of f can exhibit one of several different situations near y_1, including the following:

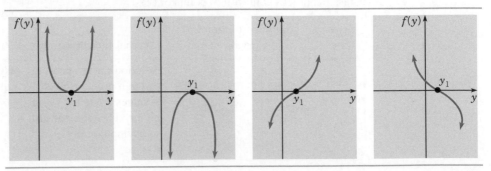

We summarize this discussion as the following theorem.

THEOREM 1.2.1	**Linearization About an Equilibrium Point.** Let y_1 be an equilibrium point of Eq. (2) and assume that f has a continuous derivative in a vicinity of y_1.

    i. If $f'(y_1) < 0$, then y_1 is an asymptotically stable equilibrium point.

    ii. If $f'(y_1) > 0$, then y_1 is an unstable equilibrium point.

    iii. If $f'(y_1) = 0$, then more information is needed to classify y_1.

Solutions and Direction Fields for $y' = f(t, y)$

Phase line diagrams allow us to infer qualitative properties of solutions of autonomous equations, that is, equations of the form $y' = f(y)$. More generally, the right side of a first order equation can depend on both the dependent and independent variables. The standard, or **normal form**, for a first order differential equation is

$$\frac{dy}{dt} = f(t, y). \tag{6}$$

Here f is a given function of the two variables t and y, sometimes referred to as the **rate function**. If the independent variable t appears explicitly in the rate function, then the equation is said to be **nonautonomous**.

A **solution** of Eq. (6) is a differentiable function $y = \phi(t)$ that satisfies the equation. This means that if we substitute $\phi(t)$ into the equation in place of the dependent variable y, the resulting equation

$$\phi'(t) = f(t, \phi(t)) \tag{7}$$

must be true for all t in the interval where $\phi(t)$ is defined. Equation (7) may be read as "at each point $(t, \phi(t))$ the slope $\phi'(t)$ of the line tangent[3] to the integral curve must be equal to $f(t, \phi(t))$" (see Figure 1.2.5).

It is not necessary to have a solution of Eq. (6) to draw direction field vectors. If a solution passes through the point (t, y), then the slope of the direction vector at that point is given by $f(t, y)$. Thus a direction field for equations of the form (6) can be constructed by evaluating f at each point of a rectangular grid consisting of at least a few hundred points. Then, at each point of the grid, a short line segment is drawn whose slope is the value of f at that point. Thus each line segment is tangent to the graph of the solution passing through that point. A direction field drawn on a fairly fine grid gives a good picture of the overall behavior of solutions of a differential equation.

Direction Fields for Autonomous Equations. Since the right side of an autonomous equation $y' = f(y)$ does not depend on t, slopes of direction field vectors for autonomous equations can vary only in the vertical direction of the ty-plane. Thus the slope of each direction field vector on a horizontal line $y = \alpha$, where α is a constant, will be $f(\alpha)$, as we now illustrate.

---

[3] Recall from calculus that the direction vector for the tangent line at $\mathbf{r}(t) = t\,\mathbf{i} + \phi(t)\,\mathbf{j}$ is $\mathbf{r}'(t) = 1\,\mathbf{i} + \phi'(t)\,\mathbf{j}$, which has slope $\phi'(t)$. Here $\mathbf{i}$ and $\mathbf{j}$ are unit vectors in the horizontal and vertical directions, respectively, of the xy-plane.

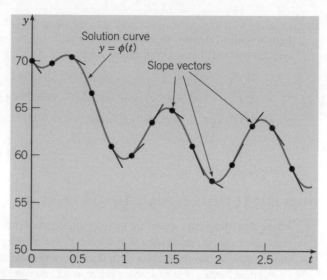

FIGURE 1.2.5 The path taken by an integral curve of a differential equation $y' = f(t, y)$ is determined by the slope vectors generated by $f(t, y)$ at each point on the path.

EXAMPLE
3

Draw a direction field for Eq. (4):

$$\frac{du}{dt} = -k(u - T_0).$$

Our task is simplified slightly if we assign numerical values to k and T_0, but the procedure is the same regardless of which values we choose. If we let $k = 1.5$ and $T_0 = 60$, then Eq. (4) becomes

$$du/dt = -1.5(u - 60) \tag{4a}$$

Suppose that we choose a value for u. Then, by evaluating the right side of Eq. (4a), we can find the corresponding value of du/dt. For instance, if $u = 70$, then $du/dt = -15$. This means that the slope of a solution $u = \phi(t)$ has the value -15 at any point where $u = 70$. We can display this information graphically in the tu-plane by drawing short line segments with slope -15 at several points on the line $u = 70$. Similarly, if $u = 50$, then $du/dt = 15$, so we draw line segments with slope 15 at several points on the line $u = 50$. We obtain Figure 1.2.6 by proceeding in the same way with other values of u. Figure 1.2.6 is an example of what is called a **direction field** or sometimes a **slope field**.

The importance of Figure 1.2.6 is that each line segment is a tangent line to the graph of a solution of Eq. (4a). Consequently, by looking at the direction field, we can visualize how solutions of Eq. (4a) vary with time. On a printed copy of a direction field we can even sketch (approximately) graphs of solutions by drawing curves that are always tangent to line segments in the direction field. Thus the general geometric behavior of the integral curves can be inferred from the direction field in Figure 1.2.6.

This approach can be applied equally well to the more general Eq. (4), where the parameters k and T_0 are unspecified positive numbers. The conclusions are essentially the same. The equilibrium solution of Eq. (4) is $u = T_0$. Solutions below the equilibrium solution increase with time, those above it decrease with time, and all other solutions approach the equilibrium solution as t becomes large.

The connection between integral curves and direction fields is an important concept for understanding how the right side of a differential equation, such as $u' = -k(u - T_0)$,

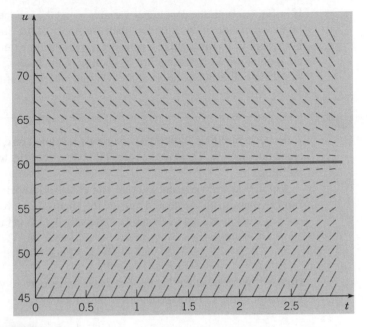

FIGURE 1.2.6   Direction field and equilibrium solution $u = 60$ for $u' = -1.5(u - 60)$.

determines the behavior of solutions and gives rise to the integral curves. However, using modern software packages, it is just as easy to plot the graphs of numerical approximations to solutions as it is to draw direction fields. We will frequently do this because the behavior of solutions of a first order equation is usually made most clear by overlaying the direction field with a representative set of integral curves, as shown in Figure 1.2.7. Such a sampling of integral curves facilitates visualization of the many other integral curves determined by the direction field generated by the right side of the differential equation.

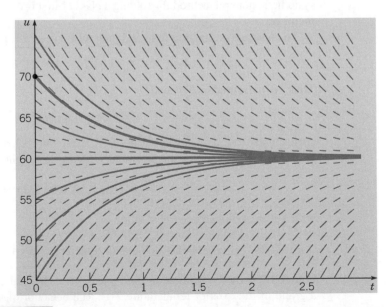

FIGURE 1.2.7   Direction field for $u' = -1.5(u - 60)$ overlaid with the integral curves shown in Figure 1.1.2.

If in Eq. (6) both the dependent variable y and the independent variable t appear explicitly on the right side of the equation, then the slopes of the direction field vectors will vary with both t and y. To illustrate, we consider the following extension of Example 1 in Section 1.1, an application of Newton's law of cooling to the heating and cooling of a building subject to periodic diurnal variation in the external air temperature.

EXAMPLE 4

Heating and Cooling of a Building

Consider a building, thought of as a partly insulated box, that is subject to external temperature fluctuations. Construct a model that describes the temperature fluctuations inside the building.

Let $u(t)$ and $T(t)$ be the internal and external temperatures, respectively, at time t. Assuming that the air inside and outside the enclosure is well mixed, we use Newton's law of cooling, just as we did in Example 1, Section 1.1, to get the differential equation

$$\frac{du}{dt} = -k\big[u - T(t)\big]. \tag{8}$$

However we now allow for the external temperature to vary with time. If we assume that the temperature of the external air is described by

$$T(t) = T_0 + A \sin \omega t, \tag{9}$$

then Eq. (8) can be written as

$$\frac{du}{dt} + ku = kT_0 + kA \sin \omega t. \tag{10}$$

In Problem 30 we ask you to verify that

$$u = T_0 + \frac{kA}{k^2 + \omega^2} (k \sin \omega t - \omega \cos \omega t) + ce^{-kt} \tag{11}$$

is a solution of Eq. (10), where c is an arbitrary real constant. In Chapter 2 we present a systematic general method for solving a class of first order equations of which Eq. (10) is a member.

To construct a direction field and integral curves for Eq. (10), we suppose that

$$k = 1.5(\text{day})^{-1}, \quad T_0 = 60°F, \quad A = 15°F, \quad \text{and} \quad \omega = 2\pi.$$

Thus t is measured in days and

$$T(t) = 60 + 15 \sin(2\pi t) \tag{12}$$

corresponds to a daily variation of 15°F above and below a mean temperature of 60°F. Inserting these values for the parameters into Eqs. (10) and (11) gives

$$\frac{du}{dt} + 1.5u = 90 + 22.5 \sin 2\pi t \tag{13}$$

with a corresponding general solution

$$u = 60 + \frac{22.5}{2.25 + \pi^2} (1.5 \sin 2\pi t - 2\pi \cos 2\pi t) + ce^{-1.5t}. \tag{14}$$

Figure 1.2.8 shows a direction field and several integral curves for Eq. (13) along with a graph of the exterior temperature $T(t)$. The behavior of solutions is a bit more complicated than those shown in Figure 1.2.7 because the right side of Eq. (13) depends on the independent variable t as well as the dependent variable u. Figure 1.2.8 shows that after approximately two days, all solutions begin to exhibit similar behavior. From the general solution (14), it is evident that for large t,

$$u(t) \approx U(t) = 60 + \frac{22.5}{2.25 + 4\pi^2}(1.5 \sin 2\pi t - 2\pi \cos 2\pi t), \qquad (15)$$

since $ce^{-1.5t} \to 0$ as $t \to \infty$. The function $U(t)$ in expression (15) is referred to as the **steady-state solution** of Eq. (13). Using trigonometric identities, we can write $U(t)$ in the form (see Problem 31)

$$U(t) = 60 + \frac{22.5}{\sqrt{2.25 + 4\pi^2}} \sin(2\pi t - \delta) \approx 60 + 3.4831 \sin(2\pi t - 1.33645), \qquad (16)$$

where $\delta = \cos^{-1}(1.5/\sqrt{2.25 + 4\pi^2})$. Comparing $U(t)$ with $T(t)$, we see that for large t the air temperature within the building varies sinusoidally at the same frequency as the external air temperature, but with a time lag of $t_{\text{lag}} = 1.33645/(2\pi) = 0.2127$ days and an amplitude of only $3.4831°F$ about a mean temperature of $60°F$. Does the qualitative behavior of the steady state solution agree with what you expect based on physical reasoning and experience?

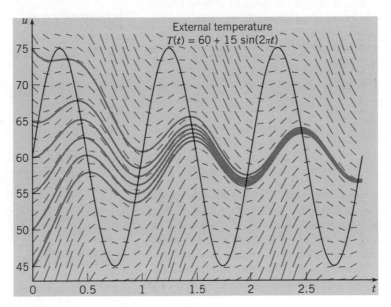

FIGURE 1.2.8 Direction field and integral curves for $u' + 1.5u = T(t)$. The variation in external temperature is described by $T(t) = 60 + 15 \sin 2\pi t$.

PROBLEMS

Phase Line Diagrams. Problems 1 through 7 involve equations of the form $dy/dt = f(y)$. In each problem, sketch the graph of $f(y)$ versus y, determine the critical (equilibrium) points, and classify each one as asymptotically stable or unstable. Draw the phase line, and sketch several graphs of solutions in the ty-plane.

1. $dy/dt = y(y - 1)(y - 2), \qquad y_0 \geq 0$

2. $dy/dt = e^y - 1, \qquad -\infty < y_0 < \infty$

3. $dy/dt = e^{-y} - 1, \qquad -\infty < y_0 < \infty$

4. $dy/dt = -2(\arctan y)/(1 + y^2), \qquad -\infty < y_0 < \infty$

5. $dy/dt = y^2(y + 1)(y - 3)$, $\qquad -\infty < y_0 < \infty$

6. $dy/dt = ay + by^2$, $\qquad a > 0$, $\quad b > 0$, $\quad y_0 \geq 0$

7. $dy/dt = ay + by^2$, $\qquad a > 0$, $\quad b > 0$, $\quad -\infty < y_0 < \infty$

Problems 8 through 13 involve equations of the form $dy/dt = f(y)$. In each problem sketch the graph of $f(y)$ versus y, determine the critical (equilibrium) points, and classify each one as asymptotically stable, unstable, or semistable. Draw the phase line, and sketch several graphs of solutions in the ty-plane.

8. $dy/dt = -k(y - 1)^2$, $\qquad k > 0$, $\quad -\infty < y_0 < \infty$

9. $dy/dt = y^2(y^2 - 1)$, $\qquad -\infty < y_0 < \infty$

10. $dy/dt = y(1 - y^2)$, $\qquad -\infty < y_0 < \infty$

11. $dy/dt = ay - b\sqrt{y}$, $\qquad a > 0$, $\quad b > 0$, $\quad y_0 \geq 0$

12. $dy/dt = y^2(4 - y^2)$, $\qquad -\infty < y_0 < \infty$

13. $dy/dt = y^2(1 - y)^2$, $\qquad -\infty < y_0 < \infty$

Direction Fields. In each of Problems 14 through 19 draw a direction field for the given differential equation. Based on the direction field, determine the behavior of y as $t \to \infty$. If this behavior depends on the initial value of y at $t = 0$, describe the dependency.

14. $y' = 3 - 2y$

15. $y' = 2y - 3$

16. $y' = 3 + 2y$

17. $y' = -1 - 2y$

18. $y' = 1 + 2y$

19. $y' = y + 2$

In each of Problems 20 through 23 draw a direction field for the given differential equation. Based on the direction field, determine the behavior of y as $t \to \infty$. If this behavior depends on the initial value of y at $t = 0$, describe this dependency. Note that in these problems the equations are not of the form $y' = ay + b$, and the behavior of their solutions is somewhat more complicated than the solutions shown in Figures 1.2.6 and 1.2.7.

20. $y' = y(4 - y)$

21. $y' = -y(5 - y)$

22. $y' = y^2$

23. $y' = y(y - 2)^2$

Consider the following list of differential equations, some of which produced the direction fields shown in Figures 1.2.9 through 1.2.14. In each of Problems 24 through 29 identify the differential equation that corresponds to the given direction field.

(a) $y' = 2y - 1$

(b) $y' = 2 + y$

(c) $y' = y - 2$

(d) $y' = y(y + 3)$

(e) $y' = y(y - 3)$

(f) $y' = 1 + 2y$

(g) $y' = -2 - y$

(h) $y' = y(3 - y)$

(i) $y' = 1 - 2y$

(j) $y' = 2 - y$

24. The direction field of Figure 1.2.9.

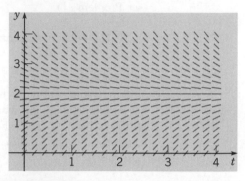

FIGURE 1.2.9 Direction field for Problem 24.

25. The direction field of Figure 1.2.10.

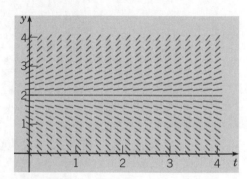

FIGURE 1.2.10 Direction field for Problem 25.

26. The direction field of Figure 1.2.11.

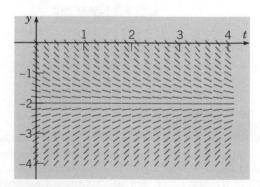

FIGURE 1.2.11 Direction field for Problem 26.

27. The direction field of Figure 1.2.12.

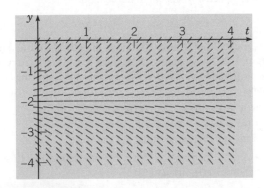

FIGURE 1.2.12 Direction field for Problem 27.

28. The direction field of Figure 1.2.13.

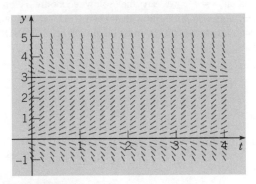

FIGURE 1.2.13 Direction field for Problem 28.

29. The direction field of Figure 1.2.14.

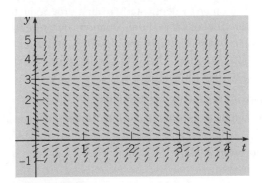

FIGURE 1.2.14 Direction field for Problem 29.

30. Verify that the function in Eq. (11) is a solution of Eq. (10).

31. Show that $A \sin \omega t + B \cos \omega t = R \sin(\omega t - \delta)$, where $R = \sqrt{A^2 + B^2}$ and δ is the angle defined by $R \cos \delta = A$ and $R \sin \delta = -B$.

Applications.

32. If in the exponential model for population growth, $dy/dt = ry$, the constant growth rate r is replaced by a growth rate $r(1 - y/K)$ that decreases linearly as the size of the population increases, we obtain the logistic model for population growth,

$$\frac{dy}{dt} = ry\left(1 - \frac{y}{K}\right), \qquad \text{(i)}$$

in which K is referred to as the *carrying capacity* of the population. Sketch the graph of $f(y)$, find the critical points, and determine whether each is asymptotically stable or unstable.

33. An equation that is frequently used to model the population growth of cancer cells in a tumor is the Gompertz equation

$$\frac{dy}{dt} = ry \ln(K/y),$$

where r and K are positive constants.

(a) Sketch the graph of $f(y)$ versus y, find the critical points, and determine whether each is asymptotically stable or unstable.

(b) For each y in $0 < y \leq K$, show that dy/dt, as given by the Gompertz equation, is never less than dy/dt, as given by the logistic equation, Eq. (i) in Problem 32.

34. In addition to the Gompertz equation (see Problem 33), another equation used to model the growth of cancerous tumors is the Bertalanffy equation

$$\frac{dV}{dt} = aV^{2/3} - bV,$$

where a and b are positive constants. This model assumes that the tumor grows at a rate proportional to surface area, while the loss of tumor mass due to cell death is proportional to the volume of the tumor. Sketch the graph of $f(V)$ versus V, find the critical points, and determine whether each is asymptotically stable or unstable.

35. A chemical of fixed concentration c_i flows into a continuously stirred tank reactor at a constant volume flow rate r_i and flows out at the same rate. While in the reactor, the chemical undergoes a simple reaction in which it disappears at a rate proportional to the concentration:

$$\frac{dc}{dt} = \frac{r_i}{V}c_i - r_i\frac{c}{V} - kc, \qquad \text{(i)}$$

where V is the volume of the reactor and k is the rate of reaction.

(a) Use the dimensionless variables

$$C = \frac{c}{c_i}, \qquad \tau = \frac{t}{V/r_i}$$

to express Eq. (i) in dimensionless form

$$\frac{dC}{d\tau} = 1 - C - \alpha C, \qquad \text{(ii)}$$

where

$$\alpha = \frac{kV}{r_i}.$$

(b) Determine the equilibrium solution of Eq. (ii), draw a phase line diagram, and determine whether the equilibrium solution is asymptotically stable or unstable. Then sketch a phase portrait with a representative set of solution curves.

36. A pond forms as water collects in a conical depression of radius a and depth h. Suppose that water flows in at a constant rate k and is lost through evaporation at a rate proportional to the surface area.

(a) Show that the volume $V(t)$ of water in the pond at time t satisfies the differential equation

$$dV/dt = k - \alpha\pi(3a/\pi h)^{2/3}V^{2/3},$$

where α is the coefficient of evaporation.

(b) Find the equilibrium depth of water in the pond. Is the equilibrium asymptotically stable?

(c) Find a condition that must be satisfied if the pond is not to overflow.

37. The Solow model of economic growth (ignoring the effects of capital stock depreciation) is

$$k' = \sigma f(k) - (n + g)k, \tag{i}$$

where k is *capital stock per unit of effective labor*, $f(k)$ is *GDP per unit of effective labor*, and $\sigma, 0 < \sigma < 1$, is the fraction of gross domestic product (GDP) devoted to investment.

The parameters n and g, growth rates of labor L and technology A, respectively, appear in the equations

$$\frac{dL}{dt} = nL, \qquad \frac{dA}{dt} = gA.$$

The product $A(t)L(t)$ is referred to as **effective labor** and the **output** Y of the economy is given by $Y = ALf(k)$. Assume that the production function $f(k)$ satisfies the following conditions:

(i) $f(0) = 0, \quad f(k) > 0$ for $k > 0$,

(ii) $f'(k) > 0, \quad f''(k) < 0$ for $k > 0$,

(iii) $\lim_{k\to 0} f'(k) = \infty, \quad \lim_{k\to\infty} f'(k) = 0$.

For example, the function $f(k) = ck^\alpha$ where $0 < \alpha < 1$ satisfies conditions (i), (ii), and (iii).

(a) Draw a phase line diagram of Eq. (i) by sketching the graphs of *actual investment* $\sigma f(k)$ and *break-even investment* $(n + g)k$ on the same set of coordinate axes. Show that Eq. (i) has an asymptotically stable equilibrium solution k^*.

(b) When $k = k^*$, we say the Solow economy is on its *balanced growth path*. Show that when $k = k^*$, the output of the economy grows at the combined growth rates of labor and technology,

$$\frac{dY}{dt} = (n + g)Y.$$

1.3 Definitions, Classification, and Terminology

In Sections 1.1 and 1.2 we gave examples to introduce you to a number of important topics in the context of first order differential equations: *mathematical modeling, solutions, integral curves, initial value problems, phase line diagrams, direction fields, equilibrium points, and concepts of stability*. Prior to embarking on an in-depth study of first order equations, we briefly step back and give you a broader view of differential equations by presenting a few important definitions, introducing some commonly used terminology, and discussing different ways that differential equations are classified. This background information will enhance your understanding of the subject in the following ways:

▶ It will provide you with an organizational framework for the subject;

▶ it will acquaint you with some of the language used to discuss the subject in a sensible manner;

▶ it will give you perspective on the subject as a whole.

We begin with a definition of a differential equation.

DEFINITION 1.3.1

Differential Equation. An equation that contains derivatives of one or more unknown functions with respect to one or more independent variables is said to be a **differential equation**.

The a priori unknown functions referred to in Definition 1.3.1 are dependent variables. When we write down a differential equation, such as Eq. (2) in Section 1.1,

$$\frac{du}{dt} = -k(u - T_0),$$

the unknown function u is considered to be a function of t, and is therefore a dependent variable. Thus Definition 1.3.1 may be alternatively expressed as "an equation that contains derivatives of one or more dependent variables with respect to one or more independent variables is said to be a **differential equation**."

Definition 1.3.1 underlies the following classifications based on (i) the number of independent variables, (ii) the number of unknown functions, and (iii) the highest order derivatives that appear in the equations.

Ordinary and Partial Differential Equations

We make a distinction between differential equations in which there is only one independent variable and differential equations in which there are two or more independent variables.

If the unknown function (or functions) depend on a single independent variable, then the only derivatives that appear in the equation are ordinary derivatives. In this case the differential equation is said to be an **ordinary differential equation** (**ODE**). For example, all of the differential equations that appear in Section 1.1 are ODEs.

If the unknown function (or functions) depend on more than one independent variable, and partial derivatives appear in the equation, then the differential equation is said to be a **partial differential equation** (**PDE**). Examples of PDEs are the three archetypal equations of mathematical physics shown in Table 1.3.1.

TABLE 1.3.1	Three archetypal PDEs of mathematical physics.

heat equation,	$\dfrac{\partial u}{\partial t} = D\dfrac{\partial^2 u}{\partial x^2},$	independent variables t and x.	(1)
wave equation,	$\dfrac{\partial^2 y}{\partial t^2} = c^2\dfrac{\partial^2 y}{\partial x^2},$	independent variables t and x.	(2)
Laplace's equation,	$\dfrac{\partial^2 V}{\partial x^2} + \dfrac{\partial^2 V}{\partial y^2} = 0,$	independent variables x and y.	(3)

In Eq. (1) $u(x, t)$ is the temperature of a metal rod at position x at time t; in Eq. (2) $y(x, t)$ is the vertical displacement from equilibrium of a horizontal vibrating string at position x at time t; in Eq. (3) $V(x, y)$ is the electric potential at the point (x, y) in a metal plate with a prescribed distribution of electric charge around the boundary. All of these equations have counterparts with $n \geq 3$ independent variables.

Systems of Differential Equations

Another classification of differential equations depends on the number of unknown functions that are involved. If there is a single function to be determined, then one equation is sufficient and is referred to as a *scalar equation*. All of the differential equations that appear in Section 1.1 are scalar ODEs. However, if there are two or more unknown functions, then a system of equations is required. Systems arise whenever there are two or more components that interact in some manner.[1] For example, the Lotka–Volterra, or predator–prey equations are important in ecological modeling. They have the form

$$dx/dt = \ \ ax - \alpha xy$$
$$dy/dt = -cy + \gamma xy. \tag{4}$$

where $x(t)$ and $y(t)$ are the respective populations of the prey and predator species. These equations provide an example of what is often one of the main problems confronting the mathematical modeler: "In terms of the variables, how do we mathematically represent the interaction of various components of the phenomenon?" The system (4) arises from the following assumptions. The prey are assumed to have an unlimited food supply, and to reproduce exponentially unless subject to predation; this exponential growth is represented in the first equation of the system (4) by the term ax. The rate of predation upon the prey is assumed to be proportional to the product of the predator and prey populations; this is represented above by $-\alpha xy$. If either x or y is zero, then there is no predation. In the second equation γxy represents the growth of the predator population. A different constant is used since the rate at which the predator population grows is not necessarily equal to the rate at which it consumes the prey. The term $-cy$ represents the loss rate of the predators due to either natural death or emigration; it leads to an exponential decay in the absence of prey. The constants $a, \alpha, c,$ and γ are based on empirical observations and depend on the particular species being studied. Systems of equations are discussed in Chapters 3, 6, and 7; in particular, the Lotka–Volterra equations are examined in Section 7.4. In some areas of applications, it is not unusual to encounter very large systems containing hundreds, or even thousands of equations.

Order

The **order** of a differential equation is the order of the highest derivative, ordinary or partial, that appears in the equation. The equations in Section 1.1 are all first order ODEs, while each of Eqs. (1), (2), and (3) is a second order PDE. The equation

$$ay'' + by' + cy = f(t), \tag{5}$$

where a, b, and c are given constants, and f is a given function, is a second order ODE. Equation (5) is a useful model of physical systems, for example, the motion of a mass attached to a spring, or the current in an electric circuit; we will consider it in detail in Chapter 4. More generally, the equation

$$F[t, u(t), u'(t), \ldots, u^{(n)}(t)] = 0 \tag{6}$$

is an ODE of the nth order. Equation (6) expresses a relation between the independent variable t and the values of the function u and its first n derivatives $u'(t), u''(t), \ldots, u^{(n)}(t)$. It is convenient and customary to write y for $u(t)$ with $y', \ldots, y^{(n)}$ standing for $u'(t), u''(t), \ldots, u^{(n)}(t)$. Thus Eq. (6) is written as

$$F\left[t, y, y', \ldots, y^{(n)}\right] = 0. \tag{7}$$

---

[1]We will frequently use the word *system* to refer to (i) a real-world group or combination of interrelated, interdependent, or interacting elements forming a collective entity, and (ii) a system of equations that model that entity. Although closely related and often identified with one another, they are not the same.

For example,

$$y''' + 2e^t y'' + yy' = t^4 \tag{8}$$

is a third order equation for $y = u(t)$. Occasionally, other letters will be used instead of t and y for the independent and dependent variables; the meaning should be clear from the context.

We assume that it is always possible to solve a given ordinary differential equation for the highest derivative. Thus we assume Eq. (7) can be written as

$$y^{(n)} = f(t, y, y'', \dots, y^{(n-1)}). \tag{9}$$

We study only equations of the form (9), although in the process of solving them, we often find it convenient to rewrite them in other forms.

Linear and Nonlinear Equations

DEFINITION 1.3.2

Linear Differential Equation. An nth order ordinary differential equation $F(t, y, y', \dots, y^{(n)}) = 0$ is said to be **linear** if it can be written in the form

$$a_0(t)y^{(n)} + a_1(t)y^{(n-1)} + \cdots + a_n(t)y = g(t).^2 \tag{10}$$

The functions $a_0, a_1, \dots, a_n$, called the **coefficients** of the equation, can depend at most on the independent variable t. Equation (10) is said to be **homogeneous** if the term $g(t)$ is zero for all t. Otherwise, the equation is **nonhomogeneous**.

Important special cases of Eq. (10) are first order linear equations, $a_0(t)y' + a_1(t)y = g(t)$, the subject of Section 2.2, and second order linear equations,

$$a_0(t)y'' + a_1(t)y' + a_2(t)y = g(t),$$

which we take up in Chapter 4.

An ODE that is not of the form (10) is a **nonlinear equation**. The distinction between a linear ODE and a nonlinear ODE hinges only on how the dependent variable y and its derivatives $y', y'', \dots, y^{(n)}$ appear in the equation: for an equation to be linear, they can appear in no other way except as designated by the form (10).

Common reasons that an ODE is nonlinear are that there are terms in the equation in which the dependent variable y or any of its derivatives

(i) are arguments of a nonlinear function, for example, terms such as $\sin y$, e^{-y}, or $\sqrt{1 + y^2}$,

(ii) appear as products, or are raised to a power other than 1, such as y^2 and yy'.

These statements also apply to equations in which there are two or more dependent variables, that is, to systems of differential equations. The presence of such terms often makes it easy to determine that an equation is nonlinear by observation. For example, Eq. (8) is nonlinear because of the term yy'. Each equation in system (4) is nonlinear because of the terms that involve the product xy of the dependent variables.

---

[2]Comparing Eq. (10) with Eq. (7), we see that an nth order ODE is linear only if

$$F\left[t, y, y', \dots, y^{(n)}\right] = a_0(t)y^{(n)} + a_1(t)y^{(n-1)} + \cdots + a_n(t)y - g(t). \tag{11}$$

If this is the case, F is said to be a **linear** function of the variables $y, y', y'', \dots, y^{(n)}$.

To show that a given equation is linear, you need only match its coefficients with Eq. (10) of appropriate order, as we show in the following example.

EXAMPLE 1

Show that

$$x^3 y''' + 3x^2 y'' + 4y = \sin(\ln x) \tag{12}$$

is a linear differential equation and state whether the equation is homogeneous or nonhomogeneous.

If, in Eq. (10), the independent variable is chosen to be x instead of t, and we set $n = 3$, $a_0(x) = x^3$, $a_1(x) = 3x^2$, $a_2(x) = 0$, $a_3(x) = 4$, and $g(x) = \sin(\ln x)$, we see that Eq. (10) reduces to $x^3 y''' + 3x^2 y'' + 4y = \sin(\ln x)$. Since $g(x)$ is not the zero function, the equation is nonhomogeneous.

Solutions

In Example 2, Section 1.1, we showed that directly substituting

$$u = T_0 + ce^{-kt}, \quad c \text{ an arbitrary constant,} \tag{13}$$

into the differential equation

$$\frac{du}{dt} = -k(u - T_0) \tag{14}$$

results in the identity

$$-kce - kt = -kce - kt, \quad -\infty < t < \infty,$$

and therefore the function in Eq. (13) is a solution of Eq. (14). The following definition generalizes this notion of a solution to nth order differential equations.

DEFINITION 1.3.3

Solution of a Differential Equation. A **solution** of the ordinary differential equation (9) on the interval $\alpha < t < \beta$ is a function ϕ such that $\phi', \phi'', \ldots, \phi^{(n)}$ exist and satisfy

$$\phi^{(n)}(t) = f\left[t, \phi(t), \phi'(t), \ldots, \phi^{(n-1)}(t)\right] \tag{15}$$

for every t in $\alpha < t < \beta$.

Thus, to determine if a given function is a solution of a differential equation, we substitute the function into the equation. If, upon substitution, the differential equation reduces to an identity, then the function is a solution. Otherwise, the function is not a solution.

EXAMPLE 2

Show that $y_1(t) = \cos t$ and $y_2(t) = \sin t$ are solutions of

$$y'' + y = 0 \tag{16}$$

on the interval $-\infty < t < \infty$.

Since $y_1'(t) = -\sin t$ and $y_1''(t) = -\cos t$, substituting $y_1(t)$ into Eq. (16) yields

$$\underbrace{-\cos t}_{y_1''(t)} + \underbrace{\cos t}_{y_1(t)} = 0$$

for all t. Therefore $y_1(t)$ is a solution of Eq. (16) on $-\infty < t < \infty$. Similarly, substituting $y_2(t)$ into Eq. (16) gives

$$\underbrace{-\sin t}_{y_2''(t)} + \underbrace{\sin t}_{y_2(t)} = 0$$

for all t, so $y_2(t)$ is also a solution of Eq. (16) on $-\infty < t < \infty$.

EXAMPLE 3

Show that $y(x) = c_1 x^2 + c_2 x^2 \ln x + \frac{1}{4} \ln x + \frac{1}{4}$, where c_1 and c_2 are arbitrary constants, is a solution of

$$x^2 y'' - 3xy' + 4y = \ln x \tag{17}$$

on the interval $0 < x < \infty$.

Subsituting $y(x)$ into the left side of Eq. (17) yields

$$x^2 \underbrace{\left(2c_1 + c_2(2 \ln x + 3) - \frac{1}{4x^2} \right)}_{y''(x)} - 3x \underbrace{\left(2c_1 x + c_2(2x \ln x + 3) + \frac{1}{4x} \right)}_{y'(x)} + 4 \underbrace{\left(c_1 x^2 + c_2 x^2 \ln x + \frac{1}{4} \ln x + \frac{1}{4} \right)}_{y(x)}$$

$$= c_1(2x^2 - 6x^2 + 4x^2) + c_2(2x^2 \ln x + 3x^2 - 6x^2 \ln x - 3x^2 + 4x^2 \ln x) - \frac{1}{4} - \frac{3}{4} + \ln x + 1$$

$$= c_1 \cdot 0 + c_2 \cdot 0 + \ln x.$$

Thus $y(x)$ satisfies Eq. (17) for all $0 < x < \infty$).

Initial Value Problems

Recall that in Section 1.1 we found solutions of certain equations by a process of direct integration. For instance, we found that the equation

$$\frac{du}{dt} = -k(u - T_0) \tag{18}$$

has the solution

$$u = T_0 + ce^{-kt}, \tag{19}$$

where c is an arbitrary constant. Each value of c corresponds to an integral curve in the tu-plane. If we want the solution that satisfies the condition

$$u(t_0) = u_0, \tag{20}$$

that is, the integral curve in the tu-plane that passes through the point (t_0, u_0), we substitute Eq. (19) into Eq. (20) to get

$$T_0 + ce^{-kt_0} = u_0. \tag{21}$$

Solving Eq. (21) for c gives

$$c = e^{kt_0}(u_0 - T_0), \tag{22}$$

and then replacing c in Eq. (19) by the right side of Eq. (22), we get

$$u = T_0 + (u_0 - T_0)e^{-k(t-t_0)}, \tag{23}$$

the solution of the initial value problem consisting of Eq. (18) and Eq. (20).

Let us now consider the simple second order equation

$$\frac{d^2y}{dt^2} = 0. \tag{24}$$

Integrating Eq. (24) twice results in two undetermined constants:

$$y(t) = c_1 t + c_2. \tag{25}$$

Values for c_1 and c_2 may be determined, for example, by requiring two initial conditions

$$y(t_0) = y_0 \quad \text{and} \quad y'(t_0) = y_1. \tag{26}$$

Substituting the solution (25) into Eqs. (26) gives

$$c_1 t_0 + c_2 = y_0 \quad \text{and} \quad c_1 = y_1.$$

Since $c_1 = y_1$, it follows that $c_2 = y_0 - t_0 y_1$. The solution of the initial value problem consisting of Eqs. (24) and (26) is therefore the straight line in the ty-plane with slope y_1 that passes through the point (t_0, y_0),

$$y = y_1(t - t_0) + y_0.$$

In general, solving an nth order ordinary differential equation results in n constants of integration $c_1, c_2, \ldots, c_n$. In applications, these constants of integration are determined by a set of auxiliary constraints, called **initial conditions**, on the solutions.

DEFINITION 1.3.4

Initial Value Problem. An **initial value problem** for an nth order differential equation

$$y^{(n)} = f(t, y, y'', \ldots, y^{(n-1)}) \tag{27}$$

on an interval I consists of Eq. (27) together with n initial conditions

$$y(t_0) = y_0, \quad y'(t_0) = y_1, \quad \ldots, \quad y^{(n-1)}(t_0) = y_{n-1} \tag{28}$$

prescribed at a point $t_0 \in I$, where $y_0, y_1, \ldots, y_{n-1}$ are given constants.

Thus $y = \phi(t)$ is a **solution of the initial value problem** (27), (28) on I if, in addition to satisfying Eq. (27) on I,

$$\phi(t_0) = y_0, \quad \phi'(t_0) = y_1, \quad \ldots, \quad \phi^{(n-1)}(t_0) = y_{n-1}.$$

EXAMPLE 4

Show that $\phi(t) = 2\cos t - 3\sin t$ is a solution of the initial value problem

$$y'' + y = 0, \qquad y(0) = 2, \quad y'(0) = -3, \tag{29}$$

on the interval $-\infty < t < \infty$.

Just as in Example 2, we substitute $\phi(t)$ into $y'' + y = 0$ to find that

$$\underbrace{-2\cos t + 3\sin t}_{\phi''(t)} + \underbrace{2\cos t - 3\sin t}_{\phi(t)} = 0$$

for all t. Therefore $\phi(t)$ is a solution of $y'' + y = 0$ on $-\infty < t < \infty$. Next we must check to see if the initial conditions specified in the initial value problem (29) are satisfied. Since

$$\phi(0) = 2\cos 0 - 3\sin 0 = 2$$

and

$$\phi'(0) = -2\sin 0 - 3\cos 0 = -3,$$

we conclude that $\phi(t)$ is a solution of the initial value problem (29).

PROBLEMS

In each of Problems 1 through 6, determine the order of the given differential equation; also state whether the equation is linear or nonlinear.

1. $t^2 \dfrac{d^2y}{dt^2} + t\dfrac{dy}{dt} + 2y = \sin t$

2. $(1 + y^2)\dfrac{d^2y}{dt^2} + t\dfrac{dy}{dt} + y = e^t$

3. $\dfrac{d^4y}{dt^4} + \dfrac{d^3y}{dt^3} + \dfrac{d^2y}{dt^2} + \dfrac{dy}{dt} + y = 1$

4. $\dfrac{dy}{dt} + ty^2 = 0$

5. $\dfrac{d^2y}{dt^2} + \sin(t + y) = \sin t$

6. $\dfrac{d^3y}{dt^3} + t\dfrac{dy}{dt} + (\cos^2 t)y = t^3$

Show that Eq. (10) can be matched to each equation in Problems 7 through 12 by a suitable choice of n, coefficients $a_0, a_1, \ldots, a_n$, and function g. In each case, state whether the equation is homogeneous or nonhomogeneous.

7. $\dfrac{dQ}{dt} = -\left(\dfrac{1}{1+t}\right)Q + 2\sin t$

8. $\dfrac{d^2y}{dt^2} = ty$

9. $x^2\dfrac{d^2y}{dx^2} - 3x\dfrac{dy}{dx} + 4y = \ln x, \quad x > 0$

10. $\dfrac{d}{dx}\left[(1 - x^2)\dfrac{d}{dx}P_n\right] + n(n + 1)P_n = 0, \quad n$ constant

11. $\dfrac{d^4y}{dt^4} + (\cos t)\dfrac{d^2y}{dt^2} + y = e^{-t}\sin t$

12. $\dfrac{d}{dx}\left[p(x)\dfrac{dy}{dx}\right] - q(x)y + \lambda r(x)y = 0, \quad \lambda$ constant

In each of Problems 13 through 20, verify that each given function is a solution of the differential equation.

13. $y'' - y = 0; \quad y_1(t) = e^t, \quad y_2(t) = \cosh t$

14. $y'' + 2y' - 3y = 0; \quad y_1(t) = e^{-3t}, \quad y_2(t) = e^t$

15. $ty' - y = t^2; \quad y = 3t + t^2$

16. $y'''' + 4y''' + 3y = t; \quad y_1(t) = t/3,$
$y_2(t) = e^{-t} + t/3$

17. $2t^2y'' + 3ty' - y = 0, \quad t > 0; \quad y_1(t) = t^{1/2},$
$y_2(t) = t^{-1}$

18. $t^2y'' + 5ty' + 4y = 0, \quad t > 0; \quad y_1(t) = t^{-2},$
$y_2(t) = t^{-2}\ln t$

19. $y'' + y = \sec t, \quad 0 < t < \pi/2;$
$y = (\cos t)\ln\cos t + t\sin t$

20. $y' - 2ty = 1; \quad y = e^{t^2}\int_0^t e^{-s^2}\,ds + e^{t^2}$

In each of Problems 21 through 24, determine the values of r for which the given differential equation has solutions of the form $y = e^{rt}$.

21. $y' + 2y = 0$

22. $y'' - y = 0$

23. $y'' + y' - 6y = 0$

24. $y''' - 3y'' + 2y' = 0$

In each of Problems 25 and 26, determine the values of r for which the given differential equation has solutions of the form $y = t^r$ for $t > 0$.

25. $t^2y'' + 4ty' + 2y = 0$

26. $t^2y'' - 4ty' + 4y = 0$

In Problems 27 through 31, verify that $y(t)$ satisfies the given differential equation. Then determine a value of the constant C so that $y(t)$ satisfies the given initial condition.

27. $y' + 2y = 0; \quad y(t) = Ce^{-2t}, \quad y(0) = 1$

28. $y' + (\sin t)y = 0; \quad y(t) = Ce^{\cos t}, \quad y(\pi) = 1$

29. $y' + (2/t)y = (\cos t)/t^2;$ $y(t) = (\sin t)/t^2 + C/t^2,$
$y(1) = \frac{1}{2}$

30. $ty' + (t + 1)y = t;$ $y(t) = (1 - 1/t) + Ce^{-t}/t,$
$y(\ln 2) = 1$

31. $2y' + ty = 2;$ $y = e^{-t^2/4} \int_0^t e^{s^2/4} \, ds + Ce^{-t^2/4},$
$y(0) = 1$

32. Verify that the function $\phi(t) = c_1 e^{-t} + c_2 e^{-2t}$ is a solution of the linear equation

$$y'' + 3y' + 2y = 0$$

for any choice of the constants c_1 and c_2. Determine c_1 and c_2 so that each of the following initial conditions is satisfied:

(a) $y(0) = -1,$ $y'(0) = 4$
(b) $y(0) = 2,$ $y'(0) = 0$

33. Verify that the function $\phi(t) = c_1 e^t + c_2 t e^t$ is a solution of the linear equation

$$y'' - 2y' + y = 0$$

for any choice of the constants c_1 and c_2. Determine c_1 and c_2 so that each of the following initial conditions is satisfied:

(a) $y(0) = 3,$ $y'(0) = 1$
(b) $y(0) = 1,$ $y'(0) = -4$

34. Verify that the function $\phi(t) = c_1 e^{-t} \cos 2t + c_2 e^{-t} \sin 2t$ is a solution of the linear equation

$$y'' + 2y' + 5y = 0$$

for any choice of the constants c_1 and c_2. Determine c_1 and c_2 so that each of the following initial conditions is satisfied:

(a) $y(0) = 1,$ $y'(0) = 1$
(b) $y(0) = 2,$ $y'(0) = 5$

© OSTILL / iStockphoto

First Order Differential Equations

This chapter deals with differential equations of first order,

$$\frac{dy}{dt} = f(t, y), \tag{1}$$

where f is a given function of two variables. Any differentiable function $y = \phi(t)$ that satisfies this equation for all t in some interval is called a solution. Our object is to develop methods for finding solutions or, if that is not possible, approximating them. Unfortunately, for an arbitrary function f, there is no general method for solving Eq. (1) in terms of elementary functions. Instead, we will describe several methods, each of which is applicable to a certain subclass of first order equations. The most important of these are separable equations (Section 2.1), linear equations (Section 2.2), and exact equations (Section 2.6). In Section 2.5 we discuss another subclass of first order equations, autonomous equations, for which geometrical methods yield valuable information about solutions. Finally, in Section 2.7 we describe other types of first order differential equations that can be transformed into separable or linear equations, and then solved. Methods for constructing numerical approximations to solutions are introduced and discussed in Chapter 8. Along the way, especially in Sections 2.3 and 2.5, we point out some of the many areas of application in which first order differential equations provide useful mathematical models.

2.1 Separable Equations

In Section 1.1 we used a process of direct integration to solve first order linear equations of the form

$$\frac{dy}{dt} = ay + b, \tag{2}$$

where a and b are constants. We will now show that this process is actually applicable to a much larger class of equations.

We will use x, rather than t, to denote the independent variable in this section for two reasons. In the first place, different letters are frequently used for the variables in a differential equation, and you should not become too accustomed to using a single pair. In particular, x often occurs as the independent variable. Further, we want to reserve t for another purpose later in this section.

The general first order equation is

$$\frac{dy}{dx} = f(x, y). \tag{3}$$

The equations we want to consider first are called separable, because the right side $f(x, y)$ has a special form.

DEFINITION 2.1.1

Separable Differential Equation. If the right side $f(x, y)$ of Eq. (3) can be written as the product of a function that depends only on x times another function that depends only on y,

$$\frac{dy}{dx} = f(x, y) = p(x)q(y), \tag{4}$$

then the equation is called **separable**.

If a differential equation is separable, Definition 2.1.1 means we can find two such functions p and q. For example, Eq. (2) written in the variables (x, y) becomes

$$\frac{dy}{dx} = f(x, y) = ay + b. \tag{5}$$

One choice for the functions p and q for Eq. (5) is $p(x) = 1$ and $q(y) = ay + b$, and therefore Eq. (5) is separable.

Once we know a differential equation is separable, we can find an expression for the solution by integration. A convenient shortcut uses the differentials dx and $dy = y'(x)\,dx$, and multiplying Eq. (4) by dx gives

$$dy = p(x)q(y)\,dx. \tag{6}$$

We assume $q(y)$ is nonzero for y values of interest, divide Eq. (6) by q, and integrate both sides to produce

$$\int q(y)^{-1}dy = \int p(x)\,dx. \tag{7}$$

Substituting our choices for p and q, we can integrate both sides of Eq. (7). For example, to solve Eq. (5), the same calculation steps can be used as in Section 1.1 [where the solution Eq. (8) is found from Eq. (2) in that section],

$$y(x) = -\frac{b}{a} + ce^{ax} \quad (c \neq 0). \tag{8}$$

In addition to Eq. (4) for any separable equation, we will use another equivalent form:

$$M(x) + N(y)\frac{dy}{dx} = 0. \tag{9}$$

The equivalence follows by choosing, for example, $M(x) = -p(x)$ and $N(y) = 1/q(y)$. This means that a **separable** equation can be written in the differential form

$$M(x)\,dx + N(y)\,dy = 0. \tag{10}$$

Then, if you wish, terms involving each variable may be separated by the equals sign. The differential form (10) is also more symmetric and tends to diminish the distinction between independent and dependent variables.

A separable equation can be solved by integrating the functions M and N. We illustrate the process by an example and then discuss it in general for Eq. (9).

EXAMPLE 1

Show that the equation

$$\frac{dy}{dx} = \frac{x^2}{1 - y^2} \tag{11}$$

is separable, and then find an equation for its integral curves.

If we write Eq. (11) as

$$-x^2 + (1 - y^2)\frac{dy}{dx} = 0, \tag{12}$$

then it has the form (9) and is therefore separable. Next, observe that the first term in Eq. (12) is the derivative of $-x^3/3$ with respect to x. Further, if we think of y as a function of x, then by the chain rule

$$\frac{d}{dx}\left(y - \frac{y^3}{3}\right) = \frac{d}{dy}\left(y - \frac{y^3}{3}\right)\frac{dy}{dx} = (1 - y^2)\frac{dy}{dx}.$$

Thus Eq. (12) can be written as

$$\frac{d}{dx}\left(-\frac{x^3}{3}\right) + \frac{d}{dx}\left(y - \frac{y^3}{3}\right) = 0,$$

or

$$\frac{d}{dx}\left(-\frac{x^3}{3} + y - \frac{y^3}{3}\right) = 0.$$

Therefore by integrating, we obtain

$$-x^3 + 3y - y^3 = c, \tag{13}$$

where c is an arbitrary constant. Equation (13) is an equation for the integral curves of Eq. (11). A direction field and several integral curves are shown in Figure 2.1.1. Any differentiable function $y = \phi(x)$ that satisfies Eq. (13) is a solution of Eq. (11). An equation of the integral curve passing through a particular point (x_0, y_0) can be found by substituting x_0 and y_0 for x and y, respectively, in Eq. (13) and determining the corresponding value of c.

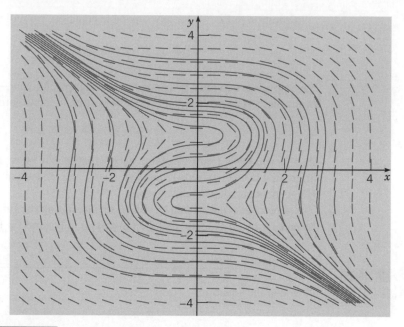

FIGURE 2.1.1 Direction field and integral curves of $y' = x^2/(1-y^2)$.

General Method for Separable Equations

Essentially the same procedure can be followed for any separable equation. Returning to Eq. (9), let H_1 and H_2 be any antiderivatives of M and N, respectively. Thus

$$H_1'(x) = M(x), \qquad H_2'(y) = N(y), \tag{14}$$

and Eq. (9) becomes

$$H_1'(x) + H_2'(y)\frac{dy}{dx} = 0. \tag{15}$$

According to the chain rule, if y is a function of x, then

$$H_2'(y)\frac{dy}{dx} = \frac{dH_2(y)}{dy}\frac{dy}{dx} = \frac{d}{dx}H_2(y). \tag{16}$$

Consequently, we can write Eq. (15) as

$$\frac{d}{dx}[H_1(x) + H_2(y)] = 0. \tag{17}$$

By integrating Eq. (17), we obtain

$$H_1(x) + H_2(y) = c, \tag{18}$$

where c is an arbitrary constant. Any differentiable function $y = \phi(x)$ that satisfies Eq. (18) is a solution of Eq. (9); in other words, Eq. (18) defines the solution implicitly rather than explicitly. In practice, Eq. (18) is usually obtained from Eq. (10) by integrating the first term with respect to x and the second term with respect to y, or equivalently, from Eq. (7). This more direct procedure is illustrated in Examples 2 and 3 below.

If, in addition to the differential equation, an initial condition

$$y(x_0) = y_0 \tag{19}$$

is prescribed, then the solution of Eq. (9) satisfying this condition is obtained by setting $x = x_0$ and $y = y_0$ in Eq. (18). This gives

$$c = H_1(x_0) + H_2(y_0). \tag{20}$$

Substituting this value of c in Eq. (18) and noting that

$$H_1(x) - H_1(x_0) = \int_{x_0}^{x} M(s)\, ds, \qquad H_2(y) - H_2(y_0) = \int_{y_0}^{y} N(s)\, ds,$$

we obtain

$$\int_{x_0}^{x} M(s)\, ds + \int_{y_0}^{y} N(s)\, ds = 0. \tag{21}$$

Equation (21) is an implicit representation of the solution of the differential equation (9) that also satisfies the initial condition (19). You should bear in mind that the determination of an explicit formula for the solution requires that Eq. (21) be solved for y as a function of x. Unfortunately, it is often impossible to do this analytically; in such cases, you can resort to numerical methods to find approximate values of y for given values of x. Alternatively, if it is possible to solve for x in terms of y, then this can often be very helpful.

EXAMPLE 2

Solve the initial value problem

$$\frac{dy}{dx} = \frac{3x^2 + 4x + 2}{2(y - 1)}, \qquad y(0) = -1, \tag{22}$$

and determine the interval in which the solution exists.

The differential equation can be written as

$$2(y - 1)\, dy = (3x^2 + 4x + 2)\, dx.$$

Integrating the left side with respect to y and the right side with respect to x gives

$$y^2 - 2y = x^3 + 2x^2 + 2x + c, \tag{23}$$

where c is an arbitrary constant. To determine the solution satisfying the prescribed initial condition, we substitute $x = 0$ and $y = -1$ in Eq. (23), obtaining $c = 3$. Hence the solution of the initial value problem is given implicitly by

$$y^2 - 2y = x^3 + 2x^2 + 2x + 3. \tag{24}$$

To obtain the solution explicitly, we must solve Eq. (24) for y in terms of x. That is a simple matter in this case, since Eq. (24) is quadratic in y, and we obtain

$$y = 1 \pm \sqrt{x^3 + 2x^2 + 2x + 4}. \tag{25}$$

Equation (25) gives two solutions of the differential equation, only one of which, however, satisfies the given initial condition. This is the solution corresponding to the minus sign in Eq. (25), so we finally obtain

$$y = \phi(x) = 1 - \sqrt{x^3 + 2x^2 + 2x + 4} \tag{26}$$

as the solution of the initial value problem (22). Note that if the plus sign is chosen by mistake in Eq. (25), then we obtain the solution of the same differential equation that satisfies the initial condition $y(0) = 3$. Finally, to determine the interval in which the solution (26) is valid, we must find the interval (containing the initial point $x = 0$) in which the quantity under the radical is positive. The only real zero of this expression is $x = -2$, so the desired interval is $x > -2$. The solution of the initial value problem and some other integral curves of the differential equation are shown in Figure 2.1.2. Observe that the boundary of the interval of validity of the solution (26) is determined by the point $(-2, 1)$ at which the tangent line is vertical.

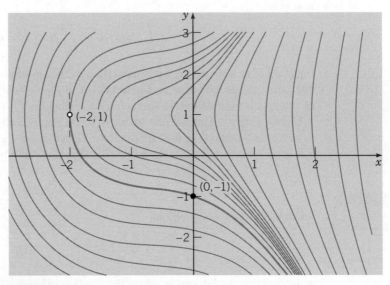

FIGURE 2.1.2 Integral curves of $y' = (3x^2 + 4x + 2)/2(y - 1)$.

EXAMPLE 3

Solve the equation

$$\frac{dy}{dx} = \frac{4x - x^3}{4 + y^3} \tag{27}$$

and draw graphs of several integral curves. Also find the solution passing through the point $(0, 1)$ and determine its interval of validity.

Rewriting Eq. (27) as

$$(4 + y^3)\, dy = (4x - x^3)\, dx,$$

integrating each side, multiplying by 4, and rearranging the terms, we obtain

$$y^4 + 16y + x^4 - 8x^2 = c, \tag{28}$$

where c is an arbitrary constant. Any differentiable function $y = \phi(x)$ that satisfies Eq. (28) is a solution of the differential equation (27). Graphs of Eq. (28) for several values of c are shown in Figure 2.1.3.

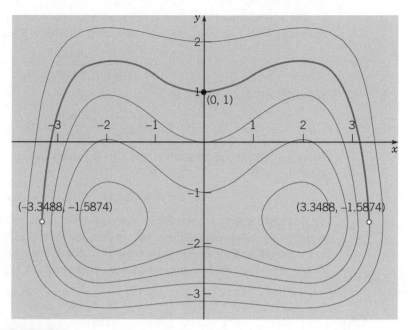

FIGURE 2.1.3 Integral curves of $y' = (4x - x^3)/(4 + y^3)$. The solution passing through $(0, 1)$ is shown by the heavy curve.

To find the particular solution passing through $(0, 1)$, we set $x = 0$ and $y = 1$ in Eq. (28) with the result that $c = 17$. Thus the solution in question is given implicitly by

$$y^4 + 16y + x^4 - 8x^2 = 17. \tag{29}$$

It is shown by the heavy curve in Figure 2.1.3. The interval of validity of this solution extends on either side of the initial point as long as the function remains differentiable. From the figure, we see that the interval ends when we reach points where the tangent line is vertical. It follows, from the differential equation (27), that these are points where $4 + y^3 = 0$, or $y = (-4)^{1/3} \cong -1.5874$. From Eq. (29), the corresponding values of x are $x \cong \pm 3.3488$. These points are marked on the graph in Figure 2.1.3.

Sometimes an equation of the form (3),

$$\frac{dy}{dx} = f(x, y),$$

has a constant solution $y = y_0$. Such a solution is usually easy to find because if $f(x, y_0) = 0$ for some value y_0 and for all x, then the constant function $y = y_0$ is a solution of the differential equation (3). For example, the equation

$$\frac{dy}{dx} = \frac{(y - 3)\cos x}{1 + 2y^2} \tag{30}$$

has the constant solution $y = 3$. Other solutions of this equation can be found by separating the variables and integrating.

The investigation of a first order nonlinear equation can sometimes be facilitated by regarding both x and y as functions of a third variable t. Then

$$\frac{dy}{dx} = \frac{dy/dt}{dx/dt}. \tag{31}$$

If the differential equation is

$$\frac{dy}{dx} = \frac{F(x, y)}{G(x, y)}, \tag{32}$$

then, by comparing numerators and denominators in Eqs. (31) and (32), we obtain the system

$$dx/dt = G(x, y), \qquad dy/dt = F(x, y). \tag{33}$$

At first sight it may seem unlikely that a problem will be simplified by replacing a single equation by a pair of equations, but, in fact, the system (33) may well be more amenable to investigation than the single equation (32). Nonlinear systems of the form (33) are introduced in Section 3.6 and discussed more extensively in Chapter 7.

Note on Explicit and Implicit Solutions

In Example 2, it was not difficult to solve explicitly for y as a function of x. However this situation is exceptional, and often it will be better to leave the solution in implicit form, as in Examples 1 and 3. Thus, in the problems below and in other sections where nonlinear equations appear, the words "solve the following differential equation" mean to find the solution explicitly if it is convenient to do so, but otherwise, to find an equation defining the solution implicitly.

PROBLEMS

In each of Problems 1 through 12, solve the given differential equation.

1. $y' = x^4/y$

2. $y' = x^2/y(1 + x^3)$

3. $y' + y^3 \sin x = 0$

4. $y' = (7x^2 - 1)/(7 + 5y)$

5. $y' = (\sin^2 2x)(\cos^2 y)$

6. $xy' = (1 - y^2)^{1/2}$

7. $yy' = (x + xy^2)e^{x^2}$

8. $\dfrac{dy}{dx} = \dfrac{x^2 + e^{-x}}{y^2 - e^y}$

9. $\dfrac{dy}{dx} = \dfrac{x^2}{1 + y^2}$

10. $\dfrac{dy}{dx} = \dfrac{\sec^2 x}{1 + y^3}$

11. $\dfrac{dy}{dx} = 4\sqrt{xy}$

12. $\dfrac{dy}{dx} = x(y - y^2)$

In each of Problems 13 through 28:

(a) Find the solution of the given initial value problem in explicit form.

(b) Plot the graph of the solution.

(c) Determine (at least approximately) the interval in which the solution is defined.

13. $y' = (1 - 12x)y^2, \qquad y(0) = -\dfrac{1}{8}$

14. $y' = (3 - 2x)/y, \qquad y(1) = -6$

15. $x\,dx + ye^{-x}dy = 0, \qquad y(0) = 1$

16. $dr/d\theta = r^2/\theta, \qquad r(1) = 2$

17. $y' = 3x/(y + x^2 y), \qquad y(0) = -7$

18. $y' = 2x/(1 + 2y), \qquad y(2) = 0$

19. $y' = 2xy^2 + 4x^3 y^2, \qquad y(1) = -2$

20. $\dfrac{dy}{dx} = x^2 e^{-3y}, \qquad y(2) = 0$

21. $\dfrac{dy}{dx} = (1 + y^2)\tan 2x, \qquad y(0) = -\sqrt{3}$

22. $y' = x(x^2 + 1)/6y^5, \qquad y(0) = -1/\sqrt[3]{2}$

23. $y' = (3x^2 - e^x)/(2y - 11), \qquad y(0) = 11$

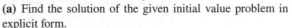

24. $x^2 y' = y - xy$, $y(1) = 2$

25. $y' = (e^{-x} - e^x)/(3 + 4y)$, $y(0) = 1$

26. $2y \dfrac{dy}{dx} = \dfrac{x}{\sqrt{x^2 - 4}}$, $y(3) = -1$

27. $\sin 2x\, dx + \cos 3y\, dy = 0$, $y(\pi/2) = \pi/3$

28. $y^2(1 - x^2)^{1/2}\, dy = \arcsin x\, dx$, $y(0) = 1$

In Problems 29 through 36, obtain the requested results by solving the given equations analytically or, if necessary, by graphing numerically generated approximations to the solutions.

29. Solve the initial value problem

$$y' = (1 + 3x^2)/(12y^2 - 12y), \qquad y(0) = 2$$

and determine the interval in which the solution is valid.
Hint: To find the interval of definition, look for points where the integral curve has a vertical tangent.

30. Solve the initial value problem

$$y' = 2x^2/(2y^2 - 6), \qquad y(1) = 0$$

and determine the interval in which the solution is valid.
Hint: To find the interval of definition, look for points where the integral curve has a vertical tangent.

31. Solve the initial value problem

$$y' = 2y^2 + xy^2, \qquad y(0) = 1$$

and determine where the solution attains its minimum value.

32. Solve the initial value problem

$$y' = (6 - e^x)/(3 + 2y), \qquad y(0) = 0$$

and determine where the solution attains its maximum value.

33. Solve the initial value problem

$$y' = 2\cos 2x/(10 + 2y), \qquad y(0) = -1$$

and determine where the solution attains its maximum value.

34. Solve the initial value problem

$$y' = 2(1 + x)(1 + y^2), \qquad y(0) = 0$$

and determine where the solution attains its minimum value.

35. Consider the initial value problem

$$y' = ty(4 - y)/3, \qquad y(0) = y_0.$$

(a) Determine how the behavior of the solution as t increases depends on the initial value y_0.
(b) Suppose that $y_0 = 0.5$. Find the time T at which the solution first reaches the value 3.98.

36. Consider the initial value problem

$$y' = ty(4 - y)/(1 + t), \qquad y(0) = y_0 > 0.$$

(a) Determine how the solution behaves as $t \to \infty$.
(b) If $y_0 = 2$, find the time T at which the solution first reaches the value 3.99.
(c) Find the range of initial values for which the solution lies in the interval $3.99 < y < 4.01$ by the time $t = 2$.

37. Solve the equation

$$\frac{dy}{dx} = \frac{ay + b}{cy + d},$$

where a, b, c, and d are constants.

2.2 Linear Equations: Method of Integrating Factors

In Section 1.1, we were able to find an explicit analytic solution of the differential equation

$$\frac{du}{dt} = -k(u - T_0) \tag{1}$$

modeling heat exchange between an object and its constant temperature surroundings. The same method can also be used to solve the differential equation

$$\frac{dp}{dt} = rp - a \tag{2}$$

for the population of field mice preyed on by owls (see Example 3 in Section 1.1), and the differential equation

$$m\frac{dv}{dt} = mg - \gamma v \tag{3}$$

for the velocity of a falling object (see Problem 8 in Section 1.1). But the method cannot be used to solve Example 4 in Section 1.2

$$\frac{du}{dt} = -k[u - T_0 - A\sin(\omega t)] \tag{4}$$

for the temperature in a building subject to a varying external temperature.

There is no general method for finding analytic solutions to all first order differential equations. What we can do is this. Given a first order differential equation, determine if it belongs to a class of equations for which we know a corresponding solution method that works for all members of that class. Then the method can be applied to solve the given equation. For this approach to be useful, we need a collection of important classes of equations and their corresponding solution methods. We have seen one such class already, the separable equations in Section 2.1; we will see another in this section; and others will be taken up in Sections 2.5 through 2.7. The class of equations that we consider here is specified by the following definition.

DEFINITION 2.2.1

A differential equation that can be written in the form

$$\frac{dy}{dt} + p(t)y = g(t) \tag{5}$$

is said to be a **first order linear equation** in the dependent variable y.

We can always find a solution to a first order linear equation, provided that a solution exists. Note that each of Eqs. (1)–(4) is a first order linear equation because each can be obtained by an appropriate choice of $p(t)$ and $g(t)$ in Eq. (5). For example, if we write Eq. (4) as

$$\frac{du}{dt} + ku = kT_0 + kA\sin(\omega t), \tag{6}$$

we see that it is a special case of Eq. (5) in which the dependent variable is u, $p(t) = k$ and $g(t) = kT_0 + kA\sin(\omega t)$.

Equation (5) is referred to as the **standard form** for a first order linear equation. The more general first order linear equation,

$$a_0(t)\frac{dy}{dt} + a_1(t)y = h(t), \tag{7}$$

can be put in the form of Eq. (5) by dividing by $a_0(t)$, provided that $a_0(t)$ is not zero. The function $p(t)$ in Eq. (5) is a **coefficient** in the equation. If $g(t) = 0$, Eq. (5) takes the form

$$\frac{dy}{dt} + p(t)y = 0, \tag{8}$$

and is said to be a **homogeneous** linear equation; otherwise, the equation is **nonhomogeneous**.

The equations

$$y' = \sin(t)\,y \quad \text{and} \quad ty' + 2y = 0$$

are linear and homogeneous, while

$$y' + \frac{2}{t}y = \frac{\cos t}{t^2} \quad \text{and} \quad (1+t^2)y' = 4ty + (1+t^2)^{-2}$$

are linear and nonhomogeneous.

The equations

$$y' + y^2 = 0 \quad \text{and} \quad y' + \sin(ty) = 1 + t$$

are not linear (we then say they are **nonlinear**), because the dependent variable y is squared in the first equation, and appears as an argument of the sine function in the second equation. Thus these equations cannot be written in the form of Eq. (5).

The method that we use to solve Eq. (5) is due to Leibniz, who was a co-inventor of calculus. It involves multiplying the equation by a certain function $\mu(t)$, chosen so that the resulting equation is readily integrable. The function $\mu(t)$ is called an **integrating factor**, and the first challenge is to determine how to find it. We will introduce this method in a simple example, and then show that the method extends to the general equation (5).

EXAMPLE 1

Solve the differential equation

$$\frac{dy}{dt} - 2y = 4 - t. \tag{9}$$

Plot the graphs of several solutions and draw a direction field. Find the particular solution whose graph contains the point $(0, -2)$. Discuss the behavior of solutions as $t \to \infty$.

The first step is to multiply Eq. (9) by a function $\mu(t)$, as yet undetermined, so that

$$\mu(t)\frac{dy}{dt} - 2\mu(t)y = \mu(t)(4 - t). \tag{10}$$

The idea now is to try to find a $\mu(t)$, so that we recognize the left side of Eq. (10) as the derivative of some particular expression. If we can, then we can integrate Eq. (10), even though we do not know the function y. To guide our choice of the integrating factor $\mu(t)$, observe that the left side of Eq. (10) contains two terms and that the first term is part of the result of differentiating the product $\mu(t)y$. Thus let us try to find $\mu(t)$, so that the left side of Eq. (10) becomes the derivative of the expression $\mu(t)y$. If we compare the left side of Eq. (10) with the differentiation formula

$$\frac{d}{dt}[\mu(t)y] = \mu(t)\frac{dy}{dt} + \frac{d\mu(t)}{dt}y, \tag{11}$$

we note that the first terms are identical for any $\mu(t)$, and that the second terms also agree, provided that we choose $\mu(t)$ to satisfy

$$\frac{d\mu(t)}{dt} = -2\mu(t). \tag{12}$$

Therefore our search for an integrating factor will be successful if we can find a solution of Eq. (12). Perhaps you can already identify a function that satisfies Eq. (12): What well-known function from calculus has a derivative that is equal to -2 times the original function? More systematically, rewrite Eq. (12) as

$$\frac{d\mu(t)/dt}{\mu(t)} = -2, \tag{13}$$

which is equivalent to

$$\frac{d}{dt} \ln |\mu(t)| = -2. \tag{14}$$

Then it follows that

$$\ln |\mu(t)| = -2t + C, \tag{15}$$

or

$$\mu(t) = ce^{-2t}. \tag{16}$$

Thus the function $\mu(t)$ given by Eq. (16) is an integrating factor for Eq. (9). Since we need just one integrating factor, we choose c to be one in Eq. (16) and use $\mu(t) = e^{-2t}$.

Now we return to Eq. (9), multiply it by the integrating factor e^{-2t}, and obtain

$$e^{-2t}\frac{dy}{dt} - 2e^{-2t}y = 4e^{-2t} - te^{-2t}. \tag{17}$$

By the choice we have made of the integrating factor, the left side of Eq. (17) is the derivative of $e^{-2t}y$, so that Eq. (17) becomes

$$\frac{d}{dt}(e^{-2t}y) = 4e^{-2t} - te^{-2t}. \tag{18}$$

By integrating both sides of Eq. (18), we obtain

$$e^{-2t}y = 4\int e^{-2t}dt - \int te^{-2t}dt + c, \tag{19}$$

and, using integration by parts on the second integral,

$$e^{-2t}y = -2e^{-2t} - \left[-\frac{1}{2}te^{-2t} - \frac{1}{4}e^{-2t} \right] + c, \tag{20}$$

where c is an arbitrary constant. Finally, by solving Eq. (20) for y, we obtain

$$y = -\frac{7}{4} + \frac{1}{2}t + ce^{2t}. \tag{21}$$

Equation (21) is referred to as the general solution of Eq. (9) because it contains all solutions of that equation. To find the solution passing through the point $(0, -2)$, we set $t = 0$ and $y = -2$ in Eq. (21), obtaining $-2 = -\frac{7}{4} + c$. Thus $c = -\frac{1}{4}$, and the desired solution is

$$y = -\frac{7}{4} + \frac{1}{2}t - \frac{1}{4}e^{2t}. \tag{22}$$

Figure 2.2.1 includes the graphs of Eq. (21) for several values of c with a direction field in the background. The solution passing through $(0, -2)$ is shown by the heavy curve. The behavior of the family of solutions (21) for large values of t is determined by the term ce^{2t}. If $c \neq 0$, then the solution grows exponentially large in magnitude, with the same sign as c itself. Thus the solutions diverge as t becomes large. The boundary between solutions that ultimately grow positively from those that ultimately grow negatively occurs when $c = 0$. If we substitute $c = 0$ into Eq. (21) and then set $t = 0$, we find that $y = -\frac{7}{4}$ is the separation point on the y-axis. Note that, for this initial value, the solution is $y = -\frac{7}{4} + \frac{1}{2}t$; it grows positively, but linearly rather than exponentially.

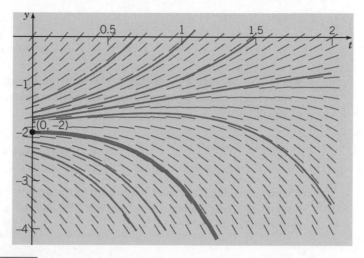

FIGURE 2.2.1 Direction field and integral curves of $y' - 2y = 4 - t$.

The Method of Integrating Factors for Solving $y' + p(t)y = g(t)$

Proceeding as in Example 1, we can apply the method of integrating factors to Eq. (5),

$$\frac{dy}{dt} + p(t)y = g(t),$$

where p and g are given functions. To determine an appropriate integrating factor, we multiply Eq. (5) by an as yet undetermined function $\mu(t)$, obtaining

$$\mu(t)\frac{dy}{dt} + p(t)\mu(t)y = \mu(t)g(t). \tag{23}$$

Following the same approach used in Example 1, we see that the left side of Eq. (23) is the derivative of the product $\mu(t)y$, provided that $\mu(t)$ satisfies the equation

$$\frac{d\mu(t)}{dt} = p(t)\mu(t). \tag{24}$$

If we assume temporarily that $\mu(t)$ is positive, then we have

$$\frac{d\mu(t)/dt}{\mu(t)} = p(t),$$

and consequently,

$$\ln \mu(t) = \int p(t)\,dt + k.$$

By choosing the arbitrary constant k to be zero, we obtain the simplest possible function for μ, namely,

$$\mu(t) = \exp \int p(t)\,dt. \tag{25}$$

Note that $\mu(t)$ is positive for all t, as we assumed. Returning to Eq. (23), we have

$$\frac{d}{dt}[\mu(t)y] = \mu(t)g(t). \tag{26}$$

Hence

$$\mu(t)y = \int \mu(t)g(t)\,dt + c, \tag{27}$$

where c is an arbitrary constant. Sometimes the integral in Eq. (27) can be evaluated in terms of elementary functions. However, this is not always possible, so the general solution of Eq. (5) is

$$y = \frac{1}{\mu(t)}\left[\int_{t_0}^{t} \mu(s)g(s)\,ds + c\right], \tag{28}$$

where t_0 is some convenient lower limit of integration. Observe that Eq. (28) involves two integrations, one to obtain $\mu(t)$ from Eq. (25) and the other to determine y from Eq. (28). In summary:

> *Use the following steps to solve any first order linear equation.*
> 1. Put the equation in standard form $y' + p(t)y = g(t)$.
> 2. Calculate the integrating factor $\mu(t) = e^{\int p(t)\,dt}$.
> 3. Multiply the equation by $\mu(t)$ and write it in the form $[\mu(t)y]' = \mu(t)g(t)$.
> 4. Integrate this equation to obtain $\mu(t)y = \int \mu(t)g(t)\,dt + c$.
> 5. Solve for y.

These steps constitute a systematic method, or algorithm, for solving any first order linear equation. The primary results of this algorithm are Eq. (25) for the integrating factor and Eq. (28) for the solution.

Note: It is helpful to memorize Eq. (25) for the integrating factor, but avoid memorizing Eq. (28). To solve a particular problem, you should understand and apply the steps of the procedure above, rather than risk errors in memorizing Eq. (28).

EXAMPLE 2

Solve the initial value problem

$$ty' + 2y = 4t^2, \tag{29}$$

$$y(1) = 2. \tag{30}$$

In order to determine $p(t)$ and $g(t)$ correctly, we must first rewrite Eq. (29) in the standard form (5). Thus we have

$$y' + \frac{2}{t}y = 4t, \tag{31}$$

so $p(t) = 2/t$ and $g(t) = 4t$. To solve Eq. (31), we first compute the integrating factor $\mu(t)$:

$$\mu(t) = \exp \int \frac{2}{t}\,dt = e^{2\ln|t|} = t^2.$$

On multiplying Eq. (31) by $\mu(t) = t^2$, we obtain

$$t^2 y' + 2ty = (t^2 y)' = 4t^3,$$

and therefore

$$t^2 y = t^4 + c,$$

where c is an arbitrary constant. It follows that

$$y = t^2 + \frac{c}{t^2} \tag{32}$$

is the general solution of Eq. (29). Integral curves of Eq. (29) for several values of c are shown in Figure 2.2.2. To satisfy the initial condition (30), it is necessary to choose $c = 1$; thus

$$y = t^2 + \frac{1}{t^2}, \qquad t > 0 \tag{33}$$

is the solution of the initial value problem (29), (30). This solution is shown by the heavy curve in Figure 2.2.2. Note that it becomes unbounded and is asymptotic to the positive y-axis as $t \to 0$ from the right. This is the effect of the infinite discontinuity in the coefficient $p(t)$ at the origin. The function $y = t^2 + (1/t^2)$ for $t < 0$ is not part of the solution of this initial value problem.

This is the first example in which the solution fails to exist for some values of t. Again, this is due to the infinite discontinuity in $p(t)$ at $t = 0$, which restricts the solution to the interval $0 < t < \infty$.

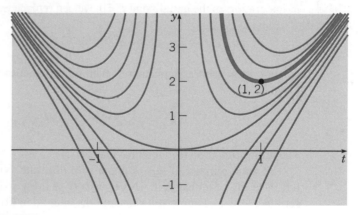

FIGURE 2.2.2   Integral curves of $ty' + 2y = 4t^2$.

Looking again at Figure 2.2.2, we see that some solutions (those for which $c > 0$) are asymptotic to the positive y-axis as $t \to 0$ from the right, while other solutions (for which $c < 0$) are asymptotic to the negative y-axis. The solution for which $c = 0$, namely, $y = t^2$, remains bounded and differentiable even at $t = 0$. If we generalize the initial condition (30) to

$$y(1) = y_0, \tag{34}$$

then $c = y_0 - 1$ and the solution (33) becomes

$$y = t^2 + \frac{y_0 - 1}{t^2}, \qquad t > 0. \tag{35}$$

As in Example 1, this is another instance where there is a critical initial value, namely, $y_0 = 1$, that separates solutions that behave in two very different ways.

EXAMPLE 3

Solve the initial value problem

$$2y' + ty = 2, \tag{36}$$

$$y(0) = 1. \tag{37}$$

First divide the differential equation (36) by two, obtaining

$$y' + \frac{t}{2}y = 1. \tag{38}$$

Thus $p(t) = t/2$, and the integrating factor is $\mu(t) = \exp(t^2/4)$. Then multiply Eq. (38) by $\mu(t)$, so that

$$e^{t^2/4}y' + \frac{t}{2}e^{t^2/4}y = e^{t^2/4}. \tag{39}$$

The left side of Eq. (39) is the derivative of $e^{t^2/4}y$, so by integrating both sides of Eq. (39), we obtain

$$e^{t^2/4}y = \int e^{t^2/4}dt + c. \tag{40}$$

The integral on the right side of Eq. (40) cannot be evaluated in terms of the usual elementary functions, so we leave the integral unevaluated. However, by choosing the lower limit of integration as the initial point $t = 0$, we can replace Eq. (40) by

$$e^{t^2/4}y = \int_0^t e^{s^2/4}ds + c, \tag{41}$$

where c is an arbitrary constant. It then follows that the general solution y of Eq. (36) is given by

$$y = e^{-t^2/4}\int_0^t e^{s^2/4}ds + ce^{-t^2/4}. \tag{42}$$

The initial condition (37) requires that $c = 1$.

The main purpose of this example is to illustrate that sometimes the solution must be left in terms of an integral. This is usually at most a slight inconvenience, rather than a serious obstacle. For a given value of t the integral in Eq. (42) is a definite integral and can be approximated to any desired degree of accuracy by using readily available numerical integrators. By repeating this process for many values of t and plotting the results, you can obtain a graph of a solution. Alternatively, you can use a numerical approximation method, such as Euler's method, discussed in Section 8.1, or others discussed in Chapter 8, that proceed directly from the differential equation and need no expression for the solution. Modern software packages such as Maple, Mathematica, and Matlab, among others, readily execute such procedures and produce graphs of solutions of differential equations.

Both of these procedures for finding solutions have advantages. For example, the analytical formula Eq. (42) contains a constant c that you can easily identify as the value $y(0)$ of the solution at $t = 0$. The solution $y(t)$ obviously depends on $y(0)$, which may be considered a **parameter** of the solution. In many applications it is important to understand the dependence of a solution on its parameters. In this example, Eq. (42) shows concisely that the solution $y(t)$ depends on the parameter $y(0)$ as a multiple of one exponential function; in other words, the parameter specifies how much of that exponential is included in the solution. Figure 2.2.3 displays graphs of Eq. (42) for several values of c. From the figure it is reasonable to conjecture that all the solutions approach a common limit as $t \to \infty$. It is interesting that the analytical formula Eq. (42) allows determination of the exact limiting value and of the rate of approach to the limiting value (see Problem 32). Alternatively, the graphs in the figure can also be found directly by numerical approximation methods. However, with that precedure, the dependence of the solution on the parameter $y(0)$ and the limiting value for large positive t are not so apparent.

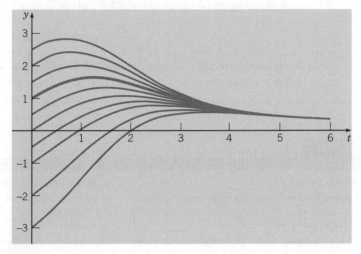

FIGURE 2.2.3   Integral curves of $2y' + ty = 2$. The heavy curve satisfies $y(0) = 1$.

For our last example, we solve Eq. (10) of Section 1.2, the differential equation from Example 4 of that section describing the heating and cooling of a building subject to external temperature variations.

EXAMPLE 4

Find the general solution of

$$\frac{du}{dt} = -k[u - T_0 - A\sin(\omega t)].$$

Solution

The standard form of this equation is

$$\frac{du}{dt} + ku = kT_0 + kA\sin(\omega t).$$

Using the integrating factor, $\mu(t) = e^{kt}$, we obtain

$$\left(e^{kt}u\right)' = kT_0 e^{kt} + kAe^{kt}\sin(\omega t). \tag{43}$$

Integrating and solving for u give

$$u = T_0 + kAe^{-kt}\int e^{kt}\sin(\omega t)\,dt + ce^{-kt}. \tag{44}$$

Setting $I = \int e^{kt}\sin(\omega t)\,dt$ and integrating by parts twice, we find that

$$I = -\frac{1}{\omega}e^{kt}\cos(\omega t) + \frac{k}{\omega^2}e^{kt}\sin(\omega t) - \frac{k^2}{\omega^2}I.$$

Solving for I, we get

$$I = \int e^{kt} \sin(\omega t)\, dt = \frac{1}{k^2 + \omega^2} e^{kt} \left[k \sin(\omega t) - \omega \cos(\omega t) \right]. \tag{45}$$

Of course, the integral in Eq. (44) can also be easily done using a computer algebra system. Substituting the result (45) into Eq. (44) then gives

$$u = T_0 + \frac{kA}{k^2 + \omega^2} \left[k \sin(\omega t) - \omega \cos(\omega t) \right] + c e^{-kt},$$

in agreement with Eq. (11) of Section 1.2. Some graphs of the solution are shown in Figure 1.2.8.

PROBLEMS

In each of Problems 1 through 12:
(a) Draw a direction field for the given differential equation.
(b) Based on an inspection of the direction field, describe how solutions behave for large t.
(c) Find the general solution of the given differential equation, and use it to determine how solutions behave as $t \to \infty$.

1. $y' + 4y = t + e^{-2t}$

2. $y' - 2y = t^2 e^{2t}$

3. $y' + y = te^{-t} + 1$

4. $y' + (1/t)y = 5 \cos 2t, \qquad t > 0$

5. $y' - 2y = 3e^t$

6. $ty' + 2y = \sin t, \qquad t > 0$

7. $y' + 2ty = 16te^{-t^2}$

8. $(1 + t^2)y' + 4ty = (1 + t^2)^{-2}$

9. $2y' + y = 3t$

10. $ty' - y = t^3 e^{-t}, \qquad t > 0$

11. $y' + y = 5 \sin 2t$

12. $2y' + y = 3t^2$

In each of Problems 13 through 20, find the solution of the given initial value problem.

13. $y' - y = 2te^{2t}, \qquad y(0) = 1$

14. $y' + 2y = te^{-2t}, \qquad y(1) = 0$

15. $ty' + 4y = t^2 - t + 1, \qquad y(1) = \frac{1}{4}, \quad t > 0$

16. $y' + (2/t)y = (\cos t)/t^2, \qquad y(\pi) = 0, \quad t > 0$

17. $y' - 2y = e^{2t}, \qquad y(0) = 2$

18. $ty' + 2y = \sin t, \qquad y(\pi/2) = 3, \quad t > 0$

19. $t^3 y' + 4t^2 y = e^{-t}, \qquad y(-1) = 0, \quad t < 0$

20. $ty' + (t + 1)y = t, \qquad y(\ln 2) = 1, \quad t > 0$

In each of Problems 21 through 23:
(a) Draw a direction field for the given differential equation. How do solutions appear to behave as t becomes large? Does the behavior depend on the choice of the initial value a? Let

a_0 be the value of a for which the transition from one type of behavior to another occurs. Estimate the value of a_0.
(b) Solve the initial value problem and find the critical value a_0 exactly.
(c) Describe the behavior of the solution corresponding to the initial value a_0.

21. $y' - \frac{1}{3}y = 3 \cos t, \qquad y(0) = a$

22. $2y' - y = e^{t/3}, \qquad y(0) = a$

23. $3y' - 2y = e^{-\pi t/2}, \qquad y(0) = a$

In each of Problems 24 through 26:
(a) Draw a direction field for the given differential equation. How do solutions appear to behave as $t \to 0$? Does the behavior depend on the choice of the initial value a? Let a_0 be the value of a for which the transition from one type of behavior to another occurs. Estimate the value of a_0.
(b) Solve the initial value problem and find the critical value a_0 exactly.
(c) Describe the behavior of the solution corresponding to the initial value a_0.

24. $ty' + (t + 1)y = 2te^{-t}, \qquad y(1) = a, \quad t > 0$

25. $ty' + 2y = (\sin t)/t, \qquad y(-\pi/2) = a, \quad t < 0$

26. $(\sin t)y' + (\cos t)y = e^t, \qquad y(1) = a, \quad 0 < t < \pi$

27. Consider the initial value problem

$$y' + \frac{1}{2}y = 2 \cos t, \qquad y(0) = -1.$$

Find the coordinates of the first local maximum point of the solution for $t > 0$.

28. Consider the initial value problem

$$y' + \frac{4}{3}y = 1 - \frac{1}{4}t, \qquad y(0) = y_0.$$

Find the value of y_0 for which the solution touches, but does not cross, the t-axis.

29. Consider the initial value problem

$$y' + \frac{1}{4}y = 3 + 2 \cos 2t, \qquad y(0) = 0.$$

(a) Find the solution of this initial value problem and describe its behavior for large t.

(b) Determine the value of t for which the solution first intersects the line $y = 12$.

30. Find the value of y_0 for which the solution of the initial value problem

$$y' - y = 1 + 3 \sin t, \qquad y(0) = y_0.$$

remains finite as $t \to \infty$.

31. Consider the initial value problem

$$y' - \frac{3}{2} y = 3t + 2e^t, \qquad y(0) = y_0.$$

Find the value of y_0 that separates solutions that grow positively as $t \to \infty$ from those that grow negatively. How does the solution that corresponds to this critical value of y_0 behave as $t \to \infty$?

32. Show that all solutions of $2y' + ty = 2$ [Eq. (36) of the text] approach a limit as $t \to \infty$, and find the limiting value. *Hint:* Consider the general solution, Eq. (42), and use L'Hôpital's rule on the first term.

33. Show that if a and λ are positive constants, and b is any real number, then every solution of the equation

$$y' + ay = be^{-\lambda t}$$

has the property that $y \to 0$ as $t \to \infty$.
Hint: Consider the cases $a = \lambda$ and $a \neq \lambda$ separately.

In each of Problems 34 through 37, construct a first order linear differential equation whose solutions have the required behavior as $t \to \infty$. Then solve your equation and confirm that the solutions do indeed have the specified property.

34. All solutions have the limit 3 as $t \to \infty$.

35. All solutions are asymptotic to the line $y = 4 - t$ as $t \to \infty$.

36. All solutions are asymptotic to the line $y = 2t - 5$ as $t \to \infty$.

37. All solutions approach the curve $y = 2 - t^2$ as $t \to \infty$.

38. Consider the initial value problem

$$y' + ay = g(t), \qquad y(t_0) = y_0.$$

Assume that a is a positive constant and that $g(t) \to g_0$ as $t \to \infty$. Show that $y(t) \to g_0/a$ as $t \to \infty$. Construct an example with a nonconstant $g(t)$ that illustrates this result.

39. Variation of Parameters. Consider the following method of solving the general linear equation of first order:

$$y' + p(t)y = g(t). \tag{i}$$

(a) If $g(t) = 0$ for all t, show that the solution is

$$y = A \exp\left[-\int p(t)\, dt\right], \tag{ii}$$

where A is a constant.

(b) If $g(t)$ is not everywhere zero, assume that the solution of Eq. (i) is of the form

$$y = A(t) \exp\left[-\int p(t)\, dt\right], \tag{iii}$$

where A is now a function of t. By substituting for y in the given differential equation, show that $A(t)$ must satisfy the condition

$$A'(t) = g(t) \exp\left[\int p(t)\, dt\right]. \tag{iv}$$

(c) Find $A(t)$ from Eq. (iv). Then substitute for $A(t)$ in Eq. (iii) and determine y. Verify that the solution obtained in this manner agrees with that of Eq. (28) in the text. This technique is known as the method of **variation of parameters**; it is discussed in detail in Section 4.7 in connection with second order linear equations.

In each of Problems 40 through 43 use the method of Problem 39 to solve the given differential equation.

40. $y' - 6y = t^6 e^{6t}$

41. $y' + (1/t)y = 3 \cos 2t, \qquad t > 0$

42. $ty' + 2y = \sin t, \qquad t > 0$

43. $2y' + y = 3t^2$

2.3 Modeling with First Order Equations

Differential equations are of interest to nonmathematicians primarily because of the possibility of using them to investigate a wide variety of problems in engineering and in the physical, biological, and social sciences. One reason for this is that mathematical models and their solutions lead to equations relating the variables and parameters in the problem. These equations often enable you to make predictions about how the natural process will behave in various circumstances. For example, all the figures in Section 2.2 show solution features that can be found by examining the parameter dependence of solution formulas. These features can be interpreted in terms of the physical behavior of the systems that the differential equations model. Furthermore, it is often easy to vary parameters in the mathematical model over wide ranges, whereas this may be very time-consuming or expensive, if not impossible, in an experimental setting. Nevertheless mathematical modeling and experiment or observation are both critically important and have somewhat complementary roles in scientific investigations. Mathematical models are validated by comparison of their predictions with experimental results. On the other hand, mathematical analyses may suggest the most promising directions for experimental exploration, and they may indicate fairly precisely what experimental data will be most helpful.

In Section 1.1 we formulated and investigated a few simple mathematical models. We begin by recapitulating and expanding on some of the conclusions reached in that section. Regardless of the specific field of application, there are three identifiable stages that are always present in the process of mathematical modeling.

▶ Construction of the Model. In this stage, you translate the physical situation into mathematical terms, often using the steps listed at the end of Section 1.1. Perhaps most critical at this stage is to *state clearly the physical principle(s) that are believed to govern the process*. For example, it has been observed that in some circumstances heat passes from a warmer to a cooler body at a rate proportional to the temperature difference, that objects move about in accordance with Newton's laws of motion, and that isolated insect populations grow at a rate proportional to the current population. Each of these statements involves a rate of change (derivative) and consequently, when expressed mathematically, leads to a differential equation. The differential equation is a mathematical model of the process.

It is important to realize that the mathematical equations are almost always only an approximate description of the actual process. For example, bodies moving at speeds comparable to the speed of light are not governed by Newton's laws, insect populations do not grow indefinitely as stated because of eventual limitations on their food supply, and heat transfer is affected by factors other than the temperature difference. Alternatively, one can adopt the point of view that the mathematical equations exactly describe the operation of a simplified or ideal physical model, which has been constructed (or imagined) so as to embody the most important features of the actual process. Sometimes, the process of mathematical modeling involves the conceptual replacement of a discrete process by a continuous one. For instance, the number of members in an insect population is an integer; however, if the population is large, it may seem reasonable to consider it to be a continuous variable and even to speak of its derivative.

▶ Analysis of the Model. Once the problem has been formulated mathematically, you are often faced with the problem of solving one or more differential equations or, failing that, of finding out as much as possible about the properties of the solution. It may happen that this mathematical problem is quite difficult, and if so, further approximations may be required at this stage to make the problem more susceptible to mathematical investigation. For example, a nonlinear equation may be approximated by a linear one, or a slowly varying coefficient may be replaced by a constant. Naturally, any such approximations must also be examined from the physical point of view to make sure that the simplified mathematical problem still reflects the essential features of the physical process under investigation. At the same time, an intimate knowledge of the physics of the problem may suggest reasonable mathematical approximations that will make the mathematical problem more amenable to analysis. *This interplay of understanding of physical phenomena and knowledge of mathematical techniques and their limitations is characteristic of applied mathematics* at its best, and it is indispensable in successfully constructing useful mathematical models of intricate physical processes.

▶ Comparison with Experiment or Observation. Finally, *having obtained the solution (or at least some information about it), you must interpret this information in the context in which the problem arose*. In particular, you should always check that the mathematical solution appears physically reasonable. If possible, calculate the values of the solution at selected points and compare them with experimentally observed values. Or ask whether the behavior of the solution after a long time is consistent with observations. Or examine the solutions corresponding to certain special values of parameters in the problem. Of course, the fact that the mathematical solution appears to be reasonable does not guarantee that it is correct. However, if the predictions of the mathematical model are seriously inconsistent with observations of the physical system it purports to describe, this suggests that errors have been made in solving the mathematical problem, that the mathematical model itself needs refinement, or that observations must be made with greater care.

The examples in this section are typical of applications in which first order differential equations arise. Subsequently, models for the evolution of the population of a particular species in a given region are treated extensively in Section 2.5.

EXAMPLE 1

Mixing

At time $t = 0$ a tank contains Q_0 pounds (lb) of salt dissolved in 100 gallons (gal) of water; see Figure 2.3.1. Assume that water containing $\frac{1}{4}$ lb of salt/gal is entering the tank at a rate of r gal/minutes (min) and that the well-stirred mixture is draining from the tank at the same rate. Set up the initial value problem that describes this flow process. Find (a) the quantity of salt $Q(t)$ in the tank at any time and (b) the limiting quantity Q_L that is present after a very long time. (c) If $r = 3$ and $Q_0 = 2Q_L$, find the time T after which the salt level is within 2% of Q_L. Also find (d) the flow rate that is required if the value of T is not to exceed 45 min.

r gal/min, $\frac{1}{4}$ lb/gal

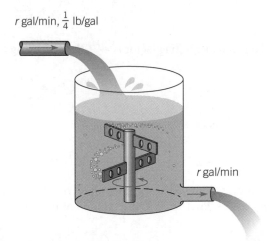

r gal/min

FIGURE 2.3.1 The water tank in Example 1.

We assume that salt is neither created nor destroyed in the tank, by chemical reactions for example. Therefore variations in the amount of salt are due solely to the flows in and out of the tank. More precisely, the *rate of change of salt in the tank*, dQ/dt, is equal to the *rate at which salt is flowing in* minus the *rate at which it is flowing out*. In symbols,

$$\frac{dQ}{dt} = \text{rate in} - \text{rate out}. \tag{1}$$

It is critically important that all terms have the same physical units so that the equation is internally consistent. The rate at which salt enters the tank is the concentration $\frac{1}{4}$ lb/gal times the flow rate r gal/min, or $(r/4)$ lb/min. To find the rate at which salt leaves the tank, we need to multiply the concentration of salt in the tank by the rate of outflow, r gal/min. Since the rates of flow in and out are equal, the volume of water in the tank remains constant at 100 gal, and since the mixture is "well stirred," the concentration throughout the tank is the same, namely, $[Q(t)/100]$ lb/gal. Therefore the rate at which salt leaves the tank is $[rQ(t)/100]$ lb/min. Thus the differential equation governing this process is

$$\frac{dQ}{dt} = \frac{r}{4} - \frac{rQ}{100}. \tag{2}$$

The initial condition is

$$Q(0) = Q_0. \tag{3}$$

Upon thinking about the problem physically, we might anticipate that eventually the mixture originally in the tank will be essentially replaced by the mixture flowing in, whose concentration is $\frac{1}{4}$ lb/gal. Consequently, we might expect that ultimately the amount of salt in the tank would be very close to 25 lb. We can also find the limiting amount $Q_L = 25$ by setting dQ/dt equal to zero in Eq. (2) and solving the resulting algebraic equation for Q.

(a) To find $Q(t)$ at any time t, note that Eq. (2) is both linear and separable. Rewriting it in the standard form for a linear equation, we have

$$\frac{dQ}{dt} + \frac{rQ}{100} = \frac{r}{4}. \tag{4}$$

Thus the integrating factor is $e^{rt/100}$. Mutiplying by this factor and integrating, we obtain

$$e^{rt/100}Q(t) = \frac{r}{4}\frac{100}{r}e^{rt/100} + c,$$

so the general solution is

$$Q(t) = 25 + ce^{-rt/100}, \tag{5}$$

where c is an arbitrary constant. To satisfy the initial condition (3), we must choose $c = Q_0 - 25$. Therefore the solution of the initial value problem (2), (3) is

$$Q(t) = 25 + (Q_0 - 25)e^{-rt/100}, \tag{6}$$

or

$$Q(t) = 25(1 - e^{-rt/100}) + Q_0e^{-rt/100}. \tag{7}$$

(b) From Eq. (6) or (7), you can see that $Q(t) \to 25$ lb as $t \to \infty$, so the limiting value Q_L is 25, confirming our physical intuition. Further $Q(t)$ approaches the limit more rapidly as r increases. In interpreting the solution (7), note that the second term on the right side is the portion of the original salt that remains at time t, while the first term gives the amount of salt in the tank due to the action of the flow processes. Plots of the solution for $r = 3$ and for several values of Q_0 are shown in Figure 2.3.2.

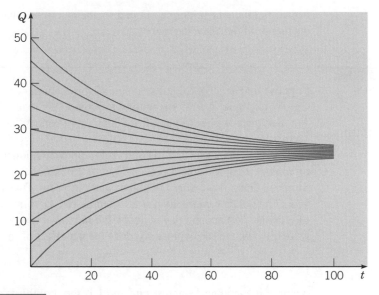

FIGURE 2.3.2 Solutions of the initial value problem (2), (3) for $r = 3$ and several values of Q_0.

(c) Now suppose that $r = 3$ and $Q_0 = 2Q_L = 50$; then Eq. (6) becomes

$$Q(t) = 25 + 25e^{-0.03t}. \tag{8}$$

Since 2% of 25 is 0.5, we wish to find the time T at which $Q(t)$ has the value 25.5. Substituting $t = T$ and $Q = 25.5$ in Eq. (8) and solving for T, we obtain

$$T = (\ln 50)/0.03 \cong 130.4 \text{ min.} \tag{9}$$

(d) To determine r so that $T = 45$, return to Eq. (6); set $t = 45$, $Q_0 = 50$, $Q(t) = 25.5$; and solve for r. The result is

$$r = (100/45)\ln 50 \cong 8.69 \text{ gal/min.} \tag{10}$$

In other words, the parameter r is specified by a requirement in the problem. We can see from Eq. (6) that generally the larger the value of r, the more rapid is the change from Q_0 to Q_L. If there were experimental data for comparison with the solution, the validity of predictions from the model could be appraised. In principle, flow rates can be measured with considerable accuracy. However the assumption of a uniform concentration of salt in the tank may be more questionable, since this may depend on how the liquid in the tank is stirred, whether the incoming flow is distributed or concentrated in one location, and perhaps the shape of the tank.

Example 1 illustrates what is called a "compartment model," for which a substance (in this case, water of different salt concentrations) flows into and out of a compartment (in this case, a tank). *The fundamental balance principle expressed by Eq. (1) can be applied in many different circumstances.* Compartment models are often used in problems involving a pollutant in a lake, or a drug in an organ of the body, among others. In such cases, the flow rates may be more challenging to determine or may vary with time. Similarly, the concentration in the compartment may be far from uniform in some cases. Finally, the rates of inflow and outflow may be different, which means that the variation of the amount of liquid in the compartment must also be taken into account. The next example illustrates the use of the balance principle (1) in a financial setting, where the "compartment" is a loan account and the "flow" is money.

EXAMPLE
2

A Car Loan

Suppose that a recent college graduate wishes to borrow $20,000 in order to purchase a new car, for example. A lender is willing to provide the loan with an annual interest rate of 8%. The borrower wishes to pay off the loan in 4 years. What monthly payment is required to do this?

This is an instance where a continuous approximation to a discrete process may be easier to analyze than the actual process. Let $S(t)$ be the balance due on the loan at any time t. Suppose that S is measured in dollars and t in years. Then dS/dt has the units of dollars per year. The balance on the loan is affected by two factors: the accumulation of interest tends to increase $S(t)$ and the payments by the borrower tend to reduce it. Based on the balance principle (1), we can express dS/dt as the net effect of these two factors. Thus we obtain

$$\frac{dS}{dt} = rS - 12k, \tag{11}$$

where r is the annual interest rate and k is the monthly payment rate. Note that k must be multiplied by 12 in Eq. (11), so that *all terms will have the same units* of dollars per year. The initial condition is

$$S(0) = S_0, \tag{12}$$

where S_0 is the amount of the loan.

For the situation stated in this example, $r = 0.08$ and $S_0 = 20,000$, so we have the initial value problem

$$\frac{dS}{dt} = 0.08S - 12k, \qquad S(0) = 20,000. \tag{13}$$

If we rewrite the differential equation as

$$S' - 0.08S = -12k,$$

then the integrating factor is $e^{-0.08t}$, and after an integration we obtain

$$e^{-0.08t}S = \frac{12}{0.08}ke^{-0.08t} + c,$$

or

$$S = 150k + ce^{0.08t}. \tag{14}$$

From the initial condition it follows that $c = 20,000 - 150k$, so the solution of the initial value problem (13) is

$$S = 20,000e^{0.08t} - 150k(e^{0.08t} - 1). \tag{15}$$

To find the monthly payment needed to pay off the loan in 4 years, we set $t = 4$, $S = 0$, and solve Eq. (15) for k. The result is

$$k = \frac{20,000}{150}\frac{e^{0.32}}{e^{0.32} - 1} = \$486.88. \tag{16}$$

The total amount paid over the life of the loan is 48 times $486.88, or $23,370.24; thus the total interest payment is $3,370.24.

The solution (15) can also be used to answer other possible questions. For example, suppose that the borrower wants to limit the monthly payment to $450. One way to do this is to extend the period of the loan beyond 4 years to T years, thereby increasing the number of payments. To find the required time period, set $k = 450$, $S = 0$, $t = T$ and solve for T, with the result that

$$T = \frac{\ln(27/19)}{0.08} \cong 4.39 \text{ years,} \tag{17}$$

or about 53 months. Equation (15) is another example of using the dependence on a parameter (in this case, k) in a solution formula to satisfy a specified requirement.

To assess the accuracy of this continuous model, we can solve the problem more precisely with a discrete-time model (see Problem 14). The comparison shows that the continuous model understates the monthly payment by only $1.38, or about 0.28%.

The approach used in Example 2 can also be applied to the more general initial value problem (11), (12), whose solution is

$$S = S_0 e^{rt} - 12\frac{k}{r}(e^{rt} - 1). \tag{18}$$

Note that (18) contains a total of three parameters—S_0, k, and r—as well as the variable t, and each can be useful in answering specified questions. The result (18) can be used in a large number of financial circumstances, including various kinds of investment plans, as well as loans and mortgages. For an investment situation, $S(t)$ is the balance in the investor's account, r is the estimated rate of return (interest, dividends, capital gains), and k is the monthly rate of deposits or withdrawals. The first term in expression (18) is the part of $S(t)$ that is due to the return accumulated on the initial amount S_0, and the second term is the part that is due to the deposit or withdrawal rate k.

The advantage of stating the problem in this general way without specific values for S_0, r, or k lies in the generality of the resulting formula (18) for $S(t)$. Using the parameter dependence in this formula, we can readily compare the results of different investment programs or different rates of return. The Problems offer other illustrations.

EXAMPLE 3

Chemicals in a Pond

Consider a pond that initially contains 10 million gal of fresh water. Stream water containing an undesirable chemical flows into the pond at the rate of 5 million gal/year, and the mixture in the pond flows out through an overflow culvert at the same rate. The concentration $\gamma(t)$ of chemical in the incoming water varies periodically with time t, measured in years, according to the expression $\gamma(t) = 2 + \sin 2t$ g/gal. Construct a mathematical model of this flow process and determine the amount of chemical in the pond at any time. Plot the solution and describe in words the effect of the variation in the incoming concentration.

Since the incoming and outgoing flows of water are the same, the amount of water in the pond remains constant at 10^7 gal. Let us denote the mass of the chemical by $Q(t)$, measured in grams. This example is another compartment model that is similar to Example 1, and the same inflow/outflow principle applies. Thus

$$\frac{dQ}{dt} = \text{rate in} - \text{rate out,}$$

where "rate in" and "rate out" refer to the rates at which the chemical flows into and out of the pond, respectively. The rate at which the chemical flows in is given by

$$\text{Rate in} = (5 \times 10^6) \text{ gal/year} \, (2 + \sin 2t) \text{ g/gal.} \tag{19}$$

The concentration of chemical in the pond is $Q(t)/10^7$ g/gal, so the rate of flow out is

$$\text{Rate out} = (5 \times 10^6) \text{ gal/year} \, [Q(t)/10^7] \text{ g/gal} = Q(t)/2 \text{ g/year.} \tag{20}$$

Thus we obtain the differential equation

$$\frac{dQ}{dt} = (5 \times 10^6)(2 + \sin 2t) - \frac{Q(t)}{2}, \tag{21}$$

where *each term has the units of grams per year*.

To make the coefficients more manageable, it is convenient to introduce a new dependent variable defined by $q(t) = Q(t)/10^6$ or $Q(t) = 10^6 \, q(t)$. This means that $q(t)$ is measured in millions of grams, or megagrams. If we make this substitution in Eq. (21), then each term contains the factor 10^6, which can be canceled. If we also transpose the term involving $q(t)$ to the left side of the equation, we finally have

$$\frac{dq}{dt} + \frac{1}{2}q = 10 + 5\sin 2t. \tag{22}$$

Originally, there is no chemical in the pond, so the initial condition is

$$q(0) = 0. \tag{23}$$

Equation (22) is linear, and although the right side is a function of time, the coefficient of q is a constant. Thus the integrating factor is $e^{t/2}$. Multiplying Eq. (22) by this factor and integrating the resulting equation, we obtain the general solution

$$q(t) = 20 - \frac{40}{17}\cos 2t + \frac{10}{17}\sin 2t + ce^{-t/2}. \tag{24}$$

The initial condition (23) requires that $c = -300/17$, so the solution of the initial value problem (22), (23) is

$$q(t) = 20 - \frac{40}{17}\cos 2t + \frac{10}{17}\sin 2t - \frac{300}{17}e^{-t/2}. \tag{25}$$

A plot of the solution (25) is shown in Figure 2.3.3, along with the line $q = 20$. The exponential term in the solution is important for small t, but it diminishes rapidly as t increases. Later, the solution consists of an oscillation, due to the $\sin 2t$ and $\cos 2t$ terms, about the constant level $q = 20$. Note that if the $\sin 2t$ term were not present in Eq. (22), then $q = 20$ would be the equilibrium solution of that equation.

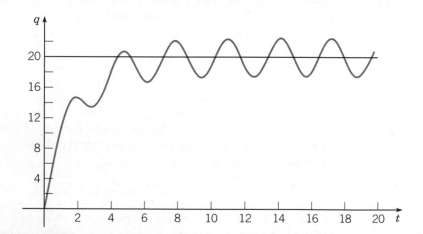

FIGURE 2.3.3 Solution of the initial value problem (22), (23).

Let us now consider the adequacy of the mathematical model itself for this problem. The model rests on several assumptions that have not yet been stated explicitly. In the first place, the amount of water in the pond is controlled entirely by the rates of flow in and out—none is lost by evaporation or by seepage into the ground, and none is gained by rainfall. The same is true of the chemical; it flows into and out of the pond, but none is absorbed by fish or other organisms living in the pond. In addition, we assume that the concentration of chemical in the pond is uniform throughout the pond. Whether the results obtained from the model are accurate depends strongly on the validity of these simplifying assumptions.

EXAMPLE 4

Escape Velocity

A body of constant mass m is projected away from the earth in a direction perpendicular to the earth's surface with an initial velocity v_0. Assuming that there is no air resistance, but taking into account the variation of the earth's gravitational field with distance, find an expression for the velocity during the ensuing motion. Also find the initial velocity that is required to lift the body to a given maximum altitude ξ above the surface of the earth, and find the least initial velocity for which the body will not return to the earth; the latter is the **escape velocity**.

Let the positive x-axis point away from the center of the earth along the line of motion, with $x = 0$ lying on the earth's surface; see Figure 2.3.4. The figure is drawn horizontally to remind you that gravity is directed toward the center of the earth, which is not necessarily downward from a perspective away from the earth's surface. The gravitational force acting on the body (i.e., its weight) is inversely proportional to the square of the distance from the center of the earth and is given by $w(x) = -k/(x + R)^2$, where k is a constant, R is the radius of the earth, and the minus sign signifies that $w(x)$ is directed in the negative x direction. We know that on the earth's surface $w(0)$ is given by $-mg$, where g is the acceleration due to gravity at sea level. Therefore $k = mgR^2$ and

$$w(x) = -\frac{mgR^2}{(R + x)^2}.\tag{26}$$

Since there are no other forces acting on the body, the equation of motion is

$$m\frac{dv}{dt} = -\frac{mgR^2}{(R + x)^2},\tag{27}$$

and the initial condition is

$$v(0) = v_0.\tag{28}$$

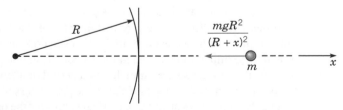

FIGURE 2.3.4 A body in the earth's gravitational field.

Unfortunately, Eq. (27) involves too many variables because it depends on t, x, and v. To remedy this situation, we can eliminate t from Eq. (27) by thinking of x, rather than t, as the independent variable. Thus we must express dv/dt in terms of dv/dx by the chain rule;

hence

$$\frac{dv}{dt} = \frac{dv}{dx}\frac{dx}{dt} = v\frac{dv}{dx},$$

and Eq. (27) is replaced by

$$v\frac{dv}{dx} = -\frac{gR^2}{(R+x)^2}. \tag{29}$$

Equation (29) is separable but not linear, so by separating the variables and integrating, we obtain

$$\frac{v^2}{2} = \frac{gR^2}{R+x} + c. \tag{30}$$

Since $x = 0$ when $t = 0$, the initial condition (28) at $t = 0$ can be replaced by the condition that $v = v_0$ when $x = 0$. Hence $c = (v_0^2/2) - gR$ and

$$v = \pm\sqrt{v_0^2 - 2gR + \frac{2gR^2}{R+x}}. \tag{31}$$

Note that Eq. (31) gives the velocity as a function of altitude rather than as a function of time. The plus sign must be chosen if the body is rising, and the minus sign if it is falling back to earth.

To determine the maximum altitude ξ that the body reaches, we set $v = 0$ and $x = \xi$ in Eq. (31) and then solve for ξ, obtaining

$$\xi = \frac{v_0^2 R}{2gR - v_0^2}. \tag{32}$$

Solving Eq. (32) for v_0, we find the initial velocity required to lift the body to the altitude ξ, namely,

$$v_0 = \sqrt{2gR\frac{\xi}{R+\xi}}. \tag{33}$$

The escape velocity v_e is then found by letting $\xi \to \infty$. Consequently,

$$v_e = \sqrt{2gR}. \tag{34}$$

The numerical value of v_e is approximately 6.9 miles (mi)/s, or 11.2 km/s.

The preceding calculation of the escape velocity neglects the effect of air resistance, so the actual escape velocity (including the effect of air resistance) is somewhat higher. On the other hand, the effective escape velocity can be significantly reduced if the body is transported a considerable distance above sea level before being launched. Both gravitational and frictional forces are thereby reduced; air resistance, in particular, diminishes quite rapidly with increasing altitude. You should keep in mind also that it may well be impractical to impart too large an initial velocity instantaneously; space vehicles, for instance, receive their initial acceleration during a period of a few minutes. Our result for the escape velocity, in terms of its dependence on the parameters of the problem, provides more insight than a result that uses specific numerical values for the parameters. For instance, (34) shows clearly that the escape velocity does not depend on the mass m of the body. Furthermore, on the moon, with a radius of about 0.27 R and with a gravitational constant of about 0.17 g, Eq. (34) reveals that the escape velocity from the moon is 0.214 times that on Earth, or about 2.4 km/s.

PROBLEMS

1. Consider a tank used in certain hydrodynamic experiments. After one experiment the tank contains 150 liter (L) of a dye solution with a concentration of 3 g/L. To prepare for the next experiment, the tank is to be rinsed with fresh water flowing in at a rate of 3 L/min, the well-stirred solution flowing out at the same rate. Find the time that will elapse before the concentration of dye in the tank reaches 2% of its original value.

2. A tank initially contains 200 L of pure water. A mixture containing a concentration of γ g/L of salt enters the tank at a rate of 4 L/min, and the well-stirred mixture leaves the tank at the same rate. Find an expression in terms of γ for the amount of salt in the tank at any time t. Also find the limiting amount of salt in the tank as $t \to \infty$.

3. A tank originally contains 160 gal of fresh water. Then water containing $\frac{1}{4}$ lb of salt per gallon is poured into the tank at a rate of 4 gal/min, and the mixture is allowed to leave at the same rate. After 8 min the process is stopped, and fresh water is poured into the tank at a rate of 6 gal/min, with the mixture again leaving at the same rate. Find the amount of salt in the tank at the end of an additional 8 min.

4. A tank with a capacity of 500 gal originally contains 200 gal of water with 100 lb of salt in solution. Water containing 1 lb of salt per gallon is entering at a rate of 3 gal/min, and the mixture is allowed to flow out of the tank at a rate of 2 gal/min. Find the amount of salt in the tank at any time prior to the instant when the solution begins to overflow. Find the concentration (in pounds per gallon) of salt in the tank when it is on the point of overflowing. Compare this concentration with the theoretical limiting concentration if the tank had infinite capacity.

5. A tank contains 100 gal of water and 50 oz of salt. Water containing a salt concentration of $\frac{1}{4}(1 + \frac{1}{2} \sin t)$ oz/gal flows into the tank at a rate of 2 gal/min, and the mixture in the tank flows out at the same rate.

(a) Find the amount of salt in the tank at any time.

(b) Plot the solution for a time period long enough so that you see the ultimate behavior of the graph.

(c) The long-time behavior of the solution is an oscillation about a certain constant level. What is this level? What is the amplitude of the oscillation?

6. Suppose that a tank containing a certain liquid has an outlet near the bottom. Let $h(t)$ be the height of the liquid surface above the outlet at time t. Torricelli's principle states that the outflow velocity v at the outlet is equal to the velocity of a particle falling freely (with no drag) from the height h.

(a) Show that $v = \sqrt{2gh}$, where g is the acceleration due to gravity.

(b) By equating the rate of outflow to the rate of change of liquid in the tank, show that $h(t)$ satisfies the equation

$$A(h)\frac{dh}{dt} = -\alpha a\sqrt{2gh}, \tag{i}$$

where $A(h)$ is the area of the cross section of the tank at height h and a is the area of the outlet. The constant α is a contraction coefficient that accounts for the observed fact that the cross section of the (smooth) outflow stream is smaller than a. The value of α for water is about 0.6.

(c) Consider a water tank in the form of a right circular cylinder that is 3 m high above the outlet. The radius of the tank is 1 m and the radius of the circular outlet is 0.1 m. If the tank is initially full of water, determine how long it takes to drain the tank down to the level of the outlet.

7. An outdoor swimming pool loses 0.05% of its water volume every day it is in use, due to losses from evaporation and from excited swimmers who splash water. A system is available to continually replace water at a rate of G gallons per day of use.

(a) Find an expression, in terms of G, for the equilibrium volume of the pool. Sketch a few graphs for the volume $V(t)$, including all possible types of solutions.

(b) If the pool volume is initially 1% above its equilibrium value, find an expression for $V(t)$.

(c) What is the replacement rate G required to maintain 12,000 gal of water in the pool?

8. Cholesterol is produced by the body for the construction of cell walls, and is also absorbed from certain foods. The blood cholesterol level is measured in units of milligrams per deciliter, or mg/dl. The net cholesterol production or destruction by the body is modeled by a rate r per day, times the difference between the body's "natural" cholesterol level (a constant) and the actual cholesterol level at any time t. The rate of absorption from food is estimated as a constant k in milligrams per deciliter per day.

(a) A person's cholesterol level at the start of a testing period is 150 mg/dl. Find an expression for the cholesterol level at any subsequent time t. If the rate r is 0.10 per day and the natural level is 100 mg/dl, find the cholesterol level of the person 10 days after the start of the testing period, in terms of k.

(b) If $k = 25$, what is the cholesterol level of this person after a long time?

(c) Suppose this person starts a low-cholesterol diet. What must the value of k be so that the long-time cholesterol level is 180 mg/dl?

9. Imagine a medieval world. In this world a Queen wants to poison a King, who has a wine keg with 500 L of his favorite wine. The Queen gives a conspirator a liquid containing 5 g/L of poison, which must be poured slowly into the keg at a rate

of 0.5 L/min. The poisoner must also remove the well-stirred mixture at the same rate, so that the keg is not suspiciously full.

(a) Find a formula for the amount of poison in the keg at any time, measured from the start of the pouring by the poisoner.
(b) A plot is hatched for the King to drink wine from the keg while he is on a hunt, where he will become so addled that his prey will surely kill him. The poisoner must pour for a time T, when the poison in the keg reaches a dangerous concentration of 0.005 g/L. Find T.
(c) The Lord High Inquisitor of the Realm never learned about differential equations. Nonetheless, knowing the basic numbers (keg size, poison concentrations, etc.), he can produce an estimate for the time T that the poisoner was at the keg. In fact, his estimate is within 2% of the exact value found in part (b). What is the Lord's estimate? In the context of differential equations, why is it so close to the exact value obtained from the solution?

10. Suppose an amount S_0 is invested at an annual rate of return r percent, compounded continuously.

(a) Find the number of years T that are required for the original amount to double in value, as a function of r.
(b) Determine T if $r = 8\%$.
(c) Find the annual percentage rate that is needed for the original investment to double in 8 years.
(d) A rough guideline, known as early as the 15th century, is the "Rule of 72": an investment doubles when rT is about 72. How accurate is it for the examples in (b) and (c)? Explain the basis for the "Rule of 72." Why do you suppose the number 72 was chosen long ago?

11. A young person with no initial capital invests k dollars per year at an annual rate of return r. Assume that investments are made continuously and that the return is compounded continuously.

(a) Determine the sum $S(t)$ accumulated at any time t.
(b) If $r = 5.5\%$, determine k so that $1 million will be available for retirement in 42 years.
(c) If $k = \$4,000$/year, determine the return rate r that must be obtained to have $1 million available in 42 years.

12. A homebuyer can afford to spend no more than $800/month on mortgage payments. Suppose that the interest rate is 9% and that the term of the mortgage is 20 years. Assume that interest is compounded continuously and that payments are also made continuously.

(a) Determine the maximum amount that this buyer can afford to borrow.
(b) Determine the total interest paid during the term of the mortgage.

13. A recent college graduate borrows $100,000 at an interest rate of 9% to purchase a condominium. Anticipating

steady salary increases, the buyer expects to make payments at a monthly rate of $800(1 + t/120)$, where t is the number of months since the loan was made.

(a) Assuming that this payment schedule can be maintained, when will the loan be fully paid?
(b) Assuming the same payment schedule, how large a loan could be paid off in exactly 20 years?

14. A Difference Equation. In this problem, we approach the loan problem in Example 2 from a discrete viewpoint. This leads to a difference equation rather than a differential equation.

(a) Let S_0 be the initial balance of the loan, and let S_n be the balance after n months. Show that

$$S_n = (1 + r)S_{n-1} - k, \quad n = 1, 2, 3, \ldots, \tag{i}$$

where r is the monthly interest rate and k is the monthly payment. In Example 2, the annual interest rate is 8%, so here we take $r = 0.08/12$.
(b) Let $R = 1 + r$, so that Eq. (i) becomes

$$S_n = RS_{n-1} - k, \quad n = 1, 2, 3, \ldots. \tag{ii}$$

Find S_1, S_2, and S_3.
(c) Use an induction argument to show that

$$S_n = R^n S_0 - \frac{R^n - 1}{R - 1} k \tag{iii}$$

for each positive integer n.
(d) Let $S_0 = 20{,}000$ and suppose that (as in Example 2) the loan is to be paid off in 48 months. Find the value of k and compare it with the result of Example 2.

15. An important tool in archeological research is radiocarbon dating, developed by the American chemist Willard F. Libby. This is a means of determining the age of certain wood and plant remains, hence of animal or human bones or artifacts found buried at the same levels. Radiocarbon dating is based on the fact that some wood or plant remains contain residual amounts of carbon-14, a radioactive isotope of carbon. This isotope is accumulated during the lifetime of the plant and begins to decay at its death. Since the half-life of carbon-14 is long (approximately 5,730 years[1]), measurable amounts of carbon-14 remain after many thousands of years. If even a tiny fraction of the original amount of carbon-14 is still present, then by appropriate laboratory measurements the *proportion* of the original amount of carbon-14 that remains can be accurately determined. In other words, if $Q(t)$ is the amount of carbon-14 at time t and Q_0 is the original amount, then the ratio $Q(t)/Q_0$ can be determined, at least if this quantity is not too small. Present measurement techniques permit the use of this method for time periods of 50,000 years or more.

(a) Assuming that Q satisfies the differential equation $Q' = -rQ$, determine the decay constant r for carbon-14.

[1] *McGraw-Hill Encyclopedia of Science and Technology*, 8th ed. (New York: McGraw-Hill, 1997), Vol. 5, p. 48.

(b) Find an expression for $Q(t)$ at any time t, if $Q(0) = Q_0$.

(c) Suppose that certain remains are discovered in which the current residual amount of carbon-14 is 50% of the original amount. Determine the age of these remains.

16. The population of mosquitoes in a certain area increases at a rate proportional to the current population, and in the absence of other factors, the population doubles each week. There are 800,000 mosquitoes in the area initially, and predators (birds, bats, etc.) eat 30,000 mosquitoes/day. Determine the population of mosquitoes in the area at any time.

 17. Suppose that a certain population has a growth rate that varies with time and that this population satisfies the differential equation

$$\frac{dy}{dt} = \frac{(0.5 + \sin t)y}{5}.$$

(a) If $y(0) = 1$, find (or estimate) the time τ at which the population has doubled. Choose other initial conditions and determine whether the doubling time τ depends on the initial population.

(b) Suppose that the growth rate is replaced by its average value $\frac{1}{10}$. Determine the doubling time τ in this case.

(c) Suppose that the term $\sin t$ in the differential equation is replaced by $\sin 2\pi t$; that is, the variation in the growth rate has a substantially higher frequency. What effect does this have on the doubling time τ?

(d) Plot the solutions obtained in parts (a), (b), and (c) on a single set of axes.

 18. Suppose that a certain population satisfies the initial value problem

$$\frac{dy}{dt} = r(t)y - k, \qquad y(0) = y_0,$$

where the growth rate $r(t)$ is given by $r(t) = (1 + \sin t)/5$, and k represents the rate of predation.

(a) Suppose that $k = \frac{1}{5}$. Plot y versus t for several values of y_0 between $\frac{1}{2}$ and 1.

(b) Estimate the critical initial population y_c below which the population will become extinct.

(c) Choose other values of k and find the corresponding y_c for each one.

(d) Use the data you have found in parts (b) and (c) to plot y_c versus k.

19. Newton's law of cooling states that the temperature of an object changes at a rate proportional to the difference between its temperature and that of its surroundings. Suppose that the temperature of a cup of coffee obeys Newton's law of cooling. If the coffee has a temperature of 200°F when freshly poured, and 1 min later has cooled to 190°F in a room at 70°F, determine when the coffee reaches a temperature of 150°F.

20. Heat transfer from a body to its surroundings by radiation, based on the Stefan–Boltzmann law, is described by the differential equation

$$\frac{du}{dt} = -\alpha(u^4 - T^4), \qquad \text{(i)}$$

where $u(t)$ is the absolute temperature of the body at time t, T is the absolute temperature of the surroundings, and α is a constant depending on the physical parameters of the body. However, if u is much larger than T, then solutions of Eq. (i) are well approximated by solutions of the simpler equation

$$\frac{du}{dt} = -\alpha u^4. \qquad \text{(ii)}$$

Suppose that a body with initial temperature 2000 K is surrounded by a medium with temperature 300 K and that $\alpha = 2.0 \times 10^{-12}$ K^{-3}/s.

(a) Determine the temperature of the body at any time by solving Eq. (ii).

(b) Plot the graph of u versus t.

(c) Find the time τ at which $u(\tau) = 600$, that is, twice the ambient temperature. Up to this time, the error in using Eq. (ii) to approximate the solutions of Eq. (i) is no more than 1%.

21. Consider a lake of constant volume V containing at time t an amount $Q(t)$ of pollutant, evenly distributed throughout the lake with a concentration $c(t)$, where $c(t) = Q(t)/V$. Assume that water containing a concentration k of pollutant enters the lake at a rate r, and that water leaves the lake at the same rate. Suppose that pollutants are also added directly to the lake at a constant rate P. Note that the given assumptions neglect a number of factors that may, in some cases, be important—for example, the water added or lost by precipitation, absorption, and evaporation; the stratifying effect of temperature differences in a deep lake; the tendency of irregularities in the coastline to produce sheltered bays; and the fact that pollutants are not deposited evenly throughout the lake but (usually) at isolated points around its periphery. The results below must be interpreted in the light of the neglect of such factors as these.

(a) If at time $t = 0$ the concentration of pollutant is c_0, find an expression for the concentration $c(t)$ at any time. What is the limiting concentration as $t \to \infty$?

(b) If the addition of pollutants to the lake is terminated ($k = 0$ and $P = 0$ for $t > 0$), determine the time interval T that must elapse before the concentration of pollutants is reduced to 50% of its original value; to 10% of its original value.

(c) Table 2.3.1 contains data[2] for several of the Great Lakes. Using these data, determine from part (b) the time T necessary to reduce the contamination of each of these lakes to 10% of the original value.

---

[2]This problem is based on R. H. Rainey, "Natural Displacement of Pollution from the Great Lakes," *Science* *155* (1967), pp. 1242–1243; the information in the table was taken from that source.

TABLE 2.3.1 Volume and flow data for the Great Lakes.

Lake	V (km$^3 \times 10^3$)	r (km^3/year)
Superior	12.2	65.2
Michigan	4.9	158
Erie	0.46	175
Ontario	1.6	209

 22. A ball with mass 0.25 kg is thrown upward with initial velocity 24 m/s from the roof of a building 26 m high. Neglect air resistance.

(a) Find the maximum height above the ground that the ball reaches.
(b) Assuming that the ball misses the building on the way down, find the time that it hits the ground.
(c) Plot the graphs of velocity and position versus time.

 23. Assume that conditions are as in Problem 22 except that there is a force due to air resistance of $|v|/30$, where the velocity v is measured in meters per second.

(a) Find the maximum height above the ground that the ball reaches.
(b) Find the time that the ball hits the ground.
(c) Plot the graphs of velocity and position versus time. Compare these graphs with the corresponding ones in Problem 22.

 24. Assume that conditions are as in Problem 22 except that there is a force due to air resistance of $v_2/1325$, where the velocity v_2 is measured in meters per second.

(a) Find the maximum height above the ground that the ball reaches.
(b) Find the time that the ball hits the ground.
(c) Plot the graphs of velocity and position versus time. Compare these graphs with the corresponding ones in Problems 22 and 23.

 25. A skydiver weighing 180 lb (including equipment) falls vertically downward from an altitude of 5,000 ft and opens the parachute after 10 s of free fall. Assume that the force of air resistance is $0.75|v|$ when the parachute is closed and $12|v|$ when the parachute is open, where the velocity v is measured in feet per second.

(a) Find the speed of the skydiver when the parachute opens.
(b) Find the distance fallen before the parachute opens.
(c) What is the limiting velocity v_L after the parachute opens?
(d) Determine how long the skydiver is in the air after the parachute opens.
(e) Plot the graph of velocity versus time from the beginning of the fall until the skydiver reaches the ground.

26. A rocket sled having an initial speed of 160 mi/h is slowed by a channel of water. Assume that, during the braking process, the acceleration a is given by $a(v) = -\mu v^2$, where v is the velocity and μ is a constant.

(a) As in Example 4 in the text, use the relation $dv/dt = v(dv/dx)$ to write the equation of motion in terms of v and x.
(b) If it requires a distance of 2200 ft to slow the sled to 16 mi/h, determine the value of μ.
(c) Find the time τ required to slow the sled to 16 mi/h.

27. A body of constant mass m is projected vertically upward with an initial velocity v_0 in a medium offering a resistance $k|v|$, where k is a constant. Neglect changes in the gravitational force.

(a) Find the maximum height x_m attained by the body and the time t_m at which this maximum height is reached.
(b) Show that if $kv_0/mg < 1$, then t_m and x_m can be expressed as

$$t_m = \frac{v_0}{g}\left[1 - \frac{1}{2}\frac{kv_0}{mg} + \frac{1}{3}\left(\frac{kv_0}{mg}\right)^2 - \cdots\right],$$

$$x_m = \frac{v_0^2}{2g}\left[1 - \frac{2}{3}\frac{kv_0}{mg} + \frac{1}{2}\left(\frac{kv_0}{mg}\right)^2 - \cdots\right].$$

(c) Show that the quantity kv_0/mg is dimensionless.

28. A body of mass m is projected vertically upward with an initial velocity v_0 in a medium offering a resistance $k|v|$, where k is a constant. Assume that the gravitational attraction of the earth is constant.

(a) Find the velocity $v(t)$ of the body at any time.
(b) Use the result of part (a) to calculate the limit of $v(t)$ as $k \to 0$, that is, as the resistance approaches zero. Does this result agree with the velocity of a mass m projected upward with an initial velocity v_0 in a vacuum?
(c) Use the result of part (a) to calculate the limit of $v(t)$ as $m \to 0$, that is, as the mass approaches zero.

29. A body falling in a relatively dense fluid, oil for example, is acted on by three forces (see Figure 2.3.5): a resistive force R, a buoyant force B, and its weight w due to gravity. The buoyant force is equal to the weight of the fluid displaced by the object. For a slowly moving spherical body of radius a, the resistive force is given by Stokes's law, $R = 6\pi\,\mu a|v|$, where v is the velocity of the body, and μ is the coefficient of viscosity of the surrounding fluid.

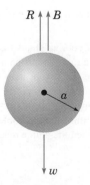

FIGURE 2.3.5 A body falling in a dense fluid.

(a) Find the limiting velocity of a solid sphere of radius a and density ρ falling freely in a medium of density ρ' and coefficient of viscosity μ.

(b) In 1910 R. A. Millikan studied the motion of tiny droplets of oil falling in an electric field. A field of strength E exerts a force Ee on a droplet with charge e. Assume that E has been adjusted, so the droplet is held stationary ($v = 0$), and that w and B are as given above. Find an expression for e. Millikan repeated this experiment many times, and from the data that he gathered he was able to deduce the charge on an electron.

30. A mass of 0.40 kg is dropped from rest in a medium offering a resistance of $0.2|v|$, where v is measured in meters per second.

(a) If the mass is dropped from a height of 25 m, find its velocity when it hits the ground.

(b) If the mass is to attain a velocity of no more than 8 m/s, find the maximum height from which it can be dropped.

(c) Suppose that the resistive force is $k|v|$, where v is measured in meters per second and k is a constant. If the mass is dropped from a height of 25 m and must hit the ground with a velocity of no more than 8 m/s, determine the coefficient of resistance k that is required.

31. Suppose that a rocket is launched straight up from the surface of the earth with initial velocity $v_0 = \sqrt{2gR}$, where R is the radius of the earth. Neglect air resistance.

(a) Find an expression for the velocity v in terms of the distance x from the surface of the earth.

(b) Find the time required for the rocket to go 240,000 mi (the approximate distance from the earth to the moon). Assume that $R = 4{,}000$ mi.

32. Let $v(t)$ and $w(t)$, respectively, be the horizontal and vertical components of the velocity of a batted (or thrown) baseball. In the absence of air resistance, v and w satisfy the equations

$$dv/dt = 0, \qquad dw/dt = -g.$$

(a) Show that

$$v = u \cos A, \qquad w = -gt + u \sin A,$$

where u is the initial speed of the ball and A is its initial angle of elevation.

(b) Let $x(t)$ and $y(t)$, respectively, be the horizontal and vertical coordinates of the ball at time t. If $x(0) = 0$ and $y(0) = h$, find $x(t)$ and $y(t)$ at any time t.

(c) Let $g = 32$ ft/s^2, $u = 125$ ft/s, and $h = 3$ ft. Plot the trajectory of the ball for several values of the angle A; that is, plot $x(t)$ and $y(t)$ parametrically.

(d) Suppose the outfield wall is at a distance L and has height H. Find a relation between u and A that must be satisfied if the ball is to clear the wall.

(e) Suppose that $L = 350$ ft and $H = 10$ ft. Using the relation in part (d), find (or estimate from a plot) the range of values of A that corresponds to an initial velocity of $u = 110$ ft/s.

(f) For $L = 350$ ft and $H = 10$ ft, find the minimum initial velocity u and the corresponding optimal angle A for which the ball will clear the wall.

33. A more realistic model (than that in Problem 32) of a baseball in flight includes the effect of air resistance. In this case, the equations of motion are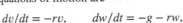

$$dv/dt = -rv, \qquad dw/dt = -g - rw,$$

where r is the coefficient of resistance.

(a) Determine $v(t)$ and $w(t)$ in terms of initial speed u and initial angle of elevation A.

(b) Find $x(t)$ and $y(t)$ if $x(0) = 0$ and $y(0) = h$.

(c) Plot the trajectory of the ball for $r = \frac{1}{5}$, $u = 125$, $h = 3$, and for several values of A. How do the trajectories differ from those in Problem 32 with $r = 0$?

(d) Assuming that $r = \frac{1}{5}$ and $h = 3$, find the minimum initial velocity u and the optimal angle A for which the ball will clear a wall that is 350 ft distant and 10 ft high. Compare this result with that in Problem 32(f).

34. Brachistochrone Problem. One of the famous problems in the history of mathematics is the brachistochrone[3] problem: to find the curve along which a particle will slide without friction in the minimum time from one given point P to another Q, the second point being lower than the first but not directly beneath it (see Figure 2.3.6). This problem was posed by Johann Bernoulli in 1696 as a challenge problem to the mathematicians of his day. Correct solutions were found by Johann Bernoulli and his brother Jakob Bernoulli and by Isaac Newton, Gottfried Leibniz, and the Marquis de L'Hôpital. The brachistochrone problem is important in the development of mathematics as one of the forerunners of the calculus of variations.

In solving this problem, it is convenient to take the origin as the upper point P and to orient the axes as shown in Figure 2.3.6. The lower point Q has coordinates (x_0, y_0). It is then

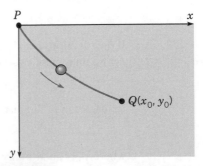

FIGURE 2.3.6 The brachistochrone.

---

[3]The word "brachistochrone" comes from the Greek words *brachistos*, meaning shortest, and *chronos*, meaning time.

possible to show that the curve of minimum time is given by a function $y = \phi(x)$ that satisfies the differential equation

$$(1 + y'^2)y = k^2, \tag{i}$$

where k^2 is a certain positive constant to be determined later.

(a) Solve Eq. (i) for y'. Why is it necessary to choose the positive square root?

(b) Introduce the new variable t by the relation

$$y = k^2 \sin^2 t. \tag{ii}$$

Show that the equation found in part (a) then takes the form

$$2k^2 \sin^2 t \, dt = dx. \tag{iii}$$

(c) Letting $\theta = 2t$, show that the solution of Eq. (iii) for which $x = 0$ when $y = 0$ is given by

$$\begin{aligned} x &= k^2(\theta - \sin\theta)/2, \\ y &= k^2(1 - \cos\theta)/2. \end{aligned} \tag{iv}$$

Equations (iv) are parametric equations of the solution of Eq. (i) that passes through (0, 0). The graph of Eqs. (iv) is called a **cycloid**.

(d) If we make a proper choice of the constant k, then the cycloid also passes through the point (x_0, y_0) and is the solution of the brachistochrone problem. Find k if $x_0 = 1$ and $y_0 = 2$.

2.4 Differences Between Linear and Nonlinear Equations

Up to now, we have been primarily concerned with showing that first order differential equations can be used to investigate many different kinds of problems in the natural sciences, and with presenting methods of solving such equations if they are either linear or separable. Now it is time to turn our attention to some more general questions about differential equations and to explore, in more detail, some important ways in which nonlinear equations differ from linear ones.

▶ **Existence and Uniqueness of Solutions.** So far, we have discussed a number of initial value problems, each of which had a solution and apparently only one solution. This raises the question of whether this is true of all initial value problems for first order equations. In other words, does every initial value problem have exactly one solution? This may be an important question even for nonmathematicians. If you encounter an initial value problem in the course of investigating some physical problem, you might want to know that it has a solution before spending very much time and effort in trying to find it. Further, if you are successful in finding one solution, you might be interested in knowing whether you should continue a search for other possible solutions or whether you can be sure that there are no other solutions. For linear equations, the answers to these questions are given by the following fundamental theorem.

THEOREM 2.4.1	If the functions p and g are continuous on an open interval $I = (\alpha, \beta)$ containing the point $t = t_0$, then there exists a unique function $y = \phi(t)$ that satisfies the differential equation $$y' + p(t)y = g(t) \tag{1}$$ for each t in I, and that also satisfies the initial condition $$y(t_0) = y_0, \tag{2}$$ where y_0 is an arbitrary prescribed initial value.

Observe that Theorem 2.4.1 states that the given initial value problem *has* a solution and also that the problem has *only one* solution. In other words, the theorem asserts both the *existence* and *uniqueness* of the solution of the initial value problem (1), (2).

Remarks.

1. Theorem 2.4.1 states that the solution exists throughout any interval I containing the initial point t_0 in which the coefficients p and g are continuous. That is, the solution can be discontinuous or fail to exist only at points where at least one of p and g is discontinuous. Such points can often be identified at a glance.

2. The interval I need not have finite length. Specifically, α could be $-\infty$ and β could be ∞.

Outline of Proof: The proof of this theorem is partly contained in the discussion in Section 2.2 leading to the formula [Eq. (27) in Section 2.2]

$$\mu(t)y = \int \mu(t)g(t)\,dt + c, \tag{3}$$

where [Eq. (25) in Section 2.2]

$$\mu(t) = \exp \int p(t)\,dt. \tag{4}$$

The derivation in Section 2.2 shows that if Eq. (1) has a solution, then it must be given by Eq. (3). By looking a little more closely at that derivation, we can also conclude that the differential equation (1) must indeed have a solution. Since p is continuous for $\alpha < t < \beta$, it follows that μ is defined in this interval and is a nonzero differentiable function. Upon multiplying Eq. (1) by $\mu(t)$, we obtain

$$[\mu(t)y]' = \mu(t)g(t). \tag{5}$$

Since both μ and g are continuous, the function μg is integrable, and Eq. (3) follows from Eq. (5). Further the integral of μg is differentiable, so y as given by Eq. (3) exists and is differentiable throughout the interval $\alpha < t < \beta$. By substituting the expression for y from Eq. (3) into either Eq. (1) or Eq. (5), you can easily verify that this expression satisfies the differential equation throughout the interval $\alpha < t < \beta$. Finally, the initial condition (2) determines the constant c uniquely, so there is only one solution of the initial value problem, thus completing the proof.

Equation (4) determines the integrating factor $\mu(t)$ only up to a multiplicative factor that depends on the lower limit of integration. If we choose this lower limit to be t_0, then

$$\mu(t) = \exp \int_{t_0}^{t} p(s)\,ds, \tag{6}$$

and it follows that $\mu(t_0) = 1$. Using the integrating factor given by Eq. (6), and choosing the lower limit of integration in Eq. (3) also to be t_0, we obtain the general solution of Eq. (1) in the form

$$y = \frac{1}{\mu(t)}\left[\int_{t_0}^{t} \mu(s)g(s)\,ds + c\right]. \tag{7}$$

To satisfy the initial condition (2), we must choose $c = y_0$. Thus the solution of the initial value problem (1), (2) is

$$y = \frac{1}{\mu(t)}\left[\int_{t_0}^{t} \mu(s)g(s)\,ds + y_0\right], \tag{8}$$

where $\mu(t)$ is given by Eq. (6).

Turning now to nonlinear differential equations, we must replace Theorem 2.4.1 by a more general theorem, such as the following.

THEOREM
2.4.2

Let the functions f and $\partial f/\partial y$ be continuous in some rectangle $\alpha < t < \beta$, $\gamma < y < \delta$ containing the point (t_0, y_0). Then, in some interval $t_0 - h < t < t_0 + h$ contained in $\alpha < t < \beta$, there is a unique solution $y = \phi(t)$ of the initial value problem

$$y' = f(t, y), \qquad y(t_0) = y_0. \tag{9}$$

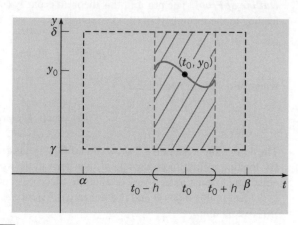

FIGURE 2.4.1 Illustration of rectangular region in Theorem 2.4.2.

Observe that the hypotheses in Theorem 2.4.2 reduce to those in Theorem 2.4.1 if the differential equation is linear. For then $f(t, y) = -p(t)y + g(t)$ and $\partial f(t, y)/\partial y = -p(t)$, so the continuity of f and $\partial f/\partial y$ is equivalent to the continuity of p and g in this case. The proof of Theorem 2.4.1 was comparatively simple because it could be based on the expression (3) that gives the solution of an arbitrary linear equation. There is no corresponding expression for the solution of the differential equation (9), so the proof of Theorem 2.4.2 is much more difficult. It is discussed in more advanced books on differential equations.

Remarks.

1. The conditions stated in Theorem 2.4.2 are sufficient to guarantee the existence of a unique solution of the initial value problem (9) in some interval $t_0 - h < t < t_0 + h$, as illustrated in Figure 2.4.1 but they are not necessary. That is, the conclusion remains true under slightly weaker hypotheses about the function f. In fact, the existence of a solution (but not its uniqueness) can be established on the basis of the continuity of f alone.

2. An important geometrical consequence of the uniqueness parts of Theorems 2.4.1 and 2.4.2 is that the graphs of two solutions cannot intersect each other. Otherwise, there would be two solutions that satisfy the initial condition corresponding to the point of intersection, in violation of Theorem 2.4.1 or 2.4.2.

We now consider some examples.

EXAMPLE
1

Use Theorem 2.4.1 to find an interval in which the initial value problem

$$ty' + 2y = 4t^2, \tag{10}$$

$$y(1) = 2 \tag{11}$$

has a unique solution.

Rewriting Eq. (10) in the standard form (1), we have, assuming $t \neq 0$,

$$y' + \frac{2}{t}y = 4t,$$

so $p(t) = 2/t$ and $g(t) = 4t$. Thus, for this equation, g is continuous for all t, while p is continuous only for $t < 0$ or for $t > 0$. The interval $t > 0$ contains the initial point; consequently, Theorem 2.4.1 guarantees that the problem (10), (11) has a unique solution on the interval $0 < t < \infty$. In Example 2 of Section 2.2, we found the solution of this initial value problem to be

$$y = t^2 + \frac{1}{t^2}, \qquad t > 0. \tag{12}$$

Now suppose that the initial condition (11) is changed to $y(-1) = 2$. Then Theorem 2.4.1 asserts the existence of a unique solution for $t < 0$. As you can readily verify, the solution is again given by Eq. (12), but now on the interval $-\infty < t < 0$.

EXAMPLE 2

Apply Theorem 2.4.2 to the initial value problem

$$\frac{dy}{dt} = \frac{3t^2 + 4t + 2}{2(y - 1)}, \qquad y(0) = -1. \tag{13}$$

Note that Theorem 2.4.1 is not applicable to this problem since the differential equation is nonlinear. To apply Theorem 2.4.2, observe that

$$f(t, y) = \frac{3t^2 + 4t + 2}{2(y - 1)}, \qquad \frac{\partial f}{\partial y}(t, y) = -\frac{3t^2 + 4t + 2}{2(y - 1)^2}.$$

Thus each of these functions is continuous everywhere except on the line $y = 1$. Consequently, a rectangle can be drawn about the initial point $(0, -1)$ in which both f and $\partial f/\partial y$ are continuous. Therefore Theorem 2.4.2 guarantees that the initial value problem has a unique solution in some interval about $t = 0$. However, even though the rectangle can be stretched infinitely far in both the positive and negative t directions, this does not necessarily mean that the solution exists for all t. Indeed, the initial value problem (13) was solved in Example 2 of Section 2.1 and the solution exists only for $t > -2$.

Now suppose we change the initial condition to $y(0) = 1$. The initial point now lies on the line $y = 1$, so no rectangle can be drawn about it within which f and $\partial f/\partial y$ are continuous. Consequently, Theorem 2.4.2 says nothing about possible solutions of this modified problem. However, if we separate the variables and integrate, as in Section 2.1, we find that

$$y^2 - 2y = t^3 + 2t^2 + 2t + c.$$

Further if $t = 0$ and $y = 1$, then $c = -1$. Finally, by solving for y, we obtain

$$y = 1 \pm \sqrt{t^3 + 2t^2 + 2t}. \tag{14}$$

Equation (14) provides two functions that satisfy the given differential equation for $t > 0$ and also satisfy the initial condition $y(0) = 1$. Thus the initial value problem consisting of the differential equation (13) with the initial condition $y(0) = 1$ does not have a unique solution. The two solutions are shown in Figure 2.4.2.

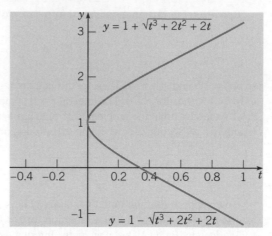

FIGURE 2.4.2　Nonunique solutions of the differential equation (13) with the initial condition $y(0) = 1$.

EXAMPLE
3

Consider the initial value problem

$$y' = y^{1/3}, \qquad y(0) = 0 \tag{15}$$

for $t \geq 0$. Apply Theorem 2.4.2 to this initial value problem and then solve the problem.

The function $f(t, y) = y^{1/3}$ is continuous everywhere, but $\partial f / \partial y$ is not defined when $y = 0$, and hence is not continuous there. Thus Theorem 2.4.2 does not apply to this problem and no conclusion can be drawn from it. However, by the remark following Theorem 2.4.2, the continuity of f does guarantee the existence of solutions, but not their uniqueness.

To understand the situation more clearly, we must actually solve the problem, which is easy to do since the differential equation is separable. Thus we have

$$y^{-1/3} dy = dt,$$

so

$$\tfrac{3}{2} y^{2/3} = t + c$$

and

$$y = \left[\tfrac{2}{3}(t + c) \right]^{3/2}.$$

The initial condition is satisfied if $c = 0$, so

$$y = \phi_1(t) = \left(\tfrac{2}{3} t \right)^{3/2}, \qquad t \geq 0 \tag{16}$$

satisfies both of Eqs. (15). On the other hand, the function

$$y = \phi_2(t) = -\left(\tfrac{2}{3} t \right)^{3/2}, \qquad t \geq 0 \tag{17}$$

is also a solution of the initial value problem. Moreover the function

$$y = \psi(t) = 0, \qquad t \geq 0 \tag{18}$$

is yet another solution. Indeed, it is not hard to show that, for any arbitrary positive t_0, the function

$$y = \chi(t) = \begin{cases} 0, & \text{if } 0 \le t < t_0, \\ \pm \left[\frac{2}{3}(t - t_0) \right]^{3/2}, & \text{if } t \ge t_0 \end{cases} \tag{19}$$

is continuous, differentiable (in particular at $t = t_0$), and is a solution of the initial value problem (15). Hence this problem has an infinite family of solutions; see Figure 2.4.3, where a few of these solutions are shown.

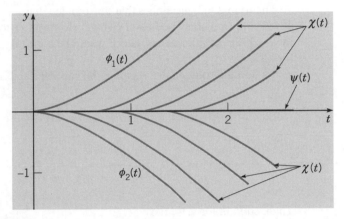

FIGURE 2.4.3 Several solutions of the initial value problem $y' = y^{1/3}$, $y(0) = 0$.

As already noted, the nonuniqueness of the solutions of the problem (15) does not contradict the existence and uniqueness theorem, since the theorem is not applicable if the initial point lies on the t-axis. If (t_0, y_0) is any point not on the t-axis, however, then the theorem guarantees that there is a unique solution of the differential equation $y' = y^{1/3}$ passing through (t_0, y_0).

▶ **Interval of Definition.** According to Theorem 2.4.1, the solution of a linear equation (1),

$$y' + p(t)y = g(t),$$

subject to the initial condition $y(t_0) = y_0$, exists throughout any interval about $t = t_0$ in which the functions p and g are continuous. Thus vertical asymptotes or other discontinuities in the solution can occur only at points of discontinuity of p or g. For instance, the solutions in Example 1 (with one exception) are asymptotic to the y-axis, corresponding to the discontinuity at $t = 0$ in the coefficient $p(t) = 2/t$, but none of the solutions has any other point where it fails to exist and to be differentiable. The one exceptional solution shows that solutions may sometimes remain continuous even at points of discontinuity of the coefficients.

On the other hand, for a nonlinear initial value problem satisfying the hypotheses of Theorem 2.4.2, the interval in which a solution exists may be difficult to determine. The solution $y = \phi(t)$ is certain to exist as long as the point $(t, \phi(t))$ remains within a region in which the hypotheses of Theorem 2.4.2 are satisfied. This is what determines the value of h in that theorem. However, since $\phi(t)$ is usually not known, it may be impossible to locate the point $(t, \phi(t))$ with respect to this region. In any case, the interval in which a solution exists may have no simple relationship to the function f in the differential equation $y' = f(t, y)$. This is illustrated by the following example.

EXAMPLE 4

Solve the initial value problem

$$y' = y^2, \qquad y(0) = 1, \tag{20}$$

and determine the interval in which the solution exists.

Theorem 2.4.2 guarantees that this problem has a unique solution since $f(t, y) = y^2$ and $\partial f / \partial y = 2y$ are continuous everywhere. However Theorem 2.4.2 does not give an interval in which the solution exists, and it would be a mistake to conclude that the solution exists for all t.

To find the solution, we separate the variables and integrate, with the result that

$$y^{-2}\, dy = dt \tag{21}$$

and

$$-y^{-1} = t + c.$$

Then, solving for y, we have

$$y = -\frac{1}{t + c}. \tag{22}$$

To satisfy the initial condition, we must choose $c = -1$, so

$$y = \frac{1}{1 - t} \tag{23}$$

is the solution of the given initial value problem. Clearly, the solution becomes unbounded as $t \to 1$; therefore the solution exists only in the interval $-\infty < t < 1$. There is no indication from the differential equation itself, however, that the point $t = 1$ is in any way remarkable. Moreover, if the initial condition is replaced by

$$y(0) = y_0, \tag{24}$$

then the constant c in Eq. (22) must be chosen to be $c = -1/y_0$, and it follows that

$$y = \frac{y_0}{1 - y_0 t} \tag{25}$$

is the solution of the initial value problem with the initial condition (24). Observe that the solution (25) becomes unbounded as $t \to 1/y_0$, so the interval of existence of the solution is $-\infty < t < 1/y_0$ if $y_0 > 0$, and is $1/y_0 < t < \infty$ if $y_0 < 0$. Figure 2.4.4 shows the solution for $y_0 > 0$. This example illustrates another feature of initial value problems for nonlinear equations; namely, the singularities of the solution may depend in an essential way on the initial conditions as well as the differential equation.

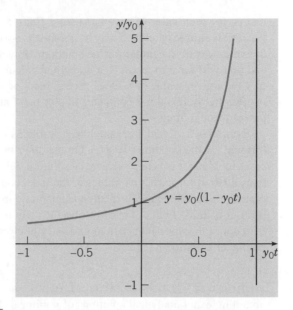

FIGURE 2.4.4 The solution (25) of the initial value problem $y' = y^2, y(0) = y_0 > 0$. Note that we plot y/y_0 versus $y_0 t$; thus the vertical asymptote is at $y_0 t = 1$.

▶ General Solution. Another way in which linear and nonlinear equations differ concerns the concept of a general solution. For a first order linear equation it is possible to obtain a solution containing one arbitrary constant, from which all possible solutions follow by specifying values for this constant. For nonlinear equations this may not be the case; even though a solution containing an arbitrary constant may be found, there may be other solutions that cannot be obtained by giving values to this constant. For instance, for the differential equation $y' = y^2$ in Example 4, the expression in Eq. (22) contains an arbitrary constant, but does not include all solutions of the differential equation. To show this, observe that the function $y = 0$ for all t is certainly a solution of the differential equation, but it cannot be obtained from Eq. (22) by assigning a value to c. In this example, we might anticipate that something of this sort might happen because, to rewrite the original differential equation in the form (21), we must require that y is not zero. However the existence of "additional" solutions is not uncommon for nonlinear equations; a less obvious example is given in Problem 14. Thus we will use the term "general solution" only when discussing linear equations.

▶ Implicit Solutions. Recall again that, for an initial value problem for a first order linear equation, Eq. (8) provides an explicit formula for the solution $y = \phi(t)$. As long as the necessary antiderivatives can be found, the value of the solution at any point can be determined merely by substituting the appropriate value of t into the equation. The situation for nonlinear equations is much less satisfactory. Usually, the best that we can hope for is to find an equation

$$F(t, y) = 0 \tag{26}$$

involving t and y that is satisfied by the solution $y = \phi(t)$. Even this can be done only for differential equations of certain particular types, of which separable equations are the most important. The equation (26) is called an **integral**, or **first integral**, of the differential equation, and (as we have already noted) its graph is an integral curve, or perhaps a family of integral curves. Equation (26), assuming it can be found, defines the solution implicitly;

that is, for each value of t we must solve Eq. (26) to find the corresponding value of y. If Eq. (26) is simple enough, it may be possible to solve it for y by analytical means and thereby obtain an explicit formula for the solution. However more often this will not be possible, and you will have to resort to a numerical calculation to determine an approximate value of y for a given value of t. Once several pairs of values of t and y have been calculated, it is often helpful to use a computer to plot them and then to sketch the integral curve that passes through them.

Examples 2, 3, and 4 are nonlinear problems in which it is easy to solve for an explicit formula for the solution $y = \phi(t)$. On the other hand, Examples 1 and 3 in Section 2.1 are cases in which it is better to leave the solution in implicit form, and to use numerical means to evaluate it for particular values of the independent variable. The latter situation is more typical; unless the implicit relation is quadratic in y, or has some other particularly simple form, it is unlikely that it can be solved exactly by analytical methods. Indeed, more often than not, it is impossible even to find an implicit expression for the solution of a first order nonlinear equation.

▶ Graphical or Numerical Construction of Integral Curves. Because of the difficulty in obtaining exact analytical solutions of nonlinear differential equations, methods that yield approximate solutions or other qualitative information about solutions are of correspondingly greater importance. We have already described, in Section 1.2, how the direction field of a differential equation can be constructed. The direction field can often show the qualitative form of solutions and can also be helpful in identifying regions of the ty-plane where solutions exhibit interesting features that merit more detailed analytical or numerical investigation. Graphical methods for first order equations are discussed further in Section 2.5. A systematic discussion of numerical methods appears in Chapter 8. However, it is not necessary to study the numerical algorithms themselves in order to use effectively one of the many software packages that generate and plot numerical approximations to solutions of initial value problems.

▶ Summary. The linear equation $y' + p(t)y = g(t)$ has several nice properties that can be summarized in the following statements:

1. Assuming that the coefficients are continuous, there is a general solution, containing an arbitrary constant, that includes all solutions of the differential equation. A particular solution that satisfies a given initial condition can be picked out by choosing the proper value for the arbitrary constant.

2. There is an expression for the solution, namely, Eq. (7) or Eq. (8). Moreover, although it involves two integrations, the expression is an explicit one for the solution $y = \phi(t)$ rather than an equation that defines ϕ implicitly.

3. The possible points of discontinuity, or singularities, of the solution can be identified (without solving the problem) merely by finding the points of discontinuity of the coefficients. If the coefficients are continuous for all t, then the solution not only exists and is continuous for all t, but it is also continuously differentiable for all t.

None of these statements is true, in general, of nonlinear equations. Although a nonlinear equation may well have a solution involving an arbitrary constant, there may also be other solutions. There is no general formula for solutions of nonlinear equations. If you are able to integrate a nonlinear equation, you are likely to obtain an equation defining solutions implicitly rather than explicitly. Finally, the singularities of solutions of nonlinear equations can usually be found only by solving the equation and examining the solution. It is likely that the singularities will depend on the initial condition as well as the differential equation.

PROBLEMS

Existence and Uniqueness of Solutions. In each of Problems 1 through 6, use Theorem 2.4.1 to determine (without solving the problem) an interval in which the solution of the given initial value problem is certain to exist.

1. $(t-3)y' + (\ln t)y = 2t, \qquad y(1) = 2$

2. $t(t-4)y' + y = 0, \qquad y(2) = 1$

3. $y' + (\tan t)y = \sin t, \qquad y(\pi) = 0$

4. $(4-t^2)y' + 2ty = 3t^2, \qquad y(-3) = 1$

5. $(4-t^2)y' + 2ty = 3t^2, \qquad y(1) = -3$

6. $(\ln t)y' + y = \cot t, \qquad y(2) = 3$

In each of Problems 7 through 12, state where in the ty-plane the hypotheses of Theorem 2.4.2 are satisfied.

7. $y' = \dfrac{t-y}{2t+5y}$

8. $y' = (1 - t^2 - y^2)^{1/2}$

9. $y' = \dfrac{\ln|ty|}{1 - t^2 + y^2}$

10. $y' = (t^2 + y^2)^{3/2}$

11. $\dfrac{dy}{dt} = \dfrac{1+t^2}{3y - y^2}$

12. $\dfrac{dy}{dt} = \dfrac{(\cot t)y}{1+y}$

13. Consider the initial value problem $y' = y^{1/3}$, $y(0) = 0$ from Example 3 in the text.

(a) Is there a solution that passes through the point $(1, 1)$? If so, find it.

(b) Is there a solution that passes through the point $(2, 1)$? If so, find it.

(c) Consider all possible solutions of the given initial value problem. Determine the set of values that these solutions attain at $t = 2$.

14. (a) Verify that both $y_1(t) = 1 - t$ and $y_2(t) = -t^2/4$ are solutions of the initial value problem

$$y' = \frac{-t + (t^2 + 4y)^{1/2}}{2}, \qquad y(2) = -1.$$

Where are these solutions valid?

(b) Explain why the existence of two solutions of the given problem does not contradict the uniqueness part of Theorem 2.4.2.

(c) Show that $y = ct + c^2$, where c is an arbitrary constant, satisfies the differential equation in part (a) for $t \geq -2c$. If $c = -1$, the initial condition is also satisfied, and the solution $y = y_1(t)$ is obtained. Show that there is no choice of c that gives the second solution $y = y_2(t)$.

Dependence of Solutions on Initial Conditions. In each of Problems 15 through 18, solve the given initial value problem and determine how the interval in which the solution exists depends on the initial value y_0.

15. $y' = -4t/y, \qquad y(0) = y_0$

16. $y' = 2ty^2, \qquad y(0) = y_0$

17. $y' + y^3 = 0, \qquad y(0) = y_0$

18. $y' = t^2/y(1 + t^3), \qquad y(0) = y_0$

In each of Problems 19 through 22, draw a direction field and plot (or sketch) several solutions of the given differential equation. Describe how solutions appear to behave as t increases and how their behavior depends on the initial value y_0 when $t = 0$.

19. $y' = ty(3 - y)$

20. $y' = y(3 - ty)$

21. $y' = -y(3 - ty)$

22. $y' = t - 1 - y^2$

Linearity Properties

23. (a) Show that $\phi(t) = e^{2t}$ is a solution of $y' - 2y = 0$ and that $y = c\phi(t)$ is also a solution of this equation for any value of the constant c.

(b) Show that $\phi(t) = 1/t$ is a solution of $y' + y^2 = 0$ for $t > 0$ but that $y = c\phi(t)$ is not a solution of this equation unless $c = 0$ or $c = 1$. Note that the equation of part (b) is nonlinear, whereas that of part (a) is linear.

24. Show that if $y = \phi(t)$ is a solution of $y' + p(t)y = 0$, then $y = c\phi(t)$ is also a solution for any value of the constant c.

25. Let $y = y_1(t)$ be a solution of

$$y' + p(t)y = 0, \qquad \text{(i)}$$

and let $y = y_2(t)$ be a solution of

$$y' + p(t)y = g(t). \qquad \text{(ii)}$$

Show that $y = y_1(t) + y_2(t)$ is also a solution of Eq. (ii).

26. (a) Show that the solution (7) of the general linear equation (1) can be written in the form

$$y = cy_1(t) + y_2(t), \qquad \text{(i)}$$

where c is an arbitrary constant. Identify the functions y_1 and y_2.

(b) Show that y_1 is a solution of the differential equation

$$y' + p(t)y = 0, \qquad \text{(ii)}$$

corresponding to $g(t) = 0$.

(c) Show that y_2 is a solution of the full linear equation (1). We see later (e.g., in Section 4.5) that solutions of higher order linear equations have a pattern similar to Eq. (i).

Discontinuous Coefficients. Linear differential equations sometimes occur in which one or both of the functions p and g have jump discontinuities. If t_0 is such a point of discontinuity, then it is necessary to solve the equation separately for $t < t_0$ and $t > t_0$. Afterward, the two solutions are matched so that y is continuous at t_0. This is accomplished by a proper choice of the arbitrary constants. Problems 27 and 28 illustrate this situation. Note in each case that it is impossible to make y' continuous at t_0: explain why, just from examining the differential equations.

27. Solve the initial value problem

$$y' + 2y = g(t), \qquad y(0) = 0,$$

where

$$g(t) = \begin{cases} 1, & 0 \leq t \leq 1, \\ 0, & t > 1. \end{cases}$$

28. Solve the initial value problem

$$y' + p(t)y = 0, \qquad y(0) = 1,$$

where

$$p(t) = \begin{cases} 2, & 0 \leq t \leq 1, \\ 1, & t > 1. \end{cases}$$

29. Consider the initial value problem

$$y' + p(t)y = g(t), \qquad y(t_0) = y_0. \tag{i}$$

(a) Show that the solution of the initial value problem (i) can be written in the form

$$y = y_0 \exp\left(-\int_{t_0}^{t} p(s)\, ds \right)$$
$$+ \int_{t_0}^{t} \exp\left(-\int_{s}^{t} p(r)\, dr \right) g(s)\, ds.$$

(b) Assume that $p(t) \geq p_0 > 0$ for all $t \geq t_0$ and that $g(t)$ is bounded for $t \geq t_0$ (i.e., there is a constant M such that $|g(t)| \leq M$ for all $t \geq t_0$). Show that the solution of the initial value problem (i) is bounded for $t \geq t_0$.

(c) Construct an example with nonconstant $p(t)$ and $g(t)$ that illustrates this result.

2.5 Autonomous Equations and Population Dynamics

In Section 1.2 we first encountered the following important class of first order equations in which the independent variable does not appear explicitly.

DEFINITION 2.5.1	**Autonomous Equation.** A differential equation that can be written as $$\frac{dy}{dt} = f(y). \tag{1}$$ is said to be **autonomous**.

We will now discuss these equations in the context of the growth or decline of the population of a given species, an important issue in fields ranging from medicine to ecology to global economics. A number of other applications are mentioned in some of the problems. Recall that in Section 2.1 we considered the special case of Eq. (1) in which the form of the right side is $f(y) = ay + b$.

Equation (1) is separable, and it can be solved using the approach discussed in Section 2.1. However, the main purpose of this section is to show how geometrical methods can be used to obtain important qualitative information about solutions directly from the differential equation, without solving the equation. Of fundamental importance in this effort are the concepts of stability and instability of solutions of differential equations. These ideas were introduced informally in Chapter 1. They are discussed further here and will be examined in greater depth and in a more general setting in Chapters 3 and 7.

▶ Exponential Growth. Let $y = \phi(t)$ be the population of the given species at time t. The simplest hypothesis concerning the variation of population is that the rate of change of y is proportional to the current value of y. For example, if the population doubles, then the number of births in a given time period should also double. Thus we have

$$dy/dt = ry, \tag{2}$$

where the constant of proportionality r is called the **rate of growth or decline**, depending on whether it is positive or negative. Here we assume that $r > 0$, so the population is growing.

Solving Eq. (2) subject to the initial condition

$$y(0) = y_0, \tag{3}$$

we obtain

$$y = y_0 e^{rt}. \tag{4}$$

Thus the mathematical model consisting of the initial value problem (2), (3) with $r > 0$ predicts that the population will grow exponentially for all time, as shown in Figure 2.5.1 for several values of y_0. Under ideal conditions, Eq. (4) has been observed to be reasonably accurate for many populations, at least for limited periods of time. However it is clear that such ideal conditions cannot continue indefinitely; eventually, limitations on space, food supply, or other resources will reduce the growth rate and bring an end to uninhibited exponential growth.

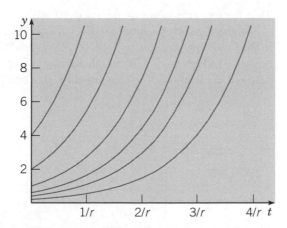

FIGURE 2.5.1 Exponential growth: y versus t for $dy/dt = ry$.

▶ Logistic Growth. To account for the fact that the growth rate actually depends on the population, we replace the constant r in Eq. (2) by a function $h(y)$ and thereby obtain the modified equation

$$dy/dt = h(y)y. \tag{5}$$

We now want to choose $h(y)$, so that $h(y) \cong r > 0$ when y is small, $h(y)$ decreases as y grows larger, and $h(y) < 0$ when y is sufficiently large. The simplest function that has these properties is $h(y) = r - ay$, where a is also a positive constant. Using this function in Eq. (5), we obtain

$$dy/dt = (r - ay)y. \tag{6}$$

Equation (6) is known as the Verhulst equation or the **logistic equation**. It is often convenient to write the logistic equation in the equivalent form

$$\frac{dy}{dt} = r\left(1 - \frac{y}{K}\right)y, \tag{7}$$

where $K = r/a$. In this equation, the constant r is referred to as the **intrinsic growth rate**, that is, the growth rate in the absence of any limiting factors. The interpretation of K will become clear shortly.

Before proceeding to investigate the solutions of Eq. (7), let us look at a specific example.

EXAMPLE 1

Consider the differential equation

$$\frac{dy}{dt} = \left(1 - \frac{y}{3}\right)y. \tag{8}$$

Without solving the equation, determine the qualitative behavior of its solutions and sketch the graphs of a representative sample of them.

As in Section 1.2, the constant solutions are of particular importance. They satisfy the algebraic equation

$$\left(1 - \frac{y}{3}\right)y = 0.$$

Thus the constant solutions are $y = \phi_1(t) = 0$ and $y = \phi_2(t) = 3$.

To visualize other solutions of Eq. (8) and to sketch their graphs quickly, we can proceed in the following way. Let $f(y) = (1 - y/3)y$ and draw the graph of $f(y)$ versus y. The graph is shown in Figure 2.5.2. Remember that $f(y_0)$ represents the slope of a line tangent to the graph of the solution of (8) passing through a point (t, y_0) in the ty-plane.

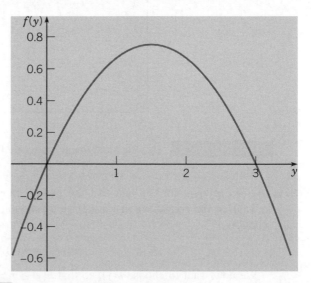

FIGURE 2.5.2   Graph of $f(y) = (1 - y/3)y$ versus y.

To sketch solutions of Eq. (8), the first step is to draw the equilibrium solutions $y = \phi_1(t) = 0$ and $y = \phi_2(t) = 3$; see the thick dashed lines in Figure 2.5.3. Then from Figure 2.5.2 note that $f(y) > 0$ for $0 < y < 3$. Thus in the ty-plane solutions are increasing (have a positive slope) for $0 < y < 3$. A few of these solutions are shown in Figure 2.5.3. These solution curves flatten out near $y = 0$ and $y = 3$ because, from Figure 2.5.2, their slopes, given by $f(y)$, are near zero there. The slopes reach a maximum at $y = \frac{3}{2}$, the vertex of the parabola. Observe also that $f(y)$ or dy/dt is increasing for $y < \frac{3}{2}$ and decreasing for $y > \frac{3}{2}$. This means that the graphs of y versus t are concave up for $y < \frac{3}{2}$ and concave down for $y > \frac{3}{2}$. In other words, solution curves have an inflection point as they cross the line $y = \frac{3}{2}$.

For $y > 3$ you can see from Figure 2.5.2 that $f(y)$, or dy/dt, is negative and decreasing. Therefore the graphs of y versus t for this range of y are decreasing and concave up. They also become flatter as they approach the equilibrium solution $y = 3$. Some of these graphs are also shown in Figure 2.5.3.

None of the other solutions can intersect the equilibrium solutions $y = 0$ and $y = 3$ at a finite time. If they did, they would violate the uniqueness part of Theorem 2.4.2, which states that only one solution can pass through any given point in the ty-plane. As such, the equilibrium solutions partition the ty-plane into disjoint regions, and the shape of the solution curves in each region is determined by the sign and slope of $f(y)$.

Finally, although we have drawn Figures 2.5.2 and 2.5.3 using a computer, very similar qualitatively correct sketches can be drawn by hand, without any computer assistance, by following the steps described in this example.

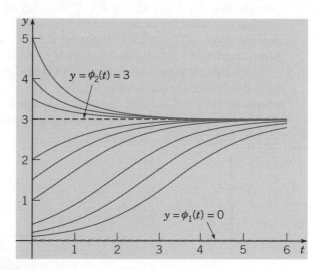

FIGURE 2.5.3 Graphs of solutions of Eq. (8): $y' = (1 - y/3)y$.

We now return to a consideration of the more general Eq. (7),

$$\frac{dy}{dt} = r\left(1 - \frac{y}{K}\right)y,$$

where r and K are positive constants. We can proceed, just as in Example 1, to draw a qualitatively correct sketch of solutions of this equation.

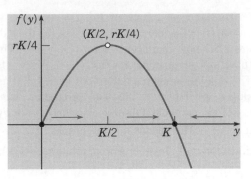

FIGURE 2.5.4 $f(y)$ versus y for $dy/dt = r(1 - y/K)y$.

To find the equilibrium solutions, we set dy/dt equal to zero and solve the resulting algebraic equation

$$r\left(1 - \frac{y}{K}\right)y = 0.$$

Thus, for Eq. (7), the equilibrium solutions are $y = \phi_1(t) = 0$ and $y = \phi_2(t) = K$.

Next we draw the graph of $f(y)$ versus y. In the case of Eq. (7), $f(y) = r(1 - y/K)y$, so the graph is the parabola shown in Figure 2.5.4. The intercepts are $(0, 0)$ and $(K, 0)$, corresponding to the critical points of Eq. (7), and the vertex of the parabola is $(K/2, rK/4)$. Observe that $dy/dt > 0$ for $0 < y < K$ [since $dy/dt = f(y)$]; therefore y is an increasing function of t when y is in this interval. This is indicated by the rightward-pointing arrows near the y-axis in Figure 2.5.4. Similarly, if $y > K$, then $dy/dt < 0$; hence y is decreasing, as indicated by the leftward-pointing arrow in Figure 2.5.4.

The y-axis, or phase line, is shown in Figure 2.5.5a. The dots at $y = 0$ and $y = K$ are the critical points, or equilibrium solutions. The arrows again indicate that y is increasing whenever $0 < y < K$ and that y is decreasing whenever $y > K$. We see from the phase line that $y = 0$ is unstable and $y = 3$ is asymptotically stable.

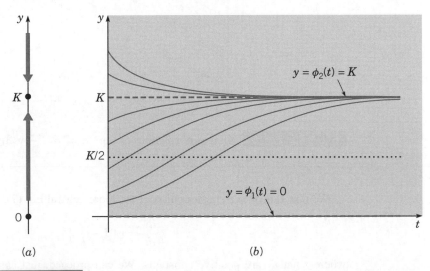

FIGURE 2.5.5 Logistic growth: $dy/dt = r(1 - y/K)y$. (*a*) The phase line. (*b*) Plots of y versus t, with equilibrium solutions shown by thick dashed lines.

Further, from Figure 2.5.4, note that if y is near zero or K, then the slope $f(y)$ is near zero, so the solution curves are relatively flat. They become steeper as the value of y leaves the neighborhood of zero or K.

To sketch the graphs of solutions of Eq. (7) in the ty-plane, we start with the equilibrium solutions $y = 0$ and $y = K$; then we draw other curves that are increasing when $0 < y < K$, decreasing when $y > K$, and flatten out as y approaches either of the values 0 or K. Thus the graphs of solutions of Eq. (7) must have the general shape shown in Figure 2.5.5b, regardless of the values of r and K.

As in Example 1, the concavity of a solution curve changes as it passes through the value $y = K/2$ corresponding to the vertex of the parabola in Figure 2.5.4. Thus each solution curve has an inflection point when $y = K/2$. Further, Figure 2.5.5b may seem to show that other solutions intersect the equilibrium solution $y = K$, but this is impossible by the uniqueness part of Theorem 2.4.2. Thus, although other solutions may be asymptotic to the equilibrium solution as $t \to \infty$, they cannot intersect it at any finite time.

Finally, observe that K is the upper bound that is approached, but not exceeded, by growing populations starting below this value. Thus it is natural to refer to K as the **saturation level**, or the **environmental carrying capacity**, for the given species.

A comparison of Figures 2.5.1 and 2.5.5b reveals that solutions of the nonlinear equation (7) are strikingly different from those of the linear equation (1), at least for large values of t. Regardless of the value of K, that is, no matter how small the nonlinear term in Eq. (7), solutions of that equation approach a finite value as $t \to \infty$, whereas solutions of Eq. (1) grow (exponentially) without bound as $t \to \infty$. Thus even a tiny nonlinear term in the differential equation has a decisive effect on the solution for large t.

Phase Line for General Autonomous Differential Equations

The same methods can be applied to the general autonomous equation (1),

$$dy/dt = f(y).$$

The **equilibrium solutions** of this equation can be found by locating the roots of $f(y) = 0$. The zeros of $f(y)$ are also called **critical points**.

We showed in Section 1.2 (see Theorem 1.2.1) that if y_1 is a critical point and if $f'(y_1) < 0$, then all nearby solutions are approaching $y = y_1$, so that y_1 is asymptotically stable. On the other hand, if y_1 is a critical point and if $f'(y_1) > 0$, then nearby solutions are departing from $y = y_1$, so y_1 is unstable.

To carry the investigation one step further, we can determine the concavity of the solution curves and the location of inflection points by finding d^2y/dt^2. From the differential equation (1), we obtain (using the chain rule)

$$\frac{d^2y}{dt^2} = \frac{d}{dt}\frac{dy}{dt} = \frac{d}{dt}f(y) = f'(y)\frac{dy}{dt} = f'(y)f(y). \tag{9}$$

The graph of y versus t is concave up when $y'' > 0$, that is, when f and f' have the same sign. Similarly, it is concave down when $y'' < 0$, which occurs when f and f' have opposite signs. The signs of f and f' can be easily identified from the graph of $f(y)$ versus y. Inflection points may occur when $f'(y) = 0$.

In many situations, it is sufficient to have the qualitative information about a solution $y = \phi(t)$ of Eq. (7) that is shown in Figure 2.5.5b. This information was obtained entirely from the graph of $f(y)$ versus y, and without solving the differential equation (7). However, if we wish to have a more detailed description of logistic growth—for example, if we wish to know the value of the population at some particular time—then we must solve Eq. (7)

subject to the initial condition (3). Provided that $y \neq 0$ and $y \neq K$, we can write Eq. (7) in the form

$$\frac{dy}{(1 - y/K)y} = r \, dt.$$

Using a partial fraction expansion on the left side, we have

$$\left(\frac{1}{y} + \frac{1/K}{1 - y/K} \right) dy = r \, dt.$$

Then, by integrating both sides, we obtain

$$\ln |y| - \ln \left| 1 - \frac{y}{K} \right| = rt + c, \tag{10}$$

where c is an arbitrary constant of integration to be determined from the initial condition $y(0) = y_0$. We have already noted that if $0 < y_0 < K$, then y remains in this interval for all time. Thus in this case we can remove the absolute value bars in Eq. (10), and by taking the exponential of both sides, we find that

$$\frac{y}{1 - y/K} = Ce^{rt}, \tag{11}$$

where $C = e^c$. In order to satisfy the initial condition $y(0) = y_0$, we must choose $C = y_0/[1 - (y_0/K)]$. Using this value for C in Eq. (11) and solving for y, we obtain

$$y = \frac{y_0 K}{y_0 + (K - y_0)e^{-rt}}. \tag{12}$$

We have derived the solution (12) under the assumption that $0 < y_0 < K$. If $y_0 > K$, then the details of dealing with Eq. (10) are only slightly different, and we leave it to you to show that Eq. (12) is also valid in this case. Finally, note that Eq. (12) also contains the equilibrium solutions $y = \phi_1(t) = 0$ and $y = \phi_2(t) = K$ corresponding to the initial conditions $y_0 = 0$ and $y_0 = K$, respectively.

All the qualitative conclusions that we reached earlier by geometrical reasoning can be confirmed by examining the solution (12). In particular, if $y_0 = 0$, then Eq. (12) requires that $y(t) = 0$ for all t. If $y_0 > 0$, and if we let $t \to \infty$ in Eq. (12), then we obtain

$$\lim_{t \to \infty} y(t) = y_0 K / y_0 = K.$$

Thus, for each $y_0 > 0$, the solution approaches the equilibrium solution $y = \phi_2(t) = K$ asymptotically as $t \to \infty$. Therefore the constant solution $\phi_2(t) = K$ is an asymptotically stable solution of Eq. (7). After a long time, the population is close to the saturation level K regardless of the initial population size, as long as it is positive. Other solutions approach the equilibrium solution more rapidly as r increases.

On the other hand, the situation for the equilibrium solution $y = \phi_1(t) = 0$ is quite different. Even solutions that start very near zero grow as t increases and, as we have seen, approach K as $t \to \infty$. So, the solution $\phi_1(t) = 0$ is an unstable equilibrium solution. This means that the only way to guarantee that the solution remains near zero is to make sure its initial value is *exactly* equal to zero.

EXAMPLE
2

The logistic model has been applied to the natural growth of the halibut population in certain areas of the Pacific Ocean.[4] Let y, measured in kilograms, be the total mass, or biomass, of the halibut population at time t. The parameters in the logistic equation are estimated to have the values $r = 0.71$/year and $K = 80.5 \times 10^6$ kg. If the initial biomass is $y_0 = 0.25K$, find the biomass two years later. Also find the time τ for which $y(\tau) = 0.75K$.

It is convenient to scale the solution (12) to the carrying capacity K; thus we write Eq. (12) in the form

$$\frac{y}{K} = \frac{y_0/K}{(y_0/K) + [1 - (y_0/K)]e^{-rt}}. \tag{13}$$

Using the data given in the problem, we find that

$$\frac{y(2)}{K} = \frac{0.25}{0.25 + 0.75e^{-1.42}} \cong 0.5797.$$

Consequently, $y(2) \cong 46.7 \times 10^6$ kg.

To find τ, we can first solve Eq. (13) for t. We obtain

$$e^{-rt} = \frac{(y_0/K)[1 - (y/K)]}{(y/K)[1 - (y_0/K)]}.$$

Hence

$$t = -\frac{1}{r} \ln \frac{(y_0/K)[1 - (y/K)]}{(y/K)[1 - (y_0/K)]}. \tag{14}$$

Using the given values of r and y_0/K and setting $y/K = 0.75$, we find that

$$\tau = -\frac{1}{0.71} \ln \frac{(0.25)(0.25)}{(0.75)(0.75)} = \frac{1}{0.71} \ln 9 \cong 3.095 \text{ years.}$$

The graphs of y/K versus t for the given parameter values and for several initial conditions are shown in Figure 2.5.6.

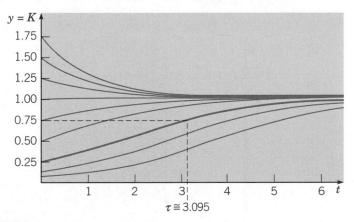

FIGURE 2.5.6 y/K versus t for population model of halibut in the Pacific Ocean.

[4]A good source of information on the population dynamics and economics involved in making efficient use of a renewable resource, with particular emphasis on fisheries, is the book by Clark listed in the references at the end of the book. The parameter values used here are given on page 53 of this book and were obtained from a study by H. S. Mohring.

▶ A Critical Threshold. We now turn to a consideration of the equation

$$\frac{dy}{dt} = -r\left(1 - \frac{y}{T}\right)y, \tag{15}$$

where r and T are given positive constants. Observe that (except for replacing the parameter K by T) this equation differs from the logistic equation (7) only in the presence of the minus sign on the right side. However, as we will see, the solutions of Eq. (15) behave very differently from those of Eq. (7).

For Eq. (15), the graph of $f(y)$ versus y is the parabola shown in Figure 2.5.7. The intercepts on the y-axis are the critical points $y = 0$ and $y = T$, corresponding to the equilibrium solutions $\phi_1(t) = 0$ and $\phi_2(t) = T$. If $0 < y < T$, then $dy/dt < 0$, and y decreases as t increases. On the other hand, if $y > T$, then $dy/dt > 0$, and y grows as t increases. Thus $\phi_1(t) = 0$ is an asymptotically stable equilibrium solution and $\phi_2(t) = T$ is an unstable one. Further $f'(y)$ is negative for $0 < y < T/2$ and positive for $T/2 < y < T$, so the graph of y versus t is concave up and concave down, respectively, in these intervals. Also, $f'(y)$ is positive for $y > T$, so the graph of y versus t is also concave up there.

Figure 2.5.8a shows the phase line (the y-axis) for Eq. (15).

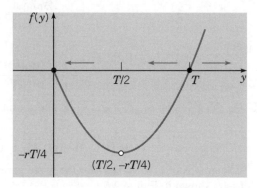

FIGURE 2.5.7 $f(y)$ versus y for $dy/dt = -r(1 - y/T)y$.

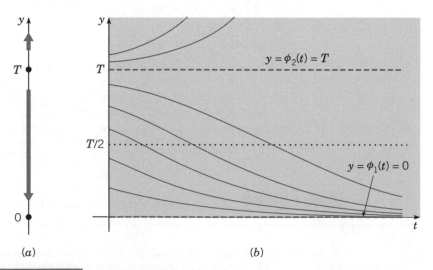

FIGURE 2.5.8 Growth with a threshold: $dy/dt = -r(1 - y/T)y$. (a) The phase line. (b) Plots of y versus t.

Solution curves of Eq. (15) can now be sketched quickly using the procedure outlined in Table 1.2.1. The result is Figure 2.5.8b, which is a qualitatively accurate sketch of solutions of Eq. (15) for any r and T. From this figure, it appears that as time increases, y either approaches zero or grows without bound, depending on whether the initial value y_0 is less than or greater than T. Thus T is a **threshold level**, below which growth does not occur.

We can confirm the conclusions that we have reached through geometrical reasoning by solving the differential equation (15). This can be done by separating the variables and integrating, just as we did for Eq. (7). However, if we note that Eq. (15) can be obtained from Eq. (7) by replacing K by T and r by $-r$, then we can make the same substitutions in the solution (12) and thereby obtain

$$y = \frac{y_0 T}{y_0 + (T - y_0)e^{rt}}, \tag{16}$$

which is the solution of Eq. (15) subject to the initial condition $y(0) = y_0$.

If $0 < y_0 < T$, then it follows from Eq. (16) that $y \to 0$ as $t \to \infty$. This agrees with our qualitative geometric analysis. If $y_0 > T$, then the denominator on the right side of Eq. (16) is zero for a certain finite value of t. We denote this value by t^* and calculate it from

$$y_0 - (y_0 - T)e^{rt^*} = 0,$$

which gives

$$t^* = \frac{1}{r} \ln \frac{y_0}{y_0 - T}. \tag{17}$$

Thus, if the initial population y_0 is above the threshold T, the threshold model predicts that the graph of y versus t has a vertical asymptote at $t = t^*$. In other words, the population becomes unbounded in a finite time, whose value depends on y_0, T, and r. The existence and location of this asymptote were not apparent from the geometric analysis, so, in this case, the explicit solution yields additional important qualitative, as well as quantitative, information.

The populations of some species exhibit the threshold phenomenon. If too few are present, then the species cannot propagate itself successfully and the population becomes extinct. However, if a population larger than the threshold level can be brought together, then further growth occurs. Of course, the population cannot become unbounded, so eventually Eq. (15) must be modified to take this into account.

Critical thresholds also occur in other circumstances. For example, in fluid mechanics, equations of the form (7) or (15) often govern the evolution of a small disturbance y in a *laminar* (or smooth) fluid flow. For instance, if Eq. (15) holds and $y < T$, then the disturbance is damped out and the laminar flow persists. However, if $y > T$, then the disturbance grows larger and the laminar flow breaks up into a turbulent one. In this case, T is referred to as the *critical amplitude*. Experimenters speak of keeping the disturbance level in a wind tunnel sufficiently low so they can study laminar flow over an airfoil, for example.

▶ Logistic Growth with a Threshold. As we mentioned in the last subsection, the threshold model (15) may need to be modified so that unbounded growth does not occur when y is above the threshold T. The simplest way to do this is to introduce another factor that will have the effect of making dy/dt negative when y is large. Thus we consider

$$\frac{dy}{dt} = -r\left(1 - \frac{y}{T}\right)\left(1 - \frac{y}{K}\right)y, \tag{18}$$

where $r > 0$ and $0 < T < K$.

The graph of $f(y)$ versus y is shown in Figure 2.5.9. There are now three critical points, $y = 0$, $y = T$, and $y = K$, corresponding to the equilibrium solutions $\phi_1(t) = 0$, $\phi_2(t) = T$,

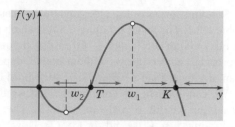

FIGURE 2.5.9 $f(y)$ versus y for $dy/dt = -r(1 - y/T)(1 - y/K)y$.

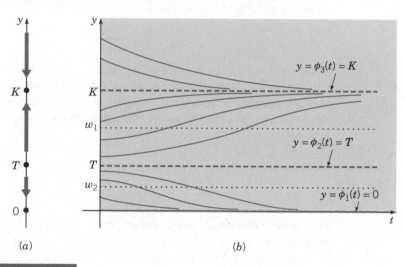

(a) (b)

FIGURE 2.5.10 Logistic growth with a threshold: $dy/dt = -r(1 - y/T)(1 - y/K)y$. (a) The phase line. (b) Plots of y versus t.

and $\phi_3(t) = K$, respectively. From Figure 2.5.9, it is clear that $dy/dt > 0$ for $T < y < K$, and consequently, y is increasing there. The reverse is true for $y < T$ and for $y > K$. Consequently, the equilibrium solutions $\phi_1(t)$ and $\phi_3(t)$ are asymptotically stable, and the solution $\phi_2(t)$ is unstable.

The phase line for Eq. (18) is shown in Figure 2.5.10a, and the graphs of some solutions are sketched in Figure 2.5.10b. Make sure that you understand the relation between these two figures, as well as the relation between Figures 2.5.9 and 2.5.10a. From Figure 2.5.10b we see that if y starts below the threshold T, then y declines to ultimate extinction. On the other hand, if y starts above T, then y eventually approaches the carrying capacity K. The inflection points on the graphs of y versus t in Figure 2.5.10b correspond to the maximum and minimum points, w_1 and w_2, respectively, on the graph of $f(y)$ versus y in Figure 2.5.9. These values can be obtained by differentiating the right side of Eq. (18) with respect to y, setting the result equal to zero, and solving for y. We obtain

$$w_{1,2} = (K + T \pm \sqrt{K^2 - KT + T^2})/3, \tag{19}$$

where the plus sign yields w_1 and the minus sign w_2.

A model of this general sort apparently describes the population of the passenger pigeon,[5] which was present in the United States in vast numbers until late in the 19th century. It was heavily hunted for food and for sport, and consequently its numbers were drastically reduced by the 1880s. Unfortunately, the passenger pigeon could apparently breed successfully only when present in a large concentration, corresponding to a relatively high threshold

[5]See, for example, Oliver L. Austin, Jr., *Birds of the World* (New York: Golden Press, 1983), pp. 143–145.

T. Although a reasonably large number of individual birds remained alive in the late 1880s, there were not enough in any one place to permit successful breeding, and the population rapidly declined to extinction. The last survivor died in 1914. The precipitous decline in the passenger pigeon population from huge numbers to extinction in a few decades was one of the early factors contributing to a concern for conservation in this country.

PROBLEMS

1. Suppose that a certain population obeys the logistic equation $dy/dt = ry[1 - (y/K)]$.

(a) If $y_0 = K/3$, find the time τ at which the initial population has doubled. Find the value of τ corresponding to $r = 0.025$ per year.

(b) If $y_0/K = \alpha$, find the time T at which $y(T)/K = \beta$, where $0 < \alpha, \beta < 1$. Observe that $T \to \infty$ as $\alpha \to 0$ or as $\beta \to 1$. Find the value of T for $r = 0.025$ per year, $\alpha = 0.1$, and $\beta = 0.9$.

2. Another equation that has been used to model population growth is the Gompertz equation

$$dy/dt = ry\ln(K/y),$$

where r and K are positive constants.

(a) Sketch the graph of $f(y)$ versus y, find the critical points, and determine whether each is asymptotically stable or unstable.

(b) For $0 \le y \le K$, determine where the graph of y versus t is concave up and where it is concave down.

(c) For each y in $0 < y \le K$, show that dy/dt, as given by the Gompertz equation, is never less than dy/dt, as given by the logistic equation.

3. **(a)** Solve the Gompertz equation

$$dy/dt = ry\ln(K/y),$$

subject to the initial condition $y(0) = y_0$.
Hint: You may wish to let $u = \ln(y/K)$.

(b) For the data given in Example 2 in the text ($r = 0.71$ per year, $K = 80.5 \times 10^6$ kg, $y_0/K = 0.25$), use the Gompertz model to find the predicted value of $y(2)$.

(c) For the same data as in part (b), use the Gompertz model to find the time τ at which $y(\tau) = 0.75K$.

4. A pond forms as water collects in a conical depression of radius a and depth h. Suppose that water flows in at a constant rate k and is lost through evaporation at a rate proportional to the surface area.

(a) Show that the volume $V(t)$ of water in the pond at time t satisfies the differential equation

$$dV/dt = k - \alpha\pi(3a/\pi h)^{2/3}V^{2/3},$$

where α is the coefficient of evaporation.

(b) Find the equilibrium depth of water in the pond. Is the equilibrium asymptotically stable?

(c) Find a condition that must be satisfied if the pond is not to overflow.

5. Consider a cylindrical water tank of constant cross section A. Water is pumped into the tank at a constant rate k and leaks out through a small hole of area a in the bottom of the tank. From Torricelli's principle in hydrodynamics (see Problem 6 in Section 2.3), it follows that the rate at which water flows through the hole is $\alpha a\sqrt{2gh}$, where h is the current depth of water in the tank, g is the acceleration due to gravity, and α is a contraction coefficient that satisfies $0.5 \le \alpha \le 1.0$.

(a) Show that the depth of water in the tank at any time satisfies the equation

$$dh/dt = (k - \alpha a\sqrt{2gh})/A.$$

(b) Determine the equilibrium depth h_e of water, and show that it is asymptotically stable. Observe that h_e does not depend on A.

Epidemics. The use of mathematical methods to study the spread of contagious diseases goes back at least to some work by Daniel Bernoulli in 1760 on smallpox. In more recent years, many mathematical models have been proposed and studied for many different diseases.[6] Problems 6 through 8 deal with a few of the simpler models and the conclusions that can be drawn from them. Similar models have also been used to describe the spread of rumors and of consumer products.

6. Suppose that a given population can be divided into two parts: those who have a given disease and can infect others, and those who do not have it but are susceptible. Let x be the proportion of susceptible individuals and y the proportion of infectious individuals; then $x + y = 1$. Assume that the disease spreads by contact between sick and well members of the population and that the rate of spread dy/dt is proportional to the number of such contacts. Further, assume that members of both groups move about freely among each other, so the number of contacts is proportional to the product of x and y. Since $x = 1 - y$, we obtain the initial value problem

$$dy/dt = \alpha y(1 - y), \qquad y(0) = y_0, \qquad \text{(i)}$$

where α is a positive proportionality factor, and y_0 is the initial proportion of infectious individuals.

---

[6]A standard source is the book by Bailey listed in the references. The models in Problems 6 through 8 are discussed by Bailey in Chapters 5, 10, and 20, respectively.

(a) Find the equilibrium points for the differential equation (i) and determine whether each is asymptotically stable, semistable, or unstable.

(b) Solve the initial value problem (i) and verify that the conclusions you reached in part (a) are correct. Show that $y(t) \to 1$ as $t \to \infty$, which means that ultimately the disease spreads through the entire population.

7. Some diseases (such as typhoid fever) are spread largely by *carriers*, individuals who can transmit the disease but who exhibit no overt symptoms. Let x and y, respectively, denote the proportion of susceptibles and carriers in the population. Suppose that carriers are identified and removed from the population at a rate β, so

$$dy/dt = -\beta y. \tag{i}$$

Suppose also that the disease spreads at a rate proportional to the product of x and y; thus

$$dx/dt = -\alpha xy. \tag{ii}$$

(a) Determine y at any time t by solving Eq. (i) subject to the initial condition $y(0) = y_0$.

(b) Use the result of part (a) to find x at any time t by solving Eq. (ii) subject to the initial condition $x(0) = x_0$.

(c) Find the proportion of the population that escapes the epidemic by finding the limiting value of x as $t \to \infty$.

8. Daniel Bernoulli's work in 1760 had the goal of appraising the effectiveness of a controversial inoculation program against smallpox, which at that time was a major threat to public health. His model applies equally well to any other disease that, once contracted and survived, confers a lifetime immunity.

Consider the cohort of individuals born in a given year ($t = 0$), and let $n(t)$ be the number of these individuals surviving t years later. Let $x(t)$ be the number of members of this cohort who have not had smallpox by year t and who are therefore still susceptible. Let β be the rate at which susceptibles contract smallpox, and let v be the rate at which people who contract smallpox die from the disease. Finally, let $\mu(t)$ be the death rate from all causes other than smallpox. Then dx/dt, the rate at which the number of susceptibles changes, is given by

$$dx/dt = -[\beta + \mu(t)]x. \tag{i}$$

The first term on the right side of Eq. (i) is the rate at which susceptibles contract smallpox, and the second term is the rate at which they die from all other causes. Also

$$dn/dt = -v\beta x - \mu(t)n, \tag{ii}$$

where dn/dt is the death rate of the entire cohort, and the two terms on the right side are the death rates due to smallpox and to all other causes, respectively.

(a) Let $z = x/n$ and show that z satisfies the initial value problem

$$dz/dt = -\beta z(1 - vz), \qquad z(0) = 1. \tag{iii}$$

Observe that the initial value problem (iii) does not depend on $\mu(t)$.

(b) Find $z(t)$ by solving Eq. (iii).

(c) Bernoulli estimated that $v = \beta = \frac{1}{8}$. Using these values, determine the proportion of 20-year-olds who have not had smallpox.

Note: On the basis of the model just described and the best mortality data available at the time, Bernoulli calculated that if deaths due to smallpox could be eliminated ($v = 0$), then approximately 3 years could be added to the average life expectancy (in 1760) of 26 years 7 months. He therefore supported the inoculation program.

9. Chemical Reactions. A second order chemical reaction involves the interaction (collision) of one molecule of a substance P with one molecule of a substance Q to produce one molecule of a new substance X; this is denoted by $P + Q \to X$. Suppose that p and q, where $p \neq q$, are the initial concentrations of P and Q, respectively, and let $x(t)$ be the concentration of X at time t. Then $p - x(t)$ and $q - x(t)$ are the concentrations of P and Q at time t, and the rate at which the reaction occurs is given by the equation

$$dx/dt = \alpha(p - x)(q - x), \tag{i}$$

where α is a positive constant.

(a) If $x(0) = 0$, determine the limiting value of $x(t)$ as $t \to \infty$ without solving the differential equation. Then solve the initial value problem and find $x(t)$ for any t.

(b) If the substances P and Q are the same, then $p = q$ and Eq. (i) is replaced by

$$dx/dt = \alpha(p - x)^2. \tag{ii}$$

If $x(0) = 0$, determine the limiting value of $x(t)$ as $t \to \infty$ without solving the differential equation. Then solve the initial value problem and determine $x(t)$ for any t.

Bifurcation Points. For an equation of the form

$$dy/dt = f(a, y), \tag{i}$$

where a is a real parameter, the critical points (equilibrium solutions) usually depend on the value of a. As a steadily increases or decreases, it often happens that at a certain value of a, called a **bifurcation point**, critical points come together, or separate, and equilibrium solutions may either be lost or gained. Bifurcation points are of great interest in many applications, because near them the nature of the solution of the underlying differential equation is undergoing an abrupt change. For example, in fluid mechanics a smooth (laminar) flow may break up and become turbulent. Or an axially loaded column may suddenly buckle and exhibit a large lateral displacement. Or, as the amount of one of the chemicals in a certain mixture is increased, spiral wave patterns of varying color may suddenly emerge in an originally quiescent fluid. Problems 10 through 12 describe three types of bifurcations that can occur in simple equations of the form (i).

10. Consider the equation

$$dy/dt = a - y^2. \tag{ii}$$

(a) Find all of the critical points for Eq. (ii). Observe that there are no critical points if $a < 0$, one critical point if $a = 0$, and two critical points if $a > 0$.

(b) Draw the phase line in each case and determine whether each critical point is asymptotically stable, semistable, or unstable.

(c) In each case, sketch several solutions of Eq. (ii) in the ty-plane.

(d) If we plot the location of the critical points as a function of a in the ay-plane, we obtain Figure 2.5.11. This is called the **bifurcation diagram** for Eq. (ii). The bifurcation at $a = 0$ is called a **saddle–node** bifurcation. This name is more natural in the context of second order systems, which are discussed in Chapter 7.

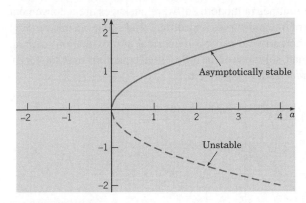

FIGURE 2.5.11 Bifurcation diagram for $y' = a - y^2$.

11. Consider the equation

$$dy/dt = ay - y^3 = y(a - y^2). \qquad \text{(iii)}$$

(a) Again consider the cases $a < 0$, $a = 0$, and $a > 0$. In each case, find the critical points, draw the phase line, and determine whether each critical point is asymptotically stable, semistable, or unstable.

(b) In each case, sketch several solutions of Eq. (iii) in the ty-plane.

(c) Draw the bifurcation diagram for Eq. (iii), that is, plot the location of the critical points versus a. For Eq. (iii), the bifurcation point at $a = 0$ is called a **pitchfork bifurcation**; your diagram may suggest why this name is appropriate.

12. Consider the equation

$$dy/dt = ay - y^2 = y(a - y). \qquad \text{(iv)}$$

(a) Again consider the cases $a < 0$, $a = 0$, and $a > 0$. In each case, find the critical points, draw the phase line, and determine whether each critical point is asymptotically stable, semistable, or unstable.

(b) In each case, sketch several solutions of Eq. (iv) in the ty-plane.

(c) Draw the bifurcation diagram for Eq. (iv). Observe that for Eq. (iv) there are the same number of critical points for $a < 0$ and $a > 0$ but that their stability has changed. For $a < 0$, the equilibrium solution $y = 0$ is asymptotically stable and $y = a$ is unstable, while for $a > 0$ the situation is reversed. Thus there has been an **exchange of stability** as a passes through the bifurcation point $a = 0$. This type of bifurcation is called a **transcritical bifurcation**.

2.6 Exact Equations and Integrating Factors

For first order equations, there are a number of integration methods that are applicable to various classes of problems. We have already discussed linear equations and separable equations. Here we consider a class of equations known as exact equations for which there is also a well-defined method of solution.

EXAMPLE 1

Solve the differential equation

$$2x + y^2 + 2xyy' = 0. \qquad \text{(1)}$$

The equation is neither linear nor separable, so the methods suitable for those types of equations are not applicable here. However observe that the function $\psi(x, y) = x^2 + xy^2$ has the property that

$$2x + y^2 = \frac{\partial \psi}{\partial x}, \qquad 2xy = \frac{\partial \psi}{\partial y}. \qquad \text{(2)}$$

Therefore the differential equation can be written as

$$2x + y^2 + 2xyy' = \frac{\partial \psi}{\partial x} + \frac{\partial \psi}{\partial y}\frac{dy}{dx} = 0. \tag{3}$$

Assuming that y is a function of x and using the multivariable chain rule, it follows that

$$\frac{\partial \psi}{\partial x} + \frac{\partial \psi}{\partial y}\frac{dy}{dx} = \frac{d\psi}{dx} = \frac{d}{dx}(x^2 + xy^2) = 0. \tag{4}$$

Therefore, by integrating with respect to x, we obtain

$$\psi(x, y) = x^2 + xy^2 = c, \tag{5}$$

where c is an arbitrary constant. Equation (5) defines solutions of Eq. (1) implicitly.

The integral curves of Eq. (1) are the level curves, or contour lines, of the function $\psi(x, y)$ given by Eq. (5). Contour plotting routines in modern software packages are a convenient way to plot a representative sample of integral curves for a differential equation, once $\psi(x, y)$ has been determined. This is an alternative to using a numerical approximation method, such as Euler's method, to approximate solutions of the differential equation (see Chapter 8). Some integral curves for Eq. (1) are shown in Figure 2.6.1.

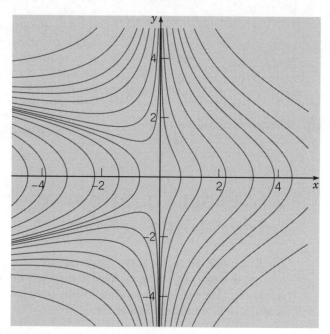

FIGURE 2.6.1 Integral curves of Eq. (1).

In solving Eq. (1), the key step was the recognition that there is a function ψ that satisfies Eqs. (2). More generally, let the differential equation

$$M(x, y) + N(x, y)y' = 0 \tag{6}$$

be given. Suppose that we can identify a function ψ such that

$$\frac{\partial \psi}{\partial x}(x, y) = M(x, y), \qquad \frac{\partial \psi}{\partial y}(x, y) = N(x, y), \tag{7}$$

and such that $\psi(x, y) = c$ defines $y = \phi(x)$ implicitly as a differentiable function of x. Then

$$M(x, y) + N(x, y)y' = \frac{\partial \psi}{\partial x} + \frac{\partial \psi}{\partial y} \frac{dy}{dx} = \frac{d}{dx} \psi[x, \phi(x)]$$

and the differential equation (6) becomes

$$\frac{d}{dx} \psi[x, \phi(x)] = 0. \tag{8}$$

In this case, Eq. (6) is said to be an **exact** differential equation. Solutions of Eq. (6), or the equivalent Eq. (8), are given implicitly by

$$\psi(x, y) = c, \tag{9}$$

where c is an arbitrary constant.

In Example 1, it was relatively easy to see that the differential equation was exact and, in fact, easy to find its solution, by recognizing the required function ψ. For more complicated equations, it may not be possible to do this so easily. A systematic way of determining whether a given differential equation is exact is provided by the following theorem.

THEOREM
2.6.1

Let the functions M, N, M_y, and N_x, where subscripts denote partial derivatives, be continuous in the rectangular[7] region R: $\alpha < x < \beta, \gamma < y < \delta$. Then Eq. (6),

$$M(x, y) + N(x, y)y' = 0,$$

is an exact differential equation in R if and only if

$$M_y(x, y) = N_x(x, y) \tag{10}$$

at each point of R. That is, there exists a function ψ satisfying Eqs. (7),

$$\psi_x(x, y) = M(x, y), \qquad \psi_y(x, y) = N(x, y),$$

if and only if M and N satisfy Eq. (10).

The proof of this theorem has two parts. First, we show that if there is a function ψ such that Eqs. (7) are true, then it follows that Eq. (10) is satisfied. Computing M_y and N_x from Eqs. (7), we obtain

$$M_y(x, y) = \psi_{xy}(x, y), \qquad N_x(x, y) = \psi_{yx}(x, y). \tag{11}$$

Since M_y and N_x are continuous, it follows that ψ_{xy} and ψ_{yx} are also continuous. This guarantees their equality by Clairaut's theorem, and Eq. (10) follows.

We now show that if M and N satisfy Eq. (10), then Eq. (6) is exact. The proof involves the construction of a function ψ satisfying Eqs. (7),

$$\psi_x(x, y) = M(x, y), \qquad \psi_y(x, y) = N(x, y).$$

We begin by integrating the first of Eqs. (7) with respect to x, holding y constant. We obtain

$$\psi(x, y) = Q(x, y) + h(y), \tag{12}$$

---

[7]It is not essential that the region be rectangular, only that it be simply connected. In two dimensions, this means that the region has no holes in its interior. Thus, for example, rectangular or circular regions are simply connected, but an annular region is not. More details can be found in most books on advanced calculus.

where $Q(x, y)$ is any differentiable function such that $\partial Q(x, y)/\partial x = M(x, y)$. For example, we might choose

$$Q(x, y) = \int_{x_0}^{x} M(s, y) \, ds, \tag{13}$$

where x_0 is some specified constant in $\alpha < x_0 < \beta$. The function h in Eq. (12) is an arbitrary differentiable function of y, playing the role of the arbitrary constant. Now we must show that it is always possible to choose $h(y)$ so that the second of Eqs. (7) is satisfied, that is, $\psi_y = N$. By differentiating Eq. (12) with respect to y and setting the result equal to $N(x, y)$, we obtain

$$\psi_y(x, y) = \frac{\partial Q}{\partial y}(x, y) + h'(y) = N(x, y).$$

Then, solving for $h'(y)$, we have

$$h'(y) = N(x, y) - \frac{\partial Q}{\partial y}(x, y). \tag{14}$$

In order for us to determine $h(y)$ from Eq. (14), the right side of Eq. (14), despite its appearance, must be a function of y only. To establish that this is true, we can differentiate the quantity in question with respect to x, obtaining

$$\frac{\partial N}{\partial x}(x, y) - \frac{\partial}{\partial x}\frac{\partial Q}{\partial y}(x, y). \tag{15}$$

By interchanging the order of differentiation in the second term of expression (15), we have

$$\frac{\partial N}{\partial x}(x, y) - \frac{\partial}{\partial y}\frac{\partial Q}{\partial x}(x, y),$$

or, since $\partial Q/\partial x = M$,

$$\frac{\partial N}{\partial x}(x, y) - \frac{\partial M}{\partial y}(x, y),$$

which is zero because of Eq. (10). Hence, despite its apparent form, the right side of Eq. (14) does not, in fact, depend on x. Then we find $h(y)$ by integrating Eq. (14), and upon substituting this function in Eq. (12), we obtain the required function $\psi(x, y)$. This completes the proof of Theorem 2.6.1.

It is possible to obtain an explicit expression for $\psi(x, y)$ in terms of integrals (see Problem 17), but in solving specific exact equations, it is usually simpler and easier just to repeat the procedure used in the preceding proof. That is, integrate $\psi_x = M$ with respect to x, including an arbitrary function of $h(y)$ instead of an arbitrary constant, and then differentiate the result with respect to y and set it equal to N. Finally, use this last equation to solve for $h(y)$. The next example illustrates this procedure.

EXAMPLE 2

Solve the differential equation

$$(y \cos x + 2xe^y) + (\sin x + x^2 e^y - 1)y' = 0. \tag{16}$$

It is easy to see that

$$M_y(x, y) = \cos x + 2xe^y = N_x(x, y),$$

so the given equation is exact. Thus there is a $\psi(x, y)$ such that

$$\psi_x(x, y) = y \cos x + 2xe^y = M(x, y),$$
$$\psi_y(x, y) = \sin x + x^2 e^y - 1 = N(x, y).$$

Integrating the first of these equations with respect to x, we obtain

$$\psi(x, y) = y \sin x + x^2 e^y + h(y). \tag{17}$$

Next, finding ψ_y from Eq. (17) and then setting the result equal to N give

$$\psi_y(x, y) = \sin x + x^2 e^y + h'(y) = \sin x + x^2 e^y - 1.$$

Thus $h'(y) = -1$ and $h(y) = -y$. The constant of integration can be omitted since any additive constant tacked onto $h(y)$ will just be absorbed into the c in $\psi(x, y) = c$. Substituting for $h(y)$ in Eq. (17) gives

$$\psi(x, y) = y \sin x + x^2 e^y - y.$$

Hence solutions of Eq. (16) are given implicitly by

$$y \sin x + x^2 e^y - y = c. \tag{18}$$

If an initial condition is prescribed, then it determines the value of c corresponding to the integral curve passing through the given initial point. For example, if the initial condition is $y(3) = 0$, then $c = 9$. Some integral curves of Eq. (16) are shown in Figure 2.6.2; the one passing through $(3, 0)$ is heavier than the others.

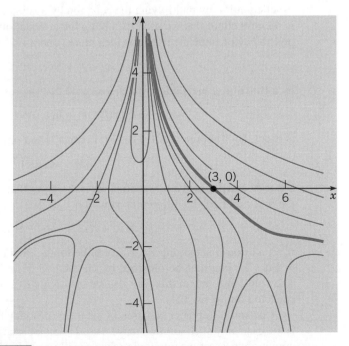

FIGURE 2.6.2 Integral curves of Eq. (16); the heavy curve is the integral curve through the initial point (3, 0).

EXAMPLE 3

Solve the differential equation

$$(3xy + y^2) + (x^2 + xy)y' = 0. \tag{19}$$

Here

$$M_y(x, y) = 3x + 2y, \qquad N_x(x, y) = 2x + y;$$

since $M_y \neq N_x$, the given equation is not exact. To see that it cannot be solved by the procedure described above, let us seek a function ψ such that

$$\psi_x(x, y) = 3xy + y^2 = M(x, y), \qquad \psi_y(x, y) = x^2 + xy = N(x, y). \tag{20}$$

Integrating the first of Eqs. (20) gives

$$\psi(x, y) = \tfrac{3}{2}x^2y + xy^2 + h(y), \tag{21}$$

where h is an arbitrary function of y only. To try to satisfy the second of Eqs. (20), we compute ψ_y from Eq. (21) and set it equal to N, obtaining

$$\tfrac{3}{2}x^2 + 2xy + h'(y) = x^2 + xy$$

or

$$h'(y) = -\tfrac{1}{2}x^2 - xy. \tag{22}$$

Since the right side of Eq. (22) depends on x as well as y, it is impossible to solve Eq. (22) for $h(y)$. Thus there is no $\psi(x, y)$ satisfying both of Eqs. (20).

▶ Integrating Factors. It is sometimes possible to convert a differential equation that is not exact into an exact equation by multiplying the equation by a suitable integrating factor. Recall that this is the procedure that we used in solving linear equations in Section 2.2. To investigate the possibility of implementing this idea more generally, let us multiply the equation

$$M(x, y) + N(x, y)y' = 0 \tag{23}$$

by a function μ and then try to choose μ so that the resulting equation

$$\mu(x, y)M(x, y) + \mu(x, y)N(x, y)y' = 0 \tag{24}$$

is exact. By Theorem 2.6.1, Eq. (24) is exact if and only if

$$(\mu M)_y = (\mu N)_x. \tag{25}$$

Since M and N are given functions, Eq. (25) states that the integrating factor μ must satisfy the first order partial differential equation

$$M\mu_y - N\mu_x + (M_y - N_x)\mu = 0. \tag{26}$$

If a function μ satisfying Eq. (26) can be found, then Eq. (24) will be exact. The solution of Eq. (24) can then be obtained by the method described in the first part of this section. The solution found in this way also satisfies Eq. (23), since the integrating factor μ can be canceled out of Eq. (24).

A partial differential equation of the form (26) may have more than one solution. If this is the case, any such solution may be used as an integrating factor of Eq. (23). This possible nonuniqueness of the integrating factor is illustrated in Example 4.

Unfortunately, Eq. (26), which determines the integrating factor μ, is ordinarily at least as hard to solve as the original equation (23). Therefore, although in principle, integrating factors are powerful tools for solving differential equations, in practice, they can be found only in special cases. The most important situations in which simple integrating factors can be found occur when μ is a function of only one of the variables x or y, instead of both. Let us determine a condition on M and N so that Eq. (23) has an integrating factor μ that depends only on x. Assuming that μ is a function only of x, we have

$$(\mu M)_y = \mu M_y, \qquad (\mu N)_x = \mu N_x + N \frac{d\mu}{dx} .$$

Thus, if $(\mu M)_y$ is to equal $(\mu N)_x$, it is necessary that

$$\frac{d\mu}{dx} = \frac{M_y - N_x}{N} \mu. \tag{27}$$

If $(M_y - N_x)/N$ is a function only of x, then there is an integrating factor μ that also depends only on x. Further $\mu(x)$ can be found by solving Eq. (27), which is both linear and separable.

A similar procedure can be used to determine a condition under which Eq. (23) has an integrating factor depending only on y; see Problem 23.

EXAMPLE 4

Find an integrating factor for the equation

$$(3xy + y^2) + (x^2 + xy)y' = 0 \tag{19}$$

and then solve the equation.

In Example 3, we showed that this equation is not exact. Let us determine whether it has an integrating factor that depends only on x. By computing the quantity $(M_y - N_x)/N$, we find that

$$\frac{M_y(x, y) - N_x(x, y)}{N(x, y)} = \frac{3x + 2y - (2x + y)}{x^2 + xy} = \frac{x + y}{x(x + y)} = \frac{1}{x} . \tag{28}$$

Thus there is an integrating factor μ that is a function only of x, and it satisfies the differential equation

$$\frac{d\mu}{dx} = \frac{\mu}{x} . \tag{29}$$

Hence

$$\mu(x) = x. \tag{30}$$

Multiplying Eq. (19) by this integrating factor, we obtain

$$(3x^2 y + xy^2) + (x^3 + x^2 y)y' = 0. \tag{31}$$

The latter equation is exact, and it is easy to show that its solutions are given implicitly by

$$x^3 y + \tfrac{1}{2} x^2 y^2 = c. \tag{32}$$

Solutions may also be readily found in explicit form since Eq. (32) is quadratic in y. Some integral curves of Eq. (19) are shown in Figure 2.6.3.

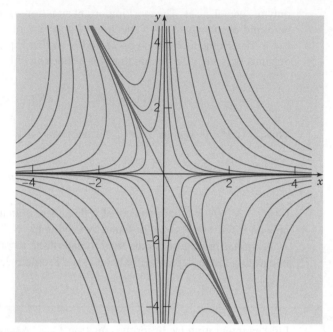

FIGURE 2.6.3 Integral curves of Eq. (19).

You may also verify that a second integrating factor of Eq. (19) is

$$\mu(x, y) = \frac{1}{xy(2x + y)},$$

and that the same solution is obtained, though with much greater difficulty, if this integrating factor is used (see Problem 32).

PROBLEMS

Exact Equations. In each of Problems 1 through 12:
(a) Determine whether the equation is exact. If it is exact, then:
(b) Solve the equation.
(c) Use a computer to draw several integral curves.

1. $(2x + 3) + (2y - 2)y' = 0$

2. $(2x + 4y) + (2x - 2y)y' = 0$

3. $(3x^2 - 2xy + 2) + (6y^2 - x^2 + 3)y' = 0$

4. $(2xy^2 + 2y) + (2x^2y + 2x)y' = 0$

5. $\dfrac{dy}{dx} = -\dfrac{4x + 2y}{2x + 3y}$

6. $\dfrac{dy}{dx} = -\dfrac{4x - 2y}{2x - 3y}$

7. $(e^x \sin y - 2y \sin x) + (e^x \cos y + 2\cos x)y' = 0$

8. $(e^x \sin y + 3y) - (3x - e^x \sin y)y' = 0$

9. $(ye^{xy} \cos 2x - 2e^{xy} \sin 2x + 2x) + (xe^{xy} \cos 2x - 3)y' = 0$

10. $(y/x + 6x) + (\ln x - 2)y' = 0, \qquad x > 0$

11. $(x \ln y + xy) + (y \ln x + xy)y' = 0; \quad x > 0, \quad y > 0$

12. $\dfrac{x}{(x^2 + y^2)^{3/2}} + \dfrac{y}{(x^2 + y^2)^{3/2}}y' = 0$

In each of Problems 13 and 14, solve the given initial value problem and determine, at least approximately, where the solution is valid.

13. $(2x - y) + (2y - x)y' = 0, \qquad y(1) = 3$

14. $(9x^2 + y - 1) - (4y - x)y' = 0, \qquad y(1) = 0$

In each of Problems 15 and 16, find the value of b for which the given equation is exact, and then solve it using that value of b.

15. $(xy^2 + bx^2y) + (x + y)x^2y' = 0$

16. $(ye^{2xy} + x) + bxe^{2xy}y' = 0$

17. Assume that Eq. (6) meets the requirements of Theorem 2.6.1 in a rectangle R and is therefore exact. Show that a

possible function $\psi(x, y)$ is

$$\psi(x, y) = \int_{x_0}^{x} M(s, y_0)\, ds + \int_{y_0}^{y} N(x, t)\, dt,$$

where (x_0, y_0) is a point in R.

18. Show that any separable equation

$$M(x) + N(y)y' = 0$$

is also exact.

Integrating Factors. In each of Problems 19 through 22:

(a) Show that the given equation is not exact but becomes exact when multiplied by the given integrating factor.
(b) Solve the equation.
(c) Use a computer to draw several integral curves.

19. $x^2y^3 + x(1 + y^2)y' = 0, \qquad \mu(x, y) = 1/xy^3$

20. $\left(\dfrac{\sin y}{y} - 2e^{-x} \sin x \right)$

$$+ \left(\dfrac{\cos y + 2e^{-x} \cos x}{y} \right) y' = 0, \qquad \mu(x, y) = ye^x$$

21. $y + (2x - ye^y)y' = 0, \qquad \mu(x, y) = y$

22. $(x + 2) \sin y + x \cos y\, y' = 0, \qquad \mu(x, y) = xe^x$

23. Show that if $(N_x - M_y)/M = Q$, where Q is a function of y only, then the differential equation

$$M + Ny' = 0$$

has an integrating factor of the form

$$\mu(y) = \exp \int Q(y)\, dy.$$

24. Show that if $(N_x - M_y)/(xM - yN) = R$, where R depends on the quantity xy only, then the differential equation

$$M + Ny' = 0$$

has an integrating factor of the form $\mu(xy)$. Find a general formula for this integrating factor.

In each of Problems 25 through 31:
(a) Find an integrating factor and solve the given equation.
(b) Use a computer to draw several integral curves.

25. $(3x^2y + 2xy + y^3) + (x^2 + y^2)y' = 0$

26. $y' = e^{2x} + y - 1$

27. $1 + (x/y - \sin y)y' = 0$

28. $y + (2xy - e^{-2y})y' = 0$

29. $e^x + (e^x \cot y + 2y \csc y)y' = 0$

30. $4\left(\dfrac{x^3}{y^2} + \dfrac{3}{y} \right) + 3\left(\dfrac{x}{y^2} + 4y \right) y' = 0$

31. $\left(3x + \dfrac{6}{y} \right) + \left(\dfrac{x^2}{y} + 3\dfrac{y}{x} \right) \dfrac{dy}{dx} = 0$

Hint: See Problem 24.

32. Use the integrating factor $\mu(x, y) = [xy(2x+y)]^{-1}$ to solve the differential equation

$$(3xy + y^2) + (x^2 + xy)y' = 0.$$

Verify that the solution is the same as that obtained in Example 4 with a different integrating factor.

2.7 Substitution Methods

In the preceding sections we developed techniques for solving three important classes of differential equations, namely, separable, linear, and exact. But the differential equations arising in many, if not most, applications do not fall into these three categories. In some cases, though, an appropriate substitution or a change of variable can be used to transform the equation into a member of one of these classes. This section focuses on two such types of equations.

Homogeneous Differential Equations

A function $f(x, y)$ is **homogeneous of degree k** if

$$f(\lambda x, \lambda y) = \lambda^k f(x, y), \tag{1}$$

for all (x, y) in its domain. For example, $f(x, y) = \dfrac{x^2 - xy + y^2}{xy}$ is homogeneous of degree 0 because

$$f(\lambda x, \lambda y) = \dfrac{\lambda^2 x^2 - (\lambda x)(\lambda y) + \lambda^2 y^2}{(\lambda x)(\lambda y)} = \dfrac{\lambda^2 \left[x^2 - xy + y^2 \right]}{\lambda^2 \left[xy \right]} = \lambda^0 f(x, y)$$

and $f(x, y) = x \ln x - x \ln y$ is homogeneous of degree 1 because

$$f(\lambda x, \lambda y) = \lambda x \left(\ln(\lambda x) - \ln(\lambda y) \right) = \lambda x \ln \left(\frac{x}{y} \right) = \lambda^1 f(x, y).$$

Accordingly, we define the following class of differential equations.

DEFINITION 2.7.1	**Homogeneous Differential Equation.** A differential equation of the form $M(x, y) + N(x, y)\frac{dy}{dx} = 0$ is **homogeneous** if $M(x, y)$ and $N(x, y)$ are homogeneous functions of the same degree k.

Generally, if $f(x, y)$ is homogeneous of degree k, then $f(x, y)$ can be expressed equivalently as $x^k \cdot f\left(1, \frac{y}{x} \right)$ and as $y^k \cdot f\left(\frac{x}{y}, 1 \right)$. Table 2.7.1 illustrates these forms for the above two functions.

TABLE 2.7.1   Equivalent forms of homogeneous functions.

$f(x, y)$	Written in the form $x^k \cdot f\left(1, \frac{y}{x} \right)$	Written in the form $y^k \cdot f\left(\frac{x}{y}, 1 \right)$
$\dfrac{x^2 - xy + y^2}{xy}$	$x^0 \cdot \dfrac{1 - \left(\frac{y}{x} \right) + \left(\frac{y}{x} \right)^2}{\left(\frac{y}{x} \right)}$	$y^0 \cdot \dfrac{\left(\frac{x}{y} \right)^2 - \left(\frac{x}{y} \right) + 1}{\left(\frac{x}{y} \right)}$
$x \ln x - x \ln y$	$x^1 \cdot \left[-\ln \left(\frac{y}{x} \right) \right]$	$y^1 \cdot \left[\left(\frac{x}{y} \right) \ln \left(\frac{x}{y} \right) \right]$

Consider a homogeneous equation

$$M(x, y) + N(x, y)\frac{dy}{dx} = 0, \tag{2}$$

where $M(x, y)$ and $N(x, y)$ are homogeneous functions of degree k. Using the above observation,

$$M(x, y) = x^k \cdot M\left(1, \frac{y}{x} \right), \qquad N(x, y) = x^k \cdot N\left(1, \frac{y}{x} \right). \tag{3}$$

Substituting (3) into (2) and simplifying, we obtain

$$\frac{dy}{dx} = -\frac{M(x, y)}{N(x, y)} = -\frac{x^k M\left(1, \frac{y}{x} \right)}{x^k N\left(1, \frac{y}{x} \right)} = -\frac{M\left(1, \frac{y}{x} \right)}{N\left(1, \frac{y}{x} \right)}. \tag{4}$$

Let us define a new variable u by $u = \frac{y}{x}$. Note that u is a function of x because y is a function of x. This is equivalent to $y = ux$, so

$$\frac{dy}{dx} = u + x\frac{du}{dx}.$$

Substituting these into (4) produces the following differential equation in terms of u and x:

$$u + x\frac{du}{dx} = \underbrace{-\frac{M(1, u)}{N(1, u)}}_{\text{A function of } u}. \tag{5}$$

Observe that Eq. (5) is separable. Indeed, simplifying further yields

$$-\left(\frac{1}{\frac{M(1,u)}{N(1,u)} + u}\right) du = \frac{1}{x} dx.$$

This equation can be solved as in Section 2.1 to obtain a function, typically implicitly defined, in terms of u and x. The solution of the original equation is then obtained by resubstituting in $u = \frac{y}{x}$.

Similarly, if

$$M(x, y) = y^k \cdot M\left(\frac{x}{y}, 1\right), \qquad N(x, y) = y^k \cdot N\left(\frac{x}{y}, 1\right) \tag{6}$$

is used instead of Eq. (3), then define a new variable v by $v = \frac{x}{y}$. Viewing y as the independent variable, note that v is a function of y because x is a function of y, so

$$\frac{dx}{dy} = v + y\frac{dv}{dy}.$$

In the same manner, substituting these into Eq. (2) would produce a separable equation in terms of v and y.

When faced with solving a homogeneous differential equation, either of these substitutions will work, but often using one form over the other will significantly reduce the algebra involved.

Remark. When solving a homogeneous differential equation, it is not necessary to first rewrite the functions involved in terms of the quantities $\frac{y}{x}$ or $\frac{x}{y}$. Rather than introducing the variable u or v in its original form of $\frac{y}{x}$ or $\frac{x}{y}$, respectively, it is more practical to make the substitution $y = ux$ or $x = vy$ directly.

EXAMPLE 1

Solve the differential equation

$$\frac{dy}{dx} = \frac{x^2 - xy + y^2}{xy}. \tag{7}$$

This equation is not separable, linear, or exact. But it can be written in the form Eq. (2), where

$$M(x, y) = -\frac{x^2 - xy + y^2}{xy}, \qquad N(x, y) = 1.$$

Both $M(x, y)$ and $N(x, y)$ are homogeneous of degree 0, so Eq. (7) is homogeneous. Following the above discussion, and using the substitution

$$y = ux, \qquad \frac{dy}{dx} = u + x\frac{du}{dx}$$

in Eq. (7) yields

$$u + x\frac{du}{dx} = \frac{x^2 - x(ux) + (ux)^2}{x(ux)} = \frac{1 - u + u^2}{u}. \tag{8}$$

Separating the variables in Eq. (8) leads to

$$\frac{u}{1 - u} du = \frac{1}{x} dx. \tag{9}$$

We solve Eq. (9) by integrating both sides to arrive at the implicitly defined solution

$$-u - \ln|1 - u| = \ln|x| + c. \tag{10}$$

The solution of Eq. (7) is then obtained by resubstituting $u = \frac{y}{x}$ into Eq. (10):

$$\frac{y}{x} + \ln|x - y| = c, \qquad x \neq 0. \tag{11}$$

Some integral curves for Eq. (7) are shown in Figure 2.7.1.

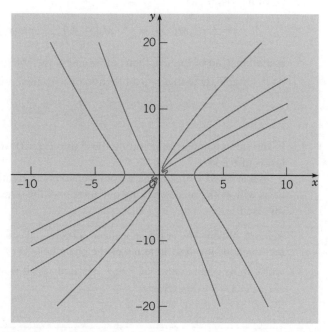

FIGURE 2.7.1   Integral curves for Eq. (7).

EXAMPLE
2

Solve the differential equation

$$\left(2x^2y + 4y^2x + 4y^3\right) \frac{dy}{dx} = xy^2. \tag{12}$$

This equation can be written in the form Eq. (2), where

$$M(x, y) = -xy^2, \qquad N(x, y) = 2x^2y + 4y^2x + 4y^3.$$

Both $M(x, y)$ and $N(x, y)$ are homogeneous of degree 3, so Eq. (12) is homogeneous. For illustrative purposes we shall use the substitution

$$x = vy, \qquad \frac{dx}{dy} = v + y\frac{dv}{dy}$$

in Eq. (12). Before doing so, we must rewrite Eq. (12) as

$$\left(2x^2y + 4y^2x + 4y^3\right) = xy^2\frac{dx}{dy}, \tag{13}$$

so that the substitution can be readily made. Doing so yields

$$\left(2(vy)^2 y + 4y^2(vy) + 4y^3\right) = (vy)y^2 \left[v + y\frac{dv}{dy}\right]. \tag{14}$$

Simplifying Eq. (14) yields

$$v^2 + 4v + 4 = vy\frac{dv}{dy}. \tag{15}$$

Separating the variables in Eq. (15) leads to

$$\frac{v}{(v+2)^2}dv = \frac{1}{y}dy. \tag{16}$$

We solve Eq. (16) by integrating both sides to arrive at the implicitly defined solution

$$\ln|v+2| + \frac{2}{v+2} = \ln|y| + c. \tag{17}$$

The solution of Eq. (12) is then obtained by resubstituting $v = \frac{x}{y}$ into Eq. (17):

$$\ln\left|\frac{x+2y}{y}\right| + \frac{2y}{x+2y} = \ln|y| + c, \qquad y \neq 0. \tag{18}$$

Remark. Looking back, had we used the substitution

$$y = ux, \qquad \frac{dy}{dx} = u + x\frac{du}{dx},$$

the algebra in Eq. (14) would have been slightly worse in that simplifying the left side would have entailed multiplying two binomials, whereas we only had to multiply a monomial times a binomial in Eq. (14) when using $x = vy$.

Bernoulli Differential Equations

A first order differential equation related to linear differential equations is the so-called *Bernoulli equation*, named after Jacob Bernoulli (1654–1705) and solved first by Leibnitz in 1696. Such an equation has the following form.

DEFINITION 2.7.2

Bernoulli Differential Equation. A differential equation of the form

$$\frac{dy}{dt} + q(t)y = r(t)y^n, \tag{19}$$

where n is any real number, is called a **Bernoulli equation**.

If $n = 0$, then Eq. (19) is linear, and if $n = 1$, then Eq. (19) is separable, linear, and homogeneous. For all other real values of n, Eq. (19) is not one of the forms studied thus far in the chapter.

To solve a Bernoulli equation when n is neither 0 nor 1, we shall make a substitution that reduces it to a linear equation that can subsequently be solved using the method of integrating factors. Specifically, we perform the following initial steps to transform Eq. (19) into a linear equation.

First divide Eq. (19) by y^n to obtain

$$y^{-n}\frac{dy}{dt} + q(t)y^{1-n} = r(t). \tag{20}$$

Define $u = y^{1-n}$, which is a function of t. Observe that

$$\frac{du}{dt} = (1-n)y^{-n}\frac{dy}{dt},$$

or equivalently,

$$y^{-n}\frac{dy}{dt} = \frac{1}{(1-n)}\frac{du}{dt}. \tag{21}$$

Substituting Eq. (21) into Eq. (20) yields

$$\frac{1}{1-n}\frac{du}{dt} + q(t)u(t) = r(t),$$

and subsequently,

$$\frac{du}{dt} + \underbrace{(1-n)q(t)}_{\text{Call this } p(t)}u(t) = \underbrace{(1-n)r(t)}_{\text{Call this } g(t)}, \tag{22}$$

which is a linear differential equation (in u).

Now solve Eq. (22) as you would any other linear differential equation. Once you obtain the solution $u(t)$, resubstitute $u(t) = y^{1-n}$ to determine the solution $y(t)$ of the original differential equation (19).

EXAMPLE 3

Solve the initial value problem

$$\frac{dy}{dt} + y = y^3, \quad y(0) = y_0, \tag{23}$$

where $-1 < y_0 < 1$. Determine the long-term behavior of the solution of Eq. (23) for such initial conditions.

To begin, divide both sides of the equation by y^3 to obtain

$$y^{-3}\frac{dy}{dt} + y^{-2} = 1.$$

Let $u = y^{-2}$ and observe that

$$\frac{du}{dt} = -2y^{-3}\frac{dy}{dt},$$

or equivalently,

$$-\frac{1}{2}\frac{du}{dt} = y^{-3}\frac{dy}{dt}. \tag{24}$$

Using the new variable u with Eq. (24) transforms the original equation into the linear equation

$$\frac{du}{dt} - 2u(t) = -2. \tag{25}$$

Solving Eq. (25) using the method of integrating factors leads to

$$u(t) = 1 + Ce^{2t}. \tag{26}$$

Resubstituting $u = y^{-2}$ into Eq. (26), we conclude that the general solution of Eq. (23) is

$$y(t) = \left(\frac{1}{1 + Ce^{2t}}\right)^{1/2}.$$ (27)

Applying the initial condition, we find that

$$C = \frac{1 - y_0^2}{y_0^2},$$

so that the solution of Eq. (23) is

$$y(t) = \frac{y_0}{\sqrt{y_0^2 + \left(1 - y_0^2\right) e^{2t}}}.$$ (28)

Solution curves of Eq. (23) for various values of y_0 are shown in Figure 2.7.2.

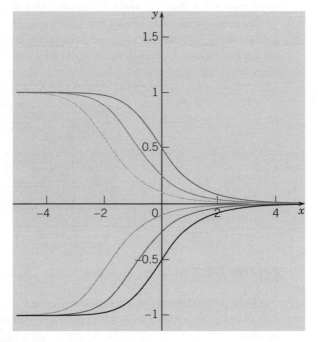

FIGURE 2.7.2 Solution curves for Eq. (23).

It appears from Figure 2.7.2 that the solution curves approach zero as $t \to \infty$, at least for the initial data used. To make this observation precise, observe that Eq. (23) is an autonomous equation that can be written as

$$\frac{dy}{dt} = y^3 - y = y(y - 1)(y + 1).$$ (29)

The equilibrium points are -1, 0, and 1 and the phase line is given by Figure 2.7.3. As such, y_0 is an asymptotically stable equilibrium point, so we conclude that indeed all solutions of Eq. (23) for which $-1 < y_0 < 1$ will tend to zero as $t \to \infty$.

FIGURE 2.7.3 Phase line for Eq. (29).

Relationships Among Classes of Equations

We have developed techniques for solving separable, linear, and exact equations, as well as transformation methods used to convert other equations (e.g., homogeneous and Bernoulli equations) into one of these types. The initial struggle you face when solving a first order differential equation is determining which of these techniques, if any, is applicable. In fact, sometimes more than one approach can be used to solve an equation.

The interrelationships among the main equation types are displayed in Figure 2.7.4. We use arrows to indicate that the type of equation listed near its tail can be transformed into the type of equation to which the arrowhead points.

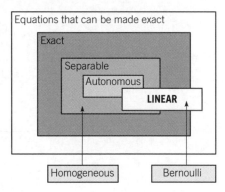

FIGURE 2.7.4 Interrelationships among equation types.

There is a collection of exercises at the end of the section that will challenge you to classify equations and to solve those that have multiple classifications using more than one method.

PROBLEMS

Homogeneous Differential Equations. In each of Problems 1 through 10:

(a) Determine if the equation is homogeneous. If it is homogeneous, then:

(b) Solve the equation.

(c) Use a computer to draw several integral curves.

1. $y\dfrac{dy}{dx} = x + 1$

2. $(y^4 + 1)\dfrac{dy}{dx} = x^4 + 1$

3. $\dfrac{3x^3 - xy^2}{3x^2 y + y^3} \cdot \dfrac{dy}{dx} = 1$

4. $x(x - 1)\dfrac{dy}{dx} = y(y + 1)$

5. $\sqrt{x^2 - y^2} + y = x\dfrac{dy}{dx}$

6. $xy\dfrac{dy}{dx} = (x + y)^2$

7. $\dfrac{dy}{dx} = \dfrac{4y - 7x}{5x - y}$

8. $x\dfrac{dy}{dx} - 4\sqrt{y^2 - x^2} = y, \qquad y > 0$

9. $\dfrac{dy}{dx} = \dfrac{y^4 + 2xy^3 - 3x^2y^2 - 2x^3y}{2x^2y^2 - 2x^3y - 2x^4}$

10. $(y + xe^{x/y})\dfrac{dy}{dx} = ye^{x/y}$

In Problems 11 and 12, solve the given initial value problem and determine, at least approximately, where the solution is valid.

11. $xy\dfrac{dy}{dx} = x^2 + y^2, \quad y(2) = 1$

12. $\dfrac{dy}{dx} = \dfrac{x+y}{x-y}, \quad y(5) = 8$

Bernoulli Differential Equations. In each of Problems 13 through 22:

(a) Write the Bernoulli equation in the proper form (19).
(b) Solve the equation.
(c) Use a computer to draw several integral curves.

13. $t\dfrac{dy}{dt} + y = t^2y^2$

14. $\dfrac{dy}{dt} = y\left(ty^3 - 1\right)$

15. $\dfrac{dy}{dt} + \dfrac{3}{t}y = t^2y^2$

16. $t^2y' + 2ty - y^3 = 0, \qquad t > 0$

17. $5(1 + t^2)\dfrac{dy}{dt} = 4ty\left(y^3 - 1\right)$

18. $3t\dfrac{dy}{dt} + 9y = 2ty^{5/3}$

19. $\dfrac{dy}{dt} = y + \sqrt{y}$

20. $y' = ry - ky^2$, $r > 0$ and $k > 0$. This equation is important in population dynamics and is discussed in detail in Section 2.5.

21. $y' = \epsilon y - \sigma y^3$, $\epsilon > 0$ and $\sigma > 0$. This equation occurs in the study of the stability of fluid flow.

22. $dy/dt = (\Gamma\cos t + T)y - y^3$, where Γ and T are constants. This equation also occurs in the study of the stability of fluid flow.

23. A differential equation of the form

$$\dfrac{dy}{dt} = A(t) + B(t)y + C(t)y^2 \qquad \text{(i)}$$

is called a **Riccati equation**. Such equations arise in optimal control theory.

(a) If y_1 is a known solution of (i), prove that the substitution $y = y_1 + v$ transforms (i) into a Bernoulli equation with $n = 2$.
(b) Solve the equation $\dfrac{dy}{dt} + 3ty = 4 - 4t^2 + y^2$, after showing that it has $y = 4t$ as a particular solution.

Mixed Practice. In each of Problems 24 through 36:

(a) List each of the following classes into which the equation falls: autonomous, separable, linear, exact, Bernoulli, homogeneous.
(b) Solve the equation. If it has more than one classification, solve it *two* different ways.

24. $(3x - y)\dfrac{dx}{dy} + (9y - 2x) = 0$

25. $1 = (3e^y - 2x)\dfrac{dy}{dx}$

26. $\dfrac{dy}{dx} - 4e^xy^2 = y$

27. $x\dfrac{dy}{dx} + (x + 1)y = x$

28. $\dfrac{dy}{dx} = \dfrac{xy^2 - \frac{1}{2}\sin 2x}{\left(1 - x^2\right)y}$

29. $\dfrac{\sqrt{x}}{y}\dfrac{dy}{dx} = 1$

30. $\left(5xy^2 + 5y\right) + \left(5x^2y + 5x\right)\dfrac{dy}{dx} = 0$

31. $2xy\dfrac{dy}{dx} + \ln x = -y^2 - 1$

32. $(2 - x)\dfrac{dy}{dx} = y + 2(2 - x)^5$

33. $x\dfrac{dy}{dx} = -\dfrac{1}{\ln x}$

34. $\dfrac{dx}{dy} = \dfrac{2xy + x^2}{3y^2 + 2xy}$

35. $4xy\dfrac{dy}{dx} = 8x^2 + 5y^2$

36. $\dfrac{dy}{dx} + y - \sqrt[4]{y} = 0$

CHAPTER SUMMARY

In this chapter we discuss a number of special solution methods for first order equations $dy/dt = f(t, y)$. The most important types of equations that can be solved analytically are **linear**, **separable**, and **exact** equations. Others, like Bernoulli and homogeneous equations, can be transformed into one of these. For equations that cannot be solved by symbolic analytic methods, it is necessary to resort to geometrical and numerical methods.

Some aspects of the qualitative theory of differential equations are also introduced in this chapter: existence and uniqueness of solutions; stability properties of equilibrium solutions of autonomous equations.

Section 2.1   Separable Equations $M(x) + N(y)dy/dx = 0$ can be solved by direct integration.

Section 2.2   Linear Equations $\frac{dy}{dt} + p(t)y = g(t)$ can be solved using the method of integrating factors.

Section 2.3   Modeling We discuss mathematical models for several types of problems that lead to either linear or separable equations: mixing tanks, compound interest, and projectile motion.

Section 2.4   Qualitative Theory We study the existence and uniqueness of solutions to initial value problems.

▶ Conditions guaranteeing existence and uniqueness of solutions are given in Theorems 2.4.1 and 2.4.2 for linear and nonlinear equations, respectively.

▶ We show examples of initial value problems where solutions are not unique or become unbounded in finite time.

Section 2.5   Qualitative Theory We investigate autonomous equations, equilibrium solutions, and their stability characteristics.

▶ Autonomous equations are of the form $dy/dt = f(y)$.

▶ Critical points (equilibrium solutions) are solutions of $f(y) = 0$.

▶ Whether an equilibrium solution is **asymptotically stable**, **semistable**, or **unstable** determines to a great extent the long-time (asymptotic) behavior of solutions.

Section 2.6   Exact Equations $M(x, y)dx + N(x, y)dy = 0$ is exact if and only if $\partial M/\partial y = \partial N/\partial x$.

▶ Direct integration of an exact equation leads to implicitly defined solutions $F(x, y) = c$, where $\partial F/\partial x = M$ and $\partial F/\partial y = N$.

▶ Some differential equations can be made exact if a special integrating factor can be found.

Section 2.7   Substitution Methods Bernoulli equations can be transformed into linear equations, and homogeneous equations can be transformed into separable equations.

PROJECTS

Project 1 Harvesting a Renewable Resource

Suppose that the population y of a certain species of fish (e.g., tuna or halibut) in a given area of the ocean is described by the logistic equation

$$dy/dt = r(1 - y/K)y.$$

If the population is subjected to harvesting at a rate $H(y, t)$ members per unit time, then the harvested population is modeled by the differential equation

$$dy/dt = r(1 - y/K)y - H(y, t). \tag{1}$$

© OSTILL / iStockphoto

Although it is desirable to utilize the fish as a food source, it is intuitively clear that if too many fish are caught, then the fish population may be reduced below a useful level and possibly even driven to extinction. The following problems explore some of the questions involved in formulating a rational strategy for managing the fishery.

Project 1 PROBLEMS

1. Constant Effort Harvesting. At a given level of effort, it is reasonable to assume that the rate at which fish are caught depends on the population y: the more fish there are, the easier it is to catch them. Thus we assume that the rate at which fish are caught is given by $H(y, t) = Ey$, where E is a positive constant, with units of 1/time, that measures the total effort made to harvest the given species of fish. With this choice for $H(y, t)$, Eq. (1) becomes

$$dy/dt = r(1 - y/K)y - Ey. \qquad (i)$$

This equation is known as the **Schaefer model** after the biologist M. B. Schaefer, who applied it to fish populations.

(a) Show that if $E < r$, then there are two equilibrium points, $y_1 = 0$ and $y_2 = K(1 - E/r) > 0$.
(b) Show that $y = y_1$ is unstable and $y = y_2$ is asymptotically stable.
(c) A sustainable yield Y of the fishery is a rate at which fish can be caught indefinitely. It is the product of the effort E and the asymptotically stable population y_2. Find Y as a function of the effort E. The graph of this function is known as the yield–effort curve.
(d) Determine E so as to maximize Y and thereby find the **maximum sustainable yield** Y_m.

2. Constant Yield Harvesting. In this problem, we assume that fish are caught at a constant rate h independent of the size of the fish population, that is, the harvesting rate $H(y, t) = h$. Then y satisfies

$$dy/dt = r(1 - y/K)y - h = f(y). \qquad (ii)$$

The assumption of a constant catch rate h may be reasonable when y is large but becomes less so when y is small.

(a) If $h < rK/4$, show that Eq. (ii) has two equilibrium points y_1 and y_2 with $y_1 < y_2$; determine these points.
(b) Show that y_1 is unstable and y_2 is asymptotically stable.
(c) From a plot of $f(y)$ versus y, show that if the initial population $y_0 > y_1$, then $y \to y_2$ as $t \to \infty$, but if $y_0 < y_1$, then y decreases as t increases. Note that $y = 0$ is not an equilibrium point, so if $y_0 < y_1$, then extinction will be reached in a finite time.
(d) If $h > rK/4$, show that y decreases to zero as t increases regardless of the value of y_0.
(e) If $h = rK/4$, show that there is a single equilibrium point $y = K/2$ and that this point is semistable. Thus the maximum sustainable yield is $h_m = rK/4$, corresponding to the equilibrium value $y = K/2$. Observe that h_m has the same value as Y_m in Problem 1(d). The fishery is considered to be overexploited if y is reduced to a level below $K/2$.

Project 2 A Mathematical Model of a Groundwater Contaminant Source

Chlorinated solvents such as trichloroethylene (TCE) are a common cause of environmental contamination[8] at thousands of government and private industry facilities. TCE and other chlorinated organics, collectively referred to as dense nonaqueous phase liquids (DNAPLs), are denser than water and only slightly soluble in water. DNAPLs tend to accumulate as a separate phase below the water table and provide a long-term source of groundwater contamination. A downstream contaminant plume is formed by the process of dissolution of DNAPL into water flowing through the source region, as shown in Figure 2.P.1.

In this project, we study a first order differential equation that describes the time-dependent rate of dissolved contaminant discharge leaving the source zone and entering the plume.[9]

[8]R. W. Falta, Rao, P. S., and N. Basu, "Assessing the Impacts of Partial Mass Depletion in DNAPL Source Zones: I. Analytical Modeling of Source Strength Functions and Plume Response," *Journal of Contaminant Hydrology* 78, 4 (2005), pp. 259–280.
[9]The output of this model can then be used as input into another mathematical model that, in turn, describes the processes of advection, adsorption, dispersion, and degradation of contaminant within the plume.

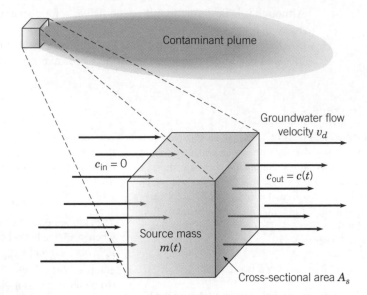

Contaminant plume

Groundwater flow
velocity v_d

$c_{\text{in}} = 0$

$c_{\text{out}} = c(t)$

Source mass
$m(t)$

Cross-sectional area A_s

FIGURE 2.P.1   Conceptual model of DNAPL source.

Parameters and variables relevant to formulating a mathematical model of contaminant discharge from the source region are defined as follows:

A_s = cross-sectional area of the source region

v_d = Darcy groundwater flow velocity[10]

$m(t)$ = total DNAPL mass in source region

$c_s(t)$ = concentration (flow averaged) of dissolved contaminant leaving the source zone

m_0 = initial DNAPL mass in source region

c_0 = source zone concentration (flow averaged) corresponding to an initial source zone mass of m_0

The equation describing the rate of DNAPL mass discharge from the source region is

$$\frac{dm}{dt} = -A_s v_d c_s(t), \tag{1}$$

whereas an algebraic relationship between $c_s(t)$ and $m(t)$ is postulated in the form of a power law,

$$\frac{c_s(t)}{c_0} = \left[\frac{m(t)}{m_0}\right]^{\gamma}, \tag{2}$$

in which $\gamma > 0$ is empirically determined. Combining Eqs. (1) and (2) (Problem 1) yields a first order differential equation

$$\frac{dm}{dt} = -\alpha m^{\gamma} \tag{3}$$

that models the dissolution of DNAPL into the groundwater flowing through the source region.

---

[10]In porous media flow, the Darcy flow velocity v_d is defined by $v_d = Q/A$, where A is a cross-sectional area available for flow and Q is the volumetric flow rate (volume/time) through A.

Project 2 PROBLEMS

1. Derive Eq. (3) from Eqs. (1) and (2) and show that $\alpha = v_d A_s c_0 / m_0^\gamma$.

2. Additional processes due to biotic and abiotic degradation contributing to source decay can be accounted for by adding a decay term to (3) that is proportional to $m(t)$,

$$m'(t) = -\alpha m^\gamma - \lambda m, \qquad \text{(i)}$$

where λ is the associated decay rate constant. Find solutions of Eq. (i) using the initial condition $m(0) = m_0$ for the following cases: (i) $\gamma = 1$, (ii) $\gamma \neq 1$ and $\lambda = 0$, (iii) $\gamma \neq 1$ and $\lambda \neq 0$. Then find expressions for $c_s(t)$ using Eq. (2).

Hint: Eq. (i) is a type of nonlinear equation known as a Bernoulli equation. A method for solving Bernoulli equations is discussed in Section 2.7.

3. Show that when $\gamma \geq 1$, the source has an infinite lifetime, but if $0 < \gamma < 1$, the source has a finite lifetime. In the latter case, find the time that the DNAPL source mass attains the value zero.

4. Assume the following values for the parameters: $m_0 = 1{,}620$ kg, $c_0 = 100$ mg/L, $A_s = 30$ m^2, $v_d = 20$ m/year, $\lambda = 0$. Use the solutions obtained in Problem 2 to plot graphs of $c_s(t)$ for each of the following cases: (i) $\gamma = 0.5$ for $0 \leq t \leq t_f$, where $c_s(t_f) = 0$, and (ii) $\gamma = 2$ for $0 \leq t \leq 100$ years.

5. Effects of Partial Source Remediation.

(a) Assume that a source remediation process results in a 90% reduction in the initial amount of DNAPL mass in the source region. Repeat Problem 4 with m_0 and c_0 in Eq. (2) replaced by $m_1 = 0.1\, m_0$ and $c_1 = (0.1)^\gamma c_0$, respectively. Compare the graphs of $c_s(t)$ in this case with the graphs obtained in Problem 4.

(b) Assume that the 90% efficient source remediation process is not applied until $t_1 = 10$ years have elapsed following the initial deposition of the contaminant. Under this scenario, plot the graphs of $c_s(t)$ using the parameters and initial conditions of Problem 4. In this case, use Eq. (2) to compute concentration for $0 \leq t < t_1$. Following remediation, use the initial condition $m(t_1) = m_1 = 0.1 m(t_1 - 0) = 0.1 \lim_{t \uparrow t_1} m(t)$ for Eq. (i) and use the following modification of Eq. (2):

$$\frac{c_s(t)}{c_1} = \left[\frac{m(t)}{m_1}\right]^\gamma, \qquad t > t_1, \qquad \text{(ii)}$$

where $c_1 = (0.1)^\gamma c(t_1 - 0) = (0.1)^\gamma \lim_{t \uparrow t_1} c(t)$ to compute concentrations for times $t > t_1$. Compare the graphs of $c_s(t)$ in this case with the graphs obtained in Problems 4 and 5(a). Can you draw any conclusions about the possible effectiveness of source remediation? If so, what are they?

Project 3 Monte Carlo Option Pricing: Pricing Financial Options by Flipping a Coin

A discrete model for change in the price of a stock over a time interval $[0, T]$ is

$$S_{n+1} = S_n + \mu S_n \Delta t + \sigma S_n \varepsilon_{n+1}\sqrt{\Delta t}, \qquad S_0 = s, \qquad \text{(1)}$$

where $S_n = S(t_n)$ is the stock price at time $t_n = n\Delta t$, $n = 0, \ldots, N-1$, $\Delta t = T/N$, μ is the annual growth rate of the stock, and σ is a measure of the stock's annual price volatility or tendency to fluctuate. Highly volatile stocks have large values for σ, for example, values ranging from 0.2 to 0.4. Each term in the sequence $\varepsilon_1, \varepsilon_2, \ldots$ takes on the value 1 or -1 depending on whether the outcome of a coin tossing experiment is heads or tails, respectively. Thus, for each $n = 1, 2, \ldots,$

$$\varepsilon_n = \begin{cases} 1 & \text{with probability} = \frac{1}{2} \\ -1 & \text{with probability} = \frac{1}{2}. \end{cases} \qquad \text{(2)}$$

A sequence of such numbers can easily be created by using one of the random number generators available in most mathematical computer software applications. Given such a sequence, the difference equation (1) can then be used to simulate a **sample path** or **trajectory** of stock prices, $\{s, S_1, S_2, \ldots, S_N\}$. The "random" terms $\sigma S_n \varepsilon_{n+1}\sqrt{\Delta t}$ on the right-hand side of (1) can be thought of as "shocks" or "disturbances" that model fluctuations in the stock price. By repeatedly simulating stock price trajectories and computing appropriate averages, it is possible to obtain estimates of the price of a **European call option**, a type of financial derivative. A statistical simulation algorithm of this type is called a **Monte Carlo method**.

A European call option is a contract between two parties, a holder and a writer, whereby, for a premium paid to the writer, the holder acquires the right (but not the obligation) to purchase the stock at a future date T (the **expiration date**) at a price K (the **strike price**) agreed upon in the contract. If the buyer elects to exercise the option on the expiration date, the writer is obligated to sell the underlying stock to the buyer at the price K. Thus the option has, associated with it, a **payoff function**

$$f(S) = \max(S - K, 0), \tag{3}$$

where $S = S(T)$ is the price of the underlying stock at the time T when the option expires (see Figure 2.P.2).

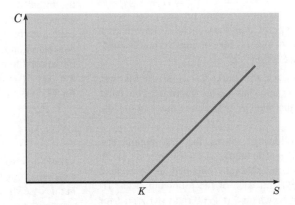

FIGURE 2.P.2   The value of a call option at expiration is $C = \max(S - K, 0)$, where K is the strike price of the option and $S = S(T)$ is the stock price at expiration.

Equation (3) is the value of the option at time T since, if $S(T) > K$, the holder can purchase, at price K, stock with market value $S(T)$ and thereby make a profit equal to $S(T) - K$ not counting the option premium. If $S(T) < K$, the holder will simply let the option expire since it would be irrational to purchase stock at a price that exceeds the market value. The option valuation problem is to determine the correct and fair price of the option at the time that the holder and writer enter into the contract.[11]

To estimate the price of a call option using a Monte Carlo method, an ensemble

$$\left\{ S_N^{(k)} = S^{(k)}(T), \quad k = 1, \dots, M \right\}$$

of M stock prices at expiration is generated using the difference equation

$$S_{n+1}^{(k)} = S_n^{(k)} + rS_n^{(k)}\Delta t + \sigma S_n^{(k)} \varepsilon_{n+1}^{(k)} \sqrt{\Delta t}, \qquad S_0^{(k)} = s. \tag{4}$$

For each $k = 1, \dots, M$, the difference equation (4) is identical to Eq. (1) except that the growth rate μ is replaced by the annual rate of interest r that it costs the writer to borrow money. Option pricing theory requires that the average value of the payoffs $\left\{ f(S_N^{(k)}), k = 1, \dots, M \right\}$ be equal to the compounded total return obtained by investing the option premium, $\hat{C}(s)$, at rate r over the life of the option,

$$\frac{1}{M} \sum_{k=1}^{M} f(S_N^{(k)}) = (1 + r\Delta t)^N \hat{C}(s). \tag{5}$$

---

[11]The 1997 Nobel Prize in Economics was awarded to Robert C. Merton and Myron S. Scholes for their work, along with Fischer Black, in developing the Black–Scholes options pricing model.

Solving (5) for $\hat{C}(s)$ yields the Monte Carlo estimate

$$\hat{C}(s) = (1 + r\Delta t)^{-N} \left\{ \frac{1}{M} \sum_{k=1}^{M} f(S_N^{(k)}) \right\} \qquad (6)$$

for the option price. Thus the Monte Carlo estimate $\hat{C}(s)$ is the present value of the average of the payoffs computed using the rules of compound interest.

Project 3 PROBLEMS

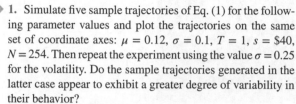

1. Simulate five sample trajectories of Eq. (1) for the following parameter values and plot the trajectories on the same set of coordinate axes: $\mu = 0.12$, $\sigma = 0.1$, $T = 1$, $s = \$40$, $N = 254$. Then repeat the experiment using the value $\sigma = 0.25$ for the volatility. Do the sample trajectories generated in the latter case appear to exhibit a greater degree of variability in their behavior?

Hint: For the ε_n's it is permissible to use a random number generator that creates normally distributed random numbers with mean 0 and variance 1.

2. Use the difference equation (4) to generate an ensemble of stock prices $S_N^{(k)} = S^{(k)}(N\Delta t)$, $k = 1, \ldots, M$ (where $T = N\Delta t$) and then use formula (6) to compute a Monte Carlo estimate of the value of a five-month call option ($T = \frac{5}{12}$ years) for the following parameter values: $r = 0.06$, $\sigma = 0.2$, and $K = \$50$. Find estimates corresponding to current stock prices of $S(0) = s = \$45$, $\$50$, and $\$55$. Use $N = 200$ time steps for each trajectory and $M \cong 10{,}000$ sample trajectories for each Monte Carlo estimate.[12] Check the accuracy of your results by comparing the Monte Carlo approximation with the value computed from the exact Black–Scholes formula

$$C(s) = \frac{s}{2} \operatorname{erfc}\left(-\frac{d_1}{\sqrt{2}}\right) - \frac{K}{2} e^{-rT} \operatorname{erfc}\left(-\frac{d_2}{\sqrt{2}}\right), \qquad (ii)$$

where

$$d_1 = \frac{1}{\sigma\sqrt{T}} \left[\ln\left(\frac{s}{k}\right) + \left(r + \frac{\sigma^2}{2}\right) T \right],$$

$$d_2 = d_1 - \sigma\sqrt{T}$$

and erfc(x) is the complementary error function,

$$\operatorname{erfc}(x) = \frac{2}{\sqrt{\pi}} \int_x^\infty e^{-t^2}\, dt.$$

3. Variance Reduction by Antithetic Variates. A simple and widely used technique for increasing the efficiency and accuracy of Monte Carlo simulations in certain situations with little additional increase in computational complexity is the method of antithetic variates. For each $k = 1, \ldots, M$, use the sequence $\left\{\varepsilon_1^{(k)}, \ldots, \varepsilon_{N-1}^{(k)}\right\}$ in Eq. (4) to simulate a payoff $f(S_N^{(k+)})$ and also use the sequence $\left\{-\varepsilon_1^{(k)}, \ldots, -\varepsilon_{N-1}^{(k)}\right\}$ in Eq. (4) to simulate an associated payoff $f(S_N^{(k-)})$. Thus the payoffs are simulated in pairs $\left\{f(S_N^{(k+)}), f(S_N^{(k-)})\right\}$. A modified Monte Carlo estimate is then computed by replacing each payoff $f(S_N^{(k)})$ in Eq. (6) by the average $[f(S_N^{(k+)}) + f(S_N^{(k-)})]/2$,

$$\hat{C}_{AV}(s) = \frac{\dfrac{1}{M} \sum_{k=1}^{M} \dfrac{f(S_N^{(k+)}) + f(S_N^{(k-)})}{2}}{(1 + r\Delta t)^N}. \qquad (iii)$$

Use the parameters specified in Problem 2 to compute several (say, 20 or so) option price estimates using Eq. (6) and an equivalent number of option price estimates using (iii). For each of the two methods, plot a histogram of the estimates and compute the mean and standard deviation of the estimates. Comment on the accuracies of the two methods.

---

[12] As a rule of thumb, you may assume that the sampling error in these Monte Carlo estimates is proportional to $1/\sqrt{M}$. Using software packages such as MATLAB that allow vector operations where all M trajectories can be simulated simultaneously greatly speeds up the calculations.

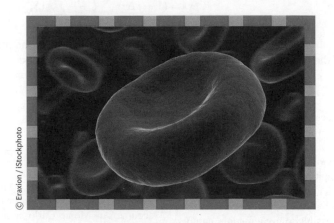

CHAPTER THREE

© Eraxion / iStockphoto

Systems of Two First Order Equations

The first two chapters of this book contain the material most essential for dealing with single first order differential equations. To proceed further, there are two natural paths that we might follow. The first is to take up the study of second order equations and the other is to consider systems of two first order equations. There are many important problems in various areas of application that lead to each of these types of problems, so both are important and we will eventually discuss both. They are also closely related to each other, as we will show. Our approach is to introduce systems of two first order equations in this chapter and to take up second order equations in Chapter 4.

There are many problem areas that involve several components linked together in some way. For example, electrical networks have this character, as do some problems in mechanics and in other fields. In these and similar cases, there are two (or more) dependent variables and the corresponding mathematical problem consists of a system of two (or more) differential equations, which can always be written as a system of first order equations. In this chapter, we consider only systems of two first order equations and we focus most of our attention on systems of the simplest kind: two first order linear equations with constant coefficients. Our goals are to show what kinds of solutions such a system may have and how the solutions can be determined and displayed graphically, so that they can be easily visualized.

3.1 Systems of Two Linear Algebraic Equations

The solution of a system of two linear differential equations with constant coefficients is directly related to the solution of an associated system of two linear algebraic equations. Consequently, we start by reviewing the properties of such linear algebraic systems.[1]

Consider the system

$$a_{11}x_1 + a_{12}x_2 = b_1, \tag{1}$$

$$a_{21}x_1 + a_{22}x_2 = b_2,$$

where $a_{11}, \ldots, a_{22}, b_1,$ and b_2 are given and x_1 and x_2 are to be determined. In matrix notation, we can write the system (1) as

$$\mathbf{Ax} = \mathbf{b}, \tag{2}$$

where

$$\mathbf{A} = \begin{pmatrix} a_{11} & a_{12} \\ a_{21} & a_{22} \end{pmatrix}, \qquad \mathbf{x} = \begin{pmatrix} x_1 \\ x_2 \end{pmatrix}, \qquad \mathbf{b} = \begin{pmatrix} b_1 \\ b_2 \end{pmatrix}. \tag{3}$$

Here $\mathbf{A}$ is a given 2×2 matrix, $\mathbf{b}$ is a given 2×1 column vector, and $\mathbf{x}$ is a 2×1 column vector to be determined.

To see what kinds of solutions the system (1) or (2) may have, it is helpful to visualize the situation geometrically. If a_{11} and a_{12} are not both zero, then the first equation in the system (1) corresponds to a straight line in the x_1x_2-plane, and similarly for the second equation. There are three distinct possibilities for two straight lines in a plane: they may intersect at a single point, they may be parallel and nonintersecting, or they may be coincident. In the first case, the system (1) or (2) is satisfied by a single pair of values of x_1 and x_2. In the second case, the system has no solutions; that is, there is no point that lies on both lines. In the third case, the system has infinitely many solutions, since every point on one line also lies on the other. The following three examples illustrate these possibilities.

EXAMPLE 1

Solve the system

$$3x_1 - x_2 = 8, \tag{4}$$

$$x_1 + 2x_2 = 5.$$

We can solve this system in a number of ways. For instance, from the first equation we have

$$x_2 = 3x_1 - 8. \tag{5}$$

Then, substituting this expression for x_2 in the second equation, we obtain

$$x_1 + 2(3x_1 - 8) = 5,$$

or $7x_1 = 21$, from which $x_1 = 3$. From Eq. (5), $x_2 = 1$. Thus the solution of the system (4) is $x_1 = 3$, $x_2 = 1$. In other words, the point $(3, 1)$ is the unique point of intersection of the two straight lines corresponding to the equations in the system (4). See Figure 3.1.1.

---

[1]We believe that much of the material in this section will be familiar to you. A more extensive discussion of linear algebra and matrices appears in Appendix A.

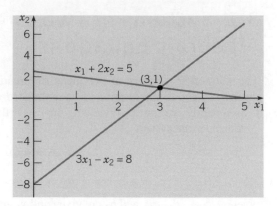

FIGURE 3.1.1   Geometrical interpretation of the system (4).

EXAMPLE 2

Solve the system

$$x_1 + 2x_2 = 1, \tag{6}$$
$$x_1 + 2x_2 = 5.$$

We can see at a glance that this system has no solution, since $x_1 + 2x_2$ cannot simultaneously take on the values 1 and 5. Proceeding more formally, as in Example 1, we can solve the first equation for x_1, with the result that $x_1 = 1 - 2x_2$. On substituting this expression for x_1 in the second equation, we obtain the false statement that $1 = 5$. Of course, you should not regard this as a demonstration that the numbers 1 and 5 are equal. Rather, you should conclude that the two equations in the system (6) are incompatible or inconsistent, and so the system has no solution. The geometrical interpretation of the system (6) is shown in Figure 3.1.2. The two lines are parallel and therefore have no points in common.

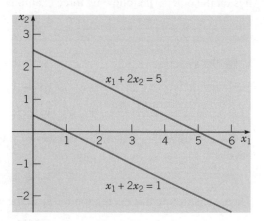

FIGURE 3.1.2   Geometrical interpretation of the system (6).

EXAMPLE 3

Solve the system

$$2x_1 + 4x_2 = 10, \tag{7}$$
$$x_1 + 2x_2 = 5.$$

Solving the second equation for x_1, we find that $x_1 = 5 - 2x_2$. Then, substituting this expression for x_1 in the first equation, we obtain $2(5 - 2x_2) + 4x_2 = 10$, or $10 = 10$. This result is true, but does not impose any restriction on x_1 or x_2. On looking at the system (7) again, we note that the two equations are multiples of each other; the first is just 2 times the second. Thus every point that satisfies one of the equations also satisfies the other. Geometrically, as shown in Figure 3.1.3, the two lines described by the equations in the system (7) are actually the same line. The system (7) has an infinite set of solutions—all of the points on this line. In other words, all values of x_1 and x_2 such that $x_1 + 2x_2 = 5$ satisfy the system (7).

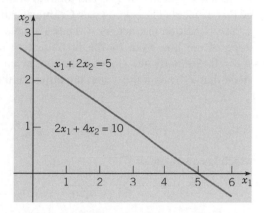

FIGURE 3.1.3 Geometrical interpretation of the system (7).

Let us now return to the system (1) and find its solution. To eliminate x_2 from the system (1), we can multiply the first equation by a_{22}, the second equation by a_{12}, and then subtract the second equation from the first. The result is

$$(a_{11}a_{22} - a_{12}a_{21})x_1 = a_{22}b_1 - a_{12}b_2, \tag{8}$$

so

$$x_1 = \frac{a_{22}b_1 - a_{12}b_2}{a_{11}a_{22} - a_{12}a_{21}} = \frac{\begin{vmatrix} b_1 & a_{12} \\ b_2 & a_{22} \end{vmatrix}}{\begin{vmatrix} a_{11} & a_{12} \\ a_{21} & a_{22} \end{vmatrix}}. \tag{9}$$

In Eq. (9), we have written the numerator and denominator of the expression for x_1 in the usual notation for 2×2 determinants.

To find a corresponding expression for x_2, we can eliminate x_1 from Eqs. (1). To do this, we multiply the first equation by a_{21}, the second by a_{11}, and then subtract the first equation from the second. We obtain

$$(a_{11}a_{22} - a_{12}a_{21})x_2 = a_{11}b_2 - a_{21}b_1, \tag{10}$$

so

$$x_2 = \frac{a_{11}b_2 - a_{21}b_1}{a_{11}a_{22} - a_{12}a_{21}} = \frac{\begin{vmatrix} a_{11} & b_1 \\ a_{21} & b_2 \end{vmatrix}}{\begin{vmatrix} a_{11} & a_{12} \\ a_{21} & a_{22} \end{vmatrix}}. \tag{11}$$

The denominator in Eqs. (9) and (11) is the **determinant of coefficients** of the system (1), or the determinant of the matrix **A**. We will denote it by det(**A**), or sometimes by Δ. Thus

$$\Delta = \det(\mathbf{A}) = \begin{vmatrix} a_{11} & a_{12} \\ a_{21} & a_{22} \end{vmatrix} = a_{11}a_{22} - a_{12}a_{21}. \tag{12}$$

As long as $\det(\mathbf{A}) \neq 0$, the expressions in Eqs. (9) and (11) give the unique values of x_1 and x_2 that satisfy the system (1). The solution of the system (1) in terms of determinants, given by Eqs. (9) and (11), is known as **Cramer's rule**.

The condition that $\det(\mathbf{A}) \neq 0$ has a simple geometric interpretation. Observe that the slope of the line given by the first equation in the system (1) is $-a_{11}/a_{12}$, as long as $a_{12} \neq 0$. Similarly, the slope of the line given by the second equation is $-a_{21}/a_{22}$, provided that $a_{22} \neq 0$. If the slopes are different, then

$$-\frac{a_{11}}{a_{12}} \neq -\frac{a_{21}}{a_{22}},$$

which is equivalent to

$$a_{11}a_{22} - a_{12}a_{21} = \det(\mathbf{A}) \neq 0.$$

Of course, if the slopes are different, then the lines intersect at a single point, whose coordinates are given by Eqs. (9) and (11). We leave it to you to consider what happens if either a_{12} or a_{22} or both are zero. Thus we have the following important result.

THEOREM 3.1.1

The system (1),

$$a_{11}x_1 + a_{12}x_2 = b_1,$$
$$a_{21}x_1 + a_{22}x_2 = b_2,$$

has a unique solution if and only if the determinant

$$\Delta = a_{11}a_{22} - a_{12}a_{21} \neq 0.$$

The solution is given by Eqs. (9) and (11). If $\Delta = 0$, then the system (1) has either no solution or infinitely many.

We now introduce two matrices of special importance, as well as some associated terminology. The 2×2 identity matrix is denoted by **I** and is defined to be

$$\mathbf{I} = \begin{pmatrix} 1 & 0 \\ 0 & 1 \end{pmatrix}. \tag{13}$$

Note that the product of **I** with any 2×2 matrix or with any 2×1 vector is just the matrix or vector itself.

For a given 2×2 matrix **A**, there may be another 2×2 matrix **B** such that $\mathbf{AB} = \mathbf{BA} = \mathbf{I}$. There may be no such matrix **B**, but if there is, then it can be shown that only one exists. The matrix **B** is called the **inverse** of **A** and is denoted by $\mathbf{B} = \mathbf{A}^{-1}$. If $\mathbf{A}^{-1}$ exists, then **A** is called **nonsingular** or **invertible**. On the other hand, if $\mathbf{A}^{-1}$ does not exist, then **A** is said to be **singular** or **noninvertible**. In Problem 37, we ask you to show that if $\mathbf{A}$ is given by Eq. (3), then $\mathbf{A}^{-1}$, when it exists, is given by

$$\mathbf{A}^{-1} = \frac{1}{\det(\mathbf{A})} \begin{pmatrix} a_{22} & -a_{12} \\ -a_{21} & a_{11} \end{pmatrix}. \tag{14}$$

It is easy to verify that Eq. (14) is correct simply by multiplying $\mathbf{A}$ and $\mathbf{A}^{-1}$ together. Equation (14) strongly suggests that $\mathbf{A}$ is nonsingular if and only if $\det(\mathbf{A}) \neq 0$, and this is, in fact, true. If $\det(\mathbf{A}) = 0$, then $\mathbf{A}$ is singular, and conversely.

We now return to the system (2). If $\mathbf{A}$ is nonsingular, multiply each side of Eq. (2) on the left by $\mathbf{A}^{-1}$. This gives

$$\mathbf{A}^{-1}\mathbf{A}\mathbf{x} = \mathbf{A}^{-1}\mathbf{b},$$

or

$$\mathbf{I}\mathbf{x} = \mathbf{A}^{-1}\mathbf{b},$$

or

$$\mathbf{x} = \mathbf{A}^{-1}\mathbf{b}. \tag{15}$$

It is straightforward to show that the result (15) agrees with Eqs. (9) and (11).

▶ Homogeneous Systems. If $b_1 = b_2 = 0$ in the system (1), then the system is said to be **homogeneous**; otherwise, it is called **nonhomogeneous**. Thus the general system of two linear homogeneous algebraic equations has the form

$$a_{11}x_1 + a_{12}x_2 = 0, \tag{16}$$

$$a_{21}x_1 + a_{22}x_2 = 0,$$

or, in matrix notation,

$$\mathbf{A}\mathbf{x} = \mathbf{0}. \tag{17}$$

For the homogeneous system (16), the corresponding straight lines must pass through the origin. Thus the lines always have at least one point in common, namely, the origin. If the two lines coincide, then every point on each line also lies on the other and the system (16) has infinitely many solutions. The two lines cannot be parallel and nonintersecting. In most applications, the solution $x_1 = 0$, $x_2 = 0$ is of little interest and it is often called the **trivial solution**. According to Eqs. (9) and (11), or Eq. (15), this is the only solution when $\det(\mathbf{A}) \neq 0$, that is, when $\mathbf{A}$ is nonsingular. Nonzero solutions occur if and only if $\det(\mathbf{A}) = 0$, that is, when $\mathbf{A}$ is singular. We summarize these results in the following theorem.

THEOREM 3.1.2	The homogeneous system (16) always has the trivial solution $x_1 = 0$, $x_2 = 0$, and this is the only solution when $\det(\mathbf{A}) \neq 0$. Nontrivial solutions exist if and only if $\det(\mathbf{A}) = 0$. In this case, unless $\mathbf{A} = \mathbf{0}$, all solutions are proportional to any nontrivial solution; in other words, they lie on a line through the origin. If $\mathbf{A} = \mathbf{0}$, then every point in the x_1x_2-plane is a solution of Eqs. (16).

The following examples illustrate the two possible cases.

EXAMPLE 4

Solve the system

$$3x_1 - x_2 = 0, \tag{18}$$

$$x_1 + 2x_2 = 0.$$

From the first equation, we have $x_2 = 3x_1$. Then, substituting into the second equation, we obtain $7x_1 = 0$, or $x_1 = 0$. Then $x_2 = 0$ also. Note that the determinant of coefficients has the value 7 (which is not zero), so this confirms the first part of Theorem 3.1.2 in this case. Figure 3.1.4 shows the two lines corresponding to the equations in the system (18).

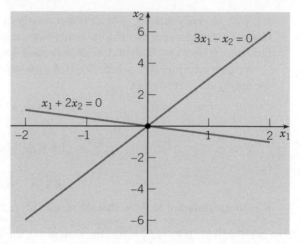

FIGURE 3.1.4 Geometrical interpretation of the system (18).

EXAMPLE
5

Solve the system

$$2x_1 + 4x_2 = 0, \tag{19}$$
$$x_1 + 2x_2 = 0.$$

From the second equation, we have $x_1 = -2x_2$. Then, from the first equation, we obtain $-4x_2 + 4x_2 = 0$, or $0 = 0$. Thus x_2 is not determined, but remains arbitrary. If $x_2 = c$, where c is an arbitrary constant, then $x_1 = -2c$. Thus solutions of the system (19) are of the form $(-2c, c)$, or $c(-2, 1)$, where c is any number. The system (19) has an infinite set of solutions, all of which are proportional to $(-2, 1)$, or to any other nontrivial solution. In the system (19) the two equations are multiples of each other and the determinant of coefficients has the value zero. See Figure 3.1.5.

Eigenvalues and Eigenvectors

The equation $\mathbf{y} = \mathbf{Ax}$, where $\mathbf{A}$ is a given 2×2 matrix, can be viewed as a transformation, or mapping, of a two-dimensional vector $\mathbf{x}$ to a new two-dimensional vector $\mathbf{y}$. For example, suppose that

$$\mathbf{A} = \begin{pmatrix} 1 & 1 \\ 4 & 1 \end{pmatrix}, \qquad \mathbf{x} = \begin{pmatrix} 1 \\ 1 \end{pmatrix}. \tag{20}$$

Then

$$\mathbf{y} = \mathbf{Ax} = \begin{pmatrix} 1 & 1 \\ 4 & 1 \end{pmatrix} \begin{pmatrix} 1 \\ 1 \end{pmatrix} = \begin{pmatrix} 2 \\ 5 \end{pmatrix}; \tag{21}$$

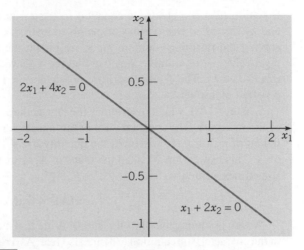

FIGURE 3.1.5 Geometrical interpretation of the system (19).

thus the original vector $\mathbf{x}$ has been transformed into the new vector $\mathbf{y}$. Similarly, if $\mathbf{A}$ is given by Eq. (20) and $\mathbf{x} = \begin{pmatrix} 2 \\ -1 \end{pmatrix}$, then

$$\mathbf{y} = \mathbf{A}\mathbf{x} = \begin{pmatrix} 1 & 1 \\ 4 & 1 \end{pmatrix} \begin{pmatrix} 2 \\ -1 \end{pmatrix} = \begin{pmatrix} 1 \\ 7 \end{pmatrix}, \tag{22}$$

and so on.

In many applications it is of particular importance to find those vectors that a given matrix transforms into vectors that are multiples of the original vectors. In other words, we want $\mathbf{y}$ to be a multiple of $\mathbf{x}$; that is, $\mathbf{y} = \lambda\mathbf{x}$, where λ is some (scalar) constant. In this case, the equation $\mathbf{y} = \mathbf{A}\mathbf{x}$ becomes

$$\mathbf{A}\mathbf{x} = \lambda\mathbf{x}. \tag{23}$$

If $\mathbf{x} = \mathbf{0}$, then Eq. (23) is true for any $\mathbf{A}$ and for any λ, so we require $\mathbf{x}$ to be a *nonzero* vector. Then, since $\mathbf{I}\mathbf{x} = \mathbf{x}$, we can rewrite Eq. (23) in the form

$$\mathbf{A}\mathbf{x} = \lambda\mathbf{I}\mathbf{x}, \tag{24}$$

or

$$(\mathbf{A} - \lambda\mathbf{I})\mathbf{x} = \mathbf{0}. \tag{25}$$

To see the elements of $\mathbf{A} - \lambda\mathbf{I}$, we write

$$\mathbf{A} - \lambda\mathbf{I} = \begin{pmatrix} a_{11} & a_{12} \\ a_{21} & a_{22} \end{pmatrix} - \begin{pmatrix} \lambda & 0 \\ 0 & \lambda \end{pmatrix} = \begin{pmatrix} a_{11} - \lambda & a_{12} \\ a_{21} & a_{22} - \lambda \end{pmatrix}. \tag{26}$$

Recall that we are looking for nonzero vectors $\mathbf{x}$ that satisfy the homogeneous system (25). By Theorem 3.1.2, nonzero solutions of this system occur if and only if the determinant of coefficients is zero. Thus we require that

$$\det(\mathbf{A} - \lambda\mathbf{I}) = \begin{vmatrix} a_{11} - \lambda & a_{12} \\ a_{21} & a_{22} - \lambda \end{vmatrix} = 0. \tag{27}$$

Writing Eq. (27) in expanded form, we obtain

$$(a_{11} - \lambda)(a_{22} - \lambda) - a_{12}a_{21} = \lambda^2 - (a_{11} + a_{22})\lambda + a_{11}a_{22} - a_{12}a_{21} = 0. \tag{28}$$

Equation (28) is a quadratic equation in λ, so it has two roots λ_1 and λ_2. The values λ_1 and λ_2 are called **eigenvalues** of the given matrix **A**. By replacing λ by λ_1 in Eq. (25) and solving the resulting equation for **x**, we obtain the **eigenvector** $\mathbf{x}_1$ corresponding to the eigenvalue λ_1. In a similar way, we find the eigenvector $\mathbf{x}_2$ that corresponds to the second eigenvalue λ_2. The eigenvectors are not determined uniquely, but only up to an arbitrary constant multiplier.

Equation (28), which determines the eigenvalues, is called the **characteristic equation** of the matrix **A**. The constant term in this equation is just the determinant of **A**. The coefficient of λ in Eq. (28) involves the quantity $a_{11} + a_{22}$, the sum of the diagonal elements of **A**. This expression is called the **trace** of **A**, or tr(**A**). Thus the characteristic equation is sometimes written as

$$\lambda^2 - \text{tr}(\mathbf{A})\lambda + \det(\mathbf{A}) = 0. \tag{29}$$

We are assuming that the elements of **A** are real numbers. Consequently, the coefficients in the characteristic equation (28) are also real. As a result, the eigenvalues λ_1 and λ_2 may be real and different, real and equal, or complex conjugates. The following examples illustrate the calculation of eigenvalues and eigenvectors in each of these cases.

EXAMPLE 6

Find the eigenvalues and eigenvectors of the matrix

$$\mathbf{A} = \begin{pmatrix} 1 & 1 \\ 4 & 1 \end{pmatrix}. \tag{30}$$

In this case, Eq. (25) becomes

$$\begin{pmatrix} 1-\lambda & 1 \\ 4 & 1-\lambda \end{pmatrix} \begin{pmatrix} x_1 \\ x_2 \end{pmatrix} = \begin{pmatrix} 0 \\ 0 \end{pmatrix}. \tag{31}$$

The characteristic equation is

$$(1-\lambda)^2 - 4 = \lambda^2 - 2\lambda - 3 = (\lambda - 3)(\lambda + 1) = 0, \tag{32}$$

so the eigenvalues are $\lambda_1 = 3$ and $\lambda_2 = -1$.

To find the eigenvector $\mathbf{x}_1$ associated with the eigenvalue λ_1, we substitute $\lambda = 3$ in Eq. (31). Thus we obtain

$$\begin{pmatrix} -2 & 1 \\ 4 & -2 \end{pmatrix} \begin{pmatrix} x_1 \\ x_2 \end{pmatrix} = \begin{pmatrix} 0 \\ 0 \end{pmatrix}. \tag{33}$$

Observe that the rows in Eq. (33) are proportional to each other (as required by the vanishing of the determinant of coefficients), so we need only consider one row of this equation. Consequently, $-2x_1 + x_2 = 0$, or $x_2 = 2x_1$, while x_1 remains arbitrary. Thus

$$\mathbf{x}_1 = \begin{pmatrix} c \\ 2c \end{pmatrix} = c \begin{pmatrix} 1 \\ 2 \end{pmatrix}, \tag{34}$$

where c is an arbitrary constant. From Eq. (34), we see that there is an infinite set of eigenvectors associated with the eigenvalue λ_1. It is usually convenient to choose one member of this set to represent the entire set. For example, in this case, we might choose

$$\mathbf{x}_1 = \begin{pmatrix} 1 \\ 2 \end{pmatrix}, \tag{35}$$

and even refer to it as *the* eigenvector corresponding to λ_1. However you should never forget that there are actually infinitely many other eigenvectors, each of which is proportional to the chosen representative.

In the same way, we can find the eigenvector $\mathbf{x}_2$ corresponding to the eigenvalue λ_2. By substituting $\lambda = -1$ in Eq. (31), we obtain

$$\begin{pmatrix} 2 & 1 \\ 4 & 2 \end{pmatrix} \begin{pmatrix} x_1 \\ x_2 \end{pmatrix} = \begin{pmatrix} 0 \\ 0 \end{pmatrix}. \tag{36}$$

Thus $x_2 = -2x_1$, so the eigenvector $\mathbf{x}_2$ is

$$\mathbf{x}_2 = \begin{pmatrix} 1 \\ -2 \end{pmatrix}, \tag{37}$$

or any vector proportional to this one.

EXAMPLE
7

Find the eigenvalues and eigenvectors of the matrix

$$\mathbf{A} = \begin{pmatrix} -\frac{1}{2} & 1 \\ -1 & -\frac{1}{2} \end{pmatrix}. \tag{38}$$

In this case, we obtain, from Eq. (25),

$$\begin{pmatrix} -\frac{1}{2} - \lambda & 1 \\ -1 & -\frac{1}{2} - \lambda \end{pmatrix} \begin{pmatrix} x_1 \\ x_2 \end{pmatrix} = \begin{pmatrix} 0 \\ 0 \end{pmatrix}. \tag{39}$$

The characteristic equation is

$$\left(-\frac{1}{2} - \lambda \right)^2 + 1 = \lambda^2 + \lambda + \frac{5}{4} = 0, \tag{40}$$

so the eigenvalues are

$$\lambda_1 = -\frac{1}{2} + i, \qquad \lambda_2 = -\frac{1}{2} - i. \tag{41}$$

For $\lambda = \lambda_1$ Eq. (39) reduces to

$$\begin{pmatrix} -i & 1 \\ -1 & -i \end{pmatrix} \begin{pmatrix} x_1 \\ x_2 \end{pmatrix} = \begin{pmatrix} 0 \\ 0 \end{pmatrix}. \tag{42}$$

Thus $x_2 = ix_1$ and the eigenvector $\mathbf{x}_1$ corresponding to the eigenvalue λ_1 is

$$\mathbf{x}_1 = \begin{pmatrix} 1 \\ i \end{pmatrix}, \tag{43}$$

or any vector proportional to this one. In a similar way, we find the eigenvector $\mathbf{x}_2$ corresponding to λ_2, namely,

$$\mathbf{x}_2 = \begin{pmatrix} 1 \\ -i \end{pmatrix}. \tag{44}$$

Observe that $\mathbf{x}_1$ and $\mathbf{x}_2$ are also complex conjugates. This will always be the case when the matrix $\mathbf{A}$ has real elements and a pair of complex conjugate eigenvalues.

EXAMPLE 8

Find the eigenvalues and eigenvectors of the matrix

$$\mathbf{A} = \begin{pmatrix} 1 & -1 \\ 1 & 3 \end{pmatrix}. \tag{45}$$

From Eq. (25), we obtain

$$\begin{pmatrix} 1-\lambda & -1 \\ 1 & 3-\lambda \end{pmatrix} \begin{pmatrix} x_1 \\ x_2 \end{pmatrix} = \begin{pmatrix} 0 \\ 0 \end{pmatrix}. \tag{46}$$

Consequently, the characteristic equation is

$$(1-\lambda)(3-\lambda) + 1 = \lambda^2 - 4\lambda + 4 = (\lambda - 2)^2 = 0, \tag{47}$$

and the eigenvalues are $\lambda_1 = \lambda_2 = 2$.

Returning to Eq. (46) and setting $\lambda = 2$, we find that

$$\begin{pmatrix} -1 & -1 \\ 1 & 1 \end{pmatrix} \begin{pmatrix} x_1 \\ x_2 \end{pmatrix} = \begin{pmatrix} 0 \\ 0 \end{pmatrix}. \tag{48}$$

Hence $x_2 = -x_1$, so there is an eigenvector

$$\mathbf{x}_1 = \begin{pmatrix} 1 \\ -1 \end{pmatrix}. \tag{49}$$

As usual, any other (nonzero) vector proportional to $\mathbf{x}_1$ is also an eigenvector.

However, in contrast to the two preceding examples, in this case there is only one distinct family of eigenvectors, which is typified by the vector $\mathbf{x}_1$ in Eq. (49). This situation is common when a matrix $\mathbf{A}$ has a repeated eigenvalue.

The following example shows that it is also possible for a repeated eigenvalue to be accompanied by two distinct eigenvectors.

EXAMPLE 9

Find the eigenvalues and eigenvectors of the matrix

$$\mathbf{A} = \begin{pmatrix} 2 & 0 \\ 0 & 2 \end{pmatrix}. \tag{50}$$

In this case, Eq. (25) becomes

$$\begin{pmatrix} 2-\lambda & 0 \\ 0 & 2-\lambda \end{pmatrix} \begin{pmatrix} x_1 \\ x_2 \end{pmatrix} = \begin{pmatrix} 0 \\ 0 \end{pmatrix}. \tag{51}$$

Thus the characteristic equation is

$$(2-\lambda)^2 = 0, \tag{52}$$

and the eigenvalues are $\lambda_1 = \lambda_2 = 2$. Returning to Eq. (51) and setting $\lambda = 2$, we obtain

$$\begin{pmatrix} 0 & 0 \\ 0 & 0 \end{pmatrix} \begin{pmatrix} x_1 \\ x_2 \end{pmatrix} = \begin{pmatrix} 0 \\ 0 \end{pmatrix}. \tag{53}$$

Thus no restriction is placed on x_1 and x_2; in other words, every nonzero vector in the plane is an eigenvector of this matrix **A**. For example, we can choose as eigenvectors

$$\mathbf{x}_1 = \begin{pmatrix} 1 \\ 0 \end{pmatrix}, \qquad \mathbf{x}_2 = \begin{pmatrix} 0 \\ 1 \end{pmatrix}, \tag{54}$$

or any other pair of nonzero vectors that are not proportional to each other.

Sometimes a matrix depends on a parameter and, in this case, its eigenvalues also depend on the parameter.

EXAMPLE 10

Consider the matrix

$$\mathbf{A} = \begin{pmatrix} 2 & \alpha \\ -1 & 0 \end{pmatrix}, \tag{55}$$

where α is a parameter. Find the eigenvalues of **A** and describe their dependence on α.
  The characteristic equation is

$$(2 - \lambda)(-\lambda) + \alpha = \lambda^2 - 2\lambda + \alpha = 0, \tag{56}$$

so the eigenvalues are

$$\lambda = \frac{2 \pm \sqrt{4 - 4\alpha}}{2} = 1 \pm \sqrt{1 - \alpha}. \tag{57}$$

Observe that, from Eq. (57), the eigenvalues are real and different when $\alpha < 1$, real and equal when $\alpha = 1$, and complex conjugates when $\alpha > 1$. As α varies, the case of equal eigenvalues occurs as a transition between the other two cases.

We will need the following result in Sections 3.3 and 3.4.

THEOREM 3.1.3

Let **A** have real or complex eigenvalues λ_1 and λ_2 such that $\lambda_1 \neq \lambda_2$, and let the corresponding eigenvectors be

$$\mathbf{x}_1 = \begin{pmatrix} x_{11} \\ x_{21} \end{pmatrix} \quad \text{and} \quad \mathbf{x}_2 = \begin{pmatrix} x_{12} \\ x_{22} \end{pmatrix}.$$

If **X** is the matrix with first and second columns taken to be $\mathbf{x}_1$ and $\mathbf{x}_2$, respectively,

$$\mathbf{X} = \begin{pmatrix} x_{11} & x_{12} \\ x_{21} & x_{22} \end{pmatrix}, \tag{58}$$

then

$$\det(\mathbf{X}) = \begin{vmatrix} x_{11} & x_{12} \\ x_{21} & x_{22} \end{vmatrix} \neq 0. \tag{59}$$

Proof

To prove this theorem, we will assume that $\det(\mathbf{X}) = 0$ and then show that this leads to a contradiction. If $\det(\mathbf{X}) = 0$, then the linear combination

$$x_{12}\mathbf{x}_1 - x_{11}\mathbf{x}_2 = \begin{pmatrix} x_{12}x_{11} - x_{11}x_{12} \\ x_{12}x_{21} - x_{11}x_{22} \end{pmatrix} = \begin{pmatrix} 0 \\ 0 \end{pmatrix}. \tag{60}$$

The first component is obviously zero, while the second component is $-\det(\mathbf{X}) = 0$.

Equation (60) implies that

$$\mathbf{x}_2 = k\mathbf{x}_1, \tag{61}$$

where k is a nonzero scalar. To show this, we consider the four cases (i) $x_{11} \neq 0$ and $x_{12} \neq 0$, (ii) $x_{11} = 0$, (iii) $x_{12} = 0$, and (iv) $x_{11} = x_{12} = 0$.

In case (i), we write Eq. (60) in the form $x_{11}\mathbf{x}_2 = x_{12}\mathbf{x}_1$ and divide through by x_{11} to get $\mathbf{x}_2 = k\mathbf{x}_1$, where $k = x_{12}/x_{11}$ is nonzero.

In case (ii), if $x_{11} = 0$, then the second component of Eq. (60) reduces to $x_{12}x_{21} = 0$. Since $\mathbf{x}_1$ is an eigenvector, it cannot be $\mathbf{0}$; consequently, its second component, x_{21}, must be nonzero. It follows that $x_{12} = 0$. Then $\mathbf{x}_1$ and $\mathbf{x}_2$ must have the forms

$$\mathbf{x}_1 = \begin{pmatrix} 0 \\ x_{21} \end{pmatrix} \quad \text{and} \quad \mathbf{x}_2 = \begin{pmatrix} 0 \\ x_{22} \end{pmatrix}, \tag{62}$$

where both x_{21} and x_{22} are nonzero, because the eigenvectors $\mathbf{x}_1$ and $\mathbf{x}_2$ are nonzero by definition. It follows that Eq. (61) holds with $k = x_{22}/x_{21}$.

Interchanging the roles of x_{11} and x_{12}, case (iii) is identical to case (ii).

Case (iv) results in Eq. (62) directly. Thus Eq. (61) also holds with $k = x_{22}/x_{21}$.

Thus we have established that the assumption $\det(\mathbf{X}) = 0$ implies Eq. (61) in which $k \neq 0$. Multiplying Eq. (61) by $\mathbf{A}$ gives

$$\mathbf{A}\mathbf{x}_2 = \mathbf{A}(k\mathbf{x}_1) = k\mathbf{A}\mathbf{x}_1 = k\lambda_1\mathbf{x}_1,$$

whereas on the other hand, we have

$$\mathbf{A}\mathbf{x}_2 = \lambda_2\mathbf{x}_2 = \lambda_2(k\mathbf{x}_1) = k\lambda_2\mathbf{x}_1.$$

Taking the difference of these two equations gives us the equation

$$k(\lambda_1 - \lambda_2)\mathbf{x}_1 = \mathbf{0} \tag{63}$$

in which $k \neq 0$, $\lambda_1 - \lambda_2 \neq 0$ since $\lambda_1 \neq \lambda_2$, and $\mathbf{x}_1 \neq \mathbf{0}$ because it is an eigenvector. Since the assumption $\det(\mathbf{X}) = 0$ leads to the contradictory statement (63), we conclude that $\det(\mathbf{X}) \neq 0$.

Remark. If you are familiar with the concepts of linear dependence and linear independence, then you will recognize that Theorem 3.1.3 states that if the eigenvalues of $\mathbf{A}$ are distinct, then the eigenvectors are linearly independent.

PROBLEMS

Solving Linear Systems. In each of Problems 1 through 12:
(a) Find all solutions of the given system of equations.
(b) Sketch the graph of each equation in the system. Are the lines intersecting, parallel, or coincident?

1. $2x_1 + 3x_2 = 7$, $-3x_1 + x_2 = -5$
2. $x_1 - 2x_2 = 10$, $2x_1 + 3x_2 = 6$
3. $x_1 + 3x_2 = 0$, $2x_1 - x_2 = 0$
4. $-x_1 + 2x_2 = 4$, $2x_1 - 4x_2 = -6$
5. $2x_1 - 3x_2 = 4$, $x_1 + 2x_2 = -5$
6. $3x_1 - 2x_2 = 0$, $-6x_1 + 4x_2 = 0$
7. $2x_1 - 3x_2 = 6$, $-4x_1 + 6x_2 = -12$

8. $4x_1 + x_2 = 0, \quad 4x_1 - 3x_2 = -12$
9. $x_1 + 4x_2 = 10, \quad 4x_1 + x_2 = 10$
10. $x_1 + x_2 = -1, \quad -x_1 + 2x_2 = 4$
11. $4x_1 - 3x_2 = 0, \quad -2x_1 + 5x_2 = 0$
12. $2x_1 + 5x_2 = 0, \quad 4x_1 + 10x_2 = 0$

Eigenvalues and Eigenvectors. In each of Problems 13 through 32, find the eigenvalues and eigenvectors of the given matrix.

13. $\mathbf{A} = \begin{pmatrix} 3 & -2 \\ 2 & -2 \end{pmatrix}$

14. $\mathbf{A} = \begin{pmatrix} 3 & -2 \\ 4 & -1 \end{pmatrix}$

15. $\mathbf{A} = \begin{pmatrix} 3 & -4 \\ 1 & -1 \end{pmatrix}$

16. $\mathbf{A} = \begin{pmatrix} 1 & -2 \\ 3 & -4 \end{pmatrix}$

17. $\mathbf{A} = \begin{pmatrix} -1 & -4 \\ 1 & -1 \end{pmatrix}$

18. $\mathbf{A} = \begin{pmatrix} \frac{5}{4} & \frac{3}{4} \\ -\frac{3}{4} & -\frac{1}{4} \end{pmatrix}$

19. $\mathbf{A} = \begin{pmatrix} -\frac{3}{2} & 1 \\ -\frac{1}{4} & -\frac{1}{2} \end{pmatrix}$

20. $\mathbf{A} = \begin{pmatrix} 2 & -1 \\ 3 & -2 \end{pmatrix}$

21. $\mathbf{A} = \begin{pmatrix} 2 & -5 \\ 1 & -2 \end{pmatrix}$

22. $\mathbf{A} = \begin{pmatrix} 6 & 3 \\ 2 & 1 \end{pmatrix}$

23. $\mathbf{A} = \begin{pmatrix} 1 & 1 \\ 4 & -2 \end{pmatrix}$

24. $\mathbf{A} = \begin{pmatrix} 2 & -\frac{5}{2} \\ \frac{9}{5} & -1 \end{pmatrix}$

25. $\mathbf{A} = \begin{pmatrix} -3 & \frac{5}{2} \\ -\frac{5}{2} & 2 \end{pmatrix}$

26. $\mathbf{A} = \begin{pmatrix} 1 & -1 \\ 5 & -3 \end{pmatrix}$

27. $\mathbf{A} = \begin{pmatrix} 1 & \frac{4}{3} \\ -\frac{9}{4} & -3 \end{pmatrix}$

28. $\mathbf{A} = \begin{pmatrix} -2 & 1 \\ 1 & -2 \end{pmatrix}$

29. $\mathbf{A} = \begin{pmatrix} 1 & 2 \\ -5 & -1 \end{pmatrix}$

30. $\mathbf{A} = \begin{pmatrix} -1 & -\frac{1}{2} \\ 2 & -3 \end{pmatrix}$

31. $\mathbf{A} = \begin{pmatrix} \frac{5}{4} & \frac{3}{4} \\ \frac{3}{4} & \frac{5}{4} \end{pmatrix}$

32. $\mathbf{A} = \begin{pmatrix} 2 & \frac{1}{2} \\ -\frac{1}{2} & 1 \end{pmatrix}$

In each of Problems 33 through 36:
(a) Find the eigenvalues of the given matrix.
(b) Describe how the nature of the eigenvalues depends on the parameter α in the matrix $\mathbf{A}$.

33. $\mathbf{A} = \begin{pmatrix} 2 & \alpha \\ 1 & -3 \end{pmatrix}$

34. $\mathbf{A} = \begin{pmatrix} 3 & 4 \\ -\alpha & 2 \end{pmatrix}$

35. $\mathbf{A} = \begin{pmatrix} 1 & 2 \\ 3 & \alpha \end{pmatrix}$

36. $\mathbf{A} = \begin{pmatrix} 1 & -\alpha \\ 2\alpha & 3 \end{pmatrix}$

37. If $\det(\mathbf{A}) \neq 0$, derive the result in Eq. (14) for $\mathbf{A}^{-1}$.
38. Show that $\lambda = 0$ is an eigenvalue of the matrix $\mathbf{A}$ if and only if $\det(\mathbf{A}) = 0$.

3.2 Systems of Two First Order Linear Differential Equations

We begin our discussion of systems of differential equations with a model of heat exchange between the air inside a greenhouse and an underground rockbed in which heat, derived from solar radiation, is accumulated and stored during the daytime. The mathematical model requires two dependent variables: (1) the air temperature in the greenhouse, and (2) the temperature of the rockbed. At night, when the air temperature outside the greenhouse is low, heat from the rockbed is used to help keep the air in the greenhouse warm, thereby reducing the cost of operating an electrical or petroleum-based heating system.

EXAMPLE 1

A Rockbed Heat Storage System

Consider the schematic diagram of the greenhouse/rockbed system in Figure 3.2.1. The rockbed, consisting of rocks ranging in size from 2 to 15 cm, is loosely packed so that air can easily pass through the void space between the rocks. The rockbed, and the underground portion of the air ducts used to circulate air through the system, are thermally insulated from the surrounding soil. Rocks are a good material for storing heat since they have a high energy-storage capacity, are inexpensive, and have a long life.

During the day, air in the greenhouse is heated primarily by solar radiation. Whenever the air temperature in the greenhouse exceeds an upper threshold value, a thermostatically

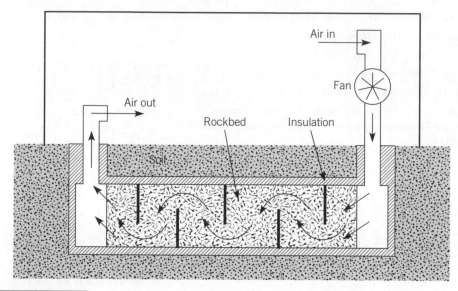

FIGURE 3.2.1 Schematic side view of greenhouse and rock storage bed.

controlled fan circulates the air through the system, thereby transferring heat to the rockbed. At night, when the air temperature in the greenhouse drops below a lower threshold value, the fan again turns on, and heat stored in the rockbed warms the circulating air.

We wish to study temperature variation in the greenhouse during the nighttime phase of the cycle. A simplified model for the system is provided by *lumped system thermal analysis*, in which we treat the physical system as if it consists of two interacting components. Assume that the air in the system is well mixed so that both the temperature of the air in the greenhouse and the temperature of the rockbed are functions of time, but not location. Let us denote the air temperature in the greenhouse by $u_1(t)$ and the temperature of the rockbed by $u_2(t)$. We will measure t in hours and temperature in degrees Celsius.

The following table lists the relevant parameters that appear in the mathematical model below. We use the subscripts 1 and 2 to indicate thermal and physical properties of the air and the rock medium, respectively.

m_1, m_2	total masses of air and rock
C_1, C_2	specific heats of air and rock
A_1, A_2	area of above-ground greenhouse enclosure and area of the air–rock interface
h_1, h_2	heat transfer coefficients across interface areas A_1 and A_2
T_a	temperature of air external to the greenhouse

The units of C_1, and C_2 are J/kg · °C, while the units of h_1 and h_2 are J/h · m^3 · °C. The area of the air–rock interface is approximately equal to the sum of the surface areas of the rocks in the rock storage bed.

Using the law of conservation of energy, we get the differential equations

$$m_1 C_1 \frac{du_1}{dt} = -h_1 A_1 (u_1 - T_a) - h_2 A_2 (u_1 - u_2), \tag{1}$$

$$m_2 C_2 \frac{du_2}{dt} = -h_2 A_2 (u_2 - u_1). \tag{2}$$

Equation (1) states that the rate of change of energy in the air internal to the system equals the rate at which heat flows across the above-ground enclosure (made of glass or polyethylene) plus the rate at which heat flows across the underground air–rock interface. In each case, the rates are proportional to the difference in temperatures of the materials on each side of the interface. The algebraic signs that multiply each term on the right are chosen so that heat flows in the direction from hot to cool. Equation (2) arises from the following reasoning. Since the rockbed is insulated around its boundary, heat can enter or leave the rockbed only across the air–rock interface. Energy conservation requires that the rate at which heat is gained or lost by the rockbed through this interface must equal the heat lost or gained by the greenhouse air through the same interface. Thus the right side of Eq. (2) is equal to the negative of the second term on the right-hand side of Eq. (1).

Dividing Eq. (1) by $m_1 C_1$ and Eq. (2) by $m_2 C_2$ gives

$$\frac{du_1}{dt} = -k_1(u_1 - T_a) - k_2(u_1 - u_2), \tag{3}$$

$$\frac{du_2}{dt} = -\epsilon k_2(u_2 - u_1), \tag{4}$$

or, rearranging terms, we get

$$\frac{du_1}{dt} = -(k_1 + k_2)u_1 + k_2 u_2 + k_1 T_a, \tag{5}$$

$$\frac{du_2}{dt} = \epsilon k_2 u_1 - \epsilon k_2 u_2, \tag{6}$$

where

$$k_1 = \frac{h_1 A_1}{m_1 C_1}, \qquad k_2 = \frac{h_2 A_2}{m_1 C_1}, \qquad \text{and} \qquad \epsilon = \frac{m_1 C_1}{m_2 C_2}. \tag{7}$$

Note that Eqs. (5) and (6) are an extension to two materials of Newton's law of cooling, introduced in Example 1 in Section 1.1. In the current application, the dimensionless parameter ϵ, the ratio of the energy storage capacity of the greenhouse air to the energy storage capacity of the rockbed, is small relative to unity[2] because $m_1 C_1$ is much smaller than $m_2 C_2$.

Let us suppose that $t = 0$ corresponds to the beginning of the nighttime phase of the cycle, and that starting values of u_1 and u_2 are specified at this time by

$$u_1(0) = u_{10}, \qquad u_2(0) = u_{20}. \tag{8}$$

The pair of differential equations (5) and (6), together with the initial conditions (8), constitute a mathematical model for the variation of air temperature in the greenhouse and the temperature of the rockbed during the nighttime phase of the cycle. Solutions of these equations for $u_1(t)$ and $u_2(t)$ can assist us in designing a solar-powered rockbed heating system for the greenhouse. In particular, the solutions can help us determine the size of the rockbed relative to the volume of the greenhouse, to determine optimal fan speeds, and to determine required rockbed temperatures satisfactory for keeping the greenhouse warm during the night.

In order to plot graphs of solutions of the initial value problem (5), (6), and (8) on a computer, it is necessary to assign numerical values to the parameters k_1, k_2, ϵ, and T_a as well as to the initial conditions u_{10} and u_{20}. Values for k_1, k_2, and ϵ may be estimated by consulting

---

[2] In mathematics, the notation $0 < \epsilon \ll 1$ is frequently used to represent the statement, "The positive number ϵ is small compared to unity."

tables of heat transfer coefficients and thermal properties of gases and building materials, whereas values for T_a, u_{10}, and u_{20} may be chosen to represent different experimental scenarios of interest. In general, T_a can vary with t, but we will assume the simplest case, in which T_a is assumed to be constant. In preliminary stages of model development, an investigator may use a number of educated guesses for the parameter values to get a feeling for how solutions behave and to see how sensitive solutions are to changes in the values of the parameters. For the time being, we choose the following values as a compromise between realism and analytical convenience:

$$k_1 = \tfrac{7}{8}, \qquad k_2 = \tfrac{3}{4}, \qquad \epsilon = \tfrac{1}{3}, \qquad \text{and} \qquad T_a = 16°C. \tag{9}$$

Substituting these values into Eqs. (5) and (6) then gives

$$\frac{du_1}{dt} = -\left(\frac{13}{8}\right) u_1 + \left(\frac{3}{4}\right) u_2 + 14, \tag{10}$$

$$\frac{du_2}{dt} = \left(\frac{1}{4}\right) u_1 - \left(\frac{1}{4}\right) u_2. \tag{11}$$

Equations (10) and (11) constitute an example of a first order **system** of differential equations. Each equation contains the unknown temperature functions, u_1 and u_2, of the two interacting components that make up the system. The equations cannot be solved separately, but must be investigated together. In dealing with systems of equations, it is most advantageous to use vector and matrix notation. This saves space, facilitates calculations, and emphasizes the similarity between systems of equations and single (scalar) equations, which we discussed in Chapters 1 and 2.

Matrix Notation, Vector Solutions, and Component Plots

We begin by rewriting Eqs. (10) and (11) in the form

$$\begin{pmatrix} du_1/dt \\ du_2/dt \end{pmatrix} = \begin{pmatrix} -\frac{13}{8} & \frac{3}{4} \\ \frac{1}{4} & -\frac{1}{4} \end{pmatrix} \begin{pmatrix} u_1 \\ u_2 \end{pmatrix} + \begin{pmatrix} 14 \\ 0 \end{pmatrix}. \tag{12}$$

Next we define the vectors **u** and **b** and the matrix **K** to be

$$\mathbf{u} = \begin{pmatrix} u_1 \\ u_2 \end{pmatrix}, \qquad \mathbf{b} = \begin{pmatrix} 14 \\ 0 \end{pmatrix}, \qquad \mathbf{K} = \begin{pmatrix} -\frac{13}{8} & \frac{3}{4} \\ \frac{1}{4} & -\frac{1}{4} \end{pmatrix}. \tag{13}$$

Then Eq. (12) takes the form

$$\frac{d\mathbf{u}}{dt} = \mathbf{K}\mathbf{u} + \mathbf{b}. \tag{14}$$

Using vector notation, the initial conditions in Eq. (8) are expressed as

$$\mathbf{u}(0) = \mathbf{u}_0 = \begin{pmatrix} u_{10} \\ u_{20} \end{pmatrix}. \tag{15}$$

Using methods discussed later in this chapter, we will be able to find all solutions of Eq. (14). But first, we want to discuss *what* we mean by a solution of this equation. Consider the vector function

$$\mathbf{u} = \begin{pmatrix} u_1(t) \\ u_2(t) \end{pmatrix} = 8e^{-t/8} \begin{pmatrix} 1 \\ 2 \end{pmatrix} - 4e^{-7t/4} \begin{pmatrix} 6 \\ -1 \end{pmatrix} + \begin{pmatrix} 16 \\ 16 \end{pmatrix}$$

$$= \begin{pmatrix} 8e^{-t/8} - 24e^{-7t/4} + 16 \\ 16e^{-t/8} + 4e^{-7t/4} + 16 \end{pmatrix}. \tag{16}$$

There are two ways that we can show $\mathbf{u}$ in Eq. (16) is a solution of Eq. (14). One way is to substitute the two components of $\mathbf{u}$, $u_1(t) = 8e^{-t/8} - 24e^{-7t/4} + 16$ and $u_2(t) = 16e^{-t/8} + 4e^{-7t/4} + 16$, into each of Eqs. (10) and (11). If, upon substitution, each equation reduces to an identity,[3] then $\mathbf{u}$ is a solution. Equivalently, we can use matrix formalism to show that substituting $\mathbf{u}$ into Eq. (14) results in a vector identity, as we now demonstrate. Substituting the right side of Eq. (16) for $\mathbf{u}$ on the left side of Eq. (14) gives

$$\frac{d\mathbf{u}}{dt} = \begin{pmatrix} -e^{-t/8} + 42e^{-7t/4} \\ -2e^{-t/8} - 7e^{-7t/4} \end{pmatrix}. \tag{17}$$

On the other hand, substituting for $\mathbf{u}$ on the right side of Eq. (14), and using the rules of matrix algebra, yield

$$\mathbf{K}\mathbf{u} + \mathbf{b} = \begin{pmatrix} -\frac{13}{8} & \frac{3}{4} \\ \frac{1}{4} & -\frac{1}{4} \end{pmatrix} \begin{pmatrix} 8e^{-t/8} - 24e^{-7t/4} + 16 \\ 16e^{-t/8} + 4e^{-7t/4} + 16 \end{pmatrix} + \begin{pmatrix} 14 \\ 0 \end{pmatrix}$$

$$= \begin{pmatrix} -13e^{-t/8} + 39e^{-7t/4} - 26 + 12e^{-t/8} + 3e^{-7t/4} + 12 \\ 2e^{-t/8} - 6e^{-7t/4} + 4 - 4e^{-t/8} - e^{-7t/4} - 4 \end{pmatrix} + \begin{pmatrix} 14 \\ 0 \end{pmatrix}$$

$$= \begin{pmatrix} -e^{-t/8} + 42e^{-7t/4} \\ -2e^{-t/8} - 7e^{-7t/4} \end{pmatrix}. \tag{18}$$

Since the right side of Eq. (17) and the last term in Eq. (18) agree, $\mathbf{u}$ in Eq. (16) is indeed a solution of $\mathbf{u}' = \mathbf{K}\mathbf{u} + \mathbf{b}$.

Note also that if we evaluate Eq. (16) at $t = 0$, we get

$$\mathbf{u}(0) = \begin{pmatrix} 0 \\ 36 \end{pmatrix}. \tag{19}$$

Thus $\mathbf{u}$ in Eq. (16) is a solution of the initial value problem (14) and (19).

The components of $\mathbf{u}$ are scalar valued functions of t, so we can plot their graphs. Plots of u_1 and u_2 versus t are called **component plots**. In Figure 3.2.2 we show the component plots of u_1 and u_2. Component plots are useful because they display the detailed dependence of u_1 and u_2 on t for a particular solution of Eq. (14). From Figure 3.2.2 we see that the temperature u_1 of the air, ostensibly heated by the rockpile, rises rapidly, in 2.3 h, from 0°C to approximately 21.6°C. During that same period, the rockpile cools from 36°C to roughly 28°C, a change of only 8°C. Thereafter, both u_1 and u_2 slowly approach the same constant limiting value as $t \to \infty$. From Eq. (16) we see that

$$\lim_{t \to \infty} \mathbf{u}(t) = \begin{pmatrix} 16 \\ 16 \end{pmatrix}, \tag{20}$$

---

[3]In this context, an identity is an equation that is true for all values of the variables, for example, $0 = 0$ or $t \sin(\omega t) = t \sin(\omega t)$. A vector identity is an equation in which corresponding components of the equation are identities, for example, $\cos(\omega t)\mathbf{i} + t \sin(\omega t)\mathbf{j} = \cos(\omega t)\mathbf{i} + t \sin(\omega t)\mathbf{j}$.

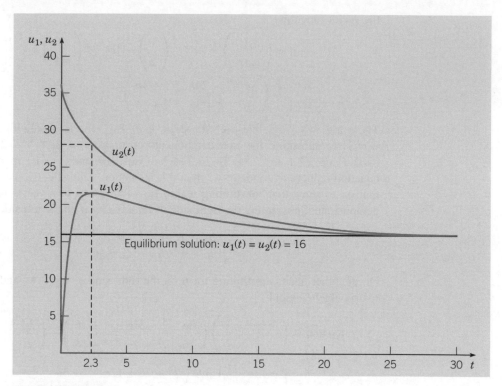

FIGURE 3.2.2   Component plots of the solutions to the initial value problems (14), (19) and (14), (21).

since all other terms in the solution contain one of the exponential factors, $e^{-t/8}$ or $e^{-7t/4}$, which tend to 0 as $t \to \infty$. Note that the constant vector $\mathbf{u} = 16\mathbf{i} + 16\mathbf{j}$ is also a solution, since substituting it into Eq. (14) results in the identity $\mathbf{0} = \mathbf{0}$. This solution satisfies the initial condition

$$\mathbf{u}(0) = \begin{pmatrix} 16 \\ 16 \end{pmatrix}. \tag{21}$$

Included in Figure 3.2.2 are plots of these constant values of u_1 and u_2. Physically, this solution corresponds to thermal equilibrium: the temperature of the greenhouse air and the temperature of the rockbed, if set initially to the outside air temperature T_a, in this case 16°C, do not change in time.

A disadvantage of this graphical approach is that we must construct another set of component plots each time we change the initial conditions for Eq. (14). Changing the values of k_1, k_2, and ϵ may require additional component plots. Fortunately, other methods of graphing solutions of Eq. (14) are available, so that we need not be overwhelmed by a dizzying array of component plots.

Geometry of Solutions: Direction Fields and Phase Portraits

We introduce here some common, and conceptually useful, terminology for systems such as Eq. (14). The variables u_1 and u_2 are often called **state variables**, since their values at any time describe the **state** of the system. For our greenhouse/rockbed system, the state variables are the temperature of the air in the greenhouse and the temperature of the rockbed. Similarly, the vector $\mathbf{u} = u_1\mathbf{i} + u_2\mathbf{j}$ is called the **state vector** of the system. The u_1u_2-plane itself is called the **state space**. If there are only two state variables, the $u_1 u_2$-plane may be called the **state plane** or, more commonly, the **phase plane**.

If $u_1(t)$ and $u_2(t)$ are the components of a solution $\mathbf{u}$ to Eq. (14), the parametric equations

$$u_1 = u_1(t), \qquad u_2 = u_2(t) \tag{22}$$

give the coordinates u_1 and u_2 of a point in the phase plane as functions of time. Each value of the parameter t determines a point $(u_1(t), u_2(t))$, and the set of all such points is a curve in the phase plane. The solution $\mathbf{u} = u_1(t)\mathbf{i} + u_2(t)\mathbf{j}$ may be thought of as a position vector of a particle moving in the phase plane. As t advances, the tip of the vector $\mathbf{u}$ traces a curve in the phase plane, called a **trajectory** or **orbit**, that graphically displays the path of the state of the system in the state space. The direction of motion along a solution trajectory is obtained by noting that for each t, a solution must satisfy the equation

$$\mathbf{u}'(t) = \mathbf{K}\mathbf{u}(t) + \mathbf{b}. \tag{23}$$

Equation (23) shows that the velocity $\mathbf{u}'(t)$ of the particle at the point with position vector $\mathbf{u}(t)$ is given by the vector $\mathbf{K}\mathbf{u}(t) + \mathbf{b}$. Thus, if we draw the vector $\mathbf{K}\mathbf{u} + \mathbf{b}$ with its tail at the point with position vector $\mathbf{u}$, it indicates the instantaneous direction of motion of the particle along a solution curve at that point, and its length tells us the speed of the solution as it passes through that point. We have done this for several values of $t \in [1, 3]$ on part of the trajectory associated with $\mathbf{u}$ in Eq. (16), as shown in Figure 3.2.3. Since these velocity vectors vary greatly in length, from the very long to the very short, we have scaled the lengths of the vectors so that they fit nicely into the graph window. Note that it is not necessary to know a solution of Eq. (23) in order to plot these velocity vectors. If a trajectory passes through the point with position vector $\mathbf{u}$, then the velocity vector that we attach to that point is obtained by simply evaluating the right side of the system (23) at $\mathbf{u}$. In other

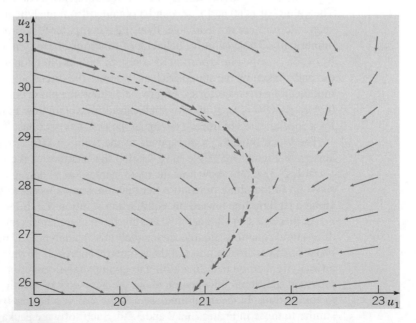

FIGURE 3.2.3   The vector field for the system (23) in the rectangle $19 \leq u_1 \leq 23$, $25.75 \leq u_2 \leq 31$ that contains the solution (16) for $1 \leq t \leq 3$.

words, the right side of Eq. (23) defines a **vector field** $\mathbf{F}(u_1, u_2)$ for the system through the relation

$$\mathbf{F}(u_1, u_2) = \mathbf{Ku} + \mathbf{b} = \left\{ -\frac{13u_1}{8} + \frac{3u_2}{4} + 14 \right\} \mathbf{i} + \left\{ \frac{u_1}{4} - \frac{u_2}{4} \right\} \mathbf{j}. \tag{24}$$

In the rectangular region shown in Figure 3.2.3, we have plotted several (uniformly scaled) velocity vectors using Eq. (23).

Direction Fields

In general, the right side of a system of first order equations defines a vector field that governs the direction and speed of motion of the solution at each point in the phase plane. Because the vectors generated by a vector field for a specific system often vary significantly in length, it is customary to scale each nonzero vector so that they all have the same length. These vectors are then referred to as **direction field vectors** for the system (14) and the resulting picture is called the **direction field**. Direction fields are easily constructed using a computer algebra system. Since direction field vectors all have the same length, we lose information about the speed of motion of solutions, but retain information about the direction of motion. Just as for the direction fields that we saw in Chapters 1 and 2 for single first order equations, we are able to infer the general behavior of trajectories, or solutions, of a system such as Eq. (14) by looking at the direction field.

Phase Portraits

Using a computer, it is just as easy to generate solution trajectories as it is to generate direction fields. A plot of a representative sample of the trajectories, *including any constant solutions*, is called a **phase portrait** of the system of equations. Each trajectory in the phase portrait is generated by plotting, in the u_1u_2-plane, the set of points with coordinates $(u_1(t), u_2(t))$ for several values of t, and then drawing a curve through the resulting points. The values of $u_1(t)$ and $u_2(t)$ can be obtained from analytical solutions of Eq. (14), or, if such solutions are not available, by using a computer to approximate solutions of Eq. (14) numerically. The qualitative behavior of solutions of a system is generally made most clear by overlaying the phase portrait for a system with its direction field. The trajectories indicate the paths taken by the state variables, whereas the direction field indicates the direction of motion along the trajectories. Figure 3.2.4 shows the phase portrait for Eq. (14) in the region $0 \leq u_1 \leq 30$, $0 \leq u_2 \leq 40$. We shall carefully explain how to draw such portraits by hand for a special cases of linear systems in the next few sections.

The black trajectory corresponds to the solution (16), for which the component plots appear in Figure 3.2.2. We have labeled the position of the state of the system at times $t = 0, 1, 2, 3, 4, 5$ to show how the rapid increase in u_1 during the first 2 to 3 hours does not show up in the phase portrait. After that, the state appears to move at a much slower rate along a straight line toward the equilibrium solution $\mathbf{u} = 16\mathbf{i} + 16\mathbf{j}$ discussed earlier.

As discussed in Chapter 1, there are three complementary approaches to the study of differential equations: geometric (or qualitative), analytic, and numeric. Direction fields and phase portraits are examples of the geometric approach. In the next section, we will present an analytic method that, in addition to giving analytic solutions to the system (12) or (14), provides additional information about the geometry of solutions and the rate of approach to equilibrium. In the meantime, we assume that you have software that can produce plots similar to those in Figures 3.2.2 and 3.2.4. Such software packages plot direction fields and compute numerical approximations to solutions, which can be displayed as phase plane trajectories and component plots. They are very useful, not only for plotting solutions of

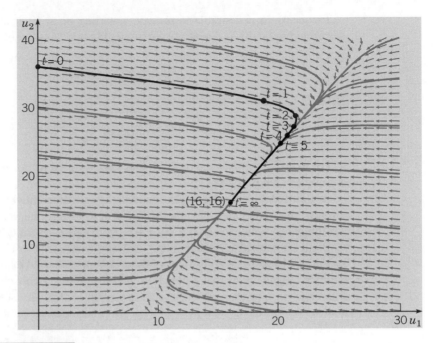

FIGURE 3.2.4   Phase portrait for the system (14).

systems that can be easily solved, but especially for investigating systems that are more difficult, if not impossible, to solve analytically.

Solutions of Two First Order Linear Equations

The system (12) or (14) belongs to a much more general class of problems, which we obtain by replacing the constant coefficients on the right side by functions of the independent variable t. Thus we obtain the general system of two first order linear differential equations

$$\begin{pmatrix} dx/dt \\ dy/dt \end{pmatrix} = \begin{pmatrix} p_{11}(t)x + p_{12}(t)y + g_1(t) \\ p_{21}(t)x + p_{22}(t)y + g_2(t) \end{pmatrix}. \tag{25}$$

Using vector notation, we can write the linear system (25) as

$$\frac{d\mathbf{x}}{dt} = \mathbf{P}(t)\mathbf{x} + \mathbf{g}(t), \tag{26}$$

where

$$\mathbf{x} = \begin{pmatrix} x \\ y \end{pmatrix}, \qquad \mathbf{P}(t) = \begin{pmatrix} p_{11}(t) & p_{12}(t) \\ p_{21}(t) & p_{22}(t) \end{pmatrix}, \qquad \text{and} \qquad \mathbf{g}(t) = \begin{pmatrix} g_1(t) \\ g_2(t) \end{pmatrix}. \tag{27}$$

Again, we refer to x and y as state variables, to $\mathbf{x}$ as the state vector, and to the xy-plane as the phase plane.

The system (25) is called a **first order linear system of dimension two** because it consists of first order equations and because its state space (the xy-plane) is two-dimensional. Further, if $\mathbf{g}(t) = \mathbf{0}$ for all t, that is, $g_1(t) = g_2(t) = 0$ for all t, then the system is said to be **homogeneous**. Otherwise, it is **nonhomogeneous**. For example, the system (12), or (14), is nonhomogeneous.

Frequently it is desirable to provide a starting point for investigation of a system by describing its configuration at a given instant in time. Thus, we often have initial conditions

$$x(t_0) = x_0, \qquad y(t_0) = y_0, \tag{28}$$

or

$$\mathbf{x}(t_0) = \mathbf{x}_0, \tag{29}$$

where $\mathbf{x}_0 = x_0\mathbf{i} + y_0\mathbf{j}$. Equations (25) and (28), or in vector form Eqs. (26) and (29), form an **initial value problem**.

A **solution** of the system (25) consists of two differentiable functions $x = \phi(t)$ and $y = \psi(t)$ that satisfy Eqs. (25) for all values of t in some interval I. In vector terminology, a solution is a vector $\mathbf{x} = \boldsymbol{\phi}(t) = \phi(t)\mathbf{i} + \psi(t)\mathbf{j}$ that satisfies Eq. (26) for all t in I.

From a graphical or visual point of view, the system (25), or (26), is relatively difficult to investigate because the right sides of these equations depend explicitly on the independent variable t. This means that a direction field for these systems changes with time. For the same reason, a phase portrait for such a system is also not useful. It is still possible, of course, to draw component plots of x versus t and y versus t for such systems.

Existence and Uniqueness of Solutions

It may be important to know that the initial value problem (25), (28), or (26), (29) has a unique solution, and there is a theorem (stated below) that asserts this is the case. This theorem is analogous to Theorem 2.4.1 that deals with initial value problems for first order linear equations. The importance of such a theorem to mathematicians is first, that it ensures that a problem you are trying to solve actually has a solution; second, that if you are successful in finding a solution, you can be sure that it is the only one; and third, that it promotes confidence in using numerical approximation methods when you are sure that there is actually a solution curve to approximate.

THEOREM 3.2.1

Let each of the functions $p_{11}, \ldots, p_{22}, g_1$, and g_2 be continuous on an open interval $I = \alpha < t < \beta$, let t_0 be any point in I, and let x_0 and y_0 be any given numbers. Then there exists a unique solution of the system (25)

$$\begin{pmatrix} dx/dt \\ dy/dt \end{pmatrix} = \begin{pmatrix} p_{11}(t)x + p_{12}(t)y + g_1(t) \\ p_{21}(t)x + p_{22}(t)y + g_2(t) \end{pmatrix}$$

that also satisfies the initial conditions (28)

$$x(t_0) = x_0, \qquad y(t_0) = y_0.$$

Further the solution exists throughout the interval I.

In some cases, the existence of solutions can be demonstrated by actually finding them, and much of this book is devoted to that goal. However, a proof of Theorem 3.2.1, in general, is too difficult to give here; it may be found in more advanced books on differential equations.

Observe that the interval of existence of the solution is the entire interval I in which the hypotheses are satisfied. Further the initial values x_0 and y_0 are completely arbitrary.

Linear Autonomous Systems

If the right side of Eq. (26) does not depend explicitly on the independent variable t, the system is said to be **autonomous**. For Eq. (26) to be autonomous, the elements of the coefficient matrix $\mathbf{P}$ and the components of the vector $\mathbf{g}$ must be constants. We will usually use the notation

$$\frac{d\mathbf{x}}{dt} = \mathbf{A}\mathbf{x} + \mathbf{b}, \tag{30}$$

where $\mathbf{A}$ is a constant matrix and $\mathbf{b}$ is a constant vector, to denote autonomous linear systems. Since the entries of $\mathbf{A}$ and the components of $\mathbf{b}$ are constants, and therefore continuous for all t, Theorem 3.2.1 implies that any solution of Eq. (30) exists and is unique on the entire t-axis. Direction fields and phase portraits are effective tools for studying Eq. (30), since the vector field $\mathbf{A}\mathbf{x} + \mathbf{b}$ does not change with time. Note that Eq. (14) for the greenhouse/rockbed system is of this type.

Recall that in Section 2.5 we found that equilibrium, or constant, solutions were of particular importance in the study of single first order autonomous equations. This is also true for autonomous systems of equations. For the linear autonomous system (30), we find the **equilibrium solutions**, or **critical points**, by setting $d\mathbf{x}/dt$ equal to zero. Hence any solution of

$$\mathbf{A}\mathbf{x} = -\mathbf{b} \tag{31}$$

is a critical point of Eq. (30). If the coefficient matrix $\mathbf{A}$ has an inverse, as we usually assume, then Eq. (31) has a single solution, namely, $\mathbf{x} = -\mathbf{A}^{-1}\mathbf{b}$. This is then the only critical point of the system (30). However, if $\mathbf{A}$ is singular, then Eq. (31) has either no solution or infinitely many.

It is important to understand that critical points are found by solving algebraic, rather than differential, equations. As we shall see later, the behavior of trajectories in the vicinity of critical points can also be determined by algebraic methods. Thus a good deal of information about solutions of autonomous systems can be found without actually solving the system.

**EXAMPLE
2**

Consider again the greenhouse/rockbed system of Example 1. Find the critical point of the system (10) and (11), and describe the behavior of the trajectories in its neighborhood.

To find the critical point, we set $du_1/dt = 0$ and $du_2/dt = 0$ in Eqs. (10) and (11) to get

$$-\tfrac{13}{8}u_1 + \tfrac{3}{4}u_2 + 14 = 0, \tag{32}$$

$$\tfrac{1}{4}u_1 - \tfrac{1}{4}u_2 = 0. \tag{33}$$

or, using matrix notation,

$$\begin{pmatrix} -\tfrac{13}{8} & \tfrac{3}{4} \\ \tfrac{1}{4} & -\tfrac{1}{4} \end{pmatrix} \begin{pmatrix} u_1 \\ u_2 \end{pmatrix} + \begin{pmatrix} 14 \\ 0 \end{pmatrix} = \mathbf{0}. \tag{34}$$

Equation (33) implies that $u_2 = u_1$. Substituting u_1 for u_2 in Eq. (32) then gives $\tfrac{7}{8}u_1 = 14$. Thus the only critical point of the system (10) and (11) is $u_1 = u_2 = 16$, in agreement with the equilibrium solution found in Eq. (21).

From the phase portrait in Figure 3.2.4, it appears that all solutions of the system (14) approach the critical point as $t \to \infty$. However, no other trajectory can actually reach the critical point in a finite time, since this would violate the uniqueness part of Theorem 3.2.1.

This behavior is borne out by the component plots in Figure 3.2.2, which indicate that $u_1 = u_2 = 16$ is a horizontal asymptote for the solutions shown there.

The qualitative behavior of solutions shown in Figure 3.2.4 is consistent with our physical intuition about how solutions of Eq. (14) should behave. After a very long time, the temperature of the rockpile and the temperature of the greenhouse air will both be essentially equal to the outside air temperature.

Transformation of a Second Order Equation to a System of First Order Equations

One or more higher order equations can always be transformed into a system of first order equations. Thus such systems can be considered as the most fundamental problem area in differential equations.

To illustrate one common way (but not the only way) of doing this transformation, let us consider the second order equation

$$y'' + p(t)y' + q(t)y = g(t), \tag{35}$$

where p, q, and g are given functions that we assume to be continuous on an interval I. First, we introduce new dependent variables x_1 and x_2 in the form of a table,

$$x_1 = y \tag{36}$$

$$x_2 = y'. \tag{37}$$

Next, we differentiate each line in the table to get

$$x_1' = y' \tag{38}$$

$$x_2' = y''. \tag{39}$$

We then use Eqs. (36) and (37) in the table, and the differential equation (35) itself, to express the right sides of Eqs. (38) and (39) in terms of the state variables x_1 and x_2. Using Eq. (37), we can replace the right side y' of Eq. (38) by x_2 to get

$$x_1' = x_2. \tag{40}$$

Solving for y'' in Eq. (35), we can write Eq. (39) as

$$x_2' = -q(t)y - p(t)y' + g(t),$$

or

$$x_2' = -q(t)x_1 - p(t)x_2 + g(t), \tag{41}$$

where we have again used Eqs. (36) and (37). Equations (40) and (41) form a system of two first order equations that is equivalent to the original Eq. (35). Using matrix notation, we can write this system as

$$\mathbf{x}' = \begin{pmatrix} 0 & 1 \\ -q(t) & -p(t) \end{pmatrix} \mathbf{x} + \begin{pmatrix} 0 \\ g(t) \end{pmatrix}. \tag{42}$$

Initial conditions for Eq. (35) are of the form

$$y(t_0) = y_0, \qquad y'(t_0) = y_1, \tag{43}$$

that is, the value of both the dependent variable and its derivative must be specified at some point t_0 in I. These initial conditions are then transferred to the state variables x_1 and x_2 by using Eqs. (36) and (37), $x_1(t_0) = y_0$ and $x_2(t_0) = y_1$, or, using vector notation,

$$\mathbf{x}(t_0) = \begin{pmatrix} y_0 \\ y_1 \end{pmatrix}. \tag{44}$$

Since the initial value problem (42), (44) is a special case of the initial value problem (26) and (29), Theorem 3.2.1 can be applied to the initial value problem (35), (43).

Remark. Once we have the solution $\mathbf{x}(t)$ of the IVP (42) and (44) we can obtain the solution $y(t)$ of the IVP (35) and (43) by extracting the first component of $\mathbf{x}(t)$.

EXAMPLE 3

Consider the differential equation

$$u'' + 0.25u' + 2u = 3\sin t. \tag{45}$$

Suppose that initial conditions

$$u(0) = 2, \qquad u'(0) = -2 \tag{46}$$

are also given. As we will show in the next chapter, this initial value problem can serve as a model for a vibrating spring–mass system in which u represents the displacement of the mass from its equilibrium position and u' represents the velocity of the mass. Transform this problem into an equivalent one for a system of first order equations. Use a computer to produce component plots of the displacement and velocity of the mass.

If we let $x_1 = u$ and $x_2 = u'$ and follow the steps leading from Eq. (35) to Eqs. (40) and (41), we obtain the system

$$x_1' = x_2, \qquad x_2' = -2x_1 - 0.25x_2 + 3\sin t. \tag{47}$$

The initial conditions (46) lead directly to

$$x_1(0) = 2, \qquad x_2(0) = -2. \tag{48}$$

The initial value problem (47) and (48) is equivalent to Eqs. (45) and (46). In matrix notation we write this initial value problem as

$$\mathbf{x}' = \begin{pmatrix} 0 & 1 \\ -2 & -0.25 \end{pmatrix} \mathbf{x} + \begin{pmatrix} 0 \\ 3\sin t \end{pmatrix}, \qquad \mathbf{x}(0) = \begin{pmatrix} 2 \\ -2 \end{pmatrix}. \tag{49}$$

Plots of $u = x_1(t)$ and $u' = x_2(t)$ are shown in Figure 3.2.5.

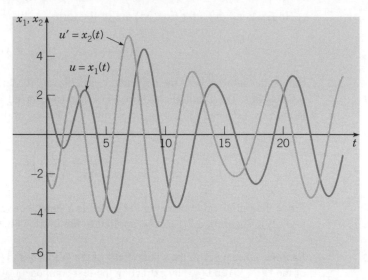

FIGURE 3.2.5 Component plots of the solution of the initial value problem (49).

PROBLEMS

Writing Systems in Matrix Form. In each of Problems 1 through 8, state whether the given system is autonomous or nonautonomous and also whether it is homogeneous or nonhomogeneous. Then write the system using matrix notation.

1. $x' = y, \quad y' = x + 4$

2. $x' = x + 2y + \sin t, \quad y' = -x + y - \cos t$

3. $x' = -2tx + y, \quad y' = 3x - y$

4. $x' = x + 2y + 4, \quad y' = -2x + y - 3$

5. $x' = 3x - y, \quad y' = x + 2y$

6. $x' = -x + ty, \quad y' = tx - y$

7. $x' = x + y + 4, \quad y' = -2x + (\sin t)y$

8. $x' = 3x - 4y, \quad y' = x + 3y$

Solutions of Linear Systems.

9. **(a)** Show that the functions $x(t) = 3e^t + 7e^{-t}$ and $y(t) = 3e^t + 21e^{-t}$ are solutions of the system

$$x' = 2x - y,$$
$$y' = 3x - 2y,$$

satisfying the initial conditions $x(0) = 10$ and $y(0) = 24$.

(b) Use a computer to plot the graphs of x and y versus t.

10. **(a)** Show that the functions

$$x(t) = e^{-t}(2\cos 2t - \sin 2t)$$

$$y(t) = -e^{-t}(2\sin 2t + \cos 2t)$$

are solutions of the system

$$x' = -x + 2y,$$
$$y' = -2x - y,$$

satisfying the initial conditions $x(0) = 2$ and $y(0) = -1$.
(b) Use a computer to plot the graphs of x and y versus t.

11. **(a)** Show that

$$\mathbf{x} = \begin{pmatrix} \sin t - t\cos t \\ t\sin t \end{pmatrix}$$

is a solution of the system

$$\mathbf{x}' = \begin{pmatrix} 0 & 1 \\ -1 & 0 \end{pmatrix}\mathbf{x} + \begin{pmatrix} 0 \\ 2\sin t \end{pmatrix}$$

satisfying the initial condition

$$\mathbf{x}(0) = \begin{pmatrix} 0 \\ 0 \end{pmatrix}.$$

(b) Use a computer to draw component plots of $\mathbf{x}$.

12. **(a)** Show that

$$\mathbf{x} = e^{-t}\begin{pmatrix} 2t - 1 \\ t - 1 \end{pmatrix} + \begin{pmatrix} 6t + 2 \\ 2t - 1 \end{pmatrix}$$

is a solution of the system

$$\mathbf{x}' = \begin{pmatrix} 1 & -4 \\ 1 & -3 \end{pmatrix} \mathbf{x} + \begin{pmatrix} 2t \\ -3 \end{pmatrix}$$

satisfying the initial condition

$$\mathbf{x}(0) = \begin{pmatrix} 1 \\ -2 \end{pmatrix}.$$

(b) Use a computer to draw component plots of **x**.

Equilibrium Solutions and Phase Portraits.

13. Find the equilibrium solution, or critical point, of Eqs. (5) and (6) in Example 1.

14. In the limiting case, $\epsilon \to 0$, Eqs. (5) and (6) of Example 1 reduce to the partially coupled system

$$\frac{du_1}{dt} = -(k_1 + k_2)u_1 + k_2 u_2 + k_1 T_a, \tag{i}$$

$$\frac{du_2}{dt} = 0. \tag{ii}$$

Thus, if initial conditions $u_1(0) = u_{10}$ and $u_2(0) = u_{20}$ are prescribed, Eq. (ii) implies that $u_2(t) = u_{20}$ for all $t \geq 0$. Therefore Eq. (i) reduces to a first order equation with one dependent variable,

$$\frac{du_1}{dt} = -(k_1 + k_2)u_1 + k_2 u_{20} + k_1 T_a. \tag{iii}$$

(a) Find the critical point (equilibrium solution) of Eq. (iii) and classify it as asymptotically stable or unstable. Then draw the phase line, and sketch several graphs of solutions in the tu_1-plane.
(b) Find the solution of Eq. (iii) subject to the initial condition $u_1(0) = u_{10}$ and use it to verify the qualitative results of part (a).
(c) What is the physical interpretation of setting $\epsilon = 0$? Give a physical interpretation of the equilibrium solution found in part (a).
(d) What do these qualitative results imply about the sizing of the rock storage pile in combination with temperatures that can be achieved in the rock storage pile during the daytime?

In each of Problems 15 through 20:
(a) Find the equilibrium solution, or critical point, of the given system.
(b) Use a computer to draw a direction field and phase portrait centered at the critical point.
(c) Describe how solutions of the system behave in the vicinity of the critical point.

15. $x' = -x + y + 1, \qquad y' = x + y - 3$

16. $x' = -x - 4y - 4, \qquad y' = x - y - 6$

17. $x' = -0.25x - 0.75y + 8, \qquad y' = 0.5x + y - 11.5$

18. $x' = -2x + y - 11, \qquad y' = -5x + 4y - 35$

19. $x' = x + y - 3, \qquad y' = -x + y + 1$

20. $x' = -5x + 4y - 35, \qquad y' = -2x + y - 11$

Second Order Differential Equations.
In Problems 21 through 24, transform the given equation into a system of first order equations.

21. $u'' + 0.5u' + 2u = 0$

22. $2u'' + 0.5u' + 8u = 6 \sin 2t$

23. $t^2 u'' + tu' + (t^2 - 0.25)u = 0$

24. $t^2 u'' + 3tu' + 5u = t^2 + 4$

In each of Problems 25 and 26, transform the given initial value problem into an initial value problem for two first order equations. Then write the system in matrix form.

25. $u'' + 0.25u' + 4u = 2 \cos 3t, \qquad u(0) = 1,$
$u'(0) = -2$

26. $tu'' + u' + tu = 0, \qquad u(1) = 1, \quad u'(1) = 0$

Applications.
Electric Circuits.   The theory of electric circuits, such as that shown in Figure 3.2.6, consisting of inductors, resistors, and capacitors, is based on Kirchhoff's laws: (1) At any node (or junction), the sum of currents flowing into that node is equal to the sum of currents flowing out of that node, and (2) the net voltage drop around each closed loop is zero. In addition to Kirchhoff's laws, we also have the relation between the current $i(t)$ in amperes through each circuit element and the voltage drop $v(t)$ in volts across the element:

$$v = Ri, \qquad R = \text{resistance in ohms};$$

$$C\frac{dv}{dt} = i, \qquad C = \text{capacitance in farads}[4];$$

$$L\frac{di}{dt} = v, \qquad L = \text{inductance in henries}.$$

Kirchhoff's laws and the current–voltage relation for each circuit element provide a system of algebraic and differential equations from which the voltage and current throughout the circuit can be determined. Problems 27 through 29 illustrate the procedure just described.

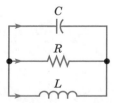

FIGURE 3.2.6   A parallel LRC circuit; see Problem 27.

---

[4]Most capacitors have capacitances measured in microfarads or picofarads. We use farad as the unit for numerical convenience.

27. Consider the circuit shown in Figure 3.2.6. Let i_1, i_2, and i_3 be the currents through the capacitor, resistor, and inductor, respectively. Likewise, let v_1, v_2, and v_3 be the corresponding voltage drops. The arrows denote the arbitrarily chosen directions in which currents and voltage drops will be taken to be positive.

(a) Applying Kirchhoff's second law to the upper loop in the circuit, show that

$$v_1 - v_2 = 0. \tag{i}$$

In a similar way, show that

$$v_2 - v_3 = 0. \tag{ii}$$

(b) Applying Kirchhoff's first law to either node in the circuit, show that

$$i_1 + i_2 + i_3 = 0. \tag{iii}$$

(c) Use the current–voltage relation through each element in the circuit to obtain the equations

$$Cv_1' = i_1, \qquad v_2 = Ri_2, \qquad Li_3' = v_3. \tag{iv}$$

(d) Eliminate v_2, v_3, i_1, and i_2 among Eqs. (i) through (iv) to obtain

$$Cv_1' = -i_3 - \frac{v_1}{R}, \qquad Li_3' = v_1. \tag{v}$$

These equations form a system of two equations for the variables v_1 and i_3.

28. Consider the circuit shown in Figure 3.2.7. Use the method outlined in Problem 27 to show that the current i through the inductor and the voltage v across the capacitor satisfy the system of differential equations

$$\frac{di}{dt} = -i - v, \qquad \frac{dv}{dt} = 2i - v.$$

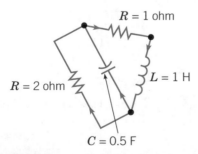

R = 1 ohm

R = 2 ohm

L = 1 H

C = 0.5 F

FIGURE 3.2.7 The circuit in Problem 28.

29. Consider the circuit shown in Figure 3.2.8. Use the method outlined in Problem 27 to show that the current i through the inductor and the voltage v across the capacitor satisfy the system of differential equations

$$L\frac{di}{dt} = -R_1 i - v, \qquad C\frac{dv}{dt} = i - \frac{v}{R_2}.$$

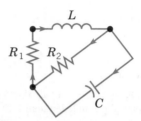

L

R_1 R_2

C

FIGURE 3.2.8 The circuit in Problem 29.

Mixing Problems.

30. Each of the tanks shown in Figure 3.2.9 contains a brine solution. Assume that Tank 1 initially contains 30 gallons (gal) of water and 55 ounces (oz) of salt, and Tank 2 initially contains 20 gal of water and 26 oz of salt. Water containing 1 oz/gal of salt flows into Tank 1 at a rate of 1.5 gal/min, and the well-stirred solution flows from Tank 1 to Tank 2 at a rate of 3 gal/min. Additionally, water containing 3 oz/gal of salt flows into Tank 2 at a rate of 1 gal/min (from the outside). The well-stirred solution in Tank 2 drains out at a rate of 4 gal/min, of which some flows back into Tank 1 at a rate of 1.5 gal/min, while the remainder leaves the system. Note that the volume of solution in each tank remains constant since the total rates of flow in and out of each tank are the same: 3 gal/min in Tank 1 and 4 gal/min in Tank 2.

(a) Denoting the amount of salt in Tank 1 and Tank 2 by $Q_1(t)$ and $Q_2(t)$, respectively, use the principle of mass balance (see Example 1 of Section 2.3) to show that

$$\frac{dQ_1}{dt} = -0.1Q_1 + 0.075Q_2 + 1.5,$$

$$\frac{dQ_2}{dt} = 0.1Q_1 - 0.2Q_2 + 3,$$

$$Q_1(0) = 55, \qquad Q_2(0) = 26.$$

(b) Write the initial value problem (i) using matrix notation.
(c) Find the equilibrium values Q_1^E and Q_2^E of the system.
(d) Use a computer to draw component plots of the initial value problem (i), and the equilibrium solutions, over the time interval $0 \leq t \leq 50$.
(e) Draw a phase portrait for the system centered at the critical point.

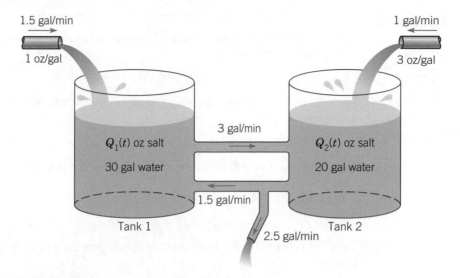

1.5 gal/min

1 oz/gal

1 gal/min

3 oz/gal

$Q_1(t)$ oz salt

30 gal water

3 gal/min

$Q_2(t)$ oz salt

20 gal water

1.5 gal/min

Tank 1

2.5 gal/min

Tank 2

FIGURE 3.2.9 The interconnected tanks in Problem 30.

31. Consider two interconnected tanks similar to those in Figure 3.2.9. Tank 1 initially contains 60 gal of water and Q_1^0 oz of salt, and Tank 2 initially contains 100 gal of water and Q_2^0 oz of salt. Water containing q_1 oz/gal of salt flows into Tank 1 at a rate of 3 gal/min. The mixture in Tank 1 flows out at a rate of 4 gal/min, of which half flows into Tank 2, while the remainder leaves the system. Water containing q_2 oz/gal of salt also flows into Tank 2 from the outside at the rate of 1 gal/min. The mixture in Tank 2 leaves the tank at a rate of 3 gal/min, of which some flows back into Tank 1 at a rate of 1 gal/min, while the rest leaves the system.

(a) Draw a diagram that depicts the flow process described above. Let $Q_1(t)$ and $Q_2(t)$, respectively, be the amount of salt in each tank at time t. Write down differential equations and initial conditions for Q_1 and Q_2 that model the flow process.
(b) Find the equilibrium values Q_1^E and Q_2^E in terms of the concentrations q_1 and q_2.
(c) Is it possible (by adjusting q_1 and q_2) to obtain $Q_1^E = 60$ and $Q_2^E = 50$ as an equilibrium state?
(d) Describe which equilibrium states are possible for this system for various values of q_1 and q_2.

3.3 Homogeneous Linear Systems with Constant Coefficients

In the preceding section, we noted that a linear autonomous system of first order differential equations has the form

$$\frac{d\mathbf{x}}{dt} = \mathbf{A}\mathbf{x} + \mathbf{b}, \tag{1}$$

where $\mathbf{A}$ and $\mathbf{b}$ are a constant matrix and a constant vector, respectively. In Example 1 of that section, we studied a greenhouse/rockbed heat storage problem modeled by the equations

$$\begin{pmatrix} du_1/dt \\ du_2/dt \end{pmatrix} = \begin{pmatrix} -\frac{13}{8} & \frac{3}{4} \\ \frac{1}{4} & -\frac{1}{4} \end{pmatrix} \begin{pmatrix} u_1 \\ u_2 \end{pmatrix} + \begin{pmatrix} 14 \\ 0 \end{pmatrix}, \tag{2}$$

which are of the form (1). Further, in Example 2, we found that the system (2) has a single critical point, or equilibrium solution, given by $\hat{u}_1 = 16$ and $\hat{u}_2 = 16$. It is often convenient to shift the origin of the phase plane, by means of a simple transformation, so that it coincides with a critical point.

EXAMPLE
1

A Rockbed
Heat Storage
System
Revisited

Consider the greenhouse/rockbed heat storage problem modeled by the system (2), subject to the initial conditions

$$u_1(0) = 0, \qquad u_2(0) = 36. \tag{3}$$

Let $x_1(t)$ and $x_2(t)$ be the deviations of $u_1(t)$ and $u_2(t)$ from their respective equilibrium values, that is,

$$x_1(t) = u_1(t) - 16, \qquad x_2(t) = u_2(t) - 16. \tag{4}$$

Find the system of differential equations and the initial conditions satisfied by $x_1(t)$ and $x_2(t)$.

Rewrite Eqs. (4) in the form

$$u_1(t) = 16 + x_1(t), \qquad u_2(t) = 16 + x_2(t). \tag{5}$$

Then, by substituting the right sides of Eqs. (5) for u_1 and u_2 in the first of Eqs. (2), we obtain

$$(16 + x_1)' = -\tfrac{13}{8}(16 + x_1) + \tfrac{3}{4}(16 + x_2) + 14$$

or

$$x_1' = -26 - \tfrac{13}{8}x_1 + 12 + \tfrac{3}{4}x_2 + 14$$

$$= -\tfrac{13}{8}x_1 + \tfrac{3}{4}x_2, \tag{6}$$

since the constant terms on the right side add to zero. Proceeding the same way with the second of Eqs. (2), we find that

$$x_2' = \tfrac{1}{4}(16 + x_1) - \tfrac{1}{4}(16 + x_2)$$

$$= \tfrac{1}{4}x_1 - \tfrac{1}{4}x_2. \tag{7}$$

If we write Eqs. (6) and (7) in vector form, we have

$$\begin{pmatrix} dx_1/dt \\ dx_2/dt \end{pmatrix} = \begin{pmatrix} -\tfrac{13}{8} & \tfrac{3}{4} \\ \tfrac{1}{4} & -\tfrac{1}{4} \end{pmatrix} \begin{pmatrix} x_1 \\ x_2 \end{pmatrix}, \tag{8}$$

or

$$\frac{d\mathbf{x}}{dt} = \begin{pmatrix} -\tfrac{13}{8} & \tfrac{3}{4} \\ \tfrac{1}{4} & -\tfrac{1}{4} \end{pmatrix} \mathbf{x}, \tag{9}$$

where $\mathbf{x}(t) = x_1(t)\mathbf{i} + x_2(t)\mathbf{j}$. In terms of x_1 and x_2, the initial conditions (3) become

$$x_1(0) = -16, \qquad x_2(0) = 20. \tag{10}$$

Observe that, by introducing the variables x_1 and x_2 defined by Eq. (4), we have transformed the nonhomogeneous system (2) into the homogeneous system (9) with the same coefficient matrix. The variables x_1 and x_2 specify the temperatures of the air and the rockbed relative to their respective equilibrium values, whereas u_1 and u_2 are the corresponding actual temperatures. The result of using x_1 and x_2 instead of u_1 and u_2 is to simplify the system (2) by eliminating the nonhomogeneous term. In geometrical language we have shifted coordinates so that the critical point $(16, 16)$ in the u_1u_2-plane is now located at the origin in the x_1x_2-plane.

On Reducing $\mathbf{x}' = \mathbf{Ax} + \mathbf{b}$ to $\mathbf{x}' = \mathbf{Ax}$

If $\mathbf{A}$ has an inverse, then the only critical, or equilibrium, point of $\mathbf{x}' = \mathbf{Ax} + \mathbf{b}$ is $\mathbf{x}_{eq} = -\mathbf{A}^{-1}\mathbf{b}$. As illustrated in Example 1, in such cases it is convenient to shift the origin of the phase plane to the critical point using the coordinate transformation

$$\mathbf{x} = \mathbf{x}_{eq} + \tilde{\mathbf{x}}, \tag{11}$$

as shown in Figure 3.3.1.

Thus $\tilde{\mathbf{x}} = \mathbf{x} - \mathbf{x}_{eq}$ represents the difference between $\mathbf{x}$ and the equilibrium state $\mathbf{x}_{eq}$. Substituting the right side of Eq. (11) for $\mathbf{x}$ in $\mathbf{x}' = \mathbf{Ax} + \mathbf{b}$ gives

$$\frac{d}{dt}\left(\mathbf{x}_{eq} + \tilde{\mathbf{x}}\right) = \mathbf{A}\left(\mathbf{x}_{eq} + \tilde{\mathbf{x}}\right) + \mathbf{b},$$

or

$$\frac{d\tilde{\mathbf{x}}}{dt} = \mathbf{A}\tilde{\mathbf{x}}, \tag{12}$$

since $d\mathbf{x}_{eq}/dt = \mathbf{0}$ and $\mathbf{Ax}_{eq} + \mathbf{b} = \mathbf{A}(-\mathbf{A}^{-1}\mathbf{b}) + \mathbf{b} = -\mathbf{Ib} + \mathbf{b} = \mathbf{0}$. Dropping the tilde, if $\mathbf{x} = \boldsymbol{\phi}(t)$ is a solution of the homogeneous system $\mathbf{x}' = \mathbf{Ax}$, then the solution of the nonhomogeneous system $\mathbf{x}' = \mathbf{Ax} + \mathbf{b}$ is given by $\mathbf{x} = \boldsymbol{\phi}(t) + \mathbf{x}_{eq} = \boldsymbol{\phi}(t) - \mathbf{A}^{-1}\mathbf{b}$.

On the other hand, if $\mathbf{A}$ is singular, there may be no critical points, and then it will not be possible to reduce the nonhomogeneous system to a homogeneous one in this manner. Similarly, if the nonhomogeneous term $\mathbf{b}$ depends on t, then it is no longer possible to eliminate it by a simple change of variables similar to Eq. (11). However, even in these cases, if the homogeneous system $\mathbf{x}' = \mathbf{Ax}$ can be solved, then the nonhomogeneous system $\mathbf{x}' = \mathbf{Ax} + \mathbf{b}$ can also be solved by well-established methods that we will discuss in Section 4.7. Thus the homogeneous system $\mathbf{x}' = \mathbf{Ax}$ is the more fundamental problem and we will focus most of our attention on it.

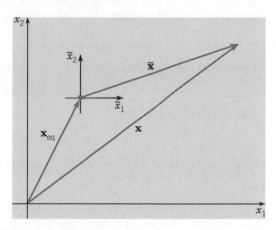

FIGURE 3.3.1 Translating the origin to the critical point $\mathbf{x}_{eq}$.

Superposition Principle and Linear Independence

Most of the remaining content of this chapter is devoted to solving linear systems of dimension two of the form

$$\mathbf{x}'(t) = \mathbf{Ax}(t) \tag{13}$$

for various cases of $\mathbf{A}$ and forming phase portraits to visualize the behavior of the solution trajectories. We first discuss how to form the general solution of (13).

Suppose $\mathbf{x}_1(t)$ is a solution of (13) and c_1 is an arbitrary constant. Then

$$\mathbf{x}(t) = c_1 \mathbf{x}_1(t) \tag{14}$$

is also a solution of (13). To show that $c_1 \mathbf{x}_1$ is a solution of (13), we must show that

$$\left(c_1 \mathbf{x}_1(t) \right)' - \mathbf{A} c_1 \mathbf{x}_1(t) \tag{14a}$$

is zero. But, we can write

$$\left(c_1 \mathbf{x}_1(t) \right)' - \mathbf{A} c_1 \mathbf{x}_1(t) = c_1 [\mathbf{x}_1'(t) - \mathbf{A} \mathbf{x}_1(t)]. \tag{14b}$$

The expression in brackets on the right side of (14b) is zero because $\mathbf{x}_1$ is a solution of (13); therefore the expression in Eq. (14a) is also zero, and $c_1 \mathbf{x}_1$ is a solution of (13) for any constant c_1.

Further, if $\mathbf{x}_1(t)$ and $\mathbf{x}_2(t)$ are both solutions of (13) and c_1 and c_2 are arbitrary constants, then the linear combination

$$\mathbf{x}(t) = c_1 \mathbf{x}_1(t) + c_2 \mathbf{x}_2(t) \tag{15}$$

is also a solution of (13) because

$$\left(c_1 \mathbf{x}_1(t) + c_2 \mathbf{x}_2(t) \right)' - \mathbf{A} \left(c_1 \mathbf{x}_1(t) + c_2 \mathbf{x}_2(t) \right) =$$
$$c_1 \underbrace{\left(\mathbf{x}_1(t)' - \mathbf{A} \mathbf{x}_1(t) \right)}_{= \mathbf{0}} + c_2 \underbrace{\left(\mathbf{x}_2(t)' - \mathbf{A} \mathbf{x}_2(t) \right)}_{= \mathbf{0}} = \mathbf{0}. \tag{16}$$

which is true because $\mathbf{x}_1$ and $\mathbf{x}_2$ are solutions of (13).

Since we did not use particular forms of the solutions $\mathbf{x}_1(t)$ and $\mathbf{x}_2(t)$, (16) holds *for all* solutions of (13). We formalize this observation as the following theorem.

THEOREM 3.3.1

Principle of Superposition. Suppose that $\mathbf{x}_1(t)$ and $\mathbf{x}_2(t)$ are solutions of Eq. (13),

$$\mathbf{x}'(t) = \mathbf{A} \mathbf{x}(t).$$

Then the expression (15)

$$\mathbf{x}(t) = c_1 \mathbf{x}_1(t) + c_2 \mathbf{x}_2(t),$$

where c_1 and c_2 are arbitrary constants, is also a solution.

This theorem expresses one of the fundamental properties of linear homogeneous systems. Starting with two specific solutions, you can immediately generate a much larger (doubly infinite, in fact) family of solutions, provided that the two solutions are distinct in the sense that we now describe.

Suppose now that there is a prescribed initial condition

$$\mathbf{x}(t_0) = \mathbf{x}_0, \tag{17}$$

where t_0 is any given value of t, and $\mathbf{x}_0$ is any given constant vector. Is it possible to choose the constants c_1 and c_2 in Eq. (15) so as to satisfy the initial condition (17)? By applying Eq. (17) in Eq. (15), we obtain

$$c_1 \mathbf{x}_1(t_0) + c_2 \mathbf{x}_2(t_0) = \mathbf{x}_0, \tag{18}$$

or, in matrix form,

$$\begin{pmatrix} x_{11}(t_0) & x_{12}(t_0) \\ x_{21}(t_0) & x_{22}(t_0) \end{pmatrix} \begin{pmatrix} c_1 \\ c_2 \end{pmatrix} = \begin{pmatrix} x_{10} \\ x_{20} \end{pmatrix}. \tag{19}$$

The notation in Eq. (19) is such that, for example, x_{12} is the first component of the vector $\mathbf{x}_2$, and so on. Thus the first subscript identifies the component of a vector and the second subscript identifies the vector itself. According to Theorem 3.1.1, system (19) can be solved uniquely for c_1 and c_2 for any values of x_{10} and x_{20} if and only if the determinant of the coefficient matrix is nonzero. It is possible to write down expressions for c_1 and c_2 that satisfy Eqs. (19), but it is usually preferable just to solve this system of equations whenever it is necessary.

The determinant

$$W[\mathbf{x}_1, \mathbf{x}_2](t) = \begin{vmatrix} x_{11}(t) & x_{12}(t) \\ x_{21}(t) & x_{22}(t) \end{vmatrix} \tag{20}$$

is called the **Wronskian determinant** or, more simply, the **Wronskian** of the two solutions $\mathbf{x}_1(t)$ and $\mathbf{x}_2(t)$. Thus, the initial value problem (13), (17) has a unique solution if and only if $W[\mathbf{x}_1, \mathbf{x}_2](t_0) \neq 0$.

The Wronskian is also closely associated with the important concept of linear independence of solutions.

DEFINITION 3.3.2

Linear Independence of Solutions. Suppose that $\mathbf{x}_1(t)$ and $\mathbf{x}_2(t)$ are solutions of Eq. (13) on an interval I. We say that $\mathbf{x}_1$ and $\mathbf{x}_2$ are **linearly dependent** if there exists a constant k such that

$$\mathbf{x}_1(t) = k\mathbf{x}_2(t), \quad \text{for all } t \text{ in } I.$$

(That is, $\mathbf{x}_1$ is a constant multiple of $\mathbf{x}_2$.) Otherwise, $\mathbf{x}_1$ and $\mathbf{x}_2$ are **linearly independent**.

Remark. Geometrically, if two vectors in the plane are linearly dependent and they are translated so their tails coincide with the origin, then the vectors lie on the same line through the origin. If they are linearly independent, this is not true. This observation will be helpful when forming phase portraits for Eq. (13).

The following theorem states how the Wronskian can be used to determine whether two solutions of Eq. (13) are linearly independent. We omit the proof of Theorem 3.3.3 because the same result in a more general context is proved in Section 6.2.

THEOREM 3.3.3

Wronskian and Linear Independence. Suppose

$$\mathbf{x}_1(t) = \begin{pmatrix} x_{11}(t) \\ x_{21}(t) \end{pmatrix} \quad \text{and} \quad \mathbf{x}_2(t) = \begin{pmatrix} x_{12}(t) \\ x_{22}(t) \end{pmatrix}$$

are solutions of Eq. (13) on an interval I. Then $\mathbf{x}_1$ and $\mathbf{x}_2$ are **linearly independent** if and only if

$$W[\mathbf{x}_1, \mathbf{x}_2](t) \neq 0, \quad \text{for all } t \text{ in } I. \tag{21}$$

Accordingly, two linearly independent solutions of Eq. (13) are often called a **fundamental set of solutions**.

If $\mathbf{x}_1(t)$ and $\mathbf{x}_2(t)$ comprise a fundamental set of Eq. (13), then the linear combination of $\mathbf{x}_1$ and $\mathbf{x}_2$ given by Eq. (15) with arbitrary coefficients c_1 and c_2 is called the **general solution** of Eq. (13).

We summarize this as the following theorem.

THEOREM 3.3.4

Suppose that $\mathbf{x}_1(t)$ and $\mathbf{x}_2(t)$ are two solutions of Eq. (13),

$$\mathbf{x}' = \mathbf{Ax},$$

and that their Wronskian is not zero. Then $\mathbf{x}_1(t)$ and $\mathbf{x}_2(t)$ form a fundamental set of solutions, and the general solution of Eq. (13) is given by Eq. (15),

$$\mathbf{x} = c_1\mathbf{x}_1(t) + c_2\mathbf{x}_2(t),$$

where c_1 and c_2 are arbitrary constants. If there is a given initial condition $\mathbf{x}(t_0) = \mathbf{x}_0$, where $\mathbf{x}_0$ is any constant vector, then this condition determines the constants c_1 and c_2 uniquely.

This theorem will also reappear in a more general setting in Chapter 6.

To begin to understand how we might solve Eq. (13), let us first consider a particularly simple example.

The Eigenvalue Method for Solving $\mathbf{x}' = \mathbf{Ax}$

EXAMPLE 2

Consider the system

$$\frac{d\mathbf{x}}{dt} = \begin{pmatrix} -1 & 0 \\ 0 & -4 \end{pmatrix} \mathbf{x}. \tag{22}$$

Find solutions of the system (22) and then find the particular solution that satisfies the initial condition

$$\mathbf{x}(0) = \begin{pmatrix} 2 \\ 3 \end{pmatrix}. \tag{23}$$

The most important feature of this system is apparent if we write it in scalar form, that is,

$$x_1' = -x_1, \qquad x_2' = -4x_2. \tag{24}$$

Each equation involves only one of the unknown variables; as a result, the two equations can be solved independently. By solving Eqs. (24), we obtain

$$x_1 = c_1 e^{-t}, \qquad x_2 = c_2 e^{-4t}, \tag{25}$$

where c_1 and c_2 are arbitrary constants. Then, by writing the solution (25) in vector form, we have

$$\mathbf{x} = \begin{pmatrix} c_1 e^{-t} \\ c_2 e^{-4t} \end{pmatrix} = c_1 \begin{pmatrix} e^{-t} \\ 0 \end{pmatrix} + c_2 \begin{pmatrix} 0 \\ e^{-4t} \end{pmatrix} = c_1 e^{-t} \begin{pmatrix} 1 \\ 0 \end{pmatrix} + c_2 e^{-4t} \begin{pmatrix} 0 \\ 1 \end{pmatrix}. \tag{26}$$

Note that this solution consists of two terms, each of which involves a vector multiplied by a certain exponential function.

To satisfy the initial conditions (23), we can set $t = 0$ in Eq. (26); then

$$\mathbf{x}(0) = c_1 \begin{pmatrix} 1 \\ 0 \end{pmatrix} + c_2 \begin{pmatrix} 0 \\ 1 \end{pmatrix}. \tag{27}$$

Consequently, we must choose $c_1 = 2$ and $c_2 = 3$. The solution of the system (13) that satisfies the initial conditions (23) is

$$\mathbf{x} = 2e^{-t} \begin{pmatrix} 1 \\ 0 \end{pmatrix} + 3e^{-4t} \begin{pmatrix} 0 \\ 1 \end{pmatrix}. \tag{28}$$

Extension to a General System

We now turn to a consideration of a general system of two first order linear homogeneous differential equations with constant coefficients. We will usually write such a system in the form (13)

$$\frac{d\mathbf{x}}{dt} = \mathbf{A}\mathbf{x},$$

where

$$\mathbf{x} = \begin{pmatrix} x_1 \\ x_2 \end{pmatrix}, \qquad \mathbf{A} = \begin{pmatrix} a_{11} & a_{12} \\ a_{21} & a_{22} \end{pmatrix}. \tag{29}$$

The elements of the matrix $\mathbf{A}$ are given real constants and the components of the vector $\mathbf{x}$ are to be determined.

To solve Eq. (13), we are guided by the form of the solution in Example 2 and assume that

$$\mathbf{x} = e^{\lambda t}\mathbf{v}, \tag{30}$$

where $\mathbf{v}$ and λ are a constant vector and a scalar, respectively, to be determined. By substituting from Eq. (30) into Eq. (13) and noting that $\mathbf{v}$ and λ do not depend on t, we obtain

$$\lambda e^{\lambda t}\mathbf{v} = \mathbf{A}e^{\lambda t}\mathbf{v}.$$

Further $e^{\lambda t}$ is never zero, so we have

$$\mathbf{A}\mathbf{v} = \lambda\mathbf{v}, \tag{31}$$

or

$$(\mathbf{A} - \lambda\mathbf{I})\mathbf{v} = \mathbf{0}, \tag{32}$$

where $\mathbf{I}$ is the 2×2 identity matrix.

We showed in Section 3.1 (see also Appendix A.4) that Eq. (32) is precisely the equation that determines the eigenvalues and eigenvectors of the matrix $\mathbf{A}$. Thus

$$\mathbf{x} = e^{\lambda t}\mathbf{v}$$

is a solution of

$$\frac{d\mathbf{x}}{dt} = \mathbf{A}\mathbf{x},$$

provided that λ is an eigenvalue and $\mathbf{v}$ is a corresponding eigenvector of the coefficient matrix $\mathbf{A}$. The eigenvalues λ_1 and λ_2 are the roots of the characteristic equation

$$\det(\mathbf{A} - \lambda\mathbf{I}) = \begin{vmatrix} a_{11} - \lambda & a_{12} \\ a_{21} & a_{22} - \lambda \end{vmatrix} = (a_{11} - \lambda)(a_{22} - \lambda) - a_{12}a_{21}$$

$$= \lambda^2 - (a_{11} + a_{22})\lambda + a_{11}a_{22} - a_{12}a_{21} = 0. \tag{33}$$

For each eigenvalue, we can solve the system (32) and thereby obtain the corresponding eigenvector $\mathbf{v}_1$ or $\mathbf{v}_2$. Recall that the eigenvectors are determined only up to an arbitrary nonzero constant multiplier.

Since the elements of $\mathbf{A}$ are real-valued, the eigenvalues may be real and different, real and equal, or complex conjugates. We will restrict our discussion in this section to the first case, and will defer consideration of the latter two possibilities until the following two sections.

Real and Different Eigenvalues

We assume now that λ_1 and λ_2 are real and different. Then, using the eigenvalues and the corresponding eigenvectors, we can write down two solutions of Eq. (13), namely,

$$\mathbf{x}_1(t) = e^{\lambda_1 t}\mathbf{v}_1, \qquad \mathbf{x}_2(t) = e^{\lambda_2 t}\mathbf{v}_2. \tag{34}$$

If $\mathbf{x}_1$ and $\mathbf{x}_2$ are given by Eqs. (34), then their Wronskian is

$$W[\mathbf{x}_1, \mathbf{x}_2](t) = \begin{vmatrix} v_{11}e^{\lambda_1 t} & v_{12}e^{\lambda_2 t} \\ v_{21}e^{\lambda_1 t} & v_{22}e^{\lambda_2 t} \end{vmatrix} = \begin{vmatrix} v_{11} & v_{12} \\ v_{21} & v_{22} \end{vmatrix} e^{(\lambda_1 + \lambda_2)t}. \tag{35}$$

The exponential function is never zero, so whether $W[\mathbf{x}_1, \mathbf{x}_2](t)$ is zero depends entirely on the determinant whose columns are the eigenvectors of the coefficient matrix $\mathbf{A}$. Since the eigenvectors

$$\mathbf{v}_1 = \begin{pmatrix} v_{11} \\ v_{21} \end{pmatrix} \qquad \text{and} \qquad \mathbf{v}_2 = \begin{pmatrix} v_{12} \\ v_{22} \end{pmatrix}$$

belong to the eigenvalues λ_1 and λ_2, respectively, and $\lambda_1 \neq \lambda_2$, we know from Theorem 3.1.3 that the determinant

$$\begin{vmatrix} v_{11} & v_{12} \\ v_{21} & v_{22} \end{vmatrix} \neq 0.$$

Therefore the Wronskian of the vectors $\mathbf{x}_1$ and $\mathbf{x}_2$ in Eq. (35) is nonzero and so $\{\mathbf{x}_1, \mathbf{x}_2\}$ is a fundamental set of (13), and the general solution of (13) is $\mathbf{x}(t) = c_1 e^{\lambda_1 t}\mathbf{v}_1 + c_2 e^{\lambda_2 t}\mathbf{v}_2$.

EXAMPLE 3

A Rockbed Heat Storage System Revisited

Consider again the greenhouse/rockbed heat storage problem from Example 1 modeled by Eq. (9), in coordinates centered at the critical point,

$$\frac{d\mathbf{x}}{dt} = \begin{pmatrix} -\frac{13}{8} & \frac{3}{4} \\ \frac{1}{4} & -\frac{1}{4} \end{pmatrix} \mathbf{x} = \mathbf{A}\mathbf{x}. \tag{36}$$

Find the general solution of this system. Then plot a direction field, a phase portrait, and several component plots of the system (36).

To solve the system (36), we let $\mathbf{x} = e^{\lambda t}\mathbf{v}$. Following the steps that lead from Eq. (30) to Eq. (31), we get the eigenvalue problem

$$(\mathbf{A} - \lambda\mathbf{I})\mathbf{v} = \begin{pmatrix} -\frac{13}{8} - \lambda & \frac{3}{4} \\ \frac{1}{4} & -\frac{1}{4} - \lambda \end{pmatrix}\begin{pmatrix} v_1 \\ v_2 \end{pmatrix} = \begin{pmatrix} 0 \\ 0 \end{pmatrix}, \tag{37}$$

for the matrix $\mathbf{A}$ of coefficients in system (36). Thus the characteristic equation is

$$\lambda^2 + \frac{15}{8}\lambda + \frac{7}{32} = 0, \tag{38}$$

so the eigenvalues are $\lambda_1 = -\frac{7}{4}$ and $\lambda_2 = -\frac{1}{8}$. Setting $\lambda = -\frac{7}{4}$ in Eq. (37) gives the system

$$\begin{pmatrix} \frac{1}{8} & \frac{3}{4} \\ \frac{1}{4} & \frac{3}{2} \end{pmatrix}\begin{pmatrix} v_1 \\ v_2 \end{pmatrix} = \begin{pmatrix} 0 \\ 0 \end{pmatrix}. \tag{39}$$

Thus $\frac{1}{4}v_1 + \frac{3}{2}v_2 = 0$, so we take the eigenvector $\mathbf{v}_1$ corresponding to the eigenvalue $\lambda_1 = -\frac{7}{4}$ to be

$$\mathbf{v}_1 = \begin{pmatrix} 6 \\ -1 \end{pmatrix}. \tag{40}$$

Similarly, corresponding to $\lambda = -\frac{1}{8}$, Eq. (37) yields

$$\begin{pmatrix} -\frac{3}{2} & \frac{3}{4} \\ \frac{1}{4} & -\frac{1}{8} \end{pmatrix}\begin{pmatrix} v_1 \\ v_2 \end{pmatrix} = \begin{pmatrix} 0 \\ 0 \end{pmatrix}. \tag{41}$$

Hence $2v_1 - v_2 = 0$, so the eigenvector can be chosen as

$$\mathbf{v}_2 = \begin{pmatrix} 1 \\ 2 \end{pmatrix}. \tag{42}$$

Observe that Eqs. (39) and (41) determine the eigenvectors only up to an arbitrary nonzero multiplicative constant. We have chosen the constants so that v_1 and v_2 have small integer components, but any vectors proportional to those given by Eqs. (40) and (42) could also be used.

The corresponding solutions of the differential equation are

$$\mathbf{x}_1(t) = e^{-7t/4}\begin{pmatrix} 6 \\ -1 \end{pmatrix}, \qquad \mathbf{x}_2(t) = e^{-t/8}\begin{pmatrix} 1 \\ 2 \end{pmatrix}. \tag{43}$$

The Wronskian of these solutions is

$$W[\mathbf{x}_1, \mathbf{x}_2](t) = \begin{vmatrix} 6e^{-7t/4} & e^{-t/8} \\ -e^{-7t/4} & 2e^{-t/8} \end{vmatrix} = 13e^{-15t/8}, \tag{44}$$

which is never zero. Hence the solutions $\mathbf{x}_1$ and $\mathbf{x}_2$ form a fundamental set, and the general solution of the system (36) is

$$\mathbf{x} = c_1\mathbf{x}_1(t) + c_2\mathbf{x}_2(t) \tag{45}$$

$$= c_1 e^{-7t/4}\begin{pmatrix} 6 \\ -1 \end{pmatrix} + c_2 e^{-t/8}\begin{pmatrix} 1 \\ 2 \end{pmatrix}, \tag{46}$$

where c_1 and c_2 are arbitrary constants. The solution satisfying the initial conditions (10), $x_1(0) = -16$ and $x_2(0) = 20$, is found by solving the system

$$\begin{pmatrix} 6 & 1 \\ -1 & 2 \end{pmatrix} \begin{pmatrix} c_1 \\ c_2 \end{pmatrix} = \begin{pmatrix} -16 \\ 20 \end{pmatrix} \tag{47}$$

for c_1 and c_2, with the result that $c_1 = -4$ and $c_2 = 8$.

To visualize the solution (46), it is helpful to consider trajectories in the $x_1 x_2$-plane for various values of the constants c_1 and c_2. First, we consider the special cases when $c_2 = 0$ and then when $c_1 = 0$. If we set $c_2 = 0$ in Eq. (46), we get $\mathbf{x} = c_1 \mathbf{x}_1(t)$ or, in scalar form,

$$x_1 = 6c_1 e^{-7t/4}, \qquad x_2 = -c_1 e^{-7t/4}.$$

By eliminating t between these two equations, we see that this solution lies on the straight line $x_2 = -x_1/6$; see Figure 3.3.2a. This is the line through the origin in the direction of the eigenvector $\mathbf{v}_1$. If we look at the solution as the trajectory of a moving particle, then the particle is in the fourth quadrant when $c_1 > 0$ and in the second quadrant when $c_1 < 0$. In either case, the particle moves toward the origin as t increases.

Next we set $c_1 = 0$ in Eq. (46) to get $\mathbf{x} = c_2 \mathbf{x}_2(t)$, or

$$x_1 = c_2 e^{-t/8}, \qquad x_2 = 2c_2 e^{-t/8}.$$

This solution lies on the line $x_2 = 2x_1$, whose direction is determined by the eigenvector $\mathbf{v}_2$. This solution is in the first quadrant when $c_2 > 0$ and in the third quadrant when $c_2 < 0$, as shown in Figure 3.3.2a. In both cases the particle moves toward the origin as t increases.

Now suppose we want to know the behavior of a solution trajectory of (36) that passes through a point *not* on one of the two pictured lines. Since the uniqueness portion of Theorem 3.2.1 guarantees that two solution curves cannot intersect, these two lines restrict the behavior of all other solution trajectories. Moreover the solution (46) is a combination of $\mathbf{x}_1(t)$ and $\mathbf{x}_2(t)$, so all solutions approach the origin as $t \to \infty$. For large t the term $c_2 \mathbf{x}_2(t)$ is dominant and the term $c_1 \mathbf{x}_1(t)$ is negligible in comparison. To see this, we write Eq. (46) in the form

$$\mathbf{x} = e^{-t/8} \left[c_1 e^{-13t/8} \begin{pmatrix} 6 \\ -1 \end{pmatrix} + c_2 \begin{pmatrix} 1 \\ 2 \end{pmatrix} \right] \cong c_2 e^{-t/8} \begin{pmatrix} 1 \\ 2 \end{pmatrix}, \quad \text{if } c_2 \neq 0 \text{ and } t \text{ is large.} \tag{48}$$

Thus all solutions for which $c_2 \neq 0$ approach the origin tangent to the line $x_2 = 2x_1$ as $t \to \infty$.

On the other hand, if you look backward in time and let $t \to -\infty$, then the term $c_1 \mathbf{x}_1(t)$ is the dominant one (unless $c_1 = 0$). This is made clear by writing Eq. (46) in the form

$$\mathbf{x} = e^{-7t/4} \left[c_1 \begin{pmatrix} 6 \\ -1 \end{pmatrix} + c_2 e^{13t/8} \begin{pmatrix} 1 \\ 2 \end{pmatrix} \right]. \tag{49}$$

Since $e^{13t/8}$ is negligible for large negative values of t, the slopes of all trajectories for which c_2 is not zero approach the limit $-1/6$ as t approaches negative infinity.

Figure 3.3.2b shows the trajectories corresponding to $\mathbf{x}_1(t)$ and $\mathbf{x}_2(t)$, as well as several other trajectories of the system (36). Notice that the straight-line solution curves containing the eigenvectors determine the principal direction of motion of all other trajectories, and the corresponding eigenvalues are the growth and decay rates of the motion along these curves. As such, you can quickly produce a rough sketch of the phase portraits once you have drawn the straight-line solutions and determined the directions of motion along them. In other words, it is a phase portrait. Observe that Figure 3.3.2b is essentially the same as Figure 3.2.4, except that now the critical point occurs at the origin.

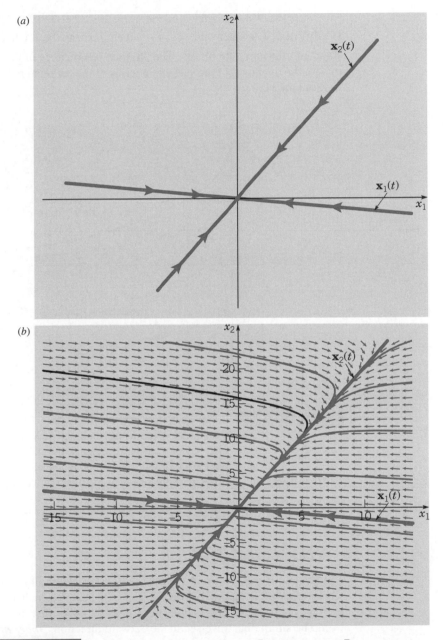

FIGURE 3.3.2   (a) The eigenvectors $\mathbf{v}_1 = (6, -1)^T$ and $\mathbf{v}_2 = (1, 2)^T$ with directions of motion indicated by the arrowheads. (b) Phase portrait for the system (36). As $t \to \infty$, solution curves approach the origin tangent to the line generated by $\mathbf{v}_2 = (1, 2)^T$. As $t \to -\infty$, solution curves move to infinity in a direction parallel to $\mathbf{v}_1 = (6, -1)^T$.

In Figure 3.3.3 we plot the components of the solution of Eq. (36) subject to the initial conditions (10),

$$\mathbf{x} = -4e^{-7t/4}\begin{pmatrix} 6 \\ -1 \end{pmatrix} + 8e^{-t/8}\begin{pmatrix} 1 \\ 2 \end{pmatrix}. \tag{50}$$

The large negative eigenvalue $\lambda_1 = -\frac{7}{4}$ governs the relatively rapid adjustment to a quasi-steady state. The eigenvalue $\lambda_2 = -\frac{1}{8}$ then governs the relatively slow decay toward the static equilibrium state $(0, 0)^T$. The physical significance of the eigenvalues and eigenvectors on the dynamics of the system, in terms of the parameters k_1, k_2, ϵ, and T_a, is explored in Problem 31.

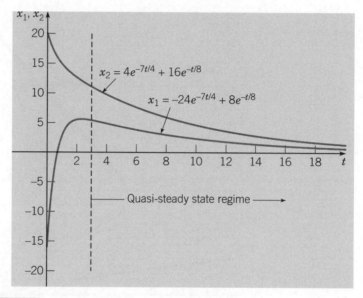

FIGURE 3.3.3 Component plots of the solution (50). The system settles into a quasi-steady state within the first few hours due to the rapid decay of $-4e^{-7t/4}(6\mathbf{i} - 1\mathbf{j})$. Subsequently, the slow decay of $\mathbf{x} \cong 8e^{-t/8}(1\mathbf{i} + 2\mathbf{j})$ to static equilibrium is governed by the eigenvalue $\lambda_2 = -\frac{1}{8}$. Note the strong resemblance between this figure and Figure 3.2.2.

Nodal Sources and Nodal Sinks

The pattern of trajectories in Figure 3.3.2*b* is typical of all second order systems $\mathbf{x}' = \mathbf{A}\mathbf{x}$ whose eigenvalues are real, different, and of the same sign. The origin is called a **node** for such a system. If the eigenvalues were positive rather than negative, then the trajectories would be similar but traversed in the outward direction. *Nodes are **asymptotically stable** if the eigenvalues are negative and **unstable** if the eigenvalues are positive.* Asymptotically stable nodes and unstable nodes are also referred to as **nodal sinks** and **nodal sources**, respectively.

The next example illustrates the qualitative behavior of a system in which the eigenvalues are real, nonzero, and of opposite signs.

Saddle Points

EXAMPLE 4

Consider the system

$$\mathbf{x}' = \begin{pmatrix} 1 & 1 \\ 4 & 1 \end{pmatrix} \mathbf{x}. \tag{51}$$

Find the general solution and draw a phase portrait.

To find solutions, we assume that $\mathbf{x} = e^{\lambda t}\mathbf{v}$ and substitute for $\mathbf{x}$ in Eq. (51). This results in the system of algebraic equations

$$\begin{pmatrix} 1 - \lambda & 1 \\ 4 & 1 - \lambda \end{pmatrix} \begin{pmatrix} v_1 \\ v_2 \end{pmatrix} = \begin{pmatrix} 0 \\ 0 \end{pmatrix}. \tag{52}$$

The characteristic equation is

$$\begin{vmatrix} 1 - \lambda & 1 \\ 4 & 1 - \lambda \end{vmatrix} = (1 - \lambda)^2 - 4 = \lambda^2 - 2\lambda - 3$$

$$= (\lambda - 3)(\lambda + 1) = 0, \tag{53}$$

so the eigenvalues are $\lambda_1 = 3$ and $\lambda_2 = -1$. If $\lambda = 3$, then the system (52) reduces to the single equation

$$-2v_1 + v_2 = 0. \tag{54}$$

Thus $v_2 = 2v_1$, and the eigenvector corresponding to $\lambda_1 = 3$ can be taken as

$$\mathbf{v}_1 = \begin{pmatrix} 1 \\ 2 \end{pmatrix}. \tag{55}$$

Similarly, corresponding to $\lambda_2 = -1$, we find that $v_2 = -2v_1$, so the eigenvector is

$$\mathbf{v}_2 = \begin{pmatrix} 1 \\ -2 \end{pmatrix}. \tag{56}$$

The corresponding solutions of the differential equation are

$$\mathbf{x}_1(t) = e^{3t} \begin{pmatrix} 1 \\ 2 \end{pmatrix}, \qquad \mathbf{x}_2(t) = e^{-t} \begin{pmatrix} 1 \\ -2 \end{pmatrix}. \tag{57}$$

The Wronskian of these solutions is

$$W[\mathbf{x}_1, \mathbf{x}_2](t) = \begin{vmatrix} e^{3t} & e^{-t} \\ 2e^{3t} & -2e^{-t} \end{vmatrix} = -4e^{2t}, \tag{58}$$

which is never zero. Hence the solutions $\mathbf{x}_1$ and $\mathbf{x}_2$ form a fundamental set, and the general solution of the system (51) is

$$\mathbf{x} = c_1\mathbf{x}_1(t) + c_2\mathbf{x}_2(t)$$

$$= c_1 e^{3t} \begin{pmatrix} 1 \\ 2 \end{pmatrix} + c_2 e^{-t} \begin{pmatrix} 1 \\ -2 \end{pmatrix}, \tag{59}$$

where c_1 and c_2 are arbitrary constants.

To visualize the solution (59) in the x_1x_2-plane, we first determine the straight-line solutions by setting $c_2 = 0$ and then $c_1 = 0$. We start with $\mathbf{x} = c_1\mathbf{x}_1(t)$ or, in scalar form,

$$x_1 = c_1e^{3t}, \qquad x_2 = 2c_1e^{3t}.$$

By eliminating t between these two equations, we see that this solution lies on the straight line $x_2 = 2x_1$; see Figure 3.3.4. This is the line through the origin in the direction of the eigenvector $\mathbf{v}_1$. If we think of the solution as the trajectory of a moving particle, then the particle is in the first quadrant when $c_1 > 0$ and in the third quadrant when $c_1 < 0$. In either case the particle departs from the origin as t increases. Next consider $\mathbf{x} = c_2\mathbf{x}_2(t)$, or

$$x_1 = c_2e^{-t}, \qquad x_2 = -2c_2e^{-t}.$$

This solution lies on the line $x_2 = -2x_1$, whose direction is determined by the eigenvector $\mathbf{v}_2$. The solution is in the fourth quadrant when $c_2 > 0$ and in the second quadrant when $c_2 < 0$, as shown in Figure 3.3.4. In both cases the particle moves toward the origin as t increases.

The solution (59) is a combination of $\mathbf{x}_1(t)$ and $\mathbf{x}_2(t)$. For large t the term $c_1\mathbf{x}_1(t)$ is dominant and the term $c_2\mathbf{x}_2(t)$ is negligible. Thus all solutions for which $c_1 \neq 0$ are asymptotic to the line $x_2 = 2x_1$ as $t \to \infty$. Similarly, all solutions for which $c_2 \neq 0$ are asymptotic to the line $x_2 = -2x_1$ as $t \to -\infty$. The graphs of several trajectories comprise the phase portrait in Figure 3.3.4.

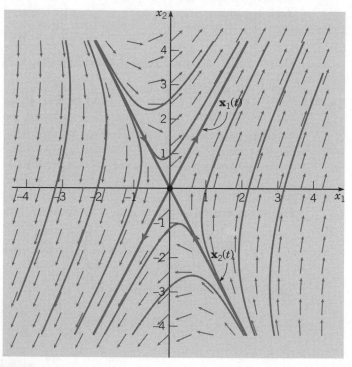

FIGURE 3.3.4 Direction field and phase portrait for the system (51); the origin is a saddle point.

You can also draw component plots of x_1 or x_2 versus t; some typical plots of x_1 versus t are shown in Figure 3.3.5, and those of x_2 versus t are similar. For certain initial conditions it follows that $c_1 = 0$ in Eq. (59) so that $x_1 = c_2e^{-t}$ and $x_1 \to 0$ as $t \to \infty$. One such graph is shown in Figure 3.3.5, corresponding to a trajectory that approaches the origin in

Figure 3.3.4. For most initial conditions, however, $c_1 \neq 0$ and x_1 is given by $x_1 = c_1 e^{3t} + c_2 e^{-t}$. Then the presence of the positive exponential term causes x_1 to grow exponentially in magnitude as t increases. Several graphs of this type are shown in Figure 3.3.5, corresponding to trajectories that depart from the neighborhood of the origin in Figure 3.3.4.

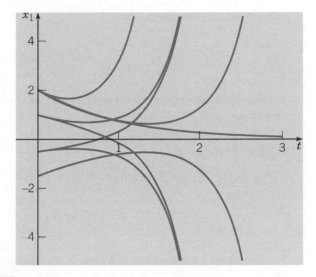

FIGURE 3.3.5   Typical component plots of x_1 versus t for the system (51).

The pattern of trajectories in Figure 3.3.4 is typical of all second order systems $\mathbf{x}' = \mathbf{A}\mathbf{x}$ for which the eigenvalues are real and of opposite signs. The origin is called a **saddle point** in this case. *Saddle points are always **unstable** because almost all trajectories depart from them as t increases.* A computer was used to plot the phase portrait in Figure 3.3.4, but you can draw a qualitatively accurate sketch of the trajectories as soon as you know the eigenvalues and eigenvectors. Phase portraits can also provide insight into the behavior of the solution trajectories of homogeneous second order differential equations with constant coefficients.

EXAMPLE 5

Consider the differential equations

$$y'' + 5y' + 6y = 0, \tag{60}$$

$$y'' - 5y' + 6y = 0. \tag{61}$$

Find the general solution and draw the phase portrait for each of them and comment on the differences.

Using the change of variable introduced in Eq. (42) of Section 3.2 with $q(t) = 6, p(t) = 5$, and $g(t) = 0$, we can rewrite Eq. (60) as the first order system

$$\mathbf{x}' = \begin{pmatrix} 0 & 1 \\ -6 & -5 \end{pmatrix} \mathbf{x}. \tag{62}$$

Assume $\mathbf{x}(t) = e^{\lambda t}\mathbf{v}$ and substitute for $\mathbf{x}$ in Eq. (62) to obtain the system of algebraic equations

$$\begin{pmatrix} -\lambda & 1 \\ -6 & -5 - \lambda \end{pmatrix} \begin{pmatrix} v_1 \\ v_2 \end{pmatrix} = \begin{pmatrix} 0 \\ 0 \end{pmatrix}. \tag{63}$$

The characteristic equation is

$$\begin{vmatrix} -\lambda & 1 \\ -6 & -5 - \lambda \end{vmatrix} = \lambda^2 + 5\lambda + 6 = (\lambda + 2)(\lambda + 3) = 0,$$

so the eigenvalues are $\lambda_1 = -2$ and $\lambda_2 = -3$. If $\lambda = -2$, then Eq. (63) reduces to the single equation

$$2v_1 + v_2 = 0.$$

Thus $v_2 = -2v_1$ and the eigenvector corresponding to $\lambda_1 = -2$ can be taken as

$$\mathbf{v}_1 = \begin{pmatrix} 1 \\ -2 \end{pmatrix}.$$

Similarly, corresponding to $\lambda_2 = -3$, we find that $v_2 = -3v_1$, so the eigenvector is

$$\mathbf{v}_2 = \begin{pmatrix} 1 \\ -3 \end{pmatrix}.$$

The corresponding solutions of Eq. (62) are

$$\mathbf{x}_1(t) = e^{-2t} \begin{pmatrix} 1 \\ -2 \end{pmatrix}, \qquad \mathbf{x}_2(t) = e^{-3t} \begin{pmatrix} 1 \\ -3 \end{pmatrix}.$$

As in Example 4, the Wronskian of these two solutions is nonzero for all t. Hence the solutions $\mathbf{x}_1$ and $\mathbf{x}_2$ form a fundamental set, so the general solution of Eq. (62) is

$$\mathbf{x}(t) = c_1\mathbf{x}_1(t) + c_2\mathbf{x}_2(t)$$

$$= c_1 e^{-2t} \begin{pmatrix} 1 \\ -2 \end{pmatrix} + c_2 e^{-3t} \begin{pmatrix} 1 \\ -3 \end{pmatrix} \tag{64}$$

$$= \begin{pmatrix} c_1 e^{-2t} + c_2 e^{-3t} \\ -2c_1 e^{-2t} - 3c_2 e^{-3t} \end{pmatrix}.$$

Correspondingly, the general solution $y(t)$ of Eq. (60) is the first component of $\mathbf{x}(t)$, namely,

$$y(t) = c_1 e^{-2t} + c_2 e^{-3t}. \tag{65}$$

Since the eigenvalues are distinct negative real numbers, the origin is a nodal sink. As such, we expect all solution trajectories of Eq. (62) to approach the origin as $t \to \infty$, and likewise for the solution curves of Eq. (65) to approach zero rapidly as $t \to \infty$. This is illustrated in the phase portrait for Eq. (62) and some typical solution curves of Eq. (65), shown in Figures 3.3.6a and 3.3.6b, respectively.

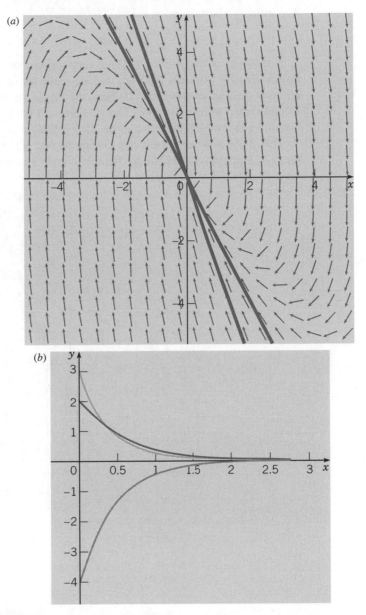

FIGURE 3.3.6 (*a*) Phase portrait for (62). (*b*) Typical solution curves of (65).

Turning our attention to (61), the system form of Eq. (61) is

$$\mathbf{x}' = \begin{pmatrix} 0 & 1 \\ -6 & 5 \end{pmatrix} \mathbf{x}. \tag{66}$$

The general solution is given by (64), but with $\mathbf{x}_1$ and $\mathbf{x}_2$ given by

$$\mathbf{x}_1(t) = e^{2t} \begin{pmatrix} 1 \\ 2 \end{pmatrix}, \qquad \mathbf{x}_2(t) = e^{3t} \begin{pmatrix} 1 \\ 3 \end{pmatrix}.$$

Accordingly, the solution of Eq. (61) is

$$y(t) = c_1 e^{2t} + c_2 e^{3t}. \tag{67}$$

Although Eq. (60) and Eq. (61) are very similar, the long-term behavior of the solution trajectories are markedly different. Indeed, since the eigenvalues associated with Eq. (66) are now distinct *positive* real numbers, the origin is classified as a nodal source and the solution trajectories move *away* from the origin. Likewise, the solution curves for Eq. (67) become unboundedly large as $t \to \infty$. This is illustrated in Figures 3.3.7a and b.

This illustrates that a mere sign change on one of the parameters in a second order differential equation can lead to drastically different end behavior.

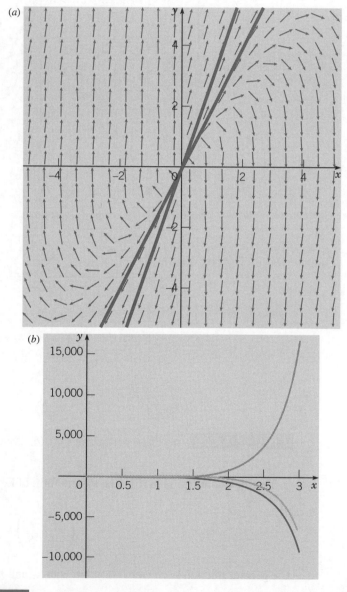

FIGURE 3.3.7 (*a*) Phase portrait for Eq. (66). (*b*) Typical solution curves of Eq. (67).

One Zero Eigenvalue

The preceding Examples 4 and 5 illustrate the cases in which the eigenvalues are of the same sign (a node) or of opposite signs (a saddle point). In both these cases the matrix $\mathbf{A}$ is nonsingular and, consequently, the origin is the only critical point of the system. Another possibility is that one eigenvalue is zero and the other is not. This situation occurs as a transition state between a node and a saddle point. If $\lambda = 0$ is an eigenvalue, then from the characteristic Eq. (25) it follows that $\det(\mathbf{A}) = 0$, so $\mathbf{A}$ is singular. Consequently, there is an infinite set of critical points, all of which lie on a line through the origin. The following example illustrates this case.

EXAMPLE 6

Consider the system

$$\mathbf{x}' = \mathbf{A}\mathbf{x} = \begin{pmatrix} -1 & 4 \\ \frac{1}{2} & -2 \end{pmatrix} \mathbf{x}. \tag{68}$$

Find the critical points. Then solve the system, draw a phase portrait, and describe how the solutions behave.

Observe that $\det(\mathbf{A}) = 0$, so $\mathbf{A}$ is singular. Solutions of $\mathbf{A}\mathbf{x} = \mathbf{0}$ satisfy $x_1 = 4x_2$, so each point on this line is a critical point of the system (68). To solve the system, we assume that $\mathbf{x} = e^{\lambda t}\mathbf{v}$ and substitute for $\mathbf{x}$ in Eq. (68). This results in the system of algebraic equations

$$\begin{pmatrix} -1 - \lambda & 4 \\ \frac{1}{2} & -2 - \lambda \end{pmatrix} \begin{pmatrix} v_1 \\ v_2 \end{pmatrix} = \begin{pmatrix} 0 \\ 0 \end{pmatrix}. \tag{69}$$

The characteristic equation is

$$(-1 - \lambda)(-2 - \lambda) - 2 = \lambda^2 + 3\lambda = 0, \tag{70}$$

so the eigenvalues are $\lambda_1 = 0$ and $\lambda_2 = -3$. For $\lambda = 0$, it follows from Eq. (69) that $v_1 = 4v_2$, so the eigenvector corresponding to λ_1 is

$$\mathbf{v}_1 = \begin{pmatrix} 4 \\ 1 \end{pmatrix}. \tag{71}$$

In a similar way we obtain the eigenvector corresponding to $\lambda_2 = -3$, namely,

$$\mathbf{v}_2 = \begin{pmatrix} -2 \\ 1 \end{pmatrix}. \tag{72}$$

Thus two solutions of the system (68) are

$$\mathbf{x}_1(t) = \begin{pmatrix} 4 \\ 1 \end{pmatrix}, \qquad \mathbf{x}_2(t) = e^{-3t} \begin{pmatrix} -2 \\ 1 \end{pmatrix}. \tag{73}$$

The Wronskian of these two solutions is $6e^{-3t}$, which is not zero, so the general solution of the system (68) is

$$\mathbf{x} = c_1 \mathbf{x}_1(t) + c_2 \mathbf{x}_2(t) = c_1 \begin{pmatrix} 4 \\ 1 \end{pmatrix} + c_2 e^{-3t} \begin{pmatrix} -2 \\ 1 \end{pmatrix}. \tag{74}$$

If initial conditions are given, they will determine appropriate values for c_1 and c_2.

The solution $\mathbf{x}_1(t)$ is independent of t, that is, it is a constant solution. If $c_2 = 0$, then $\mathbf{x}(t)$ is proportional to $\mathbf{x}_1(t)$ and thus corresponds to a point on the line determined by the eigenvector $\mathbf{v}_1$. Such a solution remains stationary for all time and its trajectory is a single point. This behavior is consistent with the fact that every point on this line is a critical point, as we noted earlier. On the other hand, if $c_1 = 0$, then $\mathbf{x}(t)$ is proportional to $\mathbf{x}_2(t)$. In this case, the trajectory approaches the origin along the line determined by the eigenvector $\mathbf{v}_2$.

A phase portrait for the system (68) is shown in Figure 3.3.8. Any solution starting at a point on the line of critical points remains fixed for all time at its starting point. A solution starting at any other point in the plane moves on a line parallel to $\mathbf{v}_2$ toward the point of intersection of this line with the line of critical points. The phase portrait in Figure 3.3.8 is typical of that for any second order system $\mathbf{x}' = \mathbf{A}\mathbf{x}$ with one zero eigenvalue and one negative eigenvalue. If the nonzero eigenvalue is positive rather than negative, then the pattern of trajectories is also similar to Figure 3.3.8, but the direction of motion is away from the line of critical points rather than toward it.

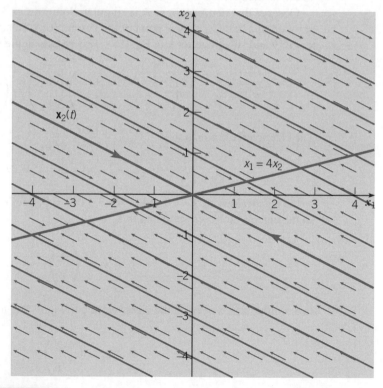

FIGURE 3.3.8 A direction field and phase portrait for the system (68).

We summarize the different phase portraits appearing in this section in Table 3.3.1.

TABLE 3.3.1	Phase portraits for $\mathbf{x}' = \mathbf{Ax}$ when $\mathbf{A}$ has distinct real eigenvalues.			
Eigenvalues		Sample Phase Portrait	Type of Critical Point	Stability
$\lambda_1 \neq \lambda_2$ Both positive			$(0, 0)$ is a **nodal source**.	Unstable
$\lambda_1 \neq \lambda_2$ Both negative			$(0, 0)$ is a **nodal sink**.	Asymptotically stable
$\lambda_1 \neq \lambda_2$ Opposite signs			$(0, 0)$ is a **saddle**.	Unstable
$\lambda_1 = 0$ and $\lambda_2 > 0$				
$\lambda_1 = 0$ and $\lambda_2 < 0$				

PROBLEMS

 General Solutions of Systems. In each of Problems 1 through 12, find the general solution of the given system of equations. Also draw a direction field and a phase portrait. Describe the behavior of the solutions as $t \to \infty$.

1. $\mathbf{x}' = \begin{pmatrix} 3 & -2 \\ 2 & -2 \end{pmatrix} \mathbf{x}$

2. $\mathbf{x}' = \begin{pmatrix} 1 & -2 \\ 3 & -4 \end{pmatrix} \mathbf{x}$

3. $\mathbf{x}' = \begin{pmatrix} 2 & -1 \\ 3 & -2 \end{pmatrix} \mathbf{x}$

4. $\mathbf{x}' = \begin{pmatrix} 1 & 1 \\ 4 & -2 \end{pmatrix} \mathbf{x}$

5. $\mathbf{x}' = \begin{pmatrix} 4 & -3 \\ 8 & -6 \end{pmatrix} \mathbf{x}$

6. $\mathbf{x}' = \begin{pmatrix} -2 & 1 \\ 1 & -2 \end{pmatrix} \mathbf{x}$

7. $\mathbf{x}' = \begin{pmatrix} \frac{5}{4} & \frac{3}{4} \\ \frac{3}{4} & \frac{5}{4} \end{pmatrix} \mathbf{x}$

8. $\mathbf{x}' = \begin{pmatrix} -\frac{3}{4} & -\frac{7}{4} \\ \frac{1}{4} & \frac{5}{4} \end{pmatrix} \mathbf{x}$

9. $\mathbf{x}' = \begin{pmatrix} -\frac{1}{4} & -\frac{3}{4} \\ \frac{1}{2} & 1 \end{pmatrix} \mathbf{x}$

10. $\mathbf{x}' = \begin{pmatrix} 5 & -1 \\ 3 & 1 \end{pmatrix} \mathbf{x}$

11. $\mathbf{x}' = \begin{pmatrix} -2 & 1 \\ -5 & 4 \end{pmatrix} \mathbf{x}$

12. $\mathbf{x}' = \begin{pmatrix} 3 & 6 \\ -1 & -2 \end{pmatrix} \mathbf{x}$

 In each of Problems 13 through 16, solve the given initial value problem. Draw component plots of x_1 and x_2 versus t. Describe the behavior of the solution as $t \to \infty$.

13. $\mathbf{x}' = \begin{pmatrix} 1 & -2 \\ 3 & -4 \end{pmatrix} \mathbf{x}, \quad \mathbf{x}(0) = \begin{pmatrix} 3 \\ 1 \end{pmatrix};$

see Problem 2.

14. $\mathbf{x}' = \begin{pmatrix} 2 & -1 \\ 3 & -2 \end{pmatrix} \mathbf{x}, \quad \mathbf{x}(0) = \begin{pmatrix} 2 \\ 5 \end{pmatrix};$

see Problem 3.

15. $\mathbf{x}' = \begin{pmatrix} 5 & -1 \\ 3 & 1 \end{pmatrix} \mathbf{x}, \quad \mathbf{x}(0) = \begin{pmatrix} 2 \\ -1 \end{pmatrix};$

see Problem 10.

16. $\mathbf{x}' = \begin{pmatrix} -2 & 1 \\ -5 & 4 \end{pmatrix} \mathbf{x}, \quad \mathbf{x}(0) = \begin{pmatrix} 1 \\ 3 \end{pmatrix};$

see Problem 11.

Phase Portraits and Component Plots. In each of Problems 17 through 24, the eigenvalues and eigenvectors of a matrix $\mathbf{A}$ are given. Consider the corresponding system $\mathbf{x}' = \mathbf{A}\mathbf{x}$. Without using a computer, draw each of the following graphs:
(a) Sketch a phase portrait of the system.
(b) Sketch the trajectory passing through the initial point $(2, 3)$.
(c) For the trajectory in part (b), sketch the component plots of x_1 versus t and of x_2 versus t on the same set of axes.

17. $\lambda_1 = -1, \quad \mathbf{v}_1 = \begin{pmatrix} -1 \\ 2 \end{pmatrix};$

$\lambda_2 = -2, \quad \mathbf{v}_2 = \begin{pmatrix} 1 \\ 2 \end{pmatrix}$

18. $\lambda_1 = 1, \quad \mathbf{v}_1 = \begin{pmatrix} -1 \\ 2 \end{pmatrix};$

$\lambda_2 = -2, \quad \mathbf{v}_2 = \begin{pmatrix} 1 \\ 2 \end{pmatrix}$

19. $\lambda_1 = -1, \quad \mathbf{v}_1 = \begin{pmatrix} -1 \\ 2 \end{pmatrix};$

$\lambda_2 = 2, \quad \mathbf{v}_2 = \begin{pmatrix} 1 \\ 2 \end{pmatrix}$

20. $\lambda_1 = 1, \quad \mathbf{v}_1 = \begin{pmatrix} 1 \\ 2 \end{pmatrix};$

$\lambda_2 = 2, \quad \mathbf{v}_2 = \begin{pmatrix} 1 \\ -2 \end{pmatrix}$

21. $\lambda_1 = 0.5, \quad \mathbf{v}_1 = \begin{pmatrix} 1 \\ 4 \end{pmatrix};$

$\lambda_2 = -0.5, \quad \mathbf{v}_2 = \begin{pmatrix} 4 \\ 1 \end{pmatrix}$

22. $\lambda_1 = -0.5, \quad \mathbf{v}_1 = \begin{pmatrix} 2 \\ 1 \end{pmatrix};$

$\lambda_2 = -0.8, \quad \mathbf{v}_2 = \begin{pmatrix} -1 \\ 2 \end{pmatrix}$

23. $\lambda_1 = 0.3, \quad \mathbf{v}_1 = \begin{pmatrix} 1 \\ -2 \end{pmatrix};$

$\lambda_2 = 0.6, \quad \mathbf{v}_2 = \begin{pmatrix} 1 \\ 3 \end{pmatrix}$

24. $\lambda_1 = 1.5, \quad \mathbf{v}_1 = \begin{pmatrix} -1 \\ 2 \end{pmatrix};$

$\lambda_2 = -1, \quad \mathbf{v}_2 = \begin{pmatrix} 3 \\ 1 \end{pmatrix}$

Second Order Equations. For Problems 25 through 30:
(a) Write as a system of first order equations.
(b) Determine the general solution of the system in (a).
(c) Determine the general solution of the second order equation.
(d) Draw a phase portrait.
(e) Classify the critical points.

25. $y'' + 7y' + 10y = 0$

26. $5y'' - 4y' = 0$

27. $y'' - 7y' + 12y = 0$

28. $2y'' + 7y' = 0$

29. $6y'' + 7y' + 2y = 0$

30. $3y'' + 11y' - 4y = 0$

Applications.

31. Obtaining exact, or approximate, expressions for eigenvalues and eigenvectors in terms of the model parameters is often useful for understanding the qualitative behavior of solutions to a dynamical system. We illustrate using Example 1 in Section 3.2.

(a) Show that the general solution of Eqs. (5) and (6) in Section 3.2 can be represented as

$$\mathbf{u} = c_1 \mathbf{x}_1(t) + c_2 \mathbf{x}_2(t) + \hat{\mathbf{u}}, \tag{i}$$

where $\hat{\mathbf{u}}$ is the equilibrium solution (see Problem 13, Section 3.2) to the system and $\{\mathbf{x}_1(t), \mathbf{x}_2(t)\}$ is a fundamental set of solutions to the homogeneous equation

$$\mathbf{x}' = \begin{pmatrix} -(k_1 + k_2) & k_2 \\ \epsilon k_2 & -\epsilon k_2 \end{pmatrix} \mathbf{x} = \mathbf{K}\mathbf{x}.$$

(b) Assuming that $0 < \epsilon \ll 1$ (i.e., ϵ is positive and small relative to unity), show that approximations to the eigenvalues of **K** are

$$\lambda_1(\epsilon) \cong -(k_1 + k_2) - \frac{\epsilon k_2^2}{k_1 + k_2} \qquad \text{and}$$

$$\lambda_2(\epsilon) \cong -\frac{\epsilon k_1 k_2}{k_1 + k_2}.$$

(c) Show that approximations to the corresponding eigenvectors are

$$\mathbf{v}_1(\epsilon) = \left(1, -\epsilon k_2/(k_1 + k_2)\right)^T \qquad \text{and}$$

$$\mathbf{v}_2(\epsilon) = \left(k_2/(k_1 + k_2), 1\right)^T.$$

(d) Use the approximations obtained in parts (b) and (c) and the equilibrium solution $\hat{\mathbf{u}}$ to write down an approximation to the general solution (i). Assuming that nominal values for k_1 and k_2 are near unity, and that $0 < \epsilon \ll 1$, sketch a qualitatively accurate phase portrait for Eqs. (5) and (6) in Section 3.2. Compare your sketch with the phase portrait in Figure 3.2.4, Section 3.2.

(e) Show that when t is large and the system is in the quasi-steady state,

$$u_1 \cong \frac{k_1}{k_1 + k_2}T_a + \frac{k_2}{k_1 + k_2}u_2,$$

$$u_2 \cong (u_{20} - T_a)\exp\left(-\frac{\epsilon k_1 k_2}{k_1 + k_2}t\right) + T_a,$$

where $u_2(0) = u_{20}$. Now let $\epsilon \to 0$. Interpret the results. Compare with the results of Problem 14 in Section 3.2.

(f) Give a physical explanation of the significance of the eigenvalues on the dynamical behavior of the solution (i). In particular, relate the eigenvalues to fast and slow temporal changes in the state of the system. What implications does the value of ϵ have with respect to the design of the greenhouse/rockbed system? Explain, giving due consideration to local climatic conditions and construction costs.

Electric Circuits. Problems 32 and 33 are concerned with the electric circuit described by the system of differential equations in Problem 29 of Section 3.2:

$$\frac{d}{dt}\begin{pmatrix} i \\ v \end{pmatrix} = \begin{pmatrix} -R_1/L & -1/L \\ 1/C & -1/CR_2 \end{pmatrix}\begin{pmatrix} i \\ v \end{pmatrix}. \tag{i}$$

32. (a) Find the general solution of Eq. (i) if $R_1 = 1$ ohm, $R_2 = \frac{3}{5}$ ohm, $L = 2$ henries, and $C = \frac{2}{3}$ farad.
(b) Show that $i(t) \to 0$ and $v(t) \to 0$ as $t \to \infty$, regardless of the initial values $i(0)$ and v_0.

33. Consider the preceding system of differential equations (i).

(a) Find a condition on R_1, R_2, C, and L that must be satisfied if the eigenvalues of the coefficient matrix are to be real and different.
(b) If the condition found in part (a) is satisfied, show that both eigenvalues are negative. Then show that $i(t) \to 0$ and $v(t) \to 0$ as $t \to \infty$, regardless of the initial conditions.

34. Dependence on a Parameter. Consider the system

$$\mathbf{x}' = \begin{pmatrix} -1 & -1 \\ -\alpha & -1 \end{pmatrix}\mathbf{x}.$$

(a) Solve the system for $\alpha = 0.5$. What are the eigenvalues of the coefficient matrix? Classify the equilibrium point at the origin as to type.
(b) Solve the system for $\alpha = 2$. What are the eigenvalues of the coefficient matrix? Classify the equilibrium point at the origin as to type.
(c) In parts (a) and (b), solutions of the system exhibit two quite different types of behavior. Find the eigenvalues of the coefficient matrix in terms of α and determine the value of α between 0.5 and 2 where the transition from one type of behavior to the other occurs.

3.4 Complex Eigenvalues

In this section we again consider a system of two linear homogeneous equations with constant coefficients

$$\mathbf{x}' = \mathbf{A}\mathbf{x}, \tag{1}$$

where the coefficient matrix **A** is real-valued. If we seek solutions of the form $\mathbf{x} = e^{\lambda t}\mathbf{v}$, then it follows, as in Section 3.3, that λ must be an eigenvalue and **v** a corresponding eigenvector of the coefficient matrix **A**. Recall that the eigenvalues λ_1 and λ_2 of **A** are the roots of the quadratic equation

$$\det(\mathbf{A} - \lambda\mathbf{I}) = 0, \tag{2}$$

and that the corresponding eigenvectors satisfy

$$(\mathbf{A} - \lambda\mathbf{I})\mathbf{v} = \mathbf{0}. \tag{3}$$

Since $\mathbf{A}$ is real-valued, the coefficients in Eq. (2) for λ are real, and any complex eigenvalues must occur in conjugate pairs. For example, if $\lambda_1 = \mu + i\nu$, where μ and ν are real, is an eigenvalue of $\mathbf{A}$, then so is $\lambda_2 = \mu - i\nu$. Before further analyzing the general system (1), we consider an example.

EXAMPLE
1

Consider the system

$$\mathbf{x}' = \begin{pmatrix} -\frac{1}{2} & 1 \\ -1 & -\frac{1}{2} \end{pmatrix} \mathbf{x}. \tag{4}$$

Find a fundamental set of solutions and display them graphically in a phase portrait and component plots.

To find a fundamental set of solutions, we assume that

$$\mathbf{x} = e^{\lambda t} \mathbf{v} \tag{5}$$

and obtain the set of linear algebraic equations

$$\begin{pmatrix} -\frac{1}{2} - \lambda & 1 \\ -1 & -\frac{1}{2} - \lambda \end{pmatrix} \begin{pmatrix} v_1 \\ v_2 \end{pmatrix} = \begin{pmatrix} 0 \\ 0 \end{pmatrix} \tag{6}$$

for the eigenvalues and eigenvectors of $\mathbf{A}$. The characteristic equation is

$$\begin{vmatrix} -\frac{1}{2} - \lambda & 1 \\ -1 & -\frac{1}{2} - \lambda \end{vmatrix} = \lambda^2 + \lambda + \frac{5}{4} = 0; \tag{7}$$

therefore the eigenvalues are $\lambda_1 = -\frac{1}{2} + i$ and $\lambda_2 = -\frac{1}{2} - i$. For $\lambda_1 = -\frac{1}{2} + i$, we obtain from Eq. (6)

$$\begin{pmatrix} -i & 1 \\ -1 & -i \end{pmatrix} \begin{pmatrix} v_1 \\ v_2 \end{pmatrix} = \begin{pmatrix} 0 \\ 0 \end{pmatrix}. \tag{8}$$

In scalar form, the first of these equations is $-iv_1 + v_2 = 0$ and the second equation is $-i$ times the first. Thus we have $v_2 = iv_1$, so the eigenvector corresponding to the eigenvalue λ_1 is

$$\mathbf{v}_1 = \begin{pmatrix} 1 \\ i \end{pmatrix} \tag{9}$$

or any multiple of this vector. A similar calculation for the second eigenvalue $\lambda_2 = -\frac{1}{2} - i$ leads to

$$\mathbf{v}_2 = \begin{pmatrix} 1 \\ -i \end{pmatrix}. \tag{10}$$

Thus the eigenvectors as well as the eigenvalues are complex conjugates. The corresponding solutions of the system (4) are

$$\mathbf{x}_1(t) = e^{(-1/2+i)t} \begin{pmatrix} 1 \\ i \end{pmatrix}, \qquad \mathbf{x}_2(t) = e^{(-1/2-i)t} \begin{pmatrix} 1 \\ -i \end{pmatrix}. \tag{11}$$

The Wronskian of $\mathbf{x}_1(t)$ and $\mathbf{x}_2(t)$ is readily calculated to be $-2ie^{-t}$, which is never zero, so these solutions form a fundamental set. However $\mathbf{x}_1$ and $\mathbf{x}_2$ are complex-valued, and for many purposes it is desirable to find a fundamental set of real-valued solutions. To do this, we start by finding the real and imaginary parts[5] of $\mathbf{x}_1(t)$. First, we use the properties of exponents to write

$$\mathbf{x}_1(t) = e^{-t/2} e^{it} \begin{pmatrix} 1 \\ i \end{pmatrix}. \tag{12}$$

Then, using Euler's formula for e^{it}, namely,

$$e^{it} = \cos t + i \sin t, \tag{13}$$

we obtain

$$\mathbf{x}_1(t) = e^{-t/2} (\cos t + i \sin t) \begin{pmatrix} 1 \\ i \end{pmatrix}. \tag{14}$$

Finally, by carrying out the multiplication indicated in Eq. (14), we find that

$$\mathbf{x}_1(t) = \begin{pmatrix} e^{-t/2} \cos t \\ -e^{-t/2} \sin t \end{pmatrix} + i \begin{pmatrix} e^{-t/2} \sin t \\ e^{-t/2} \cos t \end{pmatrix} = \mathbf{u}(t) + i\mathbf{w}(t). \tag{15}$$

The real and imaginary parts of $\mathbf{x}_1(t)$, that is,

$$\mathbf{u}(t) = \begin{pmatrix} e^{-t/2} \cos t \\ -e^{-t/2} \sin t \end{pmatrix}, \qquad \mathbf{w}(t) = \begin{pmatrix} e^{-t/2} \sin t \\ e^{-t/2} \cos t \end{pmatrix}, \tag{16}$$

are real-valued vector functions. They are also solutions of the system (4). One way to show this is simply to substitute $\mathbf{u}(t)$ and $\mathbf{w}(t)$ for $\mathbf{x}$ in Eq. (4). However, this is a very special case of a much more general result, which we will demonstrate shortly, so for the moment let us accept that $\mathbf{u}(t)$ and $\mathbf{w}(t)$ satisfy Eq. (4).

To verify that $\mathbf{u}(t)$ and $\mathbf{w}(t)$ constitute a fundamental set of solutions, we compute their Wronskian:

$$W[\mathbf{u}, \mathbf{w}](t) = \begin{vmatrix} e^{-t/2} \cos t & e^{-t/2} \sin t \\ -e^{-t/2} \sin t & e^{-t/2} \cos t \end{vmatrix}$$

$$= e^{-t}(\cos^2 t + \sin^2 t) = e^{-t}. \tag{17}$$

Since the Wronskian is never zero, it follows that $\mathbf{u}(t)$ and $\mathbf{w}(t)$ constitute a fundamental set of (real-valued) solutions of the system (4). Consequently, every solution of Eq. (4) is a linear combination of $\mathbf{u}(t)$ and $\mathbf{w}(t)$, and the general solution of Eq. (4) is

$$\mathbf{x} = c_1 \mathbf{u}(t) + c_2 \mathbf{w}(t) = c_1 \begin{pmatrix} e^{-t/2} \cos t \\ -e^{-t/2} \sin t \end{pmatrix} + c_2 \begin{pmatrix} e^{-t/2} \sin t \\ e^{-t/2} \cos t \end{pmatrix}, \tag{18}$$

where c_1 and c_2 are arbitrary constants.

The trajectories of the solutions $\mathbf{u}(t)$ and $\mathbf{w}(t)$ are shown in Figure 3.4.1. Since

$$\mathbf{u}(0) = \begin{pmatrix} 1 \\ 0 \end{pmatrix}, \qquad \mathbf{w}(0) = \begin{pmatrix} 0 \\ 1 \end{pmatrix},$$

the trajectories of $\mathbf{u}(t)$ and $\mathbf{w}(t)$ pass through the points $(1, 0)$ and $(0, 1)$, respectively. Other solutions of the system (4) are linear combinations of $\mathbf{u}(t)$ and $\mathbf{w}(t)$, and trajectories of a few of these solutions are also shown in Figure 3.4.1. Each trajectory spirals toward the origin in the clockwise direction as $t \to \infty$, making infinitely many circuits about the origin.

---

[5] You will find a summary of the necessary elementary results from complex variables in Appendix B.

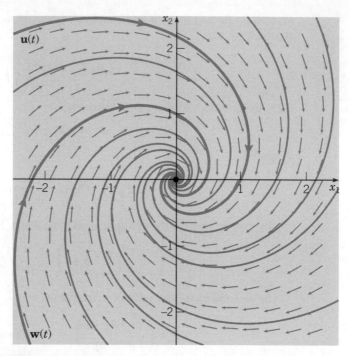

FIGURE 3.4.1   A direction field and phase portrait for the system (4).

This is due to the fact that the solutions (16) are products of decaying exponential and sine or cosine factors.

Figure 3.4.2 shows the component plots of $\mathbf{u}(t)$ and $\mathbf{w}(t)$. Note that $u_1(t)$ and $w_2(t)$ are identical, so their plots coincide, whereas $u_2(t)$ and $w_1(t)$ are negatives of each other. Each plot represents a decaying oscillation in time. Since other solutions are linear combinations of $\mathbf{u}(t)$ and $\mathbf{w}(t)$, their component plots are also decaying oscillations.

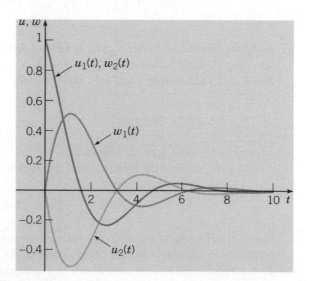

FIGURE 3.4.2   Component plots for the solutions $\mathbf{u}(t)$ and $\mathbf{w}(t)$ of the system (4).

▶ Extension to a General System. We can proceed as in Example 1 in the case of a general two-dimensional system $\mathbf{x}' = \mathbf{A}\mathbf{x}$ with complex conjugate eigenvalues. Suppose that the eigenvalues are $\lambda_1 = \mu + iv$ and $\lambda_2 = \bar{\lambda}_1 = \mu - iv$. Suppose also that $\mathbf{v}_1$ is an eigenvector corresponding to λ_1. Then λ_1 and $\mathbf{v}_1$ satisfy

$$(\mathbf{A} - \lambda_1 \mathbf{I})\mathbf{v}_1 = \mathbf{0}. \tag{19}$$

By taking the conjugate of Eq. (19) and remembering that $\mathbf{A}$, $\mathbf{I}$, and $\mathbf{0}$ are real-valued, we obtain

$$(\mathbf{A} - \bar{\lambda}_1 \mathbf{I})\bar{\mathbf{v}}_1 = \mathbf{0}. \tag{20}$$

Thus $\bar{\mathbf{v}}_1$ is an eigenvector corresponding to $\bar{\lambda}_1$, which is λ_2. Thus, for a pair of complex conjugate eigenvalues, we can always choose the eigenvectors so that they are complex conjugates as well, and we will always make this choice.

Using these eigenvalues and eigenvectors, we obtain two solutions of the system (1):

$$\mathbf{x}_1(t) = e^{(\mu+iv)t}\mathbf{v}_1, \qquad \mathbf{x}_2(t) = e^{(\mu-iv)t}\bar{\mathbf{v}}_1. \tag{21}$$

We now wish to separate $\mathbf{x}_1(t)$ into its real and imaginary parts. Recall that, by Euler's formula,

$$e^{ivt} = \cos vt + i \sin vt \tag{22}$$

and let $\mathbf{v}_1 = \mathbf{a} + i\mathbf{b}$, where $\mathbf{a}$ and $\mathbf{b}$ are real-valued. Then from Eq. (21) we have

$$\mathbf{x}_1(t) = (\mathbf{a} + i\mathbf{b})e^{\mu t}(\cos vt + i \sin vt)$$
$$= e^{\mu t}(\mathbf{a} \cos vt - \mathbf{b} \sin vt) + ie^{\mu t}(\mathbf{a} \sin vt + \mathbf{b} \cos vt). \tag{23}$$

We denote the real and imaginary parts of $\mathbf{x}_1(t)$ by

$$\mathbf{u}(t) = e^{\mu t}(\mathbf{a} \cos vt - \mathbf{b} \sin vt), \qquad \mathbf{w}(t) = e^{\mu t}(\mathbf{a} \sin vt + \mathbf{b} \cos vt), \tag{24}$$

respectively. A similar calculation, starting from $\mathbf{x}_2(t)$ in Eq. (21), leads to

$$\mathbf{x}_2(t) = \mathbf{u}(t) - i\mathbf{w}(t). \tag{25}$$

Thus the solutions $\mathbf{x}_1(t)$ and $\mathbf{x}_2(t)$ are also complex conjugates.

Next, we want to show that $\mathbf{u}(t)$ and $\mathbf{w}(t)$ are solutions of Eq. (1). Since $\mathbf{x}_1(t)$ is a solution, we can write

$$\mathbf{0} = \mathbf{x}_1' - \mathbf{A}\mathbf{x}_1 = (\mathbf{u}' + i\mathbf{w}') - \mathbf{A}(\mathbf{u} + i\mathbf{w})$$
$$= (\mathbf{u}' - \mathbf{A}\mathbf{u}) + i(\mathbf{w}' - \mathbf{A}\mathbf{w}). \tag{26}$$

A complex number (vector) is zero if and only if both its real and imaginary parts are zero, so we conclude that

$$\mathbf{u}' - \mathbf{A}\mathbf{u} = \mathbf{0}, \qquad \mathbf{w}' - \mathbf{A}\mathbf{w} = \mathbf{0}; \tag{27}$$

therefore $\mathbf{u}(t)$ and $\mathbf{w}(t)$ are (real-valued) solutions of Eq. (1).

Finally, we calculate the Wronskian of $\mathbf{u}(t)$ and $\mathbf{w}(t)$ so as to determine whether they form a fundamental set of solutions. Let $\mathbf{a} = a_1\mathbf{i} + a_2\mathbf{j}$ and $\mathbf{b} = b_1\mathbf{i} + b_2\mathbf{j}$. Then

$$W[\mathbf{u}, \mathbf{w}](t) = \begin{vmatrix} e^{\mu t}(a_1 \cos vt - b_1 \sin vt) & e^{\mu t}(a_1 \sin vt + b_1 \cos vt) \\ e^{\mu t}(a_2 \cos vt - b_2 \sin vt) & e^{\mu t}(a_2 \sin vt + b_2 \cos vt) \end{vmatrix}. \tag{28}$$

A straightforward calculation shows that

$$W[\mathbf{u}, \mathbf{w}](t) = (a_1 b_2 - a_2 b_1)e^{2\mu t}. \tag{29}$$

Assuming that $v \neq 0$, the eigenvalues $\lambda_1 = \mu + iv$ and $\lambda_2 = \mu - iv$ are not equal. Taking the determinant of the matrix formed from the corresponding eigenvectors $\mathbf{v}_1 = \mathbf{a} + i\mathbf{b}$ and $\mathbf{v}_2 = \mathbf{a} - i\mathbf{b}$, we get

$$\begin{vmatrix} a_1 + ib_1 & a_1 - ib_1 \\ a_2 + ib_2 & a_2 - ib_2 \end{vmatrix} = -2i\left(a_1 b_2 - a_2 b_1\right),$$

which is nonzero by Theorem 3.1.3, so $a_1 b_2 - a_2 b_1 \neq 0$. Consequently, the Wronskian (29) is not zero. Therefore the solutions $\mathbf{u}(t)$ and $\mathbf{w}(t)$ given by Eq. (24) form a fundamental set of solutions of the system (1). The general solution can be written as

$$\mathbf{x} = c_1 \mathbf{u}(t) + c_2 \mathbf{w}(t), \tag{30}$$

where c_1 and c_2 are arbitrary constants.

Procedure for Finding the General Solution of $x' = Ax$ When A Has Complex Eigenvalues

1. Identify the complex conjugate eigenvalues $\lambda = \mu \pm iv$.

2. Determine the eigenvector $\mathbf{v} = \begin{pmatrix} v_1 \\ v_2 \end{pmatrix}$ corresponding to $\lambda_1 = \mu + iv$ by solving

$$\left(\mathbf{A} - \lambda_1 \mathbf{I}\right)\mathbf{v} = \mathbf{0}.$$

3. Express the eigenvector $\mathbf{v}$ in the form $\mathbf{v} = \mathbf{a} + \mathbf{b}i$.

4. Write the solution $\mathbf{x}_1$ corresponding to $\mathbf{v}$ and separate it into real and imaginary parts:[6]

$$\mathbf{x}_1(t) = \underbrace{e^{\mu t}\left(\mathbf{a}\cos vt - \mathbf{b}\sin vt\right)}_{\mathbf{u}(t)} + \underbrace{ie^{\mu t}\left(\mathbf{a}\sin vt + \mathbf{b}\cos vt\right)}_{\mathbf{w}(t)}$$

It can be shown that $\mathbf{u}$ and $\mathbf{w}$ form a fundamental set of solutions for $\mathbf{x}' = \mathbf{Ax}$.

5. Then the general solution of $\mathbf{x}' = \mathbf{Ax}$ is

$$\mathbf{x}(t) = c_1 \mathbf{u}(t) + c_2 \mathbf{w}(t),$$

where c_1 and c_2 are arbitrary constants.

Spiral Points and Centers

The phase portrait in Figure 3.4.1 is typical of all two-dimensional systems $\mathbf{x}' = \mathbf{Ax}$ whose eigenvalues are complex with a negative real part. The origin is called a **spiral point** and is asymptotically stable because all trajectories approach it as t increases. Such a spiral point is often called a **spiral sink**. For a system whose eigenvalues have a positive real part, the trajectories are similar to those in Figure 3.4.1, but the direction of motion is away from the origin and the trajectories become unbounded. In this case, the origin is unstable and is often called a **spiral source**.

If the real part of the eigenvalues is zero, then there is no exponential factor in the solution and the trajectories neither approach the origin nor become unbounded. Instead, they repeatedly traverse a closed curve about the origin. An example of this behavior can be seen in Figure 3.4.3. In this case, the origin is called a **center** and is said to be **stable**, but not asymptotically stable. In all three cases, the direction of motion may be either clockwise, as in Example 1, or counterclockwise, depending on the elements of the coefficient matrix **A**.

---

[6]Rather than memorize the expressions for $\mathbf{u}(t)$ and $\mathbf{w}(t)$, once μ, v, $\mathbf{a}$, and $\mathbf{b}$ have been determined in a given problem, it is preferable to carry out the complex arithmetic used to calculate the right side of Eq. (23); then, take the real and imaginary parts for $\mathbf{u}(t)$ and $\mathbf{v}(t)$, respectively.

EXAMPLE
2

Consider the system

$$\mathbf{x}' = \begin{pmatrix} \frac{1}{2} & -\frac{5}{4} \\ 2 & -\frac{1}{2} \end{pmatrix} \mathbf{x}. \tag{31}$$

Find the general solution and the solution that satisfies the initial conditions

$$\mathbf{x}(0) = \begin{pmatrix} -1 \\ -2 \end{pmatrix}. \tag{32}$$

Draw a direction field and phase portrait for the system. Then draw component plots for the solution satisfying the initial conditions (32).

To solve the system, we assume that $\mathbf{x} = e^{\lambda t}\mathbf{v}$, substitute in Eq. (31), and thereby obtain the algebraic system $\mathbf{Av} = \lambda\mathbf{v}$, or

$$\begin{pmatrix} \frac{1}{2} - \lambda & -\frac{5}{4} \\ 2 & -\frac{1}{2} - \lambda \end{pmatrix} \begin{pmatrix} v_1 \\ v_2 \end{pmatrix} = \begin{pmatrix} 0 \\ 0 \end{pmatrix}. \tag{33}$$

The eigenvalues are found from the characteristic equation

$$\left(\frac{1}{2} - \lambda\right)\left(-\frac{1}{2} - \lambda\right) - \left(-\frac{5}{4}\right)(2) = \lambda^2 + \frac{9}{4} = 0, \tag{34}$$

so the eigenvalues are $\lambda_1 = 3i/2$ and $\lambda_2 = -3i/2$. By substituting λ_1 for λ in Eq. (33) and then solving this system, we find the corresponding eigenvector $\mathbf{v}_1$. We can find the other eigenvector $\mathbf{v}_2$ in a similar way, or we can simply take $\mathbf{v}_2$ to be the complex conjugate of $\mathbf{v}_1$. Either way, we obtain

$$\mathbf{v}_1 = \begin{pmatrix} 5 \\ 2 - 6i \end{pmatrix}, \qquad \mathbf{v}_2 = \begin{pmatrix} 5 \\ 2 + 6i \end{pmatrix}. \tag{35}$$

Hence two (complex-valued) solutions of the system (31) are

$$\mathbf{x}_1(t) = e^{3it/2} \begin{pmatrix} 5 \\ 2 - 6i \end{pmatrix}, \qquad \mathbf{x}_2(t) = e^{-3it/2} \begin{pmatrix} 5 \\ 2 + 6i \end{pmatrix}, \tag{36}$$

and the general solution of Eq. (31) can be expressed as a linear combination of $\mathbf{x}_1(t)$ and $\mathbf{x}_2(t)$ with arbitrary coefficients.

To find real-valued solutions, we can separate $\mathbf{x}_1(t)$ into its real and imaginary parts. Using Euler's formula, we have

$$\mathbf{x}_1(t) = \left[\cos(3t/2) + i\sin(3t/2)\right] \begin{pmatrix} 5 \\ 2 - 6i \end{pmatrix}$$

$$= \begin{pmatrix} 5\cos(3t/2) \\ 2\cos(3t/2) + 6\sin(3t/2) \end{pmatrix} + i \begin{pmatrix} 5\sin(3t/2) \\ 2\sin(3t/2) - 6\cos(3t/2) \end{pmatrix}. \tag{37}$$

Thus we can also write the general solution of Eq. (31) in the form

$$\mathbf{x} = c_1 \begin{pmatrix} 5\cos(3t/2) \\ 2\cos(3t/2) + 6\sin(3t/2) \end{pmatrix} + c_2 \begin{pmatrix} 5\sin(3t/2) \\ 2\sin(3t/2) - 6\cos(3t/2) \end{pmatrix}, \tag{38}$$

where c_1 and c_2 are arbitrary constants.

To satisfy the initial conditions (32), we set $t = 0$ in Eq. (38) and obtain

$$c_1 \begin{pmatrix} 5 \\ 2 \end{pmatrix} + c_2 \begin{pmatrix} 0 \\ -6 \end{pmatrix} = \begin{pmatrix} -1 \\ -2 \end{pmatrix}. \tag{39}$$

Therefore $c_1 = -\frac{1}{5}$ and $c_2 = \frac{4}{15}$. Using these values in Eq. (38), we obtain the solution that satifies the initial conditions (32). The scalar components of this solution are easily found to be

$$x_1(t) = -\cos(3t/2) + \tfrac{4}{3}\sin(3t/2), \qquad x_2(t) = -2\cos(3t/2) - \tfrac{2}{3}\sin(3t/2). \tag{40}$$

A direction field and phase portrait for the system (31) appear in Figure 3.4.3 with the heavy curve showing the trajectory passing through the point $(-1, -2)$. All of the trajectories (except the origin itself) are closed curves surrounding the origin, and each one is traversed in the counterclockwise direction, as the direction field in Figure 3.4.3 indicates. Closed trajectories correspond to eigenvalues whose real part is zero. In this case, the solution contains no exponential factor, but consists only of sine and cosine terms, and therefore is periodic. The closed trajectory is traversed repeatedly, with one period corresponding to a full circuit. The periodic nature of the solution is shown clearly by the component plots in Figure 3.4.4, which show the graphs of $x_1(t)$ and $x_2(t)$ given in Eq. (40). The period of each solution is $4\pi/3$.

As mentioned previously, when the eigenvalues are purely imaginary, the critical point at the origin is called a center. Because nearby trajectories remain near the critical point but do not approach it, the critical point is said to be stable, but not asymptotically stable.

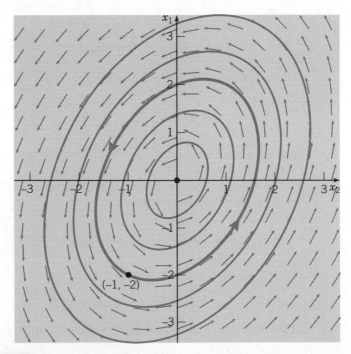

FIGURE 3.4.3   A direction field and phase portrait for the system (31); the heavy curve passes through $(-1, -2)$.

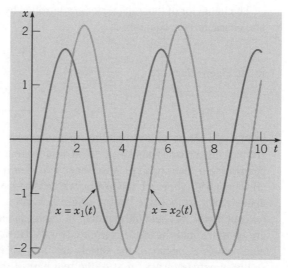

FIGURE 3.4.4 Plots of the components of the solution passing through $(-1, -2)$; they are given by Eq. (40).

A center is often a transition state between spiral points that are asymptotically stable or unstable, depending on the sign of the real part of the eigenvalues. The trajectories in Figure 3.4.3 appear to be elliptical, and it can be shown (see Problem 23) that this is always the case for centers of two-dimensional linear systems with constant coefficients.

For two-dimensional systems with real coefficients, we have now completed our description of the three main cases that can occur:

1. Eigenvalues are real and have opposite signs; $\mathbf{x} = \mathbf{0}$ is a saddle point.
2. Eigenvalues are real and have the same sign but are unequal; $\mathbf{x} = \mathbf{0}$ is a node.
3. Eigenvalues are complex with nonzero real part; $\mathbf{x} = \mathbf{0}$ is a spiral point.

Other possibilities are of less importance and occur as transitions between two of the cases just listed. For example, we have seen that purely imaginary eigenvalues occur at a transition between asymptotically stable and unstable spiral points. A zero eigenvalue occurs at the transition between a saddle point and a node, as one of the eigenvalues changes sign. Finally, real and equal eigenvalues appear when the discriminant of the characteristic equation is zero, that is, at the transition between nodes and spiral points.

EXAMPLE 3

The system

$$\mathbf{x}' = \begin{pmatrix} \alpha & 2 \\ -2 & 0 \end{pmatrix} \mathbf{x} \tag{41}$$

contains a parameter α. Describe how the solutions depend qualitatively on α; in particular, find the critical values of α at which the qualitative behavior of the trajectories in the phase plane changes markedly.

The behavior of the trajectories is controlled by the eigenvalues of the coefficient matrix. The characteristic equation is

$$\lambda^2 - \alpha\lambda + 4 = 0, \tag{42}$$

so the eigenvalues are

$$\lambda = \frac{\alpha \pm \sqrt{\alpha^2 - 16}}{2}. \tag{43}$$

From Eq. (43), it follows that the eigenvalues are complex conjugates for $-4 < \alpha < 4$ and are real otherwise. Thus two critical values are $\alpha = -4$ and $\alpha = 4$, where the eigenvalues change from real to complex, or vice versa. For $\alpha < -4$, both eigenvalues are negative, so all trajectories approach the origin, which is an asymptotically stable node. For $\alpha > 4$, both eigenvalues are positive, so the origin is again a node, this time unstable; all trajectories (except $\mathbf{x} = \mathbf{0}$) become unbounded. In the intermediate range, $-4 < \alpha < 4$, the eigenvalues are complex and the trajectories are spirals. However, for $-4 < \alpha < 0$, the real part of the eigenvalues is negative, the spirals are directed inward, and the origin is asymptotically stable, whereas for $0 < \alpha < 4$, the real part of the eigenvalues is positive and the origin is unstable. Thus $\alpha = 0$ is also a critical value where the direction of the spirals changes from inward to outward. For this value of α, the origin is a center and the trajectories are closed curves about the origin, corresponding to solutions that are periodic in time. The other critical values, $\alpha = \pm 4$, yield eigenvalues that are real and equal. In this case, the origin is again a node, but the phase portrait differs somewhat from those in Section 3.3. We take up this case in Section 3.5.

We summarize the phase portraits appearing in this section in Table 3.4.1.

TABLE 3.4.1	Phase portraits for $\mathbf{x}' = \mathbf{Ax}$ when $\mathbf{A}$ has complex eigenvalues.		
Eigenvalues	Sample Phase Portrait	Type of Critical Point	Stability
$\lambda = \mu \pm iv$ $\mu < 0$		$(0, 0)$ is a **spiral sink**.	Asymptotically stable
$\lambda = \mu \pm iv$ $\mu > 0$		$(0, 0)$ is a **spiral source**.	Unstable
$\lambda = \mu \pm iv$ $\mu = 0$		$(0, 0)$ is a **center**.	Stable

PROBLEMS

General Solutions of Systems. In each of Problems 1 through 6, express the general solution of the given system of equations in terms of real-valued functions. Also draw a direction field and a phase portrait. Describe the behavior of the solutions as $t \to \infty$.

1. $\mathbf{x}' = \begin{pmatrix} 3 & -2 \\ 4 & -1 \end{pmatrix} \mathbf{x}$ 2. $\mathbf{x}' = \begin{pmatrix} -1 & -4 \\ 1 & -1 \end{pmatrix} \mathbf{x}$

3. $\mathbf{x}' = \begin{pmatrix} 2 & -5 \\ 1 & -2 \end{pmatrix} \mathbf{x}$ 4. $\mathbf{x}' = \begin{pmatrix} 2 & -\frac{5}{2} \\ \frac{9}{5} & -1 \end{pmatrix} \mathbf{x}$

5. $\mathbf{x}' = \begin{pmatrix} 1 & -1 \\ 5 & -3 \end{pmatrix} \mathbf{x}$ 6. $\mathbf{x}' = \begin{pmatrix} 1 & 2 \\ -5 & -1 \end{pmatrix} \mathbf{x}$

In each of Problems 7 through 10, find the solution of the given initial value problem. Draw component plots of the solution and describe the behavior of the solution as $t \to \infty$.

7. $\mathbf{x}' = \begin{pmatrix} -1 & -4 \\ 1 & -1 \end{pmatrix} \mathbf{x}, \quad \mathbf{x}(0) = \begin{pmatrix} 4 \\ -3 \end{pmatrix}$;

see Problem 2.

8. $\mathbf{x}' = \begin{pmatrix} 2 & -5 \\ 1 & -2 \end{pmatrix} \mathbf{x}, \quad \mathbf{x}(0) = \begin{pmatrix} 3 \\ 2 \end{pmatrix}$;

see Problem 3.

9. $\mathbf{x}' = \begin{pmatrix} 1 & -5 \\ 1 & -3 \end{pmatrix} \mathbf{x}, \quad \mathbf{x}(0) = \begin{pmatrix} 1 \\ 1 \end{pmatrix}$

10. $\mathbf{x}' = \begin{pmatrix} -3 & 2 \\ -1 & -1 \end{pmatrix} \mathbf{x}, \quad \mathbf{x}(0) = \begin{pmatrix} 1 \\ -2 \end{pmatrix}$

Phase Portraits and Component Plots. In each of Problems 11 and 12:
(a) Find the eigenvalues of the given system.
(b) Choose an initial point (other than the origin) and sketch the corresponding trajectory in the x_1x_2-plane.
(c) For your trajectory in part (b), sketch the graphs of x_1 versus t and of x_2 versus t.

11. $\mathbf{x}' = \begin{pmatrix} \frac{3}{4} & -2 \\ 1 & -\frac{5}{4} \end{pmatrix} \mathbf{x}$ 12. $\mathbf{x}' = \begin{pmatrix} -\frac{4}{5} & 2 \\ -1 & \frac{6}{5} \end{pmatrix} \mathbf{x}$

Dependence on a Parameter. In each of Problems 13 through 20, the coefficient matrix contains a parameter α. In each of these problems:
(a) Determine the eigenvalues in terms of α.
(b) Find the critical value or values of α where the qualitative nature of the phase portrait for the system changes.
(c) Draw a phase portrait for a value of α slightly below, and for another value slightly above, each critical value.

13. $\mathbf{x}' = \begin{pmatrix} \alpha & 1 \\ -1 & \alpha \end{pmatrix} \mathbf{x}$ 14. $\mathbf{x}' = \begin{pmatrix} 0 & -5 \\ 1 & \alpha \end{pmatrix} \mathbf{x}$

15. $\mathbf{x}' = \begin{pmatrix} 2 & -5 \\ \alpha & -2 \end{pmatrix} \mathbf{x}$ 16. $\mathbf{x}' = \begin{pmatrix} \frac{5}{4} & \frac{3}{4} \\ \alpha & \frac{5}{4} \end{pmatrix} \mathbf{x}$

17. $\mathbf{x}' = \begin{pmatrix} -1 & \alpha \\ -1 & -1 \end{pmatrix} \mathbf{x}$ 18. $\mathbf{x}' = \begin{pmatrix} 3 & \alpha \\ -6 & -4 \end{pmatrix} \mathbf{x}$

19. $\mathbf{x}' = \begin{pmatrix} \alpha & 10 \\ -1 & -4 \end{pmatrix} \mathbf{x}$ 20. $\mathbf{x}' = \begin{pmatrix} 4 & \alpha \\ 8 & -6 \end{pmatrix} \mathbf{x}$

Applications.

21. Consider the electric circuit shown in Figure 3.4.5. Suppose that $R_1 = R_2 = 4$ ohms, $C = \frac{1}{2}$ farad, and $L = 8$ henries.
(a) Show that this circuit is described by the system of differential equations

$$\frac{d}{dt}\begin{pmatrix} i \\ v \end{pmatrix} = \begin{pmatrix} -\frac{1}{2} & -\frac{1}{8} \\ 2 & -\frac{1}{2} \end{pmatrix}\begin{pmatrix} i \\ v \end{pmatrix}, \tag{i}$$

where i is the current through the inductor and v is the voltage across the capacitor.
Hint: See Problem 27 of Section 3.2.
(b) Find the general solution of Eqs. (i) in terms of real-valued functions.
(c) Find $i(t)$ and $v(t)$ if $i(0) = 2$ amperes and $v(0) = 3$ volts.
(d) Determine the limiting values of $i(t)$ and $v(t)$ as $t \to \infty$. Do these limiting values depend on the initial conditions?

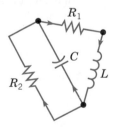

FIGURE 3.4.5   The circuit in Problem 21.

22. The electric circuit shown in Figure 3.4.6 is described by the system of differential equations

$$\frac{d}{dt}\begin{pmatrix} i \\ v \end{pmatrix} = \begin{pmatrix} 0 & 1/L \\ -1/C & -1/RC \end{pmatrix}\begin{pmatrix} i \\ v \end{pmatrix}, \tag{i}$$

where i is the current through the inductor and v is the voltage across the capacitor. These differential equations were derived in Problem 27 of Section 3.2.

(a) Show that the eigenvalues of the coefficient matrix are real and different if $L > 4R^2C$. Show that they are complex conjugates if $L < 4R^2C$.

(b) Suppose that $R = 1$ ohm, $C = \frac{1}{2}$ farad, and $L = 1$ henry. Find the general solution of the system (i) in this case.

(c) Find $i(t)$ and $v(t)$ if $i(0) = 2$ amperes and $v(0) = 1$ volt.

(d) For the circuit of part (b), determine the limiting values of $i(t)$ and $v(t)$ as $t \to \infty$. Do these limiting values depend on the initial conditions?

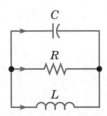

FIGURE 3.4.6 The circuit in Problem 22.

23. In this problem, we indicate how to show that the trajectories are ellipses when the eigenvalues are purely imaginary.

Consider the system

$$\begin{pmatrix} x \\ y \end{pmatrix}' = \begin{pmatrix} a_{11} & a_{12} \\ a_{21} & a_{22} \end{pmatrix} \begin{pmatrix} x \\ y \end{pmatrix}. \tag{i}$$

(a) Show that the eigenvalues of the coefficient matrix are purely imaginary if and only if

$$a_{11} + a_{22} = 0, \qquad a_{11}a_{22} - a_{12}a_{21} > 0. \tag{ii}$$

(b) The trajectories of the system (i) can be found by converting Eqs. (i) into the single equation

$$\frac{dy}{dx} = \frac{dy/dt}{dx/dt} = \frac{a_{21}x + a_{22}y}{a_{11}x + a_{12}y}. \tag{iii}$$

Use the first of Eqs. (ii) to show that Eq. (iii) is exact.

(c) By integrating Eq. (iii), show that

$$a_{21}x^2 + 2a_{22}xy - a_{12}y^2 = k, \tag{iv}$$

where k is a constant. Use Eqs. (ii) to conclude that the graph of Eq. (iv) is always an ellipse.

Hint: What is the discriminant of the quadratic form in Eq. (iv)?

3.5 Repeated Eigenvalues

We continue our consideration of two-dimensional linear homogeneous systems with constant coefficients

$$\mathbf{x}' = \mathbf{Ax} \tag{1}$$

with a discussion of the case in which the matrix $\mathbf{A}$ has a repeated eigenvalue. This case occurs when the discriminant of the characteristic equation is zero, and it is a transition state between a node and a spiral point. There are two essentially different phenomena that can occur: the repeated eigenvalue may have two independent eigenvectors, or it may have only one. The first possibility is the simpler one, so we will start with it.

EXAMPLE 1

Solve the system $\mathbf{x}' = \mathbf{Ax}$, where

$$\mathbf{A} = \begin{pmatrix} -1 & 0 \\ 0 & -1 \end{pmatrix}. \tag{2}$$

Draw a direction field, a phase portrait, and typical component plots.

To solve the system, we assume, as usual, that $\mathbf{x} = e^{\lambda t}\mathbf{v}$. Then, from Eq. (2), we obtain

$$\begin{pmatrix} -1 - \lambda & 0 \\ 0 & -1 - \lambda \end{pmatrix} \begin{pmatrix} v_1 \\ v_2 \end{pmatrix} = \begin{pmatrix} 0 \\ 0 \end{pmatrix}, \tag{3}$$

The characteristic equation is $(1 + \lambda)^2 = 0$, so the eigenvalues are $\lambda_1 = \lambda_2 = -1$. To determine the eigenvectors, we set λ first equal to λ_1 and then to λ_2 in Eq. (3). In either case, we have

$$\begin{pmatrix} 0 & 0 \\ 0 & 0 \end{pmatrix} \begin{pmatrix} v_1 \\ v_2 \end{pmatrix} = \begin{pmatrix} 0 \\ 0 \end{pmatrix}. \tag{4}$$

Thus there are no restrictions on v_1 and v_2; in other words, we can choose them arbitrarily. It is convenient to choose $v_1 = 1$ and $v_2 = 0$ for λ_1 and to choose $v_1 = 0$ and $v_2 = 1$ for λ_2. Thus we obtain two solutions of the given system:

$$\mathbf{x}_1(t) = e^{-t}\begin{pmatrix} 1 \\ 0 \end{pmatrix}, \qquad \mathbf{x}_2(t) = e^{-t}\begin{pmatrix} 0 \\ 1 \end{pmatrix}. \tag{5}$$

The general solution is

$$\mathbf{x} = c_1\mathbf{x}_1(t) + c_2\mathbf{x}_2(t). \tag{6}$$

Alternatively, you can also solve this system by starting from the scalar equations

$$x_1' = -x_1, \qquad x_2' = -x_2, \tag{7}$$

which follow directly from Eq. (2). These equations can be integrated immediately, with the result that

$$x_1(t) = c_1 e^{-t}, \qquad x_2(t) = c_2 e^{-t}. \tag{8}$$

Equations (8) are just the scalar form of Eq. (6).

A direction field and a phase portrait for the system (2) are shown in Figure 3.5.1. The trajectories lie on straight lines through the origin and (because the eigenvalues are negative) they approach the origin as $t \to \infty$. That this must be so is apparent from Eqs. (8). By eliminating t between these two equations, we find that

$$\frac{x_2(t)}{x_1(t)} = \frac{c_2}{c_1}. \tag{9}$$

Thus the ratio $x_2(t)/x_1(t)$ is a constant, so the trajectory lies on a line through the origin. The value of the constant in a particular case is determined by the initial conditions.

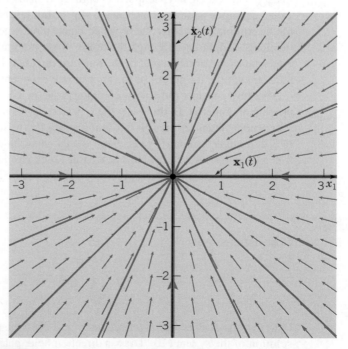

FIGURE 3.5.1 A direction field and phase portrait for the system (2).

Typical component plots are shown in Figure 3.5.2. Each graph is proportional to the graph of e^{-t}, with the proportionality constant determined by the initial condition.

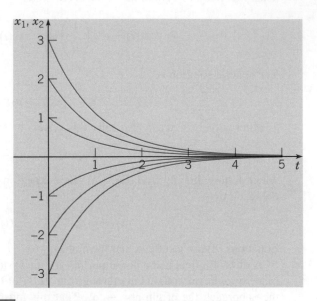

FIGURE 3.5.2 Typical component plots for the system (2).

It is possible to show that the only 2×2 matrices with a repeated eigenvalue and two independent eigenvectors are the diagonal matrices with the eigenvalues along the diagonal. Such matrices form a rather special class, since each of them is proportional to the identity matrix. The system in Example 1 is entirely typical of this class of systems. The origin is called a **proper node** or, sometimes, a **star point**. It is either asymptotically stable or unstable, according to whether the repeated eigenvalue is negative or positive. The trajectories lie along straight lines through the origin. They are traversed in the inward direction if the eigenvalues are negative and in the outward direction if they are positive. The component plots are simply the graphs of the appropriate exponential function.

We now consider the (more common) case in which repeated eigenvalues occur in a nondiagonal matrix. In this case, the repeated eigenvalue is accompanied by only a single independent eigenvector. This has implications for the solution of the corresponding system of differential equations because there is only one solution of our assumed form $\mathbf{x} = e^{\lambda t}\mathbf{v}$. To find a fundamental set of solutions, we must seek an additional solution of a different form. Before discussing the general problem of this type, we look at a relatively simple example.

EXAMPLE
2

Consider the system

$$\mathbf{x}' = \mathbf{A}\mathbf{x} = \begin{pmatrix} -\frac{1}{2} & 1 \\ 0 & -\frac{1}{2} \end{pmatrix} \mathbf{x}. \tag{10}$$

Find the eigenvalues and eigenvectors of the coefficient matrix, and then find the general solution of the system (10). Draw a direction field, phase portrait, and component plots.

Assuming that $\mathbf{x} = e^{\lambda t}\mathbf{v}$, we immediately obtain the algebraic system

$$\begin{pmatrix} -\frac{1}{2} - \lambda & 1 \\ 0 & -\frac{1}{2} - \lambda \end{pmatrix} \begin{pmatrix} v_1 \\ v_2 \end{pmatrix} = \begin{pmatrix} 0 \\ 0 \end{pmatrix}. \tag{11}$$

The characteristic equation is $(\frac{1}{2} + \lambda)^2 = 0$, so the eigenvalues are $\lambda_1 = \lambda_2 = -\frac{1}{2}$. Setting $\lambda = -\frac{1}{2}$ in Eq. (11), we have

$$\begin{pmatrix} 0 & 1 \\ 0 & 0 \end{pmatrix} \begin{pmatrix} v_1 \\ v_2 \end{pmatrix} = \begin{pmatrix} 0 \\ 0 \end{pmatrix}. \tag{12}$$

The second line in this vector equation imposes no restriction on v_1 and v_2, but the first line requires that $v_2 = 0$. So we may take the eigenvector $\mathbf{v}_1$ to be

$$\mathbf{v}_1 = \begin{pmatrix} 1 \\ 0 \end{pmatrix}. \tag{13}$$

Thus one solution of Eq. (10) is

$$\mathbf{x}_1(t) = e^{-t/2} \begin{pmatrix} 1 \\ 0 \end{pmatrix}, \tag{14}$$

but there is no second solution of the assumed form. Where should we look to find a second solution, and hence the general solution, of this system?

Let us consider the scalar equations corresponding to the vector equation (10). They are

$$x_1' = -\frac{1}{2}x_1 + x_2, \qquad x_2' = -\frac{1}{2}x_2. \tag{15}$$

Because the second of these equations does not involve x_1, we can solve this equation for x_2. The equation is linear and separable and has the solution

$$x_2 = c_2 e^{-t/2}, \tag{16}$$

where c_2 is an arbitrary constant. Then, by substituting from Eq. (16) for x_2 in the first of Eqs. (15), we obtain an equation for x_1:

$$x_1' + \frac{1}{2}x_1 = c_2 e^{-t/2}. \tag{17}$$

Equation (17) is a first order linear equation with the integrating factor $\mu(t) = e^{t/2}$. On multiplying Eq. (17) by $\mu(t)$ and integrating, we find that

$$x_1 = c_2 t e^{-t/2} + c_1 e^{-t/2}, \tag{18}$$

where c_1 is another arbitrary constant. Now writing the scalar solutions (16) and (18) in vector form, we obtain

$$\mathbf{x} = \begin{pmatrix} x_1 \\ x_2 \end{pmatrix} = e^{-t/2} \begin{pmatrix} c_1 + c_2 t \\ c_2 \end{pmatrix}$$

$$= c_1 e^{-t/2} \begin{pmatrix} 1 \\ 0 \end{pmatrix} + c_2 e^{-t/2} \left[\begin{pmatrix} 1 \\ 0 \end{pmatrix} t + \begin{pmatrix} 0 \\ 1 \end{pmatrix} \right]. \tag{19}$$

The first term on the right side of Eq. (19) is the solution we found earlier in Eq. (14). However the second term on the right side of Eq. (19) is a new solution that we did not obtain before. This solution has the form

$$\mathbf{x} = te^{-t/2}\mathbf{v} + e^{-t/2}\mathbf{w}. \tag{20}$$

Now that we know the form of the second solution, we can obtain it more directly by assuming an expression for $\mathbf{x}$ of the form (20), with unknown vector coefficients $\mathbf{v}$ and $\mathbf{w}$, and then substituting in Eq. (10). This leads to the equation

$$e^{-t/2}\mathbf{v} - \tfrac{1}{2}e^{-t/2}(\mathbf{v}t + \mathbf{w}) = e^{-t/2}\mathbf{A}(\mathbf{v}t + \mathbf{w}). \tag{21}$$

Equating coefficients of $te^{-t/2}$ and $e^{-t/2}$ on each side of Eq. (21) gives the conditions

$$\left(\mathbf{A} + \tfrac{1}{2}\mathbf{I}\right)\mathbf{v} = \mathbf{0} \tag{22}$$

and

$$\left(\mathbf{A} + \tfrac{1}{2}\mathbf{I}\right)\mathbf{w} = \mathbf{v} \tag{23}$$

for the determination of $\mathbf{v}$ and $\mathbf{w}$. Equation (22) is satisfied if $\mathbf{v}$ is the eigenvector $\mathbf{v}_1$, given by Eq. (13), associated with the eigenvalue $\lambda = -\tfrac{1}{2}$. Since $\det(\mathbf{A} + \tfrac{1}{2}\mathbf{I})$ is zero, we might expect that Eq. (23) has no solution. However, if we write out this equation in full, we have

$$\begin{pmatrix} 0 & 1 \\ 0 & 0 \end{pmatrix}\begin{pmatrix} w_1 \\ w_2 \end{pmatrix} = \begin{pmatrix} 1 \\ 0 \end{pmatrix}. \tag{24}$$

The first row of Eq. (24) requires that $w_2 = 1$, and the second row of Eq. (24) puts no restriction on either w_1 or w_2. Thus we can choose $w_1 = k$, where k is any constant. By substituting the results we have obtained in Eq. (20), we find a second solution of the system (10):

$$\mathbf{x}_2(t) = te^{-t/2}\begin{pmatrix} 1 \\ 0 \end{pmatrix} + e^{-t/2}\begin{pmatrix} 0 \\ 1 \end{pmatrix} + ke^{-t/2}\begin{pmatrix} 1 \\ 0 \end{pmatrix}. \tag{25}$$

Since the third term on the right side of Eq. (25) is proportional to $\mathbf{x}_1(t)$, we need not include it in $\mathbf{x}_2(t)$; in other words, we choose $k = 0$. You can easily verify that the Wronskian of $\mathbf{x}_1(t)$ and $\mathbf{x}_2(t)$ is

$$W[\mathbf{x}_1, \mathbf{x}_2](t) = e^{-t}, \tag{26}$$

which is never zero. Thus the general solution of the system (10) is given by a linear combination of $\mathbf{x}_1(t)$ and $\mathbf{x}_2(t)$, that is, by Eq. (19).

A direction field and phase portrait for the system (10) are shown in Figure 3.5.3. In Figure 3.5.3, the trajectories of $\mathbf{x}_1(t)$ and $\mathbf{x}_2(t)$ are indicated by the heavy curves. As $t \to \infty$, every trajectory approaches the origin tangent to the x_1-axis. This is because the dominant term in the solution (19) for large t is the $te^{-t/2}$ term in $\mathbf{x}_2(t)$. If the initial conditions are such that $c_2 = 0$, then the solution is proportional to $\mathbf{x}_1(t)$, which lies along the x_1-axis. Figure 3.5.4 shows some representative plots of x_1 versus t. The component $x_2(t)$ is a purely exponential function (without any t factor), so its graphs resemble those in Figure 3.5.2.

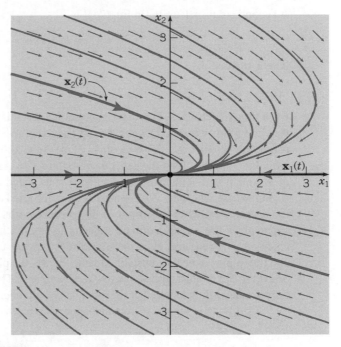

FIGURE 3.5.3 A direction field and phase portrait for the system (10).

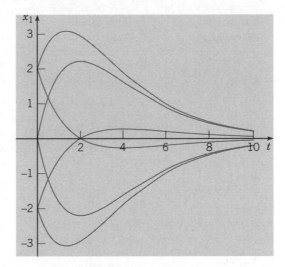

FIGURE 3.5.4 Typical plots of x_1 versus t for the system (10).

We can proceed in exactly the same way in the general case of Eq. (1). Suppose that $\lambda = \lambda_1$ is a repeated eigenvalue of the matrix $\mathbf{A}$ and that there is only one independent eigenvector $\mathbf{v}_1$. Then one solution [similar to Eq. (14)] is

$$\mathbf{x}_1(t) = e^{\lambda_1 t}\mathbf{v}_1, \tag{27}$$

where $\mathbf{v}_1$ satisfies

$$(\mathbf{A} - \lambda_1 \mathbf{I})\mathbf{v} = \mathbf{0}. \tag{28}$$

A second solution [similar to Eq. (25)] is

$$\mathbf{x}_2(t) = te^{\lambda_1 t}\mathbf{v}_1 + e^{\lambda_1 t}\mathbf{w}, \tag{29}$$

where $\mathbf{w}$ satisfies

$$(\mathbf{A} - \lambda_1\mathbf{I})\mathbf{w} = \mathbf{v}_1. \tag{30}$$

Even though $\det(\mathbf{A} - \lambda_1\mathbf{I}) = 0$, it can be shown that it is always possible to solve Eq. (30) for $\mathbf{w}$. The vector $\mathbf{w}$ is called a **generalized eigenvector** corresponding to the eigenvalue λ_1. It is possible to show that the Wronskian of $\mathbf{x}_1(t)$ and $\mathbf{x}_2(t)$ is not zero, so these solutions form a fundamental set. Thus the general solution of $\mathbf{x}' = \mathbf{A}\mathbf{x}$ in this case is a linear combination of $\mathbf{x}_1(t)$ and $\mathbf{x}_2(t)$ with arbitrary coefficients, as stated in Theorem 3.3.4.

In the case where the 2×2 matrix $\mathbf{A}$ has a repeated eigenvalue and only one eigenvector, the origin is called an **improper** or **degenerate node**. It is asymptotically stable when the eigenvalues are negative and all trajectories approach the origin as $t \to \infty$. If a positive eigenvalue is repeated, then the trajectories (except for $\mathbf{x} = \mathbf{0}$ itself) become unbounded and the improper node is unstable. In either case, the phase portrait resembles Figure 3.5.3, or Figure 3.5.5 which appears below in Example 3. The trajectories are directed inward or outward according to whether the eigenvalues are negative or positive.

We summarize the phase portraits appearing in this section in Table 3.5.1.

TABLE 3.5.1 Phase portraits for $\mathbf{x}' = \mathbf{A}\mathbf{x}$ when $\mathbf{A}$ has a single repeated eigenvalue.

Nature of $\mathbf{A}$ and Eigenvalues	Sample Phase Portrait	Type of Critical Point	Stability
$\mathbf{A} = \begin{pmatrix} \lambda & 0 \\ 0 & \lambda \end{pmatrix}$ $\lambda > 0$		$(0, 0)$ is an **unstable proper node**. *Note*: $(0, 0)$ is also called an **unstable star node**.	Unstable
$\mathbf{A} = \begin{pmatrix} \lambda & 0 \\ 0 & \lambda \end{pmatrix}$ $\lambda < 0$		$(0, 0)$ is a **stable proper node**. *Note*: $(0, 0)$ is also called a **stable star node**.	Asymptotically stable
$\mathbf{A}$ is not diagonal. $\lambda > 0$		$(0, 0)$ is an **unstable improper node**. *Note*: $(0, 0)$ is also called an **unstable degenerate node**.	Unstable
$\mathbf{A}$ is not diagonal. $\lambda < 0$		$(0, 0)$ is a **stable improper node**. *Note*: $(0, 0)$ is also called a **stable degenerate node**.	Asymptotically stable

EXAMPLE
3

Find a fundamental set of solutions of

$$x' = Ax = \begin{pmatrix} 1 & -1 \\ 1 & 3 \end{pmatrix} x \tag{31}$$

and draw a phase portrait for this system.

To solve Eq. (31), we start by finding the eigenvalues and eigenvectors of the coefficient matrix A. They satisfy the algebraic equation $(A - \lambda I) v = 0$, or

$$\begin{pmatrix} 1 - \lambda & -1 \\ 1 & 3 - \lambda \end{pmatrix} \begin{pmatrix} v_1 \\ v_2 \end{pmatrix} = \begin{pmatrix} 0 \\ 0 \end{pmatrix}. \tag{32}$$

The eigenvalues are the roots of the characteristic equation

$$\det(A - \lambda I) = \begin{vmatrix} 1 - \lambda & -1 \\ 1 & 3 - \lambda \end{vmatrix} = \lambda^2 - 4\lambda + 4 = 0. \tag{33}$$

Thus the two eigenvalues are $\lambda_1 = \lambda_2 = 2$.

To determine the eigenvectors, we must return to Eq. (32) and use the value 2 for λ. This gives

$$\begin{pmatrix} -1 & -1 \\ 1 & 1 \end{pmatrix} \begin{pmatrix} v_1 \\ v_2 \end{pmatrix} = \begin{pmatrix} 0 \\ 0 \end{pmatrix}. \tag{34}$$

Hence we obtain the single condition $v_1 + v_2 = 0$, which determines v_2 in terms of v_1, or vice versa. Thus the eigenvector corresponding to the eigenvalue $\lambda = 2$ is

$$v_1 = \begin{pmatrix} 1 \\ -1 \end{pmatrix}, \tag{35}$$

or any nonzero multiple of this vector. Consequently, one solution of the system (31) is

$$x_1(t) = e^{2t} \begin{pmatrix} 1 \\ -1 \end{pmatrix}, \tag{36}$$

but there is no second solution of the form $x = e^{\lambda t} v$.

To find a second solution, we need to assume that

$$x = t e^{2t} v + e^{2t} w, \tag{37}$$

where v and w are constant vectors. Upon substituting this expression for x in Eq. (31), we obtain

$$2 t e^{2t} v + e^{2t}(v + 2w) = A(t e^{2t} v + e^{2t} w). \tag{38}$$

Equating coefficients of $t e^{2t}$ and e^{2t} on each side of Eq. (38) gives the conditions

$$(A - 2I)v = 0 \tag{39}$$

and

$$(A - 2I)w = v \tag{40}$$

for the determination of v and w. Equation (39) is satisfied if v is an eigenvector of A corresponding to the eigenvalue $\lambda = 2$, that is, $v^T = (1, -1)$. Then Eq. (40) becomes

$$\begin{pmatrix} -1 & -1 \\ 1 & 1 \end{pmatrix} \begin{pmatrix} w_1 \\ w_2 \end{pmatrix} = \begin{pmatrix} 1 \\ -1 \end{pmatrix}. \tag{41}$$

Thus we have

$$-w_1 - w_2 = 1,$$

so if $w_1 = k$, where k is arbitrary, then $w_2 = -k - 1$. If we write

$$\mathbf{w} = \begin{pmatrix} k \\ -1 - k \end{pmatrix} = \begin{pmatrix} 0 \\ -1 \end{pmatrix} + k \begin{pmatrix} 1 \\ -1 \end{pmatrix}, \tag{42}$$

then by substituting for $\mathbf{v}$ and $\mathbf{w}$ in Eq. (37), we obtain

$$\mathbf{x} = t e^{2t} \begin{pmatrix} 1 \\ -1 \end{pmatrix} + e^{2t} \begin{pmatrix} 0 \\ -1 \end{pmatrix} + k e^{2t} \begin{pmatrix} 1 \\ -1 \end{pmatrix}. \tag{43}$$

The last term in Eq. (43) is merely a multiple of the first solution $\mathbf{x}_1(t)$ and may be ignored, but the first two terms constitute a new solution:

$$\mathbf{x}_2(t) = t e^{2t} \begin{pmatrix} 1 \\ -1 \end{pmatrix} + e^{2t} \begin{pmatrix} 0 \\ -1 \end{pmatrix}. \tag{44}$$

An elementary calculation shows that $W[\mathbf{x}_1, \mathbf{x}_2](t) = -e^{4t}$, and therefore $\mathbf{x}_1$ and $\mathbf{x}_2$ form a fundamental set of solutions of the system (31). The general solution is

$$\mathbf{x} = c_1 \mathbf{x}_1(t) + c_2 \mathbf{x}_2(t)$$

$$= c_1 e^{2t} \begin{pmatrix} 1 \\ -1 \end{pmatrix} + c_2 \left[t e^{2t} \begin{pmatrix} 1 \\ -1 \end{pmatrix} + e^{2t} \begin{pmatrix} 0 \\ -1 \end{pmatrix} \right]. \tag{45}$$

A direction field and phase portrait for the system (31) are shown in Figure 3.5.5. The trajectories of $\mathbf{x}_1(t)$ and $\mathbf{x}_2(t)$ are shown by the heavy curves. It is clear that all solutions (except the equilibrium solution $\mathbf{x} = \mathbf{0}$) become unbounded at $t \to \infty$. Thus the improper node at the origin is unstable. It is possible to show that as $t \to -\infty$, all solutions approach

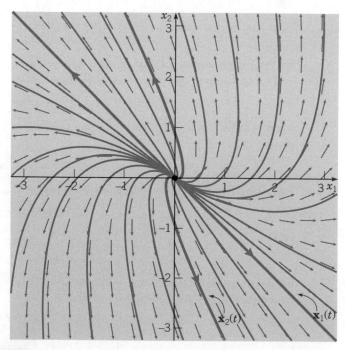

FIGURE 3.5.5   A direction field and phase portrait for the system (31).

the origin tangent to the line $x_2 = -x_1$ determined by the eigenvector. Further, as t approaches infinity, the slope of each trajectory approaches the limit -1. Some typical plots of x_1 versus t are shown in Figure 3.5.6.

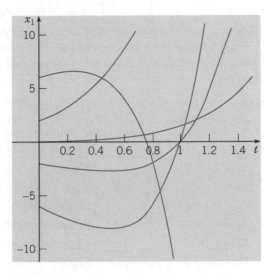

FIGURE 3.5.6   Some plots of x_1 versus t for the system (31).

▶ **Summary of Results.** This completes our investigation of the possible behavior of solutions of a two-dimensional linear homogeneous system with constant coefficients. When the coefficient matrix **A** has a nonzero determinant, there is a single equilibrium solution, or critical point, which is located at the origin. By reflecting on the possibilities explored in this section and in the two preceding ones, and by examining the corresponding figures, we can make several observations:

1. After a long time, each individual trajectory exhibits one of only three types of behavior. As $t \to \infty$, each trajectory becomes unbounded, approaches the critical point **x** = **0**, or repeatedly traverses a closed curve, corresponding to a periodic solution, that surrounds the critical point.

2. Viewed as a whole, the pattern of trajectories in each case is relatively simple. To be more specific, through each point (x_0, y_0) in the phase plane there is only one trajectory; thus the trajectories do not cross each other. Do not be misled by the figures, in which it sometimes appears that many trajectories pass through the critical point **x** = **0**. In fact, the only solution passing through the origin is the equilibrium solution **x** = **0**. The other solutions that appear to pass through the origin actually only approach this point as $t \to \infty$ or $t \to -\infty$.

3. In each case, the set of all trajectories is such that one of three situations occurs.

   a. All trajectories approach the critical point **x** = **0** as $t \to \infty$. This is the case if the eigenvalues are real and negative or complex with a negative real part. The origin is either a nodal or a spiral sink.

   b. All trajectories remain bounded but do not approach the origin as $t \to \infty$. This is the case if the eigenvalues are purely imaginary. The origin is a center.

   c. Some trajectories, and possibly all trajectories except **x** = **0**, become unbounded as $t \to \infty$. This is the case if at least one of the eigenvalues is positive or if the eigenvalues have a positive real part. The origin is a nodal source, a spiral source, or a saddle point.

The situations described in 3(a), (b), and (c) above illustrate the concepts of asymptotic stability, stability, and instability, respectively, of the equilibrium solution $\mathbf{x} = \mathbf{0}$ of the system (1),

$$\mathbf{x}' = \mathbf{A}\mathbf{x}.$$

The precise definitions of these terms are given in Section 7.1, but their basic meaning should be clear from the geometrical discussion in this section. The information that we have obtained about the system (1) is summarized in Table 3.5.2. Also see Problems 15 and 16.

TABLE 3.5.2	Stability properties of linear systems $\mathbf{x}' = \mathbf{A}\mathbf{x}$ with $\det(\mathbf{A} - \lambda\mathbf{I}) = 0$ and $\det \mathbf{A} \neq 0$.

Eigenvalues	Type of Critical Point	Stability
$\lambda_1 > \lambda_2 > 0$	Node	Unstable
$\lambda_1 < \lambda_2 < 0$	Node	Asymptotically stable
$\lambda_2 < 0 < \lambda_1$	Saddle point	Unstable
$\lambda_1 = \lambda_2 > 0$	Proper or improper node	Unstable
$\lambda_1 = \lambda_2 < 0$	Proper or improper node	Asymptotically stable
$\lambda_1, \lambda_2 = \mu \pm iv$	Spiral point	
$\mu > 0$		Unstable
$\mu < 0$		Asymptotically stable
$\lambda_1 = iv, \lambda_2 = -iv$	Center	Stable

PROBLEMS

General Solutions and Phase Portraits. In each of Problems 1 through 6, find the general solution of the given system of equations. Also draw a direction field and a phase portrait. Describe how the solutions behave as $t \to \infty$.

1. $\mathbf{x}' = \begin{pmatrix} 3 & -4 \\ 1 & -1 \end{pmatrix}\mathbf{x}$

2. $\mathbf{x}' = \begin{pmatrix} \frac{5}{4} & \frac{3}{4} \\ -\frac{3}{4} & -\frac{1}{4} \end{pmatrix}\mathbf{x}$

3. $\mathbf{x}' = \begin{pmatrix} -\frac{3}{2} & 1 \\ -\frac{1}{4} & -\frac{1}{2} \end{pmatrix}\mathbf{x}$

4. $\mathbf{x}' = \begin{pmatrix} -3 & \frac{5}{2} \\ -\frac{5}{2} & 2 \end{pmatrix}\mathbf{x}$

5. $\mathbf{x}' = \begin{pmatrix} -1 & -\frac{1}{2} \\ 2 & -3 \end{pmatrix}\mathbf{x}$

6. $\mathbf{x}' = \begin{pmatrix} 2 & \frac{1}{2} \\ -\frac{1}{2} & 1 \end{pmatrix}\mathbf{x}$

In each of Problems 7 through 12, find the solution of the given initial value problem. Draw the trajectory of the solution in the x_1x_2-plane and also draw the component plots of x_1 versus t and of x_2 versus t.

7. $\mathbf{x}' = \begin{pmatrix} 1 & -4 \\ 4 & -7 \end{pmatrix}\mathbf{x}, \qquad \mathbf{x}(0) = \begin{pmatrix} 3 \\ 2 \end{pmatrix}$

8. $\mathbf{x}' = \begin{pmatrix} -\frac{5}{2} & \frac{3}{2} \\ -\frac{3}{2} & \frac{1}{2} \end{pmatrix}\mathbf{x}, \qquad \mathbf{x}(0) = \begin{pmatrix} 3 \\ -1 \end{pmatrix}$

9. $\mathbf{x}' = \begin{pmatrix} 2 & \frac{3}{2} \\ -\frac{3}{2} & -1 \end{pmatrix}\mathbf{x}, \qquad \mathbf{x}(0) = \begin{pmatrix} 3 \\ -2 \end{pmatrix}$

10. $\mathbf{x}' = \begin{pmatrix} \frac{5}{4} & \frac{3}{4} \\ -\frac{3}{4} & -\frac{1}{4} \end{pmatrix}\mathbf{x}, \qquad \mathbf{x}(0) = \begin{pmatrix} 2 \\ 3 \end{pmatrix};$

see Problem 2.

11. $\mathbf{x}' = \begin{pmatrix} -3 & \frac{5}{2} \\ -\frac{5}{2} & 2 \end{pmatrix}\mathbf{x}, \qquad \mathbf{x}(0) = \begin{pmatrix} 3 \\ 1 \end{pmatrix};$

see Problem 4.

12. $\mathbf{x}' = \begin{pmatrix} 2 & \frac{1}{2} \\ -\frac{1}{2} & 1 \end{pmatrix}\mathbf{x}, \qquad \mathbf{x}(0) = \begin{pmatrix} 1 \\ 3 \end{pmatrix};$

see Problem 6.

13. Consider again the electric circuit in Problem 22 of Section 3.4. This circuit is described by the system of differential equations

$$\frac{d}{dt}\begin{pmatrix} i \\ v \end{pmatrix} = \begin{pmatrix} 0 & 1/L \\ -1/C & -1/RC \end{pmatrix}\begin{pmatrix} i \\ v \end{pmatrix}.$$

(a) Show that the eigenvalues are real and equal if $L = 4R^2C$.
(b) Suppose that $R = 1$ ohm, $C = 1$ farad, and $L = 4$ henries. Suppose also that $i(0) = 1$ ampere and $v(0) = 2$ volts. Find $i(t)$ and $v(t)$.

14. Trace Determinant Plane. Show that all solutions of the system

$$\mathbf{x}' = \begin{pmatrix} a & b \\ c & d \end{pmatrix}\mathbf{x}$$

approach zero as $t \to \infty$ if and only if $a + d < 0$ and $ad - bc > 0$.

15. Consider the linear system

$$\frac{dx}{dt} = a_{11}x + a_{12}y, \qquad \frac{dy}{dt} = a_{21}x + a_{22}y,$$

where a_{11}, a_{12}, a_{21}, and a_{22} are real constants. Let $p = a_{11} + a_{22}$, $q = a_{11}a_{22} - a_{12}a_{21}$, and $\Delta = p^2 - 4q$. Observe that p and q are the trace and determinant, respectively, of the coefficient matrix of the given system. Show that the critical point $(0, 0)$ is a

(a) Node if $q > 0$ and $\Delta \geq 0$;
(b) Saddle point if $q < 0$;
(c) Spiral point if $p \neq 0$ and $\Delta < 0$;
(d) Center if $p = 0$ and $q > 0$.

Hint: These conclusions can be obtained by studying the eigenvalues λ_1 and λ_2. It may also be helpful to establish, and then to use, the relations $\lambda_1\lambda_2 = q$ and $\lambda_1 + \lambda_2 = p$.

16. Continuing Problem 15, show that the critical point $(0, 0)$ is

(a) Asymptotically stable if $q > 0$ and $p < 0$;
(b) Stable if $q > 0$ and $p = 0$;
(c) Unstable if $q < 0$ or $p > 0$.

The results of Problems 15 and 16 are summarized visually in Figure 3.5.7.

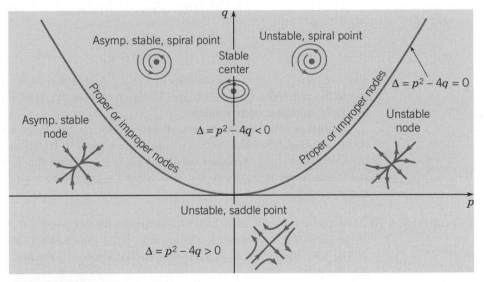

FIGURE 3.5.7 Stability diagram.

3.6 A Brief Introduction to Nonlinear Systems

In Section 3.2 we introduced the general two-dimensional first order linear system

$$\begin{pmatrix} dx/dt \\ dy/dt \end{pmatrix} = \begin{pmatrix} p_{11}(t)x + p_{12}(t)y + g_1(t) \\ p_{21}(t)x + p_{22}(t)y + g_2(t) \end{pmatrix} \tag{1}$$

or

$$\frac{d\mathbf{x}}{dt} = \mathbf{P}(t)\mathbf{x} + \mathbf{g}(t),\tag{2}$$

where

$$\mathbf{x} = \begin{pmatrix} x \\ y \end{pmatrix}, \qquad \mathbf{P}(t) = \begin{pmatrix} p_{11}(t) & p_{12}(t) \\ p_{21}(t) & p_{22}(t) \end{pmatrix}, \qquad \text{and} \qquad \mathbf{g}(t) = \begin{pmatrix} g_1(t) \\ g_2(t) \end{pmatrix}.$$

Of course, two-dimensional systems that are not of the form (1) or (2) may also occur. Such systems are said to be **nonlinear**.

A general two-dimensional first order system has the form

$$\begin{pmatrix} dx/dt \\ dy/dt \end{pmatrix} = \begin{pmatrix} f(t, x, y) \\ g(t, x, y) \end{pmatrix},\tag{3}$$

where f and g are essentially arbitrary continuous functions. Using vector notation, we have

$$\frac{d\mathbf{x}}{dt} = \mathbf{f}(t, \mathbf{x}),\tag{4}$$

where $\mathbf{x} = x\mathbf{i} + y\mathbf{j}$ and $\mathbf{f}(t, \mathbf{x}) = f(t, x, y)\mathbf{i} + g(t, x, y)\mathbf{j}$. Frequently, there will also be given initial conditions

$$x(t_0) = x_0, \qquad y(t_0) = y_0,\tag{5}$$

or

$$\mathbf{x}(t_0) = \mathbf{x}_0,\tag{6}$$

where $\mathbf{x}_0 = x_0\mathbf{i} + y_0\mathbf{j}$. Again we refer to x and y as state variables, to $\mathbf{x}$ as the state vector, and to the xy-plane as the phase plane. Equations (3) and (5), or in vector form Eqs. (4) and (6), form an **initial value problem**.

A **solution** of the system (3) consists of two differentiable functions $x = \phi(t)$ and $y = \psi(t)$ that satisfy Eqs. (3) for all values of t in some interval I. If $\phi(t)$ and $\psi(t)$ also satisfy the initial conditions (5), then they are a solution of the initial value problem (3), (5). Similar statements apply to the vector $\mathbf{x} = \boldsymbol{\phi}(t) = \phi(t)\mathbf{i} + \psi(t)\mathbf{j}$ with respect to the system (4) and initial conditions (6).

▶ Existence and Uniqueness of Solutions. To ensure the existence and uniqueness of solutions of the initial value problem (3), (5), we must place some restrictions on the functions f and g. The following theorem is analogous to Theorem 2.4.2 for first order nonlinear scalar equations.

THEOREM 3.6.1

Let each of the functions f and g and the partial derivatives $\partial f/\partial x$, $\partial f/\partial y$, $\partial g/\partial x$, and $\partial g/\partial y$ be continuous in a region R of txy-space defined by $\alpha < t < \beta$, $\alpha_1 < x < \beta_1$, $\alpha_2 < y < \beta_2$, and let the point (t_0, x_0, y_0) be in R. Then there is an interval $|t - t_0| < h$ in which there exists a unique solution of the system of differential equations (3)

$$\begin{pmatrix} dx/dt \\ dy/dt \end{pmatrix} = \begin{pmatrix} f(t, x, y) \\ g(t, x, y) \end{pmatrix}$$

that also satisfies the initial conditions (5)

$$x(t_0) = x_0, \qquad y(t_0) = y_0.$$

Note that in the hypotheses of Theorem 3.6.1, nothing is said about the partial derivatives of f and g with respect to the independent variable t. Note also that, in the conclusion, the length $2h$ of the interval in which the solution exists is not specified exactly, and in some cases it may be very short. As for first order nonlinear scalar equations, the interval of existence of solutions here may bear no obvious relation to the functions f and g, and often depends also on the initial conditions.

▶ Autonomous Systems. It is usually impossible to solve nonlinear systems exactly by analytical methods. Therefore for such systems graphical methods and numerical approximations become even more important. In Chapter 8 we discuss approximate numerical methods for such two-dimensional systems. Here we will consider systems for which direction fields and phase portraits are of particular importance. These are systems that do not depend explicitly on the independent variable t. In other words, the functions f and g in Eqs. (3) depend only on x and y and not on t. Or, in Eq. (4), the vector $\mathbf{f}$ depends only on $\mathbf{x}$ and not t. Such a system is called **autonomous**, and can be written in the form

$$\frac{dx}{dt} = f(x, y), \qquad \frac{dy}{dt} = g(x, y). \tag{7}$$

In vector notation we have

$$\frac{d\mathbf{x}}{dt} = \mathbf{f}(\mathbf{x}). \tag{8}$$

In earlier sections of this chapter, we have examined linear autonomous systems. We now want to extend the discussion to nonlinear autonomous systems.

We have seen that equilibrium, or constant, solutions are of particular importance in the study of single first order autonomous equations and of two-dimensional linear autonomous systems. We will see that the same is true for nonlinear autonomous systems of equations. To find equilibrium, or constant, solutions of the system (7), we set dx/dt and dy/dt equal to zero, and solve the resulting equations

$$f(x, y) = 0, \qquad g(x, y) = 0 \tag{9}$$

for x and y. Any solution of Eqs. (9) is a point in the phase plane that is a trajectory of an equilibrium solution. Such points are called **equilibrium points** or **critical points**. Depending on the particular forms of f and g, the nonlinear system (7) can have any number of critical points, ranging from none to infinitely many.

Sometimes the trajectories of a two-dimensional autonomous system can be found by solving a related first order differential equation. From Eqs. (7) we have

$$\frac{dy}{dx} = \frac{dy/dt}{dx/dt} = \frac{g(x, y)}{f(x, y)}, \tag{10}$$

which is a first order equation in the variables x and y. If Eq. (10) can be solved by any of the methods in Chapters 1 and 2, and if we write solutions (implicitly) as

$$H(x, y) = c, \tag{11}$$

then Eq. (11) is an equation for the trajectories of the system (7). In other words, the trajectories lie on the level curves of $H(x, y)$. Recall that there is no general way of solving Eq. (10) to obtain the function H, so this approach is applicable only in special cases.

Now let us look at some examples.

EXAMPLE
1

Consider the system

$$\frac{dx}{dt} = x - y, \qquad \frac{dy}{dt} = 2x - y - x^2. \tag{12}$$

Find a function $H(x, y)$ such that the trajectories of the system (12) lie on the level curves of H. Find the critical points and draw a phase portrait for the given system. Describe the behavior of its trajectories.

To find the critical points, we must solve the equations

$$x - y = 0, \qquad 2x - y - x^2 = 0. \tag{13}$$

From the first equation we have $y = x$; then the second equation yields $x - x^2 = 0$. Thus $x = 0$ or $x = 1$, and it follows that the critical points are $(0, 0)$ and $(1, 1)$. To determine the trajectories, note that for this system, Eq. (10) becomes

$$\frac{dy}{dx} = \frac{2x - y - x^2}{x - y}. \tag{14}$$

Equation (14) is an exact equation, as discussed in Section 2.6. Proceeding as described in that section, we find that solutions satisfy

$$H(x, y) = x^2 - xy + \tfrac{1}{2}y^2 - \tfrac{1}{3}x^3 = c, \tag{15}$$

where c is an arbitrary constant. To construct a phase portrait, you can either draw some of the level curves of $H(x, y)$, or you can plot some solutions of the system (12). In either case, you need some computer assistance to produce a plot such as that in Figure 3.6.1. The

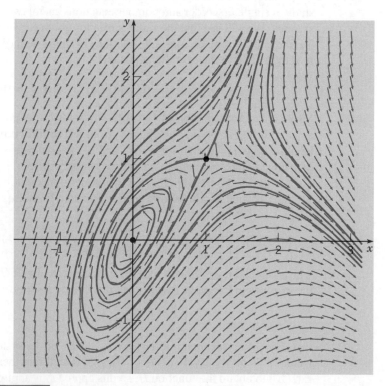

FIGURE 3.6.1   A phase portrait for the system (12).

direction of motion on the trajectories can be determined by drawing a direction field, or by evaluating dx/dt and dy/dt at one or two selected points. From Figure 3.6.1, it is clear that trajectories behave quite differently near the two critical points. Observe that there is one trajectory that departs from $(1, 1)$ as t increases from $-\infty$, loops around the other critical point (the origin), and returns to $(1, 1)$ as $t \to \infty$. Inside this loop there are trajectories that lie on closed curves surrounding $(0, 0)$. These trajectories correspond to periodic solutions that repeatedly pass through the same points in the phase plane. Trajectories that lie outside the loop ultimately appear to leave the plot window in a southeasterly direction (as $t \to \infty$) or in a northeasterly direction (as $t \to -\infty$).

It is also possible to construct component plots of particular solutions, and two are shown in Figures 3.6.2 and 3.6.3. Figure 3.6.2 provides a plot of x versus t for the solution that satisfies the initial conditions

$$x(0) = \tfrac{1}{2}, \qquad y(0) = \tfrac{1}{2}.$$

This graph confirms that the motion is periodic and enables you to estimate the period, which was not possible from the phase portrait. A plot of y versus t for this solution is similar.

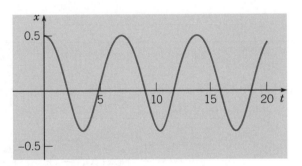

FIGURE 3.6.2 A plot of x versus t for the solution of the system (12) that passes through the point $(\tfrac{1}{2}, \tfrac{1}{2})$ at $t = 0$.

A plot of a typical unbounded solution appears in Figure 3.6.3, which shows the graph of x versus t for the solution that satisfies the initial conditions

$$x(0) = 3, \qquad y(0) = 0.$$

It appears from this figure that there may be a vertical asymptote for a value of t between 1.5 and 1.6, although you should be cautious about drawing such a conclusion from a single small plot. In Problem 22 we outline how you can show conclusively that there is an asymptote in this particular case. The existence of a vertical asymptote means that x becomes unbounded in a finite time, rather than as $t \to \infty$. A plot of y versus t for this solution is similar, except that y becomes unbounded in the negative direction. The relative magnitudes of x and y when they are both large can be determined by keeping only the most dominant terms in Eq. (15). In this way, we find that

$$y^2 \cong \tfrac{2}{3}x^3,$$

when both $|x|$ and $|y|$ are very large.

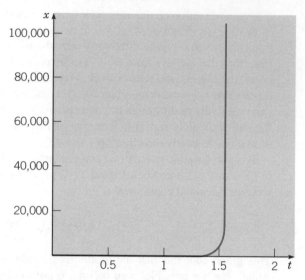

FIGURE 3.6.3 A plot of x versus t for the solution of the system (12) that passes through the point $(3, 0)$ at $t = 0$.

EXAMPLE
2

Consider the system

$$\frac{dx}{dt} = (2 + x)(y - x), \qquad \frac{dy}{dt} = (2 - x)(y + x). \tag{16}$$

Find the critical points. Draw a phase portrait and describe the behavior of the trajectories in the neighborhood of each critical point.

To find the critical points, we must solve the equations

$$(2 + x)(y - x) = 0, \qquad (2 - x)(y + x) = 0. \tag{17}$$

One way to satisfy the first equation is by choosing $x = -2$; then to satisfy the second equation, we must choose $y = 2$. Similarly, the second equation can be satisfied by choosing $x = 2$; then the first equation requires that $y = 2$. If $x \neq 2$ and $x \neq -2$, then Eqs. (17) can only be satisfied if $y - x = 0$ and $y + x = 0$. The only solution of this pair of equations is the origin. Thus the system (16) has three critical points: $(-2, 2)$, $(2, 2)$, and $(0, 0)$.

A phase portrait for Eqs. (16) is shown in Figure 3.6.4. The critical point at $(2, 2)$ attracts other trajectories in the upper-right-hand part of the phase plane. These trajectories spiral around the critical point as they approach it. All trajectories near the point $(-2, 2)$, except the critical point itself, depart from this neighborhood. Some approach the point $(2, 2)$, as we have seen, while others appear to go infinitely far away. Near the origin, there are two trajectories that approach the origin in the second and fourth quadrants. Trajectories that lie above these two trajectories approach $(2, 2)$, while those below them seem to go infinitely far away in the third quadrant.

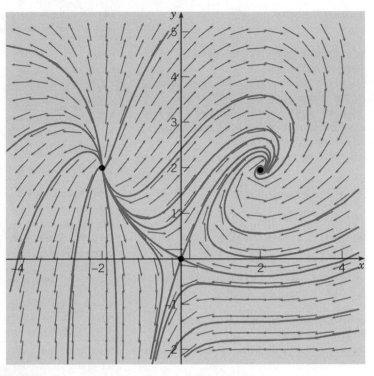

FIGURE 3.6.4   A phase portrait for the system (16).

If you look more closely at Figures 3.6.1 and 3.6.4, you will notice that, in the neighborhood of each critical point, the pattern of trajectories resembles one of the patterns found in Sections 3.3 through 3.5 for linear systems with constant coefficients. In particular, in Figure 3.6.1 it appears that the point $(1, 1)$ is a saddle point and the point $(0, 0)$ is a center. Similarly, in Figure 3.6.4 it appears that $(2, 2)$ is a spiral point, $(-2, 2)$ is a node, and $(0, 0)$ is a saddle point. This apparent relation between nonlinear autonomous systems and linear systems with constant coefficients is not accidental and we will explore it more fully in Chapter 7. In the meantime, many of the problems following this section give you a chance to draw phase portraits similar to Figures 3.6.1 and 3.6.4 and to draw conclusions about the critical points from them.

PROBLEMS

For each of the systems in Problems 1 through 6:
(a) Find an equation of the form $H(x, y) = c$ satisfied by the solutions of the given system.
(b) Without using a computer, sketch some level curves of the function $H(x, y)$.
(c) For $t > 0$, sketch the trajectory corresponding to the given initial condition and indicate the direction of motion for increasing t.

1. $dx/dt = -x, \quad dy/dt = -2y; \qquad x(0) = 4,$
$y(0) = 2$

2. $dx/dt = -x, \quad dy/dt = 2y; \qquad x(0) = 4,$
$y(0) = 2$

3. $dx/dt = -x, \quad dy/dt = 2y; \qquad x(0) = 4,$
$y(0) = 0$

4. $dx/dt = 2y, \quad dy/dt = 8x; \qquad x(0) = 2,$
$y(0) = -1$

5. $dx/dt = 2y, \quad dy/dt = 8x; \qquad x(0) = 1,$
$y(0) = -3$

6. $dx/dt = 2y$, $dy/dt = -8x$; $x(0) = 1$,
$y(0) = 2$

For each of the systems in Problems 7 through 12:
(a) Find all of the critical points.
(b) Find an equation of the form $H(x, y) = c$ satisfied by solutions of the given system.
(c) Using a computer, plot several level curves of the function H. These are trajectories of the given system. Indicate the direction of motion on each trajectory.
(d) Describe the behavior of the trajectories near each critical point.

7. $dx/dt = 2x - y$, $\quad dy/dt = x - 2y$

8. $dx/dt = -x + y$, $\quad dy/dt = x + y$

9. $dx/dt = 2x - 4y$, $\quad dy/dt = 2x - 2y$

10. $dx/dt = -x + y + x^2$, $\quad dy/dt = y - 2xy$

11. $dx/dt = 2x^2y - 3x^2 - 4y$, $\quad dy/dt = -2xy^2 + 6xy$

12. $dx/dt = 3x - x^2$, $\quad dy/dt = 2xy - 3y + 2$

For each of the systems in Problems 13 through 20:
(a) Find all the critical points.
(b) Use a computer to draw a direction field and phase portrait for the system.
(c) From the plots in part (b), describe how the trajectories behave in the vicinity of each critical point.

13. $dx/dt = x - xy$, $\quad dy/dt = y + 2xy$

14. $dx/dt = 2 - y$, $\quad dy/dt = y - x^2$

15. $dx/dt = x - x^2 - xy$, $\quad dy/dt = \frac{1}{2}y - \frac{1}{4}y^2 - \frac{3}{4}xy$

16. $dx/dt = -(x - y)(1 - x - y)$,
$dy/dt = x(2 + y)$

17. $dx/dt = y(2 - x - y)$,
$dy/dt = -x - y - 2xy$

18. $dx/dt = (2 + x)(y - x)$, $\quad dy/dt = y(2 + x - x^2)$

19. $dx/dt = -x + 2xy$, $\quad dy/dt = y - x^2 - y^2$

20. $dx/dt = y$, $\quad dy/dt = x - \frac{1}{6}x^3 - \frac{1}{5}y$

21. (a) Consider the system in Example 1. Draw a component plot of x versus t for several of the periodic solutions in the vicinity of the origin.
(b) From the plots in part (a), estimate the period and amplitude of each of the solutions. Is the period the same regardless of the amplitude? If not, how is the period related to the amplitude?

22. In this problem, we indicate how to find the asymptote suggested by Figure 3.6.3.

(a) Show that if x and y satisfy the initial conditions $x(0) = 3$ and $y(0) = 0$, then the constant c in Eq. (15) is zero. Then show that Eq. (15) can be rewritten in the form

$$(x - y)^2 + x^2 - \frac{2}{3}x^3 = 0. \tag{i}$$

(b) From the first of Eqs. (12), recall that $x - y = x'$. Use this fact with Eq. (i) to obtain the differential equation

$$(x')^2 = \frac{2}{3}x^3 - x^2, \tag{ii}$$

or

$$x' = \frac{1}{3}x\sqrt{6x - 9}. \tag{iii}$$

Why must the positive square root be chosen?
(c) Show that the solution of Eq. (iii) that satisfies the initial condition $x(0) = 3$ is

$$x = \frac{3}{2} + \frac{3}{2}\tan^2\left(\frac{1}{2}t + \frac{1}{4}\pi\right). \tag{iv}$$

Hint: In solving Eq. (iii), you may find it helpful to use the substitution $s^2 = 6x - 9$.
(d) Use the result of part (c) to show that the solution has a vertical asymptote at $t = \pi/2$. Compare this result with the graph in Figure 3.6.3.
(e) From Eq. (iv), there is another vertical asymptote at $t = -3\pi/2$. What is the significance of this asymptote?

23. A direction field and phase portrait for a certain two-dimensional dynamical system $\mathbf{x}' = \mathbf{f}(\mathbf{x})$, with critical points at $A(-1, 0)$, $B(0, 0)$, and $C(1, 0)$, are shown in Figure 3.6.5.

(a) Classify each critical point according to type (node, saddle point, spiral point, or center) and stability (unstable, stable, or asymptotically stable).
(b) Denote the solution of the initial value problem $\mathbf{x}' = \mathbf{f}(\mathbf{x})$, $\mathbf{x}(0) = \mathbf{x}_0$ by $\mathbf{x} = \boldsymbol{\phi}(t)$. For each of the following sets of initial conditions, determine $\lim_{t \to \infty} \boldsymbol{\phi}(t)$. (i) $\mathbf{x}_0 = (0, 0.1)^T$, (ii) $\mathbf{x}_0 = (0, -0.1)^T$, (iii) $\mathbf{x}_0 = (2, -1.5)^T$, (iv) $\mathbf{x}_0 = (-2, 1.5)^T$.

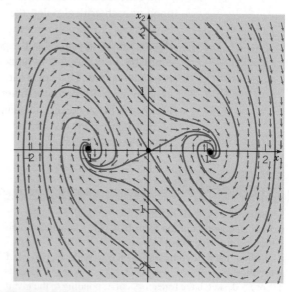

FIGURE 3.6.5 Direction field and phase portrait for Problem 23.

24. An **asymptotically stable limit cycle** is a closed trajectory C in the phase plane such that all nonclosed trajectories that start sufficiently near C, from either the inside or outside, spiral toward C as $t \to \infty$. Demonstrate numerically that the van der Pol equation (see Section 7.5),

$$u'' + 3(u^2 - 1)u' + u = 0,$$

has such a limit cycle by drawing a direction field and phase portrait for the equivalent dynamical system. Describe the speed of motion around C by viewing the component plots of $u(t)$ and $u'(t)$.

25. A model for the populations, x and y, of two species that compete for the same food supply (see Section 7.3) is

$$dx/dt = x(1 - x - y)$$
$$dy/dt = y(0.75 - y - 0.5x). \qquad \text{(i)}$$

Find all the critical points of the system (i), draw a direction field and phase portrait, and interpret the meaning of the limiting behavior, as $t \to \infty$, of x and y in terms of the populations of the two species.

CHAPTER SUMMARY

Two-dimensional systems of first order differential equations have the form

$$\frac{dx}{dt} = f(t, x, y), \qquad \frac{dy}{dt} = g(t, x, y),$$

or, using vector notation $\mathbf{x} = x\mathbf{i} + y\mathbf{j}$ and $\mathbf{f}(t, \mathbf{x}) = f(t, x, y)\mathbf{i} + g(t, x, y)\mathbf{j}$,

$$\frac{d\mathbf{x}}{dt} = \mathbf{f}(t, \mathbf{x}).$$

Section 3.1 Two-Dimensional Linear Algebra

▶ Matrix notation for a linear algebraic system of two equations in two unknowns is $\mathbf{Ax} = \mathbf{b}$.

  ▶ If $\det \mathbf{A} \neq 0$, the unique solution of $\mathbf{Ax} = \mathbf{b}$ is $\mathbf{x} = \mathbf{A}^{-1}\mathbf{b}$.

  ▶ If $\det \mathbf{A} = 0$, $\mathbf{Ax} = \mathbf{b}$ may have (i) no solution, or (ii) a straight line of solutions in the plane; in particular, if $\mathbf{b} = \mathbf{0}$ and $\mathbf{A} \neq \mathbf{0}$, the solution set is a straight line passing through the origin.

▶ **The eigenvalue problem:** $(\mathbf{A} - \lambda \mathbf{I})\mathbf{x} = \mathbf{0}$. The **eigenvalues** of $\mathbf{A}$ are solutions of the **characteristic equation** $\det(\mathbf{A} - \lambda \mathbf{I}) = 0$. An **eigenvector** for the eigenvalue λ is a nonzero solution of $(\mathbf{A} - \lambda \mathbf{I})\mathbf{x} = \mathbf{0}$. Eigenvalues may be real and different, real and equal, or complex conjugates.

Section 3.2 Systems of Two First Order Linear Equations

Variable coefficient: $\mathbf{x}' = \mathbf{P}(t)\mathbf{x} + \mathbf{g}(t)$,   Autonomous: $\mathbf{x}' = \mathbf{Ax} + \mathbf{b}$

▶ **Existence and uniqueness of solutions.** If the entries of $\mathbf{P}(t)$ and $\mathbf{g}(t)$ are continuous on I, then a unique solution to the initial value problem $\mathbf{x}' = \mathbf{P}(t)\mathbf{x} + \mathbf{g}(t)$, $\mathbf{x}(t_0) = \mathbf{x}_0$, $t_0 \in I$ exists for all $t \in I$.

▶ **Graphical techniques:** (i) component plots, and for autonomous systems, (ii) direction fields and (iii) phase portraits.

▶ **Critical points (equilibrium solutions)** of linear autonomous systems are solutions of $\mathbf{Ax} + \mathbf{b} = \mathbf{0}$.

▶ **Second order linear equations** $y'' + p(t)y' + q(t)y = g(t)$ can be transformed into systems of two first order linear equations

$$\mathbf{x}' = \begin{pmatrix} 0 & 1 \\ -q(t) & -p(t) \end{pmatrix} \mathbf{x} + \begin{pmatrix} 0 \\ g(t) \end{pmatrix},$$

where $\mathbf{x} = y\mathbf{i} + y'\mathbf{j}$.

Section 3.3   Homogeneous Systems with Constant Coefficients: $\mathbf{x}' = \mathbf{Ax}$

▶ Two solutions $\mathbf{x}_1(t)$ and $\mathbf{x}_2(t)$ to $\mathbf{x}' = \mathbf{Ax}$ form a **fundamental set of solutions** if their **Wronskian**

$$W[\mathbf{x}_1, \mathbf{x}_2](t) = \begin{vmatrix} x_{11}(t) & x_{12}(t) \\ x_{21}(t) & x_{22}(t) \end{vmatrix} = x_{11}(t)x_{22}(t) - x_{12}(t)x_{21}(t) \neq 0.$$

(In this case, $\mathbf{x}_1$ and $\mathbf{x}_2$ are **linearly independent**.)
If $\mathbf{x}_1(t)$ and $\mathbf{x}_2(t)$ are a fundamental set, then the **general solution** to $\mathbf{x}' = \mathbf{Ax}$ is $\mathbf{x} = c_1\mathbf{x}_1(t) + c_2\mathbf{x}_2(t)$, where c_1 and c_2 are arbitrary constants.

▶ When $\mathbf{A}$ has real eigenvalues $\lambda_1 \neq \lambda_2$ with corresponding eigenvectors $\mathbf{v}_1$ and $\mathbf{v}_2$,

  ▶ a general solution of $\mathbf{x}' = \mathbf{Ax}$ is $\mathbf{x} = c_1 e^{\lambda_1 t}\mathbf{v}_1 + c_2 e^{\lambda_2 t}\mathbf{v}_2$,
  ▶ if $\det(\mathbf{A}) \neq 0$, the only critical point (the origin) is (i) a **node** if the eigenvalues have the same algebraic sign, or (ii) a **saddle point** if the eigenvalues are of opposite sign.

Section 3.4   Complex Eigenvalues

▶ If the eigenvalues of $\mathbf{A}$ are $\mu \pm i\nu$, $\nu \neq 0$, with corresponding eigenvectors $\mathbf{a} \pm i\mathbf{b}$, a fundamental set of real vector solutions of $\mathbf{x}' = \mathbf{Ax}$ consists of
$\mathrm{Re}\{\exp[(\mu + i\nu)t][\mathbf{a} + i\mathbf{b}]\} = \exp(\mu t)(\cos \nu t \mathbf{a} - \sin \nu t \mathbf{b})$ and
$\mathrm{Im}\{\exp[(\mu + i\nu)t][\mathbf{a} + i\mathbf{b}]\} = \exp(\mu t)(\sin \nu t \mathbf{a} + \cos \nu t \mathbf{b})$.

▶ If $\mu \neq 0$, then the critical point (the origin) is a **spiral point**. If $\mu = 0$, then the critical points is a **center**.

Section 3.5   Repeated Eigenvalues

▶ If $\mathbf{A}$ has a single repeated eigenvalue λ, then a general solution of $\mathbf{x}' = \mathbf{Ax}$ is

  (i) $\mathbf{x} = c_1 e^{\lambda t}\mathbf{v}_1 + c_2 e^{\lambda t}\mathbf{v}_2$ if $\mathbf{v}_1$ and $\mathbf{v}_2$ are independent eigenvectors, or
  (ii) $\mathbf{x} = c_1 e^{\lambda t}\mathbf{v} + c_2 e^{\lambda t}(\mathbf{w} + t\mathbf{v})$, where $(\mathbf{A} - \lambda \mathbf{I})\mathbf{w} = \mathbf{v}$ if $\mathbf{v}$ is the only eigenvector of $\mathbf{A}$.

▶ The critical point at the origin is a **proper node** if there are two independent eigenvectors, and an **improper** or **degenerate node** if there is only one eigenvector.

Section 3.6   Nonlinear Systems

$$\text{Nonautonomous: } \mathbf{x}' = \mathbf{f}(t, \mathbf{x}), \qquad \text{Autonomous: } \mathbf{x}' = \mathbf{f}(\mathbf{x})$$

▶ Theorem 3.6.1 provides conditions that guarantee, locally in time, existence and uniqueness of solutions to the initial value problem $\mathbf{x}' = \mathbf{f}(t, \mathbf{x}), \mathbf{x}(t_0) = \mathbf{x}_0$.

▶ Examples of two-dimensional nonlinear autonomous systems suggest that locally their solutions behave much like solutions of linear systems.

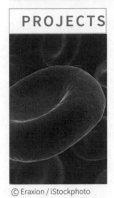

PROJECTS

Project 1 Estimating Rate Constants for an Open Two-Compartment Model

Physiological systems are often modeled by dividing them into distinct functional units or compartments. A simple two-compartment model used to describe the evolution in time of a single intravenous drug dose (or a chemical tracer) is shown in Figure 3.P.1. The central compartment, consisting of blood and extracellular water, is rapidly diffused with the drug.

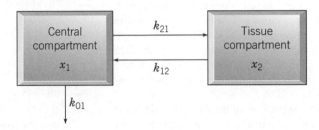

© Eraxion / iStockphoto

FIGURE 3.P.1 A two-compartment open model of a physiological system.

The second compartment, known as the tissue compartment, contains tissues that equilibriate more slowly with the drug. If x_1 is the concentration of drug in the blood and x_2 is its concentration in the tissue, the compartment model is described by the following system:

$$x_1' = -(k_{01} + k_{21})x_1 + k_{12}x_2$$
$$x_2' = k_{21}x_1 - k_{12}x_2, \tag{1}$$

or $\mathbf{x}' = \mathbf{K}\mathbf{x}$, where

$$\mathbf{K} = \begin{pmatrix} K_{11} & K_{12} \\ K_{21} & K_{22} \end{pmatrix} = \begin{pmatrix} -k_{01} - k_{21} & k_{12} \\ k_{21} & -k_{12} \end{pmatrix}. \tag{2}$$

Here, the rate constant k_{21} is the fraction per unit time of drug in the blood compartment transferred to the tissue compartment; k_{12} is the fraction per unit time of drug in the tissue compartment transferred to the blood; and k_{01} is the fraction per unit time of drug eliminated from the system.

In this project, we illustrate a method for estimating the rate constants by using time-dependent measurements of concentrations to estimate the eigenvalues and eigenvectors of the rate matrix $\mathbf{K}$ in Eq. (2) from which estimates of all rate constants can be computed.

Project 1 PROBLEMS

1. Assume that all the rate constants in Eq. (1) are positive.

(a) Show that the eigenvalues of the matrix $\mathbf{K}$ are real, distinct, and negative.
Hint: Show that the discriminant of the characteristic polynomial of $\mathbf{K}$ is positive.
(b) If λ_1 and λ_2 are the eigenvalues of $\mathbf{K}$, show that $\lambda_1 + \lambda_2 = -(k_{01} + k_{12} + k_{21})$ and $\lambda_1\lambda_2 = k_{12}k_{01}$.

2. Estimating Eigenvalues and Eigenvectors of K from Transient Concentration Data. Denote by $\mathbf{x}^*(t_k) = x_1^*(t_k)\mathbf{i} + x_2^*(t_k)\mathbf{j}$, $k = 1, 2, 3, \ldots$ measurements of the concentrations in each of the compartments. We assume that the eigenvalues of $\mathbf{K}$ satisfy $\lambda_2 < \lambda_1 < 0$. Denote the eigenvectors of λ_1 and λ_2 by

$$\mathbf{v_1} = \begin{pmatrix} v_{11} \\ v_{21} \end{pmatrix} \quad \text{and} \quad \mathbf{v_2} = \begin{pmatrix} v_{12} \\ v_{22} \end{pmatrix},$$

respectively. The solution of Eq. (1) can be expressed as

$$\mathbf{x}(t) = \alpha e^{\lambda_1 t}\mathbf{v}_1 + \beta e^{\lambda_2 t}\mathbf{v}_2, \qquad (i)$$

where α and β, assumed to be nonzero, depend on initial conditions. From Eq. (i), we note that

$$\mathbf{x}(t) = e^{\lambda_1 t}\left[\alpha\mathbf{v}_1 + \beta e^{(\lambda_2-\lambda_1)t}\mathbf{v}_2\right] \sim \alpha e^{\lambda_1 t}\mathbf{v}_1$$

$$\text{if } e^{(\lambda_2-\lambda_1)t} \sim 0. \qquad (ii)$$

(a) For values of t such that $e^{(\lambda_2-\lambda_1)t} \sim 0$, explain why the graphs of $\ln x_1(t)$ and $\ln x_2(t)$ should be approximately straight lines with slopes equal to λ_1 and intercepts equal to $\ln \alpha v_{11}$ and $\ln \alpha v_{21}$, respectively. Thus estimates of λ_1, αv_{11}, and αv_{21} may be obtained by fitting straight lines to the data $\ln x_1^*(t_n)$ and $\ln x_2^*(t_n)$ corresponding to values of t_n, where graphs of the logarithms of the data are approximately linear, as shown in Figure 3.P.2.

(b) Given that both components of the data $\mathbf{x}^*(t_n)$ are accurately represented by a sum of exponential functions of the form (i), explain how to find estimates of λ_2, βv_{12}, and βv_{22} using the residual data $\mathbf{x}_r^*(t_n) = \mathbf{x}^*(t_n) - \hat{\mathbf{v}}_1^{(\alpha)}e^{\hat{\lambda}_1 t_n}$, where estimates of λ_1 and $\alpha\mathbf{v}_1$ are denoted by $\hat{\lambda}_1$ and $\hat{\mathbf{v}}_1^{(\alpha)}$, respectively.[7]

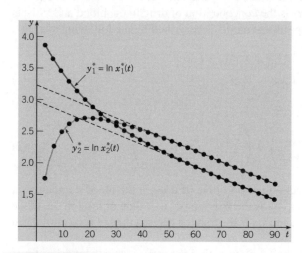

FIGURE 3.P.2 Graphs of the data $y_{1n}^* = \ln x_1^*(t_n)$ and $y_{2n}^* = \ln x_2^*(t_n)$ are approximately straight lines for values of t_n such that $e^{(\lambda_2-\lambda_1)t_n} \sim 0$.

3. Computing the Entries of K from Its Eigenvalues and Eigenvectors. Assume that the eigenvalues λ_1 and λ_2 and corresponding eigenvectors $\mathbf{v}_1$ and $\mathbf{v}_2$ of $\mathbf{K}$ are known. Show

that the entries of the matrix $\mathbf{K}$ must satisfy the following systems of equations:

$$\begin{pmatrix} v_{11} & v_{21} \\ v_{12} & v_{22} \end{pmatrix}\begin{pmatrix} K_{11} \\ K_{12} \end{pmatrix} = \begin{pmatrix} \lambda_1 v_{11} \\ \lambda_2 v_{12} \end{pmatrix} \qquad (iii)$$

and

$$\begin{pmatrix} v_{11} & v_{21} \\ v_{12} & v_{22} \end{pmatrix}\begin{pmatrix} K_{21} \\ K_{22} \end{pmatrix} = \begin{pmatrix} \lambda_1 v_{21} \\ \lambda_2 v_{22} \end{pmatrix}, \qquad (iv)$$

or, using matrix notation, $\mathbf{KV} = \mathbf{V\Lambda}$, where

$$\mathbf{V} = \begin{pmatrix} v_{11} & v_{12} \\ v_{21} & v_{22} \end{pmatrix} \quad \text{and} \quad \mathbf{\Lambda} = \begin{pmatrix} \lambda_1 & 0 \\ 0 & \lambda_2 \end{pmatrix}.$$

4. Given estimates $\hat{K}_{ij}$ of the entries of $\mathbf{K}$ and estimates $\hat{\lambda}_1$ and $\hat{\lambda}_2$ of the eigenvalues of $\mathbf{K}$, show how to obtain an estimate $\hat{k}_{01}$ of k_{01} using the relations in Problem 1(b).

5. Table 3.P.1 lists drug concentration measurements made in blood and tissue compartments over a period of 100 min. Use the method described in Problems 2 through 4 to estimate the rate coefficients k_{01}, k_{12}, and k_{21} in the system model (1). Then solve the resulting system using initial conditions from line 1 of Table 3.P.1. Verify the accuracy of your estimates by plotting the solution components and the data in Table 3.P.1 on the same set of coordinate axes.

TABLE 3.P.1 Compartment concentration measurements.

time (min)	x_1 (mg/mL)	x_2 (mg/mL)
0.000	0.623	0.000
7.143	0.374	0.113
14.286	0.249	0.151
21.429	0.183	0.157
28.571	0.145	0.150
35.714	0.120	0.137
42.857	0.103	0.124
50.000	0.089	0.110
57.143	0.078	0.098
64.286	0.068	0.087
71.429	0.060	0.077
78.571	0.053	0.068
85.714	0.047	0.060
92.857	0.041	0.053
100.000	0.037	0.047

---

[7]The procedure outlined here is called the **method of exponential peeling**. The method can be extended to cases where more than two exponential functions are required to represent the component concentrations. There must be one compartment for each exponential decay term. See, for example, D. Van Liew (1967), *Journal of Theoretical Biology* **16**, 43.

Project 2 A Blood–Brain Pharmacokinetic Model

Pharmacokinetics is the study of the time variation of drug and metabolite levels in the various fluids and tissues of the body. The discipline frequently makes use of compartment models to interpret data. In this problem, we consider a simple blood–brain compartment model (Figure 3.P.3),

$$\text{Compartment 1} \equiv \text{Blood,}$$
$$\text{Compartment 2} \equiv \text{Brain,}$$

that could be used to help estimate dosage strengths of an orally administered antidepressant drug. The rate at which the drug moves from compartment i to compartment j is denoted by the rate constant k_{ji}, while the rate at which the drug is removed from the blood is represented by the rate constant K.

A pharmaceutical company must weigh many factors in determining drug dosage parameters; of particular importance are dosage strengths that will provide flexibility to a physician in determining individual dosage regimens to conveniently maintain concentration levels at effective therapeutic values while minimizing local irritation and other adverse side effects.

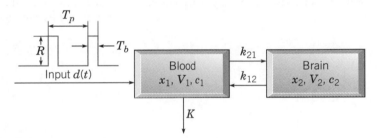

FIGURE 3.P.3 A two-compartment model for periodic drug dosages.

Assuming that the drug is rapidly absorbed into the bloodstream following its introduction into the stomach, a mathematical idealization for the dosage regimen is that of a periodic square wave

$$d(t) = \begin{cases} R, & 0 \le t \le T_b \\ 0, & T_b \le t < T_p, \end{cases}$$

where R is the rate of uptake (milligrams/hour) into the bloodstream, T_b is the time period during which the drug is absorbed into the bloodstream following oral administration, and T_p is the length of time between doses.

Project 2 PROBLEMS

1. If $x_j(t)$ represents the amount of drug (milligrams) in compartment j, $j = 1, 2$, use Figure 3.P.3 and the mass balance law

$$\frac{dx_j}{dt} = \text{compartment } j \text{ input rate}$$
$$- \text{compartment } j \text{ output rate,} \qquad \text{(i)}$$

to show that x_1 and x_2 satisfy the system

$$\frac{dx_1}{dt} = -(K + k_{21})x_1 + k_{12}x_2 + d(t)$$
$$\frac{dx_2}{dt} = k_{21}x_1 - k_{12}x_2. \qquad \text{(ii)}$$

2. If $c_i(t)$ denotes the concentration of the drug and V_i denotes the apparent volume of distribution in compartment i, use the relation $c_i = x_i/V_i$ to show that the system (ii) is transformed into

$$\frac{dc_1}{dt} = -(K + k_{21})c_1 + \frac{k_{12}V_2}{V_1}c_2 + \frac{1}{V_1}d(t)$$

$$\frac{dc_2}{dt} = \frac{V_1 k_{21}}{V_2}c_1 - k_{12}c_2. \tag{iii}$$

3. Assuming that $x_1(0) = 0$ and $x_2(0) = 0$, use the parameter values listed in the table below to perform numerical simulations of the system (iii) with the goal of recommending two different encapsulated dosage strengths $A = RT_b$ for distribution.

k_{21}	k_{12}	K	V_1	V_2	T_b
0.29/h	0.31/h	0.16/h	6 L	0.25 L	1 h

Use the following guidelines to arrive at your recommendations:

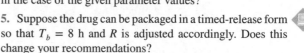 It is desirable to keep the target concentration levels in the brain as close as possible to constant levels between 10 mg/L and 30 mg/L, depending on the individual patient. The therapeutic range must be above the minimum effective concentration and below the minimum toxic concentration. For the purpose of this project, we will specify that concentration fluctuations should not exceed 25% of the average of the steady-state response.

 As a matter of convenience, a lower frequency of administration is better than a higher frequency of administration; once every 24 hours or once every 12 hours is best. Once every 9.5 hours is unacceptable and more than 4 times per day is unacceptable. Multiple doses are acceptable, that is, "take two capsules every 12 hours."

4. If a dosage is missed, explain through the simulations why it is best to skip the dose rather than to try to "catch up" by doubling the next dose, given that it is dangerous and possibly fatal to overdose on the drug. Or, does it not really matter in the case of the given parameter values?

5. Suppose the drug can be packaged in a timed-release form so that $T_b = 8$ h and R is adjusted accordingly. Does this change your recommendations?

CHAPTER FOUR

Library of Congress Prints & Photographs Division

Second Order Linear Equations

I n Chapter 3 we discussed systems of two first order equations, with primary emphasis on homogeneous linear equations with constant coefficients. In this chapter we will begin to consider second order linear equations, both homogeneous and nonhomogeneous. Since second order equations can always be transformed into a system of two first order equations, this may seem redundant. However second order equations naturally arise in many areas of application, and it is important to be able to deal with them directly. One cannot go very far in the development of fluid mechanics, heat conduction, wave motion, or electromagnetic phenomena without encountering second order linear differential equations.

4.1 Definitions and Examples

A **second order differential equation** is an equation involving the independent variable t, and an unknown function or dependent variable $y = y(t)$ along with its first and second derivatives. We will assume that it is always possible to solve for the second derivative so that the equation has the form

$$y'' = f(t, y, y'), \tag{1}$$

where f is some prescribed function. Usually, we will denote the independent variable by t since time is often the independent variable in physical problems, but sometimes we will use x instead. We will use y, or occasionally some other letter, to designate the dependent variable.

A **solution** of Eq. (1) on an interval I is a function $y = \phi(t)$, twice continuously differentiable on I, such that

$$\phi''(t) = f(t, \phi(t), \phi'(t)) \tag{2}$$

for all values of $t \in I$.

An **initial value problem** for a second order equation on an interval I consists of Eq. (1) together with two initial conditions

$$y(t_0) = y_0, \qquad y'(t_0) = y_1, \tag{3}$$

prescribed at a point $t_0 \in I$, where y_0 and y_1 are any given numbers. Thus $y = \phi(t)$ is a **solution of the initial value problem** (1), (3) on I if, in addition to satisfying Eq. (2) on I, $\phi(t_0) = y_0$ and $\phi'(t_0) = y_1$.

Remark. Observe that the initial conditions for a second order equation prescribe not only a particular point (t_0, y_0) through which the graph of the solution must pass, but also the slope $y'(t_0) = y_1$ of the graph at that point (Figure 4.1.1).

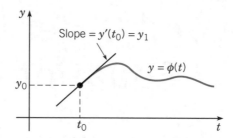

FIGURE 4.1.1   Initial conditions for a second order equation require that $y(t_0)$ and $y'(t_0)$ be prescribed.

It is reasonable to expect that two initial conditions are needed for a second order equation because, roughly speaking, two integrations are required to find a solution and each integration introduces an arbitrary constant. Presumably, two initial conditions will suffice to determine values for these two constants.

Linear Equations

The differential equation (1) is said to be **linear** if it can be written in the **standard form**

$$y'' + p(t)y' + q(t)y = g(t), \tag{4}$$

where the coefficient of y'' is equal to 1. The coefficients p, q, and g can be arbitrary functions of the independent variable t, but y, y', and y'' can appear in no other way except as designated by the form of Eq. (4).

Equation (4) is said to be **homogeneous** if the term $g(t)$ is zero for all t. Otherwise, the equation is **nonhomogeneous**, and the term $g(t)$ is referred to as the nonhomogeneous term.

A slightly more general form of a linear second order equation is

$$P(t)y'' + Q(t)y' + R(t)y = G(t). \tag{5}$$

Of course, if $P(t) \neq 0$, we can divide Eq. (5) by $P(t)$ and thereby obtain Eq. (4) with

$$p(t) = \frac{Q(t)}{P(t)}, \qquad q(t) = \frac{R(t)}{P(t)}, \qquad \text{and} \qquad g(t) = \frac{G(t)}{P(t)}. \tag{6}$$

Equation (5) is said to be a **constant coefficient** equation if P, Q, and R are constants. In this case, Eq. (5) reduces to

$$ay'' + by' + cy = g(t), \tag{7}$$

where $a \neq 0$, b, and c are given constants and we have replaced $G(t)$ by $g(t)$. Otherwise, Eq. (5) has **variable coefficients**.

Dynamical System Formulation

Recall from Section 3.2 that Eq. (1) can be converted to a system of first order equations of dimension two by introducing the state variables $x_1 = y$ and $x_2 = y'$. Then

$$x_1' = x_2,$$
$$x_2' = f(t, x_1, x_2), \tag{8}$$

or, using vector notation, $\mathbf{x}' = \mathbf{f}(t, \mathbf{x}) = x_2\mathbf{i} + f(t, x_1, x_2)\mathbf{j}$, where $\mathbf{x} = x_1\mathbf{i} + x_2\mathbf{j}$. Initial conditions for the system (8), obtained from (3), are

$$x_1(t_0) = y_0, \qquad x_2(t_0) = y_1, \tag{9}$$

or equivalently, $\mathbf{x}(t_0) = \mathbf{x}_0 = y_0\mathbf{i} + y_1\mathbf{j}$.

Thus the requirement of two initial conditions for Eq. (1) is consistent with our experience in Chapter 3 for systems such as (8). Recall, in particular, Theorem 3.6.1, which gives conditions for the existence of a unique solution of the initial value problem Eqs. (8) and (9). When we refer to the **state variables** for Eq. (1), we mean both y and y', although other choices for state variables may be used. In addition, when we refer to the **dynamical system** equivalent to Eq. (1), we mean the system of first order equations (8) expressed in terms of the state variables. Just as in Chapter 3, the evolution of the system state in time is graphically represented as a continuous **trajectory**, or **orbit**, through the phase plane or state space. It is sometimes helpful to think of an orbit as the path of a particle moving in accordance with the system of differential equations (8). The initial conditions (9) determine the starting point of the moving particle. Note that if $\phi(t)$ is a solution of Eq. (1) on I, then $\boldsymbol{\phi}(t) = \phi(t)\mathbf{i} + \phi'(t)\mathbf{j}$ is a solution of the system (8) on I (Figure 4.1.2).

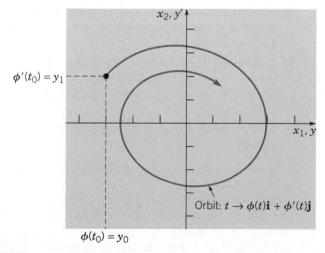

FIGURE 4.1.2 Initial conditions for the system (8) correspond to the initial conditions $y(t_0) = y_0$, $y'(t_0) = y_1$ for the second order equation $y'' = f(t, y, y')$.

Special cases of (8) corresponding to Eqs. (4) and (7) are the linear systems

$$
\begin{aligned}
x_1' &= x_2, \\
x_2' &= -q(t)x_1 - p(t)x_2 + g(t),
\end{aligned}
\tag{10}
$$

and

$$
\begin{aligned}
x_1' &= x_2, \\
x_2' &= -\frac{c}{a}x_1 - \frac{b}{a}x_2 + \frac{1}{a}g(t),
\end{aligned}
\tag{11}
$$

respectively.

If the function f on the right side of Eq. (1) is independent of t so that $f(t, y, y') = f(y, y')$, then the system (8) is **autonomous**. In this case, critical points of the system (8) are solutions of the pair of equations $x_2 = 0$ and $f(x_1, x_2) = 0$. All critical points lie on the x_1-axis with coordinates $(\bar{x}_1, 0)$, where $\bar{x}_1$ is any solution of $f(\bar{x}_1, 0) = 0$. If $f(\bar{x}_1, 0) = 0$, $\partial f/\partial x_1(\bar{x}_1, 0) \neq 0$, and $\partial f/\partial x_1$ and $\partial f/\partial x_2$ are continuous at $(\bar{x}_1, 0)$, then $(\bar{x}_1, 0)$ is an **isolated critical point**, that is, there is a neighborhood of $(\bar{x}_1, 0)$ that contains no other critical points. The linear, constant coefficient system (11) is autonomous if $g(t) = g_0$, a constant. In this case, Eq. (11) has a unique critical point at $(g_0/c, 0)$ if the coefficient c is nonzero. If $c = 0$ and $g_0 \neq 0$, then there are no critical points. Finally, if $g_0 = 0$ and $c = 0$, then the set of critical points is the entire x_1-axis.

We now present three applications arising from problems in mechanics and circuit theory. The mathematical formulation in each case consists of a second order linear equation with constant coefficients.

The Spring-Mass System

Vibrational or oscillatory behavior is observed in many mechanical, electrical, and biological systems. Understanding the motion of a mass on a spring is the first step in the investigation of more complex vibrating systems. The principles involved are common to many problems. The differential equation that describes the motion of a spring-mass system is arguably the most important equation in an introductory course in differential equations for the following reasons:

▶ It involves translating a physical description of a simple mechanical system into a prototypical mathematical model, namely, a linear, second order differential equation with constant coefficients.

▶ Understanding the behavior of solutions as parameters vary, or as external input forces are added, is fundamental to understanding the qualitative behavior of solutions of both linear, second order equations with variable coefficients and second order nonlinear equations.

▶ Mathematical properties of solutions are easily interpreted in terms of the physical system.

We consider a vertical spring of natural length l attached to a horizontal support, as shown in Figure 4.1.3a.

Next we suppose that a mass of magnitude m is attached to the lower end of the spring and slowly lowered so as to achieve its equilibrium position, as shown in Figure 4.1.3b. The mass causes an elongation L of the spring.

Our goal is to investigate motions of the mass that might be caused by an external force acting upon it, or by an initial displacement of the mass away from its equilibrium position. We consider only motions along a vertical line. Let the y-axis be vertical, with the

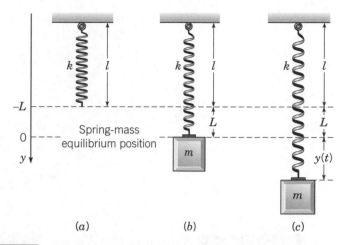

FIGURE 4.1.3 A spring-mass system with no damping or forcing.

positive direction downward, and with the origin at the equilibrium position of the mass; see Figure 4.1.3c. Then $y(t)$ denotes the displacement of the mass from its equilibrium position, and thus $y'(t)$ is its velocity. The equilibrium state of the spring-mass system corresponds to $y = 0$ and $y' = 0$, that is, both state variables are zero.

To derive an equation that describes possible motions of the mass, we need to examine the forces that may act upon it. We start with the force exerted on the mass by the spring. Denote by Δy the departure of the spring from its natural length so that if a displacement Δy from its natural length occurs, the length of the spring is $l + \Delta y$. We first assume that the force exerted by the spring on the mass is described by a function $F_s(\Delta y)$ satisfying the following properties:

▶ $F_s(0) = 0$; the spring exerts no force if $\Delta y = 0$.

▶ $F_s(\Delta y) < 0$ if $\Delta y > 0$; in an elongated state, the spring exerts a force in the upward direction.

▶ $F_s(\Delta y) > 0$ if $\Delta y < 0$; in a compressed state, the spring exerts a force in the downward direction.

Thus the direction of the force is always opposite to the displacement of the lower endpoint of the spring relative to its natural length. An example of such a function is $F_s(\Delta y) = -k\Delta y - \epsilon(\Delta y)^3$, where $k > 0$ and ϵ are constants (see Figure 4.1.4).

The spring is called a **hardening spring** if $\epsilon > 0$ and a **softening spring** if $\epsilon < 0$. If in a state of motion the maximum displacement of the spring from its natural length is small so that $\epsilon(\Delta y)^3$ is always negligible relative to $k\Delta y$, it is natural to discard the nonlinear term $\epsilon(\Delta y)^3$ and simply assume that the spring force is proportional to Δy,

$$F_s(\Delta y) = -k\Delta y. \tag{12}$$

Equation (12) is known as **Hooke's law**. It provides an excellent approximation to the behavior of real springs as long as they are not stretched or compressed too far. Unless stated otherwise, we will always assume that our springs are adequately described by Hooke's law. The **spring constant** k, sometimes referred to as the **stiffness** of the spring, is the magnitude of the spring force per unit of elongation. Thus very stiff springs have large values of k.

In the equilibrium state, there are two forces acting at the point where the mass is attached to the spring; see Figure 4.1.5.

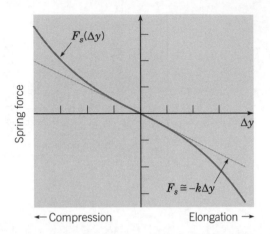

FIGURE 4.1.4   Spring force $F_s(\Delta y) = -k\Delta y - \epsilon(\Delta y)^3$ of a hardening spring ($\epsilon > 0$) as a function of departure Δy from the natural length l of the spring. A compressed spring ($\Delta y < 0$) exerts a force in the positive direction, while an elongated spring ($\Delta y > 0$) exerts a force in the negative direction. Hooke's law assumes the linear approximation $F_s(\Delta y) \cong -k\Delta y$.

FIGURE 4.1.5   Force diagram for a spring-mass system.

The gravitational force, or weight of the mass, acts downward and has magnitude mg, where g is the acceleration due to gravity. The force F_s, due to the elongated spring, acts upward, that is, in the negative direction. Since the mass is in equilibrium, the spring and gravitational forces balance each other, which means that

$$mg - kL = 0. \tag{13}$$

For a given weight $w = mg$, one can measure L and then use Eq. (13) to determine k.

In the corresponding dynamic problem, we are interested in studying the motion of the mass when it is initially displaced from equilibrium or acted on by an external force. Then $y(t)$ is related to the forces acting on the mass through Newton's law of motion,

$$my''(t) = F_{\text{net}}(t), \tag{14}$$

where y'' is the acceleration of the mass and F_{net} is the net force acting on the mass. Observe that both y and F_{net} are functions of time. In determining F_{net}, there are four separate forces that must be considered:

1. Gravitational Force. The weight $w = mg$ of the mass always acts downward.

2. **Spring Force.** The spring force F_s is assumed to be proportional to the total elongation $\Delta y = L + y$ following Hooke's law, Eq. (12) (see Figure 4.1.3c),

$$F_s = -k(L + y). \tag{15}$$

3. **Damping Force.** The damping or resistive force F_d always acts in the direction opposite to the direction of motion of the mass. This force may arise from several sources: resistance from the air or other medium in which the mass moves, internal energy dissipation due to the extension or compression of the spring, friction between the mass and the guides (if any) that constrain its motion to one dimension, or a mechanical device (dashpot; see Figure 4.1.6) that imparts a resistive force to the mass.

FIGURE 4.1.6 A damped spring-mass system.

In any case, we assume that the resistive force is proportional to the speed $|dy/dt|$ of the mass; this is usually referred to as viscous damping. If $dy/dt > 0$, then y is increasing, so the mass is moving downward. Then F_d is directed upward and is given by

$$F_d(t) = -\gamma y'(t), \tag{16}$$

where γ is a positive constant of proportionality known as the **damping constant**. On the other hand, if $dy/dt < 0$, then y is decreasing, the mass is moving upward, and F_d is directed downward. In this case, $F_d = \gamma|y'(t)|$; since $|y'(t)| = -y'(t)$, it follows that $F_d(t)$ is again given by Eq. (16). Thus, regardless of the direction of motion of the mass, the damping force is always expressed by Eq. (16).

Remark. The damping force may be rather complicated, and the assumption that it is modeled adequately by Eq. (16) may be open to question. Some dashpots do behave as Eq. (16) states, and if the other sources of dissipation are small, it may be possible to neglect them altogether or to adjust the damping constant γ to approximate them. An important benefit of the assumption (16) is that it leads to a linear (rather than a nonlinear) differential equation. In turn, this means that a thorough analysis of the system is straightforward, as we will show in the next section.

4. **External Forces or Inputs.** An applied external force $F(t)$ is directed downward or upward as $F(t)$ is positive or negative. This could be a force due to the motion of the mount to which the spring is attached, or it could be a force applied directly to the mass. Often the external force is periodic.

With $F_{\text{net}}(t) = mg + F_s(t) + F_d(t) + F(t)$, we can now rewrite Newton's law (14) as

$$
\begin{aligned}
my''(t) &= mg + F_s(t) + F_d(t) + F(t) \\
&= mg - k[L + y(t)] - \gamma y'(t) + F(t).
\end{aligned} \tag{17}
$$

Since $mg - kL = 0$ by Eq. (13), it follows that the equation of motion of the mass is a second order linear equation with constant coefficients,

$$
my''(t) + \gamma y'(t) + ky(t) = F(t), \tag{18}
$$

where the constants m, γ, and k are positive.

Remark. It is important to understand that Eq. (18) is only an approximate equation for the displacement $y(t)$. In particular, both Eqs. (12) and (16) should be viewed as approximations for the spring force and the damping force, respectively. In our derivation, we have also neglected the mass of the spring in comparison with the mass of the attached body.

The complete formulation of the vibration problem requires that we specify two initial conditions, namely, the initial position $y(0) = y_0$ and the initial velocity $y'(0) = v_0$ of the mass:

$$
y(0) = y_0, \qquad y'(0) = v_0. \tag{19}
$$

Four cases of Eq. (18) that are of particular interest are listed in the table below.

Unforced, undamped oscillator:	$my''(t) + ky(t) = 0$
Unforced, damped oscillator:	$my''(t) + \gamma y'(t) + ky(t) = 0$
Forced, undamped oscillator:	$my''(t) + ky(t) = F(t)$
Forced, damped oscillator:	$my''(t) + \gamma y'(t) + ky(t) = F(t)$

Analytical solutions and properties of unforced oscillators and forced oscillators are studied in Sections 4.4 and 4.6, respectively.

EXAMPLE 1

A mass weighing 4 pounds (lb) stretches a spring 2 in. Suppose that the mass is displaced an additional 6 in. in the positive direction and then released. The mass is in a medium that exerts a viscous resistance of 6 lb when the mass has a velocity of 3 ft/s. Under the assumptions discussed in this section formulate the initial value problem that governs the motion of the mass.

The required initial value problem consists of the differential equation (18) and initial conditions (19), so our task is to determine the various constants that appear in these equations. The first step is to choose the units of measurement. Based on the statement of the problem, it is natural to use the English rather than the metric system of units. The only time unit mentioned is the second, so we will measure t in seconds. On the other hand, both the foot and the inch appear in the statement as units of length. It is immaterial which one we use, but having made a choice, we must be consistent. To be definite, let us measure the displacement y in feet.

Since nothing is mentioned in the statement of the problem about an external force, we assume that $F(t) = 0$. To determine m, note that

$$m = \frac{w}{g} = \frac{4 \text{ lb}}{32 \text{ ft/s}^2} = \frac{1}{8} \frac{\text{lb·s}^2}{\text{ft}}.$$

The damping coefficient γ is determined from the statement that $\gamma y'$ is equal to 6 lb when y' is 3 ft/s. Therefore

$$\gamma = \frac{6 \text{ lb}}{3 \text{ ft/s}} = 2 \frac{\text{lb·s}}{\text{ft}}.$$

The spring constant k is found from the statement that the mass stretches the spring by 2 in., or 1/6 ft. Thus

$$k = \frac{4 \text{ lb}}{1/6 \text{ ft}} = 24 \frac{\text{lb}}{\text{ft}}.$$

Consequently, Eq. (18) becomes

$$\tfrac{1}{8}y'' + 2y' + 24y = 0,$$

or

$$y'' + 16y' + 192y = 0. \tag{20}$$

The initial conditions are

$$y(0) = \tfrac{1}{2}, \qquad y'(0) = 0. \tag{21}$$

The second initial condition is implied by the word "released" in the statement of the problem, which we interpret to mean that the mass is set in motion with no initial velocity.

The Linearized Pendulum

Consider the configuration shown in Figure 4.1.7, in which a mass m is attached to one end of a rigid, but weightless, rod of length L.

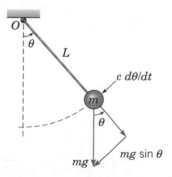

FIGURE 4.1.7 The gravitational force component $mg \sin \theta$ and damping force $c \, d\theta/dt$ are tangent to the path of the mass and act in the clockwise direction when $\theta > 0$ and $d\theta/dt > 0$.

The other end of the rod is supported at the origin O, and the rod is free to rotate in the plane of the paper. The position of the pendulum is described by the angle θ between the rod and the downward vertical direction, with the counterclockwise direction taken as positive. Since the mass is constrained to move along a circular path of radius L, the distance of the mass from its vertical downward position measured along the arc is $L\theta$. The instantaneous tangential velocity and instantaneous tangential acceleration are then given by $L\theta'$ and $L\theta''$, respectively. We consider two forces acting on the mass. The component of gravitational force tangent to the path of motion is $-mg \sin \theta$. Note that if θ is positive, this force acts in the clockwise direction; if θ is negative, then the force is counterclockwise. We also assume the presence of a damping force proportional to the angular velocity, $-c\, d\theta/dt$, where c is positive, that is always opposite to the direction of motion. This force may arise from resistance from the air or other medium in which the mass moves, friction in the pivot mechanism, or some mechanical device that imparts a resistive force to the pendulum arm or bob. The differential equation that describes the motion of the pendulum is derived by applying Newton's second law $ma = F_{\text{net}}$ along the line tangent to the path of motion,

$$mL\frac{d^2\theta}{dt^2} = -c\frac{d\theta}{dt} - mg \sin \theta, \tag{22}$$

or

$$\frac{d^2\theta}{dt^2} + \gamma\frac{d\theta}{dt} + \omega^2 \sin \theta = 0, \tag{23}$$

where $\gamma = c/mL$ and $\omega^2 = g/L$. Due to the presence of the term $\sin \theta$, Eq. (23) cannot be written in the form of Eq. (4). Thus Eq. (23) is **nonlinear**. However, if in its dynamic state the angle of deflection θ is always small, then $\sin \theta \cong \theta$ (see Figure 4.1.8.) and Eq. (23) can be approximated by the equation

$$\frac{d^2\theta}{dt^2} + \gamma\frac{d\theta}{dt} + \omega^2\theta = 0, \tag{24}$$

a constant coefficient linear equation.

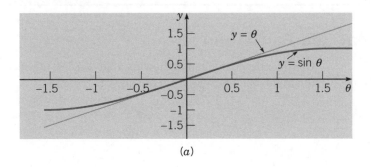

(a)

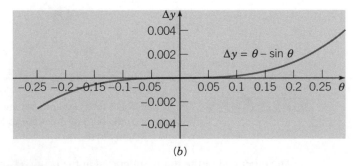

(b)

FIGURE 4.1.8 (a) Graphs of $y = \sin \theta$ and $y = \theta$. (b) The difference $|\Delta y| = |\theta - \sin \theta| < 0.0026$ for $|\theta| \leq 0.25$ radians.

The process of approximating a nonlinear equation by a linear one is called **linearization**. It is an extremely useful method for studying nonlinear systems that operate near an equilibrium point, which, in terms of state variables for the simple pendulum, corresponds to $\theta = 0$ and $\theta' = 0$.

The Series RLC Circuit

A third example of a second order linear differential equation with constant coefficients is a model of flow of electric current in the simple series circuit shown in Figure 4.1.9.

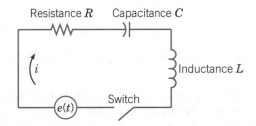

FIGURE 4.1.9 The series *RLC* circuit.

The current i, measured in amperes, is a function of time t. The resistance R (ohms), the capacitance C (farads), and the inductance L (henries) are all positive and are assumed to be known constants. The impressed voltage e (volts) is a given function of time. Another physical quantity that enters the discussion is the total charge q (coulombs) on the capacitor at time t. The relation between charge q and current i is

$$i = \frac{dq}{dt}. \tag{25}$$

The flow of current in the circuit is governed by Kirchhoff's voltage law: *in a closed circuit, the impressed voltage is equal to the algebraic sum of the voltages across the elements in the rest of the circuit.*

According to the elementary laws of electricity, we know that:

The voltage across the resistor is iR.

The voltage across the capacitor is q/C.

The voltage across the inductor is $L\,di/dt$.

Hence, by Kirchhoff's law,

$$L\frac{di}{dt} + Ri + \frac{1}{C}q = e(t). \tag{26}$$

The units have been chosen so that

1 volt = 1 ohm · 1 ampere = 1 coulomb/1 farad = 1 henry · 1 ampere/1 second.

Substituting for i from Eq. (25), we obtain the differential equation

$$Lq'' + Rq' + \frac{1}{C}q = e(t) \tag{27}$$

for the charge q. The initial conditions are

$$q(t_0) = q_0, \qquad q'(t_0) = i(t_0) = i_0. \tag{28}$$

Thus we must know the charge on the capacitor and the current in the circuit at some initial time t_0.

Alternatively, we can obtain a differential equation for the current i by differentiating Eq. (27) with respect to t, and then substituting for di/dt from Eq. (25). The result is

$$Li'' + Ri' + \frac{1}{C}i = e'(t), \tag{29}$$

with the initial conditions

$$i(t_0) = i_0, \qquad i'(t_0) = i_0'. \tag{30}$$

From Eq. (26) it follows that

$$i_0' = \frac{e(t_0) - Ri_0 - (1/C)q_0}{L}. \tag{31}$$

Hence i_0' is also determined by the initial charge and current, which are physically measurable quantities.

Remark. The most important conclusion from this discussion is that the flow of current in the circuit is described by an initial value problem of precisely the same form as the one that describes the motion of a spring-mass system or a pendulum exhibiting small amplitude oscillations. This is a good example of the unifying role of mathematics: once you know how to solve second order linear equations with constant coefficients, you can interpret the results in terms of mechanical vibrations, electric circuits, or any other physical situation that leads to the same differential equation.

PROBLEMS

In Problems 1 through 5, determine whether the differential equation is linear or nonlinear. If the equation is linear, is it homogeneous or nonhomogeneous?

1. $y'' + ty = 0$ (Airy's equation)

2. $y'' + y' + y + y^3 = 0$ (Duffing's equation)

3. $(1 - x^2)y'' - 2xy' + \alpha(\alpha + 1)y = 0$ (Legendre's equation)

4. $x^2 y'' + xy' + (x^2 - \nu^2)y = 0$ (Bessel's equation)

5. $y'' + \mu(1 - y^2)y' + y = 0$ (van der Pol's equation)

6. $y'' - ty = 1/\pi$ (Scorer's function)

7. $ax^2 y'' + bxy' + cy = d$ (Cauchy–Euler equation)

8. A mass weighing 8 lb stretches a spring 6 in. What is the spring constant for this spring?

9. A 10-kg mass attached to a vertical spring is slowly lowered to its equilibrium position. If the resulting change in the length of the spring from its rest length is 70 cm, what is the spring constant for this spring?

For each spring-mass system or electric circuit in Problems 10 through 17, write down the appropriate initial value problem based on the physical description.

10. A mass weighing 2 lb stretches a spring 6 in. The mass is pulled down an additional 3 in. and then released. Assume there is no damping.

11. A mass of 100 g stretches a spring 20 cm. The mass is set in motion from its equilibrium position with a downward velocity of 5 cm/s. Assume there is no damping.

12. A mass weighing 3 lb stretches a spring 3 in. The mass is pushed upward, contracting the spring a distance of 1 in., and then set in motion with a downward velocity of 2 ft/s. Assume there is no damping.

13. A series circuit has a capacitor of 0.25 microfarad and an inductor of 1 henry. The initial charge on the capacitor is 10^{-6} coulomb and there is no initial current.

14. A mass of 20 g stretches a spring 5 cm. Suppose that the mass is also attached to a viscous damper with a damping constant of 400 dyne·s/cm. The mass is pulled down an additional 2 cm and then released.

15. A mass weighing 16 lb stretches a spring 3 in. The mass is attached to a viscous damper with a damping constant of 2 lb·s/ft. The mass is set in motion from its equilibrium position with a downward velocity of 3 in./s.

16. A spring is stretched 10 cm by a force of 3 newtons (N). A mass of 2 kg is hung from the spring and is also attached to a viscous damper that exerts a force of 3 N when the velocity of the mass is 5 m/s. The mass is pulled down 5 cm below its equilibrium position and given an initial upward velocity of 10 cm/s.

17. A series circuit has a capacitor of 10^{-5} farad, a resistor of 3×10^2 ohms, and an inductor of 0.2 henry. The initial charge on the capacitor is 10^{-6} coulomb and there is no initial current.

18. Suppose that a mass m slides without friction on a horizontal surface. The mass is attached to a spring with spring

constant k, as shown in Figure 4.1.10, and is also subject to viscous air resistance with coefficient γ and an applied external force $F(t)$. Assuming that the force exerted on the mass by the spring obeys Hooke's law, show that the displacement $x(t)$ of the mass from its equilibrium position satisfies Eq. (18). How does the derivation of the equation of motion in this case differ from the derivation given in the text?

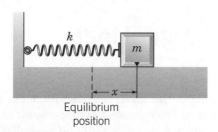

FIGURE 4.1.10 A mass attached to a spring and constrained to one-dimensional motion on a horizontal, frictionless surface.

19. Duffing's Equation
(a) For the spring-mass system shown in Figure 4.1.10, derive the differential equation for the displacement $x(t)$ of the mass from its equilibrium position using the force law $F_s(\Delta x) = -k\Delta x - \epsilon (\Delta x)^3$, where Δx represents the change in length of the spring from its natural length. Assume that damping forces and external forces are not present.
(b) Find the linearized equation for the differential equation derived in part (a) under the assumption that, in its dynamic state, the maximum displacement of the mass from its equilibrium position is small.

20. A body of mass m is attached between two springs with spring constants k_1 and k_2, as shown in Figure 4.1.11. The springs are at their rest length when the system is in the equilibrium state. Assume that the mass slides without friction but the motion is subject to viscous air resistance with coefficient γ. Find the differential equation satisfied by the displacement $x(t)$ of the mass from its equilibrium position.

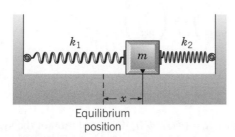

FIGURE 4.1.11 A mass attached between two springs and constrained to one-dimensional motion on a horizontal, frictionless surface.

21. A cubic block of side l and mass density ρ per unit volume is floating in a fluid of mass density ρ_0 per unit volume, where $\rho_0 > \rho$. If the block is slightly depressed and then released, it oscillates in the vertical direction. Assuming that the viscous damping of the fluid and air can be neglected, derive the differential equation of motion for this system.
Hint: Use Archimedes's principle: An object that is completely or partially submerged in a fluid is acted on by an upward (buoyant) force equal to the weight of the displaced fluid.

In Problems 22 through 26, we specify the mass, damping constant, and spring constant of an unforced spring-mass system

$$my'' + \gamma y' + ky = 0. \qquad \text{(i)}$$

Convert (i) to a planar system for the state vector $\mathbf{y} = x_1\mathbf{i} + x_2\mathbf{j} \overset{\text{def}}{=} y\mathbf{i} + y'\mathbf{j}$ and use an IVP solver to:
(a) Draw component plots of the solution of the IVP.
(b) Draw a direction field and phase portrait for the system.
(c) From the plot(s) in part (b), determine whether each critical point is asymptotically stable, stable, or unstable, and classify it as to type.

22. $m = 1$ kg, $\gamma = 0$ kg/s, $k = 1$ kg/s^2, $y(0) = 1$ m, $y'(0) = 0$ m/s (weak spring)

23. $m = 1$ kg, $\gamma = 0$ kg/s, $k = 9$ kg/s^2, $y(0) = 1$ m, $y'(0) = 0$ m/s (stiff spring)

24. $m = 1$ kg, $\gamma = 1$ kg/s, $k = 16$ kg/s^2, $y(0) = 1$ m, $y'(0) = 0$ m/s (weak damping)

25. $m = 1$ kg, $\gamma = 3$ kg/s, $k = 4$ kg/s^2, $y(0) = 1$ m, $y'(0) = 0$ m/s (strong damping)

26. $m = 1$ kg, $\gamma = -1$ kg/s, $k = 4$ kg/s^2, $y(0) = 1$ m, $y'(0) = 0$ m/s (negative damping)

27. The Linear Versus the Nonlinear Pendulum.
Convert Eq. (23) and Eq. (24) to planar systems using the state vector $\mathbf{x} = x_1\mathbf{i} + x_2\mathbf{j} = \theta\mathbf{i} + \theta'\mathbf{j}$ for each equation. For each of the following cases of parameter values and initial conditions, use an IVP solver to draw, on the same set of coordinate axes, plots of the first component of both the nonlinear and linearized system.
(a) $\gamma = 0$, $\omega^2 = 1$, $\theta(0) = \pi/8$, $\theta'(0) = 0$
(b) $\gamma = 0$, $\omega^2 = 1$, $\theta(0) = \pi/4$, $\theta'(0) = 0$

Based on the graphical output, discuss the dependence of accuracy of the linear approximation on size of initial displacement and operating time. Are the periods of the nonlinear pendulum and the linear pendulum identical?

28. (a) Numerical simulations as well as intuition based on an understanding of the origin and significance of each term in the equation suggest that solutions of the undamped spring-mass equation

$$my'' + ky = 0 \qquad \text{(i)}$$

are oscillatory. Use substitution to determine a value of r such that $y_1 = \cos rt$ and $y_2 = \sin rt$ are both solutions of

Eq. (i). Then show that $y = c_1y_1 + c_2y_2$ is also a solution of $my'' + ky = 0$ for any constants c_1 and c_2.

(b) Use the results of part (a) to find analytical solutions of the following initial value problems for $my'' + ky = 0$:

i. $m = 1$ kg, $k = 1$ kg/s^2, $y(0) = 1$ m, $y'(0) = -1$ m/s (weak spring)

ii. $m = 1$ kg, $k = 25$ kg/s^2, $y(0) = 1$ m, $y'(0) = -1$ m/s (stiff spring)

(c) Plot graphs of the two solutions obtained above on the same set of coordinate axes and discuss the amplitude and frequency of oscillations.

4.2 Theory of Second Order Linear Homogeneous Equations

In Sections 3.2 and 3.3 we stated some general properties of the solutions of systems of two first order linear equations. Some of these results were stated in the context of linear systems with constant coefficients, whereas others were stated for more general systems. Now we want to extend all of these results to systems with variable coefficients. Since a single second order equation can easily be reduced to a system of two first order equations, these results also hold, as special cases, for second order linear equations. However, our main emphasis in this chapter is on second order linear equations, so we will state and discuss these important theorems separately for such equations.

Existence and Uniqueness of Solutions

Consider the second order linear equation

$$y'' + p(t)y' + q(t)y = g(t), \tag{1}$$

where p, q, and g are continuous functions on the interval I. Initial conditions for Eq. (1) are of the form

$$y(t_0) = y_0, \qquad y'(t_0) = y_1. \tag{2}$$

By introducing the state variables $x_1 = y$ and $x_2 = y'$, we convert Eq. (1) to the first order system

$$
\begin{aligned}
x_1' &= x_2, \\
x_2' &= -q(t)x_1 - p(t)x_2 + g(t),
\end{aligned}
\tag{3}
$$

or, using matrix notation,

$$\mathbf{x}' = \begin{pmatrix} 0 & 1 \\ -q(t) & -p(t) \end{pmatrix} \mathbf{x} + \begin{pmatrix} 0 \\ g(t) \end{pmatrix}, \tag{4}$$

where

$$\mathbf{x} = \begin{pmatrix} x_1 \\ x_2 \end{pmatrix} = \begin{pmatrix} y \\ y' \end{pmatrix}. \tag{5}$$

Thus, if $y = \phi(t)$ is a solution of Eq. (1), then $\mathbf{x} = (\phi(t), \phi'(t))^T$ is a solution of the system (3) or (4). This is easily verified by setting $x_1 = \phi(t)$ and $x_2 = \phi'(t)$ in Eqs. (3),

$$\phi' = \phi'$$

and

$$\phi'' = -q(t)\phi - p(t)\phi' + g(t) \qquad \text{or} \qquad \phi''(t) + p(t)\phi'(t) + q(t)\phi(t) = g(t).$$

Conversely, if $\mathbf{x} = (\phi_1(t), \phi_2(t))^T$ is a solution of the system (3) or (4), then

$$\phi_1' = \phi_2,$$
$$\phi_2' = -q(t)\phi_1 - p(t)\phi_2 + g(t).$$

Setting $\phi_2 = \phi_1'$ in the second equation, we get

$$\phi_1''(t) = -q(t)\phi_1(t) - p(t)\phi_1'(t) + g(t) \qquad \text{or} \qquad \phi_1''(t) + p(t)\phi_1'(t) + q(t)\phi_1(t) = g(t).$$

In summary, $y = \phi(t)$ is a solution of Eq. (1) if and only if $\mathbf{x} = (\phi(t), \phi'(t))^T$ is a solution of Eq. (4). This observation allows us to pass back and forth between solutions of Eq. (1) and Eq. (4) whenever necessary. In particular, since the definition $\mathbf{x} = (y, y')^T$ relates the initial conditions (2) for Eq. (1) and the initial conditions

$$\mathbf{x}(t_0) = \mathbf{x}_0 = \begin{pmatrix} x_{10} \\ x_{20} \end{pmatrix} = \begin{pmatrix} y_0 \\ y_1 \end{pmatrix} \tag{6}$$

for Eq. (4), we can easily move back and forth between solutions of the the initial value problem (1) and (2) and solutions of the initial value problem (4) and (6).

Note that the first order system (3) is a special case of the general system of two first order linear equations

$$dx_1/dt = p_{11}(t)x_1 + p_{12}(t)x_2 + g_1(t),$$
$$dx_2/dt = p_{21}(t)x_1 + p_{22}(t)x_2 + g_2(t), \tag{7}$$

or, using matrix notation,

$$\frac{d\mathbf{x}}{dt} = \mathbf{P}(t)\mathbf{x} + \mathbf{g}(t), \tag{8}$$

where

$$\mathbf{x} = \begin{pmatrix} x_1 \\ x_2 \end{pmatrix}, \qquad \mathbf{P}(t) = \begin{pmatrix} p_{11}(t) & p_{12}(t) \\ p_{21}(t) & p_{22}(t) \end{pmatrix}, \qquad \text{and} \qquad \mathbf{g}(t) = \begin{pmatrix} g_1(t) \\ g_2(t) \end{pmatrix}. \tag{9}$$

There are three reasons why it is either convenient, or necessary, to study solutions of Eq. (1) by studying solutions of Eq. (8), of which Eq. (4) is a special case.

1. In this section we will obtain theoretical results concerning solutions of Eq. (1) as special cases of corresponding theoretical results for solutions of Eq. (8).

2. Most numerical methods require that an initial value problem for Eq. (1) be recast as an initial value problem for the first order system (3) or (4).

3. In Section 4.3 we are led to a simple and direct method for finding solutions of the constant coefficient equation $ay'' + by' + cy = 0$ by using the eigenvalue method discussed in Sections 3.3 through 3.5 to find solutions of the equivalent first order system,

$$\mathbf{x}' = \begin{pmatrix} 0 & 1 \\ -c/a & -b/a \end{pmatrix} \mathbf{x}.$$

Subsequently, we will use this direct method to find solutions of $ay'' + by' + cy = 0$.

Let us recall Theorem 3.2.1, which gives sufficient conditions for the existence of a unique solution to an initial value problem for Eq. (8). If the coefficients $p_{11}(t), \ldots, p_{22}(t)$ and the nonhomogeneous terms $g_1(t)$ and $g_2(t)$ that appear in Eq. (8) are continuous on an open interval I, and the initial condition for Eq. (8) is given by

$$\mathbf{x}(t_0) = \mathbf{x}_0 = \begin{pmatrix} x_{10} \\ x_{20} \end{pmatrix}, \tag{10}$$

where t_0 is any point in I and x_{10} and x_{20} are any given numbers, then Theorem 3.2.1 states that the initial value problem (8) and (10) has a unique solution on the interval I.

Theorem 3.2.1 applies to the initial value problem (4) and (6) as a special case. If the functions $p(t)$, $q(t)$, and $g(t)$ are continuous on an interval I, t_0 is in I, and y_1 and y_2 are any given numbers, then there is a unique solution $\mathbf{x} = (\phi_1(t), \phi_2(t))^T$ on I. The first component of this solution is therefore the unique solution $y = \phi(t) = \phi_1(t)$ of the initial value problem (1) and (2). We state this result in the following theorem.

THEOREM 4.2.1

Let $p(t)$, $q(t)$, and $g(t)$ be continuous on an open interval I, let t_0 be any point in I, and let y_0 and y_0' be any given numbers. Then there exists a unique solution $y = \phi(t)$ of the differential equation (1),

$$y'' + p(t)y' + q(t)y = g(t),$$

that also satisfies the initial conditions (2),

$$y(t_0) = y_0, \qquad y'(t_0) = y_1.$$

Further the solution exists throughout the interval I.

If the coefficients $p(t)$ and $q(t)$ are constants, and if the nonhomogeneous term $g(t)$ is not too complicated, then it is possible to solve the initial value problem (1), (2) by elementary methods. We will show you how to do this later in this chapter. Of course, the construction of a solution demonstrates the existence part of Theorem 4.2.1, although not the uniqueness part. A proof of Theorem 4.2.1 (or the more general Theorem 3.2.1) is fairly difficult, and we do not discuss it here. We will, however, accept the theorem as true and make use of it whenever necessary.

EXAMPLE 1

Find the longest interval in which the solution of the initial value problem

$$(t^2 - 3t)y'' + ty' - (t + 3)y = 0, \qquad y(1) = 2, \quad y'(1) = 1 \tag{11}$$

is certain to exist.

If the given differential equation is written in the form of Eq. (1), then $p(t) = 1/(t - 3)$, $q(t) = -(t + 3)/t(t - 3)$, and $g(t) = 0$. The only points of discontinuity of the coefficients are $t = 0$ and $t = 3$. Therefore the longest open interval, containing the initial point $t_0 = 1$, in which all the coefficients are continuous is $0 < t < 3$. Thus this is the longest interval in which Theorem 4.2.1 guarantees that the solution exists (see Figure 4.2.1).

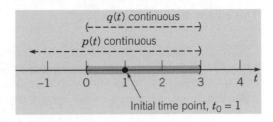

FIGURE 4.2.1 The maximum interval of existence and uniqueness for the IVP (11) guaranteed by Theorem 4.2.1 is $I = (0, 3)$.

Linear Operators and the Principle of Superposition for Linear Homogeneous Equations

For clarity, we first discuss linear operators in the context of scalar equations. Up to this point, we have used the word "linear" to classify an equation by its form. A first order equation is linear if it has the form $y' + p(t)y = g(t)$; a second order equation is linear if it has the form $y'' + p(t)y' + q(t)y = g(t)$. We now examine the property of linearity in more detail.

In an expression such as $dy/dt + p(t)y$, each of the two operations, differentiation and multiplication by a function, map y into another function. Each operation is an example of an **operator** or **transformation**. An operator maps, or associates, one function to another function. Let us use juxtaposition by D to indicate the operation of differentiation, $(Dy)(t) = dy/dt(t)$, and juxtaposition by p to indicate the operation of multiplication, $(py)(t) = p(t)y(t)$. Both of the operators D and p have the special property that they are **linear operators**. By definition, this means for all functions y_1 and y_2 for which the operations make sense (i.e., are in the domain of the operator) and all scalars c_1 and c_2,

$$D[c_1 y_1 + c_2 y_2] = c_1 Dy_1 + c_2 Dy_2 \quad \text{and} \quad p[c_1 y_1 + c_2 y_2] = c_1 p y_1 + c_2 p y_2. \tag{12}$$

Using these two operations, we can construct more elaborate linear operators. For example, if p and q are continuous functions on an interval I, we can define the second order **differential operator** L by

$$L = D^2 + pD + q \quad \text{or} \quad L = \frac{d^2}{dt^2} + p\frac{d}{dt} + q,$$

on the domain $\mathcal{D}(L)$ consisting of all continuous functions y on I such that y' and y'' exist and are continuous on I. The value of $L[y]$ at a point t is

$$L[y](t) = y''(t) + p(t)y'(t) + q(t)y(t).$$

Clearly, if $y \in \mathcal{D}(L)$, Ly is a continuous function. In terms of L, the nonhomogeneous differential equation $y'' + p(t)y' + q(t)y = g(t)$ is expressed as $L[y] = g$ and the corresponding homogeneous equation $y'' + p(t)y' + q(t)y = 0$ is expressed as $L[y] = 0$.

To see that L is linear, we need to show that $L[c_1 y_1 + c_2 y_2] = c_1 L[y_1] + c_2 L[y_2]$ for y_1 and y_2 in $\mathcal{D}(L)$ and any pair of constants c_1 and c_2. On I, we have

$$\begin{aligned}
L[c_1 y_1 + c_2 y_2] &= (c_1 y_1 + c_2 y_2)'' + p(c_1 y_1 + c_2 y_2)' + q(c_1 y_1 + c_2 y_2) \\
&= c_1 y_1'' + c_2 y_2'' + p(c_1 y_1' + c_2 y_2') + q(c_1 y_1 + c_2 y_2) \\
&= c_1(y_1'' + py_1' + qy_1) + c_2(y_2'' + py_2' + qy_2) \\
&= c_1 L[y_1] + c_2 L[y_2].
\end{aligned}$$

This proves the following theorem.

THEOREM 4.2.2

Linearity of the differential operator L. Let $L[y] = y'' + py' + qy$, where p and q are continuous functions on an interval I. If y_1 and y_2 are any twice continuously differentiable functions on I, and c_1 and c_2 are any constants, then

$$L[c_1 y_1 + c_2 y_2] = c_1 L[y_1] + c_2 L[y_2]. \tag{13}$$

The differential equations $L[y] = 0$ and $L[y] = g$ are said to be **linear** if L satisfies Eq. (13). If L is linear and y_1 and y_2 are solutions of the homogeneous equation $L[y] = 0$, then $c_1 y_1 + c_2 y_2$ is also a solution of $L[y] = 0$ for any pair of constants c_1 and c_2, since, by Eq. (13),

$$L[c_1 y_1 + c_2 y_2] = c_1 L[y_1] + c_2 L[y_2] = 0. \tag{14}$$

The property (14) is referred to as the **principle of superposition for linear homogeneous equations**. Given any pair of solutions y_1 and y_2 to $L[y] = 0$, this property enables you to enlarge the set of solutions to $L[y] = 0$ by taking linear combinations of y_1 and y_2. We state this important result in the following corollary to Theorem 4.2.2.

COROLLARY 4.2.3

Let $L[y] = y'' + py' + qy$, where p and q are continuous functions on an interval I. If y_1 and y_2 are two solutions of $L[y] = 0$, then the linear combination

$$y = c_1 y_1(t) + c_2 y_2(t) \tag{15}$$

is also a solution for any values of the constants c_1 and c_2.

If a differential operator does not satisfy the property (13), then it is **nonlinear** and the corresponding homogeneous and nonhomogeneous differential equations are said to be nonlinear. See Problem 16 for an example of a nonlinear operator.

The preceding discussion easily extends to the homogeneous system

$$\mathbf{x}' = \mathbf{P}(t)\mathbf{x}. \tag{16}$$

If the entries of $\mathbf{P}$ are continuous functions on an interval I and the components of $\mathbf{x}$ are continuously differentiable on I, we define the operator $\mathbf{K}$ by

$$\mathbf{K}[\mathbf{x}] = \mathbf{x}' - \mathbf{P}(t)\mathbf{x}. \tag{17}$$

The following theorem and corollary are parallels to Theorem 4.2.2 and Corollary 4.2.3. Their proofs are left as exercises (see Problem 27).

THEOREM 4.2.4

Let $\mathbf{K}[\mathbf{x}] = \mathbf{x}' - \mathbf{P}(t)\mathbf{x}$, where the entries of $\mathbf{P}$ are continuous functions on an interval I. If $\mathbf{x}_1$ and $\mathbf{x}_2$ are continuously differentiable vector functions on I, and c_1 and c_2 are any constants, then

$$\mathbf{K}[c_1\mathbf{x}_1 + c_2\mathbf{x}_2] = c_1\mathbf{K}[\mathbf{x}_1] + c_2\mathbf{K}[\mathbf{x}_2]. \tag{18}$$

The first order systems $\mathbf{K}[\mathbf{x}] = \mathbf{0}$ and $\mathbf{K}[\mathbf{x}] = \mathbf{g}$ are said to be linear if $\mathbf{K}$ satisfies Eq. (18). If $\mathbf{K}$ does not satisfy Eq. (18), the systems are nonlinear.

COROLLARY 4.2.5	Let $\mathbf{K}[\mathbf{x}] = \mathbf{x}' - \mathbf{P}(t)\mathbf{x}$ and suppose the entries of $\mathbf{P}$ are continuous functions on an interval I. If $\mathbf{x}_1$ and $\mathbf{x}_2$ are two solutions of $\mathbf{K}[\mathbf{x}] = \mathbf{0}$, then the linear combination

$$\mathbf{x} = c_1\mathbf{x}_1(t) + c_2\mathbf{x}_2(t) \tag{19}$$

is also a solution for any values of the constants c_1 and c_2.

Wronskians and Fundamental Sets of Solutions

Corollaries 4.2.3 and 4.2.5 enable you to enlarge the set of solutions that you have, provided that you are able to find two solutions to start with. The next question is whether the enlarged solution set is sufficient for us to satisfy a given set of initial conditions by an appropriate choice of the constants c_1 and c_2. For the constant coefficient system $\mathbf{x}' = \mathbf{A}\mathbf{x}$, this question was answered in Section 3.3. The same result also holds for the system $\mathbf{x}' = \mathbf{P}(t)\mathbf{x}$ whose coefficients are functions of t, and it can be established by the same argument. We briefly repeat the development here.

Let us suppose that $\mathbf{x}_1$ and $\mathbf{x}_2$ are solutions of Eq. (16). By Corollary 4.2.5, the linear combination

$$\mathbf{x} = c_1\mathbf{x}_1(t) + c_2\mathbf{x}_2(t)$$

is also a solution for any values of c_1 and c_2. To satisfy the initial condition (10),

$$\mathbf{x}(t_0) = \mathbf{x}_0,$$

where $\mathbf{x}_0$ is any given vector, we must choose c_1 and c_2 to satisfy

$$c_1\mathbf{x}_1(t_0) + c_2\mathbf{x}_2(t_0) = \mathbf{x}_0. \tag{20}$$

Writing out Eq. (20) more fully, we have

$$\begin{pmatrix} x_{11}(t_0) & x_{12}(t_0) \\ x_{21}(t_0) & x_{22}(t_0) \end{pmatrix} \begin{pmatrix} c_1 \\ c_2 \end{pmatrix} = \begin{pmatrix} x_{10} \\ x_{20} \end{pmatrix}. \tag{21}$$

Recall that the first subscript refers to the component of a vector, whereas the second subscript identifies the vector itself; thus, for example, x_{21} is the second component of the vector $\mathbf{x}_1$. Equation (21) has a unique solution for c_1 and c_2 regardless of the values of x_{10} and x_{20} if and only if the coefficient matrix in Eq. (21) is nonsingular, that is, if and only if the determinant of this matrix is nonzero. Since t_0 can be any point in I, the determinant must be nonzero throughout I.

As in Section 3.3, the determinant

$$W[\mathbf{x}_1, \mathbf{x}_2](t) = \begin{vmatrix} x_{11}(t) & x_{12}(t) \\ x_{21}(t) & x_{22}(t) \end{vmatrix} \tag{22}$$

is called the Wronskian of $\mathbf{x}_1$ and $\mathbf{x}_2$. Two solutions $\mathbf{x}_1$ and $\mathbf{x}_2$ whose Wronskian is not zero form a fundamental set of solutions. A linear combination of a fundamental set of solutions with arbitrary coefficients constitutes the **general solution** of the system (16). Since every possible set of initial conditions can be satisfied by a proper choice of c_1 and c_2, the general solution includes all solutions of the system (16). This last statement rests on the uniqueness part of Theorem 3.2.1, which asserts that only one solution can satisfy a given set of initial conditions. We have thus proved the following theorem.

THEOREM 4.2.6

Suppose that $\mathbf{x}_1$ and $\mathbf{x}_2$ are two solutions of Eq. (16),

$$\mathbf{x}' = \mathbf{P}(t)\mathbf{x},$$

and that their Wronskian is not zero on I. Then $\mathbf{x}_1$ and $\mathbf{x}_2$ form a fundamental set of solutions of Eq. (16), and the general solution is given by Eq. (19),

$$\mathbf{x} = c_1\mathbf{x}_1(t) + c_2\mathbf{x}_2(t),$$

where c_1 and c_2 are arbitrary constants. If there is a given initial condition $\mathbf{x}(t_0) = \mathbf{x}_0$, then this condition determines the constants c_1 and c_2 uniquely.

It is convenient to apply Theorem 4.2.6 to the system

$$\frac{d\mathbf{x}}{dt} = \begin{pmatrix} 0 & 1 \\ -q(t) & -p(t) \end{pmatrix} \mathbf{x} \tag{23}$$

in order to discuss what constitutes a general solution for the equation

$$L[y] = y'' + p(t)y' + q(t)y = 0. \tag{24}$$

Suppose that $p(t)$ and $q(t)$ are continuous on an interval I and that initial conditions $y(t_0) = b_0$ and $y'(t_0) = b_1$, $t_0 \in I$, are prescribed. Then Theorem 4.2.6 applies to the system (23) on I. If y_1 and y_2 are two solutions of Eq. (24), then the vector functions

$$\mathbf{x}_1 = \begin{pmatrix} y_1 \\ y'_1 \end{pmatrix} \qquad \text{and} \qquad \mathbf{x}_2 = \begin{pmatrix} y_2 \\ y'_2 \end{pmatrix}$$

constitute a fundamental set of solutions for Eq. (23) if

$$W[\mathbf{x}_1, \mathbf{x}_2](t) = \begin{vmatrix} y_1(t) & y_2(t) \\ y'_1(t) & y'_2(t) \end{vmatrix} = y_1(t)y'_2(t) - y'_1(t)y_2(t) \tag{25}$$

is not zero on I. If y_1 and y_2 are solutions of Eq. (24), then it is natural to define the Wronskian of y_1 and y_2 by the right-hand side of Eq. (25),

$$W[y_1, y_2](t) = \begin{vmatrix} y_1(t) & y_2(t) \\ y'_1(t) & y'_2(t) \end{vmatrix} = y_1(t)y'_2(t) - y'_1(t)y_2(t). \tag{26}$$

If $W[y_1, y_2](t)$ is not zero on I, we refer to the first component of $\mathbf{x} = c_1\mathbf{x}_1 + c_2\mathbf{x}_2$, namely,

$$y = c_1y_1(t) + c_2y_2(t),$$

as the general solution of Eq. (24). The following theorem is a parallel to Theorem 4.2.6.

THEOREM 4.2.7

Let y_1 and y_2 be two solutions of Eq. (25),

$$y'' + p(t)y' + q(t)y = 0,$$

and assume that their Wronskian, defined by Eq. (26), is not zero on I. Then y_1 and y_2 form a fundamental set of solutions, and the general solution is given by

$$y = c_1 y_1(t) + c_2 y_2(t),$$

where c_1 and c_2 are arbitrary constants. If there are given initial conditions $y(t_0) = y_0$ and $y'(t_0) = y_1$, then these conditions determine c_1 and c_2 uniquely.

Theorems 4.2.6 and 4.2.7 clarify the structure of the solutions of all systems of two linear homogeneous first order equations, and of all second order linear homogeneous equations, respectively. Assuming that the coefficients are continuous in each case, the general solution is just a linear combination of some pair of solutions whose Wronskian determinant is not zero.

In particular, since the Wronskians for sets of solutions of Eq. (23) and Eq. (24) are identical, we can easily pass back and forth between fundamental sets of solutions for both equations. If $y_1(t)$ and $y_2(t)$ form a fundamental set of solutions for Eq. (24), then

$$\mathbf{x}_1(t) = \begin{pmatrix} y_1(t) \\ y_1'(t) \end{pmatrix} \qquad \text{and} \qquad \mathbf{x}_2(t) = \begin{pmatrix} y_2(t) \\ y_2'(t) \end{pmatrix}$$

form a fundamental set of solutions for Eq. (23). If $y = c_1 y_1(t) + c_2 y_2(t)$ is a general solution of Eq. (24), then

$$\mathbf{x} = c_1 \begin{pmatrix} y_1(t) \\ y_1'(t) \end{pmatrix} + c_2 \begin{pmatrix} y_2(t) \\ y_2'(t) \end{pmatrix}$$

is a general solution of Eq. (23).

Conversely, if

$$\mathbf{x}_1(t) = \begin{pmatrix} x_{11}(t) \\ x_{21}(t) \end{pmatrix} \qquad \text{and} \qquad \mathbf{x}_2(t) = \begin{pmatrix} x_{12}(t) \\ x_{22}(t) \end{pmatrix}$$

form a fundamental set of solutions for Eq. (23), then the first components of the solution vectors, $y_1(t) = x_{11}(t)$ and $y_2(t) = x_{12}(t)$, comprise a fundamental set of solutions for Eq. (24) and the second components satisfy $x_{21}(t) = x_{11}'(t)$ and $x_{22}(t) = x_{21}'(t)$. Thus, given the general solution

$$\mathbf{x} = c_1 \begin{pmatrix} x_{11}(t) \\ x_{21}(t) \end{pmatrix} + c_2 \begin{pmatrix} x_{12}(t) \\ x_{22}(t) \end{pmatrix}$$

of Eq. (23), the general solution of Eq. (24) is $y = c_1 x_{11}(t) + c_2 x_{12}(t)$. In addition we know that $y' = c_1 x_{21}(t) + c_1 x_{22}(t)$.

We have seen that it is relatively easy to find a fundamental solution set for a system of two first order equations with constant coefficients by using the eigenvalue method discussed in Sections 3.3 through 3.5. In Section 4.3 we will use the relationship between solutions of Eq. (23) and Eq. (24) to reveal a simple and direct method for finding fundamental sets of solutions for second order linear equations with constant coefficients. However, for systems or equations with variable coefficients, the task of finding a suitable pair of solutions is usually much more challenging.

Abel's Equation for the Wronskian

The Wronskian, as defined by Eq. (22) for a system of first order equations or by Eq. (26) for a second order equation, can be calculated once we know two solutions. However it turns out that the Wronskian can also be determined directly from the differential equation(s), without knowing any solutions. The following theorem summarizes the most important properties of the Wronskian of two solutions of Eq. (16).

THEOREM
4.2.8

Abel's Theorem. The Wronskian of two solutions of the system (16),

$$\mathbf{x}' = \mathbf{P}(t)\mathbf{x},$$

is given by

$$W(t) = c\left[\exp\int \text{tr}(\mathbf{P})(t)\,dt\right] = c\exp\int [p_{11}(t) + p_{22}(t)]\,dt, \tag{27}$$

where c is a constant that depends on the pair of solutions. The Wronskian of two solutions of the second order equation (24),

$$y'' + p(t)y' + q(t)y = 0,$$

is given by

$$W(t) = c\exp\left[-\int p(t)\,dt\right], \tag{28}$$

where again c is a constant that depends on the pair of solutions.

Proof

The proof of this theorem involves the construction and solution of a differential equation satisfied by the Wronskian. We start with the Wronskian for the system (16); from Eq. (22) we have

$$W(t) = \begin{vmatrix} x_{11}(t) & x_{12}(t) \\ x_{21}(t) & x_{22}(t) \end{vmatrix} = x_{11}(t)x_{22}(t) - x_{12}(t)x_{21}(t). \tag{29}$$

Hence the derivative of $W(t)$ is

$$W'(t) = x'_{11}(t)x_{22}(t) + x_{11}(t)x'_{22}(t) - x'_{12}(t)x_{21}(t) - x_{12}(t)x'_{21}(t). \tag{30}$$

The next step is to replace all of the differentiated factors on the right side of Eq. (30) by the corresponding equation from the system of differential equations. For example, we have

$$x'_{11}(t) = p_{11}(t)x_{11}(t) + p_{12}(t)x_{21}(t)$$

and similarly for $x'_{22}(t)$, $x'_{12}(t)$, and $x'_{21}(t)$. By making these substitutions, we obtain

$$W' = (p_{11}x_{11} + p_{12}x_{21})x_{22} + x_{11}(p_{21}x_{12} + p_{22}x_{22}) \\ - (p_{11}x_{12} + p_{12}x_{22})x_{21} - x_{12}(p_{21}x_{11} + p_{22}x_{21}), \tag{31}$$

where for brevity we have omitted the independent variable t in each term. On examining the terms in Eq. (31), we observe that those involving p_{12} and p_{21} cancel. By rearranging the remaining terms, we find that

$$W' = (p_{11} + p_{22})(x_{11}x_{22} - x_{12}x_{21}),$$

or

$$\frac{dW}{dt} = [p_{11}(t) + p_{22}(t)]\, W = \text{tr}(\mathbf{P})(t)\, W. \tag{32}$$

Equation (32) is a first order separable or linear differential equation that is satisfied by the Wronskian. Its solution is readily found and is given by Eq. (27) in Theorem 4.2.8.

To establish Eq. (28), we can apply the result (27) to the system (23) that corresponds to the second order equation. For this system, $p_{11}(t) + p_{22}(t) = -p(t)$; making this substitution in Eq. (27) immediately yields Eq. (28). Equation (28) can also be derived directly from Eq. (24) by a process similar to the derivation of Eq. (27) that is given above.

EXAMPLE 2

Find the Wronskian of any pair of solutions of the system

$$\mathbf{x}' = \begin{pmatrix} 1 & 1 \\ 4 & 1 \end{pmatrix} \mathbf{x}. \tag{33}$$

The trace of the coefficient matrix is 2; therefore by Eq. (27) the Wronskian of any pair of solutions is

$$W(t) = ce^{2t}. \tag{34}$$

In Example 4 of Section 3.3, we found two solutions of the system (33) to be

$$\mathbf{x}_1(t) = \begin{pmatrix} 1 \\ 2 \end{pmatrix} e^{3t}, \qquad \mathbf{x}_2(t) = \begin{pmatrix} 1 \\ -2 \end{pmatrix} e^{-t}.$$

The Wronskian of these two solutions is

$$W(t) = \begin{vmatrix} e^{3t} & e^{-t} \\ 2e^{3t} & -2e^{-t} \end{vmatrix} = -4e^{2t}.$$

Thus for this pair of solutions, the multiplicative constant in Eq. (30) is -4. For other pairs of solutions, the multiplicative constant in the Wronskian may be different, but the exponential part will always be the same.

EXAMPLE 3

Find the Wronskian of any pair of solutions of

$$(1 - t)y'' + ty' - y = 0. \tag{35}$$

First we rewrite Eq. (35) in the standard form (24):

$$y'' + \frac{t}{1 - t} y' - \frac{1}{1 - t} y = 0. \tag{36}$$

Thus $p(t) = t/(1 - t)$ and, omitting the constant of integration, we get

$$-\int p(t)\, dt = \int \frac{t}{t - 1}\, dt = \int \left(1 + \frac{1}{t - 1}\right) dt = t + \ln|t - 1|.$$

Consequently,

$$\exp\left[-\int p(t)\,dt\right] = |t-1|e^t$$

and Theorem 4.2.8 gives

$$W(t) = c(t-1)e^t. \tag{37}$$

Note that $|t-1| = \pm(t-1)$ and that the $\pm$ sign has been incorporated into the arbitrary constant c in Eq. (37).

You can verify that $y_1(t) = t$ and $y_2(t) = e^t$ are two solutions of Eq. (35) by substituting these two functions into the differential equation. The Wronskian of these two solutions is

$$W(t) = \begin{vmatrix} t & e^t \\ 1 & e^t \end{vmatrix} = (t-1)e^t. \tag{38}$$

Thus, for these two solutions, the constant in Eq. (37) is 1.

An important property of the Wronskian follows immediately from Theorem 4.2.8. If $p_{11}(t)$ and $p_{22}(t)$ in Eq. (27) or $p(t)$ in Eq. (28) are continuous on I, then the exponential functions in these equations are always positive. If the constant c is nonzero, then the Wronskian is never zero on I. On the other hand, if $c = 0$, then the Wronskian is zero for all t in I. Thus the Wronskian is either never zero or always zero. In a particular problem, you can determine which alternative is the actual one by evaluating the Wronskian at some single convenient point. We state these conclusions in the following corollary to Theorem 4.2.8.

COROLLARY 4.2.9

If $\mathbf{x}_1(t)$ and $\mathbf{x}_2(t)$ are two solutions of the system (16) $\mathbf{x}' = \mathbf{P}(t)\mathbf{x}$, and all entries of $\mathbf{P}(t)$ are continuous on an open interval I, then the Wronskian $W[\mathbf{x}_1, \mathbf{x}_2](t)$ is either never zero or always zero in I.

If $y_1(t)$ and $y_2(t)$ are two solutions of the second order equation (24)

$$y'' + p(t)y' + q(t)y = 0,$$

where p and q are continuous on an open interval I, then the Wronskian $W[y_1, y_2](t)$ is either never zero or always zero in I.

It may seem at first that Eq. (38) in Example 3 is a counterexample to this assertion about the Wronskian because $W(t)$ given by Eq. (38) is obviously zero for $t = 1$ and nonzero otherwise. However, by writing Eq. (35) in the form (36), we see that the coefficients $p(t)$ and $q(t)$ become unbounded as $t \to 1$ and hence are discontinuous there. In any interval I not including $t = 1$, the coefficients are continuous and the Wronskian is nonzero, as claimed.

PROBLEMS

In each of Problems 1 through 8, determine the longest interval in which the given initial value problem is certain to have a unique twice differentiable solution. Do not attempt to find the solution.

1. $ty'' + 3y = t$, $y(1) = 1$, $y'(1) = 2$

2. $(t-1)y'' - 3ty' + 4y = \sin t$, $y(-2) = 2$, $y'(-2) = 1$

3. $t(t-4)y'' + 3ty' + 4y = 2$, $y(3) = 0$, $y'(3) = -1$

4. $y'' + (\cos t)y' + 3(\ln|t|)y = 0$, $y(2) = 3$, $y'(2) = 1$

5. $(x+3)y'' + xy' + (\ln|x|)y = 0$, $y(1) = 0$, $y'(1) = 1$

6. $(x-2)y'' + y' + (x-2)(\tan x)y = 0$, $y(3) = 1$, $y'(3) = 2$

7. $(1-x^2)y'' - 2xy' + (\alpha(\alpha+1) + \mu^2/(1-x^2))y = 0$, $y(0) = y_0$, $y'(0) = y_1$

8. $y'' - t/y = 1/\pi$, $y(0) = y_0$, $y'(0) = y_1$

In each of Problems 9 through 14, find the Wronskian of the given pair of functions.

9. $e^{2t}, \quad e^{-3t/2}$

10. $\cos t, \quad \sin t$

11. $e^{-2t}, \quad te^{-2t}$

12. $x, \quad xe^x$

13. $e^t \sin t, \quad e^t \cos t$

14. $\cos^2 \theta, \quad 1 + \cos 2\theta$

15. Verify that $y_1(t) = t^2$ and $y_2(t) = t^{-1}$ are two solutions of the differential equation $t^2 y'' - 2y = 0$ for $t > 0$. Then show that $c_1 t^2 + c_2 t^{-1}$ is also a solution of this equation for any c_1 and c_2.

16. Consider the differential operator T defined by $T[y] = yy'' + (y')^2$. Show that $T[y] = 0$ is a nonlinear equation by three methods below.

(a) Explain why $T[y] = 0$ cannot be put in the form of a linear equation $L[y] = y'' + py' + qy = 0$.

(b) Show that T fails to satisfy the definition of a linear operator and therefore the equation $T[y] = 0$ is a nonlinear equation.

(c) Verify that $y_1(t) = 1$ and $y_2(t) = t^{1/2}$ are solutions of the differential equation $T[y] = 0$ for $t > 0$, but $c_1 + c_2 t^{1/2}$ is not, in general, a solution of this equation.

17. Can an equation $y'' + p(t)y' + q(t)y = 0$, with continuous coefficients, have $y = \sin(t^2)$ as a solution on an interval containing $t = 0$? Explain your answer.

18. If the Wronskian W of f and g is $3e^{2t}$, and if $f(t) = e^{4t}$, find $g(t)$.

19. If the Wronskian W of f and g is $t^2 e^t$, and if $f(t) = t$, find $g(t)$.

20. If $W[f, g]$ is the Wronskian of f and g, and if $u = 2f - g$, $v = f + 2g$, find the Wronskian $W[u, v]$ of u and v in terms of $W[f, g]$.

21. If the Wronskian of f and g is $t\cos t - \sin t$, and if $u = f + 3g$, $v = f - g$, find the Wronskian of u and v.

In each of Problems 22 through 25, verify that the functions y_1 and y_2 are solutions of the given differential equation. Do they constitute a fundamental set of solutions?

22. $y'' + 4y = 0; \quad y_1(t) = \cos 2t, \quad y_2(t) = \sin 2t$

23. $y'' - 2y' + y = 0; \quad y_1(t) = e^t, \quad y_2(t) = te^t$

24. $x^2 y'' - x(x + 2)y' + (x + 2)y = 0, \quad x > 0;$
$y_1(x) = x, \quad y_2(x) = xe^x$

25. $(1 - x \cot x)y'' - xy' + y = 0, \quad 0 < x < \pi;$
$y_1(x) = x, \quad y_2(x) = \sin x$

26. Consider the equation $y'' - y' - 2y = 0$.

(a) Show that $y_1(t) = e^{-t}$ and $y_2(t) = e^{2t}$ form a fundamental set of solutions.

(b) Let $y_3(t) = -2e^{2t}$, $y_4(t) = y_1(t) + 2y_2(t)$, and $y_5(t) = 2y_1(t) - 2y_3(t)$. Are $y_3(t)$, $y_4(t)$, and $y_5(t)$ also solutions of the given differential equation?

(c) Determine whether each of the following pairs forms a fundamental set of solutions: $[y_1(t), y_3(t)]$; $[y_2(t), y_3(t)]$; $[y_1(t), y_4(t)]$; $[y_4(t), y_5(t)]$.

27. Prove Theorem 4.2.4 and Corollary 4.2.5.

Reduction of Order. Given one solution y_1 of a second order linear homogeneous equation,

$$y'' + p(t)y' + q(t)y = 0, \qquad \text{(i)}$$

a systematic procedure for deriving a second solution y_2 such that $\{y_1, y_2\}$ is a fundamental set is known as the method of **reduction of order**.

To find a second solution, assume a solution of the form $y = v(t)y_1(t)$. Substituting $y = v(t)y_1(t)$, $y' = v'(t)y_1(t) + v(t)y_1'(t)$, and $y'' = v''(t)y_1(t) + 2v'(t)y_1'(t) + v(t)y_1''(t)$ in Eq. (i) and collecting terms give

$$y_1 v'' + (2y_1' + py_1)v' + (y_1'' + py_1' + qy_1)v = 0. \qquad \text{(ii)}$$

Since y_1 is a solution of Eq. (i), the coefficient of v in Eq. (ii) is zero, so that Eq. (ii) reduces to

$$y_1 v'' + (2y_1' + py_1)v' = 0, \qquad \text{(iii)}$$

a first order equation for the function $w = v'$ that can be solved either as a first order linear equation or as a separable equation. Once v' has been found, then v is obtained by integrating w and then y is determined from $y = v(t)y_1(t)$.

This procedure is called the method of reduction of order, because the crucial step is the solution of a first order differential equation for v' rather than the original second order equation for y.

In each of Problems 28 through 38, use the method of reduction of order to find a second solution y_2 of the given differential equation such that $\{y_1, y_2\}$ is a fundamental set of solutions on the given interval.

28. $ay'' + by' + cy = 0, \quad$ where $b^2 - 4ac = 0$;
$y_1(t) = e^{-bt/(2a)}$

29. $t^2 y'' - 4ty' + 6y = 0, \quad t > 0; \qquad y_1(t) = t^2$

30. $t^2 y'' + 2ty' - 2y = 0, \quad t > 0; \qquad y_1(t) = t$

31. $t^2 y'' + 3ty' + y = 0, \quad t > 0; \qquad y_1(t) = t^{-1}$

32. $t^2 y'' - t(t + 2)y' + (t + 2)y = 0, \quad t > 0;$
$y_1(t) = t$

33. $xy'' - y' + 4x^3 y = 0, \quad x > 0; \qquad y_1(x) = \sin x^2$

34. $(x - 1)y'' - xy' + y = 0, \quad x > 1; \qquad y_1(x) = e^x$

35. $x^2 y'' - (x - 0.1875)y = 0, \quad x > 0;$
$y_1(x) = x^{1/4} e^{2\sqrt{x}}$

36. $x^2 y'' + xy' + (x^2 - 0.25)y = 0, \quad x > 0;$
$y_1(x) = x^{-1/2} \sin x$

37. The differential equation

$$xy'' - (x + N)y' + Ny = 0,$$

where N is a nonnegative integer, has been discussed by several authors.[1] One reason why it is interesting is that it has an exponential solution and a polynomial solution.

(a) Verify that one solution is $y_1(x) = e^x$.

(b) Show that a second solution has the form $y_2(x) = ce^x \int x^N e^{-x} \, dx$. Calculate $y_2(x)$ for $N = 1$ and $N = 2$; convince yourself that, with $c = -1/N!$,

$$y_2(x) = 1 + \frac{x}{1!} + \frac{x^2}{2!} + \cdots + \frac{x^N}{N!}.$$

Note that $y_2(x)$ is exactly the first $N + 1$ terms in the Taylor series about $x = 0$ for e^x, that is, for $y_1(x)$.

38. The differential equation

$$y'' + \delta(xy' + y) = 0$$

arises in the study of the turbulent flow of a uniform stream past a circular cylinder. Verify that $y_1(x) = \exp(-\delta x^2/2)$ is one solution and then find the general solution in the form of an integral.

4.3 Linear Homogeneous Equations with Constant Coefficients

In this section we study the problem of finding a fundamental set of solutions of the linear homogeneous second order differential equation with constant coefficients

$$ay'' + by' + cy = 0, \tag{1}$$

where a, b, and c are given real numbers. We assume that $a \neq 0$, since otherwise Eq. (1) is only of first order. Recall that the mathematical models derived in Section 4.1 for the unforced spring-mass system, the linearized pendulum, and the series RLC circuit were equations of this type.

Using the state variables $x_1 = y$ and $x_2 = y'$, we convert Eq. (1) into the first order system

$$\mathbf{x}' = \mathbf{A}\mathbf{x} = \begin{pmatrix} 0 & 1 \\ -c/a & -b/a \end{pmatrix} \mathbf{x}, \tag{2}$$

where

$$\mathbf{x} = \begin{pmatrix} x_1 \\ x_2 \end{pmatrix} = \begin{pmatrix} y \\ y' \end{pmatrix}.$$

As discussed in Section 4.2, $y = c_1 y_1(t) + c_2 y_2(t)$ is the general solution of Eq. (1) if and only if

$$\mathbf{x} = c_1 \begin{pmatrix} y_1(t) \\ y_1'(t) \end{pmatrix} + c_2 \begin{pmatrix} y_2(t) \\ y_2'(t) \end{pmatrix}$$

is the general solution of Eq. (2). We can find a fundamental set of vector solutions of Eq. (2) by using the eigenvalue method introduced in Sections 3.3 through 3.5. Taking the first component of each of these vector solutions then provides us with a fundamental set of solutions of Eq. (1). As a consequence, we are led to a more commonly used, simple, and direct method for finding solutions of Eq. (1).

Remark. The method described in this section for solving Eq. (1) does not apply to the linear equations $y'' + py' + qy = 0$ or $Py'' + Qy' + Ry = 0$ if the coefficients are not constants. In such cases infinite series methods or numerical methods are usually required to find either exact solutions or approximations of solutions.

---

[1] T. A. Newton, "On Using a Differential Equation to Generate Polynomials," *American Mathematical Monthly 81* (1974), pp. 592–601. Also see the references given there.

The Characteristic Equation of $ay'' + by' + cy = 0$

To find solutions of Eq. (2) using the eigenvalue method of Chapter 3, we assume that $\mathbf{x} = e^{\lambda t}\mathbf{v}$ and substitute for $\mathbf{x}$ in Eq. (2). This gives the algebraic system

$$(\mathbf{A} - \lambda\mathbf{I})\mathbf{v} = \begin{pmatrix} -\lambda & 1 \\ -c/a & -b/a - \lambda \end{pmatrix}\begin{pmatrix} v_1 \\ v_2 \end{pmatrix} = \begin{pmatrix} 0 \\ 0 \end{pmatrix}. \tag{3}$$

Since

$$\det(\mathbf{A} - \lambda\mathbf{I}) = \begin{vmatrix} -\lambda & 1 \\ -c/a & -b/a - \lambda \end{vmatrix} = \lambda^2 + \frac{b}{a}\lambda + \frac{c}{a} = \frac{1}{a}\left(a\lambda^2 + b\lambda + c\right), \tag{4}$$

the eigenvalues of $\mathbf{A}$ are the roots of

$$Z(\lambda) = a\lambda^2 + b\lambda + c = 0. \tag{5}$$

If λ is an eigenvalue of $\mathbf{A}$, then the first equation in system (3) is $-\lambda v_1 + v_2 = 0$. The corresponding eigenvector is

$$\mathbf{v} = \begin{pmatrix} 1 \\ \lambda \end{pmatrix}.$$

If λ is a root of Eq. (5), then

$$\mathbf{x} = e^{\lambda t}\mathbf{v} = \begin{pmatrix} e^{\lambda t} \\ \lambda e^{\lambda t} \end{pmatrix}$$

is a solution of Eq. (2). Consequently, the first component of $\mathbf{x}$, namely $y = e^{\lambda t}$, is a solution of Eq. (1). We summarize the preceding discussion in the following theorem.

THEOREM 4.3.1

If λ is a root of Eq. (5), $a\lambda^2 + b\lambda + c = 0$, then

    i. $y = e^{\lambda t}$ is a solution of Eq. (1), and

    ii. λ is an eigenvalue of the coefficient matrix $\mathbf{A}$ in Eq. (2) and the vector function

$$\mathbf{x} = \begin{pmatrix} e^{\lambda t} \\ \lambda e^{\lambda t} \end{pmatrix}$$

is a solution of Eq. (2).

We can arrive at Eq. (5) and conclusion (i) in Theorem 4.3.1 more directly by simply substituting $y = e^{\lambda t}$ into Eq. (1). Since $y' = \lambda e^{\lambda t}$ and $y'' = \lambda^2 e^{\lambda t}$, we find that

$$a\lambda^2 e^{\lambda t} + b\lambda e^{\lambda t} + ce^{\lambda t} = (a\lambda^2 + b\lambda + c)e^{\lambda t} = 0.$$

Since $e^{\lambda t}$ can never take on the value zero, we are led to conclude that $y = e^{\lambda t}$ is a nonzero solution of Eq. (1) if and only if λ satisfies Eq. (5). By analogy with the terminology used in solving for the eigenvalues of $\mathbf{A}$, Eq. (5) is called the **characteristic equation** (also known as the **auxiliary equation**) for the differential equation (1), and the polynomial $Z(\lambda) = a\lambda^2 + b\lambda + c$ is called the **characteristic polynomial** for Eq. (1).

Since Eq. (5) is a quadratic equation with real coefficients, it has two roots

$$\lambda_1 = \frac{-b + \sqrt{b^2 - 4ac}}{2a} \qquad \text{and} \qquad \lambda_2 = \frac{-b - \sqrt{b^2 - 4ac}}{2a}. \tag{6}$$

There will be three forms of the general solution of Eq. (2), and hence Eq. (1), depending on whether the discriminant $b^2 - 4ac$ is positive, negative, or zero:

1. If $b^2 - 4ac > 0$, the roots are real and distinct, $\lambda_1 \neq \lambda_2$.
2. If $b^2 - 4ac = 0$, the roots are real and equal, $\lambda_1 = \lambda_2$.
3. If $b^2 - 4ac < 0$, the roots are complex conjugates, $\lambda_1 = \mu + iv$, $\lambda_2 = \mu - iv$.

Case 1: Distinct Real Roots.   Since the eigenvectors corresponding to the eigenvalues λ_1 and λ_2 are

$$\mathbf{v}_1 = \begin{pmatrix} 1 \\ \lambda_1 \end{pmatrix} \qquad \text{and} \qquad \mathbf{v}_2 = \begin{pmatrix} 1 \\ \lambda_2 \end{pmatrix}, \tag{7}$$

respectively, the general solution of Eq. (2) is

$$\mathbf{x} = c_1 e^{\lambda_1 t} \begin{pmatrix} 1 \\ \lambda_1 \end{pmatrix} + c_2 e^{\lambda_2 t} \begin{pmatrix} 1 \\ \lambda_2 \end{pmatrix}. \tag{8}$$

Therefore the general solution of Eq. (1) is

$$y = c_1 e^{\lambda_1 t} + c_2 e^{\lambda_2 t}. \tag{9}$$

Case 2: Repeated Roots.   Recall from Section 3.5 that the general solution of Eq. (2), in the case of repeated roots, is

$$\mathbf{x} = c_1 e^{\lambda_1 t} \mathbf{v}_1 + c_2 e^{\lambda_1 t} (t \mathbf{v}_1 + \mathbf{w}_1),$$

where $\mathbf{v}_1$ is the eigenvector belonging to $\lambda_1 = \lambda_2 = -b/2a$ and $\mathbf{w}_1$ is any solution of $(\mathbf{A} - \lambda_1 \mathbf{I}) \mathbf{w} = \mathbf{v}_1$. As in the previous case, setting $\lambda = \lambda_1$ in Eq. (3) gives the eigenvector

$$\mathbf{v}_1 = \begin{pmatrix} 1 \\ \lambda_1 \end{pmatrix}. \tag{10}$$

Since a solution of

$$\begin{pmatrix} -\lambda_1 & 1 \\ -c/a & -b/a - \lambda_1 \end{pmatrix} \begin{pmatrix} w_1 \\ w_2 \end{pmatrix} = \begin{pmatrix} 1 \\ \lambda_1 \end{pmatrix} \tag{11}$$

is

$$\mathbf{w}_1 = \begin{pmatrix} 0 \\ 1 \end{pmatrix},$$

the general solution of Eq. (2) is

$$\mathbf{x} = c_1 e^{\lambda_1 t} \begin{pmatrix} 1 \\ \lambda_1 \end{pmatrix} + c_2 e^{\lambda_1 t} \begin{pmatrix} t \\ 1 + \lambda_1 t \end{pmatrix}. \tag{12}$$

Hence the general solution of Eq. (1) is

$$y = c_1 e^{\lambda_1 t} + c_2 t e^{\lambda_1 t}. \tag{13}$$

Case 3: Complex Conjugate Roots. In this case, the roots of Eq. (5) are conjugate complex numbers

$$\lambda_1 = \mu + iv \qquad \text{and} \qquad \lambda_2 = \mu - iv, \tag{14}$$

where

$$\mu = -\frac{b}{2a} \qquad \text{and} \qquad v = \frac{\sqrt{4ac - b^2}}{2a} \tag{15}$$

are real. As in Case 1, the corresponding eigenvectors are

$$\mathbf{v}_1 = \begin{pmatrix} 1 \\ \lambda_1 \end{pmatrix} = \begin{pmatrix} 1 \\ \mu + iv \end{pmatrix}, \qquad \mathbf{v}_2 = \begin{pmatrix} 1 \\ \lambda_2 \end{pmatrix} = \begin{pmatrix} 1 \\ \mu - iv \end{pmatrix}. \tag{16}$$

Hence two complex-valued solutions of Eq. (2) are

$$\mathbf{x}_1(t) = e^{(\mu + iv)t} \begin{pmatrix} 1 \\ \mu + iv \end{pmatrix}, \qquad \mathbf{x}_2(t) = e^{(\mu - iv)t} \begin{pmatrix} 1 \\ \mu - iv \end{pmatrix}. \tag{17}$$

We find real-valued solutions by separating $\mathbf{x}_1(t)$ into its real and imaginary parts. Using Euler's formula, we have

$$\mathbf{x}_1(t) = e^{\mu t} [\cos vt + i \sin vt] \begin{pmatrix} 1 \\ \mu + iv \end{pmatrix}$$

$$= e^{\mu t} \begin{pmatrix} \cos vt \\ \mu \cos vt - v \sin vt \end{pmatrix} + i e^{\mu t} \begin{pmatrix} \sin vt \\ \mu \sin vt + v \cos vt \end{pmatrix}. \tag{18}$$

Thus the general solution of Eq. (2), expressed in terms of real-valued functions, is

$$\mathbf{x} = c_1 e^{\mu t} \begin{pmatrix} \cos vt \\ \mu \cos vt - v \sin vt \end{pmatrix} + c_2 e^{\mu t} \begin{pmatrix} \sin vt \\ \mu \sin vt + v \cos vt \end{pmatrix}. \tag{19}$$

Taking the first component of $\mathbf{x}$ gives the general solution of Eq. (1),

$$y = c_1 e^{\mu t} \cos vt + c_2 e^{\mu t} \sin vt. \tag{20}$$

We summarize the preceding discussion relating the general solution of Eq. (1) to the general solution of the system (2) in the following theorem.

THEOREM 4.3.2

Let λ_1 and λ_2 be the roots of the characteristic equation

$$a\lambda^2 + b\lambda + c = 0.$$

Then the general solution of Eq. (1), $ay'' + by' + cy = 0$, is

$$y = c_1 e^{\lambda_1 t} + c_2 e^{\lambda_2 t}, \qquad \text{if } \lambda_1 \text{ and } \lambda_2 \text{ are real, } \lambda_1 \neq \lambda_2, \tag{21}$$

$$y = c_1 e^{\lambda_1 t} + c_2 t e^{\lambda_1 t}, \qquad \text{if } \lambda_1 = \lambda_2, \tag{22}$$

$$y = c_1 e^{\mu t} \cos vt + c_2 e^{\mu t} \sin vt, \qquad \text{if } \lambda_1 = \mu + iv \text{ and } \lambda_2 = \mu - iv. \tag{23}$$

In each case, the corresponding general solution of Eq. (2), $\mathbf{x}' = \begin{pmatrix} 0 & 1 \\ -c/a & -b/a \end{pmatrix} \mathbf{x}$, is

$$\mathbf{x} = \begin{pmatrix} x_1 \\ x_2 \end{pmatrix} = \begin{pmatrix} y \\ y' \end{pmatrix}.$$

Thus

$$\mathbf{x} = c_1 e^{\lambda_1 t} \begin{pmatrix} 1 \\ \lambda_1 \end{pmatrix} + c_2 e^{\lambda_2 t} \begin{pmatrix} 1 \\ \lambda_2 \end{pmatrix}, \qquad \text{if } \lambda_1 \text{ and } \lambda_2 \text{ are real, } \lambda_1 \neq \lambda_2, \tag{24}$$

$$\mathbf{x} = c_1 e^{\lambda_1 t} \begin{pmatrix} 1 \\ \lambda_1 \end{pmatrix} + c_2 e^{\lambda_1 t} \begin{pmatrix} t \\ 1 + \lambda_1 t \end{pmatrix}, \qquad \text{if } \lambda_1 = \lambda_2, \tag{25}$$

$$\mathbf{x} = c_1 e^{\mu t} \begin{pmatrix} \cos \nu t \\ \mu \cos \nu t - \nu \sin \nu t \end{pmatrix} + c_2 e^{\mu t} \begin{pmatrix} \sin \nu t \\ \mu \sin \nu t + \nu \cos \nu t \end{pmatrix},$$

$$\text{if } \lambda_1 = \mu + i\nu \text{ and } \lambda_2 = \mu - i\nu. \tag{26}$$

The characteristic equation can be written down without calculation, since the coefficients a, b, and c are the same as the coefficients in the differential equation. Once you know the roots of the characteristic equation, using the quadratic formula if necessary, you can immediately write down the general solution of the differential equation using one of Eqs. (21)–(23). Since these differential equations occur frequently, it is important to be able to solve them quickly and efficiently.

EXAMPLE 1

Find general solutions for each of the following differential equations:

(a) $y'' + 5y' + 6y = 0$ (b) $y'' - y' + \frac{1}{4}y = 0$

(c) $y'' + y' + y = 0$ (d) $y'' + 9y = 0$

We give the characteristic equations, the roots, and the corresponding general solutions.

(a) $\lambda^2 + 5\lambda + 6 = (\lambda + 2)(\lambda + 3) = 0, \qquad \lambda_1 = -2, \quad \lambda_2 = -3$

From Eq. (21) in Theorem 4.3.2,

$$y = c_1 e^{-2t} + c_2 e^{-3t}. \tag{27}$$

(b) $\lambda^2 - \lambda + \frac{1}{4} = (\lambda - \frac{1}{2})^2, \quad \lambda_1 = \lambda_2 = \frac{1}{2}$

From Eq. (22) in Theorem 4.3.2,

$$y = c_1 e^{t/2} + c_2 t e^{t/2}. \tag{28}$$

(c) $\lambda^2 + \lambda + 1 = 0$, $\qquad \lambda = \dfrac{-1 \pm (1-4)^{1/2}}{2} = -\dfrac{1}{2} \pm i\dfrac{\sqrt{3}}{2}$

Thus $\mu = -\frac{1}{2}$ and $\nu = \sqrt{3}/2$. From Eq. (23) in Theorem 4.3.2,

$$y = c_1 e^{-t/2} \cos(\sqrt{3}t/2) + c_2 e^{-t/2} \sin(\sqrt{3}t/2). \tag{29}$$

(d) $\lambda^2 + 9 = 0$, $\quad \lambda = \pm 3i$

Since $\mu = 0$ and $\nu = 3$, from Eq. (23) in Theorem 4.3.2,

$$y = c_1 \cos 3t + c_2 \sin 3t. \tag{30}$$

Since the real part of the conjugate complex roots is zero, no real exponential factor appears in the solution.

Initial Value Problems and Phase Portraits

Given the general solution, we can easily solve initial value problems associated with Eq. (1), and sketch phase portraits of the corresponding dynamical system (2). If $c \neq 0$, then the determinant of the matrix of coefficients in Eq. (2) is $\det \mathbf{A} = c/a \neq 0$. In this case, $(0, 0)^T$ is the only critical point, or equilibrium solution, of Eq. (2). Since the roots of the characteristic equation $a\lambda^2 + b\lambda + c = 0$ are also the eigenvalues of $\mathbf{A}$, knowledge of these roots enables us to determine immediately the type and stability properties of the equilibrium solution (see Table 3.5.1). Briefly,

▶ If the roots are real and negative, the origin is an asymptotically stable node. If the roots are real and positive, the origin is an unstable node.

▶ If the roots are real and of opposite sign, the origin is a saddle point and unstable.

▶ If the roots are complex with nonzero imaginary part, the origin is an asymptotically stable spiral point (trajectories spiral in) if the real part is negative. If the real part is positive, the origin is an unstable spiral point (trajectories spiral out).

▶ If the real part of a pair of complex roots is zero, the origin is a center (trajectories are closed curves) and is stable, but not asymptotically stable.

▶ The direction of rotation for spiral points and centers for Eq. (2) is always clockwise. To see this, we note that in order to have roots with a nonzero imaginary part, it is necessary that $b^2 - 4ac < 0$, or $ac > b^2/4 \geq 0$. Thus a and c must have the same sign. The direction field vector for Eq. (2) at the point $(1, 0)$ is $0\mathbf{i} - (c/a)\mathbf{j}$. Since the second component is negative, the direction of rotation must be clockwise.

EXAMPLE 2

Find the solution of the initial value problem

$$y'' + 5y' + 6y = 0, \qquad y(0) = 2, \quad y'(0) = 3. \tag{31}$$

Formulate the differential equation as a dynamical system, discuss the corresponding phase portrait, and draw the trajectory associated with the solution of the initial value problem.

The general solution of the differential equation is given by Eq. (27) in part (a) of Example 1. To satisfy the first initial condition, we set $t = 0$ and $y = 2$ in Eq. (27); thus c_1 and c_2 must satisfy

$$c_1 + c_2 = 2. \tag{32}$$

To use the second initial condition, we must first differentiate Eq. (27). This gives

$$y' = -2c_1e^{-2t} - 3c_2e^{-3t}. \tag{33}$$

Setting $t = 0$ and $y' = 3$ in Eq. (33), we obtain

$$-2c_1 - 3c_2 = 3. \tag{34}$$

By solving Eqs. (32) and (34), we find that $c_1 = 9$ and $c_2 = -7$. Using these values in Eq. (27), we obtain the solution

$$y = 9e^{-2t} - 7e^{-3t} \tag{35}$$

of the initial value problem (31). The graph of the solution is shown in Figure 4.3.1.

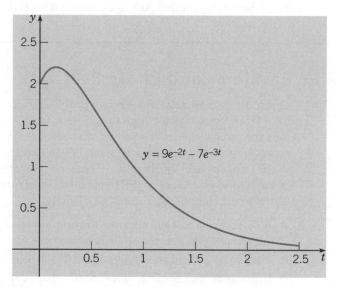

FIGURE 4.3.1 Solution of $y'' + 5y' + 6y = 0$, $y(0) = 2$, $y'(0) = 3$.

The dynamical system corresponding to the state vector $\mathbf{x} = x_1\mathbf{i} + x_2\mathbf{j} = y\mathbf{i} + y'\mathbf{j}$ is

$$\mathbf{x}' = \begin{pmatrix} 0 & 1 \\ -6 & -5 \end{pmatrix}\mathbf{x}. \tag{36}$$

From Eq. (24) in Theorem 4.3.2, the general solution of Eq. (36) is

$$\mathbf{x} = c_1e^{-2t}\begin{pmatrix} 1 \\ -2 \end{pmatrix} + c_2e^{-3t}\begin{pmatrix} 1 \\ -3 \end{pmatrix} = c_1\mathbf{x}_1(t) + c_2\mathbf{x}_2(t), \tag{37}$$

in which the first component is the general solution (27) and the second component is simply the derivative of the first component, given by Eq. (33).

The qualitative nature of the phase portrait for the system (36), drawn in Figure 4.3.2, is readily apparent from Eq. (37). The negative eigenvalues $\lambda_1 = -2$ and $\lambda_2 = -3$ of the coefficient matrix in Eq. (36) imply that the origin is a nodal sink, that trajectories approach the origin tangent to the line with direction vector $(1, -2)^T$ as $t \to \infty$. The slopes of the trajectories approach the limit -3 as t tends to negative infinity, but the trajectories are not parallel to any particular line of slope negative three. Choosing $c_1 = 9$ and $c_2 = -7$ in Eq. (37) gives the trajectory associated with the solution (35) (the heavy blue trajectory in Figure 4.3.2).

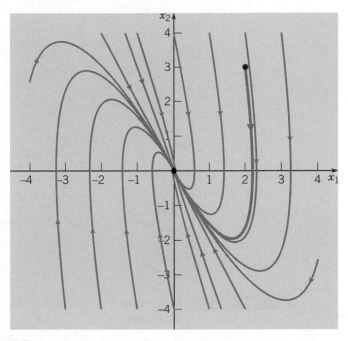

FIGURE 4.3.2   Phase portrait for the dynamical system (36) associated with Eq. (31). The heavy blue curve is the solution (35) with $y(0) = 2$ and $y'(0) = 3$. All solutions approach the origin target to the dashed line with direction vector $(1, -2)^T$ as $t \to \infty$. As $t \to -\infty$ the slopes of the trajectories approach the limit -3.

EXAMPLE 3

Find the solution of the initial value problem

$$y'' - y' + \frac{1}{4}y = 0, \qquad y(0) = 2, \quad y'(0) = \tfrac{1}{3}. \tag{38}$$

Then draw and discuss the phase portrait of $y'' - y' + 0.25y = 0$.

The general solution of the differential equation is given by Eq. (28) in Example 1. The first initial condition requires that

$$y(0) = c_1 = 2.$$

The second initial condition is satisfied by differentiating Eq. (28) and then setting $t = 0$,

$$y'(0) = \tfrac{1}{2}c_1 + c_2 = \tfrac{1}{3},$$

so $c_2 = -\tfrac{2}{3}$. Thus the solution of the initial value problem is

$$y = 2e^{t/2} - \tfrac{2}{3}te^{t/2}. \tag{39}$$

The graph of this solution is shown in Figure 4.3.3.

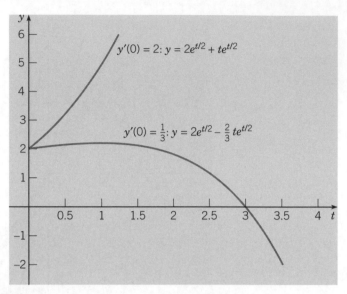

FIGURE 4.3.3 Solutions of $y'' - y' + \frac{1}{4}y = 0$, $y(0) = 2$, with $y'(0) = \frac{1}{3}$ and $y'(0) = 2$, respectively.

Let us now modify the initial value problem (38) by changing the initial slope; to be specific, let the second initial condition be $y'(0) = 2$. The solution of this modified problem is

$$y = 2e^{t/2} + te^{t/2}, \tag{40}$$

and its graph is also shown in Figure 4.3.3. The graphs shown in this figure suggest that there is a critical initial slope, with a value between $\frac{1}{3}$ and 2, that separates solutions that grow positively from those that ultimately grow negatively. By viewing the phase portrait of the dynamical system corresponding to $y'' - y' + \frac{1}{4}y = 0$, not only can we easily determine this critical slope, we can also ascertain the long-time qualitative behavior of all solutions of the differential equation.

Letting $x_1 = y$ and $x_2 = y'$, as usual, we obtain

$$\mathbf{x}' = \begin{pmatrix} 0 & 1 \\ -1/4 & 1 \end{pmatrix} \mathbf{x}. \tag{41}$$

As indicated in Eq. (25) of Theorem 4.3.2, the first component of the general solution of the system (41) is given by Eq. (28) in part (b) of Example. The second component of the general solution of (41) is given by the derivative of Eq. (28). Thus, the general solution of the system (41) is

$$\mathbf{x} = c_1 \begin{pmatrix} 1 \\ 1/2 \end{pmatrix} e^{t/2} + c_2 \left[\begin{pmatrix} 0 \\ 1 \end{pmatrix} + \begin{pmatrix} 1 \\ 1/2 \end{pmatrix} t \right] e^{t/2}. \tag{42}$$

For each choice of c_1 and c_2, a plot of the curve $t \to \langle x_1(t), x_2(t) \rangle$ yields a corresponding phase plane trajectory. The phase portrait is drawn in Figure 4.3.4. The choice $c_2 = 0$ yields $x_1 = c_1 e^{t/2}$, $x_2 = c_1 e^{t/2}/2$ corresponding to trajectories that lie on the line $x_2 = x_1/2$ (the black curve in Figure 4.3.4).

The solid and dashed heavy trajectories in Figure 4.3.4 correspond to the solutions (39) and (40) of $y'' - y' + \frac{1}{4}y = 0$, respectively. Since the slope of the straight-line

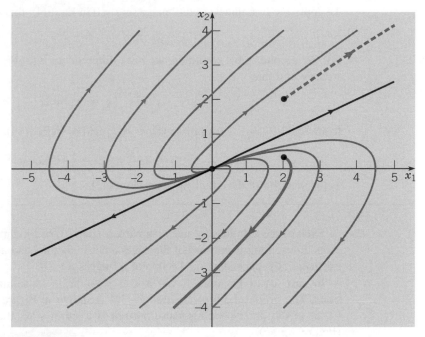

FIGURE 4.3.4 Phase portrait for the dynamical system associated with $y'' - y' + \frac{1}{4}y = 0$. The solid heavy blue curve corresponds to the solution (39) with $y(0) = 2$ and $y'(0) = \frac{1}{3}$. The dashed heavy blue curve corresponds to the solution (40) with $y(0) = 2$ and $y'(0) = 2$.

trajectories in Figure 4.3.4 is $\frac{1}{2}$, it follows that $\lim_{t\to\infty} y(t) = \infty$ for any solution of $y'' - y' + \frac{1}{4}y = 0$ such that $y'(0) > y(0)/2$ and that $\lim_{t\to\infty} y(t) = -\infty$ for any solution such that $y'(0) < y(0)/2$. It follows that if $y(0) = 2$, then $y'(0) = 1$ is a critical initial slope such that solutions with slopes greater than or equal to 1 grow positively and solutions with slopes less than 1 grow negatively. The critical point $(0, 0)$ is clearly an unstable node.

As we saw in Section 3.5, the geometrical behavior of solutions in the case of repeated roots is similar to that when the roots are real and different. If the exponents are either positive or negative, then the magnitude of the solution grows or decays accordingly; the linear factor t has little influence. However, if the repeated root is zero, then the differential equation is $y'' = 0$ and the general solution is $y = c_1 + c_2 t$. In this case, all points on the x_1-axis are critical points of the associated dynamical system.

EXAMPLE 4

Find the solution of the initial value problem

$$16y'' - 8y' + 145y = 0, \qquad y(0) = -2, \quad y'(0) = 1. \tag{43}$$

The characteristic equation is $16\lambda^2 - 8\lambda + 145 = 0$ and its roots are $\lambda = \frac{1}{4} \pm 3i$. Thus the general solution of the differential equation is

$$y = c_1 e^{t/4} \cos 3t + c_2 e^{t/4} \sin 3t. \tag{44}$$

To apply the first initial condition, we set $t = 0$ in Eq. (44); this gives

$$y(0) = c_1 = -2.$$

For the second initial condition, we must differentiate Eq. (44) and then set $t = 0$. In this way we find that

$$y'(0) = \tfrac{1}{4}c_1 + 3c_2 = 1,$$

from which $c_2 = \tfrac{1}{2}$. Using these values of c_1 and c_2 in Eq. (44), we obtain

$$y = -2e^{t/4}\cos 3t + \tfrac{1}{2}e^{t/4}\sin 3t \tag{45}$$

as the solution of the initial value problem (43).

Each of the solutions y_1 and y_2 in parts (c) and (d) of Example 1 and in Eq. (44) represents an oscillation, because of the trigonometric factors, and also either grows or decays exponentially, depending on the sign of μ (unless $\mu = 0$).

In part (c) of Example 1, we have $\mu = -\tfrac{1}{2} < 0$, so solutions are decaying oscillations. The graph of a typical solution (30) is shown in Figure 4.3.5a. The critical point $(0, 0)$ in the corresponding phase portrait is a spiral sink; trajectories rotate clockwise (Figure 4.3.5b).

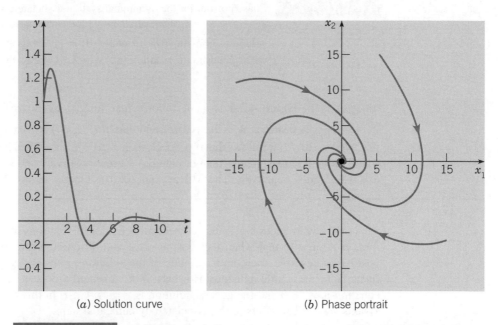

(a) Solution curve (b) Phase portrait

FIGURE 4.3.5 (a) Solutions of $y'' + y' + y = 0$, $y(0) = 1$, $y'(0) = 2$. (b) Phase plane trajectories $t \to \mathbf{x} = x_1\mathbf{i} + x_2\mathbf{j} = y\mathbf{i} + y'\mathbf{j}$ of the dynamical system equivalent to $y'' + y' + y = 0$.

If $\mu = 0$ as in part (d) of Example 1, the solutions are pure oscillations without growth or decay (Figure 4.3.6a). The origin in the phase portrait is a stable center; trajectories rotate clockwise (Figure 4.3.6b).

In Example 4, $\mu = \tfrac{1}{4} > 0$, so solutions of the differential equation in the initial value problem (45) are oscillations with exponentially growing amplitudes. The graph of the solution (46) of the given initial value problem is shown in Figure 4.3.7a. The origin in the phase portrait is a spiral source; trajectories necessarily rotate clockwise (Figure 4.3.7b).

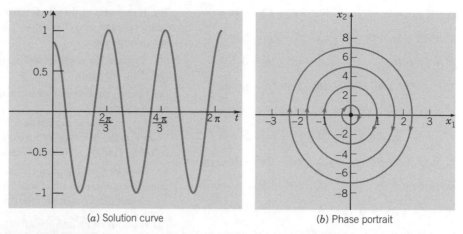

(a) Solution curve   (b) Phase portrait

FIGURE 4.3.6   (a) Solution of $y'' + 9y = 0$, $y(0) = \sqrt{3}/2$, $y'(0) = -3/2$. (b) Phase plane trajectories $t \to \mathbf{x} = x_1\mathbf{i} + x_2\mathbf{j} = y\mathbf{i} + y'\mathbf{j}$ of the dynamical system equivalent to $y'' + 9y = 0$.

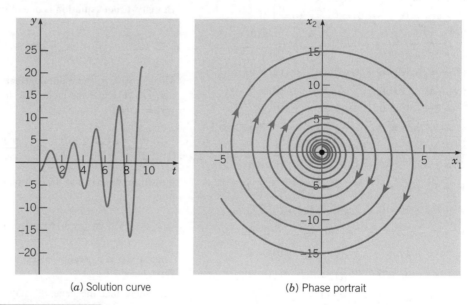

(a) Solution curve   (b) Phase portrait

FIGURE 4.3.7   (a) Solution of $16y'' - 8y' + 145y = 0$, $y(0) = -2$, $y'(0) = 1$. (b) Phase plane trajectories $t \to \mathbf{x} = x_1\mathbf{i} + x_2\mathbf{j} = y\mathbf{i} + y'\mathbf{j}$ of the dynamical system equivalent to $16y'' - 8y' + 145y = 0$.

PROBLEMS

In each of Problems 1 through 26:

(a) Find the general solution in terms of real functions.

(b) From the roots of the characteristic equation, determine whether each critical point of the corresponding dynamical system is asymptotically stable, stable, or unstable, and classify it as to type.

(c) Use the general solution obtained in part (a) to find a two-parameter family of trajectories $\mathbf{x} = x_1\mathbf{i} + x_2\mathbf{j} = y\mathbf{i} + y'\mathbf{j}$ of the corresponding dynamical system. Then sketch by hand,

or use a computer, to draw a phase portrait, including any straight-line orbits, from this family of trajectories.

1. $y'' + 2y' - 3y = 0$ 2. $y'' + 3y' + 2y = 0$

3. $y'' - 4y' + 4y = 0$ 4. $9y'' + 6y' + y = 0$

5. $y'' - 2y' + 2y = 0$ 6. $y'' - 2y' + 6y = 0$

7. $4y'' - 4y' + y = 0$ 8. $2y'' - 3y' + y = 0$

9. $6y'' - y' - y = 0$

10. $9y'' + 12y' + 4y = 0$

11. $y'' + 2y' - 8y = 0$

12. $y'' + 2y' + 2y = 0$

13. $y'' + 5y' = 0$

14. $4y'' - 9y = 0$

15. $25y'' - 20y' + 4y = 0$

16. $y'' - 4y' + 16y = 0$

17. $y'' + 6y' + 13y = 0$

18. $y'' + 2y' + 1.25y = 0$

19. $y'' - 9y' + 9y = 0$

20. $y'' - 2y' - 2y = 0$

21. $y'' + 4y' + 4y = 0$

22. $9y'' - 24y' + 16y = 0$

23. $4y'' + 9y = 0$

24. $4y'' + 9y' - 9y = 0$

25. $y'' + y' + 1.25y = 0$

26. $y'' + 4y' + 6.25y = 0$

In each of Problems 27 through 43, solve the given initial value problem. Sketch the graph of its solution and describe its behavior for increasing t.

27. $y'' + y' - 2y = 0,$ $y(0) = 1,$ $y'(0) = 1$

28. $y'' + 16y = 0,$ $y(0) = 0,$ $y'(0) = 1$

29. $9y'' - 12y' + 4y = 0,$ $y(0) = 2,$ $y'(0) = -1$

30. $y'' + 3y' + 2y = 0,$ $y(0) = 2,$ $y'(0) = -1$

31. $y'' + 4y' + 5y = 0,$ $y(0) = 1,$ $y'(0) = 0$

32. $6y'' - 5y' + y = 0,$ $y(0) = 4,$ $y'(0) = 0$

33. $y'' + 6y' + 9y = 0,$ $y(0) = 0,$ $y'(0) = 2$

34. $y'' - 2y' + 5y = 0,$ $y(\pi/2) = 0,$ $y'(\pi/2) = 2$

35. $y'' + 3y' = 0,$ $y(0) = -2,$ $y'(0) = 3$

36. $y'' + y = 0,$ $y(\pi/3) = 2,$ $y'(\pi/3) = -4$

37. $y'' + 4y' + 4y = 0,$ $y(-1) = 2,$ $y'(-1) = 1$

38. $y'' + 6y' + 3y = 0,$ $y(0) = 1,$ $y'(0) = 0$

39. $y'' + y' + 1.25y = 0,$ $y(0) = 3,$ $y'(0) = 1$

40. $2y'' + y' - 4y = 0,$ $y(0) = 0,$ $y'(0) = 1$

41. $y'' + 8y' - 9y = 0,$ $y(1) = 1,$ $y'(1) = 0$

42. $y'' + 2y' + 2y = 0,$ $y(\pi/4) = 2,$ $y'(\pi/4) = -2$

43. $4y'' - y = 0,$ $y(-2) = 1,$ $y'(-2) = -1$

44. Find a differential equation whose general solution is $y = c_1 e^{3t} + c_2 e^{-2t}$.

45. Find a differential equation whose general solution is $y = c_1 e^{-2t} + c_2 t e^{-2t}$.

46. Find a differential equation whose general solution is $y = c_1 e^{-3t} \cos 4t + c_2 e^{-3t} \sin 4t$.

In each of Problems 47 and 48, determine the values of α, if any, for which all solutions tend to zero as $t \to \infty$; also determine the values of α, if any, for which all (nonzero) solutions become unbounded as $t \to \infty$.

47. $y'' - (2\alpha - 3)y' + \alpha(\alpha - 3)y = 0$

48. $y'' + (3 - \alpha)y' - 2(\alpha - 1)y = 0$

49. If the roots of the characteristic equation are real, show that a solution of $ay'' + by' + cy = 0$ can take on the value zero at most once.

50. Consider the equation $ay'' + by' + cy = d$, where a, b, c, and d are constants.

(a) Find all equilibrium, or constant, solutions of this differential equation.

(b) Let y_e denote an equilibrium solution, and let $Y = y - y_e$. Thus Y is the deviation of a solution y from an equilibrium solution. Find the differential equation satisfied by Y.

51. Consider the equation $ay'' + by' + cy = 0$, where a, b, and c are constants with $a > 0$. Find conditions on a, b, and c such that the roots of the characteristic equation are:

(a) Real, different, and negative.

(b) Real with opposite signs.

(c) Real, different, and positive.

Cauchy–Euler Equations. A (homogeneous) second order **Cauchy–Euler equation** is one of the form

$$ax^2 y'' + bxy' + cy = 0, \qquad \text{(i)}$$

where $a \neq 0$, b, and c are constants. If $x > 0$, the substitution $t = \ln x$ and the chain rule for derivatives applied to $y(x) = y(t(x))$,

$$\frac{dy}{dx} = \frac{dy}{dt}\frac{dt}{dx} = \frac{1}{x}\frac{dy}{dt} \quad \text{and} \quad \frac{d^2y}{dx^2} = \frac{1}{x^2}\frac{d^2y}{dt^2} - \frac{1}{x^2}\frac{dy}{dt},$$

transforms Eq. (i) into the constant coefficient linear equation

$$a\frac{d^2y}{dt^2} + (b - a)\frac{dy}{dt} + cy = 0. \qquad \text{(ii)}$$

Thus if $y(t)$ is a solution of Eq. (ii), $y(\ln x)$ is a solution of Eq. (i) in $x > 0$.

52. Show that the general solution of the Cauchy–Euler equation (i) in $x > 0$ is

$$y = c_1 e^{\lambda_1 t} + c_2 e^{\lambda_2 t} = c_1 x^{\lambda_1} + c_2 x^{\lambda_2},$$
$$\text{if } \lambda_1 \text{ and } \lambda_2 \text{ are real and } \lambda_1 \neq \lambda_2,$$
$$y = c_1 e^{\lambda_1 t} + c_2 t e^{\lambda_1 t} = c_1 x^{\lambda_1} + c_2 x^{\lambda_1} \ln x,$$
$$\text{if } \lambda_1 = \lambda_2,$$
$$y = e^{\mu t}[c_1 \cos(vt) + c_2 \sin(vt)]$$
$$= x^{\mu}[c_1 \cos(v \ln x) + c_2 \sin(v \ln x)],$$
$$\text{if } \lambda_1 = \mu + iv \quad \text{and} \quad \lambda_2 = \mu - iv.$$

53. If $x < 0$, use the substitution $t = \ln(-x)$ to show that the general solutions of the Cauchy–Euler equation (i) in $x < 0$ are identical to those in Problem 52 with the exception that x is replaced by $-x$.

In each of Problems 54 through 61, find the general solution of the given Cauchy–Euler equation in $x > 0$:

54. $x^2y'' + xy' + 4y = 0$

55. $x^2y'' + 4xy' + 2y = 0$

56. $x^2y'' + 3xy' + 1.25y = 0$

57. $x^2y'' - 4xy' - 6y = 0$

58. $x^2y'' - 2y = 0$

59. $x^2y'' - 5xy' + 9y = 0$

60. $x^2y'' + 2xy' + 4y = 0$

61. $2x^2y'' - 4xy' + 6y = 0$

In each of Problems 62 through 65, find the solution of the given initial value problem. Plot the graph of the solution and describe how the solution behaves as $x \to 0$.

62. $2x^2y'' + xy' - 3y = 0, \quad y(1) = 1, \quad y'(1) = 1$

63. $4x^2y'' + 8xy' + 17y = 0, \quad y(1) = 2, \quad y'(1) = -3$

64. $x^2y'' - 5xy' + 9y = 0, \quad y(-1) = 2, \quad y'(-1) = 3$

65. $x^2y'' + 3xy' + 5y = 0, \quad y(1) = 1, \quad y'(1) = -1$

4.4 Mechanical and Electrical Vibrations

In Section 4.1 the mathematical models derived for the spring-mass system, the linearized pendulum, and the RLC circuit all turned out to be linear constant coefficient differential equations that, in the absence of a forcing function, are of the form

$$ay'' + by' + cy = 0. \tag{1}$$

To adapt Eq. (1) to a specific application merely requires interpretation of the coefficients in terms of the physical parameters that characterize the application. Using the theory and methods developed in Sections 4.2 and 4.3, we are able to solve Eq. (1) completely for all possible parameter values and initial conditions. Thus Eq. (1) provides us with an important class of problems that illustrates the linear theory described in Section 4.2 and solution methods developed in Section 4.3.

Undamped Free Vibrations

Recall that the equation of motion for the damped spring-mass system with external forcing is

$$my'' + \gamma y' + ky = F(t). \tag{2}$$

Equation (2) and the pair of conditions,

$$y(0) = y_0, \qquad y'(0) = v_0, \tag{3}$$

that specify initial position y_0 and initial velocity v_0 provide a complete formulation of the vibration problem. If there is no external force, then $F(t) = 0$ in Eq. (2).

Let us also suppose that there is no damping, so that $\gamma = 0$. This is an idealized configuration of the system, seldom (if ever) completely attainable in practice. However, if the actual damping is very small, then the assumption of no damping may yield satisfactory results over short to moderate time intervals. In this case, the equation of motion (2) reduces to

$$my'' + ky = 0. \tag{4}$$

If we divide Eq. (4) by m, it becomes

$$y'' + \omega_0^2 y = 0, \tag{5}$$

where

$$\omega_0^2 = k/m. \tag{6}$$

The characteristic equation for Eq. (5) is

$$\lambda^2 + \omega_0^2 = 0, \tag{7}$$

and the corresponding characteristic roots are $\lambda = \pm i\,\omega_0$. It follows that the general solution of Eq. (5) is

$$y = A\cos\omega_0 t + B\sin\omega_0 t. \tag{8}$$

where A and B are arbitrary constants. Substituting from Eq. (8) into the initial conditions (3) determines the integration constants A and B in terms of initial position and velocity, $A = y_0$ and $B = v_0/\omega_0$.

In discussing the solution of Eq. (5), it is convenient to rewrite Eq. (8) in the **phase-amplitude** form

$$y = R\cos(\omega_0 t - \delta). \tag{9}$$

To see the relationship between Eqs. (8) and (9), use the trigonometric identity for the cosine of the difference of the two angles, $\omega_0 t$ and δ, to rewrite Eq. (9) as

$$y = R\cos\delta\cos\omega_0 t + R\sin\delta\sin\omega_0 t. \tag{10}$$

By comparing Eq. (10) with Eq. (8), we find that A, B, R, and δ are related by the equations

$$A = R\cos\delta, \qquad B = R\sin\delta. \tag{11}$$

From these two equations, we see that (R, δ) is simply the polar coordinate representation of the point with Cartesian coordinates (A, B) (Figure 4.4.1).

Thus

$$R = \sqrt{A^2 + B^2}, \tag{12}$$

while δ satisfies

$$\cos\delta = \frac{A}{\sqrt{A^2 + B^2}}, \qquad \sin\delta = \frac{B}{\sqrt{A^2 + B^2}}. \tag{13}$$

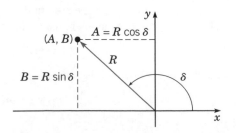

FIGURE 4.4.1 Relation between (R, δ) in Eq. (9) and (A, B) in Eq. (8).

Let arctan (B/A) be the angle that lies in the principal branch of the inverse tangent function, that is, in the interval $-\pi/2 < \hat{\delta} < \pi/2$ (Figure 4.4.2). Then the values of δ given by

$$\delta = \begin{cases} \arctan(B/A), & \text{if } A > 0, B \geq 0 \text{ (1st quadrant)} \\ \pi + \arctan(B/A), & \text{if } A < 0 \text{ (2nd or 3rd quadrant)} \\ 2\pi + \arctan(B/A), & \text{if } A > 0, B < 0 \text{ (4th quadrant)} \\ \pi/2, & \text{if } A = 0, B > 0 \\ 3\pi/2, & \text{if } A = 0, B < 0 \end{cases}$$

will lie in the interval $[0, 2\pi)$.

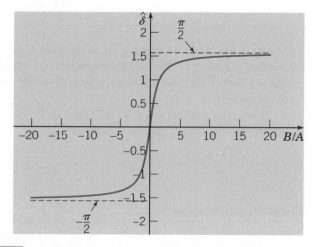

FIGURE 4.4.2   The principal branch of the arctangent function.

The graph of Eq. (9), or the equivalent Eq. (8), for a typical set of initial conditions is shown in Figure 4.4.3. The graph is a displaced cosine wave that describes a periodic, or simple harmonic, motion of the mass. The **period** of the motion is

$$T = \frac{2\pi}{\omega_0} = 2\pi \left(\frac{m}{k}\right)^{1/2}. \tag{14}$$

The circular frequency $\omega_0 = \sqrt{k/m}$, measured in radians per unit time, is called the **natural frequency** of the vibration. The maximum displacement R of the mass from equilibrium is the **amplitude** of the motion. The dimensionless parameter δ is called the **phase**, or phase

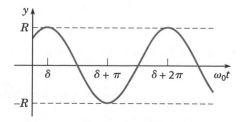

FIGURE 4.4.3   Simple harmonic motion $y = R \cos (\omega_0 t - \delta)$.

angle. The quantity δ/ω_0 measures the time shift of the wave from its normal position corresponding to $\delta = 0$.

Note that the motion described by Eq. (9) has a constant amplitude that does not diminish with time. This reflects the fact that, in the absence of damping, there is no way for the system to dissipate the energy imparted to it by the initial displacement and velocity. Further, for a given mass m and spring constant k, the system always vibrates at the same frequency ω_0, regardless of the initial conditions. However the initial conditions do help to determine the amplitude of the motion. Finally, observe from Eq. (14) that T increases as m increases, so larger masses vibrate more slowly. On the other hand, T decreases as k increases, which means that stiffer springs cause the system to vibrate more rapidly.

EXAMPLE 1

Suppose that a mass weighing 10 lb stretches a spring 2 in. If the mass is displaced an additional 2 in. and is then set in motion with an initial upward velocity of 1 ft/s, determine the position of the mass at any later time. Also determine the period, amplitude, and phase of the motion.

The spring constant is $k = 10$ lb/2 in. $= 60$ lb/ft, and the mass is $m = w/g = \frac{10}{32}$ lb·s^2/ft. Hence the equation of motion reduces to

$$y'' + 192y = 0, \tag{15}$$

and the general solution is

$$y = A\cos(8\sqrt{3}t) + B\sin(8\sqrt{3}t).$$

The solution satisfying the initial conditions $y(0) = \frac{1}{6}$ ft and $y'(0) = -1$ ft/s is

$$y = \frac{1}{6}\cos(8\sqrt{3}t) - \frac{1}{8\sqrt{3}}\sin(8\sqrt{3}t), \tag{16}$$

that is, $A = \frac{1}{6}$ and $B = -1/(8\sqrt{3})$. The natural frequency is $\omega_0 = \sqrt{192} \cong 13.856$ radians (rad)/s , so the period is $T = 2\pi/\omega_0 \cong 0.45345$ s. The amplitude R and phase δ are found from Eqs. (12) and (13). We have

$$R^2 = \frac{1}{36} + \frac{1}{192} = \frac{19}{576}, \quad \text{so} \quad R \cong 0.18162 \text{ ft.}$$

and since $A > 0$ and $B < 0$, the angle δ lies in the fourth quadrant,

$$\delta = 2\pi + \arctan(-\sqrt{3}/4) \cong 5.87455 \text{ rad.}$$

The graph of the solution (16) is shown in Figure 4.4.4.

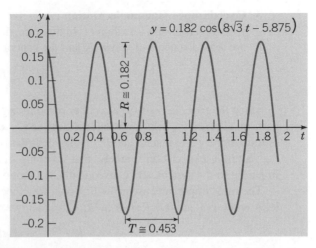

$$y = 0.182 \cos\left(8\sqrt{3}\,t - 5.875\right)$$

FIGURE 4.4.4 An undamped free vibration: $y'' + 192y = 0$, $y(0) = \frac{1}{6}$, $y'(0) = -1$.

Damped Free Vibrations

If we include the effect of damping, the differential equation governing the motion of the mass is

$$my'' + \gamma y' + ky = 0. \tag{17}$$

We are especially interested in examining the effect of variations in the damping coefficient γ for given values of the mass m and spring constant k. The roots of the corresponding characteristic equation,

$$m\lambda^2 + \gamma\lambda + k = 0, \tag{18}$$

are

$$\lambda_1, \lambda_2 = \frac{-\gamma \pm \sqrt{\gamma^2 - 4km}}{2m} = \frac{\gamma}{2m}\left(-1 \pm \sqrt{1 - \frac{4km}{\gamma^2}}\right). \tag{19}$$

There are three cases to consider, depending on the sign of the discriminant $\gamma^2 - 4km$.

1. **Underdamped Harmonic Motion** ($\gamma^2 - 4km < 0$). In this case, the roots in Eq. (19) are complex numbers $\mu \pm i\nu$ with $\mu = -\gamma/2m < 0$ and $\nu = \dfrac{(4km - \gamma^2)^{1/2}}{2m} > 0$. Hence the general solution of Eq. (18) is

$$y = e^{-\gamma t/2m}(A \cos \nu t + B \sin \nu t). \tag{20}$$

2. **Critically Damped Harmonic Motion** ($\gamma^2 - 4km = 0$). In this case, $\lambda_1 = -\gamma/2m < 0$ is a repeated root. Therefore the general solution of Eq. (17) in this case is

$$y = (A + Bt)e^{-\gamma t/2m}. \tag{21}$$

3. Overdamped Harmonic Motion ($\gamma^2 - 4km > 0$). Since m, γ, and k are positive, $\gamma^2 - 4km$ is always less than γ^2. In this case, the values of λ_1 and λ_2 given by Eq. (19) are real, distinct, and *negative*, and the general solution of Eq. (17) is

$$y = Ae^{\lambda_1 t} + Be^{\lambda_2 t}. \tag{22}$$

Since the roots in Eq. (19) are either real and negative, or complex with a negative real part, in all cases the solution y tends to zero as $t \to \infty$; this occurs regardless of the values of the arbitrary constants A and B, that is, regardless of the initial conditions. This confirms our intuitive expectation, namely, that damping gradually dissipates the energy initially imparted to the system, and consequently, the motion dies out with increasing time.

The most important case is the first one, which occurs when the damping is small. If we let $A = R \cos \delta$ and $B = R \sin \delta$ in Eq. (20), then we obtain

$$y = Re^{-\gamma t/2m} \cos(\nu t - \delta). \tag{23}$$

The displacement y lies between the curves $y = \pm Re^{-\gamma t/2m}$; hence it resembles a cosine wave whose amplitude decreases as t increases. A typical example is sketched in Figure 4.4.5. The motion is called a damped oscillation or damped vibration. The amplitude factor R depends on m, γ, k, and the initial conditions.

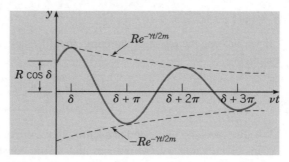

FIGURE 4.4.5   Damped vibration; $y = Re^{-\gamma t/2m} \cos(\nu t - \delta)$.

Although the motion is not periodic, the parameter ν determines the frequency with which the mass oscillates back and forth; consequently, ν is called the **quasi-frequency**. By comparing ν with the frequency ω_0 of undamped motion, we find that

$$\frac{\nu}{\omega_0} = \frac{(4km - \gamma^2)^{1/2}/2m}{\sqrt{k/m}} = \left(1 - \frac{\gamma^2}{4km}\right)^{1/2} \cong 1 - \frac{\gamma^2}{8km}. \tag{24}$$

The last approximation is valid when $\gamma^2/4km$ is small. We refer to this situation as "small damping." Thus the effect of small damping is to reduce slightly the frequency of the oscillation. By analogy with Eq. (14), the quantity $T_d = 2\pi/\nu$ is called the **quasi-period**. It is the time between successive maxima or successive minima of the position of the mass, or between successive passages of the mass through its equilibrium position while going in the same direction. The relation between T_d and T is given by

$$\frac{T_d}{T} = \frac{\omega_0}{\nu} = \left(1 - \frac{\gamma^2}{4km}\right)^{-1/2} \cong 1 + \frac{\gamma^2}{8km}, \tag{25}$$

where again the last approximation is valid when $\gamma^2/4km$ is small. Thus small damping increases the quasi-period.

Equations (24) and (25) reinforce the significance of the dimensionless ratio $\gamma^2/4km$. It is not the magnitude of γ alone that determines whether damping is large or small, but the magnitude of γ^2 compared to $4km$. When $\gamma^2/4km$ is small, then damping has a small effect on the quasi-frequency and quasi-period of the motion. On the other hand, if we want to study the detailed motion of the mass for all time, then we can *never* neglect the damping force, no matter how small.

As $\gamma^2/4km$ increases, the quasi-frequency ν decreases and the quasi-period T_d increases. In fact, $\nu \to 0$ and $T_d \to \infty$ as $\gamma \to 2\sqrt{km}$. As indicated by Eqs. (20), (21), and (22), the nature of the solution changes as γ passes through the value $2\sqrt{km}$. This value of γ is known as critical damping. The motion is said to be underdamped for values of $\gamma < 2\sqrt{km}$; while for values of $\gamma > 2\sqrt{km}$, the motion is said to be overdamped. In the critically damped and overdamped cases given by Eqs. (21) and (20), respectively, the mass creeps back to its equilibrium position but does not oscillate about it, as for small γ. Note that this analysis is consistent with the definitions of underdamped, critcally damped, and overdamped harmonic motion based on the sign of $\gamma^2 - 4km$ (see pages 245–246). Two typical examples of critically damped motion are shown in Figure 4.4.6, and the situation is discussed further in Problems 19 and 20.

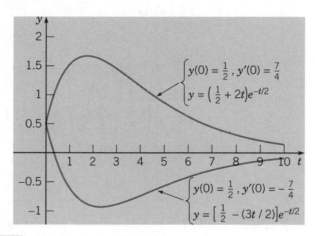

FIGURE 4.4.6   Two critically damped motions: $y'' + y' + 0.25y = 0$; $y = (A + Bt)e^{-t/2}$.

EXAMPLE 2

The motion of a certain spring-mass system is governed by the differential equation

$$y'' + 0.125y' + y = 0, \tag{26}$$

where y is measured in feet and t in seconds. If $y(0) = 2$ and $y'(0) = 0$, determine the position of the mass at any time. Find the quasi-frequency and the quasi-period, as well as the time at which the mass first passes through its equilibrium position. Find the time τ such that $|y(t)| < 0.1$ for all $t > \tau$.

The solution of Eq. (26) is

$$y = e^{-t/16}\left[A\cos\frac{\sqrt{255}}{16}t + B\sin\frac{\sqrt{255}}{16}t\right].$$

To satisfy the initial conditions, we must choose $A = 2$ and $B = 2/\sqrt{255}$; hence the solution of the initial value problem is

$$
y = e^{-t/16}\left(2\cos\frac{\sqrt{255}}{16}t + \frac{2}{\sqrt{255}}\sin\frac{\sqrt{255}}{16}t \right)
$$

$$
= \frac{32}{\sqrt{255}}e^{-t/16}\cos\left(\frac{\sqrt{255}}{16}t - \delta \right), \tag{27}
$$

where $\tan\delta = 1/\sqrt{255}$, so $\delta \cong 0.06254$. The displacement of the mass as a function of time is shown in Figure 4.4.7. For purposes of comparison, we also show the motion if the damping term is neglected.

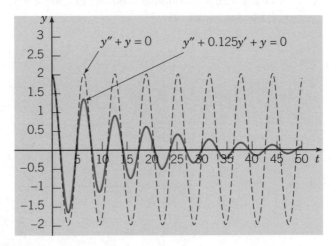

FIGURE 4.4.7 Vibration with small damping (solid curve) and with no damping (dashed curve). In each case, $y(0) = 2$ and $y'(0) = 0$.

The quasi-frequency is $v = \sqrt{255}/16 \cong 0.998$ and the quasi-period is $T_d = 2\pi/v \cong 6.295$ s. These values differ only slightly from the corresponding values (1 and 2π, respectively) for the undamped oscillation. This is also evident from the graphs in Figure 4.4.7, which rise and fall almost together. The damping coefficient is small in this example, only one-sixteenth of the critical value, in fact. Nevertheless the amplitude of the oscillation is reduced rather rapidly. Figure 4.4.8 shows the graph of the solution for $40 \leq t \leq 60$, together with the graphs of $y = \pm 0.1$. From the graph, it appears that τ is about 47.5, and by a more precise calculation we find that $\tau \cong 47.5149$ s.

To find the time at which the mass first passes through its equilibrium position, we refer to Eq. (27) and set $\sqrt{255}t/16 - \delta$ equal to $\pi/2$, the smallest positive zero of the cosine function. Then, by solving for t, we obtain

$$
t = \frac{16}{\sqrt{255}}\left(\frac{\pi}{2} + \delta \right) \cong 1.637 \text{ s}.
$$

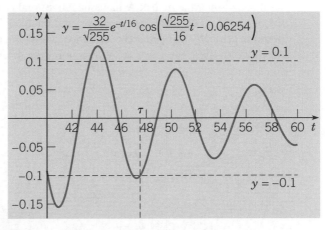

FIGURE 4.4.8 Solution of Example 2; determination of τ.

Phase Portraits for Harmonic Oscillators

The differences in the behavior of solutions of undamped and damped harmonic oscillators, illustrated by plots of displacement versus time, are completed by looking at their corresponding phase portraits. If we convert Eq. (17) to a first order system where $\mathbf{x} = x_1 \mathbf{i} + x_2 \mathbf{j} = y \mathbf{i} + y' \mathbf{j}$, we obtain

$$\mathbf{x}' = \mathbf{A}\mathbf{x} = \begin{pmatrix} 0 & 1 \\ -k/m & -\gamma/m \end{pmatrix} \mathbf{x}. \tag{28}$$

Since the eigenvalues of $\mathbf{A}$ are the roots of the characteristic equation (18), we know that the origin of the phase plane is a center, and therefore stable, for the undamped system in which $\gamma = 0$. In the underdamped case, $0 < \gamma^2 < 4km$, the origin is a spiral sink. Direction fields and phase portraits for these two cases are shown in Figure 4.4.9.

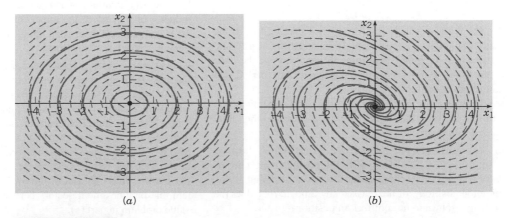

FIGURE 4.4.9 Direction field and phase portrait for (*a*) an undamped harmonic oscillator. (*b*) a damped harmonic oscillator that is underdamped.

If $\gamma_2 = 4km$, the matrix **A** has a negative, real, and repeated eigenvalue; if $\gamma^2 > 4km$, the eigenvalues of **A** are real, negative, and unequal. Thus the origin of the phase plane in both the critically damped and overdamped cases is a nodal sink. Direction fields and phase portraits for these two cases are shown in Figure 4.4.10.

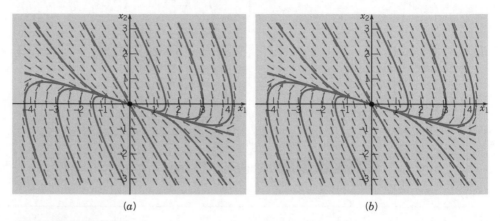

(a) (b)

FIGURE 4.4.10 Direction field and phase portrait for (a) a critically damped harmonic oscillator. (b) an overdamped harmonic oscillator.

It is clear from the phase portraits in Figure 4.4.10 that a mass can pass through the equilibrium position at most once, since trajectories either do not cross the x_2-axis, or cross it at most once, as they approach the equilibrium point. In Problem 19, you are asked to give an analytic argument of this fact.

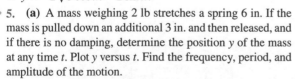

PROBLEMS

In each of Problems 1 through 4, determine ω_0, R, and δ so as to write the given expression in the form $y = R\cos(\omega_0 t - \delta)$.

1. $y = 3\cos 2t + 3\sin 2t$

2. $y = -\cos t + \sqrt{3}\sin t$

3. $y = 4\cos 3t - 2\sin 3t$

4. $y = -2\sqrt{3}\cos \pi t - 2\sin \pi t$

5. **(a)** A mass weighing 2 lb stretches a spring 6 in. If the mass is pulled down an additional 3 in. and then released, and if there is no damping, determine the position y of the mass at any time t. Plot y versus t. Find the frequency, period, and amplitude of the motion.

(b) Draw a phase portrait of the equivalent dynamical system that includes the trajectory corresponding to the initial value problem in part (a).

6. **(a)** A mass of 100 g stretches a spring 5 cm. If the mass is set in motion from its equilibrium position with a downward velocity of 10 cm/s, and if there is no damping, determine the position y of the mass at any time t. When does the mass first return to its equilibrium position?

(b) Draw a phase portrait of the equivalent dynamical system that includes the trajectory corresponding to the initial value problem in part (a).

7. A mass weighing 3 lb stretches a spring 3 in. If the mass is pushed upward, contracting the spring a distance of 1 in., and then set in motion with a downward velocity of 2 ft/s, and if there is no damping, find the position y of the mass at any time t. Determine the frequency, period, amplitude, and phase of the motion.

8. A series circuit has a capacitor of 0.25 microfarad and an inductor of 1 henry. If the initial charge on the capacitor is 10^{-6} coulomb and there is no initial current, find the charge q on the capacitor at any time t.

9. **(a)** A mass of 20 g stretches a spring 5 cm. Suppose that the mass is also attached to a viscous damper with a damping constant of 400 dyne·s/cm. If the mass is pulled down an additional 2 cm and then released, find its position y at any time t. Plot y versus t. Determine the quasi-frequency and the quasi-period. Determine the ratio of the quasi-period to the period of the corresponding undamped motion. Also find the time τ such that $|y(t)| < 0.05$ cm for all $t > \tau$.

(b) Draw a phase portrait of the equivalent dynamical system that includes the trajectory corresponding to the initial value problem in part (a).

10. A mass weighing 16 lb stretches a spring 3 in. The mass is attached to a viscous damper with a damping constant of

2 lb·s/ft. If the mass is set in motion from its equilibrium position with a downward velocity of 3 in./s, find its position y at any time t. Plot y versus t. Determine when the mass first returns to its equilibrium position. Also find the time τ such that $|y(t)| < 0.01$ in for all $t > \tau$.

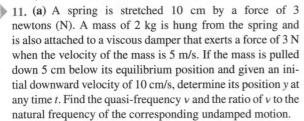

11. (a) A spring is stretched 10 cm by a force of 3 newtons (N). A mass of 2 kg is hung from the spring and is also attached to a viscous damper that exerts a force of 3 N when the velocity of the mass is 5 m/s. If the mass is pulled down 5 cm below its equilibrium position and given an initial downward velocity of 10 cm/s, determine its position y at any time t. Find the quasi-frequency ν and the ratio of ν to the natural frequency of the corresponding undamped motion.
(b) Draw a phase portrait of the equivalent dynamical system that includes the trajectory corresponding to the initial value problem in part (a).

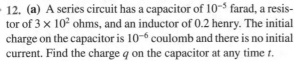

12. (a) A series circuit has a capacitor of 10^{-5} farad, a resistor of 3×10^2 ohms, and an inductor of 0.2 henry. The initial charge on the capacitor is 10^{-6} coulomb and there is no initial current. Find the charge q on the capacitor at any time t.
(b) Draw a phase portrait of the equivalent dynamical system that includes the trajectory corresponding to the initial value problem in part (a).

13. A certain vibrating system satisfies the equation $y'' + \gamma y' + y = 0$. Find the value of the damping coefficient γ for which the quasi-period of the damped motion is 50% greater than the period of the corresponding undamped motion.

14. Show that the period of motion of an undamped vibration of a mass hanging from a vertical spring is $2\pi\sqrt{L/g}$, where L is the elongation of the spring due to the mass and g is the acceleration due to gravity.

15. Show that the solution of the initial value problem

$$my'' + \gamma y' + ky = 0, \quad y(t_0) = y_0, \quad y'(t_0) = y_1$$

can be expressed as the sum $y = v + w$, where v satisfies the initial conditions $v(t_0) = y_0$, $v'(t_0) = 0$, w satisfies the initial conditions $w(t_0) = 0$, $w'(t_0) = y_1$, and both v and w satisfy the same differential equation as u. This is another instance of superposing solutions of simpler problems to obtain the solution of a more general problem.

16. Show that $A\cos\omega_0 t + B\sin\omega_0 t$ can be written in the form $r\sin(\omega_0 t - \theta)$. Determine r and θ in terms of A and B. If $R\cos(\omega_0 t - \delta) = r\sin(\omega_0 t - \theta)$, determine the relationship among R, r, δ, and θ.

17. A mass weighing 8 lb stretches a spring 1.5 in. The mass is also attached to a damper with coefficient γ. Determine the value of γ for which the system is critically damped. Be sure to give the units for γ.

18. If a series circuit has a capacitor of $C = 0.8$ microfarad and an inductor of $L = 0.2$ henry, find the resistance R so that the circuit is critically damped.

19. Assume that the system described by the equation $my'' + \gamma y' + ky = 0$ is either critically damped or overdamped. Show that the mass can pass through the equilibrium position at most once, regardless of the initial conditions.
Hint: Determine all possible values of t for which $y = 0$.

20. Assume that the system described by the equation $my'' + \gamma y' + ky = 0$ is critically damped and that the initial conditions are $y(0) = y_0$, $y'(0) = v_0$. If $v_0 = 0$, show that $y \to 0$ as $t \to \infty$ but that y is never zero. If y_0 is positive, determine a condition on v_0 that will ensure that the mass passes through its equilibrium position after it is released.

21. **Logarithmic Decrement**
(a) For the damped oscillation described by Eq. (23), show that the time between successive maxima is $T_d = 2\pi/\nu$.
(b) Show that the ratio of the displacements at two successive maxima is given by $\exp(\gamma T_d/2m)$. Observe that this ratio does not depend on which pair of maxima is chosen. The natural logarithm of this ratio is called the logarithmic decrement and is denoted by Δ.
(c) Show that $\Delta = \pi\gamma/m\nu$. Since m, ν, and Δ are quantities that can be measured easily for a mechanical system, this result provides a convenient and *practical* method for determining the damping constant of the system, which is more difficult to measure directly. In particular, for the motion of a vibrating mass in a viscous fluid, the damping constant depends on the viscosity of the fluid. For simple geometric shapes, the form of this dependence is known, and the preceding relation allows the experimental determination of the viscosity. This is one of the most accurate ways of determining the viscosity of a gas at high pressure.

22. Referring to Problem 21, find the logarithmic decrement of the system in Problem 10.

23. For the system in Problem 17, suppose that $\Delta = 3$ and $T_d = 0.3$ s. Referring to Problem 21, determine the value of the damping coefficient γ.

24. The position of a certain spring-mass system satisfies the initial value problem

$$\tfrac{3}{2}y'' + ky = 0, \qquad y(0) = 2, \quad y'(0) = v.$$

If the period and amplitude of the resulting motion are observed to be π and 3, respectively, determine the values of k and v.

25. Consider the initial value problem

$$y'' + \gamma y' + y = 0, \qquad y(0) = 2, \quad y'(0) = 0.$$

We wish to explore how long a time interval is required for the solution to become "negligible" and how this interval depends on the damping coefficient γ. To be more precise, let us seek the time τ such that $|y(t)| < 0.01$ for all $t > \tau$. Note that critical damping for this problem occurs for $\gamma = 2$.

(a) Let $\gamma = 0.25$ and determine τ, or at least estimate it fairly accurately from a plot of the solution.

(b) Repeat part (a) for several other values of γ in the interval $0 < \gamma < 1.5$. Note that τ steadily decreases as γ increases for γ in this range.

(c) Create a graph of τ versus γ by plotting the pairs of values found in parts (a) and (b). Is the graph a smooth curve?

(d) Repeat part (b) for values of γ between 1.5 and 2. Show that τ continues to decrease until γ reaches a certain critical value γ_0, after which τ increases. Find γ_0 and the corresponding minimum value of τ to two decimal places.

(e) Another way to proceed is to write the solution of the initial value problem in the form (23). Neglect the cosine factor and consider only the exponential factor and the amplitude R. Then find an expression for τ as a function of γ. Compare the approximate results obtained in this way with the values determined in parts (a), (b), and (d).

26. Consider the initial value problem

$$my'' + \gamma y' + ky = 0, \quad y(0) = y_0, \quad y'(0) = v_0.$$

Assume that $\gamma^2 < 4km$.

(a) Solve the initial value problem.

(b) Write the solution in the form $y(t) = R\exp(-\gamma t/2m)\cos(vt - \delta)$. Determine R in terms of m, γ, k, y_0, and v_0.

(c) Investigate the dependence of R on the damping coefficient γ for fixed values of the other parameters.

27. Use the differential equation derived in Problem 19 of Section 4.1 to determine the period of vertical oscillations of a cubic block floating in a fluid under the stated conditions.

28. Draw the phase portrait for the dynamical system equivalent to the differential equation considered in Example 2: $y'' + 0.125y' + y = 0$.

 29. The position of a certain undamped spring-mass system satisfies the initial value problem

$$y'' + 2y = 0, \quad y(0) = 0, \quad y'(0) = 2.$$

(a) Find the solution of this initial value problem.

(b) Plot y versus t and y' versus t on the same axes.

(c) Draw the phase portrait for the dynamical system equivalent to $y'' + 2y = 0$. Include the trajectory corresponding to the initial conditions $y(0) = 0$, $y'(0) = 2$.

30. The position of a certain spring-mass system satisfies the initial value problem

$$y'' + \tfrac{1}{4}y' + 2y = 0, \quad y(0) = 0, \quad y'(0) = 2.$$

(a) Find the solution of this initial value problem.

(b) Plot y versus t and y' versus t on the same axes.

(c) Draw the phase portrait for the dynamical system equivalent to $y'' + \tfrac{1}{4}y' + 2y = 0$. Include the trajectory corresponding to the initial conditions $y(0) = 0$, $y'(0) = 2$.

31. In the absence of damping, the motion of a spring-mass system satisfies the initial value problem

$$my'' + ky = 0, \quad y(0) = a, \quad y'(0) = b.$$

(a) Show that the kinetic energy initially imparted to the mass is $mb^2/2$ and that the potential energy initially stored in the spring is $ka^2/2$, so that initially the total energy in the system is $(ka^2 + mb^2)/2$.

(b) Solve the given initial value problem.

(c) Using the solution in part (b), determine the total energy in the system at any time t. Your result should confirm the principle of conservation of energy for this system.

32. If the restoring force of a nonlinear spring satisfies the relation

$$F_s(\Delta x) = -k\Delta x - \epsilon(\Delta x)^3,$$

where $k > 0$, then the differential equation for the displacement $x(t)$ of the mass from its equilibrium position satisfies the differential equation (see Problem 17, Section 4.1)

$$mx'' + \gamma x' + kx + \epsilon x^3 = 0.$$

Assume that the initial conditions are

$$x(0) = 0, \quad x'(0) = 1.$$

(a) Find $x(t)$ when $\epsilon = 0$ and also determine the amplitude and period of the motion.

(b) Let $\epsilon = 0.1$. Plot a numerical approximation to the solution. Does the motion appear to be periodic? Estimate the amplitude and period.

(c) Repeat part (c) for $\epsilon = 0.2$ and $\epsilon = 0.3$.

(d) Plot your estimated values of the amplitude A and the period T versus ϵ. Describe the way in which A and T, respectively, depend on ϵ.

(e) Repeat parts (c), (d), and (e) for negative values of ϵ.

4.5 Nonhomogeneous Equations; Method of Undetermined Coefficients

We now return to the nonhomogeneous equation

$$L[y] = y'' + p(t)y' + q(t)y = g(t), \tag{1}$$

where p, q, and g are given (continuous) functions on the open interval I. The equation

$$L[y] = y'' + p(t)y' + q(t)y = 0,\qquad(2)$$

in which $g(t) = 0$ and p and q are the same as in Eq. (1), is called the homogeneous equation corresponding to Eq. (1). The following two results describe the structure of solutions of the nonhomogeneous equation (1) and provide a basis for constructing its general solution.

THEOREM 4.5.1

If Y_1 and Y_2 are two solutions of the nonhomogeneous equation (1), then their difference $Y_1 - Y_2$ is a solution of the corresponding homogeneous equation (2). If, in addition, y_1 and y_2 form a fundamental set of solutions of Eq. (2), then

$$Y_1(t) - Y_2(t) = c_1 y_1(t) + c_2 y_2(t),\qquad(3)$$

where c_1 and c_2 are certain constants.

Proof

To prove this result, note that Y_1 and Y_2 satisfy the equations

$$L[Y_1](t) = g(t),\qquad L[Y_2](t) = g(t).\qquad(4)$$

Subtracting the second of these equations from the first, we have

$$L[Y_1](t) - L[Y_2](t) = g(t) - g(t) = 0.\qquad(5)$$

However, by linearity of the differential operator,

$$L[Y_1] - L[Y_2] = L[Y_1 - Y_2],$$

so Eq. (5) becomes

$$L[Y_1 - Y_2](t) = 0.\qquad(6)$$

Equation (6) states that $Y_1 - Y_2$ is a solution of Eq. (2). Finally, since all solutions of Eq. (2) can be expressed as linear combinations of a fundamental set of solutions by Theorem 4.2.7, it follows that the solution $Y_1 - Y_2$ can be so written. Hence Eq. (3) holds and the proof is complete.

THEOREM 4.5.2

The general solution of the nonhomogeneous equation (1) can be written in the form

$$\boxed{y = \phi(t) = c_1 y_1(t) + c_2 y_2(t) + Y(t),}\qquad(7)$$

where y_1 and y_2 form a fundamental set of solutions of the corresponding homogeneous equation (2), c_1 and c_2 are arbitrary constants, and Y is some specific solution of the nonhomogeneous equation (1).

Proof

The proof of Theorem 4.5.2 follows quickly from Theorem 4.5.1. Note that Eq. (3) holds if we identify Y_1 with an arbitrary solution ϕ of Eq. (1) and Y_2 with the specific solution Y. From Eq. (3) we thereby obtain

$$\phi(t) - Y(t) = c_1 y_1(t) + c_2 y_2(t),\qquad(8)$$

which is equivalent to Eq. (7). Since ϕ is an arbitrary solution of Eq. (1), the expression on the right side of Eq. (7) includes all solutions of Eq. (1); thus it is natural to call it the general solution of Eq. (1).

General Solution Strategy

Theorem 4.5.2 states that to solve the nonhomogeneous equation (1), we must do three things:

1. Find the general solution $c_1 y_1(t) + c_2 y_2(t)$ of the corresponding homogeneous equation. This solution is frequently called the **complementary solution** and may be denoted by $y_c(t)$.
2. Find some single solution $Y(t)$ of the nonhomogeneous equation. Often this solution is referred to as a **particular solution**.
3. Add together the functions found in the two preceding steps.

We have already discussed how to find $y_c(t)$, at least when the homogeneous equation (2) has constant coefficients. In the remainder of this section, we will focus on a special method of finding a particular solution $Y(t)$ of the nonhomogeneous equation (1) known as the **method of undetermined coefficients**. In Section 4.7 we present a general method known as the **method of variation of parameters**. Each method has some advantages and some possible shortcomings; these are discussed below, and again at the end of Section 4.7.

Method of Undetermined Coefficients

The method of undetermined coefficients requires that we make an initial assumption about the form of the particular solution $Y(t)$, but with the coefficients left unspecified. We then substitute the assumed expression into Eq. (1) and attempt to determine the coefficients so as to satisfy that equation. If we are successful, then we have found a solution of the differential equation (1) and can use it for the particular solution $Y(t)$. If we cannot determine the coefficients, then this means that there is no solution of the form that we assumed. In this case, we may modify the initial assumption and try again.

The main advantage of the method of undetermined coefficients is that it is straightforward to execute once the assumption is made as to the form of $Y(t)$. Its major limitation is that it is useful primarily for equations for which we can easily write down the correct form of the particular solution in advance. For this reason, this method is usually used only for problems in which the homogeneous equation has constant coefficients and the nonhomogeneous term is restricted to a relatively small class of functions. In particular, we consider only nonhomogeneous terms that consist of polynomials, exponential functions, sines, cosines, or sums or products of such functions. Despite this limitation, the method of undetermined coefficients is useful for solving many problems that have important applications. However the algebraic details may become tedious, and a computer algebra system can be very helpful in practical applications. We will illustrate the method of undetermined coefficients by several simple examples and then summarize some rules for using it.

EXAMPLE 1

Find a particular solution of

$$y'' - 3y' - 4y = 3e^{2t}. \tag{9}$$

We seek a function Y such that the combination $Y''(t) - 3Y'(t) - 4Y(t)$ is equal to $3e^{2t}$. Since the exponential function reproduces itself through differentiation, the most plausible way to achieve the desired result is to assume that $Y(t)$ is some multiple of e^{2t}, that is,

$$Y(t) = Ae^{2t},$$

where the coefficient A is yet to be determined. To find A, we calculate

$$Y'(t) = 2Ae^{2t}, \qquad Y''(t) = 4Ae^{2t},$$

and substitute for y, y', and y'' in Eq. (9). We obtain

$$(4A - 6A - 4A)e^{2t} = 3e^{2t}.$$

Hence $-6Ae^{2t}$ must equal $3e^{2t}$, so $A = -\frac{1}{2}$. Thus a particular solution is

$$Y(t) = -\frac{1}{2}e^{2t}. \tag{10}$$

EXAMPLE 2

Find a particular solution of

$$y'' - 3y' - 4y = 2\sin t. \tag{11}$$

By analogy with Example 1, let us first assume that $Y(t) = A\sin t$, where A is a constant to be determined. On substituting in Eq. (11) and rearranging the terms, we obtain

$$-5A\sin t - 3A\cos t = 2\sin t,$$

or

$$(2 + 5A)\sin t + 3A\cos t = 0. \tag{12}$$

Equation (12) can hold on an interval only if the coefficients of $\sin t$ and $\cos t$ are both zero. Thus we must have $2 + 5A = 0$ and also $3A = 0$. These contradictory requirements mean that there is no choice of the constant A that makes Eq. (12) true for all t. Thus we conclude that our assumption concerning $Y(t)$ is inadequate. The appearance of the cosine term in Eq. (12) suggests that we modify our original assumption to include a cosine term in $Y(t)$, that is,

$$Y(t) = A\sin t + B\cos t,$$

where A and B are to be determined. Then

$$Y'(t) = A\cos t - B\sin t, \qquad Y''(t) = -A\sin t - B\cos t.$$

By substituting these expressions for y, y', and y'' in Eq. (11) and collecting terms, we obtain

$$(-A + 3B - 4A)\sin t + (-B - 3A - 4B)\cos t = 2\sin t. \tag{13}$$

To satisfy Eq. (13), we must match the coefficients of $\sin t$ and $\cos t$ on each side of the equation. Thus A and B must satisfy the following pair of linear equations

$$-5A + 3B = 2, \qquad -3A - 5B = 0.$$

Hence $A = -\frac{5}{17}$ and $B = \frac{3}{17}$, so a particular solution of Eq. (11) is

$$Y(t) = -\frac{5}{17}\sin t + \frac{3}{17}\cos t.$$

The method illustrated in the preceding examples can also be used when the right side of the equation is a polynomial. Thus, to find a particular solution of

$$y'' - 3y' - 4y = 4t^2 - 1, \tag{14}$$

we initially assume that $Y(t)$ is a polynomial of the same degree as the nonhomogeneous term, that is, $Y(t) = At^2 + Bt + C$.

To summarize our conclusions up to this point:

1. If the nonhomogeneous term $g(t)$ in Eq. (1) is an exponential function $e^{\alpha t}$, then assume that $Y(t)$ is proportional to the same exponential function.

2. If $g(t)$ is $\sin \beta t$ or $\cos \beta t$, then assume that $Y(t)$ is a linear combination of $\sin \beta t$ and $\cos \beta t$.

3. If $g(t)$ is a polynomial, then assume that $Y(t)$ is a polynomial of like degree.

The same principle extends to the case where $g(t)$ is a product of any two, or all three, of these types of functions, as the next example illustrates.

EXAMPLE 3

Find a particular solution of

$$y'' - 3y' - 4y = -8e^t \cos 2t. \tag{15}$$

In this case, we assume that $Y(t)$ is the product of e^t and a linear combination of $\cos 2t$ and $\sin 2t$, that is,

$$Y(t) = Ae^t \cos 2t + Be^t \sin 2t.$$

The algebra is more tedious in this example, but it follows that

$$Y'(t) = (A + 2B)e^t \cos 2t + (-2A + B)e^t \sin 2t$$

and

$$Y''(t) = (-3A + 4B)e^t \cos 2t + (-4A - 3B)e^t \sin 2t.$$

By substituting these expressions in Eq. (15), we find that A and B must satisfy

$$10A + 2B = 8, \qquad 2A - 10B = 0.$$

Hence $A = \frac{10}{13}$ and $B = \frac{2}{13}$. Therefore a particular solution of Eq. (15) is

$$Y(t) = \frac{10}{13}e^t \cos 2t + \frac{2}{13}e^t \sin 2t.$$

Superposition Principle for Nonhomogeneous Equations

Now suppose that $g(t)$ is the sum of two terms, $g(t) = g_1(t) + g_2(t)$, and suppose that Y_1 and Y_2 are solutions of the equations

$$ay'' + by' + cy = g_1(t) \tag{16}$$

and

$$ay'' + by' + cy = g_2(t), \tag{17}$$

respectively. Then $Y_1 + Y_2$ is a solution of the differential equation

$$ay'' + by' + cy = g(t). \tag{18}$$

To prove this statement, substitute $Y_1(t) + Y_2(t)$ for y in Eq. (18) and make use of Eqs. (16) and (17). A similar conclusion holds if $g(t)$ is the sum of any finite number of terms. The practical significance of this result is that for an equation whose nonhomogeneous function $g(t)$ can be expressed as a sum, one can consider instead several simpler equations and then add together the results. The following example is an illustration of this procedure.

EXAMPLE 4

Find a particular solution of

$$y'' - 3y' - 4y = 3e^{2t} + 2\sin t - 8e^t \cos 2t. \qquad (19)$$

By splitting up the right side of Eq. (19), we obtain the three equations

$$y'' - 3y' - 4y = 3e^{2t},$$
$$y'' - 3y' - 4y = 2\sin t,$$

and

$$y'' - 3y' - 4y = -8e^t \cos 2t.$$

Solutions of these three equations have been found in Examples 1, 2, and 3, respectively. Therefore a particular solution of Eq. (19) is their sum, namely,

$$Y(t) = -\tfrac{1}{2}e^{2t} + \tfrac{3}{17}\cos t - \tfrac{5}{17}\sin t + \tfrac{10}{13}e^t \cos 2t + \tfrac{2}{13}e^t \sin 2t.$$

The procedure illustrated in these examples enables us to solve a fairly large class of problems in a reasonably efficient manner. However there is one difficulty that sometimes occurs. The next example illustrates how it arises.

EXAMPLE 5

Find a particular solution of

$$y'' - 3y' - 4y = 2e^{-t}. \qquad (20)$$

Proceeding as in Example 1, we assume that $Y(t) = Ae^{-t}$. By substituting in Eq. (20), we then obtain

$$(A + 3A - 4A)e^{-t} = 2e^{-t}. \qquad (21)$$

Since the left side of Eq. (21) is zero, there is no choice of A that satisfies this equation. Therefore there is no particular solution of Eq. (20) of the assumed form. The reason for this possibly unexpected result becomes clear if we solve the homogeneous equation

$$y'' - 3y' - 4y = 0 \qquad (22)$$

that corresponds to Eq. (20). A fundamental set of solutions of Eq. (22) is $y_1(t) = e^{-t}$ and $y_2(t) = e^{4t}$. Thus our assumed particular solution of Eq. (20) is actually a solution of the homogeneous equation (22). Consequently, it cannot possibly be a solution of the nonhomogeneous equation (20). To find a solution of Eq. (20), we must therefore consider functions of a somewhat different form.

At this stage, we have several possible alternatives. One is simply to try to guess the proper form of the particular solution of Eq. (20). Another is to solve this equation in some different way and then to use the result to guide our assumptions if this situation arises again in the future (see Problem 27 for an example of another solution method).

Still another possibility is to seek a simpler equation where this difficulty occurs and to use its solution to suggest how we might proceed with Eq. (20). Adopting the latter approach, we look for a first order equation analogous to Eq. (20). One possibility is

$$y' + y = 2e^{-t}. \tag{23}$$

If we try to find a particular solution of Eq. (23) of the form Ae^{-t}, we will fail because e^{-t} is a solution of the homogeneous equation $y' + y = 0$. However, from Section 2.2 we already know how to solve Eq. (23). An integrating factor is $\mu(t) = e^t$, and by multiplying by $\mu(t)$ and then integrating both sides, we obtain the solution

$$y = 2te^{-t} + ce^{-t}. \tag{24}$$

The second term on the right side of Eq. (24) is the general solution of the homogeneous equation $y' + y = 0$, but the first term is a solution of the full nonhomogeneous equation (23). Observe that it involves the exponential factor e^{-t} multiplied by the factor t. This is the clue that we were looking for.

We now return to Eq. (20) and assume a particular solution of the form $Y(t) = Ate^{-t}$. Then

$$Y'(t) = Ae^{-t} - Ate^{-t}, \qquad Y''(t) = -2Ae^{-t} + Ate^{-t}. \tag{25}$$

Substituting these expressions for y, y', and y'' in Eq. (20), we obtain $-5A = 2$, so $A = -\frac{2}{5}$. Thus a particular solution of Eq. (20) is

$$Y(t) = -\frac{2}{5}te^{-t}. \tag{26}$$

The outcome of Example 5 suggests a modification of the principle stated previously: if the assumed form of the particular solution duplicates a solution of the corresponding homogeneous equation, then modify the assumed particular solution by multiplying it by t. Occasionally, this modification will be insufficient to remove all duplication with the solutions of the homogeneous equation, in which case it is necessary to multiply by t a second time. For a second order equation, it will never be necessary to carry the process further than this.

Summary: Method of Undetermined Coefficients

We now summarize the steps involved in finding the solution of an initial value problem consisting of a nonhomogeneous equation of the form

$$ay'' + by' + cy = g(t), \tag{27}$$

where the coefficients a, b, and c are constants, together with a given set of initial conditions:

1. Find the general solution of the corresponding homogeneous equation.
2. Make sure that the function $g(t)$ in Eq. (27) belongs to the class of functions discussed in this section; that is, be sure it involves nothing more than exponential functions, sines, cosines, polynomials, or sums or products of such functions. If this is not the case, use the method of variation of parameters (discussed in Section 4.7).

3. If $g(t) = g_1(t) + \cdots + g_n(t)$, that is, if $g(t)$ is a sum of n terms, then form n subproblems, each of which contains only one of the terms $g_1(t), \ldots, g_n(t)$. The ith subproblem consists of the equation

$$ay'' + by' + cy = g_i(t),$$

where i runs from 1 to n.

4. For the ith subproblem, assume a particular solution $Y_i(t)$ consisting of the appropriate exponential function, sine, cosine, polynomial, or combination thereof. If there is any duplication in the assumed form of $Y_i(t)$ with the solutions of the homogeneous equation (found in Step 1), then multiply $Y_i(t)$ by t, or (if necessary) by t^2, so as to remove the duplication. See Table 4.5.1.

TABLE 4.5.1

The particular solution of $ay'' + by' + cy = g_i(t)$.

$g_i(t)$	$Y_i(t)$
$P_n(t) = a_0 t^n + a_1 t^{n-1} + \cdots + a_n$	$t^s(A_0 t^n + A_1 t^{n-1} + \cdots + A_n)$
$P_n(t)e^{\alpha t}$	$t^s(A_0 t^n + A_1 t^{n-1} + \cdots + A_n)e^{\alpha t}$
$P_n(t)e^{\alpha t} \begin{cases} \sin \beta t \\ \cos \beta t \end{cases}$	$t^s[(A_0 t^n + A_1 t^{n-1} + \cdots + A_n)e^{\alpha t} \cos \beta t + (B_0 t^n + B_1 t^{n-1} + \cdots + B_n)e^{\alpha t} \sin \beta t]$

Note. Here, s is the smallest nonnegative integer ($s = 0$, 1, or 2) that will ensure that no term in $Y_i(t)$ is a solution of the corresponding homogeneous equation. Equivalently, for the three cases, s is the number of times 0 is a root of the characteristic equation, α is a root of the characteristic equation, and $\alpha + i\beta$ is a root of the characteristic equation, respectively.

5. Find a particular solution $Y_i(t)$ for each of the subproblems. Then the sum $Y_1(t) + \cdots + Y_n(t)$ is a particular solution of the full nonhomogeneous equation (27).

6. Form the sum of the general solution of the homogeneous equation (Step 1) and the particular solution of the nonhomogeneous equation (Step 5). This is the general solution of the nonhomogeneous equation.

7. Use the initial conditions to determine the values of the arbitrary constants remaining in the general solution.

For some problems, this entire procedure is easy to carry out by hand, but in many cases it requires considerable algebra. Once you understand clearly how the method works, a computer algebra system can be of great assistance in executing the details.

EXAMPLE 6

In each of the following problems, use Table 4.5.1 to determine a suitable form for the particular solution if the method of undetermined coefficients is to be used.

(a) $y'' = 3t^3 - t$

(b) $y'' - y' - 2y = -3te^{-t} + 2\cos 4t$

(c) $y'' + 2y' + 5y = t^2 e^{-t} \sin 2t$

(d) $y'' + y = \tan t$

Solution

(a) The general solution of $y'' = 0$ is $y = c_1 1 + c_2 t$. Since $g(t) = 3t^3 - t$ is a third degree polynomial, we assume that $Y(t) = t^s \left[A_3 t^3 + A_2 t^2 + A_1 t + A_0\right]$. Here, we must take

$s = 2$ to ensure that none of the functions in the assumed form for $Y(t)$ appear in the fundamental set. Thus

$$Y(t) = A_3 t^5 + A_2 t^4 + A_1 t^3 + A_0 t^2.$$

(b) The general solution of $y'' - y' - 2y = 0$ is $y = c_1 e^{-t} + c_2 e^{2t}$. We identify two subproblems corresponding to the nonhomogeneous terms $g_1(t) = -3te^{-t}$ and $g_2(t) = 2 \cos 4t$. Since g_1 is the exponential function e^{-t} multiplied by a first degree polynomial, we assume that $Y_1(t) = t^s \left[(A_1 t + A_0) e^{-t} \right]$. Since e^{-t} is a solution of the homogeneous equation, we must take $s = 1$. Thus

$$Y_1(t) = (A_1 t^2 + A_0 t) e^{-t}.$$

The correct form for Y_2 is $Y_2(t) = t^s [B_0 \cos 4t + C_0 \sin 4t]$. Neither of the terms $\cos 4t$ or $\sin 4t$ are solutions of the homogeneous equation, so we set $s = 0$ and obtain

$$Y_2(t) = B_0 \cos 4t + C_0 \sin 4t.$$

Substituting the expression for Y_1 into $y'' - y' - 2y = -3te^{-t}$ will determine A_0 and A_1, while substituting the expression for Y_2 into $y'' - y' - 2y = 2 \cos 4t$ will determine B_0 and C_0. With these coefficients ascertained, the general solution of $y'' - y' - 2y = -3te^{-t} + 2 \cos 4t$ is $y = c_1 e^{-t} + c_2 e^{2t} + Y_1(t) + Y_2(t)$.

(c) The general solution of $y'' + 2y' + 5y = 0$ is $y = c_1 e^{-t} \cos 2t + c_2 e^{-t} \sin 2t$. In this case, $g(t) = t^2 e^{-t} \sin 2t$, that is, a second degree polynomial times $e^{-t} \sin 2t$. Since it is necessary to include both sine and cosine functions even if only one or the other is present in the nonhomogeneous expression, the correct form for $Y(t)$ is

$$Y(t) = t^s \left[(A_0 t^2 + A_1 t + A_2) e^{-t} \cos 2t + (B_0 t^2 + B_1 t + B_2) e^{-t} \sin 2t \right].$$

We must then choose $s = 1$ to ensure that none of the terms in the assumed form for $Y(t)$ are solutions of the homogeneous equation. Thus

$$Y(t) = (A_0 t^3 + A_1 t^2 + A_2 t) e^{-t} \cos 2t + (B_0 t^3 + B_1 t^2 + B_2 t) e^{-t} \sin 2t.$$

(d) The method of undetermined coefficients is not applicable to this problem since the nonhomogeneous function $g(t) = \tan t$ does not lie in the class of functions consisting of linear combinations of products of polynomials, exponential, sine, and cosine functions. However, a particular solution can be obtained by the method of variation of parameters to be discussed in Section 4.7.

The method of undetermined coefficients is self-correcting in the sense that if one assumes too little for $Y(t)$, then a contradiction is soon reached that usually points the way to the modification that is needed in the assumed form. On the other hand, if one assumes too many terms, then some unnecessary work is done and some coefficients turn out to be zero, but at least the correct answer is obtained.

PROBLEMS

In each of Problems 1 through 16, find the general solution of the given differential equation:

1. $y'' - 2y' - 3y = 3e^{2t}$

2. $y'' + 2y' + 5y = 3 \sin 2t$

3. $y'' - 2y' - 3y = -3te^{-t}$

4. $y'' + 2y' = 3 + 4 \sin 2t$

5. $y'' + 9y = t^2 e^{3t} + 6$

6. $y'' + 2y' + y = 2e^{-t}$

7. $y'' - 5y' + 4y = 2e^t$ (Compare with Problem 10 in Section 4.7.)

8. $y'' - y' - 2y = 2e^{-t}$ (Compare with Problem 11 in Section 4.7.)

9. $y'' + 2y' + y = 3e^{-t}$ (Compare with Problem 12 in Section 4.7.)

10. $4y'' - 4y' + y = 16e^{t/2}$ (Compare with Problem 13 in Section 4.7.)

11. $2y'' + 3y' + y = t^2 + 3\sin t$

12. $y'' + y = 3\sin 2t + t\cos 2t$

13. $u'' + \omega_0^2 u = \cos \omega t$, $\omega^2 \neq \omega_0^2$

14. $u'' + \omega_0^2 u = \cos \omega_0 t$

15. $y'' + y' + 4y = 2\sinh t$
Hint: $\sinh t = (e^t - e^{-t})/2$

16. $y'' - y' - 2y = \cosh 2t$
Hint: $\cosh t = (e^t + e^{-t})/2$

In each of Problems 17 through 22, find the solution of the given initial value problem.

17. $y'' + y' - 2y = 2t$, $y(0) = 0$, $y'(0) = 1$

18. $y'' + 4y = t^2 + 3e^t$, $y(0) = 0$, $y'(0) = 2$

19. $y'' - 2y' + y = te^t + 4$, $y(0) = 1$, $y'(0) = 1$

20. $y'' - 2y' - 3y = 3te^{2t}$, $y(0) = 1$, $y'(0) = 0$

21. $y'' + 4y = 3\sin 2t$, $y(0) = 2$, $y'(0) = -1$

22. $y'' + 2y' + 5y = 4e^{-t}\cos 2t$, $y(0) = 1$, $y'(0) = 0$

In each of Problems 23 through 30:
(a) Determine a suitable form for $Y(t)$ if the method of undetermined coefficients is to be used.
(b) Use a computer algebra system to find a particular solution of the given equation.

23. $y'' + 3y' = 2t^4 + t^2 e^{-3t} + \sin 3t$

24. $y'' + y = t(1 + \sin t)$

25. $y'' - 5y' + 6y = e^t\cos 2t + e^{2t}(3t + 4)\sin t$

26. $y'' + 2y' + 2y = 3e^{-t} + 2e^{-t}\cos t + 4e^{-t}t^2\sin t$

27. $y'' - 4y' + 4y = 2t^2 + 4te^{2t} + t\sin 2t$

28. $y'' + 4y = t^2\sin 2t + (6t + 7)\cos 2t$

29. $y'' + 3y' + 2y = e^t(t^2 + 1)\sin 2t + 3e^{-t}\cos t + 4e^t$

30. $y'' + 2y' + 5y = 3te^{-t}\cos 2t - 2te^{-2t}\cos t$

31. Consider the equation

$$y'' - 3y' - 4y = 2e^{-t} \qquad (i)$$

from Example 5. Recall that $y_1(t) = e^{-t}$ and $y_2(t) = e^{4t}$ are solutions of the corresponding homogeneous equation. Adapting the method of reduction of order (see the discussion preceding Problem 28 in Section 4.2), seek a solution of the nonhomogeneous equation of the form $Y(t) = v(t)y_1(t) = v(t)e^{-t}$, where $v(t)$ is to be determined.

(a) Substitute $Y(t)$, $Y'(t)$, and $Y''(t)$ into Eq. (i) and show that $v(t)$ must satisfy $v'' - 5v' = 2$.
(b) Let $w(t) = v'(t)$ and show that $w(t)$ must satisfy $w' - 5w = 2$. Solve this equation for $w(t)$.
(c) Integrate $w(t)$ to find $v(t)$ and then show that

$$Y(t) = -\tfrac{2}{5}te^{-t} + \tfrac{1}{5}c_1 e^{4t} + c_2 e^{-t}.$$

The first term on the right side is the desired particular solution of the nonhomogeneous equation. Note that it is a product of t and e^{-t}.

Nonhomogeneous Cauchy–Euler Equations. In each of Problems 32 through 35, find the general solution by using the change of variable $t = \ln x$ to transform the equation into one with constant coefficients (see the discussion preceding Problem 52 in Section 4.3).

32. $x^2 y'' - 3xy' + 4y = \ln x$

33. $x^2 y'' + 7xy' + 5y = x$

34. $x^2 y'' - 2xy' + 2y = 3x^2 + 2\ln x$

35. $x^2 y'' + xy' + 4y = \sin(\ln x)$

36. Determine the general solution of

$$y'' + \lambda^2 y = \sum_{m=1}^{N} a_m \sin m\pi t,$$

where $\lambda > 0$ and $\lambda \neq m\pi$ for $m = 1, \ldots, N$.

37. In many physical problems, the nonhomogeneous term may be specified by different formulas in different time periods. As an example, determine the solution $y = \phi(t)$ of

$$y'' + y = \begin{cases} t, & 0 \leq t \leq \pi, \\ \pi e^{\pi - t}, & t > \pi, \end{cases}$$

satisfying the initial conditions $y(0) = 0$ and $y'(0) = 1$. Assume that y and y' are also continuous at $t = \pi$. Plot the nonhomogeneous term and the solution as functions of time.
Hint: First solve the initial value problem for $t \leq \pi$; then solve for $t > \pi$, determining the constants in the latter solution from the continuity conditions at $t = \pi$.

38. Follow the instructions in Problem 37 to solve the differential equation

$$y'' + 2y' + 5y = \begin{cases} 1, & 0 \leq t \leq \pi/2, \\ 0, & t > \pi/2 \end{cases}$$

with the initial conditions $y(0) = 0$ and $y'(0) = 0$.

4.6 Forced Vibrations, Frequency Response, and Resonance

We will now investigate the situation in which a periodic external force is applied to a spring-mass system. The behavior of this simple system models that of many oscillatory systems with an external force due, for example, to a motor attached to the system. We will first consider the case in which damping is present and will look later at the idealized special case in which there is assumed to be no damping.

Forced Vibrations with Damping

Recall that the equation of motion for a damped spring-mass system with external forcing, $F(t)$, is

$$my'' + \gamma y' + ky = F(t), \tag{1}$$

where m, γ, and k are the mass, damping coefficient, and spring constant, respectively. Dividing through Eq. (1) by m puts it in the form

$$y'' + 2\delta y' + \omega_0^2 y = f(t), \tag{2}$$

where $\delta = \gamma/(2m)$, $\omega_0^2 = k/m$, and $f(t) = F(t)/m$. These definitions for δ and for ω_0 simplify important mathematical expressions that appear below as we analyze the behavior of solutions of Eq. (2).

The assumption that the external force is periodic means $f(t)$ involves a linear combination of $A\cos(\omega t)$ and $A\sin(\omega t)$ with frequency ω and amplitude A. While we could work with these forms individually, the ensuing analysis is, as we shall see, less complicated—and more informative—if we write the external force in the form of a complex-valued exponential: $f(t) = Ae^{i\omega t} = A\cos(\omega t) + iA\sin(\omega t)$, because it allows us to consider both trigonometric terms at once. Thus we wish to find the general solution of

$$y'' + 2\delta y' + \omega_0^2 y = Ae^{i\omega t}. \tag{3}$$

Note that the solutions $y_1(t)$ and $y_2(t)$ of the homogeneous equation corresponding to Eq. (3) depend on the roots λ_1 and λ_2 of the characteristic equation $\lambda^2 + 2\delta\lambda + \omega_0^2 = 0$. Note that m, γ, and k all positive imply that δ and ω_0^2 are also positive. Damped free vibrations are discussed in Section 4.4, and in Problem 51 in Section 4.3. Recall that λ_1 and λ_2 are either real and negative (when $\delta \geq \omega_0$) or are complex conjugates with a negative real part (when $0 < \delta < \omega_0$).

Because the exponent on the right-hand side of Eq. (3) is purely imaginary, its real part is zero. Consequently, the forcing function on the right-hand side of Eq. (3) cannot be a solution of the homogeneous equation. The correct form to assume for the particular solution using the method of undetermined coefficients is therefore $Y(t) = Ce^{i\omega t}$.

Substituting $Y(t)$ into Eq. (3) leads to

$$\left((i\omega)^2 + 2\delta(i\omega) + \omega_0^2\right) Ce^{i\omega t} = Ae^{i\omega t}.$$

Solving for the unknown coefficient in $Y(t)$ yields

$$C = \frac{A}{(i\omega)^2 + 2\delta(i\omega) + \omega_0^2},$$

so

$$Y(t) = \frac{1}{(i\omega)^2 + 2\delta(i\omega) + \omega_0^2} Ae^{i\omega t}. \tag{4}$$

The general solution of Eq. (3) is

$$y = y_c(t) + Y(t),$$

where $Y(t)$ is the particular solution in Eq. (4), and $y_c(t) = c_1 y_1(t) + c_2 y_2(t)$ is the general solution of the homogeneous equation with constants c_1 and c_2 depending on the initial conditions. Since the roots of $\lambda^2 + 2\delta\lambda + \omega_0^2 = 0$ are either real and negative or complex with negative real part, each of $y_1(t)$ and $y_2(t)$ contains an exponentially decaying term. As a consequence, $y_c(t) \to 0$ as $t \to 0$ and $y_c(t)$ is referred to as the **transient solution**. In many applications the transient solution is of little importance. Its primary purpose is to satisfy whatever initial conditions may be imposed. With increasing time, the energy put into the system by the initial displacement and velocity dissipates through the damping force. The motion then becomes the response of the system to the external force.

Note that without damping ($\delta = 0$), the effects of the initial conditions would persist for all time. This situation will be considered at the end of this section.

**EXAMPLE
1**

Consider the initial value problem

$$y'' + \tfrac{1}{8}y' + y = 3\cos(\omega t), \quad y(0) = 2, \quad y'(0) = 0. \tag{5}$$

Show plots of the solution for different values of the forcing frequency ω, and compare them with corresponding plots of the forcing function.

For this system we have $\delta = 1/16$ and $\omega_0 = 1$. The amplitude of the harmonic input, A, is equal to 3. The transient part of the solution of the forced problem (5) resembles the solution

$$y = e^{-t/16}\left(2\cos\frac{\sqrt{255}}{16}t + \frac{2}{\sqrt{255}}\sin\frac{\sqrt{255}}{16}t\right)$$

of the corresponding unforced problem that was discussed in Example 2 of Section 4.4. The graph of that solution was shown in Figure 4.4.7. Because the damping is relatively small, the transient solution of the problem (5) also decays fairly slowly.

Turning to the nonhomogeneous problem, since the external force is $3\cos(\omega t)$, we work with $f(t) = 3e^{i\omega t}$. From Eq. (4) we know

$$Y(t) = \frac{3}{(i\omega)^2 + i\omega/8 + 1}e^{i\omega t}. \tag{6}$$

Since $3\cos(\omega t) = \text{Re}(3e^{i\omega t})$ the particular solution for Eq. (5) is the real part of Eq. (6). To identify the real part of Eq. (6) it helps if the denominator in Eq. (6) is real-valued. To bring this about, multiply both the numerator and denominator of Eq. (6) by the complex conjugate of the denominator.

$$Y(t) = \frac{3e^{i\omega t}}{(i\omega)^2 + i\omega/8 + 1} \cdot \frac{(-i\omega)^2 - i\omega/8 + 1}{(-i\omega)^2 - i\omega/8 + 1}$$

$$= \frac{3(\cos(\omega t) + i\sin(\omega t))\left(1 - \omega^2 - i\frac{\omega}{8}\right)}{(1 - \omega^2)^2 + \frac{\omega^2}{64}}.$$

Thus, the real part of Eq. (6) is

$$Y_{\text{Re}}(t) = \text{Re}Y(t) = \frac{3}{(1 - \omega^2)^2 + \omega^2/64}\left((1 - \omega^2)\cos(\omega t) + \frac{\omega}{8}\sin(\omega t)\right). \tag{7}$$

Figures 4.6.1, 4.6.2, and 4.6.3 show the solution of the forced problem (5) for $\omega = 0.3$, $\omega = 1$, and $\omega = 2$, respectively. The graph of the corresponding forcing function is shown (as a dashed curve) in each figure.

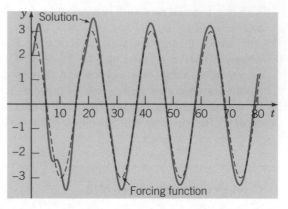

FIGURE 4.6.1 A forced vibration with damping; solution of
$y'' + 0.125y' + y = 3\cos(3t/10)$, $y(0) = 2$, $y'(0) = 0$.

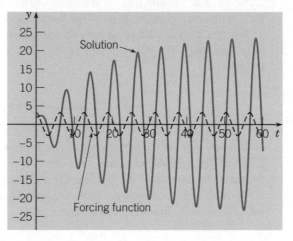

FIGURE 4.6.2 A forced vibration with damping; solution of
$y'' + 0.125y' + y = 3\cos t$, $y(0) = 2$, $y'(0) = 0$.

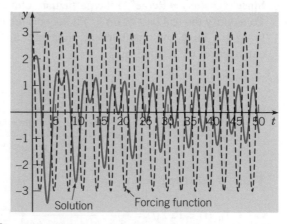

FIGURE 4.6.3 A forced vibration with damping; solution of
$y'' + 0.125y' + y = 3\cos 2t$, $y(0) = 2$, $y'(0) = 0$.

The solutions in Figures 4.6.1, 4.6.2, and 4.6.3 show three different behaviors. In each case the solution does not die out as t increases but persists indefinitely, or at least as long as the external force is applied. From Eq. (7) we see that each solution represents a steady oscillation with the same frequency as the external force. For these reasons the particular solution to a damped, harmonically forced system is called the **steady-state solution**, the **steady-state response**, the **steady-state output**, or the **forced response**.

In general, in the real-valued case with $f(t) = A\cos(\omega t)$, the real-valued particular solution is the real part of Eq. (4). Now, apply to Eq. (4) the same steps that were just applied to Eq. (6); in this way we obtain the following more general expression for the real part of Eq. (4):

$$Y_{\mathrm{Re}}(t) = \mathrm{Re}Y(t) = A\frac{(\omega_0^2 - \omega^2)\cos(\omega t) + 2\delta\omega\sin(\omega t)}{(\omega_0^2 - \omega^2)^2 + 4\delta^2\omega^2}. \tag{8}$$

The Frequency Response Function

In Example 1, even though the forcing function is a pure cosine, $\cos(\omega t)$, the forced response involves both $\sin(\omega t)$ and $\cos(\omega t)$. However, when considering the external force as a complex-valued exponential, Eq. (4) tells us that the forced output is directly proportional to the forced input:

$$\frac{Y(t)}{Ae^{i\omega t}} = \frac{1}{(i\omega)^2 + 2\delta(i\omega) + \omega_0^2}.$$

This quotient is referred to as the **frequency response** of the system. As the frequency response depends on the frequency ω (and δ and ω_0)—but not on t—it is commonly defined as

$$G(i\omega) = \frac{1}{(i\omega)^2 + 2\delta(i\omega) + \omega_0^2} = \frac{1}{(i\omega + \delta)^2 + \omega_0^2 - \delta^2}. \tag{9}$$

To continue the analysis of the frequency response function, it is convenient to represent the $G(i\omega)$ in Eq. (9) in its complex exponential form (see Figure 4.6.4),

$$G(i\omega) = |G(i\omega)|\, e^{-i\phi(\omega)} = |G(i\omega)|\,(\cos(\phi(\omega)) - i\sin(\phi(\omega))), \tag{10}$$

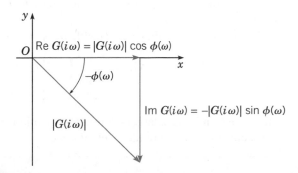

FIGURE 4.6.4 Polar coordinate representation of the frequency response function $G(i\omega)$.

where the **gain** of the frequency response is

$$|G(i\omega)| = \left(G(i\omega)\overline{G(i\omega)} \right)^{1/2} = \frac{1}{\sqrt{(\omega_0^2 - \omega^2)^2 + 4\delta^2\omega^2}} \tag{11}$$

and the **phase** of the frequency response is the angle

$$\phi(\omega) = \arccos\left(\frac{\omega_0^2 - \omega^2}{\sqrt{(\omega_0^2 - \omega^2)^2 + 4\delta^2\omega^2}} \right). \tag{12}$$

Using Eq. (10), the particular solution (4) is

$$Y(t) = G(i\omega)Ae^{i\omega t} = |G(i\omega)|\, e^{-i\phi(\omega)} Ae^{i\omega t} = A\,|G(i\omega)|\, e^{i(\omega t - \phi(\omega))}. \tag{13}$$

EXAMPLE 2

Find the gain and phase for each of the three response functions found in Example 1. Recall that $\omega_0 = 1$ and $\delta = 1/16$. The three cases are $\omega = 0.3$, $\omega = 1$, and $\omega = 2$.

Classification of the external force as low- or high-frequency is done relative to the forcing frequency ω_0. For example, the case with $\omega/\omega_0 = 0.3$ is a low-frequency force. The steady-state response is

$$Y(t) = 3\frac{0.91\cos(0.3t) + \frac{0.3}{8}\sin(0.3t)}{0.91^2 + 0.09/64}$$
$$\approx 3.29111\cos(0.3t) + 0.13562\sin(0.3t) \approx 3.2923\cos(0.3t - 0.04119).$$

That the gain is a little larger than the amplitude of the input and the phase is small are consistent with the graph shown in Figure 4.6.1.

For the comparatively high-frequency case, $\omega/\omega_0 = 2$, the particular solution is

$$Y(t) = 3\frac{-3\cos(2t) + \frac{1}{4}\sin(2t)}{9 + \frac{1}{16}}$$
$$\approx 0.99310\cos(2t) + 0.08276\sin(2t) \approx 0.99655\cos(2t - 3.0585).$$

In this case the amplitude of the steady forced response is approximately one-third the amplitude of the harmonic input and the phase between the excitation and the response is approximately π. These findings are consistent with the particular solution plotted in Figure 4.6.3.

In the third case, using Eq. (8) with $\omega/\omega_0 = 1$, the steady-state response is

$$Y(t) = \frac{3}{1/64}\left((1-1)\cos(t) + \frac{1}{8}\sin(t) \right)$$
$$= 24\sin(t) = 24\cos(t - \pi/2).$$

Here, the gain is much larger—8 times the amplitude of the harmonic input—and the phase is exactly $\pi/2$ relative to the external force.

The explicit formulas for the gain factor and phase shift given in Eqs. (11) and (12) are rather complicated. The three cases considered in Examples 1 and 2 illustrate the

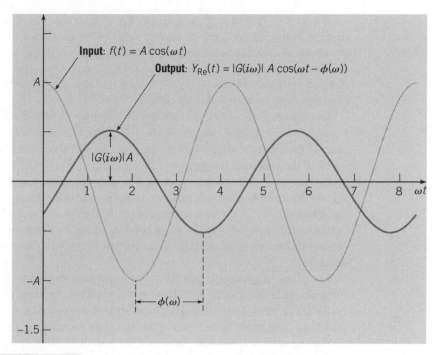

FIGURE 4.6.5 The steady-state response $Y_{\text{Re}} = |G(i\omega)|A\cos(\omega t - \phi(\omega))$ of a spring-mass system due to the harmonic input $f(t) = A\cos\omega t$.

two ways the harmonic input is modified as it passes through a spring-mass system (see Figure 4.6.5):

1. The amplitude of the output equals the amplitude of the harmonic input amplified or attenuated by the gain factor, $|G(i\omega)|$.
2. There is a phase shift in the steady-state output of magnitude $\phi(\omega)$ relative to the harmonic input.

Our next objective is to understand better how the gain $|G(i\omega)|$ and the phase shift $\phi(\omega)$ depend on the frequency of the harmonic input. For low-frequency excitation, that is, as $\omega \to 0^+$, it follows from Eq. (11) that $|G(i\omega)| \to 1/\omega_0^2 = m/k$. At the other extreme, for very high-frequency excitation, Eq. (11) implies that $|G(i\omega)| \to 0$ as $\omega \to \infty$.

The case with $\omega/\omega_0 = 1$ in Example 2 suggests that the gain can have a maximum at an intermediate value of ω. To find this maximum point, find where the derivative of $|G(i\omega)|$ with respect to ω is zero. You will find that the maximum amplitude occurs when $\omega = \omega_{\text{max}}$, where

$$\omega_{\text{max}}^2 = \omega_0^2 - 2\delta^2 = \omega_0^2 - \frac{\gamma^2}{2m^2} = \omega_0^2\left(1 - \frac{\gamma^2}{2mk}\right). \tag{14}$$

Note that $0 < \omega_{\text{max}} < \omega_0$ and, when the damping coefficient, γ, is small, ω_{max} is close to ω_0. The maximum value of the gain is

$$|G(i\omega_{\text{max}})| = \frac{m}{\gamma\omega_0\sqrt{1 - (\gamma^2/4mk)}} \approx \frac{m}{\gamma\omega_0}\left(1 + \frac{\gamma^2}{8mk}\right), \tag{15}$$

where the last expression is an approximation for small γ.

If $\gamma^2/mk > 2$, then $\omega_{\max}$, as given by Eq. (14), is imaginary. In this case, which is identified as highly damped, the maximum value of the gain occurs for $\omega = 0$, and $|G(i\omega)|$ is a monotone decreasing function of ω. Recall that critical damping occurs when $\gamma^2/mk = 4$.

For small values of γ, if follows from Eq. (15) that $|G(i\omega_{\max})| \approx m/\gamma\omega_0$. Thus, for lightly damped systems, the gain $|G(i\omega)|$ is large when $\omega/\omega_0 \approx 1$. Moreover the smaller the value of γ, the more pronounced is this effect.

Resonance is the physical tendency of solutions to periodically forced systems to have a steady-state response that oscillates with a much greater amplitude than the input. The specific frequency at which the amplitude of the steady state response has a local maximum is called the **resonant frequency** of the system.

Resonance can be an important design consideration; it can be good or bad, depending on the circumstances. It must be taken seriously in the design of structures, such as buildings and bridges, where it can produce instabilities that might lead to the catastrophic failure of the structure. On the other hand, resonance can be put to good use in the design of instruments, such as seismographs, that are intended to detect weak periodic incoming signals.

The phase angle ϕ also depends in an interesting way on ω. For ω near zero, it follows from Eq. (12) that $\cos(\phi) \approx 1$. Thus $\phi \approx 0$, and the response is nearly in phase with the excitation. That is, they rise and fall together and, in particular, they assume their respective maxima nearly together and their respective minima nearly together.

For the resonant frequency, $\omega = \omega_0$, we find that $\cos(\phi) = 0$, so $\phi = \pi/2$. In this case the response lags behind the excitation by $\pi/2$, that is, the peaks of the response occur $\pi/2$ later than the peaks of the excitation, and similarly for the valleys.

Finally, for ω very large (relative to ω_0), we have $\cos(\phi) \approx -1$. Here, $\phi \approx \pi$, so the response is nearly out of phase with the excitation. In these cases the response is minimum when the excitation is maximum, and vice versa.

We conclude this discussion of frequency response, gain, phase, and resonance by looking at typical graphs of the gain and phase. Figures 4.6.6 and 4.6.7 plot the normalized

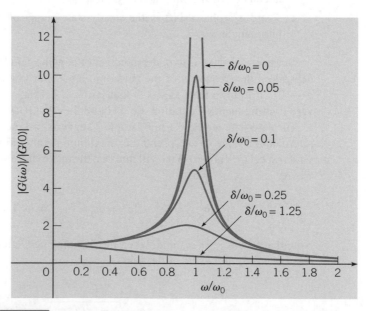

FIGURE 4.6.6 Gain function $|G(i\omega)|$ for the damped spring-mass system: $\delta/\omega_0 = \gamma/2\sqrt{mk}$.

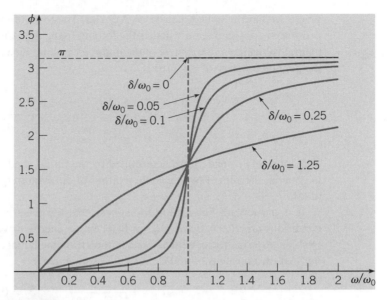

FIGURE 4.6.7 Phase function $\phi(\omega)$ for the damped spring-mass system: $\delta/\omega_0 = \gamma/2\sqrt{mk}$.

gain, $|G(i\omega)|/G(0)$, and the phase, $\phi(\omega)$, versus the normalized wavelength, ω/ω_0. With these normalizations, each frequency response curve in Figure 4.6.6 starts at height 1 when $\omega/\omega_0 = 0$. For heavily damped systems ($\gamma^2/4m > 4$), the response decreases for all $\omega > 0$. As the damping decreases, the frequency response acquires a maximum at $\omega/\omega_0 = 1$. The size of the gain increases as $\delta \to 0^+$.

In a similar way, the phase is always 0 when $\omega = 0$, $\pi/2$ when $\omega/\omega_0 = 1$, and approaches π as $\omega/\omega_0 \to \infty$, as shown in Figure 4.6.7. Notice how the transition from $\phi \approx 0$ to $\phi \approx \pi$ becomes more rapid as the damping decreases.

To conclude this introduction to the frerquency response function for damped systems, we point out how Figure 4.6.6 illustrates the usefulness of dimensionless variables. It is easy to verify that each of the quantities $|G(i\omega)|/|G(0)| = \omega^2|G(i\omega)|$, ω/ω_0, and $\delta/\omega_0 = \gamma/(2\sqrt{mk})$ is dimensionless. The importance of this observation can be seen in that the number of parameters in the problems has been reduced from the five that appear in Eq. (3)—m, γ, k, A, and ω—to the three that are in Eq. (3), namely, δ, ω_0, and ω. Thus this one family of curves, of which a few are shown in Figure 4.6.6, describes the response-versus-frequency behavior of the gain factor for all systems governed by Eq. (3). Likewise, Figure 4.6.7 shows representative curves describing the response-versus-frequency behavior of the phase shift for any solution to Eq. (3).

Forced Vibrations Without Damping

Notice that while Figures 4.6.6 and 4.6.7 include curves labeled as $\delta/\omega_0 = 0$, these curves are not governed by the formulas for $G(i\omega)$ and $\phi(\omega)$ given in this section. We conclude with a discussion of the limiting case when there is no damping.

We now assume $\gamma = 0$ in Eq. (1) so that $\delta = \gamma/2m = 0$ in Eq. (2), thereby obtaining the equation of motion of an undamped forced oscillator

$$y'' + \omega_0^2 y = A \cos \omega t, \tag{16}$$

where we have assumed that $f(t) = A\cos\omega t$. The form of the general solution of Eq. (16) is different, depending on whether the forcing frequency ω is different from or equal to the natural frequency $\omega_0 = \sqrt{k/m}$ of the unforced system. First consider the case $\omega \neq \omega_0$; then the general solution of Eq. (16) is

$$y = c_1 \cos\omega_0 t + c_2 \sin\omega_0 t + \frac{A}{(\omega_0^2 - \omega^2)}\cos\omega t. \tag{17}$$

The constants c_1 and c_2 are determined by the initial conditions. The resulting motion is, in general, the sum of two periodic motions of different frequencies (ω_0 and ω) and amplitudes.

It is particularly interesting to suppose that the mass is initially at rest, so the initial conditions are $y(0) = 0$ and $y'(0) = 0$. Then the energy driving the system comes entirely from the external force, with no contribution from the initial conditions. In this case, it turns out that the constants c_1 and c_2 in Eq. (17) are given by

$$c_1 = -\frac{A}{(\omega_0^2 - \omega^2)}, \qquad c_2 = 0, \tag{18}$$

and the solution of Eq. (16) is

$$y = \frac{A}{(\omega_0^2 - \omega^2)}(\cos\omega t - \cos\omega_0 t). \tag{19}$$

This is the sum of two periodic functions of different periods but the same amplitude. Making use of the trigonometric identities for $\cos(A \pm B)$ with $A = (\omega_0 + \omega)t/2$ and $B = (\omega_0 - \omega)t/2$, we can write Eq. (19) in the form

$$y = \left[\frac{2A}{(\omega_0^2 - \omega^2)}\sin\frac{(\omega_0 - \omega)t}{2}\right]\sin\frac{(\omega_0 + \omega)t}{2}. \tag{20}$$

If $|\omega_0 - \omega|$ is small, then $\omega_0 + \omega$ is much greater than $|\omega_0 - \omega|$. Hence $\sin((\omega_0 + \omega)t/2)$ is a rapidly oscillating function compared to $\sin((\omega_0 - \omega)t/2)$. Thus the motion is a rapid oscillation with frequency $(\omega_0 + \omega)/2$ but with a slowly varying sinusoidal amplitude

$$\frac{2A}{|\omega_0^2 - \omega^2|}\left|\sin\frac{(\omega_0 - \omega)t}{2}\right|.$$

This type of motion, possessing a periodic variation of amplitude, exhibits what is called a **beat**. For example, such a phenomenon occurs in acoustics when two tuning forks of nearly equal frequency are excited simultaneously. In this case, the periodic variation of amplitude is quite apparent to the unaided ear. In electronics, the variation of the amplitude with time is called **amplitude modulation**.

EXAMPLE 3

Solve the initial value problem

$$y'' + y = 0.5\cos 0.8t, \qquad y(0) = 0, \quad y'(0) = 0, \tag{21}$$

and plot the solution.

In this case, $\omega_0 = 1$, $\omega = 0.8$, and $A = 0.5$, so from Eq. (20) the solution of the given problem is

$$y = 2.77778 \sin 0.1t \sin 0.9t. \tag{22}$$

A graph of this solution is shown in Figure 4.6.8.

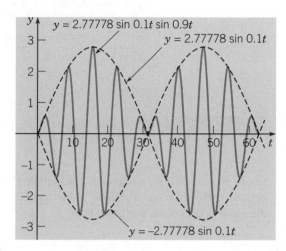

FIGURE 4.6.8 A beat; solution of $y'' + y = 0.5 \cos 0.8t$, $y(0) = 0$, $y'(0) = 0$; $y = 2.77778 \sin 0.1t \sin 0.9t$.

The amplitude variation has a slow frequency of 0.1 and a corresponding slow period of 20π. Note that a half-period of 10π corresponds to a single cycle of increasing and then decreasing amplitude. The displacement of the spring-mass system oscillates with a relatively fast frequency of 0.9, which is only slightly less than the natural frequency ω_0.

Now imagine that the forcing frequency ω is further increased, say, to $\omega = 0.9$. Then the slow frequency is halved to 0.05, and the corresponding slow half-period is doubled to 20π. The multiplier 2.7778 also increases substantially, to 5.2632. However the fast frequency is only marginally increased, to 0.95. Can you visualize what happens as ω takes on values closer and closer to the natural frequency $\omega_0 = 1$?

Now let us return to Eq. (16) and consider the case of resonance, where $\omega = \omega_0$, that is, the frequency of the forcing function is the same as the natural frequency of the system. Then the nonhomogeneous term $A \cos \omega t$ is a solution of the homogeneous equation. In this case, the solution of Eq. (16) is

$$y = c_1 \cos \omega_0 t + c_2 \sin \omega_0 t + \frac{A}{2\omega_0} t \sin \omega_0 t. \tag{23}$$

Because of the term $t \sin \omega_0 t$, the solution (23) predicts that the motion will become unbounded as $t \to \infty$ regardless of the values of c_1 and c_2; see Figure 4.6.9 for a typical example. Of course, in reality, unbounded oscillations do not occur. As soon as y becomes large, the mathematical model on which Eq. (16) is based is no longer valid, since the assumption that the spring force depends linearly on the displacement requires that y be

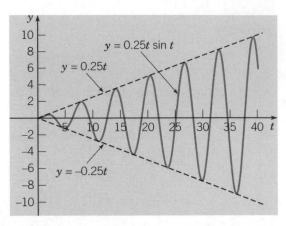

FIGURE 4.6.9 Resonance; solution of $y'' + y = 0.5 \cos t$, $y(0) = 0$, $y'(0) = 0$; $y = 0.25t \sin t$.

small. As we have seen, if damping is included in the model, the predicted motion remains bounded. However the response to the input function $A \cos \omega t$ may be quite large if the damping is small and ω is close to ω_0.

PROBLEMS

In each of Problems 1 through 4, write the given expression as a product of two trigonometric functions of different frequencies.

1. $\cos 11t - \cos 3t$ 2. $\sin 7t - \sin 4t$
3. $\cos 7\pi t + \cos 2\pi t$ 4. $\sin 9t + \sin 4t$

5. A mass weighing 4 pounds (lb) stretches a spring 1.5 in. The mass is displaced 12 in. in the positive direction from its equilibrium position and released with no initial velocity. Assuming that there is no damping and that the mass is acted on by an external force of $7 \cos 3t$ lb, formulate the initial value problem describing the motion of the mass.

6. A mass of 4 kg stretches a spring 8 cm. The mass is acted on by an external force of $8 \sin(t/2)$ newtons (N) and moves in a medium that imparts a viscous force of 4 N when the speed of the mass is 2 cm/s. If the mass is set in motion from its equilibrium position with an initial velocity of 16 cm/s, formulate the initial value problem describing the motion of the mass.

 7. **(a)** Find the solution of Problem 5.
(b) Plot the graph of the solution.
(c) If the given external force is replaced by a force $A \exp(i\omega t)$ of frequency ω, find the frequency response $G(i\omega)$, the gain $|G(i\omega)|$, and the phase $\phi(\omega) = -\arg(G(i\omega))$. Then find the value of ω for which resonance occurs.

 8. **(a)** Find the solution of the initial value problem in Problem 6.

(b) Identify the transient and steady-state parts of the solution.
(c) Plot the graph of the steady-state solution.
(d) If the given external force is replaced by a force $A \exp(i\omega t)$ of frequency ω, find the frequency response $G(i\omega)$, the gain $|G(i\omega)|$, and the phase $\phi(\omega) = -\arg(G(i\omega))$. Then find the value of ω for which the gain is maximum. Plot the graphs of $|G(i\omega)|$ and $\phi(\omega)$.

9. If an undamped spring-mass system with a mass that weighs 12 lb and a spring constant 2 lb/in. is suddenly set in motion at $t = 0$ by an external force of $15 \cos 7t$ lb, determine the position of the mass at any time and draw a graph of the displacement versus t.

10. A mass that weighs 8 lb stretches a spring 24 in. The system is acted on by an external force of $4 \sin 4t$ lb. If the mass is pulled down 6 in. and then released, determine the position of the mass at any time. Determine the first four times at which the velocity of the mass is zero.

11. A spring is stretched 6 in. by a mass that weighs 8 lb. The mass is attached to a dashpot mechanism that has a damping constant of 0.25 lb·s/ft and is acted on by an external force of $3 \cos 2t$ lb.
(a) Determine the steady-state response of this system.
(b) If the given mass is replaced by a mass m, determine the value of m for which the amplitude of the steady-state response is maximum.

12. A spring-mass system has a spring constant of 3 N/m. A mass of 2 kg is attached to the spring, and the

motion takes place in a viscous fluid that offers a resistance numerically equal to the magnitude of the instantaneous velocity. If the system is driven by an external force of $(12 \cos 3t - 8 \sin 3t)$ N, determine the steady-state response.

13. Furnish the details in determining when the gain function given by Eq. (10) is maximum, that is, show that ω_{max}^2 and $|G(i\omega_{max})|$ are given by Eqs. (14) and (15), respectively.

14. Find the solution of the initial value problem

$$y'' + y = F(t), \qquad y(0) = 0, \quad y'(0) = 0,$$

where

$$F(t) = \begin{cases} At, & 0 \le t \le \pi, \\ A(2\pi - t), & \pi < t \le 2\pi, \\ 0, & 2\pi < t. \end{cases}$$

Hint: Treat each time interval separately, and match the solutions in the different intervals by requiring that y and y' be continuous functions of t.

15. A series circuit has a capacitor of 0.25 microfarad, a resistor of 5×10^3 ohms, and an inductor of 1 henry. The initial charge on the capacitor is zero. If a 9-volt battery is connected to the circuit and the circuit is closed at $t = 0$, determine the charge on the capacitor at $t = 0.001$ s, at $t = 0.01$ s, and at any time t. Also determine the limiting charge as $t \to \infty$.

16. Consider a vibrating system described by the initial value problem

$$y'' + 0.25y' + 2y = 2 \cos \omega t,$$

$$y(0) = 0, \quad y'(0) = 2.$$

(a) Determine the steady-state part of the solution of this problem.
(b) Find the gain function $|G(i\omega)|$ of the system.
(c) Plot $|G(i\omega)|$ and $\phi(\omega) = -\arg(G(i\omega))$ versus ω.
(d) Find the maximum value of $|G(i\omega)|$ and the frequency ω for which it occurs.

17. Consider the forced but undamped system described by the initial value problem

$$y'' + y = 3 \cos \omega t, \qquad y(0) = 0, \quad y'(0) = 0.$$

(a) Find the solution $y(t)$ for $\omega \ne 1$.
(b) Plot the solution $y(t)$ versus t for $\omega = 0.7$, $\omega = 0.8$, and $\omega = 0.9$. Describe how the response $y(t)$ changes as ω varies in this interval. What happens as ω takes on values closer and closer to 1? Note that the natural frequency of the unforced system is $\omega_0 = 1$.

18. Consider the vibrating system described by the initial value problem

$$y'' + y = 3 \cos \omega t, \qquad y(0) = 1, \quad y'(0) = 1.$$

(a) Find the solution for $\omega \ne 1$.

(b) Plot the solution $y(t)$ versus t for $\omega = 0.7$, $\omega = 0.8$, and $\omega = 0.9$. Compare the results with those of Problem 17, that is, describe the effect of the nonzero initial conditions.

19. For the initial value problem in Problem 18, plot y' versus y for $\omega = 0.7$, $\omega = 0.8$, and $\omega = 0.9$, that is, draw the phase plot of the solution for these values of ω. Use a t interval that is long enough, so the phase plot appears as a closed curve. Mark your curve with arrows to show the direction in which it is traversed as t increases.

Problems 20 through 22 deal with the initial value problem

$$y'' + 0.125y' + 4y = f(t), \qquad y(0) = 2, \quad y'(0) = 0.$$

In each of these problems:
(a) Plot the given forcing function $f(t)$ versus t, and also plot the solution $y(t)$ versus t on the same set of axes. Use a t interval that is long enough, so the initial transients are substantially eliminated. Observe the relation between the amplitude and phase of the forcing term and the amplitude and phase of the response. Note that $\omega_0 = \sqrt{k/m} = 2$.
(b) Draw the phase plot of the solution, that is, plot y' versus y.

20. $f(t) = 3 \cos(t/4)$
21. $f(t) = 3 \cos 2t$
22. $f(t) = 3 \cos 6t$

23. A spring-mass system with a hardening spring (Section 4.1) is acted on by a periodic external force. In the absence of damping, suppose that the displacement of the mass satisfies the initial value problem

$$y'' + y + 0.2y^3 = \cos \omega t, y(0) = 0, \qquad y'(0) = 0.$$

(a) Let $\omega = 1$ and plot a computer-generated solution of the given problem. Does the system exhibit a beat?
(b) Plot the solution for several values of ω between $\frac{1}{2}$ and 2. Describe how the solution changes as ω increases.

24. Suppose that the system of Problem 23 is modified to include a damping term and that the resulting initial value problem is

$$y'' + 0.2y' + y + 0.2y^3 = \cos \omega t, \quad y(0) = 0, \quad y'(0) = 0.$$

(a) Plot a computer-generated solution of the given problem for several values of ω between $\frac{1}{2}$ and 2, and estimate the amplitude, say, $G_H(\omega)$, of the steady response in each case.
(b) Using the data from part (a), plot the graph of $G_H(\omega)$ versus ω. For what frequency ω is the amplitude greatest?
(c) Compare the results of parts (a) and (b) with the corresponding results for the linear spring.

4.7 Variation of Parameters

In this section we describe another method for finding a particular solution of a nonhomogeneous equation. The method, known as **variation of parameters** or **variation of constants**, is due to Lagrange and complements the method of undetermined coefficients rather well. Unlike the method of undetermined coefficients, variation of parameters is a *general method*; in principle at least, it can be applied to any linear nonhomogeneous equation or system.

It requires no detailed assumptions about the form of the solution. In this section we use this method to derive an integral representation for the particular solution of an arbitrary linear nonhomogeneous first order system of dimension 2. An analogous representation for the particular solution of an arbitrary second order linear nonhomogeneous equation then follows as a special case.

Variation of Parameters for Linear First Order Systems of Dimension 2

First consider the nonhomogeneous system

$$\mathbf{x}' = \mathbf{P}(t)\mathbf{x} + \mathbf{g}(t), \tag{1}$$

where each entry of

$$\mathbf{P}(t) = \begin{pmatrix} p_{11}(t) & p_{12}(t) \\ p_{21}(t) & p_{22}(t) \end{pmatrix} \quad \text{and} \quad \mathbf{g}(t) = \begin{pmatrix} g_1(t) \\ g_2(t) \end{pmatrix} \tag{2}$$

is continuous on an open interval I. Assume that

$$\mathbf{x}_1(t) = \begin{pmatrix} x_{11}(t) \\ x_{21}(t) \end{pmatrix} \quad \text{and} \quad \mathbf{x}_2(t) = \begin{pmatrix} x_{12}(t) \\ x_{22}(t) \end{pmatrix}$$

from a fundamental set of solutions for the homogeneous system $\mathbf{x}' = \mathbf{P}(t)\mathbf{x}$ obtained from Eq. (1) by setting $\mathbf{g}(t) = \mathbf{0}$. Thus

$$\mathbf{x}_1' = \mathbf{P}(t)\mathbf{x}_1, \qquad \mathbf{x}_2' = \mathbf{P}(t)\mathbf{x}_2, \tag{3}$$

and the Wronskian of $\mathbf{x}_1$ and $\mathbf{x}_2$ is nonzero throughout I,

$$W[\mathbf{x}_1, \mathbf{x}_2](t) = \begin{vmatrix} x_{11}(t) & x_{12}(t) \\ x_{21}(t) & x_{22}(t) \end{vmatrix} \neq 0, \quad t \in I. \tag{4}$$

Finding a particular solution of Eq. (1) via the method of variation of parameters is facilitated by introducing the convenient matrix notation

$$\mathbf{X}(t) = [\mathbf{x}_1(t), \mathbf{x}_2(t)] = \begin{pmatrix} x_{11}(t) & x_{12}(t) \\ x_{21}(t) & x_{22}(t) \end{pmatrix} \tag{5}$$

by juxtaposing the column vectors $\mathbf{x}_1$ and $\mathbf{x}_2$. The matrix (5) is referred to as a **fundamental matrix** for $\mathbf{x}' = \mathbf{P}(t)\mathbf{x}$ since the columns of $\mathbf{X}(t)$ are a fundamental set of solutions. A column-by-column comparison shows that $\mathbf{X}(t)$ satisfies the matrix differential equation

$$\mathbf{X}'(t) = [\mathbf{x}_1'(t), \mathbf{x}_2'(t)] = [\mathbf{P}(t)\mathbf{x}_1(t), \mathbf{P}(t)\mathbf{x}_2(t)] = \mathbf{P}(t)\mathbf{X}(t). \tag{6}$$

Thus Eq. (6) is equivalent to the pair of equations (3).

Recall that the general solution of $\mathbf{x}' = \mathbf{P}(t)\mathbf{x}$ has the form

$$\mathbf{x} = c_1\mathbf{x}_1(t) + c_2\mathbf{x}_2(t). \tag{7}$$

The fundamental idea behind the method of variation of parameters is to replace the constants c_1 and c_2 in Eq. (7) by functions $u_1(t)$ and $u_2(t)$, respectively,

$$\mathbf{x} = u_1(t)\mathbf{x}_1(t) + u_2(t)\mathbf{x}_2(t) = \mathbf{X}(t)\mathbf{u}(t), \tag{8}$$

where the component functions of $\mathbf{u}(t) = (u_1(t), u_2(t))^T$ must then be determined. The validity of the assumption (8) can subsequently be justified by direct verification of the resulting solution [see Problem 1(b)].

Substituting the right side of Eq. (8) into Eq. (1) yields the equation

$$\mathbf{X}'(t)\mathbf{u}(t) + \mathbf{X}(t)\mathbf{u}'(t) = \mathbf{P}(t)\mathbf{X}(t)\mathbf{u}(t) + \mathbf{g}(t),$$

which simplifies to

$$\mathbf{X}(t)\mathbf{u}'(t) = \mathbf{g}(t) \tag{9}$$

as a consequence of Eq. (6). Note that we have used the easily verified differentiation rule for the product $\mathbf{X}\mathbf{u}$, $(\mathbf{X}\mathbf{u})' = \mathbf{X}'\mathbf{u} + \mathbf{X}\mathbf{u}'$ [see Problem 1(a)]. Thus the components of $\mathbf{u}'$ satisfy the linear algebraic system of equations

$$x_{11}(t)u_1'(t) + x_{12}(t)u_2'(t) = g_1(t),$$
$$x_{21}(t)u_1'(t) + x_{22}(t)u_2'(t) = g_2(t). \tag{10}$$

Since $\det \mathbf{X}(t) = W[\mathbf{x}_1, \mathbf{x}_2](t) \neq 0$, $\mathbf{X}^{-1}(t)$ exists and is given by [see Problem 16 in Appendix A]

$$\mathbf{X}^{-1}(t) = \frac{1}{W[\mathbf{x}_1, \mathbf{x}_2](t)} \begin{pmatrix} x_{22}(t) & -x_{12}(t) \\ -x_{21}(t) & x_{11}(t) \end{pmatrix}. \tag{11}$$

In terms of $\mathbf{X}^{-1}(t)$, the solution of Eq. (9) can be represented as

$$\mathbf{u}'(t) = \mathbf{X}^{-1}(t)\mathbf{g}(t). \tag{12}$$

Thus, for $\mathbf{u}(t)$, we can select any vector from the class of vectors that satisfy Eq. (12). Since each component of $\mathbf{u}$ is determined up to an arbitrary additive constant, we denote $\mathbf{u}(t)$ by

$$\mathbf{u}(t) = \int \mathbf{X}^{-1}(t)\mathbf{g}(t)\, dt + \mathbf{c} = \begin{pmatrix} \displaystyle\int \frac{x_{22}(t)g_1(t) - x_{12}(t)g_2(t)}{W[\mathbf{x}_1, \mathbf{x}_2](t)}\, dt + c_1 \\[2ex] \displaystyle\int \frac{x_{11}(t)g_2(t) - x_{21}(t)g_1(t)}{W[\mathbf{x}_1, \mathbf{x}_2](t)}\, dt + c_2 \end{pmatrix}, \tag{13}$$

where the constant vector $\mathbf{c} = (c_1, c_2)^T$ is arbitrary. Substituting for $\mathbf{u}(t)$ in Eq. (8) gives the solution $\mathbf{x}$ of the system (1),

$$\mathbf{x} = \mathbf{X}(t)\mathbf{c} + \mathbf{X}(t) \int \mathbf{X}^{-1}(t)\mathbf{g}(t)\, dt. \tag{14}$$

Since $\mathbf{c}$ is arbitrary, any initial condition at a point t_0 can be satisfied by an appropriate choice of $\mathbf{c}$. Thus every solution of the system (1) is contained in the expression given by Eq. (14). Therefore it is the general solution of Eq. (1). Note that the first term on the right side of Eq. (14) is the general solution of the corresponding homogeneous system $\mathbf{x}' = \mathbf{P}(t)\mathbf{x}$, and the second term is a particular solution of Eq. (1).

The above results are summarized in the following theorem.

THEOREM 4.7.1

Assume that the entries of the matrices $\mathbf{P}(t)$ and $\mathbf{g}(t)$ in Eq. (2) are continuous on an open interval I and that $\mathbf{x}_1$ and $\mathbf{x}_2$ are a fundamental set of solutions of the homogeneous equation $\mathbf{x}' = \mathbf{P}(t)\mathbf{x}$ corresponding to the nonhomogeneous equation (1)

$$\mathbf{x}' = \mathbf{P}(t)\mathbf{x} + \mathbf{g}(t).$$

Then a particular solution of Eq. (1) is

$$\mathbf{x}_p(t) = \mathbf{X}(t) \int \mathbf{X}^{-1}(t)\mathbf{g}(t)\, dt, \qquad (15)$$

where the fundamental matrix $\mathbf{X}(t)$ is defined by Eq. (5). Moreover the general solution of Eq. (1) is

$$\mathbf{x}(t) = c_1\mathbf{x}_1(t) + c_2\mathbf{x}_2(t) + \mathbf{x}_p(t). \qquad (16)$$

Remark. There may be two major difficulties in using the method of variation of parameters. One is the determination of $\mathbf{x}_1(t)$ and $\mathbf{x}_2(t)$, a fundamental set of solutions of the homogeneous equation $\mathbf{x}' = \mathbf{P}(t)\mathbf{x}$. If the coefficients in that equation are not constants, these solutions are generally not easy to obtain. The other possible difficulty lies in the evaluation of the integrals appearing in Eq. (15) and this depends entirely on the nature of $\mathbf{X}^{-1}(t)$ and $\mathbf{g}(t)$.

EXAMPLE 1

Find the solution of the initial value problem

$$\mathbf{x}' = \begin{pmatrix} 1 & -4 \\ 2 & -5 \end{pmatrix}\mathbf{x} + \begin{pmatrix} 10\cos t \\ 2e^{-t} \end{pmatrix}, \qquad \mathbf{x}(0) = \begin{pmatrix} 10 \\ 4 \end{pmatrix}. \qquad (17)$$

Applying the eigenvalue method to the homogeneous equation

$$\mathbf{x}' = \begin{pmatrix} 1 & -4 \\ 2 & -5 \end{pmatrix}\mathbf{x}$$

yields the fundamental solution set

$$\mathbf{x}_1(t) = e^{-t}\begin{pmatrix} 2 \\ 1 \end{pmatrix} \qquad \text{and} \qquad \mathbf{x}_2(t) = e^{-3t}\begin{pmatrix} 1 \\ 1 \end{pmatrix}.$$

Thus the corresponding fundamental matrix and its inverse are given by

$$\mathbf{X}(t) = \begin{pmatrix} 2e^{-t} & e^{-3t} \\ e^{-t} & e^{-3t} \end{pmatrix} \qquad \text{and} \qquad \mathbf{X}^{-1}(t) = \begin{pmatrix} e^t & -e^t \\ -e^{3t} & 2e^{3t} \end{pmatrix}.$$

Since the nonhomogeneous term in the initial value problem (17) is $\mathbf{g}(t) = \begin{pmatrix} 10\cos t \\ 2e^{-t} \end{pmatrix}$, a particular solution is given by

$$\mathbf{x}_p(t) = \mathbf{X}(t) \int \mathbf{X}^{-1}(t)\mathbf{g}(t)\, dt$$

$$= \begin{pmatrix} 2e^{-t} & e^{-3t} \\ e^{-t} & e^{-3t} \end{pmatrix} \int \begin{pmatrix} e^t & -e^t \\ -e^{3t} & 2e^{3t} \end{pmatrix} \begin{pmatrix} 10\cos t \\ 2e^{-t} \end{pmatrix} dt$$

$$= \begin{pmatrix} 2e^{-t} & e^{-3t} \\ e^{-t} & e^{-3t} \end{pmatrix} \int \begin{pmatrix} 10e^t & \cos t - 2 \\ -10e^{3t} & \cos t + 4e^{2t} \end{pmatrix} dt$$

$$= \begin{pmatrix} 2e^{-t} & e^{-3t} \\ e^{-t} & e^{-3t} \end{pmatrix} \begin{pmatrix} 5e^t & \cos t + 5e^t & \sin t - 2t \\ -3e^{3t} & \cos t - e^{3t} & \sin t + 2e^{2t} \end{pmatrix}$$

$$= \begin{pmatrix} 7\cos t + 9\sin t + 2(1 - 2t)e^{-t} \\ 2\cos t + 4\sin t + 2(1 - t)e^{-t} \end{pmatrix}.$$

It follows that the general solution of the nonhomogeneous equation in Eq. (17) is

$$\mathbf{x} = c_1 \begin{pmatrix} 2e^{-t} \\ e^{-t} \end{pmatrix} + c_2 \begin{pmatrix} e^{-3t} \\ e^{-3t} \end{pmatrix} + \begin{pmatrix} 7\cos t + 9\sin t + 2(1 - 2t)e^{-t} \\ 2\cos t + 4\sin t + 2(1 - t)e^{-t} \end{pmatrix}. \tag{18}$$

The initial condition prescribed in the initial value problem (17) requires that

$$c_1 \begin{pmatrix} 2 \\ 1 \end{pmatrix} + c_2 \begin{pmatrix} 1 \\ 1 \end{pmatrix} + \begin{pmatrix} 9 \\ 4 \end{pmatrix} = \begin{pmatrix} 10 \\ 4 \end{pmatrix}$$

or

$$\begin{pmatrix} 2 & 1 \\ 1 & 1 \end{pmatrix} \begin{pmatrix} c_1 \\ c_2 \end{pmatrix} = \begin{pmatrix} 1 \\ 0 \end{pmatrix}.$$

Thus $c_1 = 1$, $c_2 = -1$, and the solution of the initial value problem (17) is

$$\mathbf{x} = \begin{pmatrix} -e^{-3t} + 7\cos t + 9\sin t + 2(2 - 2t)e^{-t} \\ -e^{-3t} + 2\cos t + 4\sin t + (3 - 2t)e^{-t} \end{pmatrix}.$$

Variation of Parameters for Linear Second Order Equations

The method of variation of parameters used to find a particular solution of Eq. (1) will now be used to find a particular solution of

$$y'' + p(t)y' + q(t)y = g(t) \tag{19}$$

by applying the method to the dynamical system equivalent to Eq. (19),

$$\mathbf{x}' = \begin{pmatrix} 0 & 1 \\ -q(t) & -p(t) \end{pmatrix} \mathbf{x} + \begin{pmatrix} 0 \\ g(t) \end{pmatrix}, \tag{20}$$

where

$$\mathbf{x}(t) = \begin{pmatrix} x_1(t) \\ x_2(t) \end{pmatrix} = \begin{pmatrix} y(t) \\ y'(t) \end{pmatrix}.$$

If $\{y_1, y_2\}$ is a fundamental set of solutions of $y'' + p(t)y' + q(t)y = 0$, then

$$\mathbf{x}_1(t) = \begin{pmatrix} y_1(t) \\ y_1'(t) \end{pmatrix} \qquad \text{and} \qquad \mathbf{x}_2(t) = \begin{pmatrix} y_2(t) \\ y_2'(t) \end{pmatrix}$$

form a fundamental set of solutions for the homogeneous system obtained from Eq. (20) by setting $g(t) = 0$. A direct method for finding a particular solution of Eq. (19) can be found by using the substitutions

$$\mathbf{X}(t) = \begin{pmatrix} y_1(t) & y_2(t) \\ y_1'(t) & y_2'(t) \end{pmatrix} \qquad \text{and} \qquad \mathbf{g}(t) = \begin{pmatrix} 0 \\ g(t) \end{pmatrix} \tag{21}$$

in the steps leading from Eq. (8) to Eq. (14). Using these substitutions in Eq. (8), the assumed form for the particular solution of Eq. (20) is

$$\mathbf{x}_p(t) = u_1(t) \begin{pmatrix} y_1(t) \\ y_1'(t) \end{pmatrix} + u_2(t) \begin{pmatrix} y_2(t) \\ y_2'(t) \end{pmatrix}. \tag{22}$$

The first component of $\mathbf{x}_p(t)$ in (22) provides us with the form for the particular solution $Y(t)$ of Eq. (19),

$$Y(t) = u_1(t)y_1(t) + u_2(t)y_2(t). \tag{23}$$

Using the expressions for $\mathbf{X}(t)$ and $\mathbf{g}(t)$ in (21) in Eq. (10), we obtain the following algebraic system for the components of $\mathbf{u}' = (u_1', u_2')$:

$$y_1(t)u_1'(t) + y_2(t)u_2'(t) = 0,$$
$$y_1'(t)u_1'(t) + y_2'(t)u_2'(t) = g(t). \tag{24}$$

The solution of Eq. (24) is

$$u_1'(t) = -\frac{y_2(t)g(t)}{W[y_1, y_2](t)}, \qquad u_2'(t) = \frac{y_1(t)g(t)}{W[y_1, y_2](t)}, \tag{25}$$

where $W[y_1, y_2](t)$ is the Wronskian of y_1 and y_2. Note that division by W is permissible since y_1 and y_2 form a fundamental set of solutions of $y'' + p(t)y' + q(t)y = 0$, and therefore their Wronskian is nonzero. By integrating Eqs. (25), we find the desired functions $u_1(t)$ and $u_2(t)$, namely,

$$u_1(t) = -\int \frac{y_2(t)g(t)}{W[y_1, y_2](t)}\, dt + c_1, \qquad u_2(t) = \int \frac{y_1(t)g(t)}{W[y_1, y_2](t)}\, dt + c_2. \tag{26}$$

If the integrals in Eqs. (26) can be evaluated in terms of elementary functions, then we substitute the results in Eq. (23), thereby obtaining the general solution of Eq. (19). More generally, the solution can always be expressed in terms of integrals, as stated in the following theorem.

| THEOREM 4.7.2 | If the functions p, q, and g are continuous on an open interval I, and if the functions y_1 and y_2 form a fundamental set of solutions of the homogeneous equation $y'' + p(t)y' + q(t)y = 0$ corresponding to the nonhomogeneous equation (19), |

$$y'' + p(t)y' + q(t)y = g(t),$$

then a particular solution of Eq. (19) is

$$Y(t) = -y_1(t) \int \frac{y_2(t)g(t)}{W[y_1, y_2](t)} \, dt + y_2(t) \int \frac{y_1(t)g(t)}{W[y_1, y_2](t)} \, dt. \qquad (27)$$

The general solution is

$$y = c_1 y_1(t) + c_2 y_2(t) + Y(t), \qquad (28)$$

as prescribed by Theorem 4.5.2.

Remarks.

i. Note that the conclusions of Theorem 4.7.2 are predicated on the assumption that the leading coefficient in Eq. (19) is 1. If, for example, a second order equation is in the form $P(t)y'' + Q(t)y' + R(t)y = G(t)$, then dividing the equation by $P(t)$ will bring it into the form of Eq. (19) with $g(t) = G(t)/P(t)$.

ii. A major advantage of the method of variation of parameters, particularly when applied to the important case of linear second order equations, is that Eq. (27) provides an expression for the particular solution $Y(t)$ in terms of an arbitrary forcing function $g(t)$. This expression is a good starting point if you wish to investigate the effect of changes to the forcing function, or if you wish to analyze the response of a system to a number of different forcing functions.

| EXAMPLE 2 | Find a particular solution of |

$$y'' + 4y = 3 \csc t. \qquad (29)$$

Observe that this problem is not a good candidate for the method of undetermined coefficients, as described in Section 4.5, because the nonhomogeneous term $g(t) = 3 \csc t$ involves a quotient with $\sin t$ in the denominator. Some experimentation should convince you that a method for finding a particular solution based on naive assumptions about its form is unlikely to succeed. For example, if we assume a particular solution of the form $Y = A \csc t$, we find that $Y'' = A \csc t + 2A \csc t \cot^2 t$, suggesting that it may be difficult or impossible to guess a correct form for Y. Thus we seek a particular solution of Eq. (29) using the method of variation of parameters.

Since a fundamental set of solutions of $y'' + 4y = 0$ is $y_1 = \cos 2t$ and $y_2 = \sin 2t$, according to Eq. (23), the form of the particular solution that we seek is

$$Y(t) = u_1(t) \cos 2t + u_2(t) \sin 2t, \qquad (30)$$

where u_1 and u_2 are to be determined. In practice, a good starting point for determining u_1 and u_2 is the system of algebraic equations (24). Substituting $y_1 = \cos 2t$, $y_2 = \sin 2t$, and

$g(t) = 3 \csc t$ into Eq. (24) gives

$$\cos 2t\, u_1'(t) + \sin 2t\, u_2'(t) = 0,$$
$$-2\sin 2t\, u_1'(t) + 2\cos 2t\, u_2'(t) = 3 \csc t. \tag{31}$$

The solution of the system (31) is

$$u_1'(t) = -\frac{3 \csc t \sin 2t}{2} = -3\cos t, \tag{32}$$

and

$$u_2'(t) = \frac{3\cos t \cos 2t}{\sin 2t} = \frac{3(1 - 2\sin^2 t)}{2 \sin t} = \frac{3}{2}\csc t - 3\sin t, \tag{33}$$

where we have used the double-angle formulas to simplify the expression for u_2'. Having obtained $u_1'(t)$ and $u_2'(t)$, we next integrate to find $u_1(t)$ and $u_2(t)$. The result is

$$u_1(t) = -3\sin t + c_1 \tag{34}$$

and

$$u_2(t) = \tfrac{3}{2}\ln|\csc t - \cot t| + 3\cos t + c_2. \tag{35}$$

On substituting these expressions in Eq. (30), we have

$$y = -3\sin t \cos 2t + \tfrac{3}{2}\ln|\csc t - \cot t|\sin 2t + 3\cos t \sin 2t$$
$$+ c_1 \cos 2t + c_2 \sin 2t.$$

Finally, by using the double-angle formulas once more, we obtain

$$y = 3\sin t + \tfrac{3}{2}\ln|\csc t - \cot t|\sin 2t + c_1 \cos 2t + c_2 \sin 2t. \tag{36}$$

The terms in Eq. (36) involving the arbitrary constants c_1 and c_2 are the general solution of the corresponding homogeneous equation, whereas the remaining terms are a particular solution of the nonhomogeneous equation (29). Thus Eq. (36) is the general solution of Eq. (29).

PROBLEMS

1. **(a)** If
$$\mathbf{X}(t) = \begin{pmatrix} x_{11}(t) & x_{12}(t) \\ x_{21}(t) & x_{22}(t) \end{pmatrix} \quad \text{and} \quad \mathbf{u}(t) = \begin{pmatrix} u_1(t) \\ u_2(t) \end{pmatrix},$$
show that $(\mathbf{X}\mathbf{u})' = \mathbf{X}'\mathbf{u} + \mathbf{X}\mathbf{u}'$.
(b) Assuming that $\mathbf{X}(t)$ is a fundamental matrix for $\mathbf{x}' = \mathbf{P}(t)\mathbf{x}$ and that $\mathbf{u}(t) = \int \mathbf{X}^{-1}(t)\mathbf{g}(t)\,dt$, use the result of part (a) to verify that $\mathbf{x}_p(t)$ given by Eq. (15) satisfies Eq. (1), $\mathbf{x}' = \mathbf{P}(t)\mathbf{x} + \mathbf{g}(t)$.

In each of Problems 2 through 5, use the method of variation of parameters to find a particular solution using the given fundamental set of solutions $\{\mathbf{x}_1, \mathbf{x}_2\}$.

2. $\mathbf{x}' = \begin{pmatrix} -3 & 1 \\ 1 & -3 \end{pmatrix}\mathbf{x} + \begin{pmatrix} 18 \\ 3t \end{pmatrix},$

$$\mathbf{x}_1 = e^{-2t}\begin{pmatrix} 1 \\ 1 \end{pmatrix}, \qquad \mathbf{x}_2 = e^{-4t}\begin{pmatrix} 1 \\ -1 \end{pmatrix}$$

3. $\mathbf{x}' = \begin{pmatrix} -1 & -2 \\ 0 & 1 \end{pmatrix}\mathbf{x} + \begin{pmatrix} e^{-t} \\ t \end{pmatrix},$

$$\mathbf{x}_1 = e^{-t}\begin{pmatrix} 1 \\ 0 \end{pmatrix}, \qquad \mathbf{x}_2 = e^{t}\begin{pmatrix} -1 \\ 1 \end{pmatrix}$$

4. $\mathbf{x}' = \begin{pmatrix} -1 & 0 \\ -1 & -1 \end{pmatrix} \mathbf{x} + \begin{pmatrix} -6 \\ 9t \end{pmatrix}$,

$\mathbf{x}_1 = e^{-t} \begin{pmatrix} 0 \\ 1 \end{pmatrix}$, $\qquad \mathbf{x}_2 = e^{-t} \begin{pmatrix} -1 \\ t \end{pmatrix}$

5. $\mathbf{x}' = \begin{pmatrix} 0 & 1 \\ -1 & 0 \end{pmatrix} \mathbf{x} + \begin{pmatrix} \cos t \\ -\sin t \end{pmatrix}$,

$\mathbf{x}_1 = \begin{pmatrix} \cos t \\ -\sin t \end{pmatrix}$, $\qquad \mathbf{x}_2 = \begin{pmatrix} \sin t \\ \cos t \end{pmatrix}$

In each of Problems 6 through 9, find the solution of the specified initial value problem:

6. The equation in Problem 2 with initial condition $\mathbf{x}(0) = (2, -1)^T$.

7. The equation in Problem 3 with initial condition $\mathbf{x}(0) = (-1, 1)^T$.

8. The equation in Problem 4 with initial condition $\mathbf{x}(0) = (1, 0)^T$.

9. The equation in Problem 5 with initial condition $\mathbf{x}(0) = (1, 1)^T$.

In each of Problems 10 through 13, use the method of variation of parameters to find a particular solution of the given differential equation. Then check your answer by using the method of undetermined coefficients.

10. $y'' - 5y' + 6y = 2e^t$ (Compare with Problem 7 in Section 4.5.)

11. $y'' - y' - 2y = 2e^{-t}$ (Compare with Problem 8 in Section 4.5.)

12. $y'' + 2y' + y = 3e^{-t}$ (Compare with Problem 9 in Section 4.5.)

13. $4y'' - 4y' + y = 16e^{t/2}$ (Compare with Problem 10 in Section 4.5.)

In each of Problems 14 through 21, find the general solution of the given differential equation. In Problems 20 and 21, g is an arbitrary continuous function.

14. $y'' + y = \tan t$, $\quad 0 < t < \pi/2$

15. $y'' + 4y = 3 \sec^2 2t$, $\quad 0 < t < \pi/4$

16. $y'' + 4y' + 4y = t^{-2} e^{-2t}$, $\quad t > 0$

17. $y'' + 4y = 3 \csc(t/2)$, $\quad 0 < t < 2\pi$

18. $4y'' + y = 2 \sec 2t$, $\quad -\pi/4 < t < \pi/4$

19. $y'' - 2y' + y = e^t/(1 + t^2)$

20. $y'' - 5y' + 6y = g(t)$

21. $y'' + 4y = g(t)$

In each of Problems 22 through 27, verify that the given functions y_1 and y_2 satisfy the corresponding homogeneous equation; then find a particular solution of the given nonhomogeneous equation. In Problems 26 and 27, g is an arbitrary continuous function.

22. $t^2 y'' - t(t + 2)y' + (t + 2)y = 2t^3$, $\quad t > 0$; $\quad y_1(t) = t$, $y_2(t) = te^t$

23. $ty'' - (1 + t)y' + y = t^2 e^{2t}$, $\quad t > 0$; $\quad y_1(t) = 1 + t$, $y_2(t) = e^t$

24. $(1 - t)y'' + ty' - y = 2(t - 1)^2 e^{-t}$, $\quad 0 < t < 1$; $y_1(t) = e^t$, $\quad y_2(t) = t$

25. $x^2 y'' + xy' + (x^2 - 0.25)y = 3x^{3/2} \sin x$, $x > 0$; $\quad y_1(x) = x^{-1/2} \sin x$, $y_2(x) = x^{-1/2} \cos x$

26. $(1 - x)y'' + xy' - y = g(x)$, $\quad 0 < x < 1$; $y_1(x) = e^x$, $\quad y_2(x) = x$

27. $x^2 y'' + xy' + (x^2 - 0.25)y = g(x)$, $\quad x > 0$; $y_1(x) = x^{-1/2} \sin x$, $\quad y_2(x) = x^{-1/2} \cos x$

In each of Problems 28 through 31, find the general solution of the nonhomogeneous Cauchy–Euler equation:

28. $t^2 y'' - 2y = 3t^2 - 1$, $\quad t > 0$

29. $x^2 y'' - 3xy' + 4y = x^2 \ln x$, $\quad x > 0$

30. $t^2 y'' - 2ty' + 2y = 4t^2$, $\quad t > 0$

31. $t^2 y'' + 7ty' + 5y = t$, $\quad t > 0$

32. Show that the solution of the initial value problem

$$L[y] = y'' + p(t)y' + q(t)y = g(t),$$
$$y(t_0) = y_0, \quad y'(t_0) = y_1$$

can be written as $y = u(t) + v(t)$, where u and v are solutions of the two initial value problems

$$L[u] = 0, \quad u(t_0) = y_0, \quad u'(t_0) = y_1,$$
$$L[v] = g(t), \quad v(t_0) = 0, \quad v'(t_0) = 0,$$

respectively. In other words, the nonhomogeneities in the differential equation and in the initial conditions can be dealt with separately. Observe that u is easy to find if a fundamental set of solutions of $L[u] = 0$ is known.

33. By choosing the lower limit of integration in Eq. (27) in the text as the initial point t_0, show that $Y(t)$ becomes

$$Y(t) = \int_{t_0}^{t} \frac{y_1(\tau)y_2(t) - y_1(t)y_2(\tau)}{y_1(\tau)y_2'(\tau) - y_1'(\tau)y_2(\tau)} g(\tau) d\tau.$$

Show that $Y(t)$ is a solution of the initial value problem

$$L[y] = g(t), \quad y(t_0) = 0, \quad y'(t_0) = 0.$$

Thus Y can be identified with v in Problem 32.

34. (a) Use the result of Problem 33 to show that the solution of the initial value problem

$$y'' + y = g(t), \quad y(t_0) = 0, \quad y'(t_0) = 0$$

is

$$y = \int_{t_0}^{t} \sin(t - s)g(s) ds.$$

(b) Use the result of Problem 32 to find the solution of the initial value problem

$$y'' + y = g(t), \quad y(0) = y_0, \quad y'(0) = y_1.$$

35. Use the result of Problem 33 to find the solution of the initial value problem

$$L[y] = (D - a)(D - b)y = g(t),$$

$$y(t_0) = 0, \quad y'(t_0) = 0,$$

where a and b are real numbers with $a \neq b$.

36. Use the result of Problem 33 to find the solution of the initial value problem

$$L[y] = [D^2 - 2\lambda D + (\lambda^2 + \mu^2)]y = g(t),$$

$$y(t_0) = 0, \quad y'(t_0) = 0.$$

Note that the roots of the characteristic equation are $\lambda \pm i\mu$.

37. Use the result of Problem 33 to find the solution of the initial value problem

$$L[y] = (D - a)^2 y = g(t),$$

$$y(t_0) = 0, \quad y'(t_0) = 0,$$

where a is any real number.

38. By combining the results of Problems 35 through 37, show that the solution of the initial value problem

$$L[y] = (D^2 + bD + c)y = g(t),$$

$$y(t_0) = 0, \quad y'(t_0) = 0,$$

where b and c are constants, has the form

$$y = \phi(t) = \int_{t_0}^{t} K(t - s) g(s) \, ds. \tag{i}$$

The function K depends only on the solutions y_1 and y_2 of the corresponding homogeneous equation and is independent of the nonhomogeneous term. Once K is determined, all nonhomogeneous problems involving the same differential operator L are reduced to the evaluation of an integral. Note also that, although K depends on both t and s, only the combination $t - s$ appears, so K is actually a function of a single variable. When we think of $g(t)$ as the input to the problem and of $\phi(t)$ as the output, it follows from Eq. (i) that the output depends on the input over the entire interval from the initial point t_0 to the current value t. The integral in Eq. (i) is called the **convolution** of K and g, and K is referred to as the **kernel**.

39. The method of reduction of order (see the discussion preceding Problem 28 in Section 4.2) can also be used for the nonhomogeneous equation

$$y'' + p(t)y' + q(t)y = g(t), \tag{i}$$

provided one solution y_1 of the corresponding homogeneous equation is known. Let $y = v(t)y_1(t)$ and show that y satisfies Eq. (i) if v is a solution of

$$y_1(t)v'' + [2y_1'(t) + p(t)y_1(t)]v' = g(t). \tag{ii}$$

Equation (ii) is a first order linear equation for v'. Solving this equation, integrating the result, and then multiplying by $y_1(t)$ lead to the general solution of Eq. (i).

In each of Problems 40 and 41, use the method outlined in Problem 39 to solve the given differential equation:

40. $ty'' - (1 + t)y' + y = t^2 e^{2t}, \quad t > 0; \quad y_1(t) = 1 + t$
(see Problem 23)

41. $(1 - t)y'' + ty' - y = 2(t - 1)^2 e^{-t}, \quad 0 < t < 1;$
$y_1(t) = e^t$ (see Problem 24)

CHAPTER SUMMARY

Section 4.1 Many simple vibrating systems are modeled by second order linear equations. Mathematical descriptions of spring-mass systems and series RLC circuits lead directly to such equations. Using a technique known as **linearization**, second order linear equations are often used as approximate models of nonlinear second order systems that operate near an equilibrium point. An example of this is the pendulum undergoing small oscillations about its downward-hanging equilibrium state.

Section 4.2

▶ The theory of second order linear equations

$$y'' + p(t)y' + q(t)y = g(t), \qquad p, q, \text{ and } g \text{ continuous on } I$$

follows from the theory of first order linear systems of dimension 2 by converting the equation into an equivalent dynamical system for $\mathbf{x} = (y, y')^T$,

$$\mathbf{x}' = \begin{pmatrix} 0 & 1 \\ -q(t) & -p(t) \end{pmatrix} \mathbf{x} + \begin{pmatrix} 0 \\ g(t) \end{pmatrix}.$$

▶ If $L[y] = y'' + py' + qy$, and y_1 and y_2 are solutions of $L[y] = 0$, then the linear combination $y = c_1y_1 + c_2y_2$ is also a solution of $L[y] = 0$ since L is a **linear operator**:

$$L[c_1y_1 + c_2y_2] = c_1L[y_1] + c_2L[y_2].$$

▶ If $\mathbf{K}[\mathbf{x}] = \mathbf{x}' - \mathbf{P}(t)\mathbf{x}$ and $\mathbf{x}_1$ and $\mathbf{x}_2$ are solutions of $\mathbf{K}[\mathbf{x}] = \mathbf{0}$, then the linear combination $\mathbf{x} = c_1\mathbf{x}_1 + c_2\mathbf{x}_2$ is also a solution of $\mathbf{K}[\mathbf{x}] = \mathbf{0}$ since $\mathbf{K}$ is a **linear operator**:

$$\mathbf{K}[c_1\mathbf{x}_1 + c_2\mathbf{x}_2] = c_1\mathbf{K}[\mathbf{x}_1] + c_2\mathbf{K}[\mathbf{x}_2].$$

▶ Two solutions y_1 and y_2 to the homogeneous equation

$$y'' + p(t)y' + q(t)y = 0, \qquad p \text{ and } q \text{ continuous on } I$$

form a **fundamental set** on I if their **Wronskian**

$$W[y_1, y_2](t) = \begin{vmatrix} y_1(t) & y_2(t) \\ y_1'(t) & y_2'(t) \end{vmatrix} = y_1(t)y_2'(t) - y_1'(t)y_2(t)$$

is nonzero for some (and hence all) $t \in I$. If y_1 and y_2 form a fundamental set of solutions to the homogeneous equation, then a **general solution** is

$$y = c_1y_1(t) + c_2y_2(t),$$

where c_1 and c_2 are arbitrary constants.

▶ Two solutions $\mathbf{x}_1$ and $\mathbf{x}_2$ to the homogeneous equation $\mathbf{x}' = \mathbf{P}(t)\mathbf{x}$, $\mathbf{P}(t)$ continuous on I form a **fundamental set** on I if their **Wronskian**

$$W[\mathbf{x}_1, \mathbf{x}_2](t) = \begin{vmatrix} x_{11}(t) & x_{12}(t) \\ x_{21}(t) & x_{22}(t) \end{vmatrix} = x_{11}(t)x_{22} - x_{12}(t)x_{21}(t)$$

is nonzero for some (and hence all) $t \in I$. If $\mathbf{x}_1$ and $\mathbf{x}_2$ form a fundamental set of solutions to the homogeneous equation, then a **general solution** is

$$\mathbf{x} = c_1\mathbf{x}_1(t) + c_2\mathbf{x}_2(t),$$

where c_1 and c_2 are arbitrary constants.

▶ $y = c_1y_1(t) + c_2y_2(t)$ is a general solution of $y'' + py' + qy = 0$ if and only if $\mathbf{x} = c_1 \begin{pmatrix} y_1 \\ y_1' \end{pmatrix} + c_2 \begin{pmatrix} y_2 \\ y_2' \end{pmatrix}$ is a general solution of $\mathbf{x}' = \begin{pmatrix} 0 & 1 \\ -q(t) & -p(t) \end{pmatrix} \mathbf{x}$.

Section 4.3   Constant Coefficient Equations

▶ The form of a general solution to $ay'' + by' + cy = 0$, $\quad a \neq 0$, depends on the roots

$$\lambda_1 = \frac{-b + \sqrt{b^2 - 4ac}}{2a} \quad \text{and} \quad \lambda_2 = \frac{-b - \sqrt{b^2 - 4ac}}{2a}$$

of the **characteristic equation** $a\lambda^2 + b\lambda + c = 0$:

▶ **Real and distinct roots.** If $b^2 - 4ac > 0$, λ_1 and λ_2 are real and distinct, and a general solution is

$$y = c_1e^{\lambda_1 t} + c_2e^{\lambda_2 t}.$$

▶ **Repeated roots.** If $b^2 - 4ac = 0$, $\lambda = \lambda_1 = \lambda_2 = -\dfrac{b}{2a}$ is a repeated root, and a general solution is

$$y = c_1 e^{\lambda_1 t} + c_2 t e^{\lambda_2 t}.$$

▶ **Complex roots.** If $b^2 - 4ac < 0$, the roots of $a\lambda^2 + b\lambda + c = 0$ are complex, $\lambda_1 = \mu + iv$ and $\lambda_2 = \mu - iv$, and a general solution of $ay'' + by' + cy = 0$ is

$$y = c_1 e^{\mu t} \cos vt + c_2 e^{\mu t} \sin vt.$$

▶ In each of the above cases, the general solution of the corresponding dynamical system $\mathbf{x}' = \mathbf{Ax} = \begin{pmatrix} 0 & 1 \\ -c/a & -b/a \end{pmatrix} \mathbf{x}$ is $\mathbf{x} = \begin{pmatrix} y \\ y' \end{pmatrix}$. The eigenvalues of $\mathbf{A}$ are also the roots of $ay'' + by' + cy = 0$; they determine the type and stability properties of the equilibrium solution $\mathbf{x} = \mathbf{0}$, which is unique if $\det \mathbf{A} = c/a \neq 0$.

Section 4.4

▶ The solution of the undamped spring-mass system $my'' + ky = 0$ can be expressed using phase-amplitude notation as $y = R \cos(\omega_0 t - \phi)$.

▶ For the damped spring-mass system $my'' + \gamma y' + ky = 0$, the motion is **overdamped** if $\gamma^2 - 4mk > 0$, **critically damped** if $\gamma^2 - 4mk = 0$, and **underdamped** if $\gamma^2 - 4mk < 0$.

Section 4.5

▶ If $\mathbf{x}_p$ is any particular solution to $\mathbf{x}' = \mathbf{P}(t)\mathbf{x} + \mathbf{g}(t)$ and $\{\mathbf{x}_1, \mathbf{x}_2\}$ is a fundamental set of solutions to the corresponding homogeneous equation, then a general solution to the nonhomogeneous equation is

$$\mathbf{x} = c_1 \mathbf{x}_1(t) + c_2 \mathbf{x}_2(t) + \mathbf{x}_p(t),$$

where c_1 and c_2 are arbitrary constants.

▶ If Y is any particular solution to $y'' + p(t)y' + q(t)y = g(t)$ and $\{y_1, y_2\}$ is a fundamental set of solutions to the corresponding homogeneous equation, then a general solution of the nonhomogeneous equation is

$$y = c_1 y_1(t) + c_2 y_2(t) + Y(t),$$

where c_1 and c_2 are arbitrary constants.

▶ **Undetermined Coefficients** $\qquad ay'' + by' + cy = g(t), \quad a \neq 0$
This special method for finding a particular solution is primarily useful if the equation has constant coefficients and $g(t)$ is a polynomial, an exponential function, a sine function, a cosine function, or a linear combination of products of these functions. In this case, the appropriate form for a particular solution, given in Table 4.5.1, can be determined.

Section 4.6

▶ The steady-state response to $my'' + \gamma y' + ky = Ae^{i\omega t}$ can be expressed in the form

$$Y = A|G(i\omega)|e^{i(\omega t - \phi(\omega))},$$

where $G(i\omega)$ is called the **frequency response** of the system and $|G(i\omega)|$ and $\phi(\omega)$ are the **gain** and the **phase** of the frequency response, respectively.

▶ Information about the steady-state response as a function of the frequency ω of the input signal is contained in $G(i\omega)$. Frequencies at which $|G(i\omega)|$ are sharply peaked are called **resonant frequencies** of the system.

Section 4.7 Variation of Parameters

▶ If $\mathbf{x}_1$ and $\mathbf{x}_2$ form a fundamental set for $\mathbf{x}' = \mathbf{P}(t)\mathbf{x}$, then a particular solution to $\mathbf{x}' = \mathbf{P}(t)\mathbf{x} + \mathbf{g}(t)$ is

$$\mathbf{x}_p(t) = \mathbf{X}(t) \int \mathbf{X}^{-1}(t)\mathbf{g}(t)\, dt,$$

where $\mathbf{X}(t) = [\mathbf{x}_1(t), \mathbf{x}_2(t)]$.

▶ Applying the last result to the nonhomogeneous second order equation, $y'' + p(t)y' + q(t)y = g(t)$, yields a particular solution

$$Y(t) = -y_1(t) \int \frac{y_2(t)g(t)}{W[y_1, y_2](t)}\, dt + y_2(t) \int \frac{y_1(t)g(t)}{W[y_1, y_2](t)}\, dt,$$

where y_1 and y_2 form a fundamental set for the corresponding homogeneous equation.

PROJECTS

Project 1 A Vibration Insulation Problem

Library of Congress Prints & Photographs Division

Passive isolation systems are sometimes used to insulate delicate equipment from unwanted vibrations. For example, in order to insulate electrical monitoring equipment from vibrations present in the floor of an industrial plant, the equipment may be placed on a platform supported by flexible mountings resting on the floor. A simple physical model for such a system is shown in Figure 4.P.1, where the mountings are modeled as an equivalent linear spring with spring constant k, the combined mass of the platform and equipment is m, and viscous damping with damping coefficient γ is assumed. Assume also that only vertical motion occurs. In this project, we use this model to illustrate an important fundamental principle in the design of a passive isolation system.

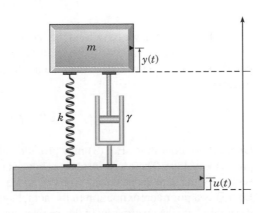

FIGURE 4.P.1 Equipment supported by flexible mountings.

Project 1 PROBLEMS

1. Denote by $y(t)$ the displacement of the platform from its equilibrium position relative to a fixed frame of reference. Relative to the same fixed reference frame, let $u(t)$ represent the displacement of the plant floor from its equilibrium position. Show that the differential equation for the motion of the platform is

$$y'' + 2\zeta\omega_0 y' + \omega_0^2 y = 2\zeta\omega_0 u' + \omega_0^2 u, \tag{i}$$

where $\omega_0 = \sqrt{k/m}$ and $\zeta = \gamma/2\sqrt{mk}$ is a dimensionless parameter known as the **viscous damping ratio**.

2. Denote by $G(i\omega)$ the frequency response of Eq. (i), that is, the ratio of the steady-state response to the harmonic input $u(t) = Ae^{i\omega t}$. The **transmissibility** T_R of the platform mounting system is then defined to be the corresponding gain function, $T_R = |G(i\omega)|$. Show that T_R can be expressed as

$$T_R = \frac{\sqrt{1 + \left(2\zeta\dfrac{\omega}{\omega_0}\right)^2}}{\sqrt{\left[1 - \left(\dfrac{\omega}{\omega_0}\right)^2\right]^2 + \left[2\zeta\dfrac{\omega}{\omega_0}\right]^2}}.$$

3. Plot the graphs of T_R versus the dimensionless ratio ω/ω_0 for $\zeta = 0.1, 0.5, 1, 2$. For what values of ω/ω_0 is $T_R = 1$? Explain why the graphs imply that it is desirable that the mountings have a low natural frequency in order to isolate the vibration source from the equipment platform, and that using low-stiffness isolators is one way to achieve this.

4. The vibrations in the floor of an industrial plant lie in the range 16–75 hertz (Hz). The combined mass of the equipment and platform is 38 kg, and the viscous damping ratio of the suspension is 0.2. Find the maximum value of the spring stiffness if the amplitude of the transmitted vibration is to be less than 10% of the amplitude of the floor vibration over the given frequency range.

5. Test the results of your design strategy for the situation described in Problem 4 by performing several numerical simulations of Eq. (i) in Problem 1 using various values of k as well as various input frequencies while holding the value of ζ fixed. For each simulation, plot the input $u(t)$ and the response $y(t)$ on the same set of coordinate axes. Do the numerical simulations support your theoretical results?

Project 2 Linearization of a Nonlinear Mechanical System

An undamped, one degree-of-freedom mechanical system with forces dependent on position is modeled by a second order differential equation

$$mx'' + f(x) = 0, \tag{1}$$

where x denotes the position coordinate of a particle of mass m and $-f(x)$ denotes the force acting on the mass. We assume that $f(0) = 0$, so $\mathbf{x} = 0\mathbf{i} + 0\mathbf{j}$ is a critical point of the equivalent dynamical system. If $f(x)$ has two derivatives at $x = 0$, then its second degree Taylor formula with remainder at $x = 0$ is

$$f(x) = f'(0)x + \frac{1}{2}f''(z)x^2, \tag{2}$$

where $0 < z < x$ and we have used the fact that $f(0) = 0$.

If the operating range of the system is such that $f''(z)x^2/2$ is always negligible compared to the linear term $f'(0)x$, then the nonlinear equation (1) may be approximated by its linearization,

$$mx'' + f'(0)x = 0. \tag{3}$$

Under these conditions Eq. (3) may provide valuable information about the motion of the physical system. There are instances, however, where nonlinear behavior of $f(x)$, represented by the remainder term $f''(z)x^2/2$ in Eq. (2), is not negligible relative to $f'(0)x$ and Eq. (3) is a poor approximation to the actual system. In such cases, it is necessary to study the nonlinear system directly. In this project, we explore some of these questions in the context of a simple nonlinear system consisting of a mass attached to a pair of identical springs and confined to motion in the horizontal direction on a frictionless surface, as shown in

Figure 4.P.2. The springs are assumed to obey Hooke's law, $F_s(\Delta y) = -k\Delta y$, where Δy is the change in length of each spring from its equilibrium length L. When the mass is at its equilibrium position $x = 0$, both springs are assumed to have length $L + h$ with $h \geq 0$. Thus in the rest state both springs are either elongated and under tension ($h > 0$) or at their natural rest length and under zero tension ($h = 0$).

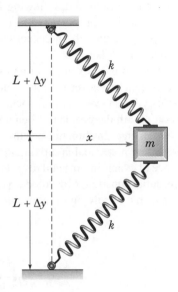

FIGURE 4.P.2   Horizontal motion of a mass attached to two identical springs.

Project 2 PROBLEMS

1. Show that the differential equation describing the motion of the mass in Figure 4.P.2 is

$$ mx'' + 2kx \left[1 - \frac{L}{\sqrt{(L+h)^2 + x^2}} \right] = 0. \qquad (i) $$

2. **(a)** Find the linearization of Eq. (i) at $x = 0$.
(b) In the case that $h > 0$, what is the natural frequency of the linearized system? Explain how the natural frequency of the linearized system depends on h and L.
(c) On the same set of coordinate axes, plot the graphs of

$$ y = f(x) = 2kx \left[1 - \frac{L}{\sqrt{(L+h)^2 + x^2}} \right], $$

its first degree Taylor polynomial $y_1 = f'(0)x$, and its third degree Taylor polynomial $y_3 = f'(0)x + f''(0)x^2/2 + f'''(0)x^3/6$.

Construct these plots for each of the following sets of parameter values:

i. $L = 1, k = 1, h = 1$
ii. $L = 1, k = 1, h = 0.5$

iii. $L = 1, k = 1, h = 0.1$
iv. $L = 1, k = 1, h = 0$

(d) Find the dependence on h of the approximate range of values of x such that $|y - y_1| \approx |y_3 - y_1| < 0.1$.
(e) In the case that $h = 0$, explain why the solution of the linearized equation obtained in Problem 2 is a poor approximation to the solution of the nonlinear equation (i) for any initial conditions other than $x(0) = 0, x'(0) = 0$.

3. Subject to the initial conditions $x(0) = 0, x'(0) = 1$, draw the graph of the numerical approximation of the solution of Eq. (1) for $0 \leq t \leq 30$. On the same set of coordinate axes, draw the graph of the solution of the linearized equation using the same initial conditions. Do this for each of the following sets of parameter values:

i. $m = 1, L = 1, k = 1,$ and $h = 1$
ii. $m = 1, L = 1, k = 1,$ and $h = 0.5$
iii. $m = 1, L = 1, k = 1,$ and $h = 0.1$

Compare the accuracy of the period length and amplitude exhibited by the solution of the linearized equation to the period length and amplitude of the solution of the nonlinear equation as the value of h decreases.

Project 3 A Spring-Mass Event Problem

A mass of magnitude m is confined to one-dimensional motion between two springs on a frictionless horizontal surface, as shown in Figure 4.P.3. The mass, which is unattached to either spring, undergoes simple inertial motion whenever the distance from the origin of the center point of the mass, x, satisfies $|x| < L$. When $x \geq L$, the mass is in contact with the spring on the right and, following Hooke's law, experiences a force in the negative x direction (to the left) proportional to $x - L$. Similarly, when $x \leq -L$, the mass is in contact with the spring on the left and experiences a force in the positive x direction (to the right) proportional to $x + L$.

This problem is an example of an **event problem** in differential equations. The events of interest occur whenever the mass initiates or terminates contact with a spring. A typical way to describe an event is to associate an event function $g(t, x)$, which may or may not depend on t, with the problem. Then an event is said to occur at time t^* if $g(t^*, x(t^*)) = 0$. For example, an event function for this problem could be $g(x) = (x - L)(x + L)$. One strategy for finding numerical approximations to solutions of an event problem is to locate the occurrence of events in time and restart the integration there so as to deal with the changes in the differential equation. Most modern commercial software packages contain ODE solvers that are capable of solving event problems.

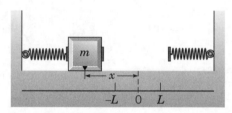

FIGURE 4.P.3   A mass bouncing back and forth between two springs.

Project 3 PROBLEMS

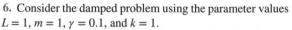

1. Assuming that both springs have spring constant k and that there is a damping force proportional to velocity x' with damping constant γ, write down the differential equation that describes the motion of the mass. Note that the expression for the force function depends on which of the three different parts of its domain x lies within.

2. The **Heaviside**, or **unit step function**, is defined by

$$u(x - c) = \begin{cases} 1, & x \geq c \\ 0, & x < c. \end{cases}$$

Use the unit step function to express the differential equation in Problem 1 in a single line.

3. Is the differential equation derived in Problems 1 and 2 linear or nonlinear? Explain why.

4. In the case that the damping constant $\gamma > 0$, find the critical points of the differential equation in Problem 2 and discuss their stability properties.

5. Consider the case of an undamped problem ($\gamma = 0$) using the parameter values $L = 1$, $m = 1$, $k = 1$ and initial condi-

tions $x(0) = 2$, $x'(0) = 0$. Find the solution $x(t)$ of the initial value problem for $0 \leq t \leq t_1^*$ and $t_1^* \leq t < t_2^*$, where t_1^* and t_2^* are the times of the first and second events. Describe the physical situation corresponding to each of these two events.

6. Consider the damped problem using the parameter values $L = 1$, $m = 1$, $\gamma = 0.1$, and $k = 1$.
(a) Use a computer to draw a direction field of the corresponding dynamical system.
(b) If you have access to computer software that is capable of solving event problems, solve for and plot the graphs of $x(t)$ and $x'(t)$ for the following sets of initial conditions:
i. $x(0) = 2$, $x'(0) = 0$
ii. $x(0) = 5$, $x'(0) = 0$

Give a physical explanation of why the limiting values of the trajectories as $t \to \infty$ depend on the initial conditions.
(c) Draw a phase portrait for the equivalent dynamical system.

7. Describe some other physical problems that could be formulated as event problems in differential equations.

Project 4 Euler–Lagrange Equations

Recall from calculus that a function $f(x)$ has a horizontal tangent at $x = a$ if $f'(a) = 0$. Such a point is called a stationary point; it may be, but does not have to be, a local maximum or minimum of $f(x)$. Similarly, for a function $f(x, y)$ of two variables, necessary conditions for (a, b) to be a stationary point of $f(x, y)$ are that $f_x(a, b) = 0$ and $f_y(a, b) = 0$. In this project, we explain how a generalization of this idea leads to several of the differential equations treated in this textbook. The generalization requires finding necessary conditions for a stationary point (a function) of a functional (a function of a function). For mechanical systems with one degree of freedom, the functional is frequently of the form

$$J(y) = \int_a^b F(x, y, y') \, dx, \tag{1}$$

where $y = y(x)$. Consider, for example, the undamped pendulum of Figure 4.P.4. A fundamental principle of classical mechanics is the **principle of stationary action**[2]: *the trajectory of an object in state space is the one that yields a stationary value for the integral with respect to time of the kinetic energy minus the potential energy*. If we denote kinetic energy by K and potential energy by V, the integral $\int_0^T (K - V) \, dt$ is called the **action integral** of the system. If $\theta(t)$ is the angular displacement of the pendulum, then the kinetic energy of the pendulum is $\frac{1}{2}mL^2(\theta')^2$ and its potential energy is $mgL(1 - \cos\theta)$. Therefore the principle of stationary action requires that $\theta(t)$ must be a function for which the action integral,

$$J(\theta) = \int_0^T \left[\frac{1}{2}mL^2(\theta')^2 - mgL(1 - \cos\theta) \right] dt, \tag{2}$$

is stationary. Note that $J(\theta)$ is of the form (1). Often a stationary point of the action integral is a minimum but can be a saddle point or even a maximum. According to the principle of stationary action, of all possible paths $t \to \langle \theta(t), \theta'(t) \rangle$ that the pendulum can follow in state space, the one observed in nature is the path for which the value of (2) is stationary. In this project, we want to deduce a characterization, which turns out to be a differential equation, of the stationary points (functions) of $J(y)$ in Eq. (1).

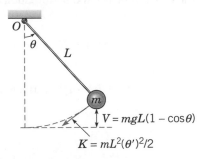

FIGURE 4.P.4 An undamped pendulum in motion has kinetic energy $K = mL^2(\theta')^2/2$ and potential energy $V = mgL(1 - \cos\theta)$.

Thus we consider the problem of finding a function $y(x)$ that is a stationary point for the functional (1) where $y(x)$ is to be chosen from a class of suitably smooth functions (two or more continuous derivatives) that satisfy the boundary conditions

$$y(a) = A, \qquad y(b) = B.$$

[2]It can be shown that the principle of stationary action is equivalent to Newton's laws of motion.

Suppose we give the name $\bar{y}(x)$ to such a stationary point. A very useful method for solving this problem is to examine the behavior of Eq. (1) for functions in a neighborhood of $\bar{y}(x)$. This is accomplished by considering small variations or perturbations to $\bar{y}(x)$ of the form $\epsilon\eta(x)$,

$$y(x) = \bar{y}(x) + \epsilon\eta(x), \tag{3}$$

where $0 < \epsilon \ll 1$ and $\eta(x)$ satisfies $\eta(a) = 0$ and $\eta(b) = 0$ so that $y(a) = A$ and $y(b) = B$ (Figure 4.P.5).

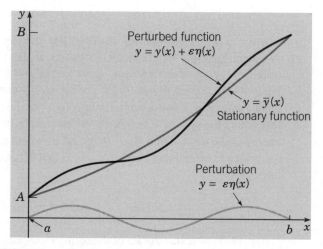

FIGURE 4.P.5   Perturbation of a function.

The parameter ϵ controls the size of the perturbation, whereas $\eta(x)$ controls the "direction" of the perturbation. If we assume that $\bar{y}(x)$ is indeed a stationary point for (1), then it is necessary that, for any fixed perturbation direction $\eta(x)$, the function

$$\phi(\epsilon) = J(\bar{y} + \epsilon\eta) = \int_a^b F(x, \bar{y} + \epsilon\eta, \bar{y}' + \epsilon\eta')\, dx \tag{4}$$

must, as a function of the single variable ϵ, be stationary at $\epsilon = 0$ (Figure 4.P.6).

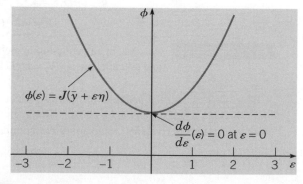

FIGURE 4.P.6   Graph of $J(\bar{y} + \epsilon\eta)$ near $\epsilon = 0$ in the case that $J(y)$ has a minimum at $\bar{y}$.

Therefore we take the derivative of $\phi(\epsilon) = J(\bar{y} + \epsilon\eta)$ in Eq. (4) at $\epsilon = 0$ and set the result equal to zero,

$$\frac{d\phi}{d\epsilon}(0) = \frac{d}{d\epsilon}J(\bar{y} + \epsilon\eta)\Big|_{\epsilon=0} = \int_a^b \left[\frac{\partial F}{\partial y}(x,\bar{y},\bar{y}')\eta + \frac{\partial F}{\partial y'}(x,\bar{y},\bar{y}')\eta'\right]dx$$

$$= \frac{\partial F}{\partial y'}(x,\bar{y},\bar{y}')\eta\Big|_{x=a}^b + \int_a^b \left[\frac{\partial F}{\partial y}(x,\bar{y},\bar{y}') - \frac{\partial^2 F}{\partial x \partial y'}(x,\bar{y},\bar{y}')\right]\eta\,dx$$

$$= \int_a^b \left[\frac{\partial F}{\partial y}(x,\bar{y},\bar{y}') - \frac{\partial^2 F}{\partial x \partial y'}(x,\bar{y},\bar{y}')\right]\eta\,dx = 0, \tag{5}$$

where we have used integration by parts and the fact that $\eta(a) = \eta(b) = 0$ to set the boundary terms that arise to zero. Since the last integral is equal to zero for any function $\eta(x)$ satisfying $\eta(a) = \eta(b) = 0$, we conclude that the part of the integrand multiplying $\eta(x)$ must be the zero function, that is,

$$\frac{\partial F}{\partial y}(x,\bar{y},\bar{y}') - \frac{\partial^2 F}{\partial x \partial y'}(x,\bar{y},\bar{y}') = 0,$$

or, reverting to the notation y instead of $\bar{y}$,

$$\frac{\partial F}{\partial y}(x,y,y') - \frac{\partial^2 F}{\partial x \partial y'}(x,y,y') = 0. \tag{6}$$

Equation (6) is known as the **Euler–Lagrange equation**[3] for the functional (1): if $J(y)$ in Eq. (1) *has a stationary function $y(x)$, then it is necessary that $y(x)$ satisfy the differential equation* (6).

EXAMPLE 1

Find the Euler–Lagrange equation for the action integral (2) of the simple undamped pendulum.

Note that the integrand in Eq. (2) is

$$F(t,\theta,\theta') = \tfrac{1}{2}mL^2(\theta')^2 - mgL(1 - \cos\theta).$$

Then $F_\theta = -mgL\sin\theta$, $F_{\theta'} = mL^2\theta'$, and $F_{\theta't} = mL^2\theta''$. Thus the Euler–Lagrange equation $F_\theta - F_{\theta't} = 0$ is $-mgL\sin\theta - mL^2\theta'' = 0$ or $\theta'' + (g/L)\sin\theta = 0$.

In the Problems that follow, several of the differential equations that appear in this textbook are derived by finding Euler–Lagrange equations of appropriate functionals.

Project 4 PROBLEMS

Problems 1 through 3 are concerned with one degree-of-freedom systems.

1. Falling Bodies. Let $y(t)$ be the height above Earth's surface of a body of mass m subject only to Earth's gravitational

acceleration g. Find the action integral and Euler–Lagrange equation for $y(t)$.

2. A Spring-Mass System. A mass m supported by a frictionless horizontal surface is attached to a spring that is, in

---

[3]The argument that allows us to conclude that the bracketed function appearing in the last integral in Eq. (5) is zero is as follows. If this function is continuous and not zero (say, positive) at some point $x = \zeta$ in the interval, then it must be positive throughout a subinterval about $x = \zeta$. If we choose an $\eta(x)$ that is positive inside this subinterval and zero outside, then the last integral in Eq. (5) will be positive— a contradiction.

turn, attached to a fixed support. Assuming that motion is restricted to one direction, let $x(t)$ represent the displacement of the mass from its equilibrium position. This position corresponds to the spring being at its rest length. If the spring is compressed, then $x < 0$ and the mass is left of its equilibrium position. If the spring is elongated, then $x > 0$ and the mass is right of its equilibrium position. Assume that the potential energy stored in the spring is given by $kx^2/2$, where the parameter k is the **spring constant** of the spring. Find the action integral and Euler–Lagrange equation for $x(t)$.

3. The Brachistochrone Problem. Find the curve $y(x)$ connecting $y(a) = A$ to $y(b) = B$ (assume $a < b$ and $B < A$), along which a particle under the influence of gravity g will slide without friction in minimum time. This is the famous Brachistochrone problem discussed in Problem 31 of Section 2.2. Bernoulli showed that the appropriate functional for the sliding time is

$$T = \int_0^T dt = \int_0^L \frac{dt}{ds}\,ds = \int_0^L \frac{ds}{v} = \frac{1}{\sqrt{2g}}\int_0^L \frac{ds}{\sqrt{y}}$$

$$= \frac{1}{\sqrt{2g}}\int_a^b \sqrt{\frac{1+y'^2}{y}}\,dx. \qquad (i)$$

(a) Justify each of the equalities in Eq. (i).
(b) From the functional on the right in Eq. (i), find the Euler–Lagrange equation for the curve $y(x)$.

Problems 4 and 5 are concerned with systems that have two degrees of freedom. The generalization of Eq. (1) to a functional with two degrees of freedom is

$$J(x,y) = \int_a^b F(t,x,y,x',y')\,dt, \qquad (7)$$

where t is a parameter, not necessarily time. Necessary conditions, stated as differential equations, for a vector-valued function $t \to \langle \bar{x}(t), \bar{y}(t)\rangle$ to be a stationary function of Eq. (7) are found by an analysis that is analogous to that leading up to Eq. (6). Consider the perturbed functional

$$\phi(\epsilon) = J(\bar{x}+\epsilon\xi, \bar{y}+\epsilon\eta)$$
$$= \int_a^b F(t,\bar{x}+\epsilon\xi, \bar{y}+\epsilon\eta, \bar{x}'+\epsilon\xi', \bar{y}'+\epsilon\eta')\,dt, \qquad (8)$$

where the perturbation vector $\langle\xi(t),\eta(t)\rangle$ satisfies

$$\langle\xi(a),\eta(a)\rangle = \langle 0,0\rangle \quad \text{and} \quad \langle\xi(b),\eta(b)\rangle = \langle 0,0\rangle. \qquad (9)$$

Thus, for $\phi(\epsilon)$ defined by Eq. (8), we compute $\phi'(0) = 0$, integrate by parts, and use the endpoint conditions (9) to arrive at

$$\int_a^b \{[F_x - F_{x't}]\xi + [F_y - F_{y't}]\eta\}\,dt = 0. \qquad (10)$$

Since Eq. (10) must hold for any ξ and η satisfying the conditions (9), we conclude that each factor multiplying ξ and η in the integrand must equal zero,

$$F_x - F_{x't} = 0$$
$$F_y - F_{y't} = 0. \qquad (11)$$

Thus the Euler–Lagrange equations for the functional (7) are given by the system (11).

4. A Two-Mass, Three-Spring System. Consider the mechanical system consisting of two masses and three springs shown in Figure 4.P.7.

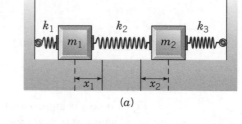

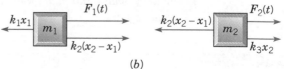

FIGURE 4.P.7　A two-mass, three-spring system.

Find the action integral for this system and the Euler–Lagrange equations for the positions $x_1(t)$ and $x_2(t)$ of the two masses.

5. The Ray Equations. In two dimensions, we consider a point sound source located at $(x, y) = (x_0, y_0)$ in a fluid medium where the sound speed $c(x, y)$ can vary with location. The initial value problem that describes the path of a ray launched from the point (x_0, y_0) with a launch angle α, measured from the horizontal, is

$$\frac{d}{ds}\left(\frac{1}{c}\frac{dx}{ds}\right) = -\frac{c_x}{c^2}, \qquad x(0) = x_0, \quad x'(0) = \cos\alpha$$
$$\frac{d}{ds}\left(\frac{1}{c}\frac{dy}{ds}\right) = -\frac{c_y}{c^2}, \qquad y(0) = y_0, \quad y'(0) = \sin\alpha, \qquad (i)$$

where $c_x = \partial c/\partial x$ and $c_y = \partial c/\partial y$. The independent variable s in Eq. (i) is the arc length of the ray measured from the source. Derive the system (i) from **Fermat's principle**: *rays, in physical space, are paths along which the transit time is minimum.* If $P : t \to \langle x(t), y(t)\rangle$, $a \le t \le b$, is a parametric representation of a ray path from (x_0, y_0) to (x_1, y_1) in the

xy-plane and sound speed is prescribed by $c(x, y)$, then the transit time functional is

$$T = \int_P \frac{ds}{c} = \int_a^b \frac{\sqrt{x'^2 + y'^2}}{c} dt, \qquad \text{(ii)}$$

where we have used the differential arc length relation

$$ds = \sqrt{x'^2 + y'^2}\, dt, \qquad \text{(iii)}$$

Find the Euler–Lagrange equations from the functional representation on the right in Eq. (ii) and use Eq. (iii) to rewrite the resulting system using arc length s as the independent variable.

6. Carry out the calculations that lead from Eq. (8) to Eq. (11) in the discussion preceding Problem 4.

CHAPTER FIVE

© Morphart Creation / Shutterstock

The Laplace Transform

A n integral transform is a relation of the form

$$F(s) = \int_{\alpha}^{\beta} K(t,s)f(t)\, dt, \tag{1}$$

which takes a given function $f(t)$ as input and outputs another function $F(s)$. The function $K(t,s)$ in Eq. (1) is called the **kernel** of the transform and the function $F(s)$ is called the **transform** of $f(t)$. It is possible that $\alpha = -\infty$ or $\beta = \infty$, or both. If $\alpha = 0$, $\beta = \infty$, and $K(t,s) = e^{-st}$, Eq. (1) takes the form

$$F(s) = \int_{0}^{\infty} e^{-st} f(t)\, dt. \tag{2}$$

In this case, $F(s)$ is called the **Laplace transform** of $f(t)$.

   In general, the parameter s can be a complex number but in most of this chapter we assume that s is real. It is customary to refer to $f(t)$ as a function, or signal, in the time or "t-domain" and $F(s)$ as its representation in the "s-domain." The Laplace transform is commonly used in engineering to study input–output relations of linear systems, to analyze feedback control systems, and to study electric circuits. One of its primary applications is to convert the problem of solving a constant coefficient linear differential equation in the t-domain into a problem involving algebraic operations in the s-domain. The general idea, illustrated by the block diagram in Figure 5.0.1, is as follows: Use the relation (2) to transform the problem for an unknown function f into a simpler problem for F, then solve this simpler problem to find F, and finally recover the desired function f from its transform F. This last step is known as "inverting the transform." The Laplace transform is particularly convenient for solving many practical engineering problems that involve mechanical or electrical systems acted on by discontinuous or impulsive forcing terms; for such problems the methods described in Chapter 4 are often rather awkward to use. In this chapter we describe how the Laplace transform method works, emphasizing problems typical of those that arise in engineering applications.

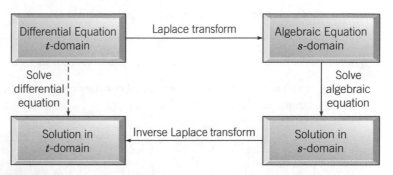

FIGURE 5.0.1 Laplace transform method for solving differential equations.

5.1 Definition of the Laplace Transform

▶ Improper Integrals. Since the Laplace transform involves an integral from zero to infinity, a knowledge of improper integrals of this type is necessary to appreciate the subsequent development of the properties of the transform. We provide a brief review of such improper integrals here. If you are already familiar with improper integrals, you may wish to skip over this review. On the other hand, if improper integrals are new to you, then you should probably consult a calculus book, where many more details and examples will be found.

An improper integral over an unbounded interval is defined as a limit of integrals over finite intervals; thus

$$\int_a^\infty f(t)\,dt = \lim_{A\to\infty} \int_a^A f(t)\,dt, \tag{3}$$

where A is a positive real number. If the integral from a to A exists for each $A > a$, and if the limit as $A \to \infty$ exists, then the improper integral is said to **converge** to that limiting value. Otherwise, the integral is said to **diverge**, or to fail to exist. The following examples illustrate both possibilities.

EXAMPLE 1

Let $f(t) = e^{ct}$, $t \geq 0$, where c is a real nonzero constant. Then

$$\int_0^\infty e^{ct}\,dt = \lim_{A\to\infty} \int_0^A e^{ct}\,dt = \lim_{A\to\infty} \frac{e^{ct}}{c}\Big|_0^A$$

$$= \lim_{A\to\infty} \frac{1}{c}(e^{cA} - 1).$$

It follows that the improper integral converges to the value $-1/c$ if $c < 0$ and diverges if $c > 0$. If $c = 0$, the integrand $f(t)$ is the constant function with value 1, and the integral again diverges.

EXAMPLE 2

Let $f(t) = 1/t$, $t \geq 1$. Then

$$\int_1^\infty \frac{dt}{t} = \lim_{A \to \infty} \int_1^A \frac{dt}{t} = \lim_{A \to \infty} \ln A.$$

Since $\lim_{A \to \infty} \ln A = \infty$, the improper integral diverges.

EXAMPLE 3

Let $f(t) = t^{-p}$, $t \geq 1$, where p is a real constant and $p \neq 1$; the case $p = 1$ was considered in Example 2. Then

$$\int_1^\infty t^{-p} \, dt = \lim_{A \to \infty} \int_1^A t^{-p} \, dt = \lim_{A \to \infty} \frac{1}{1 - p}(A^{1-p} - 1).$$

As $A \to \infty$, $A^{1-p} \to 0$ if $p > 1$, but $A^{1-p} \to \infty$ if $p < 1$. Hence $\int_1^\infty t^{-p} \, dt$ converges to the value $1/(p - 1)$ for $p > 1$ but (incorporating the result of Example 2) diverges for $p \leq 1$. These results are analogous to those for the infinite series $\sum_{n=1}^\infty n^{-p}$.

▶ The Laplace Transform

DEFINITION 5.1.1

Let f be a function on $[0, \infty)$. The Laplace transform of f is the function F defined by the integral,

$$F(s) = \int_0^\infty e^{-st} f(t) \, dt. \tag{4}$$

The domain of $F(s)$ is the set of all values of s for which the integral in Eq. (4) converges. The Laplace transform of f is denoted by both F and $\mathcal{L}\{f\}$.

Remark. We will use t (representing time) as the independent variable of functions defined by lowercase letters such as x, y, or f. The corresponding Laplace transforms will be denoted by the same letter in uppercase:

$$\mathcal{L}\{f\} = F, \qquad \mathcal{L}\{y\} = Y, \qquad \mathcal{L}\{x\} = X.$$

We will use s to denote the independent variable of the transformed function,

$$\mathcal{L}\{f\}(s) = F(s), \qquad \mathcal{L}\{y\}(s) = Y(s), \qquad \mathcal{L}\{x\}(s) = X(s).$$

The Laplace transform of the function f is the function $\mathcal{L}\{f\}$, or F, and the Laplace transform evaluated at a number s is $\mathcal{L}\{f\}(s)$, or $F(s)$. For the purposes of clarity, simplicity, or efficiency, we may depart from standard function notational conventions. For example, we may use $\mathcal{L}\{f\}$ or $\mathcal{L}\{f(t)\}$ to represent $\mathcal{L}\{f\}(s)$. Thus $\mathcal{L}\{f\}$ could be used to represent the statement "$\mathcal{L}\{f\}(s)$ where f is defined by $f(t) = t$, $t > 0$."

Before we discuss the linearity property of the Laplace transform and sufficient conditions under which the transformation in Eq. (4) makes sense, we compute the Laplace transforms of some common functions.

EXAMPLE 4

Let $f(t) = 1$, $t \geq 0$. Then, as in Example 1,

$$\mathcal{L}\{1\} = \int_0^\infty e^{-st}\, dt = -\lim_{A\to\infty} \frac{e^{-st}}{s}\bigg|_0^A = -\lim_{A\to\infty}\left(\frac{e^{-sA}}{s} - \frac{1}{s}\right) = \frac{1}{s}, \qquad s > 0.$$

EXAMPLE 5

Let $f(t) = e^{at}$, $t \geq 0$ and a real. Then, again referring to Example 1, we get

$$\mathcal{L}\{e^{at}\} = \int_0^\infty e^{-st} e^{at}\, dt = \int_0^\infty e^{-(s-a)t}\, dt = \frac{1}{s-a}, \qquad s > a.$$

EXAMPLE 6

Let $f(t) = e^{(a+bi)t}$, $t \geq 0$. As in the previous example,

$$\mathcal{L}\{e^{(a+bi)t}\} = \int_0^\infty e^{-st} e^{(a+bi)t}\, dt = \int_0^\infty e^{-(s-a-bi)t}\, dt = \frac{1}{s-a-bi}, \qquad s > a.$$

▶ Linearity of the Laplace Transform. The following theorem establishes the fact that the Laplace transform is a **linear operator**. This basic property is used frequently throughout the chapter.

THEOREM 5.1.2

Suppose that f_1 and f_2 are two functions whose Laplace transforms exist for $s > a_1$ and $s > a_2$, respectively. In addition, let c_1 and c_2 be real or complex numbers. Then, for s greater than the maximum of a_1 and a_2,

$$\boxed{\mathcal{L}\{c_1 f_1(t) + c_2 f_2(t)\} = c_1 \mathcal{L}\{f_1(t)\} + c_2 \mathcal{L}\{f_2(t)\}.} \tag{5}$$

Proof

$$\mathcal{L}\{c_1 f_1(t) + c_2 f_2(t)\} = \lim_{A\to\infty} \int_0^A e^{-st}[c_1 f_1(t) + c_2 f_2(t)]\, dt$$

$$= c_1 \lim_{A\to\infty} \int_0^A e^{-st} f_1(t)\, dt + c_2 \lim_{A\to\infty} \int_0^A e^{-st} f_2(t)\, dt$$

$$= c_1 \int_0^\infty e^{-st} f_1(t)\, dt + c_2 \int_0^\infty e^{-st} f_2(t)\, dt,$$

where the last two integrals converge if $s > \max\{a_1, a_2\}$.

Remark. The sum in Eq. (5) readily extends to an arbitrary number of terms,

$$\mathcal{L}\{c_1 f_1(t) + \cdots + c_n f_n(t)\} = c_1 \mathcal{L}\{f_1(t)\} + \cdots + c_n \mathcal{L}\{f_n(t)\}.$$

EXAMPLE 7

Find the Laplace transform of $f(t) = \sin at$, $t \geq 0$.

Since $e^{\pm iat} = \cos at \pm i \sin at$ (see Appendix B), $\sin at$ can be represented as

$$\sin at = \frac{1}{2i}(e^{iat} - e^{-iat}).$$

Using the linearity of $\mathcal{L}$ as expressed in Eq. (5) and the result from Example 6 gives

$$\mathcal{L}\{\sin at\} = \frac{1}{2i}\left[\mathcal{L}\{e^{iat}\} - \mathcal{L}\{e^{-iat}\}\right] = \frac{1}{2i}\left(\frac{1}{s - ia} - \frac{1}{s + ia}\right) = \frac{a}{s^2 + a^2}, \qquad s > 0.$$

EXAMPLE 8

Find the Laplace transform of $f(t) = 2 + 5e^{-2t} - 3\sin 4t$, $t \geq 0$.

Extending Eq. (5) to a linear combination of three terms gives

$$\mathcal{L}\{f(t)\} = 2\mathcal{L}\{1\} + 5\mathcal{L}\{e^{-2t}\} - 3\mathcal{L}\{\sin 4t\}.$$

Then, using the results from Examples 4, 5, and 7, we find that

$$\mathcal{L}\{f(t)\} = \frac{2}{s} + \frac{5}{s + 2} - \frac{12}{s^2 + 16}, \qquad s > 0.$$

If $f(t)$ contains factors that are powers of t, it may be possible to compute the Laplace transform by using integration by parts.

EXAMPLE 9

Find the Laplace transform of $f(t) = t \cos at$, $t \geq 0$.

Expressing $\cos at$ in terms of complex exponential functions, $\cos at = (e^{iat} + e^{-iat})/2$, we can write the Laplace transform of f in the form

$$\mathcal{L}\{t \cos at\} = \int_0^\infty e^{-st} t \cos at \, dt = \frac{1}{2} \int_0^\infty \left(te^{-(s-ia)t} + te^{-(s+ia)t}\right) dt.$$

Using integration by parts, we find that

$$\int_0^\infty te^{-(s-ia)t} \, dt = -\left[\frac{te^{-(s-ia)t}}{s - ia} + \frac{e^{-(s-ia)t}}{(s - ia)^2}\right]_0^\infty = \frac{1}{(s - ia)^2}, \qquad s > 0.$$

Similarly, we have

$$\int_0^\infty te^{-(s+ia)t} \, dt = \frac{1}{(s + ia)^2}, \qquad s > 0.$$

Thus

$$\mathcal{L}\{t \cos at\} = \frac{1}{2}\left[\frac{1}{(s - ia)^2} + \frac{1}{(s + ia)^2}\right] = \frac{s^2 - a^2}{(s^2 + a^2)^2}, \qquad s > 0.$$

The definitions that follow allow us to specify a large class of functions for which the Laplace transform is guaranteed to exist.

▶ Piecewise Continuous Functions

DEFINITION 5.1.3

A function f is said to be **piecewise continuous** on an interval $\alpha \leq t \leq \beta$ if the interval can be partitioned by a finite number of points $\alpha = t_0 < t_1 < \cdots < t_n = \beta$ so that:

1. f is continuous on each open subinterval $t_{i-1} < t < t_i$, and

2. f approaches a finite limit as the endpoints of each subinterval are approached from within the subinterval.

In other words, f is piecewise continuous on $\alpha \leq t \leq \beta$ if it is continuous at all but possibly finitely many points of $[\alpha, \beta]$, at each of which the function has a finite jump discontinuity. If f is piecewise continuous on $\alpha \leq t \leq \beta$ for every $\beta > \alpha$, then f is said to be piecewise continuous on $t \geq \alpha$. The graph of a piecewise continuous function is shown in Figure 5.1.1.

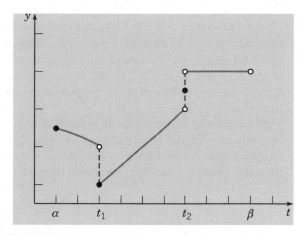

FIGURE 5.1.1   A piecewise continuous function.

We remark that continuous functions belong to the class of piecewise continuous functions and that for points of continuity the left- and right-hand limits are equal.

As shown in the next example, finding the Laplace transform of a piecewise continuous function may require expressing the right-hand side of Eq. (4) as a sum of integrals over subintervals in which the defining expressions for f change from subinterval to subinterval.

EXAMPLE 10

Find the Laplace transform of

$$f(t) = \begin{cases} e^{2t}, & 0 \leq t < 1, \\ 4, & 1 \leq t. \end{cases}$$

We observe that f is continuous on $(0, 1)$ and $(1, \infty)$. Furthermore $\lim_{t \to 1^-} f(t) = e^2$ and $\lim_{t \to 1^+} f(t) = 4$, so f has a single finite jump discontinuity at $t = 1$. Thus f is piecewise continuous on $t \geq 0$. Since the expressions defining $f(t)$ differ on each of the two subintervals $0 \leq t < 1$ and $1 \leq t$, we break the integral in Eq. (4) into two separate parts. Thus

$$
\begin{aligned}
F(s) &= \int_0^\infty e^{-st} f(t) \, dt \\
&= \int_0^1 e^{-st} e^{2t} \, dt + \int_1^\infty e^{-st} \cdot 4 \, dt \\
&= \int_0^1 e^{-(s-2)t} \, dt + 4 \lim_{A \to \infty} \int_1^A e^{-st} \, dt \\
&= -\frac{e^{-(s-2)t}}{s-2} \bigg|_{t=0}^1 - 4 \lim_{A \to \infty} \frac{e^{-st}}{s} \bigg|_{t=1}^A \\
&= \frac{1}{s-2} - \frac{e^{-(s-2)}}{s-2} - 4 \lim_{A \to \infty} \left[\frac{e^{-As}}{s} - \frac{e^{-s}}{s} \right] \\
&= \frac{1}{s-2} - \frac{e^{-(s-2)}}{s-2} + 4 \frac{e^{-s}}{s}, \qquad s > 0, s \neq 2.
\end{aligned}
$$

▶ Functions of Exponential Order. If f is piecewise continuous on the interval $a \leq t \leq A$, then it can be shown that $\int_a^A f(t) \, dt$ exists. Hence, if f is piecewise continuous for $t \geq a$, then $\int_a^A f(t) \, dt$ exists for each $A > a$. However piecewise continuity is not enough to ensure convergence of the improper integral $\int_a^\infty f(t) \, dt$, as the preceding examples show. Obviously, $f(t)$ [or $f(t)e^{-st}$ if we explicitly include e^{-st} as part of the integrand] must vanish sufficiently rapidly as $t \to \infty$.

If f cannot be integrated easily in terms of elementary functions, the definition of convergence of $\int_a^\infty f(t) \, dt$ may be difficult to apply. Frequently, the most convenient way to test the convergence or divergence of an improper integral is by using the following comparison theorem, which is analogous to a similar theorem for infinite series.

THEOREM 5.1.4

If f is piecewise continuous for $t \geq a$, if $|f(t)| \leq g(t)$ when $t \geq M$ for some positive constant M, and if $\int_M^\infty g(t) \, dt$ converges, then $\int_a^\infty f(t) \, dt$ also converges. On the other hand, if $f(t) \geq g(t) \geq 0$ for $t \geq M$, and if $\int_M^\infty g(t) \, dt$ diverges, then $\int_a^\infty f(t) \, dt$ also diverges.

The proof of this result from calculus will not be given here. It is made plausible, however, by comparing the areas represented by $\int_M^\infty g(t) \, dt$ and $\int_M^\infty |f(t)| \, dt$. The functions most useful for comparison purposes are e^{ct} and t^{-p}, which we considered in Examples 1, 2, and 3. In particular, functions that increase no faster than an exponential function as $t \to \infty$ are useful for stating the existence theorem for Laplace transforms.

DEFINITION 5.1.5

A function $f(t)$ is of **exponential order** (as $t \to +\infty$) if there exist real constants $M \geq 0, K > 0$, and a such that

$$
|f(t)| \leq K e^{at} \tag{6}
$$

when $t \geq M$.

Remark. To show that $f(t)$ is of exponential order, it suffices to show that $f(t)/e^{at}$ is bounded for all t sufficiently large.

EXAMPLE 11

Determine which of the following functions are of exponential order: (a) $f(t) = \cos at$, (b) $f(t) = t^2$, and (c) $f(t) = e^{t^2}$.

(a) Since $|\cos at| \leq 1$ for all t, inequality (6) holds if we choose $M = 0$, $K = 1$, and $a = 0$. The choice of constants is not unique. For example, $M = 10$, $K = 2$, and $a = 3$ would also suffice.

(b) By L'Hôpital's rule

$$\lim_{t \to \infty} \frac{t^2}{e^{at}} = \lim_{t \to \infty} \frac{2t}{ae^{at}} = \lim_{t \to \infty} \frac{2}{a^2 e^{at}} = 0$$

if $a > 0$. Suppose we choose $a = 1$. Then from the definition of limit, there exists an M such that $t^2/e^t \leq 0.1$, say, for all $t \geq M$. Then $|f(t)| = t^2 \leq 0.1e^t$ for all $t \geq M$. Thus $f(t)$ is of exponential order.

(c) Since

$$\lim_{t \to \infty} \frac{e^{t^2}}{e^{at}} = \lim_{t \to \infty} e^{t(t-a)} = \infty,$$

no matter how large we choose a, $f(t) = e^{t^2}$ is not of exponential order.

Existence of the Laplace Transform

The following theorem guarantees that the Laplace transform F exists if f is a piecewise continuous function of exponential order. In this chapter we deal almost exclusively with such functions.

THEOREM 5.1.6

Suppose

i. f is piecewise continuous on the interval $0 \leq t \leq A$ for any positive A, and

ii. f is of exponential order, that is, there exist real constants $M \geq 0$, $K > 0$, and a such that $|f(t)| \leq Ke^{at}$ when $t \geq M$.

Then the Laplace transform $\mathcal{L}\{f(t)\} = F(s)$, defined by Eq. (4), exists for $s > a$.

Proof

To establish this theorem, we must show that the integral in Eq. (4) converges for $s > a$. Splitting the improper integral into two parts, we have

$$F(s) = \int_0^\infty e^{-st} f(t)\, dt = \int_0^M e^{-st} f(t)\, dt + \int_M^\infty e^{-st} f(t)\, dt. \tag{7}$$

The first integral on the right side of Eq. (7) exists by hypothesis (i) of the theorem; hence the existence of $F(s)$ depends on the convergence of the second integral. By hypothesis (ii) we have, for $t \geq M$,

$$|e^{-st} f(t)| \leq Ke^{-st} e^{at} = Ke^{(a-s)t},$$

and thus, by Theorem 5.1.4, $F(s)$ exists provided that $\int_M^\infty e^{(a-s)t}\, dt$ converges. Referring to Example 1 with c replaced by $a - s$, we see that this latter integral converges when $a - s < 0$, which establishes Theorem 5.1.6.

The Laplace transforms that appear in Examples 5 through 10 all have the property of vanishing as $s \to \infty$. This property is shared by all Laplace transforms of piecewise continuous functions of exponential order as stated in the next corollary.

COROLLARY 5.1.7

If $f(t)$ satisfies the hypotheses of Theorem 5.1.6, then

$$|F(s)| \leq L/s \tag{8}$$

for some constant L as $s \to \infty$. Thus

$$\lim_{s \to \infty} F(s) = 0. \tag{9}$$

A proof of Corollary 5.1.7 is outlined in Problem 35.

Remarks

1. In general, the parameter s may be complex, and the full power of the Laplace transform becomes available only when we regard $F(s)$ as a function of a complex variable. For functions that satisfy the hypotheses of Theorem 5.1.6, the Laplace transform exists for all values of s in the set $\{s : \text{Re } s > a\}$. This follows from the proof of Theorem 5.1.6 since, if $s = \mu + i\nu$ is complex, $|e^{-st}| = |e^{-\mu t}(\cos \nu t - i \sin \nu t)| = e^{-\mu t} = e^{-\text{Re}(s)t}$.

2. Although the hypotheses of Theorem 5.1.6 are sufficient for the existence of the Laplace transform of $f(t)$, they are not necessary. In Problem 37, you are asked to compute the Laplace transform of t^p, where $p > -1$. In particular, you will show that the Laplace transform of $t^{-1/2}$ is $\sqrt{\pi/s}$.

3. Corollary 5.1.7 implies that if $\lim_{s \to \infty} F(s) \neq 0$, then F is not the Laplace transform of a piecewise continuous function of exponential order. Most of the Laplace transforms in this chapter are rational functions of s (quotients of polynomials in s). In order for such functions to be Laplace transforms of piecewise continuous functions of exponential order, Corollary 5.1.7 implies that it is necessary that the degree of the polynomial in the numerator be less than the degree of the polynomial in the denominator.

4. In Section 5.7 we will encounter a "generalized function," denoted by $\delta(t)$, such that $\mathcal{L}\{\delta(t)\} = 1$. Thus, by Corollary 5.1.7, $\delta(t)$ is not a piecewise continuous function of exponential order.

PROBLEMS

In each of Problems 1 through 4, sketch the graph of the given function. In each case, determine whether f is continuous, piecewise continuous, or neither on the interval $0 \leq t \leq 3$.

1. $f(t) = \begin{cases} t^2, & 0 \leq t \leq 1 \\ 2+t, & 1 < t \leq 2 \\ 6-t, & 2 < t \leq 3 \end{cases}$

2. $f(t) = \begin{cases} \sin(\pi t), & 0 \leq t \leq 1 \\ (t-1)^{-1}, & 1 < t \leq 2 \\ 1, & 2 < t \leq 3 \end{cases}$

3. $f(t) = \begin{cases} t^2, & 0 \leq t \leq 1 \\ 1, & 1 < t \leq 2 \\ 1-5t, & 2 < t \leq 3 \end{cases}$

4. $f(t) = \begin{cases} t, & 0 \leq t \leq 1 \\ 5-t, & 1 < t \leq 2 \\ 3, & 2 < t \leq 3 \end{cases}$

In each of Problems 5 through 12, determine whether the given function is of exponential order. If it is, find suitable values for M, K, and a in inequality (6) of Definition 5.1.5.

5. $2e^{3t}$

6. $-5e^{-6t}$

7. $e^{5t}\sin 3t$

8. t^7

9. $\cosh(t^2)$

10. $\sin(t^4)$

11. $1/(1+t)$

12. e^{t^3}

13. Find the Laplace transform of each of the following functions:
(a) t
(b) t^2
(c) t^n, where n is a positive integer

In each of Problems 14 through 17, find the Laplace transform of the given function:

14. $f(t) = \begin{cases} t, & 0 \le t \le 1 \\ 6, & 1 < t \end{cases}$

15. $f(t) = \begin{cases} 0, & 0 \le t \le 1 \\ 1, & 1 < t \le 2 \\ 0, & 2 < t \end{cases}$

16. $f(t) = \begin{cases} 0, & 0 \le t \le 1 \\ e^{-t}, & 1 < t \end{cases}$

17. $f(t) = \begin{cases} t^2, & 0 \le t \le 1 \\ 5 - t, & 1 < t \le 2 \\ 6, & 2 < t \end{cases}$

In each of Problems 18 through 21, find the Laplace transform of the given function; a and b are real constants. Recall that $\cosh bt = (e^{bt} + e^{-bt})/2$ and $\sinh bt = (e^{bt} - e^{-bt})/2$.

18. $\cosh bt$

19. $\sinh bt$

20. $e^{at}\cosh bt$

21. $e^{at}\sinh bt$

In each of Problems 22 through 24, use the facts that $\cos bt = (e^{ibt} + e^{-ibt})/2$, $\sin bt = (e^{ibt} - e^{-ibt})/2i$ to find the Laplace transform of the given function; a and b are real constants.

22. $\cos bt$

23. $e^{at}\sin bt$

24. $e^{at}\cos bt$

In each of Problems 25 through 30, using integration by parts, find the Laplace transform of the given function; n is a positive integer and a is a real constant.

25. te^{at}

26. $t\sin at$

27. $t\cosh at$

28. $t^n e^{at}$

29. $t^2\sin at$

30. $t^2\sinh at$

In each of Problems 31 through 34, determine whether the given integral converges or diverges:

31. $\int_0^\infty (t^2 + 1)^{-1}\, dt$

32. $\int_0^\infty te^{-t}\, dt$

33. $\int_1^\infty t^{-4}e^t\, dt$

34. $\int_0^\infty e^{-t}\cos t\, dt$

35. A Proof of Corollary 5.1.7
(a) Starting from (7), use the fact that $|f(t)|$ is bounded on $[0, M]$ [since $f(t)$ is piecewise continuous] to show that if $s > \max(a, 0)$, then

$$|F(s)| \le \max_{0 \le t \le M} |f(t)| \int_0^M e^{-st}\, dt + K \int_M^\infty e^{-(s-a)t}\, dt$$

$$= \max_{0 \le t \le M} |f(t)| \frac{1 - e^{-sM}}{s} + \frac{K}{s - a}e^{-(s-a)M}.$$

(b) Argue that there is a constant K_1 such that

$$\frac{Ke^{-(s-a)M}}{s - a} < \frac{K_1}{s}$$

for s sufficiently large.
(c) Conclude that for s sufficiently large, $|F(s)| \le L/s$, where

$$L = \max_{0 \le t \le M} |f(t)| + K_1.$$

36. The Gamma Function. The gamma function is denoted by $\Gamma(p)$ and defined by the integral

$$\Gamma(p + 1) = \int_0^\infty e^{-x}x^p\, dx. \qquad \text{(i)}$$

The integral converges for all $p > -1$.
(a) Show that, for $p > 0$,

$$\Gamma(p + 1) = p\Gamma(p).$$

(b) Show that $\Gamma(1) = 1$.
(c) If p is a positive integer n, show that

$$\Gamma(n + 1) = n!.$$

Since $\Gamma(p)$ is also defined when p is not an integer, this function provides an extension of the factorial function to nonintegral values of the independent variable. Note that it is also consistent to define $0! = 1$.
(d) Show that, for $p > 0$,

$$p(p + 1)\cdots(p + n - 1) = \frac{\Gamma(p + n)}{\Gamma(p)}.$$

Thus $\Gamma(p)$ can be determined for all positive values of p if $\Gamma(p)$ is known in a single interval of unit length, say, $0 < p \le 1$. It is possible to show that $\Gamma(1/2) = \sqrt{\pi}$. Find $\Gamma(3/2)$ and $\Gamma(11/2)$.

37. Consider the Laplace transform of t^p, where $p > -1$.

(a) Referring to Problem 36, show that

$$\mathcal{L}\{t^p\} = \int_0^\infty e^{-st} t^p \, dt = \frac{1}{s^{p+1}} \int_0^\infty e^{-x} x^p \, dx$$

$$= \frac{\Gamma(p+1)}{s^{p+1}}, \qquad s > 0.$$

(b) Let p be a positive integer n in (a). Show that

$$\mathcal{L}\{t^n\} = \frac{n!}{s^{n+1}}, \qquad s > 0.$$

(c) Show that

$$\mathcal{L}\{t^{-1/2}\} = \frac{2}{\sqrt{s}} \int_0^\infty e^{-x^2} \, dx, \qquad s > 0.$$

It is possible to show that

$$\int_0^\infty e^{-x^2} \, dx = \frac{\sqrt{\pi}}{2};$$

hence,

$$\mathcal{L}\{t^{-1/2}\} = \sqrt{\frac{\pi}{s}}, \qquad s > 0.$$

(d) Show that

$$\mathcal{L}\{t^{1/2}\} = \frac{\sqrt{\pi}}{2s^{3/2}}, \qquad s > 0.$$

5.2 Properties of the Laplace Transform

In the preceding section we computed the Laplace transform of several functions $f(t)$ directly from the definition

$$\mathcal{L}\{f\}(s) = \int_0^\infty e^{-st} f(t) \, dt.$$

In this section we present a number of operational properties of the Laplace transform that greatly simplify the task of obtaining explicit expressions for $\mathcal{L}\{f\}$. In addition, these properties enable us to use the transform to solve initial value problems for linear differential equations with constant coefficients.

▶ Laplace Transform of $e^{ct}f(t)$

THEOREM 5.2.1	If $F(s) = \mathcal{L}\{f(t)\}$ exists for $s > a$, and if c is a constant, then

$$\boxed{\mathcal{L}\{e^{ct}f(t)\} = F(s - c), \qquad s > a + c.} \qquad (1)$$

Proof   The proof of this theorem merely requires the evaluation of $\mathcal{L}\{e^{ct}f(t)\}$. Thus

$$\mathcal{L}\{e^{ct}f(t)\} = \int_0^\infty e^{-st} e^{ct} f(t) \, dt = \int_0^\infty e^{-(s-c)t} f(t) \, dt = F(s - c),$$

which is Eq. (1). Since $F(s)$ exists for $s > a$, $F(s - c)$ exists for $s - c > a$, or $s > a + c$.

According to Theorem 5.2.1, multiplication of $f(t)$ by e^{ct} results in a translation of the transform $F(s)$ a distance c in the positive s direction if $c > 0$. (If $c < 0$, the translation is, of course, in the negative direction.)

EXAMPLE 1

Find the Laplace transform of $g(t) = e^{-2t} \sin 4t$ and determine where it is valid.

The Laplace transform of $f(t) = \sin 4t$ is $F(s) = 4/(s^2 + 16)$, $s > 0$, (see Example 7 in Section 5.1). Using Theorem 5.2.1 with $c = -2$, the Laplace transform of g is

$$G(s) = \mathcal{L}\{e^{-2t}f(t)\} = F(s + 2) = \frac{4}{(s + 2)^2 + 16}, \qquad s > -2. \tag{2}$$

▶ **Laplace Transform of Derivatives of $f(t)$.** The usefulness of the Laplace transform, in connection with solving initial value problems, rests primarily on the fact that the transform of f' (and higher order derivatives) is related in a simple way to the transform of f. The relationship is expressed in the following theorem.

THEOREM 5.2.2

Suppose that f is continuous and f' is piecewise continuous on any interval $0 \leq t \leq A$. Suppose further that f and f' are of exponential order, with a as specified in Theorem 5.1.6. Then $\mathcal{L}\{f'(t)\}$ exists for $s > a$, and moreover

$$\boxed{\mathcal{L}\{f'(t)\} = s\mathcal{L}\{f(t)\} - f(0).} \tag{3}$$

Proof

To prove this theorem, we consider the integral

$$\int_0^A e^{-st} f'(t)\, dt.$$

If f' has points of discontinuity in the interval $0 \leq t \leq A$, let them be denoted by $t_1, t_2, \ldots, t_n$. Then we can write this integral as

$$\int_0^A e^{-st} f'(t)\, dt = \int_0^{t_1} e^{-st} f'(t)\, dt + \int_{t_1}^{t_2} e^{-st} f'(t)\, dt + \cdots + \int_{t_n}^A e^{-st} f'(t)\, dt.$$

Integrating each term on the right by parts yields

$$\int_0^A e^{-st} f'(t)\, dt = e^{-st} f(t) \Big|_0^{t_1} + e^{-st} f(t) \Big|_{t_1}^{t_2} + \cdots + e^{-st} f(t) \Big|_{t_n}^A$$

$$+ s \left[\int_0^{t_1} e^{-st} f(t)\, dt + \int_{t_1}^{t_2} e^{-st} f(t)\, dt + \cdots + \int_{t_n}^A e^{-st} f(t)\, dt \right].$$

Since f is continuous, the contributions of the integrated terms at $t_1, t_2, \ldots, t_n$ cancel. Combining the integrals gives

$$\int_0^A e^{-st} f'(t)\, dt = e^{-sA} f(A) - f(0) + s \int_0^A e^{-st} f(t)\, dt.$$

As $A \to \infty$, $e^{-sA} f(A) \to 0$ whenever $s > a$. Hence, for $s > a$,

$$\mathcal{L}\{f'(t)\} = s\mathcal{L}\{f(t)\} - f(0),$$

which establishes the theorem.

EXAMPLE
2

Verify that Theorem 5.2.2 holds for $g(t) = e^{-2t} \sin 4t$ and its derivative $g'(t) = -2e^{-2t} \sin 4t + 4e^{-2t} \cos 4t$.

The Laplace transform of $g(t)$, computed in Example 1, is $G(s) = 4/[(s + 2)^2 + 16]$. By Theorem 5.2.2

$$\mathcal{L}\{g'(t)\} = sG(s) - g(0) = \frac{4s}{(s + 2)^2 + 16}$$

since $g(0) = 0$. On the other hand, from Example 7 and Problem 22 in Section 5.1, we know that $\mathcal{L}\{\sin 4t\} = 4/(s^2 + 16)$ and $\mathcal{L}\{\cos 4t\} = s/(s^2 + 16)$. Using the linearity property of the Laplace transform and Theorem 5.2.1, it follows that

$$\mathcal{L}\{g'(t)\} = -2\mathcal{L}\{e^{-2t} \sin 4t\} + 4\mathcal{L}\{e^{-2t} \cos 4t\}$$

$$= \frac{-8}{(s + 2)^2 + 16} + \frac{4(s + 2)}{(s + 2)^2 + 16}$$

$$= \frac{4s}{(s + 2)^2 + 16}.$$

If f' and f'' satisfy the same conditions that are imposed on f and f', respectively, in Theorem 5.2.2, then it follows that the Laplace transform of f'' also exists for $s > a$ and is given by

$$\mathcal{L}\{f''(t)\} = s^2 \mathcal{L}\{f(t)\} - sf(0) - f'(0). \tag{4}$$

Indeed, provided the function f and its derivatives satisfy suitable conditions, an expression for the transform of the nth derivative $f^{(n)}$ can be derived by successive applications of Theorem 5.2.2. The result is given in the following corollary.

COROLLARY
5.2.3

Suppose that

i. the functions $f, f', \ldots, f^{(n-1)}$ are continuous and that $f^{(n)}$ is piecewise continuous on any interval $0 \le t \le A$, and

ii. $f, f', \ldots, f^{(n-1)}, f^{(n)}$ are of exponential order with a as specified in Theorem 5.1.6.

Then $\mathcal{L}\{f^{(n)}(t)\}$ exists for $s > a$ and is given by

$$\mathcal{L}\{f^{(n)}(t)\} = s^n \mathcal{L}\{f(t)\} - s^{n-1}f(0) - \cdots - sf^{(n-2)}(0) - f^{(n-1)}(0). \tag{5}$$

EXAMPLE
3

Assume that the solution of the following initial value problem satisfies the hypotheses of Corollary 5.2.3. Find its Laplace transform:

$$y'' + 2y' + 5y = e^{-t}, \qquad y(0) = 1, \quad y'(0) = -3. \tag{6}$$

Taking the Laplace transform of both sides of the differential equation in the initial value problem (6) and using the linearity property of $\mathcal{L}$ yield

$$\mathcal{L}\{y''\} + 2\mathcal{L}\{y'\} + 5\mathcal{L}\{y\} = \mathcal{L}\{e^{-t}\}.$$

Letting $Y = \mathcal{L}\{y\}$ and applying formula (5) in Corollary 5.2.3 to each of the derivative terms and using the fact that $\mathcal{L}\{e^{-t}\} = 1/(s+1)$ then yield

$$\underbrace{s^2 Y(s) - sy(0) - y'(0)}_{\mathcal{L}\{y''(t)\}(s)} + 2\underbrace{[sY(s) - y(0)]}_{\mathcal{L}\{y'(t)\}(s)} + 5Y(s) = \frac{1}{s+1}$$

or

$$s^2 Y(s) - s \cdot 1 + 3 + 2\left[sY(s) - 1\right] + 5Y(s) = \frac{1}{s+1},$$

where we have replaced $y(0)$ and $y'(0)$ using the initial conditions specified in Eqs. (6). Solving this last algebraic equation for $Y(s)$ gives

$$Y(s) = \frac{s-1}{s^2 + 2s + 5} + \frac{1}{(s+1)(s^2 + 2s + 5)} = \frac{s^2}{(s+1)(s^2 + 2s + 5)}.$$

EXAMPLE 4

In the series RLC circuit shown in Figure 5.2.1, denote the current in the circuit by $i(t)$, the total charge on the capacitor by $q(t)$, and the impressed voltage by $e(t)$. Assuming that $i(t)$ and $q(t)$ satisfy the hypotheses of Theorem 5.2.2, find the Laplace transforms of $q(t)$ and $i(t)$.

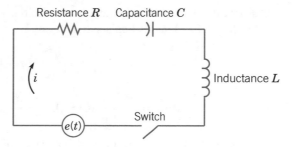

FIGURE 5.2.1 The series RLC circuit.

The relation between charge q and current i is

$$i = \frac{dq}{dt}, \tag{7}$$

while Kirchhoff's second law (*in a closed circuit the impressed voltage is equal to the algebraic sum of the voltages across the elements in the rest of the circuit*) yields the equation

$$L\frac{di}{dt} + Ri + \frac{1}{C}q = e(t). \tag{8}$$

Taking the Laplace transform of Eqs. (7) and (8) and using Theorem 5.2.2 give the system of algebraic equations

$$I(s) = sQ(s) - q(0) \tag{9}$$

and

$$L\left[sI(s) - i(0)\right] + RI(s) + \frac{1}{C}Q(s) = E(s), \tag{10}$$

where $I(s)$, $Q(s)$, and $E(s)$ are the Laplace transforms of $i(t)$, $q(t)$, and $e(t)$, respectively. Using matrix notation, Eqs. (9) and (10) may be written as

$$\begin{pmatrix} s & -1 \\ 1/C & Ls + R \end{pmatrix} \begin{pmatrix} Q(s) \\ I(s) \end{pmatrix} = \begin{pmatrix} q(0) \\ Li(0) + E(s) \end{pmatrix}. \tag{11}$$

The solution of Eq. (11) is

$$\begin{pmatrix} Q(s) \\ I(s) \end{pmatrix} = \frac{1}{Ls^2 + Rs + 1/C} \begin{pmatrix} Ls + R & 1 \\ -1/C & s \end{pmatrix} \begin{pmatrix} q(0) \\ Li(0) + E(s) \end{pmatrix}.$$

Thus

$$Q(s) = \frac{(Ls + R)q(0) + Li(0) + E(s)}{Ls^2 + Rs + 1/C} \tag{12}$$

and

$$I(s) = \frac{Lsi(0) + sE(s) - q(0)/C}{Ls^2 + Rs + 1/C}. \tag{13}$$

Note that $Q(s)$ and $I(s)$ depend on the initial current in the circuit, $i(0)$, and the initial charge on the capacitor, $q(0)$.

EXAMPLE 5

Assume that the solution of the following initial value problem satisfies the hypotheses of Corollary 5.2.3. Find its Laplace transform:

$$\frac{d^4 y}{dt^4} - y = 0, \tag{14}$$

$$y(0) = 0, \qquad \frac{dy}{dt}(0) = 0, \qquad \frac{d^2 y}{dt^2}(0) = 0, \qquad \frac{d^3 y}{dt^3}(0) = 1. \tag{15}$$

Taking the Laplace transform of both sides of the differential equation (14) and using the linearity property of $\mathcal{L}$ yield

$$\mathcal{L}\left\{ \frac{d^4 y}{dt^4} \right\} - \mathcal{L}\{y\} = 0. \tag{16}$$

The right-hand side is zero since the Laplace transform of the zero function in the t-domain is the zero function in the s-domain. Letting $Y = \mathcal{L}\{y\}$, applying formula (5) in Corollary 5.2.3 to the derivative term, and utilizing the initial conditions (15) give

$$\mathcal{L}\left\{ \frac{d^4 y}{dt^4} \right\} = s^4 Y(s) - s^3 \overbrace{y(0)}^{0} - s^2 \overbrace{\frac{dy}{dt}(0)}^{0} - s \overbrace{\frac{d^2 y}{dt^2}(0)}^{0} - \overbrace{\frac{d^3 y}{dt^3}(0)}^{1} = s^4 Y(s) - 1. \tag{17}$$

Using Eq. (17) in Eq. (16) yields the algebraic equation

$$s^4 Y(s) - 1 - Y(s) = 0,$$

the solution of which is

$$Y(s) = \frac{1}{s^4 - 1}. \tag{18}$$

▶ Laplace Transform of $t^n f(t)$.

THEOREM 5.2.4	Suppose that f is (i) piecewise continuous on any interval $0 \le t \le A$, and (ii) of exponential order with a as specified in Theorem 5.1.6. Then for any positive integer n

$$\mathcal{L}\{t^n f(t)\} = (-1)^n F^{(n)}(s), \qquad s > a. \tag{19}$$

Proof

$$F^{(n)}(s) = \frac{d^n}{ds^n} \int_0^\infty e^{-st} f(t)\, dt = \int_0^\infty \frac{\partial^n}{\partial s^n} (e^{-st}) f(t)\, dt$$

$$= \int_0^\infty (-t)^n e^{-st} f(t)\, dt = (-1)^n \int_0^\infty t^n e^{-st} f(t)\, dt$$

$$= (-1)^n \mathcal{L}\{t^n f(t)\}.$$

Thus

$$\mathcal{L}\{t^n f(t)\} = (-1)^n F^{(n)}(s).$$

The operation of interchanging the order of differentiation with respect to s and integration with respect to t is justified by a theorem from advanced calculus known as Leibniz's rule.

COROLLARY 5.2.5	For any integer $n \ge 0$

$$\mathcal{L}\{t^n\} = \frac{n!}{s^{n+1}}, \qquad s > 0. \tag{20}$$

Proof

If $f(t) = 1$, $F(s) = \mathcal{L}\{f(t)\} = 1/s$, as shown in Example 4 of Section 5.1. Then

$$F^{(n)}(s) = \frac{d^n}{ds^n} \left(\frac{1}{s} \right) = \frac{(-1)^n n!}{s^{n+1}}.$$

The result (20) follows by setting $f(t) = 1$ and $F^{(n)}(s) = \dfrac{(-1)^n n!}{s^{n+1}}$ in Eq. (19).

PROBLEMS

In each of Problems 1 through 10, find the Laplace transform of the given function. Assume that a and b are real numbers and n is a positive integer.

1. $f(t) = e^{-2t} \sin 3t$
2. $f(t) = e^{3t} \cos 2t$
3. $f(t) = t^6 - 4t^2 + 5$
4. $f(t) = t \cos 3t$
5. $f(t) = e^{-2t}(t^3 + 1)^2$
6. $f(t) = t^6 e^{5t}$
7. $f(t) = t^2 \sin bt$
8. $f(t) = t^n e^{at}$
9. $f(t) = t e^{at} \sin bt$
10. $f(t) = t e^{at} \cos bt$

11. (a) Let $F(s) = \mathcal{L}\{f(t)\}$, where $f(t)$ is piecewise continuous and of exponential order on $[0, \infty)$. Show that

$$\mathcal{L}\left\{ \int_0^t f(\tau)\, d\tau \right\} = \frac{1}{s} F(s). \tag{i}$$

Hint: Let $g_1(t) = \int_0^t f(t_1)\, dt_1$ and note that $g_1'(t) = f(t)$. Then use Theorem 5.2.2.

(b) Show that for $n \geq 2$,

$$\mathcal{L}\left\{\int_0^t \int_0^{t_n} \cdots \int_0^{t_2} f(t_1)\, dt_1 \cdots dt_n\right\} = \frac{1}{s^n} F(s). \qquad \text{(ii)}$$

In each of Problems 12 through 21, transform the given initial value problem into an algebraic equation for $Y = \mathcal{L}\{y\}$ in the s-domain. Then find the Laplace transform of the solution of the initial value problem.

12. $y'' + 2y' - 2y = 0$,
$y(0) = 2, \quad y'(0) = 1$

13. $9y'' + 12y' + 4y = 0$,
$y(0) = 2, \quad y'(0) = -1$

14. $y'' + 3y' + 2y = 0$,
$y(0) = 3, \quad y'(0) = -1$

15. $6y'' + 5y' + y = 0$,
$y(0) = 4, \quad y'(0) = 0$

16. $y'' - 2y' + 2y = t^2 e^t + 7$,
$y(0) = 1, \quad y'(0) = 1$

17. $y'' - 5y' - 6y = t^2 + 7$,
$y(0) = 1, \quad y'(0) = 0$

18. $y'' + 4y = 3e^{-2t} \sin 2t$,
$y(0) = 2, \quad y'(0) = -1$

19. $y'' + 2y' + 5y = t \cos 2t$,
$y(0) = 1, \quad y'(0) = 0$

20. $y''' + y'' + y' + y = 0$,
$y(0) = 4, \qquad y'(0) = 0, \qquad y''(0) = -2$

21. $y'''' - 6y = te^{-t}, \qquad y(0) = 0, \quad y'(0) = 0,$
$y''(0) = 0, \qquad y'''(0) = 9$

22. In Section 4.1 the differential equation for the charge on the capacitor of the RLC circuit illustrated in Figure 5.2.1 was shown to be

$$Lq'' + Rq' + \frac{1}{C}q = e(t). \qquad \text{(iii)}$$

Take the Laplace transform of Eq. (iii) to derive Eq. (12) for $Q(s)$ and then use Eq. (9) to derive Eq. (13) for $I(s)$.

In each of Problems 23 through 27, find the Laplace transform $Y(s) = \mathcal{L}\{y\}$ of the solution of the given initial value problem. A method of determining the inverse transform is developed in Section 5.5.

23. $y'' + 16y = \begin{cases} 1, & 0 \leq t < \pi, \\ 0, & \pi \leq t < \infty; \end{cases}$

$y(0) = 9, \quad y'(0) = 2$

24. $y'' + y = \begin{cases} t, & 0 \leq t < 1, \\ 0, & 1 \leq t < \infty; \end{cases}$

$y(0) = 0, \quad y'(0) = 0$

25. $y'' + 4y = \begin{cases} t, & 0 \leq t < 1, \\ 1, & 1 \leq t < \infty; \end{cases}$

$y(0) = 0, \quad y'(0) = 0$

26. A tank originally contains 640 gallons (gal) of fresh water. Then water containing $\frac{1}{2}$ pound (lb) of salt per gallon is poured into the tank at a rate of 8 gal/min, and the mixture is allowed to leave at the same rate. After 15 min the salt water solution flowing into the tank suddenly switches to fresh water flowing in at a rate of 8 gal/min, while the solution continues to leave at the same rate. Find the Laplace transform of the amount of salt $y(t)$ in the tank.

27. A damped oscillator with mass m, damping constant γ, and spring constant k is subjected to an external force $F(t) = F_0 t$ over the time interval $0 \leq t \leq T$. The external force is then removed at time T. Find the Laplace transform of the displacement $y(t)$ of the mass, assuming that the oscillator is initially in the equilibrium state.

28. The Laplace transforms of certain functions can be found conveniently from their Taylor series expansions.
(a) Using the Taylor series for $\sin t$,

$$\sin t = \sum_{n=0}^{\infty} \frac{(-1)^n t^{2n+1}}{(2n+1)!},$$

and assuming that the Laplace transform of this series can be computed term by term, verify that

$$L\{\sin t\} = \frac{1}{s^2 + 1}, \qquad s > 1.$$

(b) Let

$$f(t) = \begin{cases} (\sin t)/t, & t \neq 0, \\ 1, & t = 0. \end{cases}$$

Find the Taylor series for f about $t = 0$. Assuming that the Laplace transform of this function can be computed term by term, verify that

$$L\{f(t)\} = \arctan(1/s), \qquad s > 1.$$

(c) The Bessel function of the first kind of order zero, J_0, has the Taylor series

$$J_0(t) = \sum_{n=0}^{\infty} \frac{(-1)^n t^{2n}}{2^{2n}(n!)^2}.$$

Assuming that the following Laplace transforms can be computed term by term, verify that

$$L\{J_0(t)\} = (s^2 + 1)^{-1/2}, \qquad s > 1,$$

and

$$L\{J_0(\sqrt{t})\} = s^{-1}e^{-1/4s}, \qquad s > 0.$$

29. For each of the following initial value problems, use Theorem 5.2.4 to find the differential equation satisfied by $Y(s) = \mathcal{L}\{\phi(t)\}$, where $y = \phi(t)$ is the solution of the given initial value problem.

(a) $y'' - ty = 0;$ $y(0) = 1,$ $y'(0) = 0$ (Airy's equation)
(b) $(1 - t^2)y'' - 2ty' + \alpha(\alpha + 1)y = 0;$ $y(0) = 0,$ $y'(0) = 1$
(Legendre's equation) Note that the differential equation for $Y(s)$ is of first order in part (a), but of second order in part (b). This is due to the fact that t appears at most to the first power in the equation of part (a), whereas it appears to the second power in that of part (b). This illustrates that the Laplace transform is not often useful in solving differential equations with variable coefficients, unless all the coefficients are at most linear functions of the independent variable.

5.3 The Inverse Laplace Transform

In the preceding section we showed how the linearity of $\mathcal{L}$ and Corollary 5.2.3 provide the tools to convert linear differential equations with constant coefficients into algebraic equations in the s-domain. For example, applying these tools to the initial value problem in Example 3,

$$y'' + 2y' + 5y = e^{-t}, \qquad y(0) = 1, \quad y'(0) = -3, \tag{1}$$

led to the algebraic equation

$$(s^2 + 2s + 5)Y(s) - s + 1 = \frac{1}{s + 1}. \tag{2}$$

The solution of Eq. (2),

$$Y(s) = \frac{s^2}{(s + 1)(s^2 + 2s + 5)},$$

is presumably the Laplace transform of the solution of the initial value problem (1). Determining the function $y = \phi(t)$ corresponding to the transform $Y(s)$ is the main difficulty in solving initial value problems by the transform technique. This problem is known as the inversion problem for the Laplace transform. The following theorem, which we state without proof, allows us to define the notion of the inverse Laplace transform.

▶ **Existence of the Inverse Transform**

THEOREM 5.3.1	If $f(t)$ and $g(t)$ are piecewise continuous and of exponential order on $[0, \infty)$ and $F = G$ where $F = \mathcal{L}\{f\}$ and $G = \mathcal{L}\{g\}$, then $f(t) = g(t)$ at all points where both f and g are continuous. In particular, if f and g are continuous on $[0, \infty)$, then $f(t) = g(t)$ for all $t \in [0, \infty)$.

Remark. If $F = G$ and f and g are piecewise continuous on $[0, \infty)$, we identify f with g even though they may differ at a finite number of points or even at an infinite number of points such as $t = 1, 2, 3, \ldots$ but they cannot be different over any interval of positive length on the t-axis. For example, the following two functions, which have different values at $t = 1$,

$$f(t) = \begin{cases} 1, & 0 \le t < 1, \\ 0, & 1 \le t, \end{cases} \quad \text{and} \quad g(t) = \begin{cases} 1, & 0 \le t \le 1, \\ 0, & 1 < t, \end{cases}$$

both have the same Laplace transform, $(1 - e^{-s})/s$. An infinite number of functions can be constructed, each having $(1 - e^{-s})/s$ as its transform. However, in a practical sense, these functions are essentially the same. This "lack of uniqueness" is relatively harmless and causes no problems in the application of Laplace transforms.

Theorem 5.3.1 justifies the following definition.

DEFINITION 5.3.2

If $f(t)$ is piecewise continuous and of exponential order on $[0, \infty)$ and $\mathcal{L}\{f(t)\} = F(s)$, then we call f the **inverse Laplace transform**[1] of F, and denote it by

$$f = \mathcal{L}^{-1}\{F\}.$$

The fact that there is a one-to-one correspondence (essentially) between functions and their Laplace transforms suggests the compilation of a table, such as Table 5.3.1, giving the transforms of functions frequently encountered, and vice versa. The entries in the second column of Table 5.3.1 are the transforms of those in the first column. Perhaps more important, the functions in the first column are the inverse transforms of those in the second column. Thus, for example, if the transform of the solution of a differential equation is known, the solution itself can often be found merely by looking it up in the table. Some of the entries in Table 5.3.1 have been produced in examples and theorems, or appear as problems, in Sections 5.1 and 5.2. Others will be developed later in the chapter. The third column of the table indicates where the derivation of the given transforms may be found. Although Table 5.3.1 is sufficient for the examples and problems in this book, much larger tables are also available. Powerful computer algebra systems can also be used to find both Laplace transforms and inverse Laplace transforms. There also exist numerical algorithms for approximating inverse Laplace transforms that cannot be found using either a table or a computer algebra system.

EXAMPLE 1

Determine $\mathcal{L}^{-1}\{F\}$, where

(a) $F(s) = \dfrac{4}{s^2 + 16}$, (b) $F(s) = \dfrac{6}{(s + 2)^4}$, (c) $F(s) = \dfrac{s + 1}{s^2 + 2s + 5}$.

(a) Using line 5 of Table 5.3.1 gives

$$\mathcal{L}^{-1}\left\{\frac{4}{s^2 + 16}\right\} = \mathcal{L}^{-1}\left\{\frac{4}{s^2 + 4^2}\right\} = \sin 4t.$$

---

[1]There is a general formula for the inverse Laplace transform that requires integration in the complex s-plane, but its use requires a knowledge of the theory of functions of a complex variable, and we do not consider it in this book. It is possible, however, to develop many important properties of the Laplace transform, and to solve many interesting problems, without the use of complex variables.

| TABLE 5.3.1 | Elementary Laplace transforms. |

	$f(t) = \mathcal{L}^{-1}\{F(s)\}$	$F(s) = \mathcal{L}\{f(t)\}$	Notes
1.	1	$\dfrac{1}{s}, \quad s > 0$	Sec. 5.1; Ex. 4
2.	e^{at}	$\dfrac{1}{s-a}, \quad s > a$	Sec. 5.1; Ex. 5
3.	$t^n, \quad n = \text{positive integer}$	$\dfrac{n!}{s^{n+1}}, \quad s > 0$	Sec. 5.2; Cor. 5.2.5
4.	$t^p, \quad p > -1$	$\dfrac{\Gamma(p+1)}{s^{p+1}}, \quad s > 0$	Sec. 5.1; Prob. 37
5.	$\sin at$	$\dfrac{a}{s^2 + a^2}, \quad s > 0$	Sec. 5.1; Ex. 7
6.	$\cos at$	$\dfrac{s}{s^2 + a^2}, \quad s > 0$	Sec. 5.1; Prob. 22
7.	$\sinh at$	$\dfrac{a}{s^2 - a^2}, \quad s > |a|$	Sec. 5.1; Prob. 19
8.	$\cosh at$	$\dfrac{s}{s^2 - a^2}, \quad s > |a|$	Sec. 5.1; Prob. 18
9.	$e^{at}\sin bt$	$\dfrac{b}{(s-a)^2 + b^2}, \quad s > a$	Sec. 5.1; Prob. 23
10.	$e^{at}\cos bt$	$\dfrac{s-a}{(s-a)^2 + b^2}, \quad s > a$	Sec. 5.1; Prob. 24
11.	$t^n e^{at}, \quad n = \text{positive integer}$	$\dfrac{n!}{(s-a)^{n+1}}, \quad s > a$	Sec. 5.2; Prob. 8
12.	$u_c(t)$	$\dfrac{e^{-cs}}{s}, \quad s > 0$	Sec. 5.5; Eq. (4)
13.	$u_c(t)f(t - c)$	$e^{-cs}F(s)$	Sec. 5.5; Eq. (6)
14.	$e^{ct}f(t)$	$F(s - c)$	Sec. 5.2; Thm. 5.2.1
15.	$\displaystyle\int_0^t f(t - \tau)g(\tau)\,d\tau$	$F(s)G(s)$	Sec. 5.6; Thm. 5.8.3
16.	$\delta(t - c)$	e^{-cs}	Sec. 5.7; Eq. (14)
17.	$f^{(n)}(t)$	$s^n F(s) - s^{n-1}f(0)$ $- \cdots - f^{(n-1)}(0)$	Sec. 5.2; Cor. 5.2.3
18.	$t^n f(t)$	$(-1)^n F^{(n)}(s)$	Sec. 5.2; Thm. 5.2.4

(b) Using either line 11 or lines 3 and 14 of Table 5.3.1, we find that

$$\mathcal{L}^{-1}\left\{\frac{6}{(s+2)^4}\right\} = \mathcal{L}^{-1}\left\{\frac{3!}{s^4}\Big|_{s\to s+2}\right\} = e^{-2t}\mathcal{L}^{-1}\left\{\frac{3!}{s^4}\right\} = e^{-2t}t^3.$$

(c) Completing the square in the denominator, $s^2 + 2s + 5 = (s+1)^2 + 4$, explicitly reveals the translation $s \to s + 1$. Then using either lines 6 and 14 or line 10 of Table 5.3.1 gives

$$\mathcal{L}^{-1}\left\{\frac{s+1}{s^2 + 2s + 5}\right\} = \mathcal{L}^{-1}\left\{\frac{s+1}{(s+1)^2 + 2^2}\right\} = \mathcal{L}^{-1}\left\{\frac{s}{s^2 + 2^2}\Big|_{s\to s+1}\right\}$$

$$= e^{-t}\mathcal{L}^{-1}\left\{\frac{s}{s^2 + 2^2}\right\} = e^{-t}\cos 2t.$$

▶ Linearity of $\mathcal{L}^{-1}$. A table of Laplace transforms, such as Table 5.3.1, facilitates the task of finding the inverse Laplace transform of transforms that can easily be put into the form of those transforms of basic functions that appear in the table. However, it is not possible to create a table large enough to include all of the transforms that occur in applications. The next theorem states that $\mathcal{L}^{-1}$ is a linear operator. This property extends $\mathcal{L}^{-1}$ to all linear combinations of transforms that appear in the table.

**THEOREM
5.3.3**

Assume that $f_1 = \mathcal{L}^{-1}\{F_1\}$ and $f_2 = \mathcal{L}^{-1}\{F_2\}$ are piecewise continuous and of exponential order on $[0, \infty)$. Then for any constants c_1 and c_2,

$$\mathcal{L}^{-1}\{c_1 F_1 + c_2 F_2\} = c_1 \mathcal{L}^{-1}\{F_1\} + c_2 \mathcal{L}^{-1}\{F_2\} = c_1 f_1 + c_2 f_2.$$

Proof

Using the linearity of $\mathcal{L}$, we have

$$\mathcal{L}\{c_1 f_1 + c_2 f_2\} = c_1 F_1 + c_2 F_2.$$

Since $c_1 f_1 + c_2 f_2$ is piecewise continuous and of exponential order on $[0, \infty)$, the result follows from Definition 5.3.2.

Remark. Equality in the conclusion of Theorem 5.3.3 is in the sense discussed in the remark immediately following Theorem 5.3.1.

Theorem 5.3.3 extends to a linear combination of n Laplace transforms:

$$\mathcal{L}^{-1}\{c_1 F_1 + \cdots + c_n F_n\} = c_1 \mathcal{L}^{-1}\{F_1\} + \cdots + c_n \mathcal{L}^{-1}\{F_n\}.$$

**EXAMPLE
2**

Determine $\mathcal{L}^{-1}\left\{ \dfrac{2}{(s+2)^4} + \dfrac{3}{s^2+16} + \dfrac{5(s+1)}{s^2+2s+5} \right\}$.

Using the linearity of $\mathcal{L}^{-1}$ and the results of Example 1 gives

$$\mathcal{L}^{-1}\left\{ \frac{2}{(s+2)^4} + \frac{3}{s^2+16} + \frac{5(s+1)}{s^2+2s+5} \right\}$$

$$= \frac{1}{3}\mathcal{L}^{-1}\left\{ \frac{6}{(s+2)^4} \right\} + \frac{3}{4}\mathcal{L}^{-1}\left\{ \frac{4}{s^2+16} \right\} + 5\mathcal{L}^{-1}\left\{ \frac{s+1}{s^2+2s+5} \right\}$$

$$= \frac{1}{3}e^{-2t}t^3 + \frac{3}{4}\sin 4t + 5e^{-t}\cos 2t.$$

**EXAMPLE
3**

Find the inverse Laplace transform of $F(s) = \dfrac{3s+1}{s^2-4s+20}$.

By completing the square, the denominator can be written as $(s-2)^2 + 16$. This explicitly exposes the translation $s \to s-2$ in the s-domain. The same translation is introduced in the numerator by rewriting it as $3s + 1 = 3(s-2) + 7$. Thus

$$F(s) = \frac{3s+1}{s^2-4s+20} = \frac{3(s-2)+7}{(s-2)^2+16} = 3\frac{s-2}{(s-2)^2+4^2} + \frac{7}{4}\frac{4}{(s-2)^2+4^2}.$$

Using the linearity of $\mathcal{L}^{-1}$ and using lines 9 and 10 in Table 5.3.1, we find that

$$\mathcal{L}^{-1}\left\{\frac{3s+1}{s^2-4s+20}\right\}$$

$$= 3\mathcal{L}^{-1}\left\{\frac{s-2}{(s-2)^2+4^2}\right\} + \frac{7}{4}\mathcal{L}^{-1}\left\{\frac{4}{(s-2)^2+4^2}\right\}$$

$$= 3e^{2t}\cos 4t + \frac{7}{4}e^{2t}\sin 4t.$$

Partial Fractions

Most of the Laplace transforms that arise in the study of differential equations are rational functions of s, that is, functions of the form

$$F(s) = \frac{P(s)}{Q(s)}, \tag{3}$$

where $Q(s)$ is a polynomial of degree n and $P(s)$ is a polynomial of degree less than n. For example, the Laplace transform found in Example 3, Section 5.2,

$$Y(s) = \frac{s^2}{(s+1)(s^2+2s+5)}, \tag{4}$$

is of the form (3) with $P(s) = s^2$ and $Q(s) = (s+1)(s^2+2s+5)$. Recall from algebra (and calculus) that the method of partial fractions involves the decomposition of rational functions into an equivalent sum of simpler rational functions that have numerators which are of degree 1 or zero and denominators of degree 1 or 2, possibly raised to an integer power. The inverse Laplace transforms of these simpler rational functions can then often be found in a Laplace transform table.

EXAMPLE 4

The partial fraction expansion of the rational function (4),

$$Y(s) = \frac{s^2}{(s+1)(s^2+2s+5)} = \frac{1}{4}\frac{1}{s+1} + \frac{3}{4}\frac{s+1}{(s+1)^2+4} - \frac{2}{(s+1)^2+4},$$

can be directly verified by placing the terms on the right over the common denominator $(s+1)(s^2+2s+5) = (s+1)[(s+1)^2+4]$. Then, using the linearity of $\mathcal{L}^{-1}$ and lines 2, 9, and 10 in Table 5.3.1, we find that

$$\mathcal{L}^{-1}\{Y(s)\} = \mathcal{L}^{-1}\left\{\frac{s^2}{(s+1)(s^2+2s+5)}\right\}$$

$$= \frac{1}{4}\mathcal{L}^{-1}\left\{\frac{1}{s+1}\right\} + \frac{3}{4}\mathcal{L}^{-1}\left\{\frac{s+1}{(s+1)^2+4}\right\} - \mathcal{L}^{-1}\left\{\frac{2}{(s+1)^2+4}\right\}$$

$$= \frac{1}{4}e^{-t} + \frac{3}{4}e^{-t}\cos 2t - e^{-t}\sin 2t.$$

The last expression is the solution of the initial value problem posed in Example 3, Section 5.2, $y'' + 2y' + 5y = e^{-t}$, $y(0) = 1$, $y'(0) = -3$.

EXAMPLE
5

The partial fraction expansion of $Y(s)$ found in Example 5, Section 5.2,

$$Y(s) = \frac{1}{s^4 - 1} = \frac{1}{4}\frac{1}{s-1} - \frac{1}{4}\frac{1}{s+1} - \frac{1}{2}\frac{1}{s^2+1},$$

is verified by placing the terms on the right over a common denominator. Using the linearity of $\mathcal{L}^{-1}$ and lines 2 and 5 of Table 5.3.1, we then find that $\mathcal{L}^{-1}\{Y(s)\}$ is given by

$$\mathcal{L}^{-1}\left\{\frac{1}{s^4-1}\right\} = \frac{1}{4}\mathcal{L}^{-1}\left\{\frac{1}{s-1}\right\} - \frac{1}{4}\mathcal{L}^{-1}\left\{\frac{1}{s+1}\right\} - \frac{1}{2}\mathcal{L}^{-1}\left\{\frac{1}{s^2+1}\right\}$$

$$= \frac{1}{4}e^t - \frac{1}{4}e^{-t} - \frac{1}{2}\sin t.$$

Substitution shows that the last expression satisfies the initial value problem (14) and (15) in Example 5, Section 5.2.

We now review systematic methods for finding partial fraction decompositions of rational functions. Each expansion problem can be solved by considering the following three cases that arise:

1. **Nonrepeated Linear Factors.** If the denominator $Q(s)$ in Eq. (3) admits the factorization

$$Q(s) = (s - s_1)(s - s_2) \cdots (s - s_n),$$

where $s_1, s_2, \ldots, s_n$ are distinct, then $F(s)$ can be expanded as

$$F(s) = \frac{P(s)}{Q(s)} = \frac{a_1}{s - s_1} + \frac{a_2}{s - s_2} + \cdots + \frac{a_n}{s - s_n}, \tag{5}$$

where the a_j are constants that need to be determined.

2. **Repeated Linear Factors.** If any root s_j of $Q(s)$ is of multiplicity k, that is, the factor $s - s_j$ appears exactly k times in the factorization of $Q(s)$, then the jth term in the right-hand side of Eq. (5) must be changed to

$$\frac{a_{j_1}}{s - s_j} + \frac{a_{j_2}}{(s - s_j)^2} + \cdots + \frac{a_{j_k}}{(s - s_j)^k}, \tag{6}$$

where the constants $a_{j_1}, \ldots, a_{j_k}$ need to be determined.

3. **Quadratic Factors.** Complex conjugate roots of $Q(s)$, $\mu_j + i\nu_j$ and $\mu_j - i\nu_j$ where $\nu_j \neq 0$, give rise to quadratic factors of $Q(s)$ of the form

$$\left[s - (\mu_j + i\nu_j)\right]\left[s - (\mu_j - i\nu_j)\right] = (s - \mu_j)^2 + \nu_j^2.$$

If the roots $\mu_j + i\nu_j$ and $\mu_j - i\nu_j$ are of multiplicity k, that is, k is the highest power of $(s - \mu_j)^2 + \nu_j^2$ that divides $Q(s)$, then the partial fraction expansion of $F(s)$ must include the terms

$$\frac{a_{j_1}(s - \mu_j) + b_{j_1}\nu_j}{(s - \mu_j)^2 + \nu_j^2} + \frac{a_{j_2}(s - \mu_j) + b_{j_2}\nu_j}{\left[(s - \mu_j)^2 + \nu_j^2\right]^2} + \cdots + \frac{a_{j_k}(s - \mu_j) + b_{j_k}\nu_j}{\left[(s - \mu_j)^2 + \nu_j^2\right]^k}, \tag{7}$$

where the constants $a_{j_1}, b_{j_1}, \ldots, a_{j_k}, b_{j_k}$ need to be determined.

In the following examples, we illustrate two methods commonly used to find linear systems of algebraic equations for the undetermined coefficients.

EXAMPLE 6

Find the partial fraction decomposition of $F(s) = (s - 2)/(s^2 + 4s - 5)$ and compute $\mathcal{L}^{-1}\{F\}$.

Since $s^2 + 4s - 5 = (s - 1)(s + 5)$, we seek the partial fraction decomposition

$$\frac{s - 2}{s^2 + 4s - 5} = \frac{a_1}{s - 1} + \frac{a_2}{s + 5}. \tag{8}$$

The terms on the right-hand side of Eq. (8) can be combined by using the common denominator $(s - 1)(s + 5) = s^2 + 4s - 5$,

$$\frac{s - 2}{s^2 + 4s - 5} = \frac{a_1(s + 5) + a_2(s - 1)}{s^2 + 4s - 5}. \tag{9}$$

Equality in Eq. (9) holds if and only if the numerators of the rational functions are equal,

$$s - 2 = a_1(s + 5) + a_2(s - 1). \tag{10}$$

Equation (10) is also easily obtained by multiplying both sides of Eq. (8) by $s^2 + 4s - 5 = (s + 5)(s - 1)$ and canceling those factors common to both numerator and denominator.

Matching Polynomial Coefficients. Since two polynomials are equal if and only if coefficients of corresponding powers of s are equal, we rewrite Eq. (10) in the form $s - 2 = (a_1 + a_2)s + 5a_1 - a_2$. Matching coefficients of $s^0 = 1$ and $s^1 = s$ yields the pair of equations

$$s^0 : -2 = 5a_1 - a_2$$
$$s^1 : \ \ 1 = a_1 + a_2.$$

The solution of this system is $a_1 = -\frac{1}{6}$ and $a_2 = \frac{7}{6}$, so

$$\frac{s - 2}{s^2 + 4s - 5} = -\frac{1}{6}\frac{1}{s - 1} + \frac{7}{6}\frac{1}{s + 5}.$$

Matching Function Values. The functions of s that appear on both sides of Eq. (10) must be equal for all values of s. Since there are two unknowns to be determined, a_1 and a_2, we evaluate Eq. (10) at two different values of s to obtain a system of two independent equations for a_1 and a_2. Although there are many possibilities, choosing values of s corresponding to zeros of $Q(s)$ often yields a particularly simple system,

$$s = 1 : -1 = 6a_1$$
$$s = -5 : -7 = -6a_2.$$

Thus $a_1 = -\frac{1}{6}$ and $a_2 = \frac{7}{6}$ as before. If we choose to evaluate Eq. (10) at $s = 0$ and $s = -1$ instead, we obtain the more complicated system of equations

$$s = 0 : -2 = 5a_1 - a_2$$
$$s = -1 : -3 = 4a_1 - 2a_2$$

with the same solution obtained above. It follows that

$$\mathcal{L}^{-1}\left\{ \frac{s - 2}{s^2 + 4s - 5} \right\} = -\frac{1}{6}e^t + \frac{7}{6}e^{-5t}.$$

EXAMPLE 7

Determine $\mathcal{L}^{-1}\left\{\dfrac{s^2 + 20s + 31}{(s+2)^2(s-3)}\right\}$.

Since $s + 2$ is a linear factor of multiplicity 2 and $s - 3$ is a linear factor of multiplicity 1 in the denominator, the appropriate form for the partial fraction decomposition is

$$\frac{s^2 + 20s + 31}{(s+2)^2(s-3)} = \frac{a}{s+2} + \frac{b}{(s+2)^2} + \frac{c}{s-3}.$$

Multiplying both sides by $(s+2)^2(s-3)$ and canceling factors common to numerator and denominator yield the equation

$$s^2 + 20s + 31 = a(s+2)(s-3) + b(s-3) + c(s+2)^2. \tag{11}$$

Matching Function Values. If we choose to match function values in Eq. (11) at $s = -2$, $s = 3$, and $s = 0$, we obtain the system

$$\begin{aligned}
s = -2: \ -5 &= \qquad\ -5b \\
s = 3: \ 100 &= \qquad\qquad\qquad 25c \\
s = 0: \ \ 31 &= -6a - 3b + \ \ 4c,
\end{aligned}$$

which has the solution $a = -3$, $b = 1$, and $c = 4$.

Matching Polynomial Coefficients. If Eq. (11) is written in the form

$$s^2 + 20s + 31 = (a+c)s^2 + (-a+b+4c)s - 6a - 3b + 4c$$

and we match polynomial coefficients, we get the system

$$\begin{aligned}
s^0: \ 31 &= -6a - 3b + 4c \\
s^1: \ 20 &= \ -a + \ \ b + 4c \\
s^2: \ \ 1 &= \quad a \qquad\ + \ c,
\end{aligned}$$

which has the same solution, $a = -3$, $b = 1$, and $c = 4$.

By either method

$$\frac{s^2 + 20s + 31}{(s+2)^2(s-3)} = -\frac{3}{s+2} + \frac{1}{(s+2)^2} + \frac{4}{s-3}. \tag{12}$$

Applying $\mathcal{L}^{-1}$ to the partial fraction decomposition in Eq. (12) then yields

$$\mathcal{L}^{-1}\left\{\frac{s^2 + 20s + 31}{(s+2)^2(s-3)}\right\} = -3\mathcal{L}^{-1}\left\{\frac{1}{s+2}\right\} + \mathcal{L}^{-1}\left\{\frac{1}{(s+2)^2}\right\} + 4\mathcal{L}^{-1}\left\{\frac{1}{s-3}\right\}$$

$$= -3e^{-2t} + te^{-2t} + 4e^{3t}.$$

EXAMPLE 8

Determine $\mathcal{L}^{-1}\left\{\dfrac{14s^2 + 70s + 134}{(2s+1)(s^2 + 6s + 13)}\right\}$.

Since the discriminant of the quadratic factor $s^2 + 6s + 13$ is negative, its roots are complex, so we complete the square, writing it as $(s+3)^2 + 2^2$. Thus we assume a partial fraction decomposition of the form

$$\frac{14s^2 + 70s + 134}{(2s+1)(s^2 + 6s + 13)} = \frac{a(s+3) + b \cdot 2}{(s+3)^2 + 2^2} + \frac{c}{2s+1}.$$

Using $a(s + 3) + b \cdot 2$ for the linear factor in the numerator on the right instead of $as + b$ anticipates table entries having the forms

$$\frac{(s - \beta)}{(s - \beta)^2 + \alpha^2} \quad \text{and} \quad \frac{\alpha}{(s - \beta)^2 + \alpha^2}$$

with $\alpha = 2$ and $\beta = -3$. Multiplying both sides by $(2s + 1)(s^2 + 6s + 13) = (2s + 1)$ $[(s + 3)^2 + 2^2]$ yields the equation

$$14s^2 + 70s + 134 = [a(s + 3) + b \cdot 2](2s + 1) + c\left[(s + 3)^2 + 2^2\right]. \tag{13}$$

Evaluating Eq. (13) at $s = 0$, $s = -3$, and $s = -\frac{1}{2}$ gives the system

$$s = 0 : 134 = 3a + 2b + 13c$$
$$s = -3 : 50 = -10b + 4c$$
$$s = -\frac{1}{2} : \frac{205}{2} = \frac{41}{4}c,$$

which has the solution $c = 10$, $b = -1$, $a = 2$. Thus

$$\frac{14s^2 + 70s + 134}{(2s + 1)(s^2 + 6s + 13)} = 2\frac{(s + 3)}{(s + 3)^2 + 2^2} - \frac{2}{(s + 3)^2 + 2^2} + \frac{10}{2s + 1},$$

so

$$\mathcal{L}^{-1}\left\{\frac{14s^2 + 70s + 134}{(2s + 1)(s^2 + 6s + 13)}\right\}$$

$$= 2\mathcal{L}^{-1}\left\{\frac{(s + 3)}{(s + 3)^2 + 2^2}\right\} - \mathcal{L}^{-1}\left\{\frac{2}{(s + 3)^2 + 2^2}\right\} + 5\mathcal{L}^{-1}\left\{\frac{1}{s + \frac{1}{2}}\right\}$$

$$= 2e^{-3t}\cos 2t - e^{-3t}\sin 2t + 5e^{-t/2}.$$

Thus the appropriate form for the partial fraction decomposition of $P(s)/Q(s)$ is based on the linear and irreducible quadratic factors, counting multiplicities, of $Q(s)$ and is determined by the rules in the cases discussed above. In many textbook problems, where $Q(s)$ has rational roots, for instance, a computer algebra system can be used to find inverse transforms directly without need of a partial fraction decomposition. In real-world applications where the degree of $Q(s)$ is greater than or equal to 3, computer algorithms are available that approximate partial fraction decompositions. A polynomial root finder can also be used to assist in finding a partial fraction decomposition of $P(s)/Q(s)$.

PROBLEMS

In each of Problems 1 through 8, find the unknown constants in the given partial fraction expansion:

1. $\dfrac{3s - 40}{(s + 3)(s - 4)} = \dfrac{a}{s + 3} + \dfrac{b}{s - 4}$

2. $\dfrac{3s + 4}{(s + 2)^2} = \dfrac{a_1}{s + 2} + \dfrac{a_2}{(s + 2)^2}$

3. $\dfrac{-3s^2 + 32 - 14s}{(s + 4)(s^2 + 4)} = \dfrac{a}{s + 4} + \dfrac{bs + c \cdot 2}{s^2 + 4}$

4. $\dfrac{3s^2 + 2s + 2}{s^3} = \dfrac{a_1}{s} + \dfrac{a_2}{s^2} + \dfrac{a_3}{s^3}$

5. $\dfrac{3s^2 - 8s + 5}{(s + 1)(s^2 - 2s + 5)} = \dfrac{a}{s + 1} + \dfrac{b(s - 1) + c \cdot 2}{(s - 1)^2 + 4}$

6. $\dfrac{-2s^3 - 8s^2 + 8s + 6}{(s+3)(s+1)s^2} = \dfrac{a_1}{s} + \dfrac{a_2}{s^2} + \dfrac{b}{s+3} + \dfrac{c}{s+1}$

7. $\dfrac{8s^3 - 15s - s^5}{(s^2+1)^3} = \dfrac{a_1 s + b_1}{s^2+1} + \dfrac{a_2 s + b_2}{(s^2+1)^2} + \dfrac{a_3 s + b_3}{(s^2+1)^3}$

8. $\dfrac{s^3 + 3s^2 + 3s + 1}{(s^2+2s+5)^2} = \dfrac{a_1(s+1) + b_1 \cdot 2}{(s+1)^2 + 4}$
   $\qquad\qquad + \dfrac{a_2(s+1) + b_2 \cdot 2}{[(s+1)^2 + 4]^2}$

In each of Problems 9 through 24, use the linearity of $\mathcal{L}^{-1}$, partial fraction expansions, and Table 5.3.1 to find the inverse Laplace transform of the given function:

9. $\dfrac{30}{s^2 + 25}$

10. $\dfrac{4}{(s-3)^3}$

11. $\dfrac{2}{s^2 + 3s - 4}$

12. $\dfrac{3s}{s^2 - s - 6}$

13. $\dfrac{5s + 25}{s^2 + 10s + 74}$

14. $\dfrac{6s - 3}{s^2 - 4}$

15. $\dfrac{2s + 1}{s^2 - 2s + 2}$

16. $\dfrac{9s^2 - 12s + 28}{s(s^2 + 4)}$

17. $\dfrac{1 - 2s}{s^2 + 4s + 5}$

18. $\dfrac{2s - 3}{s^2 + 2s + 10}$

19. $3\dfrac{3s + 2}{(s-2)(s+2)(s+1)}$

20. $2\dfrac{s^3 - 2s^2 + 8}{s^2(s-2)(s+2)}$

21. $12\dfrac{s^2 - 7s + 28}{(s^2 - 8s + 25)(s+3)}$

22. $2\dfrac{s^3 + 3s^2 + 4s + 3}{(s^2+1)(s^2+4)}$

23. $\dfrac{s^3 - 2s^2 - 6s - 6}{(s^2 + 2s + 2)s^2}$

24. $\dfrac{s^2 + 3}{(s^2 + 2s + 2)^2}$

In each of Problems 25 through 28, use a computer algebra system to find the inverse Laplace transform of the given function:

25. $\dfrac{s^3 - 2s^2 - 6s - 6}{s^4 + 4s^3 + 24s^2 + 40s + 100}$

26. $\dfrac{s^3 - 3s^2 - 6s - 6}{s^7 - 6s^6 + 10s^5}$

27. $\dfrac{s^3 - 2s^2 - 6s - 6}{s^8 - 2s^7 - 2s^6 + 16s^5 - 20s^4 - 8s^3 + 56s^2 - 64s + 32}$

28. $\dfrac{s^3 - 2s^2 - 6s - 6}{s^7 - 5s^6 + 5s^5 - 25s^4 + 115s^3 - 63s^2 + 135s - 675}$

5.4 Solving Differential Equations with Laplace Transforms

The block diagram in Figure 5.0.1 shows the main steps used to solve initial value problems by the method of Laplace transforms. The mathematical steps required to carry out each stage in the process, presented in Sections 5.1, 5.2, and 5.3, are summarized below.

1. Using the linearity of $\mathcal{L}$, its operational properties (e.g., knowing how derivatives transform), and a table of Laplace transforms if necessary, the initial value problem for a linear constant coefficient differential equation is transformed into an algebraic equation in the s-domain.

2. Solving the algebraic equation gives the Laplace transform, say, $Y(s)$, of the solution of the initial value problem. This step is illustrated by Examples 3, 4, and 5 in Section 5.2.

3. The solution of the initial value problem, $y(t) = \mathcal{L}^{-1}\{Y(s)\}$, is found by using partial fraction decompositions, the linearity of $\mathcal{L}^{-1}$, and a table of Laplace transforms. Partial fraction expansions and need for a table can be avoided by using a computer algebra system or other advanced computer software functions to evaluate $\mathcal{L}^{-1}\{Y(s)\}$.

In this section we present examples ranging from first order equations to systems of equations that illustrate the entire process.

EXAMPLE 1

Find the solution of the initial value problem

$$y' + 2y = \sin 4t, \qquad y(0) = 1. \tag{1}$$

Applying the Laplace transform to both sides of the differential equation gives

$$sY(s) - y(0) + 2Y(s) = \frac{4}{s^2 + 16},$$

where we have used line 17 of Table 5.3.1 to transform $y'(t)$. Substituting for $y(0)$ from the initial condition and solving for $Y(s)$, we obtain

$$Y(s) = \frac{1}{s+2} + \frac{4}{(s+2)(s^2+16)} = \frac{s^2+20}{(s+2)(s^2+16)}.$$

The partial fraction expansion of $Y(s)$ is of the form

$$\frac{s^2+20}{(s+2)(s^2+16)} = \frac{a}{s+2} + \frac{bs+c\cdot 4}{s^2+16}, \tag{2}$$

where the linear factor $bs + c \cdot 4$ in the numerator on the right anticipates a table entry of the form $\dfrac{\mu}{s^2+\mu^2}$. Multiplying both sides of Eq. (2) by $(s+2)(s^2+16)$ gives the equation

$$s^2 + 20 = a(s^2 + 16) + (bs + c \cdot 4)(s + 2). \tag{3}$$

Evaluating Eq. (3) at $s = -2$, $s = 0$, and $s = 2$ yields the system

$$\begin{aligned}
s = -2 &: 24 = 20a \\
s = 0 &: 20 = 16a \quad\;\; + \quad 8c \\
s = 2 &: 24 = 20a + 8b + \;\; 16c.
\end{aligned} \tag{4}$$

The solution of the system (4) is $a = \frac{6}{5}$, $b = -\frac{1}{5}$, and $c = \frac{1}{10}$. Therefore the partial fraction expansion of Y is

$$Y(s) = \frac{6}{5}\frac{1}{s+2} - \frac{1}{5}\frac{s}{s^2+16} + \frac{1}{10}\frac{4}{s^2+16}.$$

It follows that the solution of the given initial value problem is

$$y(t) = \frac{6}{5}e^{-2t} - \frac{1}{5}\cos 4t + \frac{1}{10}\sin 4t.$$

EXAMPLE 2

Find the solution of the differential equation

$$y'' + y = e^{-t}\cos 2t, \tag{5}$$

satisfying the initial conditions

$$y(0) = 2, \qquad y'(0) = 1. \tag{6}$$

Taking the Laplace transform of the differential equation, we get

$$s^2 Y(s) - sy(0) - y'(0) + Y(s) = \frac{s+1}{(s+1)^2 + 4}.$$

Substituting for $y(0)$ and $y'(0)$ from the initial conditions and solving for $Y(s)$, we obtain

$$Y(s) = \frac{2s+1}{s^2+1} + \frac{s+1}{(s^2+1)\left[(s+1)^2+4\right]}. \tag{7}$$

Instead of using a common denominator to combine terms on the right side of (7) into a single rational function, we choose to compute their inverse Laplace transforms separately. The partial fraction expansion of $(2s+1)/(s^2+1)$ can be written down by observation

$$\frac{2s+1}{s^2+1} = 2\frac{s}{s^2+1} + \frac{1}{s^2+1}.$$

Thus

$$\mathcal{L}^{-1}\left\{\frac{2s+1}{s^2+1}\right\} = 2\mathcal{L}^{-1}\left\{\frac{s}{s^2+1}\right\} + \mathcal{L}^{-1}\left\{\frac{1}{s^2+1}\right\} = 2\cos t + \sin t. \tag{8}$$

The form of the partial fraction expansion for the second term on the right-hand side of Eq. (7) is

$$\frac{s+1}{(s^2+1)[(s+1)^2+4]} = \frac{as+b}{s^2+1} + \frac{c(s+1)+d\cdot 2}{(s+1)^2+4}. \tag{9}$$

Multiplying both sides of Eq. (9) by $(s^2+1)[(s+1)^2+4]$ and canceling factors common to denominator and numerator on the right yield

$$s+1 = (as+b)[(s+1)^2+4] + [c(s+1)+d\cdot 2](s^2+1).$$

Expanding and collecting coefficients of like powers of s then give the polynomial equation

$$s+1 = (a+c)s^3 + (2a+b+c+2d)s^2 + (5a+2b+c)s + 5b+c+2d.$$

Comparing coefficients of like powers of s, we have

$$\begin{aligned}
s^0: &\quad\quad 5b+c+2d = 1,\\
s^1: &\; 5a+2b+c \quad\quad = 1,\\
s^2: &\; 2a+\;\; b+c+2d = 0,\\
s^3: &\quad a \quad\quad +c \quad\quad = 0.
\end{aligned}$$

Consequently, $a = \frac{1}{10}$, $b = \frac{3}{10}$, $c = -\frac{1}{10}$, and $d = -\frac{1}{5}$, from which it follows that

$$\frac{s+1}{(s^2+1)\left[(s+1)^2+4\right]} = \frac{1}{10}\frac{s}{s^2+1} + \frac{3}{10}\frac{1}{s^2+1} - \frac{1}{10}\frac{s+1}{(s+1)^2+4} - \frac{1}{5}\frac{2}{(s+1)^2+4}.$$

Thus

$$\mathcal{L}^{-1}\left\{\frac{s+1}{(s^2+1)\left[(s+1)^2+4\right]}\right\} = \frac{1}{10}\cos t + \frac{3}{10}\sin t - \frac{1}{10}e^{-t}\cos 2t - \frac{1}{5}e^{-t}\sin 2t. \tag{10}$$

The solution of the given initial value problem follows by summing the results in Eqs. (8) and (10),

$$y(t) = \frac{21}{10}\cos t + \frac{13}{10}\sin t - \frac{1}{10}e^{-t}\cos 2t - \frac{1}{5}e^{-t}\sin 2t.$$

Characteristic Polynomials and Laplace Transforms of Differential Equations

Consider the general second order linear equation with constant coefficients

$$ay'' + by' + cy = f(t), \tag{11}$$

with initial conditions prescribed by

$$y(0) = y_0, \qquad y'(0) = y_1. \tag{12}$$

Taking the Laplace transform of Eq. (11), we obtain

$$a[s^2 Y(s) - sy(0) - y'(0)] + b[sY(s) - y(0)] + cY(s) = F(s) \tag{13}$$

or

$$(as^2 + bs + c)Y(s) - (as + b)y(0) - ay'(0) = F(s), \tag{14}$$

where $F(s)$ is the transform of $f(t)$. By solving Eq. (14) for $Y(s)$, we find that

$$Y(s) = \frac{(as + b)y(0) + ay'(0)}{as^2 + bs + c} + \frac{F(s)}{as^2 + bs + c}. \tag{15}$$

Some important observations can be made about the steps leading from Eq. (11) to Eq. (15). First, observe that the polynomial $Z(s) = as^2 + bs + c$ that multiplies $Y(s)$ in Eq. (14) and subsequently appears in the denominator on the right side of Eq. (15) is precisely the characteristic polynomial associated with Eq. (11). Second, note that the coefficient $as + b$ of $y(0)$ in the numerator in the first term on the right-hand side of Eq. (15) can be obtained by canceling the constant c in $Z(s)$ and then dividing by s,

$$\frac{as^2 + bs + \cancel{c}}{s} = as + b.$$

Similarly, the coefficient a of $y'(0)$ in the numerator in the first term on the right side of Eq. (15) can be obtained by canceling the constant in $as + b$ and again dividing by s,

$$\frac{as + \cancel{b}}{s} = a.$$

Observance of this pattern allows us to pass directly from Eq. (11) to Eq. (15). Substituting for $y(0)$ and $y'(0)$ from the initial conditions then gives

$$Y(s) = \frac{(as + b)y_0 + ay_1}{as^2 + bs + c} + \frac{F(s)}{as^2 + bs + c}.$$

Since the use of a partial fraction expansion of $Y(s)$ to determine $y(t)$ requires us to factor the polynomial $Z(s) = as^2 + bs + c$, the use of Laplace transforms does not avoid the necessity of finding roots of the characteristic equation.

The pattern described above extends easily to nth order linear equations with constant coefficients,

$$a_n y^{(n)} + a_{n-1} y^{(n-1)} + \cdots + a_1 y' + a_0 y = f(t). \tag{16}$$

Since the characteristic polynomial associated with Eq. (16) is

$$Z(s) = a_n s^n + a_{n-1} s^{n-1} + \cdots + a_1 s + a_0,$$

the Laplace transform $Y(s) = \mathcal{L}\{y(t)\}$ is given by

$$Y(s) = \frac{\left[a_n s^{n-1} + \cdots + a_1\right] y(0) + \cdots + \left[a_n s + a_{n-1}\right] y^{(n-2)}(0) + a_n y^{(n-1)}(0)}{a_n s^n + a_{n-1} s^{n-1} + \cdots + a_1 s + a_0}$$

$$+ \frac{F(s)}{a_n s^n + a_{n-1} s^{n-1} + \cdots + a_1 s + a_0}.$$

EXAMPLE 3

Find the solution of the differential equation

$$y'''' + 2y'' + y = 0 \tag{17}$$

subject to the initial conditions

$$y(0) = 1, \quad y'(0) = -1, \quad y''(0) = 0, \quad y'''(0) = 2. \tag{18}$$

Since the characteristic polynomial associated with Eq. (17) is $s^4 + 2s^2 + 1$, $Y(s)$ is given by

$$Y(s) = \frac{(s^3 + 2s)y(0) + (s^2 + 2)y'(0) + sy''(0) + y'''(0)}{s^4 + 2s^2 + 1}. \tag{19}$$

Substituting the prescribed values for the initial conditions then gives

$$Y(s) = \frac{s^3 - s^2 + 2s}{s^4 + 2s^2 + 1} = \frac{s^3 - s^2 + 2s}{(s^2 + 1)^2}. \tag{20}$$

The form for the partial fraction expansion of $Y(s)$ is

$$\frac{s^3 - s^2 + 2s}{(s^2 + 1)^2} = \frac{as + b}{s^2 + 1} + \frac{cs + d}{(s^2 + 1)^2}. \tag{21}$$

Thus the undetermined coefficients must be selected to satisfy

$$s^3 - s^2 + 2s = as^3 + bs^2 + (a + c)s + b + d.$$

Equating coefficients of like powers of s gives $a = 1$, $b = -1$, $a + c = 2$, and $b + d = 0$. The latter two equations require that $c = 1$ and $d = 1$ so that

$$Y(s) = \frac{s - 1}{s^2 + 1} + \frac{s + 1}{(s^2 + 1)^2}. \tag{22}$$

It is clear that $\mathcal{L}^{-1}\{(s - 1)/(s^2 + 1)\} = \cos t - \sin t$ but the second term on the right requires some consideration. From line 18 in Table 5.3.1, we know that

$$\mathcal{L}\{t \sin t\} = -\frac{d}{ds}\frac{1}{s^2 + 1} = \frac{2s}{(s^2 + 1)^2},$$

so

$$\mathcal{L}^{-1}\left\{\frac{s}{(s^2 + 1)^2}\right\} = \frac{1}{2}t \sin t.$$

Similarly,

$$\mathcal{L}\{t \cos t\} = -\frac{d}{ds}\frac{s}{s^2 + 1} = \frac{s^2 - 1}{(s^2 + 1)^2} = \frac{(s^2 + 1) - 2}{(s^2 + 1)^2} = \frac{1}{s^2 + 1} - 2\frac{1}{(s^2 + 1)^2}$$

so that

$$\mathcal{L}^{-1}\left\{\frac{1}{(s^2 + 1)^2}\right\} = \frac{1}{2}\sin t - \frac{1}{2}t \cos t.$$

It follows that the solution of the given initial value problem is

$$y(t) = \mathcal{L}^{-1}\left\{\frac{s - 1}{s^2 + 1}\right\} + \mathcal{L}^{-1}\left\{\frac{s + 1}{(s^2 + 1)^2}\right\} = \left(1 - \frac{1}{2}t\right)\cos t - \frac{1}{2}(1 - t)\sin t.$$

The same result can also be obtained by using a computer algebra system to invert the Laplace transform (22).

▶ Laplace Transforms of Systems of Differential Equations. The Laplace transform method easily extends to systems of equations. Consider the initial value problem

$$y_1' = a_{11}y_1 + a_{12}y_2 + f_1(t), \qquad y_1(0) = y_{10},$$
$$y_2' = a_{21}y_1 + a_{22}y_2 + f_2(t), \qquad y_2(0) = y_{20}. \tag{23}$$

Taking the Laplace transform of each equation in the system (23) gives

$$sY_1 - y_1(0) = a_{11}Y_1 + a_{12}Y_2 + F_1(s),$$
$$sY_2 - y_2(0) = a_{21}Y_1 + a_{22}Y_2 + F_2(s). \tag{24}$$

The pair of equations (24) may be rewritten in the form

$$(s - a_{11})Y_1 - a_{12}Y_2 = y_{10} + F_1(s),$$
$$-a_{21}Y_1 + (s - a_{22})Y_2 = y_{20} + F_2(s), \tag{25}$$

where we have also substituted for the initial conditions specified in Eqs. (23). Using matrix notation, the algebraic system (25) is conveniently expressed by

$$(s\mathbf{I} - \mathbf{A})\mathbf{Y} = \mathbf{y}_0 + \mathbf{F}(s), \tag{26}$$

where

$$\mathbf{Y} = \begin{pmatrix} Y_1 \\ Y_2 \end{pmatrix}, \qquad \mathbf{A} = \begin{pmatrix} a_{11} & a_{12} \\ a_{21} & a_{22} \end{pmatrix}, \qquad \mathbf{y}_0 = \begin{pmatrix} y_{10} \\ y_{20} \end{pmatrix}, \qquad \mathbf{F}(s) = \begin{pmatrix} F_1(s) \\ F_2(s) \end{pmatrix}$$

and $\mathbf{I}$ is the 2×2 identity matrix. It follows that

$$\mathbf{Y} = (s\mathbf{I} - \mathbf{A})^{-1}\mathbf{y}_0 + (s\mathbf{I} - \mathbf{A})^{-1}\mathbf{F}(s), \tag{27}$$

where, using the formula for the inverse of a 2×2 matrix given by Eq. (14) in Section 3.1,

$$(s\mathbf{I} - \mathbf{A})^{-1} = \frac{1}{|s\mathbf{I} - \mathbf{A}|} \begin{pmatrix} s - a_{22} & a_{12} \\ a_{21} & s - a_{11} \end{pmatrix}.$$

We note that $Z(s) = |s\mathbf{I} - \mathbf{A}| = s^2 - (a_{11} + a_{22})s + a_{11}a_{22} - a_{12}a_{21}$ is the characteristic polynomial of $\mathbf{A}$. Inverting each component of $\mathbf{Y}$ in Eq. (27) then yields the solution of Eqs. (23).

EXAMPLE
4

Use the Laplace transform to solve the system

$$y_1' = -3y_1 + 4y_2 + \sin t,$$
$$y_2' = -2y_1 + 3y_2 + t \tag{28}$$

subject to the initial conditions

$$y_1(0) = 0, \qquad y_2(0) = 1. \tag{29}$$

Taking the Laplace transform of each equation in the system (28) and substituting the initial conditions (29) yield the pair of algebraic equations

$$sY_1 - 0 = -3Y_1 + 4Y_2 + 1/(s^2 + 1),$$
$$sY_2 - 1 = -2Y_1 + 3Y_2 + 1/s^2.$$

Rewriting this system in the matrix form (26) gives

$$\begin{pmatrix} s+3 & -4 \\ 2 & s-3 \end{pmatrix} \begin{pmatrix} Y_1 \\ Y_2 \end{pmatrix} = \begin{pmatrix} 0 \\ 1 \end{pmatrix} + \begin{pmatrix} 1/(s^2+1) \\ 1/s^2 \end{pmatrix} = \begin{pmatrix} 1/(s^2+1) \\ (s^2+1)/s^2 \end{pmatrix}. \tag{30}$$

The solution of Eq. (30) is

$$\mathbf{Y}(s) = \frac{1}{s^2(s^2-1)(s^2+1)} \begin{pmatrix} 4s^4+s^3+5s^2+4 \\ s^5+3s^4+2s^3+4s^2+s+3 \end{pmatrix}. \tag{31}$$

Inverting the partial fraction decomposition of each component in the vector (31) then yields

$$y_1(t) = \tfrac{7}{2} e^t - 3e^{-t} + \tfrac{3}{2} \sin t - \tfrac{1}{2} \cos t - 4t,$$

and

$$y_2(t) = \tfrac{7}{2} e^t - \tfrac{3}{2} e^{-t} + \sin t - 3t - 1.$$

A computer algebra system can be used to facilitate the calculations.

The method of Laplace transforms can also be used to solve coupled systems of mixed order equations without necessarily converting them to a system of first order equations. The primary requirement is that the equations have constant coefficients.

EXAMPLE 5

Use the Laplace transform to solve the system

$$x'' + y' + 2x = 0,$$
$$2x' - y' = \cos t, \tag{32}$$

subject to the initial conditions

$$x(0) = 0, \qquad x'(0) = 0, \qquad y(0) = 0. \tag{33}$$

Letting $X(s) = \mathcal{L}\{x(t)\}$ and $Y(s) = \mathcal{L}\{y(t)\}$ and taking the Laplace transform of each equation in the system (32) give

$$s^2 X(s) - sx(0) - x'(0) + sY(s) - y(0) + 2X(s) = 0$$
$$2sX(s) - 2x(0) - sY(s) + y(0) = \frac{s}{s^2+1}. \tag{34}$$

Substituting the prescribed initial values for $x(0)$, $x'(0)$, and $y(0)$, we write the resultant system in the form

$$(s^2+2)X + sY = 0$$
$$2sX - sY = \frac{s}{s^2+1}. \tag{35}$$

Solving for X, we find

$$X(s) = \frac{s}{(s^2+1)\left[(s+1)^2+1\right]} = \frac{1}{5} \frac{s}{s^2+1} + \frac{2}{5} \frac{1}{s^2+1} - \frac{1}{5} \frac{s+1}{(s+1)^2+1} - \frac{3}{5} \frac{1}{(s+1)^2+1},$$

where the last expression is the partial fraction expansion for X. Thus

$$x(t) = \tfrac{1}{5} \cos t + \tfrac{2}{5} \sin t - \tfrac{1}{5} e^{-t} \cos t - \tfrac{3}{5} e^{-t} \sin t.$$

Solving the second equation in system (35) for Y gives $Y(s) = 2X(s) - 1/(s^2 + 1)$, so $y(t) = \mathcal{L}^{-1}\{Y(s)\} = 2\mathcal{L}^{-1}\{X(s)\} - \sin t = 2x(t) - \sin t$, that is,

$$y(t) = \tfrac{2}{5} \cos t - \tfrac{1}{5} \sin t - \tfrac{2}{5} e^{-t} \cos t - \tfrac{6}{5} e^{-t} \sin t.$$

PROBLEMS

In each of Problems 1 through 13, use the Laplace transform to solve the given initial value problem:

1. $y'' - 4y' - 12y = 0;$ $\quad y(0) = 8,\quad y'(0) = 0$

2. $y'' + 3y' + 2y = t;$ $\quad y(0) = 1,\quad y'(0) = 0$

3. $y'' - 8y' + 25y = 0;$ $\quad y(0) = 0,\quad y'(0) = 3$

4. $y'' - 4y' + 4y = 0;$ $\quad y(0) = 1,\quad y'(0) = 1$

5. $y'' - 2y' + 4y = 0;$ $\quad y(0) = 2,\quad y'(0) = 0$

6. $y'' + 4y' + 29y = e^{-2t} \sin 5t;$ $\quad y(0) = 5,\quad y'(0) = -2$

7. $y'' + \omega^2 y = \cos 2t,\quad \omega^2 \neq 4;$
$y(0) = 1,\quad y'(0) = 0$

8. $y'' - 2y' + 2y = \cos t;$ $\quad y(0) = 1,\quad y'(0) = 0$

9. $y'' - 2y' + 2y = e^{-t};$ $\quad y(0) = 0,\quad y'(0) = 1$

10. $y'' + 2y' + y = 18e^{-t};$ $\quad y(0) = 7,\quad y'(0) = -2$

11. $y^{(4)} - 4y''' + 6y'' - 4y' + y = 0;$ $\quad y(0) = 0,$
$y'(0) = 1,\quad y''(0) = 0,\quad y'''(0) = 1$

12. $y^{(4)} - y = 0;$ $\quad y(0) = 1,\quad y'(0) = 0,$
$y''(0) = 1,\quad y'''(0) = 0$

13. $y^{(4)} - 9y = 0;$ $\quad y(0) = 1,\quad y'(0) = 0,$
$y''(0) = -3,\quad y'''(0) = 0$

In each of Problems 14 through 19, use the Laplace transform to solve the given initial value problem:

14. $\mathbf{y}' = \begin{pmatrix} -5 & 1 \\ -9 & 5 \end{pmatrix} \mathbf{y},\quad \mathbf{y}(0) = \begin{pmatrix} 1 \\ 0 \end{pmatrix}$

15. $\mathbf{y}' = \begin{pmatrix} 5 & -2 \\ 6 & -2 \end{pmatrix} \mathbf{y},\quad \mathbf{y}(0) = \begin{pmatrix} 1 \\ 0 \end{pmatrix}$

16. $\mathbf{y}' = \begin{pmatrix} 4 & -4 \\ 5 & -4 \end{pmatrix} \mathbf{y},\quad \mathbf{y}(0) = \begin{pmatrix} 1 \\ 0 \end{pmatrix}$

17. $\mathbf{y}' = \begin{pmatrix} 0 & 6 \\ -6 & 0 \end{pmatrix} \mathbf{y},\quad \mathbf{y}(0) = \begin{pmatrix} 5 \\ 4 \end{pmatrix}$

18. $\mathbf{y}' = \begin{pmatrix} -4 & -1 \\ 1 & -2 \end{pmatrix} \mathbf{y},\quad \mathbf{y}(0) = \begin{pmatrix} 1 \\ 0 \end{pmatrix}$

19. $\mathbf{y}' = \begin{pmatrix} 2 & -64 \\ 1 & -14 \end{pmatrix} \mathbf{y},\quad \mathbf{y}(0) = \begin{pmatrix} 0 \\ 1 \end{pmatrix}$

In each of Problems 20 through 24, use a computer algebra system to assist in solving the given initial value problem by the method of Laplace transforms:

20. $\mathbf{y}' = \begin{pmatrix} -4 & -1 \\ 1 & -2 \end{pmatrix} \mathbf{y} + \begin{pmatrix} 2e^t \\ \sin 2t \end{pmatrix},\quad \mathbf{y}(0) = \begin{pmatrix} 1 \\ 2 \end{pmatrix}$

21. $\mathbf{y}' = \begin{pmatrix} 5 & -1 \\ 1 & 3 \end{pmatrix} \mathbf{y} + \begin{pmatrix} e^{-t} \\ 2e^t \end{pmatrix},\quad \mathbf{y}(0) = \begin{pmatrix} -3 \\ 2 \end{pmatrix}$

22. $\mathbf{y}' = \begin{pmatrix} -1 & -5 \\ 1 & 3 \end{pmatrix} \mathbf{y} + \begin{pmatrix} 3 \\ 5\cos t \end{pmatrix},\quad \mathbf{y}(0) = \begin{pmatrix} 1 \\ -1 \end{pmatrix}$

23. $\mathbf{y}' = \begin{pmatrix} -2 & 1 \\ 1 & -2 \end{pmatrix} \mathbf{y} + \begin{pmatrix} 0 \\ \sin t \end{pmatrix},\quad \mathbf{y}(0) = \begin{pmatrix} 0 \\ 0 \end{pmatrix}$

24. $\mathbf{y}' = \begin{pmatrix} 0 & 1 & -1 \\ 1 & 0 & 1 \\ 1 & 1 & 0 \end{pmatrix} \mathbf{y} + \begin{pmatrix} 0 \\ -e^{-t} \\ e^t \end{pmatrix},\quad \mathbf{y}(0) = \begin{pmatrix} 1 \\ 2 \\ 3 \end{pmatrix}$

25. Use the Laplace transform to solve the system

$$x'' - y'' + x - 4y = 0,$$
$$x' + y' = \cos t,$$

$$x(0) = 0,\quad x'(0) = 1,\quad y(0) = 0,\quad y'(0) = 2.$$

26. A radioactive substance R_1 having decay rate k_1 disintegrates into a second radioactive substance R_2 having decay rate $k_2 \neq k_1$. Substance R_2 disintegrates into R_3, which is

stable. If $m_i(t)$ represents the mass of substance R_i at time t, $i = 1, 2, 3$, the applicable equations are

$$m_1' = -k_1 m_1$$
$$m_2' = k_1 m_1 - k_2 m_2$$
$$m_3' = k_2 m_2.$$

Use the Laplace transform to solve this system under the conditions

$$m_1(0) = m_0, \qquad m_2(0) = 0, \qquad m_3(0) = 0.$$

5.5 Discontinuous Functions and Periodic Functions

In Section 5.4 we presented the general procedure used to solve initial value problems by means of the Laplace transform. Some of the most interesting elementary applications of the transform method occur in the solution of linear differential equations with discontinuous, periodic, or impulsive forcing functions. Equations of this type frequently arise in the analysis of the flow of current in electric circuits or the vibrations of mechanical systems. In this section and the following ones, we develop some additional properties of the Laplace transform that are useful in the solution of such problems. Unless a specific statement is made to the contrary, all functions appearing below are assumed to be piecewise continuous and of exponential order, so that their Laplace transforms exist, at least for s sufficiently large.

▶ The Unit Step Function. To deal effectively with functions having jump discontinuities, it is very helpful to introduce a function known as the **unit step function** or **Heaviside function**. This function is defined by

$$u(t) = \begin{cases} 0, & t < 0, \\ 1, & t \geq 0. \end{cases} \tag{1}$$

In applications, the Heaviside function often represents a force, voltage, current, or signal that is turned on at time $t = 0$, and left on thereafter. Translations of the Heaviside function are used to turn such functions on at times other than 0. For a real number c, we define

$$u_c(t) = \begin{cases} 0, & t < c, \\ 1, & t \geq c. \end{cases} \tag{2}$$

The graph of $y = u_c(t)$ is shown at the top of Figure 5.5.1. It is often necessary to turn a signal or function on at time $t = c$ and then turn it off at time $t = d > c$. This can be accomplished by using an **indicator function** $u_{cd}(t)$ for the interval $[c, d)$ defined by

$$u_{cd}(t) = u_c(t) - u_d(t) = \begin{cases} 0, & t < c \quad \text{or} \quad t \geq d, \\ 1, & c \leq t < d. \end{cases} \tag{3}$$

The graph of $y = u_{cd}(t)$ is shown at the bottom of Figure 5.5.1. Note that $u_{0d}(t) = u_0(t) - u_d(t) = 1 - u_d(t)$ for $t \geq 0$ has a negative step at $t = d$ and is used to turn off, at time $t = d$, a function that is initially turned on at $t = 0$.

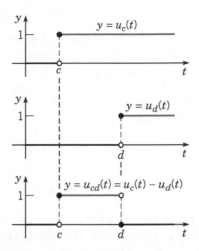

FIGURE 5.5.1 The function $u_{cd}(t) = u_c(t) - u_d(t)$ is used to model the indicator function for the intervals $[c, d)$, $(c, d]$, (c, d), and $[c, d]$.

Remark. Note that the definition of $u_c(t)$ at the point where the jump discontinuity occurs is immaterial. In fact, $u_c(t)$ need not be defined at all at $t = c$. This is consistent with comments regarding piecewise continuous functions that appear in the remark immediately following Theorem 5.3.1. For example, we will use the model $\hat{f}(t) = 3u_{24}(t) = 3[u_2(t) - u_4(t)]$ to describe the function

$$f(t) = \begin{cases} 0, & t < 2, \\ 3, & 2 < t < 4, \\ 0, & 4 < t, \end{cases}$$

even though f is not defined at $t = 2$ and $t = 4$ while $\hat{f}(2) = 3$ and $\hat{f}(4) = 0$.

EXAMPLE 1

Use the unit step function to give a representation of the piecewise continuous function

$$f(t) = \begin{cases} t, & 0 < t < 2, \\ 1, & 2 \le t < 3, \\ e^{-2t}, & 3 \le t. \end{cases}$$

This can be accomplished by:

    (i) turning on the function t at $t = 0$ and turning it off at $t = 2$,

    (ii) turning on the constant function 1 at $t = 2$ and then turning it off at $t = 3$, and

    (iii) turning on e^{-2t} at $t = 3$ and leaving it on.

Thus

$$f(t) = tu_{02}(t) + 1u_{23}(t) + e^{-2t}u_3(t)$$

$$= t[1 - u_2(t)] + 1[u_2(t) - u_3(t)] + e^{-2t}u_3(t)$$

$$= t - (t - 1)u_2(t) + (e^{-2t} - 1)u_3(t), \qquad t > 0.$$

EXAMPLE
2

Use unit step functions to represent the function described by the graph in Figure 5.5.2.

The increasing portion of the triangular pulse is a straight-line segment with slope 1 that passes through the point $(1, 0)$ in the ty-plane. It is therefore described by $t - 1$ for $1 \leq t \leq 2$. The decreasing portion of the pulse is a straight-line segment with slope -1 that passes through the point $(3, 0)$ in the ty-plane. It is described by $3 - t$ for $2 \leq t \leq 3$. Using indicator functions, we have

$$
\begin{aligned}
f(t) &= (t - 1)u_{12}(t) + (3 - t)u_{23}(t) \\
&= (t - 1)[u_1(t) - u_2(t)] + (3 - t)[u_2(t) - u_3(t)] \\
&= (t - 1)u_1(t) - 2(t - 2)u_2(t) + (t - 3)u_3(t).
\end{aligned}
$$

Using the definition of the unit step function and noting where the step function changes value, the last line can also be written in the form

$$
f(t) = \begin{cases}
0, & t < 1, \\
t - 1, & 1 \leq t < 2, \\
3 - t, & 2 \leq t < 3, \\
0, & 3 \leq t,
\end{cases}
$$

thus confirming the representation in terms of unit step functions.

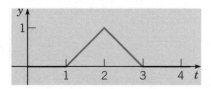

FIGURE 5.5.2   A triangular pulse.

▶ **The Laplace Transform of the Unit Step Function.** The Laplace transform of u_c, with $c \geq 0$,

$$
\mathcal{L}\{u_c(t)\} = \frac{e^{-cs}}{s}, \qquad s > 0, \tag{4}
$$

since for $s > 0$,

$$
\begin{aligned}
\mathcal{L}\{u_c(t)\} &= \int_0^\infty e^{-st} u_c(t)\, dt = \int_c^\infty e^{-st}\, dt = \lim_{A \to \infty} \int_c^A e^{-st}\, dt \\
&= \lim_{A \to \infty} \left(\frac{e^{-cs}}{s} - \frac{e^{-sA}}{s} \right) = \frac{e^{-cs}}{s}.
\end{aligned}
$$

In particular, note that $\mathcal{L}\{u(t)\} = \mathcal{L}\{u_0(t)\} = 1/s$, which is identical to the Laplace transform of the function $f(t) = 1$. This is because $f(t) = 1$ and $u(t)$ are identical functions on $[0, \infty)$.

For the indicator function $u_{cd}(t) = u_c(t) - u_d(t)$, the linearity of $\mathcal{L}$ gives

$$\mathcal{L}\{u_{cd}(t)\} = \mathcal{L}\{u_c(t)\} - \mathcal{L}\{u_d(t)\} = \frac{e^{-cs} - e^{-ds}}{s}, \qquad s > 0. \tag{5}$$

▶ **Laplace Transforms of Time-Shifted Functions.** For a given function f defined for $t \geq 0$, we will often want to consider the related function g defined by

$$y = g(t) = \begin{cases} 0, & t < c, \\ f(t - c), & t \geq c, \end{cases}$$

which represents a translation of f a distance c in the positive t direction (see Figure 5.5.3). In terms of the unit step function, we can write $g(t)$ in the convenient form

$$g(t) = u_c(t)f(t - c).$$

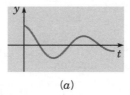 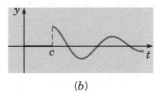

(a)   (b)

FIGURE 5.5.3 A translation of the given function. (a) $y = f(t)$; (b) $y = u_c(t)f(t - c)$.

The unit step function is particularly important in transform use because of the following relation between the transform of $f(t)$ and that of its translation $u_c(t)f(t - c)$.

THEOREM 5.5.1

If $F(s) = \mathcal{L}\{f(t)\}$ exists for $s > a \geq 0$, and if c is a nonnegative constant, then

$$\mathcal{L}\{u_c(t)f(t - c)\} = e^{-cs}\mathcal{L}\{f(t)\} = e^{-cs}F(s), \qquad s > a. \tag{6}$$

Conversely, if $f(t) = \mathcal{L}^{-1}\{F(s)\}$, then

$$u_c(t)f(t - c) = \mathcal{L}^{-1}\{e^{-cs}F(s)\}. \tag{7}$$

Theorem 5.5.1 simply states that the translation of $f(t)$ a distance c in the positive t direction corresponds to the multiplication of $F(s)$ by e^{-cs}.

Proof

To prove Theorem 5.5.1, it is sufficient to determine the transform of $u_c(t)f(t - c)$:

$$\mathcal{L}\{u_c(t)f(t - c)\} = \int_0^\infty e^{-st}u_c(t)f(t - c)\,dt$$

$$= \int_c^\infty e^{-st}f(t - c)\,dt.$$

Introducing a new integration variable $\xi = t - c$, we have

$$\mathcal{L}\{u_c(t)f(t-c)\} = \int_0^\infty e^{-(\xi+c)s}f(\xi)\,d\xi = e^{-cs}\int_0^\infty e^{-s\xi}f(\xi)\,d\xi$$

$$= e^{-cs}F(s).$$

Thus Eq. (6) is established; Eq. (7) follows by taking the inverse transform of both sides of Eq. (6).

A simple example of this theorem occurs if we take $f(t) = 1$. Recalling that $\mathcal{L}\{1\} = 1/s$, we immediately have $\mathcal{L}\{u_c(t)\} = e^{-cs}/s$ from Eq. (6). This result agrees with that of Eq. (4).

Remark. In order to use Theorem 5.5.1, the translations of u and f have to be identical, that is, their product must be of the form $f(t-c)u_c(t) = f(t-c)u(t-c)$. If, as is usually the case, a term involving a step function is of the form $g(t)u_c(t)$, a systematic way to deduce f is simply to set $f(t-c) = g(t)$ and then substitute $t + c$ for t in the resulting equation to obtain $f(t+c-c) = g(t+c)$ or $f(t) = g(t+c)$.

EXAMPLE 3

Find the Laplace transform of the function

$$f(t) = \begin{cases} t, & 0 < t < 2, \\ 1, & 2 \le t < 3, \\ e^{-2t}, & 3 \le t. \end{cases}$$

In Example 1, we found the following representation of f in terms of step functions:

$$f(t) = t - (t-1)u_2(t) + (e^{-2t} - 1)u_3(t), \qquad t > 0.$$

If we set $f_1(t) = t$, then $\mathcal{L}\{f_1(t)\} = 1/s^2$. Following the above remark, we set $f_2(t-2) = t - 1$ and substitute $t + 2$ for t so that $f_2(t) = t + 1$, and $\mathcal{L}\{f_2(t)\} = 1/s^2 + 1/s = (1 + s)/s^2$. Similarly, set $f_3(t-3) = e^{-2t} - 1$ and substitute $t + 3$ for t to obtain $f_3(t) = e^{-2(t+3)} - 1 = e^{-6}e^{-2t} - 1$. Thus $\mathcal{L}\{f_3(t)\} = e^{-6}/(s+2) - 1/s$. Using the linearity of $\mathcal{L}$ and Theorem 5.5.1 gives

$$\mathcal{L}\{f(t)\} = \mathcal{L}\{f_1(t)\} - \mathcal{L}\{f_2(t-2)u_2(t)\} + \mathcal{L}\{f_3(t-3)u_2(t)\}$$

$$= \mathcal{L}\{f_1(t)\} - e^{-2s}\mathcal{L}\{f_2(t)\} + e^{-3s}\mathcal{L}\{f_3(t)\}$$

$$= \frac{1}{s^2} - e^{-2s}\frac{1+s}{s^2} + e^{-3s}\left[\frac{e^{-6}}{s+2} - \frac{1}{s}\right].$$

EXAMPLE 4

Find the Laplace transform of

$$f(t) = \begin{cases} 0, & t < 1, \\ t - 1, & 1 \le t < 2, \\ 3 - t, & 2 \le t < 3, \\ 0, & 3 \le t. \end{cases}$$

This is the triangular pulse of Example 2, which can be represented by

$$f(t) = (t-1)u_1(t) - 2(t-2)u_2(t) + (t-3)u_3(t),$$

and is already nearly in the form to which Theorem 5.5.1 is directly applicable. If we set $f_1(t-1) = t-1$, $f_2(t-2) = 2(t-2)$, and $f_3(t-3) = t-3$, we see that $f_1(t) = t$, $f_2(t) = 2t$, and $f_3(t) = t$. It follows that

$$\mathcal{L}\{f(t)\} = \frac{e^{-s}}{s^2} - \frac{2e^{-2s}}{s^2} + \frac{e^{-3s}}{s^2}.$$

EXAMPLE 5

Find the inverse transform of

$$F(s) = \frac{1 - e^{-2s}}{s^2}.$$

From the linearity of the inverse transform, we have

$$f(t) = \mathcal{L}^{-1}\{F(s)\} = \mathcal{L}^{-1}\left\{\frac{1}{s^2}\right\} - \mathcal{L}^{-1}\left\{\frac{e^{-2s}}{s^2}\right\}$$

$$= t - u_2(t)(t-2).$$

The function f may also be written as

$$f(t) = \begin{cases} t, & 0 \le t < 2, \\ 2, & t \ge 2. \end{cases}$$

▶ Periodic Functions.

DEFINITION 5.5.2

A function f is said to be **periodic with period** $T > 0$ if

$$f(t + T) = f(t)$$

for all t in the domain of f.

A periodic function can be defined by indicating the length of its period and specifying its values over a single period. For example, the periodic function shown in Figure 5.5.4 can be expressed as

$$f(t) = \begin{cases} 1 - t, & 0 \le t < 1, \\ 0, & 1 \le t < 2, \end{cases} \quad \text{and } f(t) \text{ has period 2.} \tag{8}$$

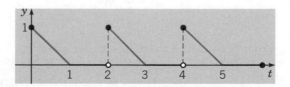

FIGURE 5.5.4 The periodic function described in Eq. (8).

More generally, in discussing a periodic function $f(t)$, it is convenient to introduce a **window function** $f_T(t)$ defined by

$$f_T(t) = f(t)[1 - u_T(t)] = \begin{cases} f(t), & 0 \le t \le T, \\ 0, & \text{otherwise,} \end{cases} \tag{9}$$

and its Laplace transform $F_T(s)$ given by

$$F_T(s) = \int_0^\infty e^{-st} f_T(t)\, dt = \int_0^T e^{-st} f(t)\, dt. \tag{10}$$

The window function specifies values of $f(t)$ over a single period. A replication of $f_T(t)$ shifted k periods to the right, that is, a distance kT, can be represented by

$$f_T(t - kT)u_{kT}(t) = \begin{cases} f(t - kT), & kT \le t < (k+1)T, \\ 0, & \text{otherwise.} \end{cases} \tag{11}$$

Summing the time-shifted replications $f_T(t - kT)u_{kT}(t)$, $k = 0, \ldots, n - 1$, gives $f_{nT}(t)$, the periodic extension of $f_T(t)$ to the interval $[0, nT]$,

$$f_{nT}(t) = \sum_{k=0}^{n-1} f_T(t - kT)u_{kT}(t). \tag{12}$$

Thus $f_{nT}(t)$ consists of exactly n periods of the window function $f_T(t)$, as shown in Figure 5.5.5.

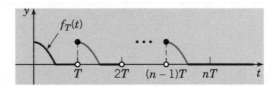

FIGURE 5.5.5 The periodic extension $f_{nT}(t)$ of the window function $f_T(t)$ to the interval $[0, nT]$, $f_{nT}(t) = \sum_{k=0}^{n-1} f_T(t - kT)u_{kT}(t)$.

The entire periodic function is then represented by

$$f(t) = \sum_{n=0}^\infty f_T(t - nT)u_{nT}(t). \tag{13}$$

THEOREM 5.5.3

If f is periodic with period T and is piecewise continuous on $[0, T]$, then

$$\mathcal{L}\{f(t)\} = \frac{F_T(s)}{1 - e^{-sT}} = \frac{\int_0^T e^{-st} f(t)\, dt}{1 - e^{-sT}}. \tag{14}$$

Proof

From Theorem 5.5.1 and Eq. (10), we know that for each $k \ge 0$,

$$\mathcal{L}\{f_T(t - kT)u_{kT}(t)\} = e^{-kTs}\mathcal{L}\{f_T(t)\} = e^{-kTs}F_T(s).$$

Using the linearity of $\mathcal{L}$, the Laplace transform of $f_{nT}(t)$ in (12) is therefore

$$F_{nT}(s) = \int_0^{nT} e^{-st}f(t)\,dt = \sum_{k=0}^{n-1} \mathcal{L}\{f_T(t-kT)u_{kT}(t)\}$$

$$= \sum_{k=0}^{n-1} e^{-kTs}F_T(s) = F_T(s)\sum_{k=0}^{n-1}\left(e^{-sT}\right)^k = F_T(s)\frac{1-(e^{-sT})^n}{1-e^{-sT}}, \qquad (15)$$

where the last equality arises from the formula for the sum of the first n terms of a geometric series. Since $e^{-sT} < 1$ for $sT > 0$, it follows from (15) that

$$F(s) = \lim_{n\to\infty}\int_0^{nT} e^{-sT}f(t)\,dt = \lim_{n\to\infty}F_T(s)\frac{1-(e^{-sT})^n}{1-e^{-sT}} = \frac{F_T(s)}{1-e^{-sT}}.$$

EXAMPLE 6

Find the Laplace transform of the periodic function described by the graph in Figure 5.5.6. The sawtooth waveform can be expressed as

$$f(t) = \begin{cases} t, & 0 \le t < 1, \\ 0, & 1 \le t < 2, \end{cases} \qquad \text{and } f(t) \text{ has period 2.}$$

Using $T = 2$, $F_T(s) = \displaystyle\int_0^2 e^{-st}f(t)\,dt = \int_0^1 e^{-st}t\,dt = \dfrac{1-e^{-s}}{s^2} - \dfrac{e^{-s}}{s}$, and Theorem 5.5.3 give

$$F(s) = \frac{1-e^{-s}}{s^2\left(1-e^{-2s}\right)} - \frac{e^{-s}}{s\left(1-e^{-2s}\right)}.$$

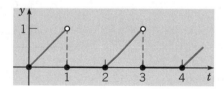

FIGURE 5.5.6 A periodic sawtooth signal.

EXAMPLE 7

Find the inverse Laplace transform of

$$F(s) = \frac{1-e^{-s}}{s\left(1-e^{-2s}\right)}. \qquad (16)$$

According to Eq. (14), the presence of $1-e^{-2s}$ in the denominator suggests that the inverse transform is a periodic function of period 2 and that $F_2(s) = (1-e^{-s})/s$. Since $\mathcal{L}^{-1}\{1/s\} = 1$ and $\mathcal{L}^{-1}\{e^{-s}/s\} = u_1(t)$, it follows that

$$f_2(t) = 1 - u_1(t) = \begin{cases} 1, & 0 \le t < 1, \\ 0, & 1 \le t < 2. \end{cases}$$

Hence

$$f(t) = \begin{cases} 1, & 0 \le t < 1, \\ 0, & 1 \le t < 2, \end{cases} \qquad \text{and } f(t) \text{ has period 2.}$$

The graph of $f(t)$, a periodic square wave, is shown in Figure 5.5.7.

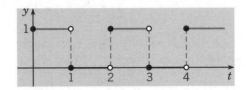

FIGURE 5.5.7   A periodic square wave.

Alternatively, note that $F(s)$ in Eq. (16) can be expressed as

$$F(s) = \frac{1}{s(1 + e^{-s})} = \frac{1}{s}\left(1 - e^{-s} + e^{-2s} - e^{-3s} + \cdots\right),$$

where the last equality arises from using the geometric series representation of $1/(1 + e^{-s})$. A term-by-term inversion of this infinite series using the linearity of $\mathcal{L}^{-1}$ yields the time domain representation

$$f(t) = 1 - u_1(t) + u_2(t) - u_3(t) + \cdots = u_{01}(t) + u_{23}(t) + \cdots$$

of the square wave shown in Figure 5.5.7.

PROBLEMS

In each of Problems 1 through 6, sketch the graph of the given function on the interval $t \ge 0$:

1. $u_1(t) + 3u_3(t) - 7u_4(t)$

2. $(t - 3)u_2(t) - (t - 2)u_3(t)$

3. $f(t - \pi)u_\pi(t)$, where $f(t) = t^2$

4. $f(t - 4)u_4(t)$, where $f(t) = \sin t$

5. $f(t - 1)u_2(t)$, where $f(t) = 2t$

6. $(t - 1)u_1(t) - 2(t - 2)u_2(t) + (t - 3)u_3(t)$

In each of Problems 7 through 12, find the Laplace transform of the given function:

7. $f(t) = \begin{cases} 0, & t < 9 \\ (t - 9)^s, & t \ge 9 \end{cases}$

8. $f(t) = \begin{cases} 0, & t < 3 \\ t^2 - 6t + 18, & t \ge 3 \end{cases}$

9. $f(t) = \begin{cases} 0, & t < \pi \\ t - \pi, & \pi \le t < 2\pi \\ 0, & t \ge 2\pi \end{cases}$

10. $f(t) = u_1(t) + 3u_5(t) - 4u_8(t)$

11. $f(t) = (t - 4)u_3(t) - (t - 3)u_4(t)$

12. $f(t) = t - u_1(t)(t - 1)$, $t \ge 0$

In each of Problems 13 through 18, find the inverse Laplace transform of the given function:

13. $F(s) = \dfrac{5!e^{-s}}{(s - 5)^6}$

14. $F(s) = \dfrac{e^{-2s}}{s^2 + s - 2}$

15. $F(s) = \dfrac{2(s - 1)e^{-2s}}{s^2 - 2s + 2}$

16. $F(s) = \dfrac{8e^{-7s}}{s^2 - 64}$

17. $F(s) = \dfrac{(s - 2)e^{-s}}{s^2 - 4s + 3}$

18. $F(s) = \dfrac{e^{-s} + e^{-2s} - e^{-3s} - e^{-4s}}{s}$

In each of Problems 19 through 21, find the Laplace transform of the given function:

19. $f(t) = \begin{cases} 1, & 0 \le t < 1 \\ 0, & t \ge 1 \end{cases}$

20. $f(t) = \begin{cases} 1, & 0 \le t < 1 \\ 0, & 1 \le t < 2 \\ 1, & 2 \le t < 3 \\ 0, & t \ge 3 \end{cases}$

21. $f(t) = 1 - u_1(t) + \cdots + u_{2n}(t) - u_{2n+1}(t) = 1 + \sum_{k=1}^{2n+1}(-1)^k u_k(t)$

In each of Problems 22 through 24, find the Laplace transform of the periodic function:

22. $f(t) = \begin{cases} 1, & 0 \leq t < 1, \\ -1, & 1 \leq t < 2; \end{cases}$ and $f(t)$ has period 2.

See Figure 5.5.8.

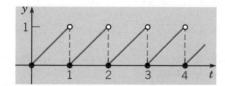

FIGURE 5.5.8 A periodic square wave.

23. $f(t) = t, 0 \leq t < 1$; and $f(t)$ has period 1. See Figure 5.5.9.

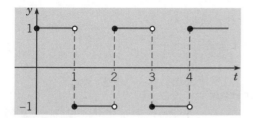

FIGURE 5.5.9 A sawtooth wave.

24. $f(t) = \sin t, 0 \leq t < \pi$; and $f(t)$ has period π. See Figure 5.5.10.

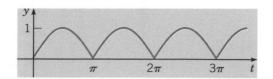

FIGURE 5.5.10 A rectified sine wave.

25. (a) If $f(t) = 1 - u_1(t)$, find $\mathcal{L}\{f(t)\}$; compare with Problem 19. Sketch the graph of $y = f(t)$.
(b) Let $g(t) = \int_0^t f(\xi)\,d\xi$, where the function f is defined in part (a). Sketch the graph of $y = g(t)$ and find $\mathcal{L}\{g(t)\}$.
(c) Let $h(t) = g(t) - u_1(t)g(t-1)$, where g is defined in part (b). Sketch the graph of $y = h(t)$ and find $\mathcal{L}\{h(t)\}$.

26. Consider the function p defined by

$$p(t) = \begin{cases} t, & 0 \leq t < 1, \\ 2 - t, & 1 \leq t < 2; \end{cases} \quad \text{and } p(t) \text{ has period 2.}$$

(a) Sketch the graph of $y = p(t)$.
(b) Find $\mathcal{L}\{p(t)\}$ by noting that p is the periodic extension of the function h in Problem 25(c) and then using the result of Theorem 5.5.3.
(c) Find $\mathcal{L}\{p(t)\}$ by noting that

$$p(t) = \int_0^t f(\xi)\,d\xi,$$

where f is the function in Problem 22, and then using Theorem 5.2.2.

5.6 Differential Equations with Discontinuous Forcing Functions

In this section we turn our attention to some examples in which the nonhomogeneous term, or forcing function, of a differential equation is modeled by a discontinuous function. This is often done even if in the actual physical system the forcing function is continuous, but changes rapidly over a very short time interval. In many applications, systems are often tested by subjecting them to discontinuous forcing functions. For example, engineers frequently wish to know how a system responds to a step input, a common situation where the input to a system suddenly changes from one constant level to another.

Although not immediately obvious, solutions of the constant coefficient equation,

$$ay'' + by' + cy = g(t),$$

are, in fact, continuous whenever the input g is piecewise continuous. To understand how discontinuous inputs can yield continuous outputs, we consider the simple initial value problem

$$y''(t) = u_c(t), \qquad y(0) = 0, \quad y'(0) = 0, \tag{1}$$

where $c > 0$. Integrating both sides of Eq. (1) from 0 to t and using the initial condition $y'(0) = 0$ yield

$$y'(t) - y'(0) = \int_0^t u_c(\tau)\, d\tau = \begin{cases} 0, & 0 \le t < c, \\ t - c, & t \ge c, \end{cases} \tag{2}$$

or

$$y'(t) = (t - c)u_c(t). \tag{3}$$

Integrating both sides of Eq. (3) from 0 to t then yields

$$y(t) - y(0) = \int_0^t (\tau - c)u_c(\tau)\, d\tau = \begin{cases} 0, & 0 \le t < c, \\ \dfrac{(t-c)^2}{2}, & t \ge c, \end{cases} \tag{4}$$

or since $y(0) = 0$,

$$y(t) = \frac{(t-c)^2}{2}u_c(t). \tag{5}$$

The graphs of $y''(t)$, $y'(t)$, and $y(t)$ are shown in Figure 5.6.1.

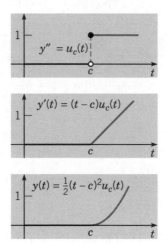

FIGURE 5.6.1   The smoothing effect of integration.

While $y''(t)$ has a jump discontinuity of magnitude 1 at $t = c$, $y'(t)$ is continuous at $t = c$, as can be seen from Eq. (3), where $\lim_{t \to c} y'(t) = 0$. From Eq. (5) we see that $y(t)$ is not only continuous for all values of $t > 0$, but its derivative $y'(t)$ exists and is continuous for all values of $t > 0$. Each integration yields a function that is smoother than the function being integrated. The smoothing effect of integration generalizes to variable coefficient equations and higher order equations. Consider

$$y'' + p(t)y' + q(t)y = g(t), \tag{6}$$

where p and q are continuous on some interval $\alpha < t < \beta$, but g is only piecewise continuous there. If $y = \psi(t)$ is a solution of Eq. (6), then ψ and ψ' are continuous on $\alpha < t < \beta$, but ψ'' has jump discontinuities at the same points as g. Similar remarks apply to higher order equations; the highest derivative of the solution appearing in the differential equation has jump discontinuities at the same points as the forcing function, but the solution itself and its lower order derivatives are continuous even at those points.

EXAMPLE
1

Describe the qualitative nature of the solution of the initial value problem

$$y'' + 4y = g(t), \tag{7}$$

$$y(0) = 0, \qquad y'(0) = 0, \tag{8}$$

where

$$g(t) = \begin{cases} 0, & 0 \le t < 5, \\ (t-5)/5, & 5 \le t < 10, \\ 1, & t \ge 10, \end{cases} \tag{9}$$

and then find the solution.

In this example, the forcing function has the graph shown in Figure 5.6.2 and is known as ramp loading.

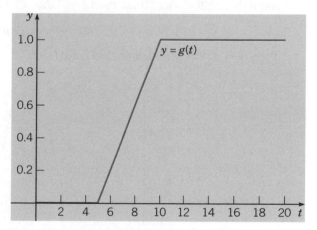

FIGURE 5.6.2   Ramp loading: $y = g(t)$ from Eq. (9).

It is relatively easy to identify the general form of the solution. For $t < 5$, the solution is simply $y = 0$. On the other hand, for $t > 10$ the solution has the form

$$y = c_1 \cos 2t + c_2 \sin 2t + \tfrac{1}{4}. \tag{10}$$

The constant $\tfrac{1}{4}$ is a particular solution of the nonhomogeneous equation, whereas the other two terms comprise the general solution of the corresponding homogeneous equation. Thus the solution (10) is a simple harmonic oscillation about $y = \tfrac{1}{4}$. Similarly, in the intermediate range $5 < t < 10$, the solution is an oscillation about a certain linear function. In an engineering context, for example, we might be interested in knowing the amplitude of the eventual steady oscillation.

To solve the problem, it is convenient to write

$$g(t) = \frac{u_5(t)(t-5) - u_{10}(t)(t-10)}{5}, \tag{11}$$

as you may verify. Then we take the Laplace transform of the differential equation and use the initial conditions, thereby obtaining

$$(s^2 + 4)Y(s) = \frac{e^{-5s} - e^{-10s}}{5s^2},$$

or

$$Y(s) = \frac{(e^{-5s} - e^{-10s})H(s)}{5}, \qquad (12)$$

where

$$H(s) = \frac{1}{s^2(s^2 + 4)}. \qquad (13)$$

Thus the solution of the initial value problem (7), (8), (9) is

$$y = \phi(t) = \frac{u_5(t)h(t-5) - u_{10}(t)h(t-10)}{5}, \qquad (14)$$

where $h(t)$ is the inverse transform of $H(s)$. The partial fraction expansion of $H(s)$ is

$$H(s) = \frac{1/4}{s^2} - \frac{1/4}{s^2 + 4}, \qquad (15)$$

and it then follows from lines 3 and 5 of Table 5.3.1 that

$$h(t) = \tfrac{1}{4}t - \tfrac{1}{8}\sin 2t. \qquad (16)$$

The graph of $y = \phi(t)$ is shown in Figure 5.6.3.

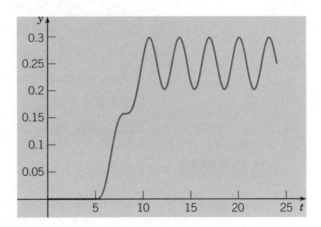

FIGURE 5.6.3   Solution of the initial value problem (7), (8), (9).

Observe that it has the qualitative form that we indicated earlier. To find the amplitude of the eventual steady oscillation, it is sufficient to locate one of the maximum or minimum points for $t > 10$. Setting the derivative of the solution (14) equal to zero, we find that the first maximum is located approximately at (10.642, 0.2979), so the amplitude of the oscillation is approximately 0.0479. Note that in this example the forcing function g is continuous but g' is discontinuous at $t = 5$ and $t = 10$. It follows that the solution ϕ and its first two derivatives are continuous everywhere, but ϕ''' has discontinuities at $t = 5$ and at $t = 10$ that match the discontinuities in g' at those points.

EXAMPLE
2

Resonance. Solve the initial value problem

$$y'' + \pi^2 y = f(t), \qquad y(0) = 0, \quad y'(0) = 0, \qquad (17)$$

where $f(t)$ is the square wave in Example 7, Section 5.5.

Taking the Laplace transform of the differential equation in the initial value problem (17) and using the initial conditions give

$$(s^2 + \pi^2)Y(s) = \frac{1}{s(1 + e^{-s})}, \tag{18}$$

where we have used the representation $1/s(1 + e^{-s})$ for the Laplace transform of $f(t)$. Thus

$$Y(s) = \frac{1}{s(s^2 + \pi^2)} \frac{1}{1 + e^{-s}}.$$

The partial fraction expansion

$$\frac{1}{s(s^2 + \pi^2)} = \frac{1}{\pi^2} \left[\frac{1}{s} - \frac{s}{s^2 + \pi^2} \right]$$

gives

$$\mathcal{L}^{-1} \left\{ \frac{1}{s(s^2 + \pi^2)} \right\} = \frac{1}{\pi^2} [1 - \cos \pi t].$$

Then, remembering the geometric series representation $1/(1 + e^{-s}) = \sum_{k=0}^{\infty}(-1)^k e^{-ks}$, we find that

$$y(t) = \frac{1}{\pi^2} \sum_{k=0}^{\infty} (-1)^k \{1 - \cos[\pi(t - k)]\} u_k(t). \tag{19}$$

Graphs of y and the square wave input are shown in Figure 5.6.4. The graph of y suggests that the system is exhibiting resonance, a fact that is not immediately apparent from the representation (19). However the trigonometric identity $\cos \pi(t - k) = \cos \pi t \cos \pi k = (-1)^k \cos \pi t$ allows us to express the sum (19) as

$$y(t) = \frac{1}{\pi^2} \sum_{k=0}^{\infty} \left[(-1)^k - \cos \pi t \right] u_k(t). \tag{20}$$

If we evaluate Eq. (20) at the integers $t = n$, then using the facts that $u_k(n) = 0$ if $k > n$ and $\cos \pi n = (-1)^n$, we find that

$$y(n) = -\frac{n}{\pi^2} \quad \text{if } n \text{ is an even integer,}$$

and

$$y(n) = \frac{n + 1}{\pi^2} \quad \text{if } n \text{ is an odd integer.}$$

The graphs of $y(n)$ versus n for $n = 1, 3, 5, \ldots$ and $y(n)$ versus n for $n = 0, 2, 4, \ldots$, form upper and lower envelopes for the graph of $y(t)$, $t \geq 0$, respectively, and are shown as dashed lines in Figure 5.6.4. From the graphs of y and the square wave input f, we see that $f(t) = 1$, a force in the positive y direction, during time intervals when $y(t)$ is increasing and $f(t) = 0$ during time intervals when $y(t)$ is decreasing. This is what a child does on a swing to increase the amplitude of the swing's motion, that is, push with your feet when moving forward, and relaxing when moving backward. Thus phase synchronization between input $f(t)$ and response $y(t)$ gives rise to the phenomenon of resonance.

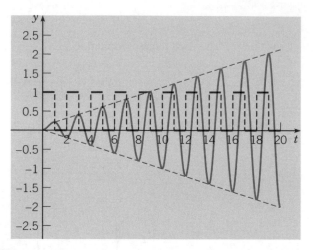

FIGURE 5.6.4 Resonance (solid blue) due to a square wave (black) input.

PROBLEMS

In each of Problems 1 through 13, find the solution of the given initial value problem. Draw the graphs of the solution and of the forcing function; explain how they are related.

1. $y'' + y = f(t);$ $\quad y(0) = 5,$ $\quad y'(0) = 3;$

$$f(t) = \begin{cases} 1, & 0 \le t < \pi/2 \\ 0, & \pi/2 \le t < \infty \end{cases}$$

2. $y'' + 2y' + 2y = h(t);$ $\quad y(0) = 5,$ $\quad y'(0) = 4;$

$$h(t) = \begin{cases} 1, & \pi \le t < 2\pi \\ 0, & 0 \le t < \pi \text{ and } t \ge 2\pi \end{cases}$$

3. $y'' + 4y = \sin t - u_{2\pi}(t)\sin(t - 2\pi);$

$y(0) = 0,$ $\quad y'(0) = 0$

4. $y'' + 4y = \sin t + u_\pi(t)\sin(t - \pi);$

$y(0) = 0,$ $\quad y'(0) = 0$

5. $y'' + 3y' + 2y = f(t);$ $\quad y(0) = 0,$ $\quad y'(0) = 0;$

$$f(t) = \begin{cases} 1, & 0 \le t < 10 \\ 0, & t \ge 10 \end{cases}$$

6. $y'' + 3y' + 2y = u_2(t);$ $\quad y(0) = 6,$ $\quad y'(0) = 6$

7. $y'' + y = u_{3\pi}(t);$ $\quad y(0) = 1,$ $\quad y'(0) = 0$

8. $y'' + y' + 1.25y = t - u_{\pi/2}(t)(t - \pi/2);$

$y(0) = 0,$ $\quad y'(0) = 0$

9. $y'' + y = g(t);$ $\quad y(0) = 6,$ $\quad y'(0) = 8;$

$$g(t) = \begin{cases} t/2, & 0 \le t < 6 \\ 3, & t \ge 6 \end{cases}$$

10. $y'' + y' + 1.25y = g(t);$ $\quad y(0) = 0,$ $\quad y'(0) = 0;$

$$g(t) = \begin{cases} \sin t, & 0 \le t < \pi \\ 0, & t \ge \pi \end{cases}$$

11. $y'' + 4y = u_\pi(t) - u_{3\pi}(t);$ $\quad y(0) = 3,$ $\quad y'(0) = 7$

12. $y^{(4)} - y = u_1(t) - u_2(t);$ $\quad y(0) = 12,$

$y'(0) = 7,$ $\quad y''(0) = 2,$ $\quad y'''(0) = -9$

13. $y^{(4)} + 5y'' + 4y = 1 - u_\pi(t);$ $\quad y(0) = 0,$

$y'(0) = 0,$ $\quad y''(0) = 0,$ $\quad y'''(0) = 0$

14. Find an expression involving $u_c(t)$ for a function f that ramps up from zero at $t = t_0$ to the value h at $t = t_0 + k$.

15. Find an expression involving $u_c(t)$ for a function g that ramps up from zero at $t = t_0$ to the value h at $t = t_0 + k$ and then ramps back down to zero at $t = t_0 + 2k$.

16. A certain spring-mass system satisfies the initial value problem

$$u'' + 0.25u' + u = kg(t), \qquad u(0) = 0, \quad u'(0) = 0,$$

where $g(t)$ is equal to 1 over the time interval $\frac{3}{2} < t < \frac{5}{2}$, is zero elsewhere, and $k > 0$ is a parameter.
(a) Sketch the graph of $g(t)$.
(b) Solve the initial value problem.
(c) Plot the solution for $k = \frac{1}{2}$, $k = 1$, and $k = 2$. Describe the principal features of the solution and how they depend on k.
(d) Find, to two decimal places, the smallest value of k for which the solution $u(t)$ reaches the value 2.
(e) Suppose $k = 2$. Find the time τ after which $|u(t)| < 0.1$ for all $t > \tau$.

17. Modify the problem in Example 1 of this section by replacing the given forcing function $g(t)$ by

$$f(t) = \frac{u_5(t)(t-5) - u_{5+k}(t)(t-5-k)}{k}.$$

(a) Sketch the graph of $f(t)$ and describe how it depends on k. For what value of k is $f(t)$ identical to $g(t)$ in the example?
(b) Solve the initial value problem

$$y'' + 4y = f(t), \qquad y(0) = 0, \quad y'(0) = 0.$$

(c) The solution in part (b) depends on k, but for sufficiently large t the solution is always a simple harmonic oscillation about $y = \frac{1}{4}$. Try to decide how the amplitude of this eventual oscillation depends on k. Then confirm your conclusion by plotting the solution for a few different values of k.

18. Consider the initial value problem

$$y'' + \tfrac{1}{3}y' + 4y = f_k(t), \qquad y(0) = 0, \quad y'(0) = 0,$$

where

$$f_k(t) = \begin{cases} 1/2k, & 4-k \le t < 4+k \\ 0, & 0 \le t < 4-k \ \text{ and } \ t \ge 4+k \end{cases}$$

and $0 < k < 4$.
(a) Sketch the graph of $f_k(t)$. Observe that the area under the graph is independent of k. If $f_k(t)$ represents a force, this means that the product of the magnitude of the force and the time interval during which it acts does not depend on k.
(b) Write $f_k(t)$ in terms of the unit step function and then solve the given initial value problem.
(c) Plot the solution for $k = 2$, $k = 1$, and $k = \frac{1}{2}$. Describe how the solution depends on k.

Resonance and Beats. In Section 4.6 we observed that an undamped harmonic oscillator (such as a spring-mass system) with a sinusoidal forcing term experiences resonance if the frequency of the forcing term is the same as the natural frequency. If the forcing frequency is slightly different from the natural frequency, then the system exhibits a beat.

In Problems 19 through 23, we explore the effect of some nonsinusoidal periodic forcing functions:

19. Consider the initial value problem

$$y'' + y = f(t), \qquad y(0) = 0, \quad y'(0) = 0,$$

where

$$f(t) = 1 + 2\sum_{k=1}^{n}(-1)^k u_{k\pi}(t).$$

(a) Find the solution of the initial value problem.
(b) Let $n = 15$ and plot the graph of $f(t)$ and $y(t)$ for $0 \le t \le 60$ on the same set of coordinate axes. Describe the solution and explain why it behaves as it does.
(c) Investigate how the solution changes as n increases. What happens as $n \to \infty$?

20. Consider the initial value problem

$$y'' + 0.1y' + y = f(t), \qquad y(0) = 0, \quad y'(0) = 0,$$

where $f(t)$ is the same as in Problem 19.
(a) Find the solution of the initial value problem.
(b) Plot the graph of $f(t)$ and $y(t)$ on the same set of coordinate axes. Use a large enough value of n and a long enough t-interval so that the transient part of the solution has become negligible and the steady state is clearly shown.
(c) Estimate the amplitude and frequency of the steady-state part of the solution.
(d) Compare the results of part (b) with those from Section 4.6 for a sinusoidally forced oscillator.

21. Consider the initial value problem

$$y'' + y = g(t), \qquad y(0) = 0, \quad y'(0) = 0,$$

where

$$g(t) = 1 + \sum_{k=1}^{n}(-1)^k u_{k\pi}(t).$$

(a) Draw the graph of $g(t)$ on an interval such as $0 \le t \le 6\pi$. Compare the graph with that of $f(t)$ in Problem 19(a).
(b) Find the solution of the initial value problem.
(c) Let $n = 15$ and plot the graph of the solution for $0 \le t \le 60$. Describe the solution and explain why it behaves as it does. Compare it with the solution of Problem 19.
(d) Investigate how the solution changes as n increases. What happens as $n \to \infty$?

22. Consider the initial value problem

$$y'' + 0.1y' + y = g(t), \qquad y(0) = 0, \quad y'(0) = 0,$$

where $g(t)$ is the same as in Problem 21.
(a) Plot the graph of the solution. Use a large enough value of n and a long enough t-interval so that the transient part of the solution has become negligible and the steady state is clearly shown.
(b) Estimate the amplitude and frequency of the steady-state part of the solution.

(c) Compare the results of part (b) with those from Problem 20 and from Section 4.6 for a sinusoidally forced oscillator.

23. Consider the initial value problem

$$y'' + y = f(t), \qquad y(0) = 0, \quad y'(0) = 0,$$

where

$$f(t) = 1 + 2 \sum_{k=1}^{n} (-1)^k u_{11k/4}(t).$$

Observe that this problem is identical to Problem 19 except that the frequency of the forcing term has been increased somewhat.

(a) Find the solution of this initial value problem.
(b) Let $n \geq 33$ and plot $y(t)$ and $f(t)$ for $0 \leq t \leq 90$ or longer on the same set of coordinate axes. Your plot should show a clearly recognizable beat.
(c) From the graph in part (b), estimate the "slow period" and the "fast period" for this oscillator.
(d) For a sinusoidally forced oscillator, it was shown in Section 4.6 that the "slow frequency" is given by $|\omega - \omega_0|/2$, where ω_0 is the natural frequency of the system and ω is the forcing frequency. Similarly, the "fast frequency" is $(\omega + \omega_0)/2$. Use these expressions to calculate the "fast period" and the "slow period" for the oscillator in this problem. How well do the results compare with your estimates from part (c)?

5.7 Impulse Functions

In some applications, it is necessary to deal with phenomena of an impulsive nature—for example, voltages or forces of large magnitude that act over very short time intervals. Such problems often lead to differential equations of the form

$$ay'' + by' + cy = g(t), \tag{1}$$

where $g(t)$ is large during a short time interval $t_0 \leq t < t_0 + \epsilon$ and is otherwise zero.

One measure of the strength of the input function is obtained by defining the integral $I(\epsilon)$ as

$$I(\epsilon) = \int_{t_0}^{t_0 + \epsilon} g(t)\, dt. \tag{2}$$

Or, since $g(t) = 0$ outside of the interval $[t_0, t_0 + \epsilon)$,

$$I(\epsilon) = \int_{-\infty}^{\infty} g(t)\, dt. \tag{3}$$

For example, the initial value problem for a mass-spring system at equilibrium during $0 \leq t < t_0$, suddenly set into motion by striking the mass with a hammer at time t_0, is

$$my'' + \gamma y' + ky = g(t), \qquad y(0) = 0, \quad y'(0) = 0, \tag{4}$$

where the nonhomogeneous term $g(t)$ represents a force of large magnitude and short duration, as shown in Figure 5.7.1. However its detailed behavior may be difficult or impossible to ascertain.

In applications where $g(t)$ represents a force, $I(\epsilon)$ is referred to as the total **impulse** of the force over the time interval $[t_0, t_0 + \epsilon)$ and has the physical units of momentum (force × time). Prior to time t_0 the mass has zero momentum, but during the time interval $[t_0, t_0 + \epsilon)$ an amount of momentum equal to $I(\epsilon)$ is transferred to the mass. The idea that we wish to quantify here is that the dominant contribution to the system response $y(t)$ for times $t \geq t_0 + \epsilon$ is primarily determined by the magnitude of the impulse $I(\epsilon)$ rather than the detailed behavior of the forcing function $g(t)$. This is illustrated by the following example.

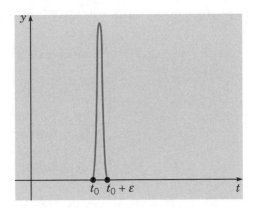

FIGURE 5.7.1 A force of large magnitude active during the short time interval $[t_0, t_0 + \epsilon)$.

EXAMPLE 1

Let I_0 be a real number and let ϵ be a small positive constant. Suppose that $t_0 = 0$ and that $g(t)$ is given by $g(t) = I_0 \delta_\epsilon(t)$, where

$$\delta_\epsilon(t) = \frac{u_0(t) - u_\epsilon(t)}{\epsilon} = \begin{cases} \dfrac{1}{\epsilon}, & 0 \le t < \epsilon, \\ 0, & t < 0 \quad \text{or} \quad t \ge \epsilon, \end{cases} \tag{5}$$

where u_0 and u_ϵ are unit step functions. Graphs of δ_ϵ for various small values of ϵ are shown in Figure 5.7.2. Since the area beneath the graph of $\delta_\epsilon(t)$ is equal to 1 for each $\epsilon > 0$, Eq. (3) implies that $I(\epsilon) = I_0$ for each $\epsilon > 0$.

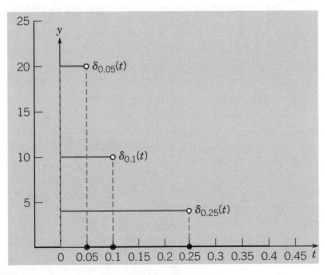

FIGURE 5.7.2 Graphs of $\delta_\epsilon(t)$ for $\epsilon = 0.25, 0.1$, and 0.05.

A model for a simple, undamped, mass-spring system ($m = 1$, $\gamma = 0$, $k = 1$) in a state of equilibrium at $t = 0$, and subjected to the force $I_0 \delta_\epsilon(t)$ prescribed in Eq. (5), is

$$y'' + y = I_0 \delta_\epsilon(t), \qquad y(0) = 0, \quad y'(0) = 0. \tag{6}$$

The response, easily obtained by using Laplace transforms, is

$$y_\epsilon(t) = \frac{I_0}{\epsilon}\{u_0(t)[1 - \cos(t)] - u_\epsilon(t)[1 - \cos(t - \epsilon)]\}$$

$$= \begin{cases} 0, & t < 0, \\ \dfrac{I_0}{\epsilon}[1 - \cos(t)], & 0 \le t < \epsilon, \\ \dfrac{I_0}{\epsilon}[\cos(t - \epsilon) - \cos(t)], & t \ge \epsilon. \end{cases} \tag{7}$$

From Eq. (7),

$$y_0(t) = \lim_{\epsilon \to 0} y_\epsilon(t) = u_0(t)I_0 \sin(t) = \begin{cases} 0, & t < 0, \\ I_0 \sin(t), & t \ge 0. \end{cases} \tag{8}$$

The result for $t = 0$ comes from the second line on the right side of Eq. (7), where $y_\epsilon(0) = 0$ for each $\epsilon > 0$. Consequently, the limit is zero. For any fixed $t > 0$, we use the third line on the right side of Eq. (7) since as ϵ tends to zero, it eventually must be less than t. Then by L'Hôpital's rule,

$$\lim_{\epsilon \to 0} I_0 \frac{\cos(t - \epsilon) - \cos(t)}{\epsilon} = I_0 \lim_{\epsilon \to 0} \sin(t - \epsilon) = I_0 \sin t. \tag{9}$$

The graphs of y_ϵ for $I_0 = 1, 2$ and $\epsilon = 0.1, 0.01, 0$ in Figure 5.7.3 show that the response is relatively insensitive to variation in small values of ϵ. Figure 5.7.3 also illustrates that the amplitude of the response is determined by I_0, a fact readily apparent in the form of expressions (7) and (8).

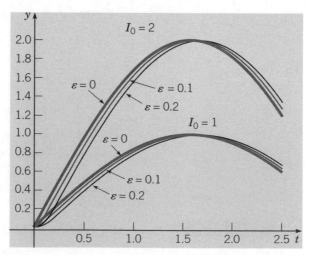

FIGURE 5.7.3 Response of $y'' + y = I_0\delta_\epsilon(t)$, $y(0) = 0$, $y'(0) = 0$ for $I_0 = 1, 2$, and $\epsilon = 0.2$ (black), 0.1 (light blue), 0 (dark blue).

Remark. From Eq. (8) we note that $\lim_{t \to 0^-} y_0(t) = \lim_{\substack{t \to 0 \\ t < 0}} y_0(t) = 0$ and $\lim_{t \to 0^+} y_0(t) = \lim_{\substack{t \to 0 \\ t > 0}} y_0(t) = 0$, so $y_0(t)$ is continuous at $t = 0$, and the first initial condition in the initial value problem (6) is satisfied. The sense in which the second initial condition is satisfied is found by considering

$$y_0'(t) = \begin{cases} 0, & t < 0, \\ I_0 \cos t, & t > 0. \end{cases} \tag{10}$$

Since $\lim_{t \to 0^-} y_0'(t) = 0$ but $\lim_{t \to 0^+} y_0(t) = \lim_{t \to 0^+} I_0 \cos t = I_0$, $y_0'(t)$ is not continuous at $t = 0$. Therefore it is necessary to interpret the second initial condition, $y'(0) = 0$, as being satisfied in the sense that $\lim_{t \to 0^-} y'(t) = 0$.

▶ **Definition of the Unit Impulse Function.** It is somewhat tedious first to solve an initial value problem using $\delta_\epsilon(t - t_0)$ to model an input pulse and then to compute the limiting behavior of the solution as $\epsilon \to 0$. This type of analysis can usually be avoided by introducing an idealized object, the **unit impulse function** δ, that imparts an impulse of magnitude 1 at $t = t_0$, but is zero for all values of t other than t_0. The properties that define the "function" δ, made precise in the following statements, derive from the limiting behavior of δ_ϵ as $\epsilon \to 0$:

1.

$$\delta(t - t_0) = \lim_{\epsilon \to 0} \delta_\epsilon(t - t_0) = 0, \qquad t \neq t_0, \tag{11}$$

2. for any function continuous on an interval $a \leq t_0 < b$ containing t_0,

$$\int_a^b f(t)\delta(t - t_0)\,dt = \lim_{\epsilon \to 0} \int_a^b f(t)\delta_\epsilon(t - t_0)\,dt = f(t_0). \tag{12}$$

The first property is due to the fact that $\delta_\epsilon(t - t_0)$ is nonzero on the interval $[t_0, t_0 + \epsilon)$, so any $t \neq t_0$ must lie outside the interval for sufficiently small values of ϵ. To understand the second property, note that if $\epsilon < b - t_0$, $[t_0, t_0 + \epsilon) \subset [a, b)$ and therefore

$$\int_a^b f(t)\delta_\epsilon(t - t_0)\,dt = \frac{1}{\epsilon} \int_{t_0}^{t_0+\epsilon} f(t)\,dt = \frac{1}{\epsilon} \cdot \epsilon \cdot f(t^*) = f(t^*), \tag{13}$$

for some number t^* with $t_0 \leq t^* \leq t_0 + \epsilon$. The second equality in Eq. (13) follows from the mean value theorem for integrals. Since $t_0 \leq t^* \leq t_0 + \epsilon$, $t^* \to t_0$ as $\epsilon \to 0$. Consequently, $f(t^*) \to f(t_0)$ as $\epsilon \to 0$ since f is continuous at t_0. Note that if $f(t) = 1$, Eqs. (11) and (12) imply that

$$\int_a^b \delta(t - t_0)\,dt = \begin{cases} 1, & \text{if } a \leq t_0 < b, \\ 0, & \text{if } t_0 \notin [a, b). \end{cases}$$

There is no ordinary function of the kind studied in elementary calculus that satisfies both of Eqs. (11) and (12). The "function" δ, defined by those equations, is an example of what are known as generalized functions; it is usually called the Dirac **delta function**.

▶ The Laplace Transform of $\delta(t - t_0)$. Since e^{-st} is a continuous function of t for all t and s, Eq. (12) gives

$$\mathcal{L}\{\delta(t - t_0)\} = \int_0^\infty e^{-st}\delta(t - t_0)\,dt = e^{-st_0} \tag{14}$$

for all $t_0 \geq 0$. In the important case when $t_0 = 0$,

$$\mathcal{L}\{\delta(t)\} = \int_0^\infty e^{-st}\delta(t)\,dt = 1. \tag{15}$$

It is often convenient to introduce the delta function when working with impulse problems and to operate formally on it as though it were a function of the ordinary kind. This is illustrated in the examples below. It is important to realize, however, that the ultimate justification of such procedures must rest on a careful analysis of the limiting operations involved. Such a rigorous mathematical theory has been developed, but we do not discuss it here.

EXAMPLE
2

In the initial value problem (6) of Example 1, we replace the model of the input pulse δ_ϵ, by δ,

$$y'' + y = I_0\delta(t), \qquad y(0) = 0, \quad y'(0) = 0.$$

Since $\mathcal{L}\{\delta(t)\} = 1$, Laplace transforming the initial value problem gives $(s^2 + 1)Y(s) = I_0$ and therefore $Y(s) = I_0/(s^2 + 1)$. Thus $y(t) = \mathcal{L}^{-1}\{Y(s)\} = I_0 \sin t$.

Although the initial condition $y(0) = 0$ is satisfied by this solution, the second initial condition is not since $y'(0) = I_0$. However, if we more properly write[2] the solution as $y(t) = u_0(t)I_0 \sin t$, both initial conditions are satisfied in the sense that $\lim_{t\to 0^-} y(t) = \lim_{t\to 0^-} y'(t) = 0$, as discussed in the remark following Example 1. Physically speaking, the delta impulse delivered to the system causes a discontinuity in the velocity $y'(t)$ at $t = 0$; before the impulse $y'(t) = 0$, after the impulse $y'(t) = I_0 \cos t$.

EXAMPLE
3

Find the solution of the initial value problem

$$2y'' + y' + 2y = \delta(t - 5), \tag{16}$$

$$y(0) = 0, \qquad y'(0) = 0. \tag{17}$$

To solve the given problem, we take the Laplace transform of the differential equation and use the initial conditions, obtaining

$$(2s^2 + s + 2)Y(s) = e^{-5s}.$$

[2]To see this, note that the solution to $y'' + y = I_0\delta(t - t_0)$, $y(t_0) = 0$, $y'(t_0) = 0$, where $t_0 > 0$ is $y(t; t_0) = u_{t_0}(t)I_0 \sin(t - t_0)$; let $t_0 \to 0$ to get $y(t; 0) = u_0(t)I_0 \sin t = u(t)I_0 \sin(t)$, where we have used the definitions for $u(t)$ and $u_0(t)$ given by Eqs. (1) and (2) in Section 5.5.

Thus

$$Y(s) = \frac{e^{-5s}}{2s^2 + s + 2} = \frac{e^{-5s}}{2} \frac{1}{\left(s + \frac{1}{4}\right)^2 + \frac{15}{16}}. \tag{18}$$

By Theorem 5.2.1 or from line 9 of Table 5.3.1,

$$\mathcal{L}^{-1}\left\{\frac{1}{\left(s + \frac{1}{4}\right)^2 + \frac{15}{16}}\right\} = \frac{4}{\sqrt{15}} e^{-t/4} \sin \frac{\sqrt{15}}{4} t. \tag{19}$$

Hence, by Theorem 5.5.1, we have

$$y = \mathcal{L}^{-1}\{Y(s)\} = \frac{2}{\sqrt{15}} u_5(t) e^{-(t-5)/4} \sin \frac{\sqrt{15}}{4} (t - 5), \tag{20}$$

which is the formal solution of the given problem. It is also possible to write y in the form

$$y = \begin{cases} 0, & t < 5, \\[2mm] \dfrac{2}{\sqrt{15}} e^{-(t-5)/4} \sin \dfrac{\sqrt{15}}{4} (t - 5), & t \geq 5. \end{cases} \tag{21}$$

The graph of the solution (20), or (21), is shown in Figure 5.7.4.

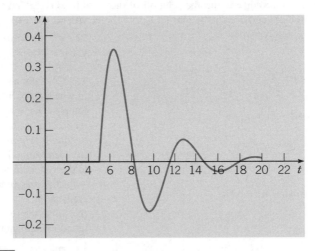

FIGURE 5.7.4   Solution of the initial value problem (16), (17).

Since the initial conditions at $t = 0$ are homogeneous and no external excitation occurs until $t = 5$, there is no response in the interval $0 < t < 5$. The impulse at $t = 5$ produces a decaying oscillation that persists indefinitely. The response is continuous at $t = 5$ despite the singularity in the forcing function at that point. However the first derivative of the solution has a jump discontinuity at $t = 5$, and the second derivative has an infinite discontinuity there. This is required by the differential equation (16), since a singularity on one side of the equation must be balanced by a corresponding singularity on the other side.

▶ $\delta(t - t_0)$ **as the Derivative of** $u(t - t_0)$. In view of the definition of $\delta(t - t_0)$ given in Eqs. (11) and (12), it follows that either

$$\int_{-\infty}^{t} \delta(\tau - t_0)\,d\tau = \begin{cases} 0, & t < t_0, \\ 1, & t > t_0, \end{cases} \tag{22}$$

or

$$\int_{-\infty}^{t} \delta(\tau - t_0)\,d\tau = u(t - t_0) \tag{23}$$

for all $t \neq t_0$. Formally differentiating both sides of Eq. (23) with respect to t and treating $\delta(t - t_0)$ as an ordinary function such that the indefinite integral on the left-hand side of Eq. (23) actually satisfies the fundamental theorem of calculus, we find that

$$\delta(t - t_0) = u'(t - t_0), \tag{24}$$

that is, the delta function is the derivative of the unit step function. In the context of the theory of **generalized functions**, or **distributions**, this is indeed true in a rigorous sense.

We note that the statement in Eq. (24) is consistent with an equivalent statement in the s-domain. If we formally apply the result $\mathcal{L}\{f'(t)\} = sF(s) - f(0)$ to $u'(t - t_0)$ for $t_0 > 0$, we find that $\mathcal{L}\{u'(t - t_0)\} = \mathcal{L}\{u'_{t_0}(t)\} = se^{-st_0}/s - u_{t_0}(0) = e^{-st_0} = \mathcal{L}\{\delta(t - t_0)\}$, in agreement with Eq. (14). Letting $t_0 \to 0$ extends the result to $\mathcal{L}\{u'(t)\} = 1 = \mathcal{L}\{\delta(t)\}$.

PROBLEMS

In each of Problems 1 through 12, find the solution of the given initial value problem and draw its graph:

1. $y'' + 2y' + 2y = \delta(t - \pi)$;
   $y(0) = 0, \quad y'(0) = 1$

2. $y'' + 4y = \delta(t - \pi) - \delta(t - 2\pi)$;
   $y(0) = 0, \quad y'(0) = 0$

3. $y'' + 3y' + 2y = \delta(t - 5) + u_{10}(t)$;
   $y(0) = 0, \quad y'(0) = \frac{1}{2}$

4. $y'' - y = -20\delta(t - 3)$;
   $y(0) = 4, \quad y'(0) = 3$

5. $y'' + 2y' + 3y = \sin t + \delta(t - 3\pi)$;
   $y(0) = 0, \quad y'(0) = 0$

6. $y'' + 4y = \delta(t - 4\pi)$;
   $y(0) = \frac{1}{2}, \quad y'(0) = 0$

7. $y'' + y = \delta(t - 2\pi)\cos t$;
   $y(0) = 0, \quad y'(0) = 1$

8. $y'' + 4y = 2\delta(t - \pi/4)$;
   $y(0) = 0, \quad y'(0) = 0$

9. $y'' + y = u_{\pi/2}(t) + 3\delta(t - 3\pi/2) - u_{2\pi}(t)$;
   $y(0) = 0, \quad y'(0) = 0$

10. $2y'' + y' + 6y = \delta(t - \pi/6)\sin t$;
    $y(0) = 0, \quad y'(0) = 0$

11. $y'' + 2y' + 2y = \cos t + \delta(t - \pi/2)$;
    $y(0) = 0, \quad y'(0) = 0$

12. $y^{(4)} - y = \delta(t - 1)$; $y(0) = 0, \quad y'(0) = 0,$
    $y''(0) = 0, \quad y'''(0) = 0$

13. Consider again the system in Example 3 of this section, in which an oscillation is excited by a unit impulse at $t = 5$. Suppose that it is desired to bring the system to rest again after exactly one cycle—that is, when the response first returns to equilibrium moving in the positive direction.
    (a) Determine the impulse $k\delta(t - t_0)$ that should be applied to the system in order to accomplish this objective. Note that $-k$ is the magnitude of the impulse and t_0 is the time of its application.
    (b) Solve the resulting initial value problem and plot its solution to confirm that it behaves in the specified manner.

14. Consider the initial value problem

    $$y'' + \gamma y' + y = \delta(t - 1), \qquad y(0) = 0, \quad y'(0) = 0,$$

    where γ is the damping coefficient (or resistance).
    (a) Let $\gamma = \frac{1}{2}$. Find the solution of the initial value problem and plot its graph.
    (b) Find the time t_1 at which the solution attains its maximum value. Also find the maximum value y_1 of the solution.
    (c) Let $\gamma = \frac{1}{4}$ and repeat parts (a) and (b).
    (d) Determine how t_1 and y_1 vary as γ decreases. What are the values of t_1 and y_1 when $\gamma = 0$?

15. Consider the initial value problem

$$y'' + \gamma y' + y = k\delta(t - 1), \quad y(0) = 0, \quad y'(0) = 0,$$

where $|k|$ is the magnitude of an impulse at $t = 1$ and γ is the damping coefficient (or resistance).
(a) Let $\gamma = \frac{1}{5}$. Find the value of k for which the response has a peak value of 6; call this value k_1.
(b) Repeat part (a) for $\gamma = \frac{1}{10}$.
(c) Determine how k_1 varies as γ decreases. What is the value of k_1 when $\gamma = 0$?

16. Consider the initial value problem

$$y'' + y = f_k(t), \qquad y(0) = 0, \quad y'(0) = 0,$$

where $f_k(t) = [u_{3-k}(t) - u_{3+k}(t)]/2k$ with $0 < k \le 1$.
(a) Find the solution $y = \phi(t, k)$ of the initial value problem.
(b) Calculate $\lim_{k \to 0} \phi(t, k)$ from the solution found in part (a).
(c) Observe that $\lim_{k \to 0} f_k(t) = \delta(t - 3)$. Find the solution $\phi_0(t)$ of the given initial value problem, with $f_k(t)$ replaced by $\delta(t - 3)$. Is it true that $\phi_0(t) = \lim_{k \to 0} \phi(t, k)$?
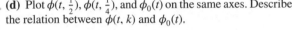
(d) Plot $\phi(t, \frac{1}{2})$, $\phi(t, \frac{1}{4})$, and $\phi_0(t)$ on the same axes. Describe the relation between $\phi(t, k)$ and $\phi_0(t)$.

Problems 17 through 22 deal with the effect of a sequence of impulses on an undamped oscillator. Suppose that

$$y'' + y = f(t), \qquad y(0) = 0, \quad y'(0) = 0.$$

For each of the following choices for $f(t)$:
(a) Try to predict the nature of the solution without solving the problem.
(b) Test your prediction by finding the solution and drawing its graph.
(c) Determine what happens after the sequence of impulses ends.

17. $f(t) = \sum_{k=1}^{20} \delta(t - k\pi)$

18. $f(t) = \sum_{k=1}^{20} (-1)^{k+1} \delta(t - k\pi)$

19. $f(t) = \sum_{k=1}^{20} \delta(t - k\pi/2)$

20. $f(t) = \sum_{k=1}^{20} (-1)^{k+1} \delta(t - k\pi/2)$

21. $f(t) = \sum_{k=1}^{15} \delta[t - (2k - 1)\pi]$

22. $f(t) = \sum_{k=1}^{40} (-1)^{k+1} \delta(t - 15k/4)$

23. The position of a certain lightly damped oscillator satisfies the initial value problem

$$y'' + 0.1y' + y = \sum_{k=1}^{20} (-1)^{k+1} \delta(t - k\pi),$$

$$y(0) = 0, \quad y'(0) = 0.$$

Observe that, except for the damping term, this problem is the same as Problem 18.
(a) Try to predict the nature of the solution without solving the problem.
(b) Test your prediction by finding the solution and drawing its graph.
(c) Determine what happens after the sequence of impulses ends.

24. Proceed as in Problem 23 for the oscillator satisfying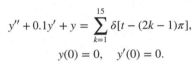

$$y'' + 0.1y' + y = \sum_{k=1}^{15} \delta[t - (2k - 1)\pi],$$

$$y(0) = 0, \quad y'(0) = 0.$$

Observe that, except for the damping term, this problem is the same as Problem 21.

25. (a) By the method of variation of parameters, show that the solution of the initial value problem

$$y'' + 2y' + 2y = f(t), \qquad y(0) = 0, \quad y'(0) = 0$$

is

$$y = \int_0^t e^{-(t-\tau)} f(\tau) \sin(t - \tau) \, d\tau.$$

(b) Show that if $f(t) = \delta(t - \pi)$, then the solution of part (a) reduces to

$$y = u_\pi(t) e^{-(t-\pi)} \sin(t - \pi).$$

(c) Use a Laplace transform to solve the given initial value problem with $f(t) = \delta(t - \pi)$ and confirm that the solution agrees with the result of part (b).

5.8 Convolution Integrals and Their Applications

Consider the initial value problem

$$y'' + y = g(t), \qquad y(0) = 0, \quad y'(0) = 0, \tag{1}$$

where the input $g(t)$ is assumed to be piecewise continuous and of exponential order for $t \ge 0$. Using the method of variation of parameters (see Section 4.7), we can express the solution of Eq. (1) in the integral form

$$y(t) = \int_0^t \sin(t - \tau)g(\tau)\, d\tau. \tag{2}$$

The integral operation involving $\sin t$ and $g(t)$ that appears on the right-hand side of Eq. (2), termed a convolution integral, is an example of the type of operation that arises naturally in representing the response or output $y(t)$ of a linear constant coefficient equation to an input function $g(t)$ in the t-domain. Other places where convolution integrals arise are in various applications in which the behavior of the system at time t depends on not only its state at time t but also its past history. Systems of this kind are sometimes called hereditary systems and occur in such diverse fields as neutron transport, viscoelasticity, and population dynamics. A formal definition of the operation exhibited in Eq. (2) follows.

▶ Definition and Properties of Convolution.

DEFINITION 5.8.1	Let $f(t)$ and $g(t)$ be piecewise continuous functions on $[0, \infty)$. The **convolution of f and g** is defined by $$h(t) = \int_0^t f(t - \tau)g(\tau)\, d\tau. \tag{3}$$

The integral in Eq. (3) is known as a **convolution integral**. It is conventional to emphasize that the convolution integral can be thought of as a "generalized product" by writing

$$h(t) = (f * g)(t). \tag{4}$$

The convolution $f * g$ has several of the properties of ordinary multiplication. These properties are summarized in the following theorem.

THEOREM 5.8.2	$f * g = g * f$	(commutative law)	(5)
	$f * (g_1 + g_2) = f * g_1 + f * g_2$	(distributive law)	(6)
	$(f * g) * h = f * (g * h)$	(associative law)	(7)
	$f * 0 = 0 * f = 0.$		(8)
Proof	The proof of Eq. (5) begins with the definition of $f * g$,		

$$(f * g)(t) = \int_0^t f(t - \tau)g(\tau)\, d\tau.$$

Changing the variable of integration from τ to $u = t - \tau$, we get

$$(f * g)(t) = -\int_t^0 f(u)g(t - u)\, du = \int_0^t g(t - u)f(u)\, du = (g * f)(t).$$

The proofs of properties (6) and (7) are left as exercises. Equation (8) follows from the fact that $f(t - \tau) \cdot 0 = 0$.

There are other properties of ordinary multiplication that the convolution integral does not have. For example, it is not true, in general, that $f * 1$ is equal to f. To see this, note that

$$(f * 1)(t) = \int_0^t f(t - \tau) \cdot 1 \, d\tau = \int_0^t f(u) \, du,$$

where $u = t - \tau$. Thus, if $(f * 1)(t) = f(t)$, it must be the case that $f'(t) = f(t)$ and therefore $f(t) = ce^t$. Since $f(0) = (f * 1)(0) = 0$, it follows that $c = 0$. Therefore the only f satisfying $(f * 1)(t) = f(t)$ is the zero function. Similarly, $f * f$ is not necessarily nonnegative. See Problem 2 for an example.

▶ The Convolution Theorem. Let us again consider the initial value problem (1). By Theorem 4.2.1, Eq. (2) is the unique solution of the initial value problem (1) for $t \geq 0$. On the other hand, Laplace transforming the initial value problem (1) and solving for $Y(s) = \mathcal{L}\{y(t)\}$ show that

$$Y(s) = \frac{1}{1 + s^2} G(s),$$

where $G = \mathcal{L}\{g\}$. By Theorem 5.3.1, there is one and only one continuous version of the inverse Laplace transform of $Y(s)$. We must therefore conclude that

$$\int_0^t \sin(t - \tau) g(\tau) \, d\tau = \mathcal{L}^{-1}\left\{ \frac{1}{1 + s^2} G(s) \right\}, \tag{9}$$

or equivalently,

$$\mathcal{L}\left\{ \int_0^t \sin(t - \tau) g(\tau) \, d\tau \right\} = \frac{1}{1 + s^2} G(s) = \mathcal{L}\{\sin t\} \mathcal{L}\{g(t)\}. \tag{10}$$

Equality between the Laplace transform of the convolution of two functions and the product of the Laplace transforms of the two functions involved in the convolution operation, as expressed in Eq. (10), is an instance of the following general result.

THEOREM 5.8.3	**Convolution Theorem**. If $F(s) = \mathcal{L}\{f(t)\}$ and $G(s) = \mathcal{L}\{g(t)\}$ both exist for $s > a \geq 0$, then $$H(s) = F(s)G(s) = \mathcal{L}\{h(t)\}, \qquad s > a, \tag{11}$$ where $$h(t) = \int_0^t f(t - \tau) g(\tau) \, d\tau = \int_0^t f(\tau) g(t - \tau) \, d\tau. \tag{12}$$
Proof	We note first that if $$F(s) = \int_0^\infty e^{-s\xi} f(\xi) \, d\xi$$ and $$G(s) = \int_0^\infty e^{-s\tau} g(\tau) \, d\tau,$$

then

$$F(s)G(s) = \int_0^\infty e^{-s\xi} f(\xi)\, d\xi \int_0^\infty e^{-s\tau} g(\tau)\, d\tau. \tag{13}$$

Since the integrand of the first integral does not depend on the integration variable of the second, we can write $F(s)G(s)$ as an iterated integral,

$$F(s)G(s) = \int_0^\infty e^{-s\tau} g(\tau) \left[\int_0^\infty e^{-s\xi} f(\xi)\, d\xi \right] d\tau$$

$$= \int_0^\infty g(\tau) \left[\int_0^\infty e^{-s(\xi+\tau)} f(\xi)\, d\xi \right] d\tau. \tag{14}$$

The latter integral can be put into a more convenient form by introducing a change of variable. Let $\xi = t - \tau$, for fixed τ, so that $d\xi = dt$. Further $\xi = 0$ corresponds to $t = \tau$ and $\xi = \infty$ corresponds to $t = \infty$. Then, the integral with respect to ξ in Eq. (14) is transformed into one with respect to t:

$$F(s)G(s) = \int_0^\infty g(\tau) \left[\int_\tau^\infty e^{-st} f(t - \tau)\, dt \right] d\tau. \tag{15}$$

The iterated integral on the right side of Eq. (15) is carried out over the shaded wedge-shaped region extending to infinity in the $t\tau$-plane shown in Figure 5.8.1.

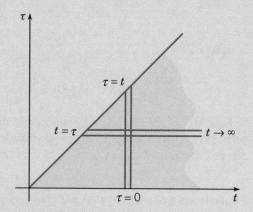

FIGURE 5.8.1 The iterated integrals in Eqs. (14) and (15) are carried out over this wedge-shaped region. Two description of this region are $\{(t,\tau); 0 \le \tau < t, 0 \le t < \infty\}$ and $\{(t,\tau); \tau \le t < \infty, 0 \le \tau < \infty\}$.

Assuming that the order of integration can be reversed, we finally obtain

$$F(s)G(s) = \int_0^\infty e^{-st} \left[\int_0^t f(t - \tau) g(\tau)\, d\tau \right] dt, \tag{16}$$

or

$$F(s)G(s) = \int_0^\infty e^{-st} h(t)\, dt$$

$$= \mathcal{L}\{h(t)\}, \tag{17}$$

where $h(t)$ is defined by Eq. (3).

EXAMPLE 1

Test Theorem 5.8.3 by finding the convolution of $f(t) = t$ and $g(t) = e^{-2t}$ in two different ways:

(i) by direct evaluation of the convolution integral (3) in Definition 5.8.1, and

(ii) by directly computing the inverse Laplace transform of $F(s)G(s)$, the product of the Laplace transforms of $f(t)$ and $g(t)$.

By using integration by parts, the convolution of $f(t) = t$ and $g(t) = e^{-2t}$ is found to be

$$(f * g)(t) = \int_0^t (t - \tau)e^{-2\tau} \, d\tau$$

$$= -\frac{1}{2}(t - \tau)e^{-2\tau}\Big|_{\tau=0}^{\tau=t} - \frac{1}{2}\int_0^t e^{-2\tau} \, d\tau$$

$$= \frac{t}{2} - \frac{1}{4} + \frac{1}{4}e^{-2t}. \tag{18}$$

On the other hand, since $F(s) = 1/s^2$ and $G(s) = 1/(s+2)$,

$$F(s)G(s) = \frac{1}{s^2(s+2)}.$$

Using partial fractions, we can express $F(s)G(s)$ in the form

$$F(s)G(s) = \frac{1}{2}\frac{1}{s^2} - \frac{1}{4}\frac{1}{s} + \frac{1}{4}\frac{1}{s+2}.$$

It readily follows that

$$\mathcal{L}^{-1}\{F(s)G(s)\} = \frac{t}{2} - \frac{1}{4} + \frac{1}{4}e^{-2t},$$

in agreement with Eq. (18).

The convolution theorem can sometimes be effectively used to compute the inverse Laplace transform of a function in the s-domain when it may not be readily apparent how to proceed using partial fraction expansions along with a table of Laplace transforms. For example, see Problems 11 and 12.

▶ **Free and Forced Responses of Input–Output Problems.** Consider the problem consisting of the differential equation

$$ay'' + by' + cy = g(t), \tag{19}$$

where a, b, and c are real constants and g is a given function, together with the initial conditions

$$y(0) = y_0, \qquad y'(0) = y_1. \tag{20}$$

The initial value problem (19), (20) is often referred to as an **input–output problem**. The coefficients a, b, and c describe the properties of some physical system, $g(t)$ is the input to the system, and the values y_0 and y_1 describe the initial state of the system. The solution of Eqs. (19) and (20), the total response, can be separated into two parts: the forced response and the free response. We now show how representations of these responses are found in both the s-domain and the t-domain.

By taking the Laplace transform of Eq. (19) and using the initial conditions (20), we obtain

$$(as^2 + bs + c)Y(s) - (as + b)y_0 - ay_1 = G(s).$$

If we set

$$H(s) = \frac{1}{as^2 + bs + c}, \tag{21}$$

then we can express $Y(s)$ as

$$Y(s) = H(s)\left[(as + b)y_0 + ay_1\right] + H(s)G(s). \tag{22}$$

Consequently,

$$y(t) = \mathcal{L}^{-1}\left\{H(s)\left[(as + b)y_0 + ay_1\right]\right\} + \int_0^t h(t - \tau)g(\tau)\,d\tau, \tag{23}$$

where we have used the convolution theorem to represent $\mathcal{L}^{-1}\left\{H(s)G(s)\right\}$ as a convolution integral. Observe that the first term on the right-hand side of Eq. (23) is the solution of the initial value problem

$$ay'' + by' + cy = 0, \qquad y(0) = y_0, \quad y'(0) = y_1, \tag{24}$$

obtained from Eqs. (19) and (20) by setting $g(t)$ equal to zero. The solution of the initial value problem (24) is called the **free response** of the system in the t-domain. Its representation in the s-domain is the first term on the right-hand side of Eq. (22.) The second term on the right-hand side of Eq. (23) is the solution of

$$ay'' + by' + cy = g(t), \qquad y(0) = 0, \quad y'(0) = 0, \tag{25}$$

in which the initial values y_0 and y_1 are each replaced by zero. The solution of the initial value problem (25) is called the **forced response** of the system. Its counterpart in the s-domain is given by the second term on the right-hand side of Eq. (22), the two representations being related by Theorem 5.8.3.

Using the time-domain methods of Chapter 4, an alternative representation of the free response is

$$y(t) = \alpha_1 y_1(t) + \alpha_2 y_2(t),$$

where $y_1(t)$ and $y_2(t)$ comprise a fundamental set of solutions of $ay'' + by' + cy = 0$, and α_1 and α_2 are determined by solving the linear system

$$\alpha_1 y_1(0) + \alpha_2 y_2(0) = y_0,$$
$$\alpha_1 y_1'(0) + \alpha_2 y_2'(0) = y_1.$$

Thus

$$\alpha_1 = \frac{y_0 y_2'(0) - y_1 y_2(0)}{y_1(0)y_2'(0) - y_1'(0)y_2(0)} \qquad \text{and} \qquad \alpha_2 = \frac{y_1 y_1(0) - y_0 y_1'(0)}{y_1(0)y_2'(0) - y_1'(0)y_2(0)}. \tag{26}$$

By the uniqueness of solutions of the initial value problem (Theorem 4.2.1) and the one-to-one correspondence between piecewise continuous functions of exponential order and their Laplace transforms (Theorem 5.3.1), we must conclude that, with the values of α_1 and α_2 specified in Eqs. (26),

$$\mathcal{L}\{\alpha_1 y_1(t) + \alpha_2 y_2(t)\} = H(s)\left[(as + b)y_0 + ay_1\right].$$

Results of the above discussion are summarized in Table 5.8.1.

TABLE 5.8.1 The total response of $ay'' + by' + cy = g(t)$, $y(0) = y_0$, $y'(0) = y_1$ as a sum of the free response and the forced response in both the s-domain and the t-domain.

	Total Response		Free Response		Forced Response
s-domain:	$Y(s)$	$=$	$H(s)[(as+b)y_0 + ay_1]$	$+$	$H(s)G(s)$
t-domain:	$y(t)$	$=$	$\alpha_1 y_1(t) + \alpha_2 y_2(t)$	$+$	$\int_0^t h(t-\tau)g(\tau)\,d\tau$

▶ Transfer Functions and Impulse Responses. There are many applications where the dominant component of the total response is the forced response and the free response is of little importance. For example, consider an automobile suspension system consisting of a quarter of the automobile's mass, a wheel, a spring, and a shock absorber. If the input to the system arises from the vertical motion of the wheel rolling over a rough surface, it is intuitively clear that the initial vertical displacement and velocity of the automobile's mass are of little importance and it is the forced response that best represents the smoothness of the ride. Mathematically speaking, if the real parts of the roots of the characteristic equation, $a\lambda^2 + b\lambda + c = 0$, for $ay'' + by' + cy = 0$ are negative, due to damping, for example, then the free response of the system is transient and often undetectable after a short period of time; it is the forced response that is of primary interest. Unless otherwise stated, when we refer to the output of an input–output system in either the t-domain or the s-domain, we mean the forced response.

DEFINITION 5.8.4	The **transfer function** of the input–output problem (19), (20) is the ratio of the forced response to the input in the s-domain. Equivalently, the transfer function is the factor in the equation for $Y(s)$ multiplying the Laplace transform of the input, $G(s)$.

Obviously, the transfer function for the input–output problem (19), (20) is $H(s)$ defined in Eq. (21),

$$H(s) = \frac{1}{as^2 + bs + c}.$$

Note that the transfer function $H(s)$ completely characterizes the system (19) since it contains all of the information concerning the coefficients of the system (19). On the other hand, $G(s)$ depends only on the external excitation $g(t)$ that is applied to the system.

The s-domain forced response $H(s)G(s)$ reduces to the transfer function $H(s)$ when the input $G(s) = 1$. This means that the corresponding input in the time domain is $g(t) = \delta(t)$ since $\mathcal{L}\{\delta(t)\} = 1$. Consequently, the inverse Laplace transform of the transfer function, $h(t) = \mathcal{L}^{-1}\{H(s)\}$, is the solution of the initial value problem

$$ay'' + by' + cy = \delta(t), \qquad y(0) = 0, \quad y'(0) = 0, \tag{27}$$

obtained from the initial value problem (25) by replacing $g(t)$ by $\delta(t)$. Thus $h(t)$ is the response of the system to a unit impulse applied at $t = 0$ under the condition that all initial conditions are zero. It is natural to call $h(t)$ the **impulse response** of the system. The forced response in the t-domain is the convolution of the impulse response and the forcing function. The block diagrams in Figure 5.8.2 represent the mapping of inputs to outputs in both the t-domain and s-domain.

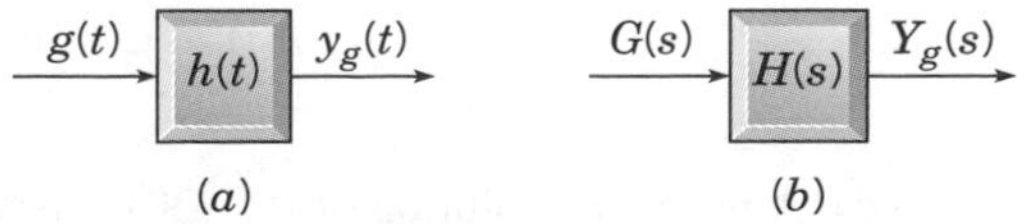

$$\xrightarrow{g(t)} \boxed{h(t)} \xrightarrow{y_g(t)} \qquad \xrightarrow{G(s)} \boxed{H(s)} \xrightarrow{Y_g(s)}$$

$$(a) \qquad\qquad\qquad (b)$$

FIGURE 5.8.2   Block diagram of forced response of $ay'' + by' + cy = g(t)$ in (a) the t-domain and (b) the s-domain.

In the t-domain, the system is characterized by the impulse response $h(t)$, as shown in Figure 5.8.2a. The response or output $y_g(t)$ due to the input $g(t)$ is represented by the convolution integral

$$y_g(t) = \int_0^t h(t - \tau)g(\tau)\,d\tau. \tag{28}$$

In the Laplace or s-domain, the system is characterized by the transfer function $H(s) = \mathcal{L}\{h(t)\}$. The output $Y_g(s)$ due to the input $G(s)$ is represented by the product of $H(s)$ and $G(s)$,

$$Y_g(s) = H(s)G(s). \tag{29}$$

The representation (29) results from taking the Laplace transform of Eq. (28) and using the convolution theorem. In summary, to obtain the system output in the t-domain:

1. Find the transfer function $H(s)$.
2. Find the Laplace transform of the input, $G(s)$.
3. Construct the output in the s-domain, a simple algebraic operation $Y_g(s) = H(s)G(s)$.
4. Compute the output in the t-domain, $y_g(t) = \mathcal{L}^{-1}\{Y_g(s)\}$.

EXAMPLE 2

Consider the input–output system

$$y'' + 2y' + 5y = g(t), \tag{30}$$

where $g(t)$ is any piecewise continuous function of exponential order.

1. Find the transfer function and the impulse response.
2. Using a convolution integral to represent the forced response, find the general solution of Eq. (30).
3. Find the total response if the initial state of the system is prescribed by $y(0) = 1$, $y'(0) = -3$.
4. Compute the forced response when $g(t) = t$.

   (i) The transfer function $H(s)$ is the Laplace transform of the solution of

   $$y'' + 2y' + 5y = \delta(t), \qquad y(0) = 0, \quad y'(0) = 0.$$

   Taking the Laplace transform of the differential equation and using the zero-valued initial conditions give $(s^2 + 2s + 5)Y(s) = 1$. Thus the transfer function is

   $$H(s) = Y(s) = \frac{1}{s^2 + 2s + 5}.$$

   To find the impulse response, $h(t)$, we compute the inverse transform of $H(s)$. Completing the square in the denominator of $H(s)$ yields

   $$H(s) = \frac{1}{(s + 1)^2 + 4} = \frac{1}{2}\frac{2}{(s + 1)^2 + 4}.$$

   Thus

   $$h(t) = \mathcal{L}^{-1}\{H(s)\} = \frac{1}{2}e^{-t}\sin 2t.$$

   (ii) A particular solution of Eq. (30) is provided by the forced response

   $$y_g(t) = \int_0^t h(t - \tau)g(\tau)\,d\tau = \frac{1}{2}\int_0^t e^{-(t-\tau)}\sin 2(t - \tau)\,g(\tau)\,d\tau.$$

The characteristic equation of $y'' + 2y' + 5y = 0$ is $\lambda^2 + 2\lambda + 5 = 0$; it has the complex solutions $\lambda = -1 \pm 2i$. Therefore a fundamental set of real-valued solutions of $y'' + 2y' + 5y = 0$ is $\{e^{-t}\cos 2t, e^{-t}\sin 2t\}$. It follows that the general solution of Eq. (30) is

$$y(t) = c_1 e^{-t} \cos 2t + c_2 e^{-t} \sin 2t + \frac{1}{2} \int_0^t e^{-(t-\tau)} \sin 2(t - \tau) g(\tau) \, d\tau.$$

(iii) The free response is obtained by requiring the complementary solution $y_c(t) = c_1 e^{-t}\cos 2t + c_2 e^{-t}\sin 2t$ to satisfy the initial conditions $y(0) = 1$ and $y'(0) = -3$. This leads to the system $c_1 = 1$ and $-c_1 + 2c_2 = -3$, so $c_2 = -1$. Consequently, the total response is represented by

$$y(t) = e^{-t}\cos 2t - e^{-t}\sin 2t + \frac{1}{2} \int_0^t e^{-(t-\tau)} \sin 2(t - \tau) g(\tau) \, d\tau.$$

Remark. It is worth noting that the particular solution represented by the forced response does not affect the values of the constants c_1 and c_2 since $y_g(0) = 0$ and $y'_g(0) = 0$.

(iv) If $g(t) = t$, the forced response may be calculated either by evaluating the convolution integral

$$y_g(t) = \frac{1}{2} \int_0^t \tau e^{-(t-\tau)} \sin 2(t - \tau) \, d\tau$$

or by computing the inverse Laplace transform of the product of $H(s)$ and the Laplace transform of t, $1/s^2$. We choose the latter method. Using partial fractions, we find that

$$Y_g(s) = \frac{1}{s^2[(s+1)^2 + 4]} = \frac{1}{5s^2} - \frac{2}{25s} + \frac{2}{25} \frac{s+1}{(s+1)^2 + 4} - \frac{3}{50} \frac{2}{(s+1)^2 + 4},$$

and therefore

$$y_g(t) = \frac{1}{5}t - \frac{2}{25} + \frac{2}{25}e^{-t}\cos 2t - \frac{3}{50}e^{-t}\sin 2t.$$

PROBLEMS

1. Establish the distributive and associative properties of the convolution integral.
(a) $f * (g_1 + g_2) = f * g_1 + f * g_2$
(b) $f * (g * h) = (f * g) * h$

2. Show, by means of the example $f(t) = \sin t$, that $f * f$ is not necessarily nonnegative.

In each of Problems 3 through 6, find the Laplace transform of the given function:

3. $f(t) = \int_0^t (t - \tau)^4 \cos 6\tau \, d\tau$

4. $f(t) = \int_0^t e^{-(t-\tau)} \sin \tau \, d\tau$

5. $f(t) = \int_0^t (t - \tau)e^\tau \, d\tau$

6. $f(t) = \int_0^t \sin(t - \tau)\cos \tau \, d\tau$

In each of Problems 7 through 12, find the inverse Laplace transform of the given function by using the convolution theorem:

7. $F(s) = \dfrac{1}{s^3(s^2 + 1)}$

8. $F(s) = \dfrac{s}{(s + 1)(s^2 + 4)}$

9. $F(s) = \dfrac{1}{(s + 3)^4(s^2 + 4)}$

10. $F(s) = \dfrac{G(s)}{s^2 + 1}$

11. $F(s) = \dfrac{1}{(s^2 + 1)^2}$

12. $F(s) = \dfrac{5s}{(s^2 + 25)^2}$

13. (a) If $f(t) = t^m$ and $g(t) = t^n$, where m and n are positive integers, show that

$$(f * g)(t) = t^{m+n+1} \int_0^1 u^m (1 - u)^n \, du.$$

(b) Use the convolution theorem to show that

$$\int_0^1 u^m (1 - u)^n \, du = \frac{m! \, n!}{(m + n + 1)!}.$$

(c) Extend the result of part (b) to the case where m and n are positive numbers but not necessarily integers.

In each of Problems 14 through 21, express the total response of the given initial value problem using a convolution integral to represent the forced response:

14. $y'' + \omega^2 y = g(t); \qquad y(0) = 0, \quad y'(0) = 1$

15. $y'' + 6y' + 25y = \sin \alpha t;$
$y(0) = 0, \quad y'(0) = 0$

16. $4y'' + 4y' + 17y = g(t);$
$y(0) = 0, \quad y'(0) = 0$

17. $y'' + y' + 1.25y = 1 - u_\pi(t);$
$y(0) = 1, \quad y'(0) = -1$

18. $y'' + 4y' + 4y = g(t);$
$y(0) = 2, \quad y'(0) = -3$

19. $y'' + 3y' + 2y = \cos \alpha t;$
$y(0) = 1, \quad y'(0) = 0$

20. $y^{(4)} - 16y = g(t); \qquad y(0) = 0,$
$y'(0) = 0, \quad y''(0) = 0, \quad y'''(0) = 0$

21. $y^{(4)} + 17y'' + 16y = g(t); \qquad y(0) = 2,$
$y'(0) = 0, \quad y''(0) = 0, \quad y'''(0) = 0$

22. Unit Step Responses. The unit step response of a system is the output $y(t)$ when the input is the unit step function $u(t)$ and all initial conditions are zero.
(a) Show that the derivative of the unit step response is the impulse response.
(b) Plot and graph on the same set of coordinates the unit step response of

$$y'' + 2\delta y' + y = u(t)$$

for $\delta = 0.2, 0.3, 0.4, 0.5, 0.6, 0.8,$ and 1.

23. Consider the equation

$$\phi(t) + \int_0^t k(t - \xi)\phi(\xi) \, d\xi = f(t),$$

in which f and k are known functions, and ϕ is to be determined. Since the unknown function ϕ appears under an integral sign, the given equation is called an **integral equation**; in particular, it belongs to a class of integral equations known as Volterra integral equations. Take the Laplace transform of the given integral equation and obtain an expression for $\mathcal{L}\{\phi(t)\}$ in terms of the transforms $\mathcal{L}\{f(t)\}$ and $\mathcal{L}\{k(t)\}$ of the given functions f and k. The inverse transform of $\mathcal{L}\{\phi(t)\}$ is the solution of the original integral equation.

24. Consider the Volterra integral equation (see Problem 23)

$$\phi(t) + \int_0^t (t - \xi)\phi(\xi) \, d\xi = \sin 2t. \qquad \text{(i)}$$

(a) Solve the integral equation (i) by using the Laplace transform.
(b) By differentiating Eq. (i) twice, show that $\phi(t)$ satisfies the differential equation

$$\phi''(t) + \phi(t) = -4 \sin 2t.$$

Show also that the initial conditions are

$$\phi(0) = 0, \qquad \phi'(0) = 2.$$

(c) Solve the initial value problem in part (b) and verify that the solution is the same as the one in part (a).

In each of Problems 25 through 27:
(a) Solve the given Volterra integral equation by using the Laplace transform.
(b) Convert the integral equation into an initial value problem, as in Problem 24(b).
(c) Solve the initial value problem in part (b) and verify that the solution is the same as the one in part (a).

25. $\phi(t) + \displaystyle\int_0^t (t - \xi)\phi(\xi) \, d\xi = 1$

26. $\phi(t) - \displaystyle\int_0^t (t - \xi)\phi(\xi) \, d\xi = 1$

27. $\phi(t) + 2\displaystyle\int_0^t \cos(t - \xi)\phi(\xi) \, d\xi = e^{-t}$

There are also equations, known as **integro-differential equations**, in which both derivatives and integrals of the unknown function appear.

In each of Problems 28 through 30:
(a) Solve the given integro-differential equation by using the Laplace transform.
(b) By differentiating the integro-differential equation a sufficient number of times, convert it into an initial value problem.
(c) Solve the initial value problem in part (b) and verify that the solution is the same as the one in part (a).

28. $\phi'(t) + \displaystyle\int_0^t (t - \xi)\phi(\xi) \, d\xi = t, \qquad \phi(0) = 0$

29. $\phi'(t) - \dfrac{1}{2}\displaystyle\int_0^t (t-\xi)^2 \phi(\xi)\,d\xi = -t, \qquad \phi(0) = 1$

30. $\phi'(t) + \phi(t) = \displaystyle\int_0^t \sin(t-\xi)\phi(\xi)\,d\xi, \qquad \phi(0) = 1$

31. The Tautochrone. A problem of interest in the history of mathematics is that of finding the tautochrone[3]—the curve down which a particle will slide freely under gravity alone, reaching the bottom in the same time regardless of its starting point on the curve. This problem arose in the construction of a clock pendulum whose period is independent of the amplitude of its motion. The tautochrone was found by Christian Huygens in 1673 by geometrical methods, and later by Leibniz and Jakob Bernoulli using analytical arguments. Bernoulli's solution (in 1690) was one of the first occasions on which a differential equation was explicitly solved.

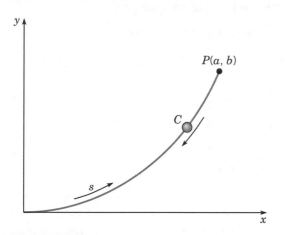

FIGURE 5.8.3 The tautochrone.

The geometric configuration is shown in Figure 5.8.3. The starting point $P(a, b)$ is joined to the terminal point $(0, 0)$ by the arc C. Arc length s is measured from the origin, and $f(y)$ denotes the rate of change of s with respect to y:

$$f(y) = \frac{ds}{dy} = \left[1 + \left(\frac{dx}{dy} \right)^2 \right]^{1/2}. \tag{i}$$

Then it follows from the principle of conservation of energy that the time $T(b)$ required for a particle to slide from P to the origin is

$$T(b) = \frac{1}{\sqrt{2g}} \int_0^b \frac{f(y)}{\sqrt{b-y}}\,dy. \tag{ii}$$

(a) Assume that $T(b) = T_0$, a constant, for each b. By taking the Laplace transform of Eq. (ii) in this case and using the convolution theorem, show that

$$F(s) = \sqrt{\frac{2g}{\pi}} \frac{T_0}{\sqrt{s}}; \tag{iii}$$

then show that

$$f(y) = \frac{\sqrt{2g}}{\pi} \frac{T_0}{\sqrt{y}}. \tag{iv}$$

Hint: See Problem 37 of Section 5.1.
(b) Combining Eqs. (i) and (iv), show that

$$\frac{dx}{dy} = \sqrt{\frac{2\alpha - y}{y}}, \tag{v}$$

where $\alpha = gT_0^2/\pi^2$.
(c) Use the substitution $y = 2\alpha \sin^2(\theta/2)$ to solve Eq. (v), and show that

$$x = \alpha(\theta + \sin\theta), \quad y = \alpha(1 - \cos\theta). \tag{vi}$$

Equations (vi) can be identified as parametric equations of a cycloid. Thus the tautochrone is an arc of a cycloid.

5.9 Linear Systems and Feedback Control

The development of methods and theory of feedback control is one of the most important achievements of modern engineering. Instances of feedback control systems are common: thermostatic control of heating and cooling systems, automobile cruise control, airplane autopilots, industrial robots, and so on. In this section we discuss the application of the Laplace transform to input–output problems and feedback control of systems that can be modeled by linear differential equations with constant coefficients. The subject area is extensive. Thus we limit our presentation to the introduction of only a few important methods and concepts.

[3]The word "tautochrone" comes from the Greek words *tauto*, meaning same, and *chronos*, meaning time.

▶ Feedback Control. Control systems are classified into two general categories: *open-loop* and *closed-loop* systems. An **open-loop** control system is one in which the control action is independent of the output. An example of an open-loop system is a toaster that is controlled by a timer. Once a slice of bread is inserted and the timer is set (the input), the toasting process continues until the timer expires. The input is not affected by the degree to which the slice of bread is toasted (the output). A **closed-loop** control system is one in which the control action depends on the output in some manner, for example, a home heating and cooling system controlled by a thermostat. A block diagram of a simple open-loop control system with transfer function $H(s)$ is shown in Figure 5.9.1a. The output is specified by $Y(s) = H(s)F(s)$, where $F(s)$ represents the control input. In Figure 5.9.1b, a feedback control loop has been added to the open-loop system. The output is used as an input to an element, represented by a transfer function $G(s)$, called a **controller**.[4] The output of the controller $U(s) = G(s)Y(s)$ is then added to (positive feedback) or subtracted from (negative feedback) the external input $F(s)$ at the junction point represented by the circle at the left side of Figure 5.9.1b.[5] In the case of negative feedback, as indicated by the minus sign at the junction point in Figure 5.9.1b, this yields an algebraic equation

$$Y(s) = H(s)\,[F(s) - U(s)] = H(s)\,[F(s) - G(s)Y(s)] \tag{1}$$

in which $Y(s)$ appears on both sides of the equal sign. Solving Eq. (1) for $Y(s)$ yields

$$Y(s) = \frac{H(s)F(s)}{1 + G(s)H(s)}. \tag{2}$$

Thus the transfer function for the closed-loop system is

$$H_G(s) = \frac{Y(s)}{F(s)} = \frac{H(s)}{1 + G(s)H(s)}. \tag{3}$$

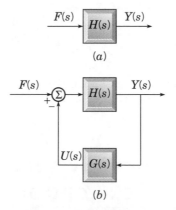

FIGURE 5.9.1   (*a*) Open-loop system. (*b*) Closed-loop system.

[4]The physical hardware associated with the quantity being controlled is often called the **plant**. The plant is usually considered to be fixed and unalterable. The elements added to implement control are referred to as the **controller**. The combined entity of the plant and the controller is called the **closed-loop system**, or simply the **system**.

[5]Note that there is some imprecision in this convention since if the transfer function of the controller were represented by $-G(s)$ instead of $G(s)$, the effect of adding and subtracting the output of the controller would be reversed. In the end, what is really important is achieving a desired performance in the closed-loop system.

It is often desirable to introduce explicitly a parameter associated with the **gain** (output amplification or attenuation) of a transfer function. For example, if the open-loop transfer function $H(s)$ is multiplied by a gain parameter K, then Eq. (3) becomes

$$H_G(s) = \frac{KH(s)}{1 + KG(s)H(s)}. \tag{4}$$

Block diagrams can be used to express relationships between inputs and outputs of components of more complex systems from which the overall transfer function may be deduced by algebraic operations. For example, from the block diagram in Figure 5.9.2 we have

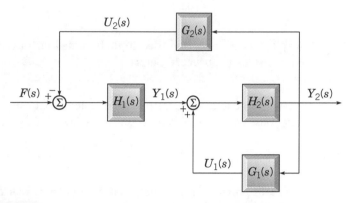

FIGURE 5.9.2 Block diagram of a feedback control system with both positive and negative feedback loops.

$$Y_1(s) = H_1(s)[F(s) - U_2(s)]$$
$$Y_2(s) = H_2(s)[Y_1(s) + U_1(s)]$$
$$U_1(s) = G_1(s)Y_2(s)$$
$$U_2(s) = G_2(s)Y_2(s).$$

Eliminating $U_1(s)$, $U_2(s)$, and $Y_1(s)$ and solving for $Y_2(s)$ yield

$$Y_2(s) = \frac{H_2(s)H_1(s)F(s)}{1 - H_2(s)G_1(s) + H_2(s)H_1(s)G_2(s)}.$$

Thus the overall transfer function of the system described by Figure 5.9.2 is

$$H(s) = \frac{Y_2(s)}{F(s)} = \frac{H_2(s)H_1(s)}{1 - H_2(s)G_1(s) + H_2(s)H_1(s)G_2(s)}. \tag{5}$$

If the transfer functions for each of the components are rational functions, then the closed-loop transfer function will also be a rational function. For example, if $G_1(s) = 1/(s + 1)$, $G_2(s) = 1/(s^2 + 2s + 2)$, $H_1(s) = 1/(s + 2)$, and $H_2(s) = 1/(s + 3)$, then Eq. (5) can be written as

$$H(s) = \frac{\dfrac{1}{(s + 3)(s + 2)}}{1 - \dfrac{1}{(s + 1)(s + 3)} + \dfrac{1}{(s^2 + 2s + 2)(s + 3)(s + 2)}}$$

$$= \frac{s^3 + 3s^2 + 4s + 2}{s^5 + 8s^4 + 24s^3 + 36s^2 + 29s + 9}. \tag{6}$$

▶ Poles, Zeros, and Stability. There are typically several application-dependent performance characteristics that a closed-loop system must satisfy. Achieving these characteristics is part of the design problem of control theory. However, independent of these characteristics, all feedback control systems must satisfy the conditions that we now define for stability.

DEFINITION 5.9.1

A function $f(t)$ defined on $0 \leq t < \infty$ is said to be **bounded** if there is a number M such that

$$|f(t)| \leq M \qquad \text{for all } t \geq 0.$$

DEFINITION 5.9.2

A system is said to be **bounded-input bounded-output (BIBO) stable** if every bounded input results in a bounded output.

We wish to characterize BIBO stable systems that can be represented by rational transfer functions. We consider transfer functions having the general form

$$H(s) = \frac{P(s)}{Q(s)} = \frac{b_m s^m + b_{m-1} s^{m-1} + \cdots + b_0}{s^n + a_{n-1} s^{n-1} + \cdots + a_0}, \tag{7}$$

where the polynomial in the numerator is $P(s) = b_m s^m + b_{m-1} s^{m-1} + \cdots + b_0$ and the polynomial in the denominator is $Q(s) = s^n + a_{n-1} s^{n-1} + \cdots + a_0$. The transfer function (7) is said to have m zeros and n poles. The reason for this is clear if we represent $H(s)$ as

$$H(s) = b_m \frac{(s - \zeta_1)(s - \zeta_2) \cdots (s - \zeta_m)}{(s - \lambda_1)(s - \lambda_2) \cdots (s - \lambda_n)},$$

where we have expressed the numerator polynomial $P(s)$ and denominator polynomial $Q(s)$ in their factored forms

$$P(s) = b_m(s - \zeta_1)(s - \zeta_2) \cdots (s - \zeta_m)$$

and

$$Q(s) = (s - \lambda_1)(s - \lambda_2) \cdots (s - \lambda_n).$$

The m values of s, namely, $\zeta_1, \zeta_2, \ldots, \zeta_m$, that make $P(s)$ zero are called the **zeros** of $H(s)$. The n values of s, that is, $\lambda_1, \lambda_2, \ldots, \lambda_n$, that make $Q(s)$ zero are called the **poles** of $H(s)$. The zeros and poles may be real or complex numbers. Since the coefficients of the polynomials are assumed to be real numbers, complex poles or zeros must occur in complex conjugate pairs. We assume that all transfer functions have more poles than zeros, that is, $n > m$. Such transfer functions are called **strictly proper**. If $H(s)$ is strictly proper, then $\lim_{s \to \infty} H(s) = 0$. Recall from Corollary 5.1.7 that this is necessary in order for $H(s)$ to be the Laplace transform of a piecewise continuous function of exponential order.

THEOREM 5.9.3

An input–output system with a strictly proper transfer function

$$H(s) = \frac{P(s)}{Q(s)} = \frac{b_m s^m + b_{m-1} s^{m-1} + \cdots + b_0}{s^n + a_{n-1} s^{n-1} + \cdots + a_0}$$

is BIBO stable if and only if all of the poles have negative real parts.

Proof   The output $Y(s)$ is related to input $F(s)$ in the s-domain by

$$Y(s) = H(s)F(s).$$

The convolution theorem gives the corresponding relation in the time domain,

$$y(t) = \int_0^t h(t-\tau)f(\tau)\,d\tau.$$

If all of the poles of $H(s)$ have negative real parts, the partial fraction expansion of $H(s)$ is a linear combination of terms of the form

$$\frac{1}{(s-\alpha)^m}, \qquad \frac{(s-\alpha)}{\left[(s-\alpha)^2 + \beta^2\right]^m}, \qquad \frac{\beta}{\left[(s-\alpha)^2 + \beta^2\right]^m},$$

where $m \geq 1$ and $\alpha < 0$. Consequently, $h(t)$ will consist only of a linear combination of terms of the form

$$t^k e^{\alpha t}, \qquad t^k e^{\alpha t}\cos\beta t, \qquad t^k e^{\alpha t}\sin\beta t, \tag{8}$$

where $k \leq m - 1$. Assuming that there exists a constant M_1 such that $|f(t)| \leq M_1$ for all $t \geq 0$, it follows that

$$|y(t)| = \left| \int_0^t h(t-\tau)f(\tau)\,d\tau \right| \leq \int_0^t |h(t-\tau)|\,|f(\tau)|\,d\tau$$

$$\leq M_1 \int_0^t |h(t-\tau)|\,d\tau = M_1 \int_0^t |h(\tau)|\,d\tau \leq M_1 \int_0^\infty |h(\tau)|\,d\tau.$$

Integrating by parts shows that integrals of any terms of the form exhibited in the list (8) are bounded by $k!/|\alpha|^{k+1}$. Consequently, the output $y(t)$ is bounded, which proves the first part of the theorem.

    To prove the second part, we now suppose that every bounded input results in a bounded response. We first show that this implies that $\int_0^\infty |h(\tau)|\,d\tau$ is bounded. Consider the bounded input defined by

$$f(t-\tau) = \begin{cases} 1, & \text{if } h(\tau) \geq 0 \text{ and } \tau < t \\ -1, & \text{if } h(\tau) < 0 \text{ and } \tau < t \\ 0, & \text{if } t < \tau. \end{cases}$$

Then

$$y(t) = \int_0^t h(\tau)f(t-\tau)\,d\tau = \int_0^t |h(\tau)|\,d\tau.$$

Since $y(t)$ is bounded, there exists a constant M_2 such that $|y(t)| \leq M_2$ for all $t \geq 0$. Then from the preceding equation we have

$$|y(t)| = \left| \int_0^t |h(\tau)|\,d\tau \right| = \int_0^t |h(\tau)|\,d\tau \leq M_2$$

for all $t \geq 0$. Thus $\int_0^t |h(\tau)|\,d\tau$ is bounded for all $t \geq 0$, and consequently, $\int_0^\infty |h(\tau)|\,d\tau$ is also bounded, as was to be shown. Since $H(s)$ is the Laplace transform of $h(t)$,

$$H(s) = \int_0^\infty e^{-st}h(t)\,dt.$$

If $\text{Re}(s) \geq 0$, then $|e^{-st}| = e^{-\text{Re}(s)t} \leq 1$ for all $t \geq 0$, so

$$|H(s)| = \left| \int_0^\infty e^{-st} h(t)\, dt \right| \leq \int_0^\infty |e^{-st}|\, |h(t)|\, dt \leq \int_0^\infty |h(t)|\, dt \leq M_2.$$

Since the rational function $|H(s)|$ is bounded on the set $\{s : \text{Re}(s) \geq 0\}$, all poles of $H(s)$ must lie in the left half of the complex s-plane.

▶ Root Locus Analysis. Theorem 5.9.3 points out the importance of pole locations with regard to the linear system stability problem. For rational transfer functions, the mathematical problem is to determine whether all of the roots of the polynomial equation

$$s^n + a_{n-1}s^{n-1} + \cdots + a_0 = 0 \tag{9}$$

have negative real parts.[6] Given numerical values of the polynomial coefficients, powerful computer programs can then be used to find the roots. It is often the case that one or more of the coefficients of the polynomial depend on a parameter. Then it is of major interest to understand how the locations of the roots in the complex s-plane change as the parameter is varied. The graph of all possible roots of Eq. (9) relative to some particular parameter is called a **root locus**. The design technique based on this graph is called the **root locus method** of analysis.

EXAMPLE 1

Suppose the open-loop transfer function for a plant is given by

$$H(s) = \frac{K}{s^2 + 2s + 2},$$

and that it is desired to synthesize a closed-loop system with negative feedback using the controller

$$G(s) = \frac{1}{s + 0.1}.$$

Since the poles of $H(s)$ are $s = -1 \pm i$ while $G(s)$ has a single pole at $s = -0.1$, Theorem 5.9.3 implies that open-loop input–output systems described by $H(s)$ and $G(s)$ are BIBO stable. The transfer function for the closed-loop system corresponding to Figure 5.9.1b is

$$H_G(s) = \frac{H}{1 + GH} = \frac{K(s + 0.1)}{s^3 + 2.1s^2 + 2.2s + K + 0.2}.$$

The root locus diagram for $H_G(s)$ as K varies from 1 to 10 is shown in Figure 5.9.3.

---

[6]A polynomial with real coefficients and roots that are either negative or pairwise conjugate with negative real parts is called a **Hurwitz polynomial**.

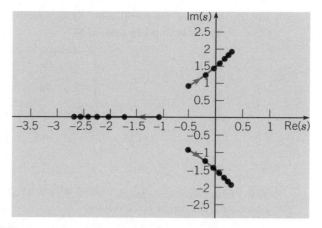

FIGURE 5.9.3 The root locus of the polynomial equation $s^3 + 2.1s^2 + 2.2s + K + 0.2 = 0$ as K varies from 1 to 10, as indicated by the blue arrows.

The value of K for which the complex roots cross the imaginary axis is found by assuming that for that value of K the roots of the polynomial, α and $\pm i\beta$, must satisfy the equation

$$(s - \alpha)(s^2 + \beta^2) = s^3 - \alpha s^2 + \beta^2 s - \alpha \beta^2 = s^3 + 2.1s^2 + 2.2s + K + 0.2.$$

Matching the coefficients of the last two polynomials on the right,

$$\alpha = -2.1, \qquad \beta^2 = 2.2, \qquad -\alpha\beta^2 = K + 0.2,$$

yields $K = 4.42$. Thus, for $K < 4.42$, the real parts of all poles of $H_G(s)$ are negative and the closed-loop system is BIBO stable. For $K = 4.42$ two poles are pure imaginary, and for $K > 4.42$ the real parts of two of the poles are positive. Thus the closed-loop system is unstable, that is, not BIBO stable, for gains $K \geq 4.42$.

▶ The Routh Stability Criterion. In view of the importance of knowing polynomial root locations for the purpose of establishing BIBO stability, a valuable tool is a criterion for establishing whether the roots of a polynomial have negative real parts in terms of the coefficients of the polynomial.[7] For example, the real parts of the roots of

$$Q(s) = s^2 + a_1 s + a_0$$

are negative if and only if $a_1 > 0$ and $a_0 > 0$, a result easily obtained by comparing with a quadratic function of s with roots $\alpha \pm i\beta$,

$$(s - \alpha + i\beta)(s - \alpha - i\beta) = s^2 - 2\alpha s + \alpha^2 + \beta^2.$$

Thus $a_1 = -2\alpha$ and $a_0 = \alpha^2 + \beta^2$. Clearly, the real parts of the roots are negative, that is, $\alpha < 0$, if and only if $a_1 > 0$ and $a_0 > 0$.

We now present, without proof, a classical result for nth degree polynomials that provides a test for determining whether all of the roots of the polynomial (9) have negative real parts in terms of the coefficients of the polynomial. It is useful if the degree n of Eq. (9)

---

[7]More generally, such a criterion is valuable for answering stability questions about linear constant coefficient systems, $\mathbf{x}' = \mathbf{Ax}$, discussed in Chapter 6. The original formulation of polynomial root location problems dates at least back to Cauchy in 1831.

is not too large and the coefficients depend on one or more parameters. The criterion is applied using a **Routh table** defined by

$$
\begin{array}{c|cccc}
s^n & 1 & a_{n-2} & a_{n-4} & \cdots \\
s^{n-1} & a_{n-1} & a_{n-3} & a_{n-5} & \cdots \\
\cdot & b_1 & b_2 & b_3 & \cdots, \\
\cdot & c_1 & c_2 & c_3 & \cdots \\
\cdot & \cdot & \cdot & \cdot &
\end{array}
$$

where $1, a_{n-1}, a_{n-2}, \ldots, a_0$ are the coefficients of the polynomial (9) and $b_1, b_2, \ldots,$ $c_1, c_2, \ldots$ are defined by the quotients

$$
b_1 = -\frac{\begin{vmatrix} 1 & a_{n-2} \\ a_{n-1} & a_{n-3} \end{vmatrix}}{a_{n-1}}, \qquad b_2 = -\frac{\begin{vmatrix} 1 & a_{n-4} \\ a_{n-1} & a_{n-5} \end{vmatrix}}{a_{n-1}}, \qquad \ldots.
$$

$$
c_1 = -\frac{\begin{vmatrix} a_{n-1} & a_{n-3} \\ b_1 & b_2 \end{vmatrix}}{b_1}, \qquad c_2 = -\frac{\begin{vmatrix} a_{n-1} & a_{n-5} \\ b_1 & b_3 \end{vmatrix}}{b_1}, \qquad \ldots.
$$

The table is continued horizontally and vertically until only zeros are obtained. Any row of the table may be multiplied by a positive constant before the next row is computed without altering the properties of the table.

THEOREM 5.9.4

The Routh Criterion. All of the roots of the polynomial equation

$$
s^n + a_{n-1}s^{n-1} + \cdots + a_0 = 0
$$

have negative real parts if and only if the elements of the first column of the Routh table have the same sign. Otherwise, the number of roots with positive real parts is equal to the number of changes of sign.

EXAMPLE 2

The Routh table for the denominator polynomial in the transfer function

$$
H_G(s) = \frac{K(s+0.1)}{s^3 + 2.1s^2 + 2.2s + K + 0.2}
$$

of Example 1 is

$$
\begin{array}{c|ccc}
s^3 & 1 & 2.2 & 0 \\
s^2 & 2.1 & K+0.2 & 0 \\
s & \dfrac{4.42-K}{2.1} & 0 & \\
s^0 & K+0.2 & &
\end{array}
$$

No sign changes occur in the first column if $K < 4.42$ and $K > -0.2$. Thus we conclude that the closed-loop system is BIBO stable if $-0.2 < K < 4.42$. If $K > 4.42$, there are two roots with positive real parts. If $K < -0.2$, there is one root with positive real part.

PROBLEMS

1. Find the transfer function of the system shown in Figure 5.9.4.

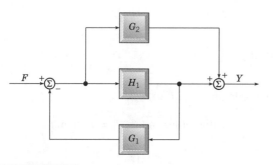

FIGURE 5.9.4 Block diagram for Problem 1.

2. Find the transfer function of the system shown in Figure 5.9.5.

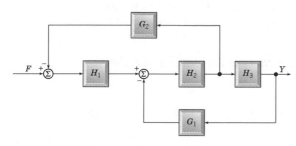

FIGURE 5.9.5 Block diagram for Problem 2.

3. If $h(t)$ is any one of the functions, $t^k e^{\alpha t}$, $t^k e^{\alpha t} \cos \beta t$, or $t^k e^{\alpha t} \sin \beta t$ and $\alpha < 0$, show that

$$\int_0^\infty |h(t)| \, dt \leq \frac{k!}{|\alpha|^{k+1}}.$$

4. Show that, if all of the real roots of a polynomial are negative and the real parts of all of the complex roots are neg-

ative, then the coefficients of the polynomial are necessarily positive. (Thus if any coefficients are negative or zero, the polynomial must have roots with nonnegative real parts. This provides a useful initial check on whether there may be roots in the right half plane.)

5. Use Routh's criterion to find necessary and sufficient conditions on a_2, a_1, and a_0 that guarantee all roots of $s^3 + a_2 s^2 + a_1 s + a_0$ lie in the left half plane.

For each of the characteristic functions in Problems 6 through 11, use Routh's stability criterion to determine the number of roots with positive real parts. Confirm your answers by using a computer to find the roots.

6. $s^3 + 5s^2 + 2s + 13$

7. $s^3 + 5s^2 + 9s + 5$

8. $s^4 + 8s^3 + 20s^2 + s - 34$

9. $s^4 + 5s^3 - 5s^2 - 35s + 34$

10. $s^4 + 12s^3 + 79s^2 + 205s + 221$

11. $s^4 + 8s^3 + 24s^2 + 32s + 100$

In each of Problems 12 through 15, use the Routh criterion to find the range of K for which all the poles of the transfer functions are in the left half of the complex plane. Then use a computer program to verify your answers by plotting the poles in the s-plane for various values of K.

12. $\dfrac{K(s+1)}{s^3 + 5s^2 + (K-6)s + K}$

13. $\dfrac{K(s+1)(s+2)}{s^3 + 4s^2 + 6s + 4 + K}$

14. $\dfrac{K(s+1)}{s^3 + 3s^2 + 14s + 8 + K}$

15. $\dfrac{K(s^2 + 2s + 2)}{s^4 + 4s^3 + 14s^2 + 20s + 16 + K}$

CHAPTER SUMMARY

For constant coefficient linear differential equations, the Laplace transform converts initial value problems in the t-domain to algebraic problems in the s-domain. The Laplace transform is used to study the input–output behavior of linear systems, feedback control systems, and electric circuits.

Section 5.1　The Laplace transform is an **integral transform**.

▶ If f is **piecewise continuous** and of **exponential order**, the Laplace transform is defined by

$$\mathcal{L}\{f\}(s) = F(s) = \int_0^\infty e^{-st} f(t)\, dt.$$

The **domain** of F is the set of all s for which the **improper integral** converges.

▶ $\mathcal{L}$ is a **linear operator**, $\mathcal{L}\{c_1 f_1 + c_2 f_2\} = c_1 \mathcal{L}\{f_1\} + c_2 \mathcal{L}\{f_2\}$.

Section 5.2　Some properties of $\mathcal{L}$: if $\mathcal{L}\{f(t)\} = F(s)$, then

▶ $\mathcal{L}\{e^{at} f(t)\}(s) = F(s - a)$,

▶ $\mathcal{L}\{f^{(n)}(t)\}(s) = s^n F(s) - s^{n-1} f(0) - \cdots - f^{(n-1)}(0)$,

▶ $\mathcal{L}\{t^n f(t)\}(s) = (-1)^n F^{(n)}(s)$.

▶ If f is piecewise continuous and of exponential order, $\lim_{s \to \infty} F(s) = 0$.

Section 5.3　If f is piecewise continuous and of exponential order and $\mathcal{L}\{f\} = F$, then f is the **inverse transform** of F and is denoted by $f = \mathcal{L}^{-1}\{F\}$.

▶ $\mathcal{L}^{-1}$ is a **linear operator**, $\mathcal{L}^{-1}\{c_1 F_1 + c_2 F_2\} = c_1 \mathcal{L}^{-1}\{F_1\} + c_2 \mathcal{L}^{-1}\{F_2\}$.

▶ Inverse transforms of many functions can be found using (i) the linearity of $\mathcal{L}^{-1}$, (ii) **partial fraction decompositions**, and (iii) a table of Laplace transforms. Labor can be greatly reduced, or eliminated, by using a computer algebra system.

Section 5.4　The Laplace transform is used to solve initial value problems for linear constant coefficient differential equations and systems of linear constant coefficient differential equations.

Section 5.5　Discontinuous functions are modeled using the **unit step function**

$$u_c(t) = \begin{cases} 0, & t < c, \\ 1, & t \ge c. \end{cases}$$

▶ $\mathcal{L}\{u_c(t) f(t - c)\} = e^{-cs} \mathcal{L}\{f(t)\} = e^{-cs} F(s)$.

▶ If f is **periodic with period** T and is piecewise continuous on $[0, T]$, then

$$\mathcal{L}\{f(t)\} = \frac{\int_0^T e^{-st} f(t)\, dt}{1 - e^{-sT}}.$$

Section 5.6　The Laplace transform is convenient for solving constant coefficient differential equations and systems with discontinuous forcing functions.

Section 5.7　Large magnitude inputs of short duration are modeled by the **unit impulse**, or **Dirac delta function**, defined by

$$\delta(t - t_0) = 0 \quad \text{if} \quad t \ne t_0,$$

and for any function f that is continuous at t_0:

$$\int_a^b f(t) \delta(t - t_0)\, dt = f(t_0) \quad \text{if} \quad a \le t_0 < b$$

▶ $\mathcal{L}\{\delta(t - t_0)\} = e^{-st_0}$.

▶ $u'(t - t_0) = \delta(t - t_0)$ in a generalized sense.

Section 5.8 The **convolution** of f and g is defined by $(f * g)(t) = \int_0^t f(t - \tau)g(\tau)\,d\tau$.

▶ **The Convolution Theorem.** $\mathcal{L}\{(f * g)(t)\} = F(s)G(s)$.

▶ The **transfer function** for $ay'' + by' + cy = g(t)$ is $H(s) = 1/(as^2 + bs + c)$ and the corresponding **impulse response** is $h(t) = \mathcal{L}^{-1}\{H(s)\}$.

▶ The **forced response** for $ay'' + by' + cy = g(t)$ is $H(s)G(s)$ in the s-domain and $\int_0^t h(t - \tau)g(\tau)\,d\tau$ in the t-domain.

Section 5.9 Transfer functions for many feedback control systems modeled by linear constant coefficient differential equations are of the form

$$H(s) = \frac{P(s)}{Q(s)} = \frac{b_m s^m + b_{m-1} s^{m-1} + \cdots + b_0}{s^n + a_{n-1} s^{n-1} + \cdots + a_0},$$

where $m < n$ (**strictly rational transfer functions**). The roots of $Q(s)$ are the **poles** of the transfer function.

▶ An input–output system with a strictly rational transfer function is **bounded-input bounded-output stable** if and only if all of the poles have negative real parts.

▶ The **Routh criterion** can be used to determine whether the real parts of the roots of a polynomial are negative in terms of the coefficients of the polynomial.

PROJECTS

Project 1 An Electric Circuit Problem

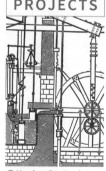

© Morphart Creation /
Shutterstock

Assume the following values for the elements in the electric circuit shown in Figure 5.P.1:

$$R = 0.01\text{ ohm}$$
$$L_1 = 1\text{ henry}$$
$$L_2 = 1\text{ henry}$$
$$C = 1\text{ farad}$$

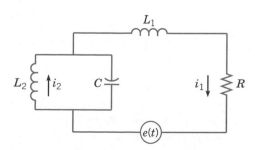

FIGURE 5.P.1

Project 1 PROBLEMS

1. Find $G(s)$ such that the Laplace transform of the charge on the capacitor, $Q(s) = \mathcal{L}\{q(t)\}$, can be expressed as $Q(s) = G(s)E(s)$, where $E(s) = \mathcal{L}\{e(t)\}$ is the Laplace transform of the impressed voltage. Assume that at time $t = 0$ the charge on the capacitor is zero and the currents i_1 and i_2 are zero.

2. Suppose that the impressed voltage is prescribed by the square wave

$$e(t) = \begin{cases} 1, & 0 \leq t < 0.6\pi, \\ -1, & 0.6\pi \leq t < 1.2\pi; \end{cases}$$

and $e(t)$ has period 1.2π.

Assuming zero initial conditions as stated in Problem 1, find an expression for $q(t)$ and plot the graphs of $e(t)$ and $q(t)$

on the same set of coordinate axes. Explain the behavior of $q(t)$.

3. Suppose that the impressed voltage is prescribed by the square wave

$$e(t) = \begin{cases} 1, & 0 \leq t < \pi/\sqrt{2}, \\ -1, & \pi/\sqrt{2} \leq t < \sqrt{2}\pi; \end{cases}$$

and $e(t)$ has period $\sqrt{2}\pi$.

Assuming zero initial conditions as stated in Problem 1, find an expression for $q(t)$ and plot the graphs of $e(t)$ and $q(t)$ on the same set of coordinate axes. Explain the behavior of $q(t)$.

Project 2 The Watt Governor, Feedback Control, and Stability

In the latter part of the 18th century, James Watt designed and built a steam engine with a rotary output motion (see Figure 5.P.2).

It was highly desirable to maintain a uniform rotational speed for powering various types of machinery, but fluctuations in steam pressure and work load on the engine caused the rotational speed to vary. At first, the speed was controlled manually by using a throttle valve to vary the flow of steam to the engine inlet. Then, using principles observed in a device for controlling the speed of the grinding stone in a wind-driven flour mill, Watt designed a **flyball** or **centrifugal governor**, based on the motion of a pair of centrifugal pendulums, to regulate the angular velocity of the steam engine's flywheel. A sketch of the essential components of the governor and its mechanical linkage to the throttle valve is shown in Figure 5.P.3.

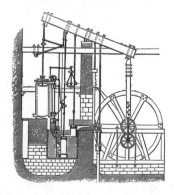

FIGURE 5.P.2   The Watt steam engine, 1781. © Morphart Creation/Shutterstock.

To understand how the mechanism automatically controls the angular velocity of the steam engine's flywheel and drive shaft assembly, assume that the engine is operating under a constant load at a desired equilibrium speed or operating point. The engine, via a belt and pulley assembly, also rotates a straight vertical shaft to which a pair of flyballs are connected. If the load on the engine is decreased or the steam pressure driving the engine increases, the engine speed increases. A corresponding increase in the rotational speed of the flyball shaft simultaneously causes the flyballs to swing outward due to an increase in centrifugal force. This motion, in turn, causes a sliding collar on the vertical shaft to move downward. A lever arm connected to the sliding collar on one end and a throttle control rod on the other end then partially closes the throttle valve, the steam flow to the engine is reduced, and the engine returns to its operating point. Adjustable elements in the linkage allow a desirable engine speed to be set during the startup phase of operation.

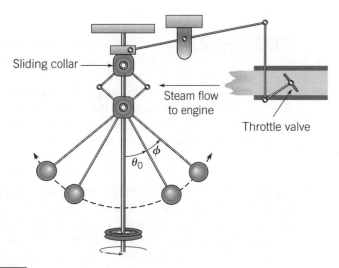

FIGURE 5.P.3   The centrifugal governor.

A Problem with Instability. Although some governors on some steam engines were successful in maintaining constant rotational speed, others would exhibit an anomalous rhythmic oscillation in which the system appeared to "hunt" unsuccessfully about its equilibrium value for a constant speed. The amplitude of oscillations would increase until limited by mechanical constraints on the motion of the flyballs or the throttle valve.

The purpose of this project is to develop a relatively simple mathematical model of the feedback control system and to gain an understanding of the underlying cause of the unstable behavior. We will need a differential equation for the angular velocity of the steam engine flywheel and drive shaft and a differential equation for the angle of deflection between the flyball connecting arms and the vertical shaft about which the flyballs revolve. Furthermore the equations need to be coupled to account for the mechanical linkage between the governor and the throttle valve.

The Flyball Motion. The model for the flyball governor follows from taking into account all of the forces acting on the flyball and applying Newton's law, $ma = F$. The angle between the flyball connecting arm and the vertical shaft about which the flyballs revolve will be denoted by θ. Assuming that the angular velocity of the vertical shaft and the rotational speed of the engine have the same value, Ω, there is a centrifugal acceleration acting on the flyballs in the outward direction of magnitude $\Omega^2 L \sin \theta$ (see Figure 5.P.4).

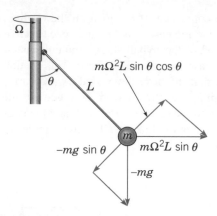

FIGURE 5.P.4 The angle of deflection of the centrifugal pendulum is determined by the opposing components of gravitational force and centrifugal force.

Recall from calculus that the magnitude of this acceleration is the curvature of the motion, $1/L \sin \theta$, times the square of the tangential velocity, $\Omega^2 L^2 \sin^2 \theta$. Taking into account the force due to gravitational acceleration and assuming a damping force of magnitude $\gamma \theta'$, we obtain the following equation for θ:

$$mL\theta'' = -\gamma \theta' - mg \sin \theta + m\Omega^2 L \sin \theta \cos \theta. \tag{1}$$

Equation (1) results from equating components of inertial forces to impressed forces parallel to the line tangent to the circular arc along which the flyball moves in the vertical plane determined by the pendulum arm and the vertical shaft (Problem 1).

Angular Velocity of the Steam Engine Flywheel and Drive Shaft. The equation for the rotational speed of the flywheel and drive shaft assembly of the steam engine is assumed to be

$$J\frac{d\Omega}{dt} = -\beta \Omega + \tau, \tag{2}$$

where J is the moment of inertia of the flywheel, the first term on the right is the torque due to the load, and the second term on the right is the steam-generated torque referred to the drive shaft.

Linearization About the Operating Point. In order to use the feedback control concepts of Section 5.9, it is necessary to determine the linear equations that are good approximations to Eqs. (1) and (2) when θ and Ω are near their equilibrium operating points, a mathematical technique known as **linearization**. The equilibrium operating point of the steam engine will be denoted by Ω_0, the equilibrium angle that the flyball connecting arm makes with the vertical will be denoted by θ_0, and the equilibrium torque delivered to the engine drive shaft will be denoted by τ_0. Note that in the equilibrium state, corresponding to $\theta'' = 0$, $\theta' = 0$, and $\Omega' = 0$, Eqs. (1) and (2) imply that

$$g = L\Omega_0^2 \cos \theta_0 \tag{3}$$

and

$$\tau_0 = \beta \Omega_0. \tag{4}$$

To linearize Eqs. (1) and (2) about θ_0, Ω_0, and τ_0, we assume that $\theta = \theta_0 + \phi$, $\Omega = \Omega_0 + y$, and $\tau = \tau_0 + u$, where ϕ, y, and u are perturbations that are small relative to θ_0, Ω_0, and τ_0, respectively. Note that ϕ represents the **error** in deflection of the flyball connecting arm from its desired value θ_0, in effect, measuring or sensing the error in the rotational speed of the engine. If $\phi > 0$, the engine is rotating too rapidly and must be slowed down; if $\phi < 0$, the engine is rotating too slowly and must be sped up. Substituting the expressions for θ, Ω, and τ into Eqs. (1) and (2), retaining only linear terms in ϕ, y, and u, and making use of Eqs. (3) and (4) yield

$$\phi'' + 2\delta\phi' + \omega_0^2\phi = \alpha_0 y \tag{5}$$

and

$$Jy' = -\beta y + u, \tag{6}$$

where $\delta = \gamma/2mL$, $\omega_0^2 = \Omega_0^2 \sin^2 \theta_0$, and $\alpha_0 = \Omega_0 \sin 2\theta_0$ (Problem 3).

The Closed-Loop System. Regarding the error in rotational speed, y, as the input, the transfer function associated with Eq. (5) is easily seen to be

$$G(s) = \frac{\alpha_0}{s^2 + 2\delta s + \omega_0^2}.$$

Thus $\Phi(s) = G(s)Y(s)$, where $\Phi(s) = \mathcal{L}\{\phi(t)\}$ and $Y(s) = \mathcal{L}\{y(t)\}$. Regarding u as the input, the transfer function associated with Eq. (6) is

$$H(s) = \frac{1}{Js + \beta}$$

so that $Y(s) = H(s)U(s)$, where $U(s) = \mathcal{L}\{u(t)\}$. The closed-loop system is synthesized by subtracting $\Phi(s)$ at the summation point, as shown in Figure 5.P.5.

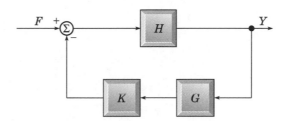

FIGURE 5.P.5   Block diagram of the feedback control system corresponding to the Watt governor linearized about the operating point.

Thus a positive error in deflection of the flyball connecting arm causes a decrease in steam-generated torque and vice versa. A proportional gain constant K has been inserted into the feedback loop to model the sensitivity of the change in steam-generated torque to the error $\Phi(s)$ in the deflection of the flyball connecting arm. Physically, K can be altered by changing the location of the pivot point of the lever arm that connects the sliding collar on the governor to the vertical rod attached to the throttle valve. Note that any external input affecting engine speed is represented by $F(s)$.

Project 2 PROBLEMS

In the following problems, we ask the reader to supply some of the details left out of the above discussion, to analyze the closed-loop system for stability properties, and to conduct a numerical simulation of the nonlinear system.

1. Work out the details leading to Eq. (1).

2. Give physical explanations for the meaning of Eqs. (3) and (4).

3. Derive the linearized system (5) and (6).

4. Show that the transfer function of the closed-loop system linearized about the operating point is

$$H_K(s) = \frac{s^2 + 2\delta s + \omega_0^2}{(Js + \beta)(s^2 + 2\delta s + \omega_0^2) + K}.$$

5. Use the Routh criterion to show that if the gain factor K is sufficiently large, $H_K(s)$ will have two poles with positive real parts and the corresponding closed-loop system is therefore unstable. Derive an expression for K_c, that value of K at which the real parts of a pair of conjugate complex poles of $H_K(s)$ are equal to 0.

6. The Nonlinear Feedback Control System. Using the relations (3) and (4), show that Eqs. (1) and (2) can be expressed in terms of ϕ and y as

$$mL\phi'' = -\gamma\phi - m\Omega_0^2 L\cos(\theta_0)\sin(\theta_0 + \phi)$$
$$+ m(\Omega_0 + y)^2 L\sin(\theta_0 + \phi)\cos(\theta_0 + \phi) \qquad \text{(i)}$$

and

$$Jy' = -\beta y - K\phi, \qquad \text{(ii)}$$

where the negative feedback loop has been incorporated into the nonlinear system.

7. Simulations. Consider the following parameter values expressed in SI units:

$$m = 12, \quad L = \tfrac{1}{2}, \quad g = 9.8, \quad \gamma = 0.01,$$
$$J = 400, \quad \beta = 20, \quad \theta_0 = \pi/6,$$

with Ω_0 determined by the relation (3).

(a) Using the above parameter values, construct a root locus plot of the poles of $H_K(s)$ as K varies over an interval containing K_c. Verify that a pair of poles cross from the left half plane to the right half plane as K increases through K_c.

(b) Conduct computer simulations of the system (i), (ii) in Problem 6 using the above parameter values. Do this for various values of K less than K_c and greater than K_c while experimenting with different values of $y(0) \neq 0$ to represent departures from the equilibrium operating point. For simplicity, you may always assume that $\phi(0) = \phi'(0) = 0$. Plot graphs of the functions $\phi(t)$ and $y(t)$ generated from the simulations. Note that if $\phi(t)$ wanders outside the interval $[-\theta_0, \pi - \theta_0]$, the results are nonphysical. (Why?) Are the results of your simulations consistent with your theoretical predictions? Discuss the results of your computer experiments addressing such issues as whether the feedback control strategy actually works, good choices for K, stability and instability, and the "hunting" phenomenon discussed above.

CHAPTER SIX

© InterNetwork Media / Photodisc / Getty Images

Systems of First Order Linear Equations

I n this chapter we build on elementary theory and solution techniques for first order linear systems introduced in a two-dimensional setting in Chapters 3 and 4. Science and engineering applications that possess even a modest degree of complexity often lead to systems of differential equations of dimension $n > 2$. The language, concepts, and tools of linear algebra, combined with certain elements of calculus, are essential for the study of these systems. The required mathematical background from matrices and linear algebra is presented in Appendix A. A reader with little or no previous exposure to this material will find it necessary to study some or all of Appendix A before proceeding to Chapter 6. Alternatively, the reader may take up the study of Chapter 6 straightaway, drawing on necessary results from Appendix A as needed.

A large portion of Chapter 6 generalizes the eigenvalue method first presented in Chapter 3 to constant coefficient linear systems of dimension $n > 2$. Since carrying out eigenvalue calculations by hand for matrices of dimension greater than 2 is time-consuming, tedious, and susceptible to error, we strongly recommend using a computer or a calculator to perform the required calculations, once an understanding of the general theory and methodology has been acquired.

6.1 Definitions and Examples

First order linear systems of dimension 2,

$$x' = p_{11}(t)x + p_{12}(t)y + g_1(t),$$
$$y' = p_{21}(t)x + p_{22}(t)y + g_2(t), \tag{1}$$

were first introduced in Section 3.2. We now discuss the general mathematical framework for first order linear systems of dimension n, followed by additional examples from science and engineering. The matrix algebra required to understand the general framework is presented in Appendix A.1.

▶ **Matrix-Valued Functions.** The principal mathematical objects involved in the study of linear systems of differential equations are **matrix-valued functions**, or simply **matrix functions**. These objects are vectors or matrices whose elements are functions of t. We write

$$\mathbf{x}(t) = \begin{pmatrix} x_1(t) \\ \vdots \\ x_n(t) \end{pmatrix}, \qquad \mathbf{P}(t) = \begin{pmatrix} p_{11}(t) & \cdots & p_{1n}(t) \\ \vdots & & \vdots \\ p_{n1}(t) & \cdots & p_{nn}(t) \end{pmatrix},$$

respectively.

The matrix $\mathbf{P} = \mathbf{P}(t)$ is said to be continuous at $t = t_0$ or on an interval $I = (\alpha, \beta)$ if each element of $\mathbf{P}$ is a continuous function at the given point or on the given interval. Similarly, $\mathbf{P}(t)$ is said to be differentiable if each of its elements is differentiable, and its derivative $d\mathbf{P}/dt$ is defined by

$$\frac{d\mathbf{P}}{dt} = \left(\frac{dp_{ij}}{dt} \right). \tag{2}$$

In other words, each element of $d\mathbf{P}/dt$ is the derivative of the corresponding element of $\mathbf{P}$. Similarly, the integral of a matrix function is defined as

$$\int_a^b \mathbf{P}(t)\, dt = \left(\int_a^b p_{ij}(t)\, dt \right). \tag{3}$$

Many of the results of elementary calculus extend easily to matrix functions, in particular,

$$\frac{d}{dt}(\mathbf{CP}) = \mathbf{C}\frac{d\mathbf{P}}{dt}, \quad \text{where } \mathbf{C} \text{ is a constant matrix,} \tag{4}$$

$$\frac{d}{dt}(\mathbf{P} + \mathbf{Q}) = \frac{d\mathbf{P}}{dt} + \frac{d\mathbf{Q}}{dt}, \tag{5}$$

$$\frac{d}{dt}(\mathbf{PQ}) = \mathbf{P}\frac{d\mathbf{Q}}{dt} + \frac{d\mathbf{P}}{dt}\mathbf{Q}. \tag{6}$$

In Eqs. (4) and (6), care must be taken in each term to avoid interchanging the order of multiplication. The definitions expressed by Eqs. (2) and (3) also apply as special cases to vectors.

First Order Linear Systems: General Framework

The general form of a first order linear system of dimension n is

$$
\begin{aligned}
x_1' &= p_{11}(t)x_1 + p_{12}(t)x_2 + \cdots + p_{1n}(t)x_n + g_1(t), \\
x_2' &= p_{21}(t)x_1 + p_{22}(t)x_2 + \cdots + p_{2n}(t)x_n + g_2(t), \\
&\;\;\vdots \\
x_n' &= p_{n1}(t)x_1 + p_{n2}(t)x_2 + \cdots + p_{nn}(t)x_n + g_n(t),
\end{aligned}
\tag{7}
$$

or, using matrix notation,

$$
\mathbf{x}' = \mathbf{P}(t)\mathbf{x} + \mathbf{g}(t),
\tag{8}
$$

where

$$
\mathbf{P}(t) = \begin{pmatrix} p_{11}(t) & \cdots & p_{1n}(t) \\ \vdots & & \vdots \\ p_{n1}(t) & \cdots & p_{nn}(t) \end{pmatrix}
$$

is referred to as the **matrix of coefficients** of the system (7) and

$$
\mathbf{g}(t) = \begin{pmatrix} g_1(t) \\ \vdots \\ g_n(t) \end{pmatrix}
$$

is referred to as the **nonhomogeneous** term of the system. We will assume that $\mathbf{P}(t)$ and $\mathbf{g}(t)$ are continuous on an interval $I = (\alpha, \beta)$. If $\mathbf{g}(t) = \mathbf{0}$ for all $t \in I$, then the system (7) or (8) is said to be **homogeneous**; otherwise, the system is said to be **nonhomogeneous**. The function $\mathbf{g}(t)$ is often referred to as the **input**, or **forcing function**, to the system (8) and provides a means for modeling interaction between the physical system represented by $\mathbf{x}' = \mathbf{P}(t)\mathbf{x}$ and the world external to the system.

The system (8) is said to have a solution on the interval I if there exists a vector

$$
\mathbf{x} = \boldsymbol{\phi}(t)
\tag{9}
$$

with n components that is differentiable at all points in the interval I and satisfies Eq. (8) at all points in this interval. In addition to the system of differential equations, there may also be given an initial condition of the form

$$
\mathbf{x}(t_0) = \mathbf{x}_0,
\tag{10}
$$

where t_0 is a specified value of t in I and $\mathbf{x}_0$ is a given constant vector with n components. The system (8) and the initial condition (10) together form an **initial value problem**.

The initial value problem (8), (10) generalizes the framework for two-dimensional systems presented in Chapter 3 to systems of dimension n. A solution (9) can be viewed as a set of parametric equations in an n-dimensional space. For a given value of t, Eq. (9) gives values of the coordinates $x_1, x_2, \ldots, x_n$ of a point $\mathbf{x}$ in n-dimensional space, $\mathbf{R}^n$. As t changes, the coordinates in general also change. The collection of points corresponding to $\alpha < t < \beta$ forms a curve in the space. As with two-dimensional systems, it is often helpful to think of the curve as the trajectory, or path, of a particle moving in accordance with the system (8). The initial condition (10) determines the starting point of the moving particle. The components of $\mathbf{x}$ are again referred to as **state variables** and the vector $\mathbf{x} = \boldsymbol{\phi}(t)$ is referred to as the **state of the system** at time t. The initial condition (10) prescribes the state of the system at time t_0. Given the initial condition (10), the differential equation (8) is a rule for advancing the state of the system through time.

▶ Linear *n*th Order Equations. Single equations of higher order can always be transformed into systems of first order equations. This is usually required if a numerical approach is planned, because almost all codes for generating numerical approximations to solutions of differential equations are written for systems of first order equations. An *n*th order linear differential equation in standard form is given by

$$\frac{d^n y}{dt^n} + p_1(t)\frac{d^{n-1} y}{dt^{n-1}} + \cdots + p_{n-1}(t)\frac{dy}{dt} + p_n(t)y = g(t). \tag{11}$$

We will assume that the functions $p_1, \ldots, p_n$ and g are continuous real-valued functions on some interval $I = (\alpha, \beta)$. Since Eq. (11) involves the *n*th derivative of y with respect to t, it will, so to speak, require n integrations to solve Eq. (11). Each of these integrations introduces an arbitrary constant. Hence we can expect that, to obtain a unique solution, it is necessary to specify n initial conditions,

$$y(t_0) = y_0, \quad y'(t_0) = y_1, \quad \ldots, \quad y^{(n-1)}(t_0) = y_{n-1}. \tag{12}$$

To transform Eq. (11) into a system of n first order equations, we introduce the variables $x_1, x_2, \ldots, x_n$ defined by

$$x_1 = y, \quad x_2 = y', \quad x_3 = y'', \quad \ldots, \quad x_n = y^{(n-1)}. \tag{13}$$

It then follows immediately that

$$
\begin{aligned}
x_1' &= x_2, \\
x_2' &= x_3, \\
&\ \ \vdots \\
x_{n-1}' &= x_n,
\end{aligned}
\tag{14}
$$

and, from Eq. (11),

$$x_n' = -p_n(t)x_1 - p_{n-1}(t)x_2 - \cdots - p_1(t)x_n + g(t). \tag{15}$$

Using matrix notation, the system (14), (15) can be written as

$$\mathbf{x}' = \begin{pmatrix} 0 & 1 & 0 & 0 & \cdots & 0 \\ 0 & 0 & 1 & 0 & \cdots & 0 \\ \vdots & \vdots & \vdots & \vdots & & \vdots \\ 0 & 0 & 0 & 0 & \cdots & 1 \\ -p_n(t) & -p_{n-1}(t) & -p_{n-2}(t) & -p_{n-3}(t) & \cdots & -p_1(t) \end{pmatrix} \mathbf{x} + \begin{pmatrix} 0 \\ 0 \\ \vdots \\ 0 \\ g(t) \end{pmatrix}, \tag{16}$$

where $\mathbf{x} = (x_1, \ldots, x_n)^T$. Using the definitions for the state variables in the list (13), the initial condition for Eq. (16) is expressed by

$$\mathbf{x}(t_0) = \begin{pmatrix} y_0 \\ y_1 \\ \vdots \\ y_{n-1} \end{pmatrix}. \tag{17}$$

Applications Modeled by First Order Linear Systems

In previous chapters we have primarily encountered first order systems of dimension 2, although systems of higher dimension appear in some of the projects. We now present several additional examples of applications that illustrate how higher dimensional linear systems can arise.

▶ Coupled Mass-Spring Systems. Interconnected systems of masses and springs are often used as a starting point in modeling flexible mechanical structures or other systems idealized as an assortment of elastically coupled bodies (for an example, see Project 1 at the end of this chapter). As a consequence of Newton's second law of motion, the mathematical description of the dynamics results in a coupled system of second order equations. Using the same technique that was demonstrated above to convert a single higher order equation into a system of first order equations, we show how a system of second order equations is easily converted to a first order system. Consider two masses, m_1 and m_2, connected to three springs in the arrangement shown in Figure 6.1.1a.

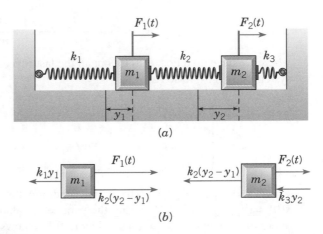

(a)

(b)

FIGURE 6.1.1 (a) A two-mass, three-spring system. (b) Free-body diagrams of the forces acting on each of the masses.

The masses are constrained to move only in the horizontal direction on a frictionless surface under the influence of external forces $F_1(t)$ and $F_2(t)$. The springs, assumed to obey Hooke's law, have spring constants k_1, k_2, and k_3. Displacements of m_1 and m_2 from their equilibrium positions are denoted by y_1 and y_2, respectively. For simplicity, we neglect the effects of friction. Aided by the free-body diagrams in Figure 6.1.1b, the following differential equations result from equating the inertial force of each mass (mass times acceleration) to the totality of external forces acting on the mass,

$$
\begin{aligned}
m_1 \frac{d^2 y_1}{dt^2} &= k_2(y_2 - y_1) - k_1 y_1 + F_1(t) \\
&= -(k_1 + k_2)y_1 + k_2 y_2 + F_1(t), \\
m_2 \frac{d^2 y_2}{dt^2} &= -k_3 y_2 - k_2(y_2 - y_1) + F_2(t) \\
&= k_2 y_1 - (k_2 + k_3)y_2 + F_2(t).
\end{aligned}
\tag{18}
$$

For example, if $y_2 > y_1 > 0$ as shown in Figure 6.1.1a, then the force ky_1 exerted on m_1 by the left most spring points in the negative direction since that spring is in an elongated state, while the force $k_2(y_2 - y_1)$ exerted on m_1 by the middle spring points in the positive direction since that spring is also in an elongated state. The second equation follows by applying analogous reasoning to m_2. Considering instantaneous configurations other than $y_2 > y_1 > 0$ yields the same set of equations. Specifying the displacement and velocity of each mass at time $t = 0$ provides initial conditions for the system (18),

$$
y_1(0) = y_{10}, \quad y_2(0) = y_{20}, \quad y_1'(0) = v_{10}, \quad y_2'(0) = v_{20}.
\tag{19}
$$

The second order system (18) is subsumed within the framework of first order systems by introducing the state variables

$$x_1 = y_1, \quad x_2 = y_2, \quad x_3 = y_1', \quad x_4 = y_2'.$$

Then

$$
\begin{aligned}
x_1' &= x_3, \\
x_2' &= x_4, \\
x_3' &= -\frac{k_1 + k_2}{m_1} x_1 + \frac{k_2}{m_1} x_2 + \frac{1}{m_1} F_1(t), \\
x_4' &= \frac{k_2}{m_2} x_1 - \frac{k_2 + k_3}{m_2} x_2 + \frac{1}{m_2} F_2(t),
\end{aligned}
\tag{20}
$$

where we have used Eqs. (18) to obtain the last two equations. Using matrix notation, the system (20) is expressed as

$$
\mathbf{x}' = \begin{pmatrix}
0 & 0 & 1 & 0 \\
0 & 0 & 0 & 1 \\
-(k_1 + k_2)/m_1 & k_2/m_1 & 0 & 0 \\
k_2/m_2 & -(k_2 + k_3)/m_2 & 0 & 0
\end{pmatrix} \mathbf{x} + \begin{pmatrix}
0 \\
0 \\
F_1(t)/m_1 \\
F_2(t)/m_2
\end{pmatrix},
\tag{21}
$$

where $\mathbf{x} = (x_1, x_2, x_3, x_4)^T$. The initial condition for the first order system (21) is then $\mathbf{x}(0) = (y_{10}, y_{20}, v_{10}, v_{20})^T$. Using the parameter values $m_1 = m_2 = k_1 = k_2 = k_3 = 1$ and under the condition of zero inputs, Figure 6.1.2 shows component plots of the solution of Eq. (21) subject to the initial condition $\mathbf{x}(0) = (-3, 2, 0, 0)^T$, that is, the left mass is pulled three units to the left and released with zero initial velocity, while the right mass is simultaneously pulled two units to the right and released with zero initial velocity.

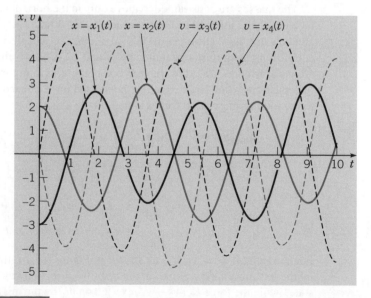

FIGURE 6.1.2   The response (displacements and velocities of the masses) of the system (21) subject to the initial conditions $x_1(0) = -3$, $x_2(0) = 2$, $x_3(0) = 0$, $x_4(0) = 0$; $m_1 = m_2 = k_1 = k_2 = k_3 = 1$ and $F_1(t) = F_2(t) = 0$.

Solutions of Eq. (21) are analyzed in Section 6.4 using the eigenvalue method.

▶ Compartment Models. A frequently used modeling paradigm idealizes the physical, biological, or economic system as a collection of subunits, or **compartments**, that exchange contents (matter, energy, capital, etc.) with one another. Compartment models were first encountered in Section 2.3, where the models had only one compartment; now we extend that modeling process to situations with multiple compartments. The amount of material in each compartment is represented by a component of a state vector and differential equations are used to describe the transport of material between compartments. For example, a model for lead in the human body views the body as composed of blood (Compartment 1), soft tissue (Compartment 2), and bones (Compartment 3), as shown in Figure 6.1.3.

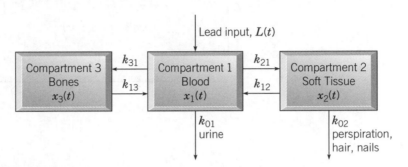

FIGURE 6.1.3   A three-compartment model of lead in the human body.

Lead enters the blood from the environment through the gastrointestinal tract and lungs and is taken up rapidly by the liver and kidneys. It is absorbed less rapidly by other soft tissues, and very slowly by the bones. Lead is then eliminated from the blood via urine, feces, skin, hair, and nails.

Denote the amount of lead in Compartment i by $x_i(t)$ and the rate at which lead moves from Compartment i to Compartment j by the rate constant k_{ji}. In this example, the amount of lead is measured in micrograms (1 microgram $= 10^{-6}$ gram) and time is measured in days. The rate constants have units of 1/day. Differential equations describing the exchange of lead between compartments are obtained by applying the mass balance law,

$$\frac{dx}{dt} = \text{input rate} - \text{output rate}, \tag{22}$$

to each compartment. For example, the lead input rate to Compartment 3 is obtained by multiplying the amount of lead in Compartment 1, x_1, by the rate constant k_{31}. The resulting input rate is $k_{31}x_1$. Similarly, the lead output rate from Compartment 3 is $k_{13}x_3$. Application of the balance law (22) to Compartment 3 then yields the differential equation

$$x_3' = k_{31}x_1 - k_{13}x_3$$

that describes the rate of change of lead with respect to time in Compartment 3.

If the rate at which lead leaves the body through urine is denoted by k_{01}, the rate of loss via perspiration, hair, and nails is denoted by k_{02}, and the exposure level to lead in the environment is denoted by L (with units of microgram/day), the principle of mass balance applied to each of the three compartments results in the following system of equations:

$$
\begin{aligned}
x_1' &= (L + k_{12}x_2 + k_{13}x_3) - (k_{21} + k_{31} + k_{01})x_1, \\
x_2' &= k_{21}x_1 - (k_{02} + k_{12})x_2, \\
x_3' &= k_{31}x_1 - k_{13}x_3.
\end{aligned}
\tag{23}
$$

Note that these equations assume that there is no transfer of lead between soft tissue and bones.

We consider a case in which the amount of lead in each compartment is initially zero but the lead-free body is then subjected to a constant level of exposure over a 365-day time period. This is followed by complete removal of the source to see how quickly the amount of lead in each compartment decreases. In Problem 12, Section 6.6, you are asked to solve the initial value problem for the system (23) that yields the component plots shown in Figure 6.1.4.

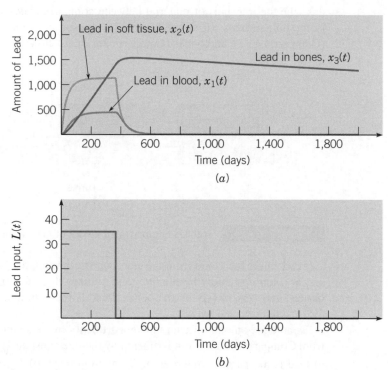

FIGURE 6.1.4 (a) Amounts of lead in body compartments during and after an exposure period of one year. (b) Lead input to Compartment 1 (blood).

Note that once the source is removed, lead is eliminated from the blood and soft tissues fairly rapidly, but persists in the bones for a much longer time. Mathematically, the slow decay in the amount of lead in the bones is due to the fact that the rate constant k_{13} is very small relative to the other rate constants.

▶ Linear Control Systems. There are many physical, biological, and economic systems in which it is desirable to **control**, or **steer**, the state of the system to some desired value or along some desired path in the state space. A standard mathematical model for linear control systems consists of the pair of equations

$$\mathbf{x}' = \mathbf{A}\mathbf{x} + \mathbf{B}\mathbf{u}(t) \tag{24}$$

$$\mathbf{y} = \mathbf{C}\mathbf{x}, \tag{25}$$

where $\mathbf{A}$ is an $n \times n$ **system matrix**, $\mathbf{B}$ is an $n \times m$ **input matrix**, and $\mathbf{C}$ is an $r \times n$ **output matrix**. Equation (24) is referred to as the **plant equation**, whereas the linear algebraic equation (25), is referred to as the **output equation**. The output $\mathbf{y}$, a vector function with r components, allows for the possibility that some components of $\mathbf{x}$ may not be directly observable, or for the possibility that only certain linear combinations of the state variables are observed or measured. The $m \times 1$ vector function $\mathbf{u}(t)$ is the **plant input**. Note that $\mathbf{B}\mathbf{u}(t)$, an $n \times 1$ vector, is simply a nonhomogeneous term in Eq. (24).

A common type of control problem is to choose or design the input $\mathbf{u}(t)$ in order to achieve some desired objective. As an example, consider again the vibration problem consisting of two masses and three springs shown in Figure 6.1.1. Given an initial state (initial position and velocity of each mass), suppose the objective is to bring the entire system to equilibrium during a specified time interval $[0, T]$ by applying a suitable forcing function only to the mass on the left. We may write the system of equations (21) in the form

$$\mathbf{x}' = \begin{pmatrix} 0 & 0 & 1 & 0 \\ 0 & 0 & 0 & 1 \\ -(k_1 + k_2)/m_1 & k_2/m_1 & 0 & 0 \\ k_2/m_2 & -(k_2 + k_3)/m_2 & 0 & 0 \end{pmatrix} \mathbf{x} + \begin{pmatrix} 0 \\ 0 \\ 1 \\ 0 \end{pmatrix} u(t), \tag{26}$$

where we have set $u(t) = F_1(t)/m_1$ and $F_2(t) = 0$. The system and input matrices in Eq. (26) are

$$\mathbf{A} = \begin{pmatrix} 0 & 0 & 1 & 0 \\ 0 & 0 & 0 & 1 \\ -(k_1 + k_2)/m_1 & k_2/m_1 & 0 & 0 \\ k_2/m_2 & -(k_2 + k_3)/m_2 & 0 & 0 \end{pmatrix} \quad \text{and} \quad \mathbf{B} = \begin{pmatrix} 0 \\ 0 \\ 1 \\ 0 \end{pmatrix},$$

respectively. If we assume that the entire state of the system is observable, then the output matrix is $\mathbf{C} = \mathbf{I}_4$ and $\mathbf{y} = \mathbf{I}_4 \mathbf{x} = \mathbf{x}$. Given an initial state $\mathbf{x}(0) = \mathbf{x}_0$, the control problem then is to specify an acceleration $u(t) = F_1(t)/m_1$ that is to be applied to mass m_1 over the time interval $0 \leq t \leq T$ so that $\mathbf{x}(T) = \mathbf{0}$. Obviously, an essential first step is to ascertain whether the desired objective can be achieved. A general answer to this question is provided, in Project 3 at the end of this chapter, in the context of an analogous system of three masses and four springs. Using the methods presented in the aforementioned project, an input function $u(t)$ that drives the system (26) from the initial state $\mathbf{x}(0) = (-3, 2, 0, 0)^T$ to equilibrium over the time interval $0 \leq t \leq 10$ is shown in Figure 6.1.5.

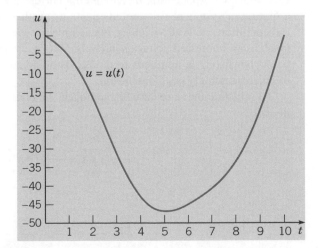

FIGURE 6.1.5 The acceleration $u(t)$, $0 \leq t \leq T$ applied to the left mass in Figure 6.1.1 drives the system from its initial state $\mathbf{x}(0) = (-3, 2, 0, 0)$ to equilibrium at time T (see Figure 6.1.6); $m_1 = m_2 = k_1 = k_2 = k_3 = 1$, $u(t) = F_1(t)/m_1$, $F_2(t) = 0$, and $T = 10$.

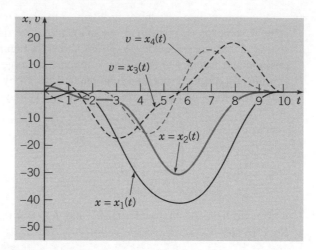

FIGURE 6.1.6 The response (displacements and velocities of the masses) of the system (26) subject to the initial conditions $x_1(0) = -3$, $x_2(0) = 2$, $x_3(0) = 0$, $x_4(0) = 0$, $m_1 = m_2 = k_1 = k_2 = k_3 = 1$ and the control input $u(t)$ shown in Figure 6.1.5.

Graphs of the resulting mass displacements and velocities are shown in Figure 6.1.6. Although the initial motion is similar to that of the unforced response shown in Figure 6.1.2, subsequent motion is greatly modified as the input acceleration forces the system to a state of rest.

▶ **The State Variable Approach to Circuit Analysis.** Energy storage elements in an electrical network are the capacitors and the inductors. The energy in a charged capacitor is due to the separation of charge on its plates. The energy in an inductor is stored in its magnetic field. The state variables of the circuit are the set of variables associated with the energy of the energy storage elements, namely, the voltage of each capacitor and the current in each inductor. Application of Kirchhoff's current law and Kirchhoff's voltage law yields a first order system of differential equations for these state variables. Then, given the initial conditions of these variables, the complete response of the circuit to a forcing function (such as an impressed voltage source) is completely determined. Obviously, electrical networks consisting of hundreds or thousands of storage elements lead to first order systems of correspondingly large dimension.

To illustrate the state variable approach, we consider the electric circuit shown in Figure 6.1.7.

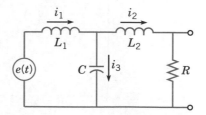

FIGURE 6.1.7 An electrical network composed of two loops.

There are two state variables, i_1 and i_2, representing the currents passing through each of the two inductors, and one state variable, v, representing the voltage on the capacitor. We therefore anticipate a coupled system of three first order differential equations describing this network.

Summing the voltages across each of the circuit elements in the left loop of the network yields

$$L_1 i_1' + v = e(t). \tag{27}$$

Similarly, summing the voltages across each of the circuit elements in the right loop of the network yields

$$L_2 i_2' + R i_2 - v = 0. \tag{28}$$

Note that adding Eq. (27) to Eq. (28) gives the sum of the voltage drops around the outer loop,

$$L_1 i_1' + L_2 i_2' + R i_2 = e(t),$$

a redundant equation since it can be obtained from Eqs. (27) and (28). Next, Kirchhoff's current law applied to the node connecting L_1, L_2, and C yields

$$i_1 = i_2 + i_3$$

or $i_3 = i_1 - i_2$. Since the voltage on the capacitor is given by $v = q/C$, where q is the charge on the capacitor, it follows that

$$C v' = q' = i_3 = i_1 - i_2. \tag{29}$$

Equations (27) through (29) are efficiently represented in matrix notation by

$$
\begin{pmatrix} i_1' \\ i_2' \\ v' \end{pmatrix} =
\begin{pmatrix} 0 & 0 & -1/L_1 \\ 0 & -R/L_2 & 1/L_2 \\ 1/C & -1/C & 0 \end{pmatrix}
\begin{pmatrix} i_1 \\ i_2 \\ v \end{pmatrix} +
\begin{pmatrix} 1/L_1 \\ 0 \\ 0 \end{pmatrix} e(t), \tag{30}
$$

where we have expressed the system in a form analogous to Eq. (24). Given initial conditions for each of the state variables, the complete response of the electrical network shown in Figure 6.1.7 to the input $e(t)$ can be found by solving the system (30) subject to the prescribed initial conditions. The output voltage across the resistor, $v_R = R i_2$, can be conveniently expressed in the form of the output equation (25),

$$
v_R = (0 \quad R \quad 0) \begin{pmatrix} i_1 \\ i_2 \\ v \end{pmatrix}.
$$

If we assume zero initial conditions for the state variables and circuit parameter values $L_1 = \frac{3}{2}, L_2 = \frac{1}{2}, C = \frac{4}{3}$, and $R = 1$, Figure 6.1.8 shows the output voltage $v_R(t) = R i_R(t)$ across the resistor due to a harmonic input $e(t) = \sin \omega t$ with $\omega = \frac{1}{2}$ (Figure 6.1.8a) and $\omega = 2$ (Figure 6.1.8b).

At the lower frequency the amplitude of the steady-state response is approximately equal to 1, whereas at the higher frequency the steady-state response is greatly attenuated. With the given parameter values, the circuit acts as a **low-pass filter**. Low-frequency input signals $e(t)$ pass through the circuit easily, while high-frequency signals are greatly attenuated. In Section 6.6, we find the amplitude of the steady-state output $v_R(t)$ of the circuit for all frequencies.

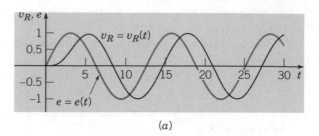

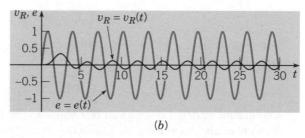

FIGURE 6.1.8 Output voltage $v_R(t) = Ri_R(t)$ across the resistor of the circuit in Figure 6.1.7 due to the input $e(t) = \sin \omega t$: (a) $\omega = \frac{1}{2}$, (b) $\omega = 2$.

PROBLEMS

1. If $\mathbf{A}(t) = \begin{pmatrix} e^t & 2e^{-t} & e^{2t} \\ 5e^t & e^{-t} & -e^{2t} \\ -e^t & 3e^{-t} & 5e^{2t} \end{pmatrix}$

and

$\mathbf{B}(t) = \begin{pmatrix} 2e^t & e^{-t} & 4e^{2t} \\ -e^t & 3e^{-t} & e^{2t} \\ 4e^t & -e^{-t} & -e^{2t} \end{pmatrix}$, find:

(a) $\mathbf{A} + 3\mathbf{B}$ **(b)** $\mathbf{AB}$

(c) $d\mathbf{A}/dt$ **(d)** $\int_0^1 \mathbf{A}(t)\,dt$

2. Verify that $\mathbf{x} = e^{-t} \begin{pmatrix} 6 \\ -8 \\ -4 \end{pmatrix} + 2e^{2t} \begin{pmatrix} 0 \\ 1 \\ -1 \end{pmatrix}$ satisfies

$\mathbf{x}' = \begin{pmatrix} 1 & 1 & 1 \\ 2 & 1 & -1 \\ 0 & -1 & 1 \end{pmatrix} \mathbf{x}$.

3. Verify that $\mathbf{\Psi} = \begin{pmatrix} e^t & e^{-2t} & e^{3t} \\ -4e^t & -e^{-2t} & 2e^{3t} \\ -e^t & -e^{-2t} & e^{3t} \end{pmatrix}$ satisfies the matrix differential equation

$\mathbf{\Psi}' = \begin{pmatrix} 1 & -1 & 4 \\ 3 & 2 & -1 \\ 2 & 1 & -1 \end{pmatrix} \mathbf{\Psi}$.

In each of Problems 4 through 9, transform the equation into an equivalent first order system:

4. $y^{(4)} + 6y''' + 3y = t$

5. $ty''' + (\sin t)y'' + 8y = \cos t$

6. $t(t-1)y^{(4)} + e^t y'' + 4t^2 y = 0$

7. $y''' + ty'' + t^2 y' + t^2 y = \ln t$

8. $(x-4)y^{(4)} + (x+1)y'' + (\tan x)y = 0$

9. $(x^2-2)y^{(6)} + x^2 y'' + 3y = 0$

10. Derive the differential equations for $x_1(t)$ and $x_2(t)$ in the system (23) by applying the balance law (22) to the compartment model illustrated in Figure 6.1.3.

11. Determine the matrix $\mathbf{K}$ and input $\mathbf{g}(t)$ if the system (23) is to be expressed using matrix notation, $\mathbf{x}' = \mathbf{Kx} + \mathbf{g}(t)$.

12. Find a system of first order linear differential equations for the four state variables of the circuit shown in Figure 6.1.9.

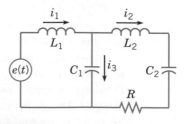

FIGURE 6.1.9

13. An initial amount α of a tracer (such as a dye or a radioactive isotope) is injected into Compartment 1 of the two-compartment system shown in Figure 6.1.10. At time $t > 0$, let $x_1(t)$ and $x_2(t)$ denote the amount of tracer in Compartment 1 and Compartment 2, respectively. Thus under the conditions stated, $x_1(0) = \alpha$ and $x_2(0) = 0$. The amounts are related to the corresponding concentrations $\rho_1(t)$ and $\rho_2(t)$ by the equations

$$x_1 = \rho_1 V_1 \qquad \text{and} \qquad x_2 = \rho_2 V_2, \qquad \text{(i)}$$

where V_1 and V_2 are the constant respective volumes of the compartments.

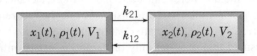

FIGURE 6.1.10 A closed two-compartment system.

The differential equations that describe the exchange of tracer between the compartments are

$$\frac{dx_1}{dt} = -k_{21}\rho_1 + k_{12}\rho_2$$

$$\frac{dx_2}{dt} = k_{21}\rho_1 - k_{12}\rho_2, \qquad \text{(ii)}$$

or, using the relations in (i),

$$\frac{dx_1}{dt} = -L_{21}x_1 + L_{12}x_2$$

$$\frac{dx_2}{dt} = L_{21}x_1 - L_{12}x_2, \qquad \text{(iii)}$$

where $L_{21} = k_{21}/V_1$ is the **fractional turnover rate** of Compartment 1 with respect to 2 and $L_{12} = k_{12}/V_2$ is the **fractional turnover rate** of Compartment 2 with respect to 1.
(a) Use Eqs. (iii) to show that

$$\frac{d}{dt}[x_1(t) + x_2(t)] = 0$$

and therefore $x_1(t) + x_2(t) = \alpha$ for all $t \geq 0$, that is, the tracer is conserved.
(b) Use the eigenvalue method to find the solution of the system (iii) subject to the initial conditions $x_1(0) = \alpha$ and $x_2(0) = 0$.
(c) What are the limiting values $\bar{x}_1 = \lim_{t\to\infty} x_1(t)$ and $\bar{x}_2 = \lim_{t\to\infty} x_2(t)$? Explain how the rate of approach to the equilibrium point $(\bar{x}_1, \bar{x}_2)$ depends on L_{12} and L_{21}.
(d) Give a qualitative sketch of the phase portrait for the system (iii).
(e) Plot the graphs of $\bar{x}_1/\alpha$ and $\bar{x}_2/\alpha$ as a function of $L_{21}/L_{12} \geq 0$ on the same set of coordinates and explain the meaning of the graphs.

14. (a) Using matrix notation, show that the system of second order equations (18) for the displacements $\mathbf{y} = (y_1, y_2)^T$ of the masses can be written in the form

$$\mathbf{y}'' = \mathbf{K}\mathbf{y} + \mathbf{f}(t).$$

(b) Under what conditions on k_1, k_2, k_3, m_1, and m_2 is $\mathbf{K}$ a symmetric matrix?

15. Consider the plant equation (26) for the control system consisting of two masses and three springs. Determine a suitable output matrix $\mathbf{C}$ for an output equation (25) if only the displacement of m_2 is observable.

6.2 Basic Theory of First Order Linear Systems

The theory presented in this section generalizes to higher dimensions the theory for two-dimensional linear systems introduced in Sections 3.2 and 4.2. The extension of this theory requires definitions for linearly dependent and linearly independent sets of vectors (see Appendix A.2). We will also use the determinant of an $n \times n$ matrix (i) to determine if its columns are linearly dependent or linearly independent and (ii) to test whether a system of n linear algebraic equations in n unknowns has a unique solution (see Theorems A.3.6 and A.3.7 in Appendix A.3).

Using the matrix notation of Section 6.1, the first order linear system of dimension n is

$$\mathbf{x}' = \mathbf{P}(t)\mathbf{x} + \mathbf{g}(t). \qquad \text{(1)}$$

Conditions under which a unique solution to the initial value problem associated with Eq. (1) exists are provided by the following generalization of Theorem 3.2.1.

THEOREM
6.2.1

(Existence and Uniqueness for First Order Linear Systems). If $\mathbf{P}(t)$ and $\mathbf{g}(t)$ are continuous on an open interval $I = (\alpha, \beta)$, then there exists a unique solution $\mathbf{x} = \boldsymbol{\phi}(t)$ of the initial value problem

$$\mathbf{x}' = \mathbf{P}(t)\mathbf{x} + \mathbf{g}(t), \qquad \mathbf{x}(t_0) = \mathbf{x}_0, \tag{2}$$

where t_0 is any point in I, and $\mathbf{x}_0$ is any constant vector with n components. Moreover the solution exists throughout the interval I.

The proof of this theorem, too difficult to include here, can be found in more advanced books on differential equations. However, just as in the two-dimensional case, the theorem is easy to apply. If the functions $p_{11}, p_{12}, \ldots, p_{nn}, g_1, \ldots, g_n$ are continuous on the interval I containing the point t_0, then we are assured that one and only one solution exists on I, irrespective of the vector $\mathbf{x}_0$ of prescribed initial values. A commonly occurring special case of the initial value problem (2) is

$$\mathbf{x}' = \mathbf{A}\mathbf{x}, \qquad \mathbf{x}(0) = \mathbf{x}_0, \tag{3}$$

where $\mathbf{A}$ is a constant $n \times n$ matrix. Since the coefficients of $\mathbf{A}$ are continuous for all values of t, Theorem 6.2.1 guarantees that a solution exists and is unique on the entire t-axis. The study of solutions of the constant coefficient initial value problem (3), based on eigenvalue methods, is taken up in the next section.

In this section we discuss properties and structure of solutions of the homogeneous equation

$$\mathbf{x}' = \mathbf{P}(t)\mathbf{x} \tag{4}$$

obtained from Eq. (1) by setting $\mathbf{g}(t) = \mathbf{0}$. The solution of the nonhomogeneous equation, $\mathbf{x}' = \mathbf{P}(t)\mathbf{x} + \mathbf{g}(t)$, is dealt with in Section 6.6.

We use the notation

$$\mathbf{x}_1(t) = \begin{pmatrix} x_{11}(t) \\ x_{21}(t) \\ \vdots \\ x_{n1}(t) \end{pmatrix}, \quad \ldots, \quad \mathbf{x}_k(t) = \begin{pmatrix} x_{1k}(t) \\ x_{2k}(t) \\ \vdots \\ x_{nk}(t) \end{pmatrix}, \quad \ldots$$

to designate specific solutions of the system (4). Note that $x_{ij}(t)$ refers to the ith component of the jth solution $\mathbf{x}_j(t)$. An expression of the form

$$c_1\mathbf{x}_1 + \cdots + c_k\mathbf{x}_k,$$

where $c_1, \ldots, c_k$ are arbitrary constants, is called a **linear combination** of solutions.

The following result, a generalization of Theorem 3.3.1, is the principal distinguishing property of linear systems. This property is not shared by nonlinear systems.

THEOREM
6.2.2

(Principle of Superposition). If $\mathbf{x}_1, \mathbf{x}_2, \ldots, \mathbf{x}_k$ are solutions of the homogeneous linear system

$$\mathbf{x}' = \mathbf{P}(t)\mathbf{x} \tag{5}$$

on the interval $I = (\alpha, \beta)$, then the linear combination

$$c_1\mathbf{x}_1 + c_2\mathbf{x}_2 + \cdots + c_k\mathbf{x}_k$$

is also a solution of Eq. (5) on I.

Proof

Let $\mathbf{x} = c_1\mathbf{x}_1 + c_2\mathbf{x}_2 + \cdots + c_k\mathbf{x}_k$. The result follows from the linear operations of matrix multiplication and differentiation:

$$\begin{aligned} \mathbf{P}(t)\mathbf{x} &= \mathbf{P}(t)[c_1\mathbf{x}_1 + \cdots + c_k\mathbf{x}_k] \\ &= c_1\mathbf{P}(t)\mathbf{x}_1 + \cdots + c_k\mathbf{P}(t)\mathbf{x}_k \\ &= c_1\mathbf{x}_1' + \cdots + c_k\mathbf{x}_k' = \mathbf{x}'. \end{aligned}$$

We use Theorem 6.2.2 to enlarge a finite set of solutions $\{\mathbf{x}_1, \mathbf{x}_2, \ldots, \mathbf{x}_k\}$ to a k-fold infinite family of solutions $c_1\mathbf{x}_1 + c_2\mathbf{x}_2 + \cdots + c_k\mathbf{x}_k$ parameterized by $c_1, \ldots, c_k$. In the discussion that follows, we show that all solutions of Eq. (5) are contained in an n-parameter family $c_1\mathbf{x}_1 + c_2\mathbf{x}_2 + \cdots + c_n\mathbf{x}_n$, provided that the n solutions $\mathbf{x}_1, \ldots, \mathbf{x}_n$ are distinct in a sense made precise by the following definition.

DEFINITION
6.2.3

The n vector functions $\mathbf{x}_1, \ldots, \mathbf{x}_n$ are said to be **linearly independent on an interval** I if the only constants $c_1, c_2, \ldots, c_n$ such that

$$c_1\mathbf{x}_1(t) + \cdots + c_n\mathbf{x}_n(t) = \mathbf{0} \tag{6}$$

for all $t \in I$ are $c_1 = c_2 = \cdots = c_n = 0$. If there exist constants $c_1, c_2, \ldots, c_n$, *not all zero*, such that Eq. (6) is true for all $t \in I$, the vector functions are said to be **linearly dependent** on I.

EXAMPLE
1

Show that the vector functions

$$\mathbf{x}_1(t) = \begin{pmatrix} e^{-2t} \\ 0 \\ -e^{-2t} \end{pmatrix} \quad \text{and} \quad \mathbf{x}_2(t) = \begin{pmatrix} e^t \\ e^t \\ e^t \end{pmatrix}$$

are linearly independent on $I = (-\infty, \infty)$.

To prove independence, we assume that

$$c_1\mathbf{x}_1(t) + c_2\mathbf{x}_2(t) = \mathbf{0} \tag{7}$$

is true for all $t \in I$. Setting $t = 0$ in Eq. (7) leads to the requirement

$$c_1 \begin{pmatrix} 1 \\ 0 \\ -1 \end{pmatrix} + c_2 \begin{pmatrix} 1 \\ 1 \\ 1 \end{pmatrix} = \begin{pmatrix} 0 \\ 0 \\ 0 \end{pmatrix},$$

which is equivalent to the system of linear equations

$$c_1 + c_2 = 0,$$
$$c_2 = 0, \tag{8}$$
$$-c_1 + c_2 = 0.$$

Since the only solution of the system (8) is $c_1 = c_2 = 0$, we conclude that the vector functions $\mathbf{x}_1$ and $\mathbf{x}_2$ are linearly independent on any interval containing $t = 0$ and, in particular, are linearly independent on $I = (-\infty, \infty)$.

EXAMPLE 2

Show that the vector functions

$$\mathbf{x}_1(t) = \begin{pmatrix} 1+t \\ t \\ 1-t \end{pmatrix}, \qquad \mathbf{x}_2(t) = \begin{pmatrix} 3 \\ t+2 \\ t \end{pmatrix}, \qquad \text{and} \qquad \mathbf{x}_3(t) = \begin{pmatrix} 1-2t \\ 2-t \\ 3t-2 \end{pmatrix}$$

are linearly dependent on $I = (-\infty, \infty)$.

We begin by assuming that

$$c_1\mathbf{x}_1(t) + c_2\mathbf{x}_2(t) + c_3\mathbf{x}_3(t) = \mathbf{0} \quad \text{for all } t \in I, \tag{9}$$

and attempt to determine constants c_1, c_2, and c_3, not all of which are zero, such that condition (9) is a true statement. Candidate values for c_1, c_2, and c_3 are found by evaluating the equation in condition (9) at a particular value of $t \in I$. For example, choosing $t = 0$ yields the system of equations

$$c_1 \begin{pmatrix} 1 \\ 0 \\ 1 \end{pmatrix} + c_2 \begin{pmatrix} 3 \\ 2 \\ 0 \end{pmatrix} + c_3 \begin{pmatrix} 1 \\ 2 \\ -2 \end{pmatrix} = \begin{pmatrix} 0 \\ 0 \\ 0 \end{pmatrix}, \tag{10}$$

which is equivalent to the system of linear equations

$$c_1 + 3c_2 + c_3 = 0,$$
$$2c_2 + 2c_3 = 0, \tag{11}$$
$$c_1 - 2c_3 = 0.$$

Using Gaussian elimination, we find the general solution of the system (11) to be

$$\mathbf{c} = \alpha \begin{pmatrix} 2 \\ -1 \\ 1 \end{pmatrix},$$

where α is arbitrary. Setting $\alpha = 1$ gives us $c_1 = 2$, $c_2 = -1$, and $c_3 = 1$. Thus Eq. (10) is true for these values of c_1, c_2, and c_3, that is, the three vectors $(1, 0, 1)^T$, $(3, 2, 0)^T$, and $(1, 2, -2)^T$ are linearly dependent. We still need to verify that the statement (9) is true using the values $c_1 = 2$, $c_2 = -1$, and $c_3 = 1$. Since

$$2 \begin{pmatrix} 1+t \\ t \\ 1-t \end{pmatrix} - 1 \begin{pmatrix} 3 \\ t+2 \\ t \end{pmatrix} + 1 \begin{pmatrix} 1-2t \\ 2-t \\ 3t-2 \end{pmatrix} = \begin{pmatrix} 2+2t-3+1-2t \\ 2t-t-2+2-t \\ 2-2t-t+3t-2 \end{pmatrix} = \begin{pmatrix} 0 \\ 0 \\ 0 \end{pmatrix}$$

for all $t \in I$, we conclude that $\mathbf{x}_1, \mathbf{x}_2$, and $\mathbf{x}_3$ are linearly dependent vector functions on I.

In Section 3.3 the Wronskian of two vector functions with two components was defined. We now extend the definition to sets of n vector functions, each with n components.

<table>
<tr><td>DEFINITION
6.2.4</td><td>

Let $\mathbf{x}_1, \ldots, \mathbf{x}_n$ be n solutions of the homogeneous linear system of differential equations $\mathbf{x}' = \mathbf{P}(t)\mathbf{x}$ and let $\mathbf{X}(t)$ be the $n \times n$ matrix whose jth column is $\mathbf{x}_j(t)$, $j = 1, \ldots, n$,

$$\mathbf{X}(t) = \begin{pmatrix} x_{11}(t) & \cdots & x_{1n}(t) \\ \vdots & & \vdots \\ x_{n1}(t) & \cdots & x_{nn}(t) \end{pmatrix}. \tag{12}$$

The **Wronskian** $W = W[\mathbf{x}_1, \ldots, \mathbf{x}_n]$ of the n solutions $\mathbf{x}_1, \ldots, \mathbf{x}_n$ is defined by

$$W[\mathbf{x}_1, \ldots, \mathbf{x}_n](t) = \det \mathbf{X}(t). \tag{13}$$

</td></tr>
</table>

The next theorem shows how W is used to test whether a set of n solutions of $\mathbf{x}' = \mathbf{P}(t)\mathbf{x}$ is linearly independent or linearly dependent on an interval I.

<table>
<tr><td>THEOREM
6.2.5</td><td>

Let $\mathbf{x}_1, \ldots, \mathbf{x}_n$ be solutions of $\mathbf{x}' = \mathbf{P}(t)\mathbf{x}$ on an interval $I = (\alpha, \beta)$ in which $\mathbf{P}(t)$ is continuous.

(i) If $\mathbf{x}_1, \ldots, \mathbf{x}_n$ are linearly independent on I, then $W[\mathbf{x}_1, \ldots, \mathbf{x}_n](t) \neq 0$ at every point in I,

(ii) If $\mathbf{x}_1, \ldots, \mathbf{x}_n$ are linearly dependent on I, then $W[\mathbf{x}_1, \ldots, \mathbf{x}_n](t) = 0$ at every point in I.

</td></tr>
<tr><td>Proof</td><td>

Assume first that $\mathbf{x}_1, \ldots, \mathbf{x}_n$ are linearly independent on I. We then want to show that $W[\mathbf{x}_1, \ldots, \mathbf{x}_n](t) \neq 0$ throughout I. To do this, we assume the contrary, that is, there is a point $t_0 \in I$ such that $W[\mathbf{x}_1, \ldots, \mathbf{x}_n](t_0) = 0$. This means that the column vectors $\{\mathbf{x}_1(t_0), \ldots, \mathbf{x}_n(t_0)\}$ are linearly dependent (Theorem A.3.6) so that there exist constants $\hat{c}_1, \ldots, \hat{c}_n$, not all zero, such that $\hat{c}_1\mathbf{x}_1(t_0) + \cdots + \hat{c}_n\mathbf{x}_n(t_0) = \mathbf{0}$. Then Theorem 6.2.2 implies that $\boldsymbol{\phi}(t) = \hat{c}_1\mathbf{x}_1(t) + \cdots + \hat{c}_n\mathbf{x}_n(t)$ is a solution of $\mathbf{x}' = \mathbf{P}(t)\mathbf{x}$ that satisfies the initial condition $\mathbf{x}(t_0) = \mathbf{0}$. The zero solution also satisfies the same initial value problem. The uniqueness part of Theorem 6.2.1 therefore implies that $\boldsymbol{\phi}$ is the zero solution, that is, $\boldsymbol{\phi}(t) = \hat{c}_1\mathbf{x}_1(t) + \cdots + \hat{c}_n\mathbf{x}_n(t) = \mathbf{0}$ for every $t \in (\alpha, \beta)$, contradicting our original assumption that $\mathbf{x}_1, \ldots, \mathbf{x}_n$ are linearly independent on I. This proves (i).

To prove (ii), assume that $\mathbf{x}_1, \ldots, \mathbf{x}_n$ are linearly dependent on I. Then there exist constants $\alpha_1, \ldots, \alpha_n$, not all zero, such that $\alpha_1\mathbf{x}_1(t) + \cdots + \alpha_n\mathbf{x}_n(t) = \mathbf{0}$ for every $t \in I$. Consequently, for each $t \in I$, the vectors $\mathbf{x}_1(t), \ldots, \mathbf{x}_n(t)$ are linearly dependent. Thus $W[\mathbf{x}_1, \ldots, \mathbf{x}_n](t) = 0$ at every point in I (Theorem A.3.6).

</td></tr>
</table>

Remark. The conclusion of Theorem 6.2.5 holds only for solutions of a system $\mathbf{x}' = \mathbf{P}(t)\mathbf{x}$ on an interval I where $\mathbf{P}(t)$ is continuous. For example, the vectors

$$\mathbf{x}_1 = \begin{pmatrix} t^2 \\ 2t \end{pmatrix} \quad \text{and} \quad \mathbf{x}_2 = \begin{pmatrix} t|t| \\ 2|t| \end{pmatrix}$$

although linearly independent on $I = (-1, 1)$, satisfy $W[\mathbf{x}_1, \mathbf{x}_2](0) = 0$ on I. Therefore, $\mathbf{x}_1$ and $\mathbf{x}_2$ do not satisfy the hypothesis of Theorem 6.2.5.

It follows from Theorem 6.2.5 that $\mathbf{x}_1, \ldots, \mathbf{x}_n$ are linearly independent solutions of $\mathbf{x}' = \mathbf{P}(t)\mathbf{x}$ on I if and only if $W[\mathbf{x}_1, \ldots, \mathbf{x}_n](t) \neq 0$ for every $t \in I$. The next theorem shows that all solutions of $\mathbf{x}' = \mathbf{P}(t)\mathbf{x}$ are contained in the n-fold infinite family $c_1\mathbf{x}_1 + \cdots + c_n\mathbf{x}_n$, provided that $\mathbf{x}_1, \ldots, \mathbf{x}_n$ are linearly independent on I.

THEOREM 6.2.6

Let $\mathbf{x}_1, \ldots, \mathbf{x}_n$ be solutions of

$$\mathbf{x}' = \mathbf{P}(t)\mathbf{x} \tag{14}$$

on the interval $\alpha < t < \beta$ such that, for some point $t_0 \in (\alpha, \beta)$, the Wronskian is nonzero, $W[\mathbf{x}_1, \ldots, \mathbf{x}_n](t_0) \neq 0$. Then each solution $\mathbf{x} = \boldsymbol{\phi}(t)$ of Eq. (14) can be expressed as a linear combination of $\mathbf{x}_1, \ldots, \mathbf{x}_n$,

$$\boldsymbol{\phi}(t) = \hat{c}_1\mathbf{x}_1(t) + \cdots + \hat{c}_n\mathbf{x}_n(t), \tag{15}$$

where the constants $\hat{c}_1, \ldots, \hat{c}_n$ are uniquely determined.

Proof

Let $\boldsymbol{\phi}(t)$ be a given solution of Eq. (14). If we set $\mathbf{x}_0 = \boldsymbol{\phi}(t_0)$, then the vector function $\boldsymbol{\phi}$ is a solution of the initial value problem

$$\mathbf{x}' = \mathbf{P}(t)\mathbf{x}, \qquad \mathbf{x}(t_0) = \mathbf{x}_0. \tag{16}$$

By the principle of superposition, the linear combination $\boldsymbol{\psi}(t) = c_1\mathbf{x}_1(t) + \cdots + c_n\mathbf{x}_n(t)$ is also a solution of (14) for any choice of constants $c_1, \ldots, c_n$. The requirement $\boldsymbol{\psi}(t_0) = \mathbf{x}_0$ leads to the linear algebraic system

$$\mathbf{X}(t_0)\mathbf{c} = \mathbf{x}_0, \tag{17}$$

where $\mathbf{X}(t)$ is defined by Eq. (12). Since $W[\mathbf{x}_1, \ldots, \mathbf{x}_n](t_0) \neq 0$, the linear algebraic system (17) has a unique solution (see Theorem A.3.7) that we denote by $\hat{c}_1, \ldots, \hat{c}_n$. Thus the particular member $\hat{\boldsymbol{\psi}}(t) = \hat{c}_1\mathbf{x}_1(t) + \cdots + \hat{c}_n\mathbf{x}_n(t)$ of the n-parameter family represented by $\boldsymbol{\psi}(t)$ also satisfies the initial value problem (16). By the uniqueness part of Theorem 6.2.1, it follows that $\boldsymbol{\phi} = \hat{\boldsymbol{\psi}} = \hat{c}_1\mathbf{x}_1 + \cdots + \hat{c}_n\mathbf{x}_n$. Since $\boldsymbol{\phi}$ is arbitrary, the result holds (with different constants, of course) for every solution of Eq. (14).

Remark. It is customary to call the n-parameter family

$$c_1\mathbf{x}_1(t) + \cdots + c_n\mathbf{x}_n(t)$$

the **general solution** of Eq. (14) if $W[\mathbf{x}_1, \ldots, \mathbf{x}_n](t_0)$ is nonzero for some $t_0 \in (\alpha, \beta)$. Theorem 6.2.6 guarantees that the general solution includes all possible solutions of Eq. (14). Any set of solutions $\mathbf{x}_1, \ldots, \mathbf{x}_n$ of Eq. (14), which are linearly independent on an interval $\alpha < t < \beta$, is said to be a **fundamental set of solutions** for that interval.

EXAMPLE 3

If

$$\mathbf{A} = \begin{pmatrix} 0 & -1 & 2 \\ 2 & -3 & 2 \\ 3 & -3 & 1 \end{pmatrix},$$

show that

$$\mathbf{x}_1(t) = e^{-2t}\begin{pmatrix} 1 \\ 0 \\ -1 \end{pmatrix}, \qquad \mathbf{x}_2(t) = e^{-t}\begin{pmatrix} 1 \\ 1 \\ 0 \end{pmatrix}, \qquad \text{and} \qquad \mathbf{x}_3(t) = e^{t}\begin{pmatrix} 1 \\ 1 \\ 1 \end{pmatrix}$$

form a fundamental set for

$$\mathbf{x}' = \mathbf{A}\mathbf{x}. \tag{18}$$

Then solve the initial value problem consisting of Eq. (18) subject to the initial condition $\mathbf{x}(0) = (1, -2, 1)^T$.

Substituting each of $\mathbf{x}_1$, $\mathbf{x}_2$, and $\mathbf{x}_3$ into Eq. (18) and verifying that the equation reduces to an identity, show that $\mathbf{x}_1$, $\mathbf{x}_2$, and $\mathbf{x}_3$ are solutions of Eq. (18). In the next section we will show you how to find $\mathbf{x}_1$, $\mathbf{x}_2$, and $\mathbf{x}_3$ yourself. Since

$$W[\mathbf{x}_1, \mathbf{x}_2, \mathbf{x}_3](0) = \begin{vmatrix} 1 & 1 & 1 \\ 0 & 1 & 1 \\ -1 & 0 & 1 \end{vmatrix} = 1 \neq 0,$$

it follows from Theorem 6.2.6 that $\mathbf{x}_1$, $\mathbf{x}_2$, and $\mathbf{x}_3$ form a fundamental set for Eq. (18) and that the general solution is $\mathbf{x} = c_1\mathbf{x}_1(t) + c_2\mathbf{x}_2(t) + c_3\mathbf{x}_3(t)$. Substituting the general solution into the initial condition $\mathbf{x}(0) = (1, -2, 1)^T$ yields the system

$$\begin{pmatrix} 1 & 1 & 1 \\ 0 & 1 & 1 \\ -1 & 0 & 1 \end{pmatrix} \begin{pmatrix} c_1 \\ c_2 \\ c_3 \end{pmatrix} = \begin{pmatrix} 1 \\ -2 \\ 1 \end{pmatrix}. \tag{19}$$

Solving Eq. (19) by Gaussian elimination gives $c_1 = 3$, $c_2 = -6$, and $c_3 = 4$. Thus the solution of Eq. (18) satisfying $\mathbf{x}(0) = (1, -2, 1)^T$ is

$$\mathbf{x} = 3e^{-2t} \begin{pmatrix} 1 \\ 0 \\ -1 \end{pmatrix} - 6e^{-t} \begin{pmatrix} 1 \\ 1 \\ 0 \end{pmatrix} + 4e^{t} \begin{pmatrix} 1 \\ 1 \\ 1 \end{pmatrix}.$$

The next theorem states that the system (4) always has at least one fundamental set of solutions.

THEOREM 6.2.7

Let

$$\mathbf{e}_1 = \begin{pmatrix} 1 \\ 0 \\ 0 \\ \vdots \\ 0 \end{pmatrix}, \quad \mathbf{e}_2 = \begin{pmatrix} 0 \\ 1 \\ 0 \\ \vdots \\ 0 \end{pmatrix}, \quad \ldots, \quad \mathbf{e}_n = \begin{pmatrix} 0 \\ 0 \\ 0 \\ \vdots \\ 1 \end{pmatrix};$$

further let $\mathbf{x}_1, \ldots, \mathbf{x}_n$ be solutions of $\mathbf{x}' = \mathbf{P}(t)\mathbf{x}$ that satisfy the initial conditions

$$\mathbf{x}_1(t_0) = \mathbf{e}_1, \quad \ldots, \quad \mathbf{x}_n(t_0) = \mathbf{e}_n,$$

respectively, where t_0 is any point in $\alpha < t < \beta$. Then $\mathbf{x}_1, \ldots, \mathbf{x}_n$ form a fundamental set of solutions of $\mathbf{x}' = \mathbf{P}(t)\mathbf{x}$.

Proof	Existence and uniqueness of the solutions $\mathbf{x}_1, \ldots, \mathbf{x}_n$ are guaranteed by Theorem 6.2.1, and it is easy to see that $W[\mathbf{x}_1, \ldots, \mathbf{x}_n](t_0) = \det \mathbf{I}_n = 1$. It then follows from Theorem 6.2.6 that $\mathbf{x}_1, \ldots, \mathbf{x}_n$ form a fundamental set of solutions for $\mathbf{x}' = \mathbf{P}(t)\mathbf{x}$.

Once one fundamental set of solutions has been found, other sets can be generated by forming (independent) linear combinations of the first set. For theoretical purposes, the set given by Theorem 6.2.7 is usually the simplest.

To summarize, any set of n linearly independent solutions of the system (4) forms a fundamental set of solutions. Under the conditions given in this section, such fundamental sets always exist, and every solution of the system (4) can be represented as a linear combination of any fundamental set of solutions.

▶ **Linear nth Order Equations.** Recall from Section 6.1 that by introducing the variables

$$x_1 = y, \quad x_2 = y', \quad x_3 = y'', \quad \ldots, \quad x_n = y^{(n-1)}, \tag{20}$$

the initial value problem for the linear nth order equation,

$$\frac{d^n y}{dt^n} + p_1(t)\frac{d^{n-1}y}{dt^{n-1}} + \cdots + p_{n-1}(t)\frac{dy}{dt} + p_n(t)y = g(t), \tag{21}$$

$$y(t_0) = y_0, \quad y'(t_0) = y_1, \quad \ldots, \quad y^{(n-1)}(t_0) = y_{n-1}, \tag{22}$$

can be expressed as an initial value problem for a first order system,

$$\mathbf{x}' = \begin{pmatrix} 0 & 1 & 0 & 0 & \cdots & 0 \\ 0 & 0 & 1 & 0 & \cdots & 0 \\ \vdots & \vdots & \vdots & \vdots & & \vdots \\ 0 & 0 & 0 & 0 & \cdots & 1 \\ -p_n(t) & -p_{n-1}(t) & -p_{n-2}(t) & -p_{n-3}(t) & \cdots & -p_1(t) \end{pmatrix} \mathbf{x} + \begin{pmatrix} 0 \\ 0 \\ \vdots \\ 0 \\ g(t) \end{pmatrix}, \tag{23}$$

$$\mathbf{x}(t_0) = \begin{pmatrix} y_0 \\ y_1 \\ \vdots \\ y_{n-1} \end{pmatrix}, \tag{24}$$

where $\mathbf{x} = (x_1, \ldots, x_n)^T$. Theorem 6.2.1 then provides sufficient conditions for the existence and uniqueness of a solution to the initial value problem (21), (22). These conditions are stated in the following corollary to Theorem 6.2.1.

COROLLARY 6.2.8	If the functions $p_1(t), p_2(t), \ldots, p_n(t)$, and $g(t)$ are continuous on the open interval $I = (\alpha, \beta)$, then there exists exactly one solution $y = \phi(t)$ of the differential equation (21) that also satisfies the initial conditions (22). This solution exists throughout the interval I.

Proof   Under the stated conditions, the matrix of coefficients and the nonhomogeneous term in Eq. (23) are continuous on I. By Theorem 6.2.1, a unique solution $\mathbf{x} = \boldsymbol{\phi}(t) = (\phi_1(t), \dots, \phi_n(t))^T$ to the initial value problem (23), (24) exists throughout I. The definitions for the state variables in the list (20) then show that $y = \phi(t) = \phi_1(t)$ is the unique solution of the initial value problem (21), (22).

We now restrict our attention to the homogeneous equation associated with Eq. (21),

$$\frac{d^n y}{dt^n} + p_1(t)\frac{d^{n-1}y}{dt^{n-1}} + \cdots + p_{n-1}(t)\frac{dy}{dt} + p_n(t)y = 0, \tag{25}$$

and the corresponding homogeneous system associated with Eq. (23)

$$\mathbf{x}' = \begin{pmatrix} 0 & 1 & 0 & 0 & \cdots & 0 \\ 0 & 0 & 1 & 0 & \cdots & 0 \\ \vdots & \vdots & \vdots & \vdots & & \vdots \\ 0 & 0 & 0 & 0 & \cdots & 1 \\ -p_n(t) & -p_{n-1}(t) & -p_{n-2}(t) & -p_{n-3}(t) & \cdots & -p_1(t) \end{pmatrix} \mathbf{x}. \tag{26}$$

Using the relations (20), the scalar functions $y_1, \dots, y_n$ are solutions of Eq. (25) if and only if the n vectors $\mathbf{x}_1 = (y_1, y_1', \dots, y_1^{(n-1)})^T, \dots, \mathbf{x}_n = (y_n, y_n', \dots, y_n^{(n-1)})^T$ are solutions of Eq. (26). In accordance with Eq. (13), we will define the Wronskian of the scalar functions $y_1, \dots, y_n$ by

$$W[y_1, \dots, y_n](t) = \begin{vmatrix} y_1(t) & y_2(t) & \cdots & y_n(t) \\ y_1'(t) & y_2'(t) & \cdots & y_n'(t) \\ \vdots & \vdots & & \vdots \\ y_1^{(n-1)}(t) & y_2^{(n-1)}(t) & & y_n^{(n-1)}(t) \end{vmatrix}. \tag{27}$$

Theorem 6.2.6 then allows us to conclude that if $y_1, \dots, y_n$ are solutions of Eq. (25) on an interval $I = (\alpha, \beta)$ and

$$W[y_1, \dots, y_n](t_0) \neq 0$$

for some $t_0 \in (\alpha, \beta)$, then each solution $y = \phi(t)$ of Eq. (25) can be written as a linear combination of $y_1, \dots, y_n$,

$$\phi(t) = \hat{c}_1 y_1(t) + \cdots + \hat{c}_n y_n(t),$$

where the constants $\hat{c}_1, \dots, \hat{c}_n$ are uniquely determined. We state this result in the following corollary to Theorem 6.2.6.

COROLLARY
6.2.9

Let $y_1, \ldots, y_n$ be solutions of

$$\frac{d^n y}{dt^n} + p_1(t)\frac{d^{n-1} y}{dt^{n-1}} + \cdots + p_{n-1}(t)\frac{dy}{dt} + p_n(t)y = 0 \qquad (28)$$

on an interval $I = (\alpha, \beta)$ in which $p_1, \ldots, p_n$ are continuous. If for some point $t_0 \in I$ these solutions satisfy

$$W[y_1, \ldots, y_n](t_0) \neq 0,$$

then each solution $y = \phi(t)$ of Eq. (28) can be expressed as a linear combination of $y_1, \ldots, y_n$,

$$\phi(t) = \hat{c}_1 y_1(t) + \cdots + \hat{c}_n y_n(t),$$

where the constants $\hat{c}_1, \ldots, \hat{c}_n$ are uniquely determined.

The terminology for solutions of the nth order scalar equation is identical to that used for solutions of $\mathbf{x}' = \mathbf{P}(t)\mathbf{x}$. A set of solutions $y_1, \ldots, y_n$ such that $W[y_1, \ldots, y_n](t_0) \neq 0$ for some $t_0 \in (\alpha, \beta)$ is called a **fundamental set of solutions** for Eq. (28) and the n-parameter family represented by the linear combination

$$y = c_1 y_1(t) + \cdots + c_n y_n(t),$$

where $c_1, \ldots, c_n$ are arbitrary constants, is called the **general solution** of Eq. (28). Corollary 6.2.9 guarantees that each solution of Eq. (28) corresponds to some member of this n-parameter family of solutions.

EXAMPLE
4

Show that $y_1(x) = x$, $y_2(x) = x^{-1}$, and $y_3(x) = x^2$ form a fundamental set of solutions for

$$x^3 y''' + x^2 y'' - 2xy' + 2y = 0 \qquad (29)$$

on $I = (0, \infty)$.

Substituting each of y_1, y_2, and y_3 into Eq. (29) shows that they are solutions on the given interval. To verify that the three functions are linearly independent on I, we compute the Wronskian

$$W[y_1, y_2, y_3](x) = \begin{vmatrix} x & x^{-1} & x^2 \\ 1 & -x^{-2} & 2x \\ 0 & 2x^{-3} & 2 \end{vmatrix} = -6x^{-1}.$$

Since $W[y_1, y_2, y_3](x) < 0$ on I, Corollary 6.2.9 implies that x, x^{-1}, and x^2 form a fundamental set for Eq. (29). Note that it is only necessary to confirm that $W[y_1, y_2, y_3]$ is nonzero at one point in I; for example, showing that $W[y_1, y_2, y_3](1) = -6$ would have sufficed.

PROBLEMS

In each of Problems 1 through 6, determine intervals in which solutions are sure to exist:

1. $y^{(4)} + 5y''' + 4y = t$

2. $ty''' + (\sin t)y'' + 4y = \cos t$

3. $t(t-1)y^{(4)} + e^t y'' + 7t^2 y = 0$

4. $y''' + ty'' + 5t^2 y' + 2t^3 y = \ln t$

5. $(x-1)y^{(4)} + (x+5)y'' + (\tan x)y = 0$

6. $(x^2 - 25)y^{(6)} + x^2 y'' + 5y = 0$

7. Consider the vectors

$$\mathbf{x}_1(t) = \begin{pmatrix} e^t \\ 2e^t \\ -e^t \end{pmatrix}, \qquad \mathbf{x}_2(t) = \begin{pmatrix} e^{-t} \\ -2e^{-t} \\ e^{-t} \end{pmatrix},$$

$$\mathbf{x}_3(t) = \begin{pmatrix} 2e^{4t} \\ 2e^{4t} \\ -8e^{4t} \end{pmatrix}$$

and let $\mathbf{X}(t)$ be the matrix whose columns are the vectors $\mathbf{x}_1(t), \mathbf{x}_2(t), \mathbf{x}_3(t)$. Compare the amounts of work between the following two methods for obtaining $W[\mathbf{x}_1, \mathbf{x}_2, \mathbf{x}_3](0)$:
Method 1: First find $|\mathbf{X}(t)|$ and then set $t = 0$.
Method 2: Evaluate $\mathbf{X}(t)$ at $t = 0$ and then find $|\mathbf{X}(0)|$.

8. Determine whether

$$\mathbf{x}_1(t) = e^{-t} \begin{pmatrix} 1 \\ 0 \\ -1 \end{pmatrix}, \qquad \mathbf{x}_2(t) = e^{-t} \begin{pmatrix} 1 \\ -6 \\ 1 \end{pmatrix},$$

$$\mathbf{x}_3(t) = e^{2t} \begin{pmatrix} 0 \\ 16 \\ -6 \end{pmatrix}$$

form a fundamental set of solutions for $\mathbf{x}' = \begin{pmatrix} 0 & 1 & 1 \\ 1 & 0 & 1 \\ 1 & 1 & 0 \end{pmatrix} \mathbf{x}$.

9. Determine whether

$$\mathbf{x}_1(t) = e^{-t} \begin{pmatrix} 1 \\ 0 \\ -1 \end{pmatrix}, \qquad \mathbf{x}_2(t) = e^{-t} \begin{pmatrix} 1 \\ -4 \\ 1 \end{pmatrix},$$

$$\mathbf{x}_3(t) = e^{8t} \begin{pmatrix} 2 \\ 1 \\ 2 \end{pmatrix}$$

form a fundamental set of solutions for $\mathbf{x}' = \begin{pmatrix} 3 & 2 & 4 \\ 2 & 0 & 2 \\ 4 & 2 & 3 \end{pmatrix} \mathbf{x}$.

10. In Section 4.2 it was shown that if $\mathbf{x}_1$ and $\mathbf{x}_2$ are solutions of

$$\mathbf{x}' = \begin{pmatrix} p_{11}(t) & p_{12}(t) \\ p_{21}(t) & p_{22}(t) \end{pmatrix} \mathbf{x},$$

on an interval I, then the Wronskian W of $\mathbf{x}_1$ and $\mathbf{x}_2$ satisfies the differential equation $W' = (p_{11} + p_{22})W$. A generalization of that proof shows that if $\mathbf{x}_1, \dots, \mathbf{x}_n$ are solutions of Eq. (4) on I, then the Wronskian of $\mathbf{x}_1, \dots, \mathbf{x}_n$, denoted by W, satisfies the differential equation

$$W' = (p_{11} + p_{22} + \cdots + p_{nn})W = \mathrm{tr}(\mathbf{P}(t))W. \qquad \text{(i)}$$

(a) Explain why Eq. (i) also implies that W is either identically zero or else never vanishes on I in accordance with Theorem 6.2.5.
(b) If $y_1, \dots, y_n$ are solutions of Eq. (28), find a counterpart to Eq. (i) satisfied by $W = W[y_1, \dots, y_n](t)$.

In each of Problems 11 through 16, verify that the given functions are solutions of the differential equations, and determine their Wronskian:

11. $y''' + y' = 0$, 1, $\cos t$, $\sin t$
12. $y^{(4)} + y'' = 0$, 1, t, $\cos t$, $\sin t$
13. $y''' + 4y'' - 4y' - 16y = 0$, e^{2t}, e^{-2t}, e^{-4t}
14. $y^{(4)} + 6y''' + 9y'' = 0$, 1, t, e^{-3t}, te^{-3t}
15. $xy''' - y'' = 0$, 1, x, x^3
16. $x^3 y''' + x^2 y'' - 2xy' + 2y = 0$, x, x^2, $1/x$
17. Verify that the differential operator defined by

$$L[y] = y^{(n)} + p_1(t)y^{(n-1)} + \cdots + p_n(t)y$$

is a linear operator. That is, show that

$$L[c_1 y_1 + c_2 y_2] = c_1 L[y_1] + c_2 L[y_2],$$

where y_1 and y_2 are n times differentiable functions and c_1 and c_2 are arbitrary constants. Hence show that if $y_1, y_2, \dots, y_n$ are solutions of $L[y] = 0$, then the linear combination $c_1 y_1 + \cdots + c_n y_n$ is also a solution of $L[y] = 0$.

6.3 Homogeneous Linear Systems with Constant Coefficients

In Chapter 3 the eigenvalue method was used to find fundamental solution sets for linear constant coefficient systems of dimension 2,

$$\mathbf{x}' = \begin{pmatrix} a_{11} & a_{12} \\ a_{21} & a_{22} \end{pmatrix} \mathbf{x}. \qquad \text{(1)}$$

In this section we extend the eigenvalue method to the system

$$\mathbf{x}' = \mathbf{A}\mathbf{x}, \tag{2}$$

where $\mathbf{A}$ is a real constant $n \times n$ matrix. As in Chapter 3, we assume solutions of the form

$$\mathbf{x} = e^{\lambda t}\mathbf{v}, \tag{3}$$

where the scalar λ and the constant $n \times 1$ vector $\mathbf{v}$ are to be determined. The steps leading to the eigenvalue problem are identical to the two-dimensional case. Substituting from Eq. (3) into Eq. (2), we find that

$$\lambda e^{\lambda t}\mathbf{v} = e^{\lambda t}\mathbf{A}\mathbf{v}, \tag{4}$$

where we have used the fact that $\mathbf{x}' = \lambda e^{\lambda t}\mathbf{v}$. Since $e^{\lambda t}$ is nonzero, Eq. (4) reduces to

$$\lambda\mathbf{v} = \mathbf{A}\mathbf{v},$$

or

$$(\mathbf{A} - \lambda\mathbf{I}_n)\mathbf{v} = \mathbf{0}, \tag{5}$$

where $\mathbf{I}_n$ is the $n \times n$ identity matrix. Given a square matrix $\mathbf{A}$, recall that the problem of

  i. finding values of λ for which Eq. (5) has nontrivial solution vectors $\mathbf{v}$, and
  ii. finding the corresponding nontrivial solutions,

is known as the **eigenvalue problem** for $\mathbf{A}$ (see Appendix A.4). We distinguish the following three cases:

1. $\mathbf{A}$ has a complete set of n linearly independent eigenvectors and all of the eigenvalues of $\mathbf{A}$ are real,
2. $\mathbf{A}$ has a complete set of n linearly independent eigenvectors and one or more complex conjugate pairs of eigenvalues,
3. $\mathbf{A}$ is defective, that is, there are one or more eigenvalues of $\mathbf{A}$ for which the geometric multiplicity is less than the algebraic multiplicity (see Appendix A.4).

In the rest of this section, we analyze the first case, whereas the second and third cases are dealt with in Sections 6.4 and 6.7, respectively.

The Matrix **A** Is Nondefective with Real Eigenvalues

THEOREM 6.3.1

Let $(\lambda_1, \mathbf{v}_1), \ldots, (\lambda_n, \mathbf{v}_n)$ be eigenpairs for the real, $n \times n$ constant matrix $\mathbf{A}$. Assume that the eigenvalues $\lambda_1, \ldots, \lambda_n$ are real and that the corresponding eigenvectors $\mathbf{v}_1, \ldots, \mathbf{v}_n$ are linearly independent. Then

$$\left\{ e^{\lambda_1 t}\mathbf{v}_1, \ldots, e^{\lambda_n t}\mathbf{v}_n \right\} \tag{6}$$

is a fundamental set of solutions to $\mathbf{x}' = \mathbf{A}\mathbf{x}$ on the interval $(-\infty, \infty)$. The general solution of $\mathbf{x}' = \mathbf{A}\mathbf{x}$ is therefore given by

$$\mathbf{x}(t) = c_1 e^{\lambda_1 t}\mathbf{v}_1 + \cdots + c_n e^{\lambda_n t}\mathbf{v}_n, \tag{7}$$

where $c_1, \ldots, c_n$ are arbitrary constants.

Remark. The eigenvalues need not be distinct. All that is required is that for each eigenvalue λ_j, the geometric multiplicity equals the algebraic multiplicity.

Proof	Let

$$\mathbf{x}_1(t) = e^{\lambda_1 t}\mathbf{v}_1, \dots, \mathbf{x}_n(t) = e^{\lambda_n t}\mathbf{v}_n.$$

We have $\mathbf{x}'_j = e^{\lambda_j t}\lambda_j\mathbf{v}_j = e^{\lambda_j t}\mathbf{A}\mathbf{v}_j = \mathbf{A}\mathbf{x}_j$, so $\mathbf{x}_j$ is a solution of $\mathbf{x}' = \mathbf{A}\mathbf{x}$ for each $j = 1, \dots, n$. To show that $\mathbf{x}_1, \dots, \mathbf{x}_n$ form a fundamental set of solutions, we evaluate the Wronskian,

$$W[\mathbf{x}_1, \dots, \mathbf{x}_n](t) = \det[e^{\lambda_1 t}\mathbf{v}_1, \dots, e^{\lambda_n t}\mathbf{v}_n] = e^{(\lambda_1 + \dots + \lambda_n)t}\det[\mathbf{v}_1, \dots, \mathbf{v}_n]. \tag{8}$$

The exponential function is never zero and since the eigenvectors $\mathbf{v}_1, \dots, \mathbf{v}_n$ are linearly independent, the determinant in the last term is nonzero. Therefore the Wronskian is nonzero and the result follows.

EXAMPLE
1

Find the general solution of

$$\mathbf{x}' = \begin{pmatrix} -4/5 & -1/5 & 4/5 \\ -1/5 & -4/5 & -4/5 \\ 2/5 & -2/5 & 3/5 \end{pmatrix}\mathbf{x}. \tag{9}$$

The characteristic polynomial of the matrix $\mathbf{A}$ of coefficients is

$$|\mathbf{A} - \lambda\mathbf{I}| = \begin{vmatrix} -4/5 - \lambda & -1/5 & 4/5 \\ -1/5 & -4/5 - \lambda & -4/5 \\ 2/5 & -2/5 & 3/5 - \lambda \end{vmatrix} = -\lambda^3 - \lambda^2 + \lambda + 1 = -(\lambda + 1)^2(\lambda - 1),$$

so the eigenvalues of $\mathbf{A}$ are $\lambda_1 = -1$ and $\lambda_2 = 1$ with algebraic multiplicities 2 and 1, respectively. To find the eigenvector(s) belonging to λ_1, we set $\lambda = -1$ in $(\mathbf{A} - \lambda\mathbf{I})\mathbf{v} = \mathbf{0}$. This gives the linear algebraic system

$$\begin{pmatrix} 1/5 & -1/5 & 4/5 \\ -1/5 & 1/5 & -4/5 \\ 2/5 & -2/5 & 8/5 \end{pmatrix}\begin{pmatrix} v_1 \\ v_2 \\ v_3 \end{pmatrix} = \begin{pmatrix} 0 \\ 0 \\ 0 \end{pmatrix}. \tag{10}$$

Using elementary row operations, we reduce the system (10) to

$$\begin{pmatrix} 1 & -1 & 4 \\ 0 & 0 & 0 \\ 0 & 0 & 0 \end{pmatrix}\begin{pmatrix} v_1 \\ v_2 \\ v_3 \end{pmatrix} = \begin{pmatrix} 0 \\ 0 \\ 0 \end{pmatrix}.$$

The only constraint on the components of $\mathbf{v}$ is $v_1 - v_2 + 4v_3 = 0$. Setting $v_2 = a_1$ and $v_3 = a_2$, where a_1 and a_2 are arbitrary constants, and then solving for v_1 give $v_1 = a_1 - 4a_2$. Consequently, the general solution of (10) can be represented by

$$\mathbf{v} = \begin{pmatrix} a_1 - 4a_2 \\ a_1 \\ a_2 \end{pmatrix} = a_1\begin{pmatrix} 1 \\ 1 \\ 0 \end{pmatrix} + a_2\begin{pmatrix} -4 \\ 0 \\ 1 \end{pmatrix}. \tag{11}$$

First, setting $a_1 = 2$ and $a_2 = 0$ and then setting $a_1 = 2$ and $a_2 = 1$ yield a pair of linearly independent eigenvectors associated with $\lambda_1 = -1$,

$$\mathbf{v}_1 = \begin{pmatrix} 2 \\ 2 \\ 0 \end{pmatrix} \quad \text{and} \quad \mathbf{v}_2 = \begin{pmatrix} -2 \\ 2 \\ 1 \end{pmatrix}.$$

Remark. Any choices for a_1 and a_2 that produce a pair of linearly independent eigenvectors for λ_1 would suffice. For example, we could have first set $a_1 = 1$ and $a_2 = 0$ and then set $a_1 = 0$ and $a_2 = 1$. Note that the choices for a_1 and a_2 used above lead to eigenvectors $\mathbf{v}_1$ and $\mathbf{v}_2$ that are orthogonal ($\mathbf{v}_1 \cdot \mathbf{v}_2 = 0$).

To find the eigenvector associated with $\lambda_2 = 1$, we set $\lambda = 1$ in $(\mathbf{A} - \lambda \mathbf{I})\mathbf{v} = \mathbf{0}$ to obtain the system

$$\begin{pmatrix} -9/5 & -1/5 & 4/5 \\ -1/5 & -9/5 & -4/5 \\ 2/5 & -2/5 & -2/5 \end{pmatrix} \begin{pmatrix} v_1 \\ v_2 \\ v_3 \end{pmatrix} = \begin{pmatrix} 0 \\ 0 \\ 0 \end{pmatrix}. \tag{12}$$

Elementary row operations reduce the system (12) to

$$\begin{pmatrix} 1 & 0 & -1/2 \\ 0 & 1 & 1/2 \\ 0 & 0 & 0 \end{pmatrix} \begin{pmatrix} v_1 \\ v_2 \\ v_3 \end{pmatrix} = \begin{pmatrix} 0 \\ 0 \\ 0 \end{pmatrix}. \tag{13}$$

Equation (12) is therefore equivalent to the two equations $v_1 - v_3/2 = 0$ and $v_2 + v_3/2 = 0$. The general solution of this pair of equations can be expressed as

$$\mathbf{v} = b_1 \begin{pmatrix} 1 \\ -1 \\ 2 \end{pmatrix}, \tag{14}$$

where b_1 is arbitrary. Choosing $b_1 = 1$ yields the eigenvector

$$\mathbf{v}_3 = \begin{pmatrix} 1 \\ -1 \\ 2 \end{pmatrix}$$

belonging to the eigenvalue $\lambda_2 = 1$. The general solution of Eq. (9) is therefore given by

$$\mathbf{x}(t) = c_1 e^{-t} \begin{pmatrix} 2 \\ 2 \\ 0 \end{pmatrix} + c_2 e^{-t} \begin{pmatrix} -2 \\ 2 \\ 1 \end{pmatrix} + c_3 e^{t} \begin{pmatrix} 1 \\ -1 \\ 2 \end{pmatrix}. \tag{15}$$

To help understand the qualitative behavior of all solutions of Eq. (9), we introduce the subset S of $\mathbf{R}^3$ spanned by the eigenvectors $\mathbf{v}_1$ and $\mathbf{v}_2$,

$$S = \{\mathbf{v} : \mathbf{v} = a_1 \mathbf{v}_1 + a_2 \mathbf{v}_2, \ -\infty < a_1, a_2 < \infty\}. \tag{16}$$

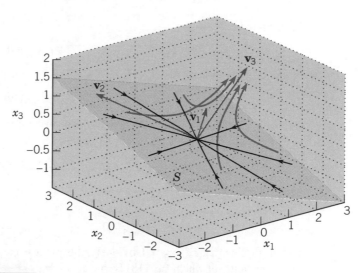

FIGURE 6.3.1 Solution trajectories for the system (9).

Geometrically, S corresponds to a plane passing through the origin. This is shown in Figure 6.3.1. From Eq. (15) we see that if $\mathbf{x}(0) \in S$, then $c_3 = 0$, and consequently, $\mathbf{x}(t) \in S$ for all $t \geq 0$. Furthermore $\mathbf{x}(t) \to (0, 0, 0)^T$ as $t \to \infty$ due to the fact that the eigenvalue $\lambda_1 = -1 < 0$. The straight-line trajectories lying in S in Figure 6.3.1 correspond to solutions $\mathbf{x}(t)$ of Eq. (9) where the initial conditions are such that $\mathbf{x}(0) = \mathbf{x}_0 \in S$. Also shown in Figure 6.3.1 are the graphs of four trajectories (heavy curves) represented by Eq. (15) in which c_3 is chosen to be slightly positive. For each of these trajectories, the initial condition $\mathbf{x}(0) \notin S$, and it is clear from Eq. (15) and Figure 6.3.1 that each trajectory in this case must asymptotically approach the line passing through the origin with direction vector $\mathbf{v}_3$, parametrically expressed by $\mathbf{x} = \tau \mathbf{v}_3, -\infty < \tau < \infty$, as $t \to \infty$.

▶ Real and Distinct Eigenvalues. A frequently occurring special case for which the general solution of $\mathbf{x}' = \mathbf{A}\mathbf{x}$ is always of the form (7) is given by the following corollary to Theorem 6.3.1.

COROLLARY 6.3.2	Suppose that the matrix $\mathbf{A}$ has n eigenpairs $(\lambda_1, \mathbf{v}_1), \dots, (\lambda_n, \mathbf{v}_n)$ with the property that the eigenvalues $\lambda_1, \dots, \lambda_n$ are real and distinct. Then $$\left\{ e^{\lambda_1 t} \mathbf{v}_1, \dots, e^{\lambda_n t} \mathbf{v}_n \right\}$$ form a fundamental solution set for the homogeneous system $\mathbf{x}' = \mathbf{A}\mathbf{x}$.
Proof	Since the eigenvalues are distinct, the eigenvectors $\mathbf{v}_1, \dots, \mathbf{v}_n$ are linearly independent. To see why this is so, note that $$(\mathbf{A} - \lambda_i \mathbf{I}_n)\mathbf{v}_j = \mathbf{A}\mathbf{v}_j - \lambda_i \mathbf{I}_n \mathbf{v}_j = \lambda_j \mathbf{v}_j - \lambda_i \mathbf{v}_j = \begin{cases} \mathbf{0}, & i = j \\ (\lambda_j - \lambda_i)\mathbf{v}_j, & i \neq j. \end{cases} \qquad (17)$$

Multiplying

$$c_1\mathbf{v}_1 + \cdots + c_n\mathbf{v}_n = \mathbf{0} \tag{18}$$

by the product matrix $(\mathbf{A} - \lambda_{n-1}\mathbf{I}_n) \cdots (\mathbf{A} - \lambda_1\mathbf{I}_n)$ and using Eqs. (17) yield the equation

$$c_n(\lambda_n - \lambda_1) \cdots (\lambda_n - \lambda_{n-1})\mathbf{v}_n = \mathbf{0},$$

Since the eigenvalues are distinct and each eigenvector is not the zero vector, this implies $c_n = 0$. Next, multiplying

$$c_1\mathbf{v}_1 + \cdots + c_{n-1}\mathbf{v}_{n-1} = \mathbf{0}$$

by the product matrix $(\mathbf{A} - \lambda_{n-2}\mathbf{I}_n) \cdots (\mathbf{A} - \lambda_1\mathbf{I}_n)$ and again using Eqs. (17) yield the equation

$$c_{n-1}(\lambda_{n-1} - \lambda_1) \cdots (\lambda_{n-1} - \lambda_{n-2})\mathbf{v}_{n-1} = \mathbf{0},$$

which implies that $c_{n-1} = 0$. Obviously, this process can be continued to show that the only constants for which Eq. (18) is true are $c_1 = c_2 = \cdots = c_n = 0$. Thus the hypothesis of Theorem (6.3.1) is satisfied and the result follows.

EXAMPLE 2

Find the general solution of

$$\mathbf{x}' = \begin{pmatrix} -1 & -1 & 1 & 1 \\ -3 & -4 & -3 & 6 \\ 0 & -3 & -2 & 3 \\ -3 & -5 & -3 & 7 \end{pmatrix} \mathbf{x}. \tag{19}$$

The characteristic polynomial of the matrix of coefficients is

$$\begin{vmatrix} -1-\lambda & -1 & 1 & 1 \\ -3 & -4-\lambda & -3 & 6 \\ 0 & -3 & -2-\lambda & 3 \\ -3 & -5 & -3 & 7-\lambda \end{vmatrix} = \lambda^4 - 5\lambda^2 + 4 = (\lambda + 2)(\lambda + 1)(\lambda - 1)(\lambda - 2).$$

Thus the eigenvalues $\lambda_1 = -2, \lambda_2 = -1, \lambda_3 = 1, \lambda_4 = 2$ are distinct and Corollary 6.3.2 is applicable. The respective eigenvectors are found to be

$$\mathbf{v}_1 = \begin{pmatrix} 1 \\ 0 \\ -1 \\ 0 \end{pmatrix}, \quad \mathbf{v}_2 = \begin{pmatrix} 1 \\ 1 \\ 0 \\ 1 \end{pmatrix}, \quad \mathbf{v}_3 = \begin{pmatrix} 1 \\ 0 \\ 1 \\ 1 \end{pmatrix}, \quad \mathbf{v}_4 = \begin{pmatrix} 0 \\ 1 \\ 0 \\ 1 \end{pmatrix}.$$

Of course, the eigenvalues and eigenvectors can be found easily by using a computer or calculator. A fundamental set of solutions of Eq. (19) is therefore

$$\left\{ e^{-2t}\begin{pmatrix} 1 \\ 0 \\ -1 \\ 0 \end{pmatrix}, \quad e^{-t}\begin{pmatrix} 1 \\ 1 \\ 0 \\ 1 \end{pmatrix}, \quad e^{t}\begin{pmatrix} 1 \\ 0 \\ 1 \\ 1 \end{pmatrix}, \quad e^{2t}\begin{pmatrix} 0 \\ 1 \\ 0 \\ 1 \end{pmatrix} \right\}$$

and the general solution is

$$\mathbf{x} = c_1 e^{-2t} \begin{pmatrix} 1 \\ 0 \\ -1 \\ 0 \end{pmatrix} + c_2 e^{-t} \begin{pmatrix} 1 \\ 1 \\ 0 \\ 1 \end{pmatrix} + c_3 e^{t} \begin{pmatrix} 1 \\ 0 \\ 1 \\ 1 \end{pmatrix} + c_4 e^{2t} \begin{pmatrix} 0 \\ 1 \\ 0 \\ 1 \end{pmatrix}.$$

From the general solution we can deduce the behavior of solutions for large t: (i) if $c_3 = c_4 = 0$, $\lim_{t \to \infty} \mathbf{x}(t) = (0, 0, 0, 0)^T$; (ii) if $c_3 \neq 0$ and $c_4 = 0$, $\mathbf{x}(t)$ asymptotically approaches the line passing through the origin in $\mathbf{R}^4$ with direction vector $\mathbf{v}_3$; (iii) if c_3 and c_4 are both nonzero, then as t approaches infinity the direction of the trajectory approaches the direction of the eigenvector $\mathbf{v}_4$.

▶ Symmetric Matrices. Even though the matrix in Example 1 has an eigenvalue ($\lambda = -1$) with algebraic multiplicity 2, we were able to find two linearly independent eigenvectors $\mathbf{v}_1$ and $\mathbf{v}_2$. As a consequence, we were able to construct the general solution. In general, given a matrix $\mathbf{A}$, the geometric multiplicity of each eigenvalue is less than or equal to its algebraic multiplicity. If for one or more eigenvalues of $\mathbf{A}$ the geometric multiplicity is less than the algebraic multiplicity, then Eq. (2) will not have a fundamental set of solutions of the form (6). However, if $\mathbf{A}$ belongs to the class of real and symmetric matrices, then Eq. (2) will always have a fundamental set of solutions of the form (6) with the general solution given by Eq. (7). This is a consequence of the following properties of the eigenvalues and eigenvectors of a real symmetric matrix $\mathbf{A}$ (see Appendix A.4):

1. All the eigenvalues $\lambda_1, \dots, \lambda_n$ of $\mathbf{A}$ are real.
2. $\mathbf{A}$ has a complete set of n real and linearly independent eigenvectors $\mathbf{v}_1, \dots, \mathbf{v}_n$. Furthermore eigenvectors corresponding to different eigenvalues are orthogonal to one another and all eigenvectors belonging to the same eigenvalue can be chosen to be orthogonal to one another.

The following continuous time, discrete space model of particle diffusion in one dimension leads to a first order system of linear differential equations $\mathbf{x}' = \mathbf{A}\mathbf{x}$, where $\mathbf{A}$ is a symmetric matrix.

EXAMPLE
3

Diffusion on a One-Dimensional Lattice with Reflecting Boundaries. Consider particles that can occupy any of n equally spaced points lying along the real line (a one-dimensional lattice), as shown in Figure 6.3.2.

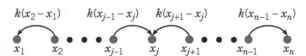

FIGURE 6.3.2 Diffusion on a one-dimensional lattice.

If $x_j(t)$ is the number of particles residing at the jth lattice point at time t, we make the following assumptions governing the movement of particles from site to site:

i. particle transitions to site j are permitted only from nearest-neighbor sites, and
ii. particles move from more populated sites to less populated sites with the rate of transition proportional to the difference between the numbers of particles at adjacent sites.

The number of particles at each interior point of the lattice is a discrete variable. To model this with a differential equation we approximate each discrete variable with a continuously differentiable variable, as we did in the population models in Chapter 2. Then the differential equation describing the rate of change in the number of particles at the jth interior point of the lattice is

$$\frac{dx_j}{dt} = k(x_{j-1} - x_j) + k(x_{j+1} - x_j) = k(x_{j-1} - 2x_j + x_{j+1}), \quad j = 2, \ldots, n-1, \qquad (20)$$

where k is a rate constant. The rate equations for the numbers of particles at the left and right endpoints are

$$\frac{dx_1}{dt} = k(x_2 - x_1) \qquad (21)$$

and

$$\frac{dx_n}{dt} = k(x_{n-1} - x_n), \qquad (22)$$

respectively. The left and right endpoints are referred to as reflecting boundaries since Eqs. (21) and (22) do not permit particles to escape from the set of lattice points. This will be made evident in the discussion that follows.

Using Eqs. (20) through (22), the system describing diffusion on a lattice consisting of $n = 3$ points is expressed in matrix notation as

$$\mathbf{x}' = k\mathbf{A}\mathbf{x}, \qquad (23)$$

where $\mathbf{x} = (x_1, x_2, x_3)^T$ and

$$\mathbf{A} = \begin{pmatrix} -1 & 1 & 0 \\ 1 & -2 & 1 \\ 0 & 1 & -1 \end{pmatrix}. \qquad (24)$$

The eigenvalues of the symmetric matrix $\mathbf{A}$ are $\lambda_1 = 0$, $\lambda_2 = -1$, and $\lambda_3 = -3$, and the corresponding eigenvectors are

$$\mathbf{v}_1 = \begin{pmatrix} 1 \\ 1 \\ 1 \end{pmatrix}, \qquad \mathbf{v}_2 = \begin{pmatrix} 1 \\ 0 \\ -1 \end{pmatrix}, \qquad \mathbf{v}_3 = \begin{pmatrix} 1 \\ -2 \\ 1 \end{pmatrix}. \qquad (25)$$

The eigenvectors are mutually orthogonal and are frequently referred to as **normal modes**. Plots of the components of the eigenvectors are shown in Figure 6.3.3.

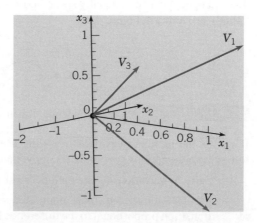

FIGURE 6.3.3   The eigenvectors, or normal modes, of the symmetric matrix A.

Thus the general solution of Eq. (23) is

$$\mathbf{x}(t) = c_1 \begin{pmatrix} 1 \\ 1 \\ 1 \end{pmatrix} + c_2 e^{-kt} \begin{pmatrix} 1 \\ 0 \\ -1 \end{pmatrix} + c_3 e^{-3kt} \begin{pmatrix} 1 \\ -2 \\ 1 \end{pmatrix}. \tag{26}$$

From Eq. (26), it is clear that

$$\lim_{t \to \infty} \mathbf{x}(t) = c_1 \mathbf{v}_1 = \begin{pmatrix} c_1 \\ c_1 \\ c_1 \end{pmatrix}, \tag{27}$$

that is, all solutions approach an equilibrium state in which the numbers of particles at each of the lattice sites are identical, that is, a uniform distribution. The components on both sides of Eq. (23) may be summed by multiplying both sides of Eq. (23) (on the left) by the row vector $(1, 1, 1)$. This yields the equation $(d/dt)(x_1 + x_2 + x_3) = 0$ and therefore $x_1(t) + x_2(t) + x_3(t)$ is constant for $t > 0$, a statement that the total number of particles in the system is conserved. Since $\lim_{t \to 0}[x_1(t) + x_2(t) + x_3(t)] = x_{10} + x_{20} + x_{30}$, the initial total number of particles, and Eq. (27) implies that $\lim_{t \to \infty}[x_1(t) + x_2(t) + x_3(t)] = 3c_1$, it follows that $3c_1 = x_{10} + x_{20} + x_{30}$. Consequently, $c_1 = (x_{10} + x_{20} + x_{30})/3$, the average value of the initial total number of particles.

By choosing $\mathbf{c} = (1, 1, 0)^T$ in the general solution (26), corresponding to the initial condition $\mathbf{x}(0) = (2, 1, 0)^T$, decay toward equilibrium that involves only the eigenvectors $\mathbf{v}_1$ and $\mathbf{v}_2$ is illustrated in Figure 6.3.4a, where we have also set the rate constant $k = 1$. In this case,

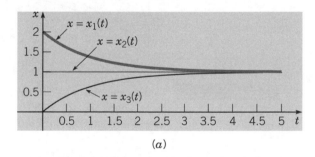

(a)

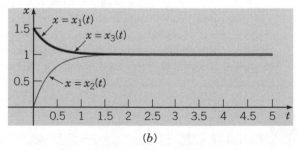

(b)

FIGURE 6.3.4 (a) Component plots of Eq. (26) with $\mathbf{c} = (1, 1, 0)^T$ corresponding to the initial condition $\mathbf{x}(0) = (2, 1, 0)^T$. ($b$) Component plots of Eq. (26) with $\mathbf{c} = \left(1, 0, \frac{1}{2}\right)^T$ corresponding to the initial condition $\mathbf{x}(0) = \left(\frac{3}{2}, 0, \frac{3}{2}\right)$. In both cases, $k = 1$.

the decay rate is controlled by the eigenvalue $\lambda_2 = -1$. Furthermore the components of $\mathbf{x}(t)$ are antisymmetric about the equilibrium solution due to the fact that $\mathbf{v}_2$ is antisymmetric about its middle component (see Figure 6.3.3).

Similarly, by choosing $\mathbf{c} = \left(1, 0, \frac{1}{2}\right)^T$ in (26), corresponding to the initial condition $\mathbf{x}(0) = \left(\frac{3}{2}, 0, \frac{3}{2}\right)^T$, decay toward equilibrium that involves only the eigenvectors $\mathbf{v}_1$ and $\mathbf{v}_3$ is shown in Figure 6.3.4b, where again $k = 1$. In this case, the rate of decay toward equilibrium is faster since it is controlled by the eigenvalue $\lambda_3 = -3$. Note also that the components of $\mathbf{x}(t)$ are symmetric about the equilibrium solution due to the fact that $\mathbf{v}_3$ is symmetric about its middle component (see Figure 6.3.3).

In Figure 6.3.5, we show component plots of (26) with $\mathbf{c} = \left(1, \frac{3}{2}, \frac{1}{2}\right)^T$, corresponding to the initial condition $\mathbf{x}(0) = (3, 0, 0)^T$ in which all of the particles are initially located at the first lattice point, again with $k = 1$. In this case, all three eigenvectors are required to represent the solution of the initial value problem.

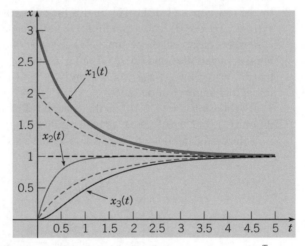

FIGURE 6.3.5   Component plots of Eq. (26) with $\mathbf{c} = \left(1, \frac{3}{2}, \frac{1}{2}\right)^T$ corresponding to the initial condition $\mathbf{x}(0) = (3, 0, 0)^T$. The dashed curves are the solutions shown in Figure 6.3.4a; $k = 1$.

The contribution to the solution from $e^{-3t}\mathbf{v}_3$ decays rapidly, while the long-term decay rate toward equilibrium is controlled by the eigenvalue $\lambda_2 = -1$. Since the component $x_2(t) = 1 - e^{-3t}$ contains no e^{-t} term, it converges more quickly than $x_1(t)$ and $x_2(t)$. In the long run, the component plots closely match those shown in Figure 6.3.4a since the total numbers of initial particles used in each of the two initial value problems are the same.

PROBLEMS

In each of Problems 1 through 8, find the general solution of the given system of equations:

1. $x_1' = -4x_1 + x_2$
    $x_2' = x_1 - 5x_2 + x_3$
    $x_3' = x_2 - 4x_3$

2. $x_1' = x_1 + 4x_2 + 4x_3$
    $x_2' = 3x_2 + 2x_3$
    $x_3' = 2x_2 + 3x_3$

3. $x_1' = 2x_1 - 4x_2 + 2x_3$
$x_2' = -4x_1 + 2x_2 - 2x_3$
$x_3' = 2x_1 - 2x_2 - x_3$

4. $x_1' = -2x_1 + 2x_2 - x_3$
$x_2' = -2x_1 + 3x_2 - 2x_3$
$x_3' = -2x_1 + 4x_2 - 3x_3$

5. $\mathbf{x}' = \begin{pmatrix} 1 & 1 & 6 \\ 1 & 6 & 1 \\ 6 & 1 & 1 \end{pmatrix} \mathbf{x}$

6. $\mathbf{x}' = \begin{pmatrix} 3 & 2 & 4 \\ 2 & 0 & 2 \\ 4 & 2 & 3 \end{pmatrix} \mathbf{x}$

7. $\mathbf{x}' = \begin{pmatrix} 1 & 1 & 1 \\ 2 & 1 & -1 \\ -8 & -5 & -3 \end{pmatrix} \mathbf{x}$

8. $\mathbf{x}' = \begin{pmatrix} 1 & -1 & 4 \\ 3 & 2 & -1 \\ 2 & 1 & -1 \end{pmatrix} \mathbf{x}$

In each of Problems 9 through 12, solve the given initial value problem and plot the graph of the solution in $\mathbf{R}^3$. Describe the behavior of the solution as $t \to \infty$.

9. $\mathbf{x}' = \begin{pmatrix} 1 & 1 & 2 \\ 0 & 2 & 2 \\ -1 & 1 & 3 \end{pmatrix} \mathbf{x}, \qquad \mathbf{x}(0) = \begin{pmatrix} 11 \\ 1 \\ 5 \end{pmatrix}$

10. $\mathbf{x}' = \begin{pmatrix} 0 & 0 & -1 \\ 2 & 0 & 0 \\ -1 & 2 & 4 \end{pmatrix} \mathbf{x}, \qquad \mathbf{x}(0) = \begin{pmatrix} 7 \\ 5 \\ 5 \end{pmatrix}$

11. $\mathbf{x}' = \begin{pmatrix} -1 & 0 & 3 \\ 0 & -2 & 0 \\ 3 & 0 & -1 \end{pmatrix} \mathbf{x}, \qquad \mathbf{x}(0) = \begin{pmatrix} 2 \\ -1 \\ -2 \end{pmatrix}$

12. $\mathbf{x}' = \begin{pmatrix} 1/2 & -1 & -3/2 \\ 3/2 & -2 & -3/2 \\ -2 & 2 & 1 \end{pmatrix} \mathbf{x}, \qquad \mathbf{x}(0) = \begin{pmatrix} 2 \\ 1 \\ 1 \end{pmatrix}$

13. Using the rate equations (20) through (22), express, in matrix notation, the system of differential equations that describe diffusion on a one-dimensional lattice with reflecting boundaries consisting of $n = 4$ lattice sites. Find the general solution of the resulting system and plot the components of each of the eigenvectors of the matrix $\mathbf{A}$ on the same set of coordinate axes. Which eigenvalue controls the long-term rate of decay toward equilibrium?

14. **Diffusion on a One-dimensional Lattice with an Absorbing Boundary.** Consider a one-dimensional lattice consisting of $n = 4$ lattice points, as shown in Figure 6.3.6.

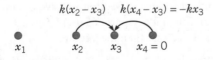

$$k(x_2 - x_3) \qquad k(x_4 - x_3) = -kx_3$$

$x_1 \qquad x_2 \qquad x_3 \quad x_4 = 0$

FIGURE 6.3.6 An absorbing boundary at the right endpoint.

Assume that the left endpoint is a reflecting boundary so that the rate equation for the number of particles $x_1(t)$ occupying that site is given by Eq. (21). Further assume that the sites 2 and 3 are interior points with rate equations given by Eq. (20) but with the condition that $x_4(t) = 0$ for all $t \geq 0$.

(a) Find the system of differential equations that describe the rate equations for $\mathbf{x} = (x_1, x_2, x_3)^T$ and express the equations in matrix notation, $\mathbf{x}' = \mathbf{A}\mathbf{x}$.

(b) Site 4 is referred to as an absorbing boundary because particles that land on that site from site 3 are removed from the diffusion process. Show that

$$\frac{d}{dt}(x_1 + x_2 + x_3) = -kx_3$$

and explain the meaning of this equation.

(c) Find the eigenvalues and eigenvectors for the matrix $\mathbf{A}$ in part (a) and plot the components of the eigenvectors. Then compare the eigenvalues and eigenvectors with those found in Example 3 of this section.

(d) Find the general solution of the system of equations found in part (a). Explain the asymptotic behavior of $\mathbf{x}(t)$ as $t \to \infty$.

15. Find constant 3×1 vectors $\mathbf{u}_1$ and $\mathbf{u}_2$ such that the solution of the initial value problem

$$\mathbf{x}' = \begin{pmatrix} 2 & -4 & -3 \\ 3 & -5 & -3 \\ -2 & 2 & 1 \end{pmatrix} \mathbf{x}, \qquad \mathbf{x}(0) = \mathbf{x}_0$$

tends to $(0, 0, 0)^T$ as $t \to \infty$ for any $\mathbf{x}_0 \in S$, where

$$S = \{\mathbf{u} : \mathbf{u} = a_1\mathbf{u}_1 + a_2\mathbf{u}_2, \ -\infty < a_1, a_2 < \infty\}.$$

In $\mathbf{R}^3$, plot solutions to the initial value problem for several different choices of $\mathbf{x}_0 \in S$ overlaying the trajectories on a graph of the plane determined by $\mathbf{u}_1$ and $\mathbf{u}_2$. Describe the behavior of solutions as $t \to \infty$ if $\mathbf{x}_0 \notin S$.

16. Find constant 4×1 vectors $\mathbf{u}_1$, $\mathbf{u}_2$, and $\mathbf{u}_3$ such that the solution of the initial value problem

$$\mathbf{x}' = \begin{pmatrix} 1 & 5 & 3 & -5 \\ 2 & 3 & 2 & -4 \\ 0 & -1 & -2 & 1 \\ 2 & 4 & 2 & -5 \end{pmatrix} \mathbf{x}, \qquad \mathbf{x}(0) = \mathbf{x}_0$$

tends to $(0, 0, 0, 0)^T$ as $t \to \infty$ for any $\mathbf{x}_0 \in S$, where

$$S = \{\mathbf{u} : \mathbf{u} = a_1\mathbf{u}_1 + a_2\mathbf{u}_2 + a_3\mathbf{u}_3, -\infty < a_1, a_2, a_3 < \infty\}.$$

17. A radioactive substance R_1 having decay rate k_1 disintegrates into a second radioactive substance R_2 having decay rate k_2. Substance R_2 disintegrates into R_3, which is stable. If $m_i(t)$ represents the mass of substance R_i at time t, $i = 1, 2, 3$,

the applicable equations are

$$m_1' = -k_1 m_1$$
$$m_2' = k_1 m_1 - k_2 m_2$$
$$m_3' = k_2 m_2.$$

Draw a block diagram of a compartment model of the overall reaction. Label the directed arrows that represent the mass flows between compartments with the appropriate radioactive decay rate constants. Use the eigenvalue method to solve the above system under the conditions

$$m_1(0) = m_0, \qquad m_2(0) = 0, \qquad m_3(0) = 0.$$

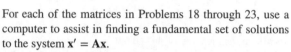 For each of the matrices in Problems 18 through 23, use a computer to assist in finding a fundamental set of solutions to the system $\mathbf{x}' = \mathbf{Ax}$.

18. $\mathbf{A} = \begin{pmatrix} -5 & 1 & -4 & -1 \\ 0 & -3 & 0 & 0 \\ 1 & -1 & 0 & 1 \\ 2 & -1 & 2 & -2 \end{pmatrix}$

19. $\mathbf{A} = \begin{pmatrix} 2 & 2 & 0 & -1 \\ 2 & -1 & 0 & 2 \\ 0 & 0 & 3 & 0 \\ -1 & 2 & 0 & 2 \end{pmatrix}$

20. $\mathbf{A} = \begin{pmatrix} 1 & 8 & 5 & 3 \\ 2 & 16 & 10 & 6 \\ 5 & -14 & -11 & -3 \\ -1 & -8 & -5 & -3 \end{pmatrix}$

21. $\mathbf{A} = \begin{pmatrix} -2 & 2 & 0 & -2 \\ -1 & 3 & -1 & 1 \\ -2 & -2 & -4 & 2 \\ -7 & 1 & -7 & 3 \end{pmatrix}$

22. $\mathbf{A} = \begin{pmatrix} -5 & -2 & -1 & 2 & 3 \\ 0 & -3 & 0 & 0 & 0 \\ 1 & 0 & -1 & 0 & -1 \\ 2 & 1 & 0 & -4 & -2 \\ -3 & -2 & -1 & 2 & 1 \end{pmatrix}$

23. $\mathbf{A} = \begin{pmatrix} 0 & -3 & -2 & 3 & 2 \\ 8 & 6 & 4 & -8 & -16 \\ -8 & -8 & -6 & 8 & 16 \\ 8 & 7 & 4 & -9 & -16 \\ -3 & -5 & -3 & 5 & 7 \end{pmatrix}$

6.4 Nondefective Matrices with Complex Eigenvalues

In this section we again consider a system of n linear homogeneous equations with constant coefficients,

$$\mathbf{x}' = \mathbf{Ax}, \tag{1}$$

where the coefficient matrix $\mathbf{A}$ is real-valued, nondefective, and has one or more complex eigenvalues. Since $\mathbf{A}$ is real, the coefficients of the characteristic equation

$$p(\lambda) = |\mathbf{A} - \lambda \mathbf{I}_n| = 0 \tag{2}$$

are real. Consequently, complex eigenvalues must occur in conjugate pairs. If $\lambda = \mu + i\nu$ and $(\lambda, \mathbf{v})$ is an eigenpair of $\mathbf{A}$, then so is $(\bar{\lambda}, \bar{\mathbf{v}})$, where $\bar{\lambda}$ and $\bar{\mathbf{v}}$ are the complex conjugates of λ and $\mathbf{v}$, respectively. It follows that the corresponding complex conjugates

$$\mathbf{u}(t) = e^{\lambda t}\mathbf{v}, \qquad \bar{\mathbf{u}}(t) = e^{\bar{\lambda}t}\bar{\mathbf{v}} \tag{3}$$

are solutions of Eq. (1). By an argument identical to that in Section 3.4, we obtain real-valued solutions of Eq. (1) by taking the real and imaginary parts of $\mathbf{u}(t)$ or $\bar{\mathbf{u}}(t)$. If we let $\mathbf{v} = \mathbf{a} + i\mathbf{b}$, where $\mathbf{a}$ and $\mathbf{b}$ are real constant $n \times 1$ vectors, then the vectors

$$\mathbf{x}_1(t) = \text{Re } \mathbf{u}(t) = e^{\mu t}(\mathbf{a} \cos \nu t - \mathbf{b} \sin \nu t)$$

$$\mathbf{x}_2(t) = \text{Im } \mathbf{u}(t) = e^{\mu t}(\mathbf{a} \sin \nu t + \mathbf{b} \cos \nu t) \tag{4}$$

are real-valued solutions of Eq. (1). It is possible to show that $\mathbf{x}_1$ and $\mathbf{x}_2$ are linearly independent solutions (see Problem 11).

If all of the eigenvectors of $\mathbf{A}$, real and complex, are linearly independent, then a fundamental set of real solutions of Eq. (1) is formed with solutions of the form (4) associated with complex eigenvalues and solutions of the form $e^{\lambda_j t}\mathbf{v}_j$ associated with real eigenvalues. For example, suppose that $\lambda_1 = \mu + iv$, $\lambda_2 = \mu - iv$, and that $\lambda_3, \ldots, \lambda_n$ are all real and distinct. Let the corresponding eigenvectors be $\mathbf{v}_1 = \mathbf{a} + i\mathbf{b}$, $\mathbf{v}_2 = \mathbf{a} - i\mathbf{b}$, $\mathbf{v}_3, \ldots, \mathbf{v}_n$. Then the general solution of Eq. (1) is

$$\mathbf{x} = c_1\mathbf{x}_1(t) + c_2\mathbf{x}_2(t) + c_3 e^{\lambda_3 t}\mathbf{v}_3 + \cdots + c_n e^{\lambda_n t}\mathbf{v}_n,$$

where $\mathbf{x}_1(t)$ and $\mathbf{x}_2(t)$ are given by Eqs. (4). We emphasize that this analysis applies only if the coefficient matrix $\mathbf{A}$ in Eq. (1) is real, for it is only then that complex eigenvalues and eigenvectors occur in conjugate pairs.

EXAMPLE 1

Find a fundamental set of real-valued solutions of the system

$$\mathbf{x}' = \begin{pmatrix} -4 & 5 & -3 \\ -17/3 & 4/3 & 7/3 \\ 23/3 & -25/3 & -4/3 \end{pmatrix} \mathbf{x} \tag{5}$$

and describe the solution trajectories.

To find a fundamental set of solutions, we assume that $\mathbf{x} = e^{\lambda t}\mathbf{v}$ and obtain the set of linear algebraic equations

$$\begin{pmatrix} -4 - \lambda & 5 & -3 \\ -17/3 & 4/3 - \lambda & 7/3 \\ 23/3 & -25/3 & -4/3 - \lambda \end{pmatrix} \mathbf{v} = \mathbf{0}. \tag{6}$$

The characteristic equation of the matrix of coefficients in Eq. (5) is

$$\begin{vmatrix} -4 - \lambda & 5 & -3 \\ -17/3 & 4/3 - \lambda & 7/3 \\ 23/3 & -25/3 & -4/3 - \lambda \end{vmatrix} = -(\lambda^3 + 4\lambda^2 + 69\lambda + 130)$$

$$= -(\lambda + 2)(\lambda^2 + 2\lambda + 65)$$

$$= -(\lambda + 2)(\lambda + 1 - 8i)(\lambda + 1 + 8i)$$

and therefore the eigenvalues are $\lambda_1 = -2$, $\lambda_2 = -1 + 8i$, and $\lambda_3 = -1 - 8i$. Substituting $\lambda = -2$ into Eq. (6) and using elementary row operations yield the reduced system

$$\begin{pmatrix} 1 & 0 & -1 \\ 0 & 1 & -1 \\ 0 & 0 & 0 \end{pmatrix} \mathbf{v} = \mathbf{0}.$$

The general solution of this system is $\mathbf{v} = a_1(1, 1, 1)^T$ and taking $a_1 = 1$ gives the eigenvector associated with $\lambda_1 = -2$,

$$\mathbf{v}_1 = \begin{pmatrix} 1 \\ 1 \\ 1 \end{pmatrix}.$$

Substituting $\lambda = -1 + 8i$ into Eq. (6) gives the system

$$
\begin{pmatrix}
-3 - 8i & 5 & -3 \\
-17/3 & 7/3 - 8i & 7/3 \\
23/3 & -25/3 & 1/3 - 8i
\end{pmatrix}
\begin{pmatrix}
v_1 \\
v_2 \\
v_3
\end{pmatrix}
=
\begin{pmatrix}
0 \\
0 \\
0
\end{pmatrix}.
$$

Using complex arithmetic and elementary row operations, we reduce this system to

$$
\begin{pmatrix}
1 & 0 & (1-i)/2 \\
0 & 1 & (1+i)/2 \\
0 & 0 & 0
\end{pmatrix}
\begin{pmatrix}
v_1 \\
v_2 \\
v_3
\end{pmatrix}
=
\begin{pmatrix}
0 \\
0 \\
0
\end{pmatrix}.
$$

The general solution is represented by

$$
\mathbf{v} = a_1
\begin{pmatrix}
1/2 \\
i/2 \\
-(1+i)/2
\end{pmatrix},
$$

where a_1 is an arbitrary constant. Taking $a_1 = 2$ gives the eigenvector belonging to λ_2,

$$
\mathbf{v}_2 =
\begin{pmatrix}
1 \\
i \\
-1 - i
\end{pmatrix}
= \mathbf{a} + i\mathbf{b},
$$

where

$$
\mathbf{a} =
\begin{pmatrix}
1 \\
0 \\
-1
\end{pmatrix}
\qquad \text{and} \qquad
\mathbf{b} =
\begin{pmatrix}
0 \\
1 \\
-1
\end{pmatrix}.
\tag{7}
$$

The eigenvector belonging to $\lambda_3 = -1 - 8i = \bar{\lambda}_2$ is the complex conjugate of $\mathbf{v}_2$,

$$
\mathbf{v}_3 = \mathbf{a} - \mathbf{b}i =
\begin{pmatrix}
1 \\
-i \\
-1 + i
\end{pmatrix}.
$$

Hence a fundamental set of solutions of the system (5) contains

$$
\mathbf{x}_1(t) = e^{-2t}
\begin{pmatrix}
1 \\
1 \\
1
\end{pmatrix},
\qquad
\mathbf{u}(t) = e^{(-1+8i)t}
\begin{pmatrix}
1 \\
i \\
-1 - i
\end{pmatrix},
$$

and

$$
\bar{\mathbf{u}}(t) = e^{(-1-8i)t}
\begin{pmatrix}
1 \\
-i \\
-1 + i
\end{pmatrix}.
$$

To obtain a fundamental set of real-valued solutions, we must find the real and imaginary parts of either $\mathbf{u}$ or $\bar{\mathbf{u}}$. A direct calculation using complex arithmetic gives

$$\mathbf{u}(t) = e^{-t}(\cos 8t + i \sin 8t) \begin{pmatrix} 1 \\ i \\ -1 - i \end{pmatrix}$$

$$= e^{-t} \begin{pmatrix} \cos 8t \\ -\sin 8t \\ -\cos 8t + \sin 8t \end{pmatrix} + i e^{-t} \begin{pmatrix} \sin 8t \\ \cos 8t \\ -\sin 8t - \cos 8t \end{pmatrix}.$$

Hence

$$\mathbf{x}_2(t) = e^{-t} \begin{pmatrix} \cos 8t \\ -\sin 8t \\ -\cos 8t + \sin 8t \end{pmatrix} \quad \text{and} \quad \mathbf{x}_3(t) = e^{-t} \begin{pmatrix} \sin 8t \\ \cos 8t \\ -\sin 8t - \cos 8t \end{pmatrix}$$

are real-valued solutions of Eq. (5). To verify that $\mathbf{x}_1(t)$, $\mathbf{x}_2(t)$, and $\mathbf{x}_3(t)$ are linearly independent, we compute their Wronskian at $t = 0$,

$$W[\mathbf{x_1}, \mathbf{x}_2, \mathbf{x}_3](0) = \begin{vmatrix} 1 & 1 & 0 \\ 1 & 0 & 1 \\ 1 & -1 & -1 \end{vmatrix} = 3.$$

Since $W[\mathbf{x}_1, \mathbf{x}_2, \mathbf{x}_3](0) \neq 0$, it follows that $\mathbf{x}_1(t)$, $\mathbf{x}_2(t)$, and $\mathbf{x}_3(t)$ form a fundamental set of (real-valued) solutions of the system (5). Therefore the general solution of Eq. (5) is

$$\mathbf{x} = c_1 e^{-2t} \begin{pmatrix} 1 \\ 1 \\ 1 \end{pmatrix} + c_2 e^{-t} \begin{pmatrix} \cos 8t \\ -\sin 8t \\ -\cos 8t + \sin 8t \end{pmatrix} + c_3 e^{-t} \begin{pmatrix} \sin 8t \\ \cos 8t \\ -\sin 8t - \cos 8t \end{pmatrix} \tag{8}$$

or, using $\mathbf{a}$ and $\mathbf{b}$ defined in Eqs. (7),

$$\mathbf{x} = c_1 e^{-2t} \mathbf{v}_1 + c_2 e^{-t}(\mathbf{a} \cos 8t - \mathbf{b} \sin 8t) + c_3 e^{-t}(\mathbf{a} \sin 8t + \mathbf{b} \cos 8t). \tag{9}$$

Plots of the components of $\mathbf{x}$ corresponding to $c_1 = 10$, $c_2 = 30$, and $c_3 = 20$ are shown in Figure 6.4.1.

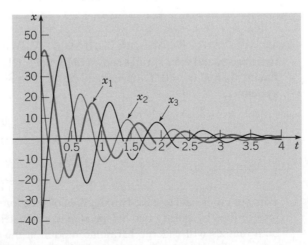

FIGURE 6.4.1   Plots of the components of the solution vector (8) using $c_1 = 10$, $c_2 = 30$, and $c_3 = 20$.

Geometric understanding of the solution (8) or (9) is facilitated by rewriting Eq. (9) in the form

$$\mathbf{x} = c_1 e^{-2t}\mathbf{v}_1 + e^{-t}(c_2 \cos 8t + c_3 \sin 8t)\mathbf{a} + e^{-t}(c_3 \cos 8t - c_2 \sin 8t)\mathbf{b}. \qquad (10)$$

Since the real parts of all eigenvalues are negative and appear in the exponential (decay) factors multiplying each term in Eqs. (8) through (10), all solutions approach $\mathbf{0}$ as $t \to \infty$. If $c_1 \neq 0$, since $\lambda_1 = -2$ is less than $\operatorname{Re}\lambda_2 = \operatorname{Re}\lambda_3 = -1$, solutions decay toward $\mathbf{0}$ in the direction parallel to $\mathbf{v}_1$ at a faster rate than their simultaneous spiral toward $\mathbf{0}$ in the plane S determined by $\mathbf{a}$ and $\mathbf{b}$,

$$S = \{\mathbf{v}: \mathbf{v} = d_1\mathbf{a} + d_2\mathbf{b}, \ -\infty < d_1, d_2 < \infty\}.$$

Solution trajectories decay toward the plane S parallel to the direction $\mathbf{v}_1$ while simultaneously spiraling toward $\mathbf{0}$ in the plane S. A typical solution trajectory for this case is shown in Figure 6.4.2. If $c_1 = 0$, solutions start in S and spiral toward $\mathbf{0}$ as $t \to \infty$, remaining in S all the while.

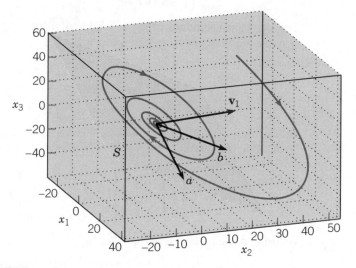

FIGURE 6.4.2 A typical solution trajectory for Eq. (5) that is not in S at time $t = 0$ spirals toward S as $t \to \infty$. The lengths of $\mathbf{v}_1$, $\mathbf{a}$, and $\mathbf{b}$ have been scaled to enhance visualization.

▶ Natural Frequencies and Principal Modes of Vibration. Consider again the system of two masses and three springs shown in Figure 6.1.1. If we assume that there are no external forces, then $F_1(t) = 0$, $F_2(t) = 0$ and Eq. (21) in Section 6.1 reduces to the homogeneous system

$$\mathbf{x}' = \begin{pmatrix} 0 & 0 & 1 & 0 \\ 0 & 0 & 0 & 1 \\ -(k_1+k_2)/m_1 & k_2/m_1 & 0 & 0 \\ k_2/m_2 & -(k_2+k_3)/m_2 & 0 & 0 \end{pmatrix}\mathbf{x}. \qquad (11)$$

This system is said to have **two degrees of freedom** since the configuration of the system is described by exactly two independent coordinates, x_1 and x_2. Under certain conditions, both masses will undergo harmonic motion at the same frequency. In such a case, both masses will attain their maximum displacements at the same times even if they do not both move in the same direction. When this kind of motion occurs, the frequency is called a **natural frequency** of the system, and the motion is called a **principal mode of vibration**. In general, the number of natural frequencies and principal modes possessed by a vibrating system of masses is equal to the number of degrees of freedom of the system. Thus, for the two degrees of freedom system under discussion, there will be two natural frequencies and two principal modes of vibration. The principal mode of vibration corresponding to the lowest natural frequency is referred to as the **first mode**. The principal mode of vibration corresponding to the next higher frequency is called the **second mode** and so on if there are more than two degrees of freedom. The following example illustrates the important relationship between eigenvalues and eigenvectors and natural frequencies and principal modes of vibration.

EXAMPLE 2

Suppose that $m_1 = 2$, $m_2 = \frac{9}{4}$, $k_1 = 1$, $k_2 = 3$, $k_3 = \frac{15}{4}$ in Eq. (11),

$$\mathbf{x}' = \begin{pmatrix} 0 & 0 & 1 & 0 \\ 0 & 0 & 0 & 1 \\ -2 & \frac{3}{2} & 0 & 0 \\ \frac{4}{3} & -3 & 0 & 0 \end{pmatrix} \mathbf{x} = \mathbf{Ax}. \tag{12}$$

Find and describe the principal modes of vibration and the associated natural frequencies for the spring-mass system described by Eq. (12).

Keep in mind that x_1 and x_2 are the positions of m_1 and m_2, relative to their equilibrium positions, and that x_3 and x_4 are their corresponding velocities. Employing the eigenvalue method of solution, we assume, as usual, that $\mathbf{x} = e^{\lambda t}\mathbf{v}$, where $(\lambda, \mathbf{v})$ must be an eigenpair of the matrix $\mathbf{A}$. It is possible, though a bit tedious, to find the eigenvalues and eigenvectors of $\mathbf{A}$ by hand, but it is easy with appropriate computer software. The characteristic polynomial of $\mathbf{A}$ is

$$\lambda^4 + 5\lambda^2 + 4 = (\lambda^2 + 1)(\lambda^2 + 4),$$

so the eigenvalues are $\lambda_1 = i$, $\lambda_2 = -i$, $\lambda_3 = 2i$, and $\lambda_4 = -2i$. The corresponding eigenvectors are

$$\mathbf{v}_1 = \begin{pmatrix} 3 \\ 2 \\ 3i \\ 2i \end{pmatrix}, \qquad \mathbf{v}_2 = \begin{pmatrix} 3 \\ 2 \\ -3i \\ -2i \end{pmatrix}, \qquad \mathbf{v}_3 = \begin{pmatrix} 3 \\ -4 \\ 6i \\ -8i \end{pmatrix}, \qquad \mathbf{v}_4 = \begin{pmatrix} 3 \\ -4 \\ -6i \\ 8i \end{pmatrix}.$$

The complex-valued solutions $e^{it}\mathbf{v}_1$ and $e^{-it}\mathbf{v}_2$ are complex conjugates, so two real-valued solutions can be found by finding the real and imaginary parts of either of them. For instance, we have

$$e^{it}\mathbf{v}_1 = (\cos t + i\sin t)\begin{pmatrix} 3 \\ 2 \\ 3i \\ 2i \end{pmatrix} = \begin{pmatrix} 3\cos t \\ 2\cos t \\ -3\sin t \\ -2\sin t \end{pmatrix} + i\begin{pmatrix} 3\sin t \\ 2\sin t \\ 3\cos t \\ 2\cos t \end{pmatrix} = \mathbf{x}_1(t) + i\mathbf{x}_2(t).$$

In a similar way, we obtain

$$e^{2it}\mathbf{v}_3 = (\cos 2t + i\sin 2t)\begin{pmatrix} 3 \\ -4 \\ 6i \\ -8i \end{pmatrix} = \begin{pmatrix} 3\cos 2t \\ -4\cos 2t \\ -6\sin 2t \\ 8\sin 2t \end{pmatrix} + i\begin{pmatrix} 3\sin 2t \\ -4\sin 2t \\ 6\cos 2t \\ -8\cos 2t \end{pmatrix} = \mathbf{x}_3(t) + i\mathbf{x}_4(t).$$

We leave it to you to verify that $\mathbf{x}_1$, $\mathbf{x}_2$, $\mathbf{x}_3$, and $\mathbf{x}_4$ are linearly independent and therefore form a fundamental set of solutions. Thus the general solution of Eq. (12) is

$$\mathbf{x} = c_1\begin{pmatrix} 3\cos t \\ 2\cos t \\ -3\sin t \\ -2\sin t \end{pmatrix} + c_2\begin{pmatrix} 3\sin t \\ 2\sin t \\ 3\cos t \\ 2\cos t \end{pmatrix} + c_3\begin{pmatrix} 3\cos 2t \\ -4\cos 2t \\ -6\sin 2t \\ 8\sin 2t \end{pmatrix} + c_4\begin{pmatrix} 3\sin 2t \\ -4\sin 2t \\ 6\cos 2t \\ -8\cos 2t \end{pmatrix}, \tag{13}$$

where c_1, c_2, c_3, and c_4 are arbitrary constants.

The state space for this system is four-dimensional, and each solution, obtained by a particular set of values for $c_1, \ldots, c_4$ in Eq. (13), corresponds to a trajectory in this space. Since each solution, given by Eq. (13), is periodic with period 2π, each trajectory is a closed curve. No matter where the trajectory starts in $\mathbf{R}^4$ at $t = 0$, it returns to that point at $t = 2\pi$, $t = 4\pi$, and so forth, repeatedly traversing the same curve in each time interval of length 2π. Thus, even though we cannot graph solutions in $\mathbf{R}^4$, we infer from Eq. (13) that the trajectories are closed curves in $\mathbf{R}^4$.

Note that the only two vibration frequencies present in Eq. (13) are $\omega_1 = \text{Im}\,\lambda_1 = 1$ and $\omega_2 = \text{Im}\,\lambda_3 = 2$ radians per second. The corresponding periods are 2π and π. In accordance with the defining property of the first mode of vibration given above, both masses will vibrate at the lowest frequency, $\omega_1 = 1$, if we choose $c_3 = c_4 = 0$ in Eq. (13). The first pure mode of vibration occurs for any c_1 and c_2, not both zero, whenever $x_1(0) = 3c_1$, $x_2(0) = 2c_1$, $x_3(0) = 3c_2$, and $x_4(0) = 2c_2$. The displacements of m_1 and m_2 are then given by

$$x_1(t) = 3(c_1\cos t + c_2\sin t) = 3A_1\cos(t - \phi_1)$$

and

$$x_2(t) = 2(c_1\cos t + c_2\sin t) = 2A_1\cos(t - \phi_1),$$

where A_1 and ϕ_1 are determined by the relationships $A_1 = \sqrt{c_1^2 + c_2^2}$, $A_1\cos\phi_1 = c_1$, and $A_1\sin\phi_1 = c_2$. Thus $x_2 = \frac{2}{3}x_1$, that is, the displacement of the right-hand mass is always in the same direction as the displacement of the left-hand mass, but only $\frac{2}{3}$ as much. In particular, the amplitude of the motion of m_2 is $\frac{2}{3}$ that of the amplitude of motion of m_1. Thus the motions of both masses are exactly in phase with one another, albeit with different amplitudes. Component plots of the displacements of m_1 and m_2 using $c_1 = 1$, $c_2 = 0$, $c_3 = 0$, and $c_4 = 0$ are shown in Figure 6.4.3. Time histories of the positions of the masses are depicted by phantom images.

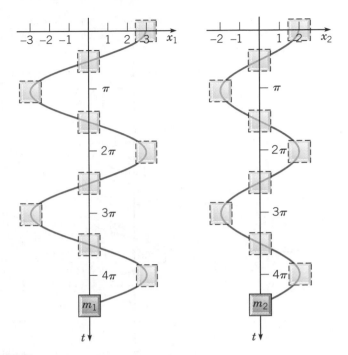

FIGURE 6.4.3 The first principal mode of vibration; the motions of m_1 and m_2 are in phase but have different amplitudes.

The second pure mode of vibration, corresponding to the frequency $\omega_2 = 2$, arises by choosing $c_1 = c_2 = 0$ in Eq. (13). This mode can be realized by choosing initial conditions

$$x_1(0) = 3c_3, \quad x_2(0) = -4c_3, \quad x_3(0) = 6c_4, \quad x_4(0) = -8c_4$$

for any c_3 and c_4, not both zero. The displacements of m_1 and m_2 in this case are given by

$$x_1(t) = 3(c_3 \cos 2t + c_4 \sin 2t) = 3A_2 \cos(2t - \phi_2)$$

and

$$x_2(t) = -4(c_3 \cos 2t + c_4 \sin 2t) = -4A_2 \cos(2t - \phi_2),$$

where $A_2 = \sqrt{c_3^2 + c_4^2}$, $A_2 \cos \phi_2 = c_3$, and $A_2 \sin \phi_2 = c_4$. Consequently, $x_2(t) = -\frac{4}{3}x_1$, that is, the displacement of m_2 is always in a direction opposite to the direction of displacement of m_1 and greater by a factor of $\frac{4}{3}$. The motions of the masses are 180 degrees out of phase with each other and have different amplitudes. Component plots of the displacements of m_1 and m_2 using $c_1 = 0$, $c_2 = 0$, $c_3 = 1$, and $c_4 = 0$ are shown in Figure 6.4.4.

Each of the pure modes of vibration is realized by one or the other special sets of initial conditions discussed above. Equation (13) shows that the motion of m_1 and m_2 for any other choice of initial conditions will be a superposition of both principal modes of vibration. For example, if initially m_1 is at equilibrium and m_2 is released with zero velocity from a position two units to the right of its equilibrium position so that the initial condition is given by $\mathbf{x}(0) = (0, 2, 0, 0)^T$, then $c_1 = \frac{1}{3}$, $c_2 = 0$, $c_3 = -\frac{1}{3}$ and $c_4 = 0$. Thus both modes of vibration are required to describe the motion. Component plots of the displacements of m_1 and m_2 for this case are shown in Figure 6.4.5. Note that the period of the vibrations is 2π.

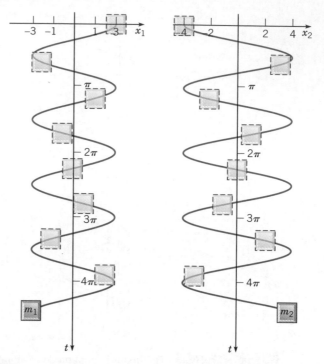

FIGURE 6.4.4 The second principal mode of vibration; the motions of m_1 and m_2 are 180 degrees out of phase and have different amplitudes.

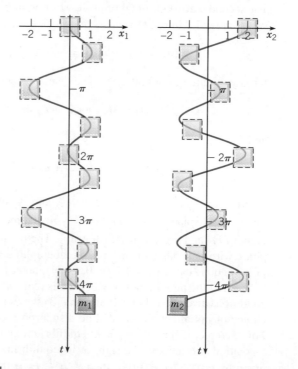

FIGURE 6.4.5 Component plots of the solution of Eq. (12) corresponding to the initial condition $\mathbf{x}(0) = (0, 2, 0, 0)^T$.

PROBLEMS

In each of Problems 1 through 8, express the general solution of the given system of equations in terms of real-valued functions:

1. $x_1' = -2x_1 + 2x_2 + x_3$
$x_2' = -2x_1 + 2x_2 + 2x_3$
$x_3' = 2x_1 - 3x_2 - 3x_3$

2. $x_1' = 2x_1 - 4x_2 - x_3$
$x_2' = x_1 + x_2 + 3x_3$
$x_3' = 3x_1 - 4x_2 - 2x_3$

3. $x_1' = -2x_2 - x_3$
$x_2' = x_1 - x_2 + x_3$
$x_3' = x_1 - 2x_2 - 2x_3$

4. $x_1' = -4x_1 + 2x_2 - x_3$
$x_2' = -6x_1 - 3x_3$
$x_3' = \frac{8}{3}x_2 - 2x_3$

5. $\mathbf{x}' = \begin{pmatrix} -7 & 6 & -6 \\ -9 & 5 & -9 \\ 0 & -1 & -1 \end{pmatrix} \mathbf{x}$

6. $\mathbf{x}' = \begin{pmatrix} 4/3 & 4/3 & -11/3 \\ -16/3 & -1/3 & 14/3 \\ 3 & -2 & -2 \end{pmatrix} \mathbf{x}$

7. $\mathbf{x}' = \begin{pmatrix} 1 & 1 & 1 \\ 2 & 1 & -1 \\ -8 & -5 & -3 \end{pmatrix} \mathbf{x}$

8. $\mathbf{x}' = \begin{pmatrix} 1 & -1 & 4 \\ 3 & 2 & -1 \\ 2 & 1 & -1 \end{pmatrix} \mathbf{x}$

9. (a) Find constant 3×1 vectors $\mathbf{a}$ and $\mathbf{b}$ such that the solution of the initial value problem

$$\mathbf{x}' = \begin{pmatrix} 3/4 & 29/4 & -11/2 \\ -3/4 & 3/4 & -5/2 \\ 5/4 & 11/4 & -5/2 \end{pmatrix} \mathbf{x}, \quad \mathbf{x}(0) = \mathbf{x}_0 \in S$$

is a closed curve lying entirely in S, where

$$S = \{\mathbf{u} : \mathbf{u} = a_1 \mathbf{a} + a_2 \mathbf{b}, \ -\infty < a_1, a_2 < \infty\}.$$

(b) In $\mathbf{R}^3$, plot solutions to the initial value problem for several different choices of $\mathbf{x}_0 \in S$ overlaying the trajectories on a graph of the plane determined by $\mathbf{a}$ and $\mathbf{b}$.

(c) In $\mathbf{R}^3$, plot the solution to the initial value problem for a choice of $\mathbf{x}_0 \notin S$. Describe the long-time behavior of the trajectory.

10. (a) Find constant 4×1 vectors $\mathbf{a}$ and $\mathbf{b}$ such that the solution of the initial value problem

$$\mathbf{x}' = \begin{pmatrix} -2 & -1 & 4 & 2 \\ -19 & -6 & 6 & 16 \\ -9 & -1 & 1 & 6 \\ -5 & -3 & 6 & 5 \end{pmatrix} \mathbf{x}, \quad \mathbf{x}(0) = \mathbf{x}_0 \in S$$

is a closed curve lying entirely in S, where

$$S = \{\mathbf{u} : \mathbf{u} = a_1 \mathbf{a} + a_2 \mathbf{b}, \ -\infty < a_1, a_2 < \infty\}.$$

(b) Describe the long-time behavior of any solutions for which $\mathbf{x}_0 \notin S$.

11. In this problem, we indicate how to show that $\mathbf{x}_1(t)$ and $\mathbf{x}_2(t)$, as given by Eqs. (4), are linearly independent. Let $\lambda_1 = \mu + i\nu$ and $\bar{\lambda}_1 = \mu - i\nu$ be a pair of conjugate eigenvalues of the coefficient matrix $\mathbf{A}$ of Eq. (1); let $\mathbf{v}_1 = \mathbf{a} + i\mathbf{b}$ and $\bar{\mathbf{v}}_1 = \mathbf{a} - i\mathbf{b}$ be the corresponding eigenvectors. Recall from Corollary 6.3.2 that if the eigenvalues of a real matrix are real and distinct, then the eigenvectors are linearly independent. The proof of Corollary 6.3.2 holds equally well if some of the eigenvalues are complex. Thus if $\nu \neq 0$ so that $\lambda_1 \neq \bar{\lambda}_1$, then $\mathbf{v}_1$ and $\bar{\mathbf{v}}_1$ are linearly independent.

(a) First we show that $\mathbf{a}$ and $\mathbf{b}$ are linearly independent. Consider the equation $c_1 \mathbf{a} + c_2 \mathbf{b} = \mathbf{0}$. Express $\mathbf{a}$ and $\mathbf{b}$ in terms of $\mathbf{v}_1$ and $\bar{\mathbf{v}}_1$, and then show that $(c_1 - ic_2)\mathbf{v}_1 + (c_1 + ic_2)\bar{\mathbf{v}}_1 = \mathbf{0}$.

(b) Show that $c_1 - ic_2 = 0$ and $c_1 + ic_2 = 0$ and then that $c_1 = 0$ and $c_2 = 0$. Consequently, $\mathbf{a}$ and $\mathbf{b}$ are linearly independent.

(c) To show that $\mathbf{x}_1(t)$ and $\mathbf{x}_2(t)$ are linearly independent, consider the equation $c_1 \mathbf{x}_1(t_0) + c_2 \mathbf{x}_2(t_0) = \mathbf{0}$, where t_0 is an arbitrary point. Rewrite this equation in terms of $\mathbf{a}$ and $\mathbf{b}$, and then proceed, as in part (b), to show that $c_1 = 0$ and $c_2 = 0$. Hence $\mathbf{x}_1(t)$ and $\mathbf{x}_2(t)$ are linearly independent at the point t_0. Therefore they are linearly independent at every point and on every interval.

12. Consider the two-mass, three-spring system of Example 2. Instead of solving the system of four first order equations, we indicate here how to proceed directly from the system of two second order equations given in Eq. (18) of Section 6.1.

(a) Show that using the parameter values of Example 2, $m_1 = 2$, $m_2 = \frac{9}{4}$, $k_1 = 1$, $k_2 = 3$, $k_3 = \frac{15}{4}$ and assuming that $F_1(t) = 0$ and $F_2(t) = 0$, Eqs. (18) in Section 6.1 can be written in the form

$$\mathbf{y}'' = \begin{pmatrix} -2 & 3/2 \\ 4/3 & -3 \end{pmatrix} \mathbf{y} = \mathbf{B}\mathbf{y}.$$

(b) Assume that $\mathbf{y} = e^{\omega t}\mathbf{u}$ and show that

$$(\mathbf{B} - \omega^2 \mathbf{I})\mathbf{u} = \mathbf{0}.$$

Note that ω^2 is an eigenvalue of **B** corresponding to the eigenvector **u**.

(c) Find the eigenvalues and eigenvectors of **B**.

(d) Write down the expressions for y_1 and y_2. There should be four arbitrary constants in these expressions.

(e) By differentiating the results from part (d), write down expressions for y_1' and y_2'. Your results from parts (d) and (e) should agree with Eq. (13) in the text.

13. Consider the two-mass, three-spring system whose equations of motion are Eqs. (11) in the text. Let $m_1 = m_2 = 1$ and $k_1 = k_2 = k_3 = 1$.

(a) Using the values of the parameters given in the previous statement, find the eigenvalues and eigenvectors of the matrix **A**.

(b) Write down the general solution of the system.

(c) For each fundamental mode, draw graphs of x_1 versus t and x_2 versus t and describe the fundamental modes of vibration.

(d) Consider the initial conditions $\mathbf{x}(0) = (-1, 3, 0, 0)^T$. Evaluate the arbitrary constants in the general solution in part (b) and plot graphs of x_1 versus t and x_2 versus t.

(e) Consider other initial conditions of your own choice, and plot graphs of x_1 versus t and x_2 versus t in each case.

14. Consider the two-mass, three-spring system whose equations of motion are Eqs. (11) in the text. Let $m_1 = m_2 = m$ and $k_1 = k_2 = k_3 = k$. Find the general solution in terms of the parameter $\omega = \sqrt{k/m}$. In terms of ω, what are the prin-

cipal modes of vibration and the corresponding natural frequencies?

For each of the matrices in Problems 15 through 18, use a computer to assist in finding a fundamental set of real solutions to the system $\mathbf{x}' = \mathbf{A}\mathbf{x}$.

15. $\mathbf{A} = \begin{pmatrix} -3 & 6 & 2 & -2 \\ 2 & -3 & -6 & 2 \\ -4 & 8 & 3 & -4 \\ 2 & -2 & -6 & 1 \end{pmatrix}$

16. $\mathbf{A} = \begin{pmatrix} -3 & -4 & 5 & 9 \\ -2 & -5 & 4 & 12 \\ -2 & 0 & -1 & 2 \\ 0 & -2 & 2 & 3 \end{pmatrix}$

17. $\mathbf{A} = \begin{pmatrix} -3 & -5 & 8 & 14 \\ -6 & -8 & 11 & 27 \\ -6 & -4 & 7 & 17 \\ 0 & -2 & 2 & 4 \end{pmatrix}$

18. $\mathbf{A} = \begin{pmatrix} 0 & 3 & 0 & -2 \\ -1/2 & 1 & -3 & -5/2 \\ 0 & 3 & -5 & -3 \\ 1 & 3 & 0 & -3 \end{pmatrix}$

6.5 Fundamental Matrices and the Exponential of a Matrix

Fundamental Matrices

The structure of the solutions of linear differential equations can be further illuminated by introducing the idea of a fundamental matrix. Suppose that $\mathbf{x}_1(t), \ldots, \mathbf{x}_n(t)$ form a fundamental set of solutions for the equation

$$\mathbf{x}' = \mathbf{P}(t)\mathbf{x} \tag{1}$$

on some interval $\alpha < t < \beta$. Then the matrix

$$\mathbf{X}(t) = [\mathbf{x}_1(t), \mathbf{x}_2(t), \ldots, \mathbf{x}_n(t)] = \begin{pmatrix} x_{11}(t) & \cdots & x_{1n}(t) \\ \vdots & & \vdots \\ x_{n1}(t) & \cdots & x_{nn}(t) \end{pmatrix}, \tag{2}$$

whose columns are the vectors $\mathbf{x}_1(t), \ldots, \mathbf{x}_n(t)$, is said to be a **fundamental matrix** for the system (1). Note that a fundamental matrix is nonsingular since its columns are linearly independent vectors (see Theorems 6.2.5, 6.2.6 and the remark following Theorem 6.2.6).

EXAMPLE 1

Find a fundamental matrix for the system

$$\mathbf{x}' = \begin{pmatrix} 1 & 1 \\ 4 & 1 \end{pmatrix} \mathbf{x}. \tag{3}$$

Note that Eq. (3) is the same system of differential equations considered in Example 4 of Section 3.3.

Using the eigenvalue method, we find the pair of linearly independent solutions

$$\mathbf{x}_1(t) = \begin{pmatrix} e^{3t} \\ 2e^{3t} \end{pmatrix}, \qquad \mathbf{x}_2(t) = \begin{pmatrix} e^{-t} \\ -2e^{-t} \end{pmatrix},$$

of Eq. (3). Thus a fundamental matrix for the system (3) is

$$\mathbf{X}(t) = \begin{pmatrix} e^{3t} & e^{-t} \\ 2e^{3t} & -2e^{-t} \end{pmatrix}. \tag{4}$$

The solutions obtained from this fundamental matrix are consistent with the general solution obtained in Example 4 of Section 3.3.

The solution of an initial value problem can be written very compactly in terms of a fundamental matrix. The general solution of Eq. (1) is

$$\mathbf{x} = c_1 \mathbf{x}_1(t) + \cdots + c_n \mathbf{x}_n(t) \tag{5}$$

or, in terms of $\mathbf{X}(t)$,

$$\mathbf{x} = \mathbf{X}(t)\mathbf{c}, \tag{6}$$

where $\mathbf{c}$ is a constant vector with arbitrary components $c_1, \ldots, c_n$. For an initial value problem consisting of the differential equation (1) and the initial condition

$$\mathbf{x}(t_0) = \mathbf{x}_0, \tag{7}$$

where t_0 is a given point in $\alpha < t < \beta$ and $\mathbf{x}_0$ is a given initial vector, it is only necessary to choose the vector $\mathbf{c}$ in Eq. (6) so as to satisfy the initial condition (7). Hence $\mathbf{c}$ must satisfy

$$\mathbf{X}(t_0)\mathbf{c} = \mathbf{x}_0. \tag{8}$$

Therefore, since $\mathbf{X}(t_0)$ is nonsingular,

$$\mathbf{c} = \mathbf{X}^{-1}(t_0)\mathbf{x}_0$$

and

$$\mathbf{x} = \mathbf{X}(t)\mathbf{X}^{-1}(t_0)\mathbf{x}_0 \tag{9}$$

is the solution of the initial value problem (1), (7). We emphasize, however, that to solve a given initial value problem, one would ordinarily solve Eq. (8) by row reduction and then substitute for $\mathbf{c}$ in Eq. (6), rather than compute $\mathbf{X}^{-1}(t_0)$ and use Eq. (9).

Recall that each column of the fundamental matrix $\mathbf{X}$ is a solution of Eq. (1). It follows that $\mathbf{X}$ satisfies the matrix differential equation

$$\mathbf{X}' = \mathbf{P}(t)\mathbf{X}. \tag{10}$$

This relation is readily confirmed by comparing the two sides of Eq. (10) column by column,

$$\underbrace{[\mathbf{x}_1', \dots, \mathbf{x}_n']}_{\mathbf{X}'} = \underbrace{[\mathbf{P}(t)\mathbf{x}_1, \dots, \mathbf{P}(t)\mathbf{x}_n]}_{\mathbf{P}(t)\mathbf{X}}.$$

Sometimes it is convenient to make use of the special fundamental matrix, denoted by $\boldsymbol{\Phi}(t)$, whose columns are the vectors $\mathbf{x}_1(t), \dots, \mathbf{x}_n(t)$ designated in Theorem 6.2.7. Besides the differential equation (1), for each $j = 1, \dots, n$, these vectors satisfy the initial condition

$$\mathbf{x}_j(t_0) = \mathbf{e}_j = \begin{pmatrix} 0 \\ \vdots \\ 0 \\ 1 \\ 0 \\ \vdots \\ 0 \end{pmatrix} \leftarrow j\text{th position.}$$

Thus $\boldsymbol{\Phi}(t)$ has the property that

$$\boldsymbol{\Phi}(t_0) = \begin{pmatrix} 1 & 0 & \cdots & 0 \\ 0 & 1 & \cdots & 0 \\ \vdots & \vdots & \ddots & \vdots \\ 0 & 0 & \cdots & 1 \end{pmatrix} = \mathbf{I}_n. \tag{11}$$

We will always reserve the symbol $\boldsymbol{\Phi}$ to denote the fundamental matrix satisfying the initial condition (11) and use $\mathbf{X}$ when an arbitrary fundamental matrix is intended. In terms of $\boldsymbol{\Phi}(t)$, the solution of the initial value problem (1), (7) is even simpler in appearance; since $\boldsymbol{\Phi}^{-1}(t_0) = \mathbf{I}_n$, it follows from Eq. (9) that

$$\mathbf{x} = \boldsymbol{\Phi}(t)\mathbf{x}_0. \tag{12}$$

Remark. Although the fundamental matrix $\boldsymbol{\Phi}(t)$ is often more complicated than $\mathbf{X}(t)$, it is especially helpful if the same system of differential equations is to be solved repeatedly subject to many different initial conditions. This corresponds to a given physical system that can be started from many different initial states. If the fundamental matrix $\boldsymbol{\Phi}(t)$ has been determined, then the solution for each set of initial conditions can be found simply by matrix multiplication, as indicated by Eq. (12). The matrix $\boldsymbol{\Phi}(t)$ thus represents a transformation of the initial conditions $\mathbf{x}_0$ into the solution $\mathbf{x}(t)$ at an arbitrary time t. Comparing Eqs. (9) and (12) makes it clear that $\boldsymbol{\Phi}(t) = \mathbf{X}(t)\mathbf{X}^{-1}(t_0)$.

EXAMPLE
2

For the system (3) of Example 1, find the fundamental matrix $\boldsymbol{\Phi}$ such that $\boldsymbol{\Phi}(0) = \mathbf{I}_2$.

The columns of $\boldsymbol{\Phi}$ are solutions of Eq. (3) that satisfy the initial conditions

$$\mathbf{x}_1(0) = \begin{pmatrix} 1 \\ 0 \end{pmatrix}, \qquad \mathbf{x}_2(0) = \begin{pmatrix} 0 \\ 1 \end{pmatrix}.$$

Since the general solution of Eq. (3) is

$$\mathbf{x} = c_1 e^{3t} \begin{pmatrix} 1 \\ 2 \end{pmatrix} + c_2 e^{-t} \begin{pmatrix} 1 \\ -2 \end{pmatrix},$$

we can find the solution satisfying the first set of these initial conditions by choosing $c_1 = c_2 = \frac{1}{2}$; similarly, we obtain the solution satisfying the second set of initial conditions by choosing $c_1 = \frac{1}{4}$ and $c_2 = -\frac{1}{4}$. Hence

$$\mathbf{\Phi}(t) = \begin{pmatrix} (e^{3t} + e^{-t})/2 & (e^{3t} - e^{-t})/4 \\ e^{3t} - e^{-t} & (e^{3t} + e^{-t})/2 \end{pmatrix}.$$

Equivalently, $\mathbf{\Phi}(t)$ can be found by computing the matrix product

$$\mathbf{\Phi}(t) = \mathbf{X}(t)\mathbf{X}^{-1}(0) = \begin{pmatrix} e^{3t} & e^{-t} \\ 2e^{3t} & -2e^{-t} \end{pmatrix} \begin{pmatrix} 1/2 & 1/4 \\ 1/2 & -1/4 \end{pmatrix}$$

$$= \begin{pmatrix} (e^{3t} + e^{-t})/2 & (e^{3t} - e^{-t})/4 \\ e^{3t} - e^{-t} & (e^{3t} + e^{-t})/2 \end{pmatrix}.$$

Note that the elements of $\mathbf{\Phi}(t)$ are more complicated than those of the fundamental matrix $\mathbf{X}(t)$ given by Eq. (4); however, it is now easy to determine the solution corresponding to any set of initial conditions.

The Matrix Exponential Function $e^{\mathbf{A}t}$

Recall that the solution of the scalar initial value problem

$$x' = ax, \qquad x(0) = x_0, \tag{13}$$

where a is a constant, is

$$x = x_0 e^{at}.$$

Now consider the corresponding initial value problem for an $n \times n$ system, namely,

$$\mathbf{x}' = \mathbf{A}\mathbf{x}, \qquad \mathbf{x}(0) = \mathbf{x}_0, \tag{14}$$

where $\mathbf{A}$ is a constant matrix. Applying the results of this section to the problem (14), we can write its solution as

$$\mathbf{x} = \mathbf{\Phi}(t)\mathbf{x}_0, \tag{15}$$

where $\mathbf{\Phi}(0) = \mathbf{I}_n$. Comparing the problems (13) and (14), and their solutions, suggests that the matrix $\mathbf{\Phi}(t)$ might have an exponential character. We now explore this possibility.

The scalar exponential function e^{at} can be represented by the power series

$$e^{at} = 1 + at + \frac{1}{2!}a^2 t^2 + \frac{1}{3!}a^3 t^3 + \cdots = \sum_{k=0}^{\infty} \frac{a^k t^k}{k!}, \tag{16}$$

which converges for all t. By analogy with the power series (16), we now define the symbolic expression $e^{\mathbf{A}t}$.

**DEFINITION
6.5.1**

Let $\mathbf{A}$ be an $n \times n$ constant matrix. The **matrix exponential function**, denoted by $e^{\mathbf{A}t}$, is defined to be

$$e^{\mathbf{A}t} = \mathbf{I}_n + \mathbf{A}t + \frac{1}{2!}\mathbf{A}^2 t^2 + \frac{1}{3!}\mathbf{A}^3 t^3 + \cdots = \sum_{k=0}^{\infty} \mathbf{A}^k \frac{t^k}{k!}. \qquad (17)$$

Remark. The powers of $\mathbf{A}$ that appear in Eq. (17) symbolically represent matrix products, with $\mathbf{A}^0 = \mathbf{I}_n$ by convention, $\mathbf{A}^2 = \mathbf{A}\mathbf{A}$, $\mathbf{A}^3 = \mathbf{A}\mathbf{A}\mathbf{A}$, and so forth. Since $\mathbf{A}$ is an $n \times n$ matrix, each term in the series is an $n \times n$ matrix, as is any partial sum $\mathbf{S}_N(t) = \sum_{k=0}^{N} \mathbf{A}^k t^k / k!$. Given any $n \times n$ matrix $\mathbf{A}$, it can be shown that each elemental sequence (i.e., the sequence associated with each entry) in the sequence of $n \times n$ matrices $\mathbf{S}_N(t)$ converges for all t as $N \to \infty$. Thus $e^{\mathbf{A}t}$ is an $n \times n$ matrix equal to the limit of the convergent sequence of partial sums, $\lim_{N \to \infty} \mathbf{S}_N(t)$. Furthermore the convergence is sufficiently rapid that all operations performed on the series (17), such as term-by-term differentiation, are justified.

**EXAMPLE
3**

Use Definition 6.5.1 to find $e^{\mathbf{A}t}$ if

$$\mathbf{A} = \begin{pmatrix} 1 & 1 \\ 0 & 1 \end{pmatrix}.$$

Noting that

$$\mathbf{A}^2 = \begin{pmatrix} 1 & 2 \\ 0 & 1 \end{pmatrix}, \quad \mathbf{A}^3 = \begin{pmatrix} 1 & 3 \\ 0 & 1 \end{pmatrix}, \quad \ldots, \quad \mathbf{A}^k = \begin{pmatrix} 1 & k \\ 0 & 1 \end{pmatrix}, \ldots$$

and using Eq. (17), we find that

$$\begin{aligned}
e^{\mathbf{A}t} &= \begin{pmatrix} 1 & 0 \\ 0 & 1 \end{pmatrix} + \frac{t}{1!}\begin{pmatrix} 1 & 1 \\ 0 & 1 \end{pmatrix} + \frac{t^2}{2!}\begin{pmatrix} 1 & 2 \\ 0 & 1 \end{pmatrix} + \cdots + \frac{t^k}{k!}\begin{pmatrix} 1 & k \\ 0 & 1 \end{pmatrix} + \cdots \\
&= \begin{pmatrix} \sum_{k=0}^{\infty} t^k/k! & \sum_{k=1}^{\infty} kt^k/k! \\ 0 & \sum_{k=0}^{\infty} t^k/k! \end{pmatrix} = \begin{pmatrix} \sum_{k=0}^{\infty} t^k/k! & t\sum_{k=0}^{\infty} t^k/k! \\ 0 & \sum_{k=0}^{\infty} t^k/k! \end{pmatrix} \\
&= \begin{pmatrix} e^t & te^t \\ 0 & e^t \end{pmatrix},
\end{aligned}$$

where we have used the series (16) with $a = 1$.

In general, it is not possible to identify the entries of $e^{\mathbf{A}t}$ in terms of elementary functions from their infinite series representations, even if the infinite series can be found. However one family of matrices for which the entries on the right-hand side of Eq. (17) are identifiable from their infinite series representation is the class of diagonal matrices, say,

$$
\mathbf{D} = \begin{pmatrix} \lambda_1 & 0 & \cdots & 0 \\ 0 & \lambda_2 & \cdots & 0 \\ \vdots & \vdots & \ddots & \vdots \\ 0 & 0 & \cdots & \lambda_n \end{pmatrix}. \tag{18}
$$

In this case, the powers of $\mathbf{D}$ are easy to calculate,

$$
\mathbf{D}^2 = \begin{pmatrix} \lambda_1^2 & 0 & \cdots & 0 \\ 0 & \lambda_2^2 & \cdots & 0 \\ \vdots & \vdots & \ddots & \vdots \\ 0 & 0 & \cdots & \lambda_n^2 \end{pmatrix}, \quad \ldots, \quad \mathbf{D}^k = \begin{pmatrix} \lambda_1^k & 0 & \cdots & 0 \\ 0 & \lambda_2^k & \cdots & 0 \\ \vdots & \vdots & \ddots & \vdots \\ 0 & 0 & \cdots & \lambda_n^k \end{pmatrix}, \quad \ldots.
$$

It follows that

$$
e^{\mathbf{D}t} = \sum_{k=0}^{\infty} \mathbf{D}^k \frac{t^k}{k!} = \begin{pmatrix} \sum_{k=0}^{\infty} \lambda_1^k t^k/k! & 0 & \cdots & 0 \\ 0 & \sum_{k=0}^{\infty} \lambda_2^k t^k/k! & \cdots & 0 \\ \vdots & \vdots & & \vdots \\ 0 & 0 & \cdots & \sum_{k=0}^{\infty} \lambda_n^k t^k/k! \end{pmatrix}
$$

$$
= \begin{pmatrix} e^{\lambda_1 t} & 0 & \cdots & 0 \\ 0 & e^{\lambda_2 t} & \cdots & 0 \\ \vdots & \vdots & & \vdots \\ 0 & 0 & \vdots & e^{\lambda_n t} \end{pmatrix}. \tag{19}
$$

The next theorem shows the equivalency of $e^{\mathbf{A}t}$ and the fundamental matrix $\mathbf{\Phi}(t)$ for the system $\mathbf{x}' = \mathbf{A}\mathbf{x}$.

THEOREM 6.5.2

If $\mathbf{A}$ is an $n \times n$ constant matrix, then

$$
\boxed{e^{\mathbf{A}t} = \mathbf{\Phi}(t).} \tag{20}
$$

Consequently, the solution to the initial value problem $\mathbf{x}' = \mathbf{A}\mathbf{x}$, $\mathbf{x}(0) = \mathbf{x}_0$, is $\mathbf{x} = e^{\mathbf{A}t}\mathbf{x}_0$.

Proof

Differentiating the series (17) term by term, we obtain

$$
\frac{d}{dt} e^{\mathbf{A}t} = \frac{d}{dt} \left(\mathbf{I}_n + \mathbf{A}t + \frac{1}{2!}\mathbf{A}^2 t^2 + \frac{1}{3!}\mathbf{A}^3 t^3 + \cdots \right)
$$

$$
= \left(\mathbf{A} + \frac{1}{1!}\mathbf{A}^2 t + \frac{1}{2!}\mathbf{A}^3 t^2 + \cdots \right)
$$

$$
= \mathbf{A} \left(\mathbf{I}_n + \mathbf{A}t + \frac{1}{2!}\mathbf{A}^2 t^2 + \cdots \right)
$$

$$
= \mathbf{A} e^{\mathbf{A}t}.
$$

Thus $e^{\mathbf{A}t}$ satisfies the differential equation

$$\frac{d}{dt}e^{\mathbf{A}t} = \mathbf{A}e^{\mathbf{A}t}. \tag{21}$$

Further, when $t = 0$, $e^{\mathbf{A}t}$ satisfies the initial condition

$$e^{\mathbf{A}t}\Big|_{t=0} = \mathbf{I}_n. \tag{22}$$

The fundamental matrix $\boldsymbol{\Phi}$ satisfies the same initial value problem as $e^{\mathbf{A}t}$, namely,

$$\boldsymbol{\Phi}' = \mathbf{A}\boldsymbol{\Phi}, \qquad \boldsymbol{\Phi}(0) = \mathbf{I}_n.$$

By the uniqueness part of Theorem 6.2.1 (extended to matrix differential equations), we conclude that $e^{\mathbf{A}t}$ and the fundamental matrix $\boldsymbol{\Phi}(t)$ are identical. Thus we can write the solution of the initial value problem (14) in the form

$$\mathbf{x} = e^{\mathbf{A}t}\mathbf{x}_0.$$

Equations (21) and (22) show that two properties of e^{at}, namely, $d(e^{at})/dt = ae^{at}$ and $e^{a\cdot 0} = 1$, generalize to $e^{\mathbf{A}t}$. The following theorem summarizes some additional properties of e^{at} that are also shared by $e^{\mathbf{A}t}$.

THEOREM 6.5.3

Let $\mathbf{A}$ and $\mathbf{B}$ be $n \times n$ constant matrices and t, τ be real or complex numbers. Then,

(a) $e^{\mathbf{A}(t+\tau)} = e^{\mathbf{A}t}e^{\mathbf{A}\tau}$.

(b) $\mathbf{A}$ commutes with $e^{\mathbf{A}t}$, that is, $\mathbf{A}e^{\mathbf{A}t} = e^{\mathbf{A}t}\mathbf{A}$.

(c) $(e^{\mathbf{A}t})^{-1} = e^{-\mathbf{A}t}$.

(d) $e^{(\mathbf{A}+\mathbf{B})t} = e^{\mathbf{A}t}e^{\mathbf{B}t}$ if $\mathbf{AB} = \mathbf{BA}$, that is, if $\mathbf{A}$ and $\mathbf{B}$ commute.

Proof

(a) From the series (17),

$$
\begin{aligned}
\mathbf{A}e^{\mathbf{A}t} &= \mathbf{A}\left(\mathbf{I}_n + \mathbf{A}t + \frac{1}{2!}\mathbf{A}^2 t^2 + \frac{1}{3!}\mathbf{A}^3 t^3 + \cdots\right) \\
&= \left(\mathbf{A} + \mathbf{A}^2 t + \frac{1}{2!}\mathbf{A}^3 t^2 + \frac{1}{3!}\mathbf{A}^4 t^3 + \cdots\right) \\
&= \left(\mathbf{I}_n + \mathbf{A}t + \frac{1}{2!}\mathbf{A}^2 t^2 + \frac{1}{3!}\mathbf{A}^3 t^3 + \cdots\right)\mathbf{A} = e^{\mathbf{A}t}\mathbf{A}.
\end{aligned}
$$

(b) Let $\boldsymbol{\Phi}(t) = e^{(\mathbf{A}+\mathbf{B})t}$ and $\hat{\boldsymbol{\Phi}}(t) = e^{\mathbf{A}t}e^{\mathbf{B}t}$. Then $\boldsymbol{\Phi}$ is the unique solution to

$$\boldsymbol{\Phi}' = (\mathbf{A} + \mathbf{B})\boldsymbol{\Phi}, \qquad \boldsymbol{\Phi}(0) = \mathbf{I}_n. \tag{23}$$

Since we are assuming that $\mathbf{A}$ and $\mathbf{B}$ commute, it follows that $e^{\mathbf{A}t}\mathbf{B} = \mathbf{B}e^{\mathbf{A}t}$. The argument makes use of the series representation (17) for $e^{\mathbf{A}t}$ and is identical to that used in the proof of part (b). Then

$$\hat{\boldsymbol{\Phi}}' = \mathbf{A}e^{\mathbf{A}t}e^{\mathbf{B}t} + e^{\mathbf{A}t}\mathbf{B}e^{\mathbf{B}t} = \mathbf{A}e^{\mathbf{A}t}e^{\mathbf{B}t} + \mathbf{B}e^{\mathbf{A}t}e^{\mathbf{B}t} = (\mathbf{A} + \mathbf{B})e^{\mathbf{A}t}e^{\mathbf{B}t} = (\mathbf{A} + \mathbf{B})\hat{\boldsymbol{\Phi}}.$$

Since $\hat{\boldsymbol{\Phi}}(0) = \mathbf{I}_n$, $\hat{\boldsymbol{\Phi}}$ is also a solution of the initial value problem (23). The result follows from the uniqueness part of Theorem 6.2.1.

(c) Multiplying the series for $e^{\mathbf{A}t}$ and $e^{\mathbf{A}\tau}$ in a manner analogous to multiplying polynomial expressions and regrouping terms appropriately yield

$$e^{\mathbf{A}t}e^{\mathbf{A}\tau} = \left(\mathbf{I}_n + \mathbf{A}t + \frac{1}{2!}\mathbf{A}^2 t^2 + \frac{1}{3!}\mathbf{A}^3 t^3 + \cdots\right)\left(\mathbf{I}_n + \mathbf{A}\tau + \frac{1}{2!}\mathbf{A}^2 \tau^2 + \frac{1}{3!}\mathbf{A}^3 \tau^3 + \cdots\right)$$

$$= \mathbf{I}_n + \mathbf{A}(t+\tau) + \frac{1}{2!}\mathbf{A}^2(t+\tau)^2 + \frac{1}{3!}\mathbf{A}^3(t+\tau)^3 + \cdots$$

$$= e^{\mathbf{A}(t+\tau)}.$$

(d) Setting $\tau = -t$ in result (c) gives $e^{\mathbf{A}(t+(-t))} = e^{\mathbf{A}t}e^{\mathbf{A}(-t)}$, or $\mathbf{I}_n = e^{\mathbf{A}t}e^{-\mathbf{A}t}$. Thus $e^{-\mathbf{A}t} = (e^{\mathbf{A}t})^{-1}$.

Methods for Constructing $e^{\mathbf{A}t}$

Given an $n \times n$ constant matrix $\mathbf{A}$, we now discuss relationships between different representations of the fundamental matrix $e^{\mathbf{A}t}$ for the system

$$\mathbf{x}' = \mathbf{A}\mathbf{x} \qquad (24)$$

that provide practical methods for its construction.

If a fundamental set of solutions, $\{\mathbf{x}_1(t), \mathbf{x}_2(t), \mathbf{x}_3(t)\}$, where

$$\mathbf{x}_1(t) = \begin{pmatrix} x_{11}(t) \\ x_{21}(t) \\ \vdots \\ x_{n1}(t) \end{pmatrix}, \quad \mathbf{x}_2(t) = \begin{pmatrix} x_{12}(t) \\ x_{22}(t) \\ \vdots \\ x_{n2}(t) \end{pmatrix}, \quad \ldots, \quad \mathbf{x}_n(t) = \begin{pmatrix} x_{1n}(t) \\ x_{2n}(t) \\ \vdots \\ x_{nn}(t) \end{pmatrix},$$

of Eq. (24) is available, then by Theorem 6.5.2,

$$e^{\mathbf{A}t} = \mathbf{X}(t)\mathbf{X}^{-1}(0), \qquad (25)$$

where

$$\mathbf{X}(t) = [\mathbf{x}_1(t), \mathbf{x}_2(t), \ldots, \mathbf{x}_n(t)] = \begin{pmatrix} x_{11}(t) & \cdots & x_{1n}(t) \\ \vdots & & \vdots \\ x_{n1}(t) & \cdots & x_{nn}(t) \end{pmatrix}.$$

EXAMPLE 4

Find $e^{\mathbf{A}t}$ for the system

$$\mathbf{x}' = \begin{pmatrix} 1 & -1 \\ 1 & 3 \end{pmatrix} = \mathbf{A}\mathbf{x}. \qquad (26)$$

Then find the inverse $(e^{\mathbf{A}t})^{-1}$ of $e^{\mathbf{A}t}$.

A fundamental set of solutions to Eq. (26) was found in Example 3 of Section 3.5. Since the characteristic equation of $\mathbf{A}$ is $(\lambda - 2)^2$, the only eigenvalue of $\mathbf{A}$, $\lambda_1 = 2$, has algebraic multiplicity 2. However the geometric multiplicity of λ_1 is 1 since the only eigenvector that can be found is $\mathbf{v}_1 = (1, -1)^T$. Thus one solution of Eq. (26) is $\mathbf{x}_1(t) = e^{2t}(1, -1)^T$. A second linearly independent solution,

$$\mathbf{x}_2(t) = te^{2t}\begin{pmatrix} 1 \\ -1 \end{pmatrix} + e^{2t}\begin{pmatrix} 0 \\ -1 \end{pmatrix} = \begin{pmatrix} te^{2t} \\ -(1+t)e^{2t} \end{pmatrix},$$

is found using the method of generalized eigenvectors for systems of dimension 2 discussed in Section 3.5. Thus a fundamental matrix for Eq. (26) is

$$\mathbf{X}(t) = [\mathbf{x}_1(t), \mathbf{x}_2(t)] = \begin{pmatrix} e^{2t} & te^{2t} \\ -e^{2t} & -(1+t)e^{2t} \end{pmatrix}.$$

Using (25) gives

$$e^{\mathbf{A}t} = \mathbf{X}(t)\mathbf{X}^{-1}(0)$$

$$= \begin{pmatrix} e^{2t} & te^{2t} \\ -e^{2t} & -(1+t)e^{2t} \end{pmatrix} \begin{pmatrix} 1 & 0 \\ -1 & -1 \end{pmatrix}$$

$$= \begin{pmatrix} (1-t)e^{2t} & -te^{2t} \\ te^{2t} & (1+t)e^{2t} \end{pmatrix}. \tag{27}$$

By part (c) of Theorem 6.5.3, the inverse of $e^{\mathbf{A}t}$ is obtained by replacing t by $-t$ in the matrix (27),

$$(e^{\mathbf{A}t})^{-1} = e^{-\mathbf{A}t} = \begin{pmatrix} (1+t)e^{-2t} & te^{-2t} \\ -te^{-2t} & (1-t)e^{-2t} \end{pmatrix}.$$

▶ $e^{\mathbf{A}t}$ When **A** is Nondefective. In the case that **A** has n linearly independent eigenvectors $\{\mathbf{v}_1, \ldots, \mathbf{v}_n\}$, then, by Theorem 6.3.1, a fundamental set for Eq. (24) is $\{e^{\lambda_1 t}\mathbf{v}_1, e^{\lambda_2 t}\mathbf{v}_2, \ldots, e^{\lambda_n t}\mathbf{v}_n\}$. In this case,

$$\mathbf{X}(t) = \begin{bmatrix} e^{\lambda_1 t}\mathbf{v}_1, e^{\lambda_2 t}\mathbf{v}_2, \ldots, e^{\lambda_n t}\mathbf{v}_n \end{bmatrix}$$

$$= \begin{pmatrix} e^{\lambda_1 t}v_{11} & e^{\lambda_2 t}v_{12} & \cdots & e^{\lambda_n t}v_{1n} \\ e^{\lambda_1 t}v_{21} & e^{\lambda_2 t}v_{22} & \cdots & e^{\lambda_n t}v_{2n} \\ \vdots & \vdots & & \vdots \\ e^{\lambda_1 t}v_{n1} & e^{\lambda_2 t}v_{n2} & \cdots & e^{\lambda_n t}v_{nn} \end{pmatrix} = \mathbf{T}e^{\mathbf{D}t}, \tag{28}$$

where

$$\mathbf{T} = [\mathbf{v}_1, \mathbf{v}_2, \ldots, \mathbf{v}_n] = \begin{pmatrix} v_{11} & v_{12} & \cdots & v_{1n} \\ v_{21} & v_{22} & \cdots & v_{2n} \\ \vdots & \vdots & & \vdots \\ v_{n1} & v_{n2} & \cdots & v_{nn} \end{pmatrix}$$

and $e^{\mathbf{D}t}$ is defined in Eq. (19). The matrix **T** is nonsingular since its columns are the linearly independent eigenvectors of **A**. Substituting $\mathbf{T}e^{\mathbf{D}t}$ for $\mathbf{X}(t)$ in Eq. (25) and noting that $\mathbf{X}(0) = \mathbf{T}$ and $\mathbf{X}^{-1}(0) = \mathbf{T}^{-1}$ give us

$$e^{\mathbf{A}t} = \underbrace{\mathbf{T}e^{\mathbf{D}t}}_{\mathbf{X}(t)} \underbrace{\mathbf{T}^{-1}}_{\mathbf{X}^{-1}(0)}. \tag{29}$$

Thus, in the case that **A** has n linearly independent eigenvectors, Eq. (29) expresses $e^{\mathbf{A}t}$ directly in terms of the eigenpairs of **A**. Note that if **A** is real, $e^{\mathbf{A}t}$ must be real and therefore the right-hand sides of Eqs. (25) and (29) must produce real matrix functions even if some of the eigenvalues and eigenvectors are complex [see Problem 23 for an alternative derivation of Eq. (29)].

EXAMPLE 5

Consider again the system of differential equations

$$\mathbf{x}' = \mathbf{A}\mathbf{x},$$

where $\mathbf{A}$ is given by Eq. (3). Use Eq. (29) to obtain $e^{\mathbf{A}t}$.

Since the eigenpairs of $\mathbf{A}$ are $(\lambda_1, \mathbf{v}_1)$ and $(\lambda_2, \mathbf{v}_2)$ with $\lambda_1 = 3$, $\lambda_2 = -1$, $\mathbf{v}_1 = (1, 2)^T$, and $\mathbf{v}_2 = (1, -2)^T$,

$$\mathbf{T} = \begin{pmatrix} 1 & 1 \\ 2 & -2 \end{pmatrix}, \qquad \mathbf{D} = \begin{pmatrix} 3 & 0 \\ 0 & -1 \end{pmatrix}, \qquad \text{and} \qquad e^{\mathbf{D}t} = \begin{pmatrix} e^{3t} & 0 \\ 0 & e^{-t} \end{pmatrix}.$$

Therefore

$$e^{\mathbf{A}t} = \underbrace{\begin{pmatrix} 1 & 1 \\ 2 & -2 \end{pmatrix}}_{\mathbf{T}} \underbrace{\begin{pmatrix} e^{3t} & 0 \\ 0 & e^{-t} \end{pmatrix}}_{e^{\mathbf{D}t}} \underbrace{\begin{pmatrix} 1/2 & 1/4 \\ 1/2 & -1/4 \end{pmatrix}}_{T^{-1}} = \begin{pmatrix} (e^{3t} + e^{-t})/2 & (e^{3t} - e^{-t})/4 \\ e^{3t} - e^{-t} & (e^{3t} + e^{-t})/2 \end{pmatrix},$$

in agreement with the results of Example 2. This is expected since in Example 2, $\mathbf{X}(t) = \mathbf{T}e^{\mathbf{D}t}$ and $\mathbf{X}^{-1}(0) = \mathbf{T}^{-1}$.

▶ **Using the Laplace Transform to Find $e^{\mathbf{A}t}$.** Recall that the method of Laplace transforms was applied to constant coefficient, first order linear systems of dimension 2 in Section 5.4. The method generalizes in a natural way to first order linear systems of dimension n. Here, we apply the method to the matrix initial value problem

$$\mathbf{\Phi}' = \mathbf{A}\mathbf{\Phi}, \qquad \mathbf{\Phi}(0) = \mathbf{I}_n, \tag{30}$$

which is equivalent to n initial value problems for the individual columns of $\mathbf{\Phi}$, each a first order system with coefficient matrix $\mathbf{A}$ and initial condition prescribed by the corresponding column of $\mathbf{I}_n$. We denote the Laplace transform of $\mathbf{\Phi}(t)$ by

$$\hat{\mathbf{\Phi}}(s) = \mathcal{L}\left\{\mathbf{\Phi}(t)\right\}(s) = \int_0^\infty e^{-st}\mathbf{\Phi}(t)\,dt.$$

Taking the Laplace transform of the differential equation in the initial value problem (30) yields

$$s\hat{\mathbf{\Phi}} - \mathbf{\Phi}(0) = \mathbf{A}\hat{\mathbf{\Phi}},$$

or

$$(s\mathbf{I}_n - \mathbf{A})\hat{\mathbf{\Phi}}(s) = \mathbf{I}_n,$$

where we have used the initial condition in (30) and rearranged terms. It follows that

$$\hat{\mathbf{\Phi}}(s) = (s\mathbf{I}_n - \mathbf{A})^{-1}. \tag{31}$$

We can then recover $\mathbf{\Phi}(t) = e^{\mathbf{A}t}$ by taking the inverse Laplace transform of the expression on the right-hand side of Eq. (31),

$$e^{\mathbf{A}t} = \mathcal{L}^{-1}\{(s\mathbf{I}_n - \mathbf{A})^{-1}\}(t). \tag{32}$$

Provided that we can find the inverse Laplace transform of each entry of the matrix $(s\mathbf{I}_n - \mathbf{A})^{-1}$, the inversion formula (32) will yield $e^{\mathbf{A}t}$ whether $\mathbf{A}$ is defective or nondefective.

EXAMPLE
6

Find $e^{\mathbf{A}t}$ for the system

$$\mathbf{x}' = \begin{pmatrix} 4 & 6 & 6 \\ 1 & 3 & 2 \\ -1 & -5 & -2 \end{pmatrix} \mathbf{x} = \mathbf{A}\mathbf{x}.$$

Then find the inverse $(e^{\mathbf{A}t})^{-1} = e^{-\mathbf{A}t}$.

The eigenvalues of $\mathbf{A}$ are $\lambda_1 = 1$ with algebraic multiplicity 1 and $\lambda_2 = 2$ with algebraic multiplicity 2 and geometric multiplicity 1, so $\mathbf{A}$ is defective. Thus we attempt to find $e^{\mathbf{A}t}$ using the Laplace transform. The matrix $s\mathbf{I}_3 - \mathbf{A}$ is given by

$$s\mathbf{I}_3 - \mathbf{A} = \begin{pmatrix} s-4 & -6 & -6 \\ -1 & s-3 & -2 \\ 1 & 5 & s+2 \end{pmatrix}.$$

Using a computer algebra system, we find that

$$(s\mathbf{I}_3 - \mathbf{A})^{-1} = \frac{1}{(s-1)(s-2)^2} \begin{pmatrix} s^2-s+4 & 6(s-3) & 6(s-1) \\ s & s^2-2s-2 & 2(s-1) \\ -(s+2) & -(5s-14) & (s-6)(s-1) \end{pmatrix}.$$

and then taking the inverse Laplace transform gives

$$e^{\mathbf{A}t} = \begin{pmatrix} (6t-3)e^{2t}+4e^t & 6(2-t)e^{2t}-12e^t & 6te^{2t} \\ (2t-1)e^{2t}+e^t & (4-2t)e^{2t}-3e^t & 2te^{2t} \\ (3-4t)e^{2t}-3e^t & (-9+4t)e^{2t}+9e^t & (1-4t)e^{2t} \end{pmatrix}.$$

Replacing t by $-t$ in the last result easily provides us with the inverse of $e^{\mathbf{A}t}$,

$$e^{-\mathbf{A}t} = \begin{pmatrix} -(6t+3)e^{-2t}+4e^{-t} & 6(2+t)e^{-2t}-12e^{-t} & -6te^{-2t} \\ -(2t+1)e^{-2t}+e^{-t} & (4+2t)e^{-2t}-3e^{-t} & -2te^{-2t} \\ (3+4t)e^{-2t}-3e^{-t} & -(9+4t)e^{-2t}+9e^{-t} & (1+4t)e^{-2t} \end{pmatrix}.$$

PROBLEMS

In each of Problems 1 through 14, find a fundamental matrix for the given system of equations. In each case, also find the fundamental matrix $e^{\mathbf{A}t}$.

1. $\mathbf{x}' = \begin{pmatrix} 3 & -2 \\ 2 & -2 \end{pmatrix} \mathbf{x}$

2. $\mathbf{x}' = \begin{pmatrix} -3 & 2 \\ 1/2 & -3 \end{pmatrix} \mathbf{x}$

3. $\mathbf{x}' = \begin{pmatrix} 3 & -4 \\ 1 & -1 \end{pmatrix} \mathbf{x}$

4. $\mathbf{x}' = \begin{pmatrix} 1/2 & -1/4 \\ 1 & -1/2 \end{pmatrix} \mathbf{x}$

5. $\mathbf{x}' = \begin{pmatrix} 1 & -5/2 \\ 1/2 & -1 \end{pmatrix} \mathbf{x}$

6. $\mathbf{x}' = \begin{pmatrix} -1 & -4 \\ 1 & -1 \end{pmatrix} \mathbf{x}$

7. $\mathbf{x}' = \begin{pmatrix} 5 & -1 \\ 3 & 1 \end{pmatrix} \mathbf{x}$

8. $\mathbf{x}' = \begin{pmatrix} 1 & -1 \\ 5 & -3 \end{pmatrix} \mathbf{x}$

9. $\mathbf{x}' = \begin{pmatrix} 2 & -1 \\ 3 & -2 \end{pmatrix} \mathbf{x}$

10. $\mathbf{x}' = \begin{pmatrix} 1/2 & 1/2 \\ 2 & -1 \end{pmatrix} \mathbf{x}$

11. $\mathbf{x}' = \begin{pmatrix} -3 & 4 \\ -1 & -2 \end{pmatrix} \mathbf{x}$

12. $\mathbf{x}' = \begin{pmatrix} -3 & 5/2 \\ -5/2 & 2 \end{pmatrix} \mathbf{x}$

13. $\mathbf{x}' = \begin{pmatrix} 1 & 1 & 1 \\ 2 & 1 & -1 \\ -8 & -5 & -3 \end{pmatrix} \mathbf{x}$ 14. $\mathbf{x}' = \begin{pmatrix} 1 & -1 & 4 \\ 3 & 2 & -1 \\ 2 & 1 & -1 \end{pmatrix} \mathbf{x}$

15. Solve the initial value problem

$$\mathbf{x}' = \begin{pmatrix} -3 & -9 \\ 1 & -3 \end{pmatrix} \mathbf{x}, \qquad \mathbf{x}(0) = \begin{pmatrix} 4 \\ 1 \end{pmatrix}$$

by using the fundamental matrix $e^{\mathbf{A}t}$ found in Problem 6.

16. Solve the initial value problem

$$\mathbf{x}' = \begin{pmatrix} 2 & -1 \\ 3 & -2 \end{pmatrix} \mathbf{x}, \qquad \mathbf{x}(0) = \begin{pmatrix} 2 \\ -1 \end{pmatrix}$$

by using the fundamental matrix $e^{\mathbf{A}t}$ found in Problem 9.

In each of Problems 17 through 20, use the method of Laplace transforms to find the fundamental matrix $e^{\mathbf{A}t}$:

17. $\mathbf{x}' = \begin{pmatrix} -4 & -1 \\ 1 & -2 \end{pmatrix} \mathbf{x}$ **18.** $\mathbf{x}' = \begin{pmatrix} 5 & -1 \\ 1 & 3 \end{pmatrix} \mathbf{x}$

19. $\mathbf{x}' = \begin{pmatrix} -1 & -5 \\ 1 & 3 \end{pmatrix} \mathbf{x}$ **20.** $\mathbf{x}' = \begin{pmatrix} 0 & 1 & -1 \\ 1 & 0 & 1 \\ 1 & 1 & 0 \end{pmatrix} \mathbf{x}$

21. Consider an oscillator satisfying the initial value problem

$$u'' + \omega^2 u = 0, \qquad u(0) = u_0, \quad u'(0) = v_0. \qquad \text{(i)}$$

(a) Let $x_1 = u$, $x_2 = u'$, and transform Eqs. (i) into the form

$$\mathbf{x}' = \mathbf{A}\mathbf{x}, \qquad \mathbf{x}(0) = \mathbf{x}_0. \qquad \text{(ii)}$$

(b) By using the series (17), show that

$$e^{\mathbf{A}t} = \mathbf{I}_2 \cos \omega t + \mathbf{A} \frac{\sin \omega t}{\omega}. \qquad \text{(iii)}$$

(c) Find the solution of the initial value problem (ii).

22. The matrix of coefficients for the system of differential equations describing the radioactive decay process in Problem 17, Section 6.3, is

$$\mathbf{A} = \begin{pmatrix} -k_1 & 0 & 0 \\ k_1 & -k_2 & 0 \\ 0 & k_2 & 0 \end{pmatrix}.$$

Use a computer algebra system to find the fundamental matrix $e^{\mathbf{A}t}$ and use the result to solve the initial value problem $\mathbf{m}' = \mathbf{A}\mathbf{m}$, $\mathbf{m}(0) = (m_0, 0, 0)^T$, where the components of $\mathbf{m} = (m_1, m_2, m_3)^T$ are the amounts of each of the three substances R_1, R_2, and R_3, the first two of which are radioactive, while the third is stable.

23. Assume that the real $n \times n$ matrix $\mathbf{A}$ has n linearly independent eigenvectors $\mathbf{v}_1, \ldots, \mathbf{v}_n$ corresponding to the (possibly repeated and possibly complex) eigenvalues $\lambda_1, \ldots, \lambda_n$. If

$\mathbf{T}$ is the matrix whose columns are the eigenvectors of $\mathbf{A}$,

$$\mathbf{T} = [\mathbf{v}_1, \ldots, \mathbf{v}_n],$$

it is shown in Appendix A.4 that

$$\mathbf{T}^{-1}\mathbf{A}\mathbf{T} = \mathbf{D} = \begin{pmatrix} \lambda_1 & 0 & \cdots & 0 \\ 0 & \lambda_2 & \cdots & 0 \\ \vdots & & & \vdots \\ 0 & 0 & \cdots & \lambda_n \end{pmatrix}, \qquad \text{(i)}$$

that is, $\mathbf{A}$ is similar to the diagonal matrix $\mathbf{D} = \text{diag}(\lambda_1, \ldots, \lambda_n)$.

(a) Use the relation in Eq. (i) to show that

$$\mathbf{A}^n = \mathbf{T}\mathbf{D}^n\mathbf{T}^{-1}$$

for each $n = 0, 1, 2, \ldots$.

(b) Use the results in part (a), Eq. (17) in Definition 6.5.1, and Eq. (19) to show that

$$e^{\mathbf{A}t} = \mathbf{T}e^{\mathbf{D}t}\mathbf{T}^{-1}.$$

24. The Method of Successive Approximations. Consider the initial value problem

$$\mathbf{x}' = \mathbf{A}\mathbf{x}, \qquad \mathbf{x}(0) = \mathbf{x}_0, \qquad \text{(i)}$$

where $\mathbf{A}$ is a constant matrix and $\mathbf{x}_0$ is a prescribed vector.

(a) Assuming that a solution $\mathbf{x} = \boldsymbol{\phi}(t)$ exists, show that it must satisfy the integral equation

$$\boldsymbol{\phi}(t) = \mathbf{x}_0 + \int_0^t \mathbf{A}\boldsymbol{\phi}(s)\, ds. \qquad \text{(ii)}$$

(b) Start with the initial approximation $\boldsymbol{\phi}^{(0)} = \mathbf{x}_0$. Substitute this expression for $\boldsymbol{\phi}(t)$ in the right-hand side of Eq. (ii) and obtain a new approximation $\boldsymbol{\phi}^{(1)}(t)$. Show that

$$\boldsymbol{\phi}^{(1)}(t) = (\mathbf{I}_n + \mathbf{A}t)\mathbf{x}_0. \qquad \text{(iii)}$$

(c) Repeat the process and thereby obtain a sequence of approximations $\boldsymbol{\phi}^{(0)}, \boldsymbol{\phi}^{(1)}, \boldsymbol{\phi}^{(2)}, \ldots, \boldsymbol{\phi}^{(n)}, \ldots$. Use an inductive argument to show that

$$\boldsymbol{\phi}^{(n)}(t) = \left(\mathbf{I}_n + \mathbf{A}t + \mathbf{A}^2 \frac{t^2}{2!} + \cdots + \mathbf{A}^n \frac{t^n}{n!} \right) \mathbf{x}_0. \qquad \text{(iv)}$$

(d) Let $n \to \infty$ and show that the solution of the initial value problem (i) is

$$\boldsymbol{\phi}(t) = e^{\mathbf{A}t}\mathbf{x}_0. \qquad \text{(v)}$$

6.6 Nonhomogeneous Linear Systems

Variation of Parameters

In this section we turn to the nonhomogeneous system

$$\mathbf{x}' = \mathbf{P}(t)\mathbf{x} + \mathbf{g}(t), \qquad \text{(1)}$$

where the $n \times n$ matrix $\mathbf{P}(t)$ and the $n \times 1$ vector $\mathbf{g}(t)$ are continuous for $\alpha < t < \beta$. Assume that a fundamental matrix $\mathbf{X}(t)$ for the corresponding homogeneous system

$$\mathbf{x}' = \mathbf{P}(t)\mathbf{x} \qquad \text{(2)}$$

has been found. We use the method of variation of parameters to construct a particular solution, and hence the general solution, of the nonhomogeneous system (1).

Since the general solution of the homogeneous system (2) is $\mathbf{X}(t)\mathbf{c}$, it is natural to proceed as in Section 4.7 and to seek a solution of the nonhomogeneous system (1) by replacing the constant vector $\mathbf{c}$ by a vector function $\mathbf{u}(t)$. Thus we assume that

$$\mathbf{x} = \mathbf{X}(t)\mathbf{u}(t), \tag{3}$$

where $\mathbf{u}(t)$ is a vector to be determined. Upon differentiating $\mathbf{x}$ as given by Eq. (3) and requiring that Eq. (1) be satisfied, we obtain

$$\mathbf{X}'(t)\mathbf{u}(t) + \mathbf{X}(t)\mathbf{u}'(t) = \mathbf{P}(t)\mathbf{X}(t)\mathbf{u}(t) + \mathbf{g}(t). \tag{4}$$

Since $\mathbf{X}(t)$ is a fundamental matrix, $\mathbf{X}'(t) = \mathbf{P}(t)\mathbf{X}(t)$ so that the terms involving $\mathbf{u}(t)$ drop out; hence Eq. (4) reduces to

$$\mathbf{X}(t)\mathbf{u}'(t) = \mathbf{g}(t). \tag{5}$$

Recall that $\mathbf{X}(t)$ is nonsingular on any interval where $\mathbf{P}$ is continuous. Hence $\mathbf{X}^{-1}(t)$ exists, and therefore

$$\mathbf{u}'(t) = \mathbf{X}^{-1}(t)\mathbf{g}(t). \tag{6}$$

Thus for $\mathbf{u}(t)$ we can select any vector from the class of vectors that satisfy Eq. (6); these vectors are determined only up to an arbitrary additive constant vector. Therefore we denote $\mathbf{u}(t)$ by

$$\mathbf{u}(t) = \int \mathbf{X}^{-1}(t)\mathbf{g}(t)\, dt + \mathbf{c}, \tag{7}$$

where the constant vector $\mathbf{c}$ is arbitrary. If the integral in Eq. (7) can be evaluated, then the general solution of the system (1) is found by substituting for $\mathbf{u}(t)$ from Eq. (7) in Eq. (3). However, even if the integral cannot be evaluated, we can still write the general solution of Eq. (1) in the form

$$\mathbf{x} = \mathbf{X}(t)\mathbf{c} + \mathbf{X}(t) \int_{t_1}^{t} \mathbf{X}^{-1}(s)\mathbf{g}(s)\, ds, \tag{8}$$

where t_1 is any point in the interval (α, β). Observe that the first term on the right side of Eq. (8) is the general solution of the corresponding homogeneous system (2), and the second term is a particular solution of Eq. (1) (see Problem 1).

Now let us consider the initial value problem consisting of the differential equation (1) and the initial condition

$$\mathbf{x}(t_0) = \mathbf{x}_0. \tag{9}$$

We can find the solution of this problem most conveniently if we choose the lower limit of integration in Eq. (8) to be the initial point t_0. Then the general solution of the differential equation is

$$\mathbf{x} = \mathbf{X}(t)\mathbf{c} + \mathbf{X}(t) \int_{t_0}^{t} \mathbf{X}^{-1}(s)\mathbf{g}(s)\, ds. \tag{10}$$

For $t = t_0$, the integral in Eq. (10) is zero, so the initial condition (9) is also satisfied if we choose

$$\mathbf{c} = \mathbf{X}^{-1}(t_0)\mathbf{x}_0. \tag{11}$$

Therefore

$$\mathbf{x} = \mathbf{X}(t)\mathbf{X}^{-1}(t_0)\mathbf{x}_0 + \mathbf{X}(t) \int_{t_0}^{t} \mathbf{X}^{-1}(s)\mathbf{g}(s)\, ds. \tag{12}$$

is the solution of the given initial value problem. Again, although it is helpful to use $\mathbf{X}^{-1}$ to write the solutions (8) and (12), it is usually better in particular cases to solve the necessary equations by row reduction than to calculate $\mathbf{X}^{-1}(t)$ and substitute into Eqs. (8) and (12).

The solution (12) takes a slightly simpler form if we use the fundamental matrix $\boldsymbol{\Phi}(t)$ satisfying $\boldsymbol{\Phi}(t_0) = \mathbf{I}_n$. In this case, we have

$$\mathbf{x} = \boldsymbol{\Phi}(t)\mathbf{x}_0 + \boldsymbol{\Phi}(t) \int_{t_0}^{t} \boldsymbol{\Phi}^{-1}(s)\mathbf{g}(s)\, ds. \tag{13}$$

▶ The Case of Constant P. If the coefficient matrix $\mathbf{P}(t)$ in Eq. (1) is a constant matrix, $\mathbf{P}(t) = \mathbf{A}$, it is natural and convenient to use the fundamental matrix $\boldsymbol{\Phi}(t) = e^{\mathbf{A}t}$ to represent solutions to

$$\mathbf{x}' = \mathbf{A}\mathbf{x} + \mathbf{g}(t). \tag{14}$$

Since $(e^{\mathbf{A}t})^{-1} = e^{-\mathbf{A}t}$, the general solution (10) takes the form

$$\mathbf{x} = e^{\mathbf{A}t}\mathbf{c} + e^{\mathbf{A}t} \int_{t_0}^{t} e^{-\mathbf{A}s}\mathbf{g}(s)\, ds. \tag{15}$$

If an initial condition is prescribed at $t = t_0$ as in Eq. (9), then $\mathbf{c} = e^{-\mathbf{A}t_0}\mathbf{x}_0$, and from Eq. (15), we get

$$\mathbf{x} = e^{\mathbf{A}(t-t_0)}\mathbf{x}_0 + e^{\mathbf{A}t} \int_{t_0}^{t} e^{-\mathbf{A}s}\mathbf{g}(s)\, ds, \tag{16}$$

where we have used the property $e^{\mathbf{A}t}e^{-\mathbf{A}t_0} = e^{\mathbf{A}(t-t_0)}$. If the initial condition is prescribed at $t = t_0 = 0$, Eq. (16) reduces to

$$\mathbf{x} = e^{\mathbf{A}t}\mathbf{x}_0 + e^{\mathbf{A}t} \int_{0}^{t} e^{-\mathbf{A}s}\mathbf{g}(s)\, ds. \tag{17}$$

EXAMPLE 1

Use the method of variation of parameters to find the solution of the initial value problem

$$\mathbf{x}' = \begin{pmatrix} -3 & 4 \\ -2 & 3 \end{pmatrix}\mathbf{x} + \begin{pmatrix} \sin t \\ t \end{pmatrix} = \mathbf{A}\mathbf{x} + \mathbf{g}(t), \qquad \mathbf{x}(0) = \begin{pmatrix} 0 \\ 1 \end{pmatrix}. \tag{18}$$

The calculations may be performed by hand. However it is highly recommended to use a computer algebra system to facilitate the operations. The eigenvalues of $\mathbf{A}$ are $\lambda_1 = 1$ and $\lambda_2 = -1$ with eigenvectors $\mathbf{v}_1 = (1, 1)^T$ and $\mathbf{v}_2 = (2, 1)^T$, respectively. A fundamental matrix for the homogeneous equation $\mathbf{x}' = \mathbf{A}\mathbf{x}$ is therefore

$$\mathbf{X}(t) = \begin{pmatrix} e^t & 2e^{-t} \\ e^t & e^{-t} \end{pmatrix}$$

and it follows that

$$e^{\mathbf{A}t} = \mathbf{X}(t)\mathbf{X}^{-1}(0) = \begin{pmatrix} 2e^{-t} - e^t & 2e^t - 2e^{-t} \\ e^{-t} - e^t & 2e^t - e^{-t} \end{pmatrix}.$$

A particular solution of the differential equation (18) is

$$\mathbf{x}_p(t) = e^{\mathbf{A}t} \int_0^t e^{-\mathbf{A}s}\mathbf{g}(s)\,ds = \begin{pmatrix} -4t + \frac{3}{2}e^t - e^{-t} - \frac{1}{2}\cos t + \frac{3}{2}\sin t \\ -1 - 3t + \frac{3}{2}e^t - \frac{1}{2}e^{-t} + \sin t \end{pmatrix}.$$

Thus the solution of the initial value problem is

$$\mathbf{x} = e^{\mathbf{A}t}\mathbf{x}_0 + \mathbf{x}_p(t)$$

$$= \begin{pmatrix} 2e^t - 2e^{-t} \\ 2e^t - e^{-t} \end{pmatrix} + \begin{pmatrix} -4t + \frac{3}{2}e^t - e^{-t} - \frac{1}{2}\cos t + \frac{3}{2}\sin t \\ -1 - 3t + \frac{3}{2}e^t - \frac{1}{2}e^{-t} + \sin t \end{pmatrix}$$

$$= \begin{pmatrix} \frac{7}{2}e^t - 3e^{-t} + \frac{3}{2}\sin t - \frac{1}{2}\cos t - 4t \\ \frac{7}{2}e^t - \frac{3}{2}e^{-t} + \sin t - 3t - 1 \end{pmatrix}.$$

Note that this is in agreement with the solution found using the method of Laplace transforms in Example 4 of Section 5.4.

Undetermined Coefficients and Frequency Response

The method of undetermined coefficients, discussed in Section 4.5, can be used to find a particular solution of

$$\mathbf{x}' = \mathbf{A}\mathbf{x} + \mathbf{g}(t)$$

if $\mathbf{A}$ is an $n \times n$ constant matrix and the entries of $\mathbf{g}(t)$ consist of polynomials, exponential functions, sines and cosines, or finite sums and products of these functions. The methodology described in Section 4.5 extends in a natural way to these types of problems and is discussed in the exercises (see Problems 14–16). In the next example, we illustrate the method of undetermined coefficients in the special but important case of determining the frequency response and gain function for a first order, constant coefficient linear system when there is a single, or one-dimensional, input and the real parts of the eigenvalues of the system matrix are all negative. In Project 2 at the end of this chapter, an analogous problem arises in the analysis of the response of tall buildings to earthquake-induced seismic vibrations.

EXAMPLE 2

Consider the circuit shown in Figure 6.6.1 that was discussed in Section 6.1. Using the circuit parameter values $L_1 = \frac{3}{2}$, $L_2 = \frac{1}{2}$, $C = \frac{4}{3}$, and $R = 1$, find the frequency response and plot a graph of the gain function for the output voltage $v_R = Ri_2(t)$ across the resistor in the circuit.

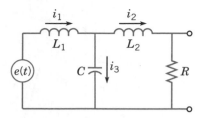

FIGURE 6.6.1 An electric circuit with input $e = e(t)$ and output $v_R = Ri_2(t)$.

By setting $\mathbf{x} = (i_1 \quad i_2 \quad v)^T$, we may express Eq. (30) in Section 6.1 as

$$\mathbf{x}' = \mathbf{A}\mathbf{x} + \mathbf{B}e(t), \tag{19}$$

where

$$\mathbf{A} = \begin{pmatrix} 0 & 0 & -1/L_1 \\ 0 & -R/L_2 & 1/L_2 \\ 1/C & -1/C & 0 \end{pmatrix} \quad \text{and} \quad \mathbf{B} = \begin{pmatrix} 1/L_1 \\ 0 \\ 0 \end{pmatrix}.$$

The frequency response is found in a manner analogous to that for finding the frequency response of a damped spring-mass system in Section 4.6. For the given values of the parameters, the eigenvalues of the system matrix $\mathbf{A}$ are $\lambda_1 = -1$, $\lambda_2 = -\frac{1}{2} + i\sqrt{3}/2$, and $\lambda_3 = \bar{\lambda}_2 = -1/2 - i\sqrt{3}/2$.[1] Since the real parts of the eigenvalues are all negative, the transient part of any solution will die out, leaving, in the long term, only the steady-state response of the system to the harmonic input $e(t) = e^{i\omega t}$. The frequency response for the state variables is then found by assuming a steady-state solution of the form

$$\mathbf{x} = \mathbf{G}(i\omega)e^{i\omega t}, \tag{20}$$

where $\mathbf{G}(i\omega)$ is a 3×1 frequency response vector. Substituting the right-hand side of Eq. (20) for $\mathbf{x}$ in Eq. (19) and canceling the $e^{i\omega t}$, which appears in every term, yield the equation

$$i\omega\mathbf{G} = \mathbf{A}\mathbf{G} + \mathbf{B}. \tag{21}$$

Solving Eq. (21) for $\mathbf{G}$, we get

$$\mathbf{G}(i\omega) = -(\mathbf{A} - i\omega\mathbf{I}_3)^{-1}\mathbf{B}. \tag{22}$$

It follows from Eq. (22) that the frequency response of the output voltage $v_R = Ri_2(t)$ is given by

$$RG_2(i\omega) = (0 \quad R \quad 0)\,\mathbf{G}(i\omega).$$

Using a computer, we can solve Eq. (21) for each ω on a sufficiently fine grid. The gain function $R|G_2(i\omega)|$ evaluated on such a grid is shown in Figure 6.6.2.

---

[1]The characteristic polynomial of $\mathbf{A}$ can be shown to be $p(\lambda) = -\left[\lambda^3 + \frac{R}{L_2}\lambda^2 + \frac{1}{C}\left(\frac{1}{L_1} + \frac{1}{L_2}\right)\lambda + \frac{R}{CL_1L_2}\right]$. The Routh stability criterion, discussed in Section 5.9, can then be used to show that the real parts of all the roots of p are negative whenever $L_1, L_2, R,$ and C are positive.

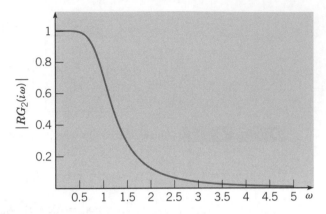

FIGURE 6.6.2   The gain function $R|G_2(i\omega)|$ of the output voltage $v_R = Rv(t)$ across the resistor in the electric circuit shown in Figure 6.6.1.

With the specified values of the circuit parameters, the circuit in Figure 6.6.1 is a **low-pass filter**, that is, a circuit offering easy passage to low-frequency signals and difficult passage to high-frequency signals.

PROBLEMS

1. Assuming that $\mathbf{X}(t)$ is a fundamental matrix for $\mathbf{x}' = \mathbf{P}(t)\mathbf{x}$, show directly that

$$\mathbf{x}_p = \mathbf{X}(t) \int_{t_1}^{t} \mathbf{X}^{-1}(s)\mathbf{g}(s)\,ds$$

is a particular solution of Eq. (1), $\mathbf{x}' = \mathbf{P}(t)\mathbf{x} + \mathbf{g}(t)$.

 In each of Problems 2 through 9, find the general solution of the given system. Calculations are greatly facilitated by using a computer algebra system.

2. $\mathbf{x}' = \begin{pmatrix} 2 & -1 \\ 3 & -2 \end{pmatrix}\mathbf{x} + \begin{pmatrix} e^t \\ t \end{pmatrix}$

3. $\mathbf{x}' = \begin{pmatrix} 1 & \sqrt{3} \\ \sqrt{3} & -1 \end{pmatrix}\mathbf{x} + \begin{pmatrix} e^t \\ \sqrt{3}e^{-t} \end{pmatrix}$

4. $\mathbf{x}' = \begin{pmatrix} 2 & -5 \\ 1 & -2 \end{pmatrix}\mathbf{x} + \begin{pmatrix} -\cos t \\ \sin t \end{pmatrix}$

5. $\mathbf{x}' = \begin{pmatrix} 1 & 1 \\ 4 & -2 \end{pmatrix}\mathbf{x} + \begin{pmatrix} e^{-2t} \\ -2e^t \end{pmatrix}$

6. $\mathbf{x}' = \begin{pmatrix} 0 & -1 & 1 \\ 0 & 2 & 0 \\ -2 & -1 & 3 \end{pmatrix}\mathbf{x} + \begin{pmatrix} 1 \\ t \\ e^{-t} \end{pmatrix}$

7. $\mathbf{x}' = \begin{pmatrix} -1/2 & 1/2 & -1/2 \\ -1 & -2 & 1 \\ 1/2 & 1/2 & -3/2 \end{pmatrix}\mathbf{x} + \begin{pmatrix} 1 \\ t \\ 11e^{-3t} \end{pmatrix}$

8. $\mathbf{x}' = \begin{pmatrix} -4 & 1 & 3 \\ 0 & -2 & 0 \\ -2 & 1 & 1 \end{pmatrix}\mathbf{x} + \begin{pmatrix} 3t \\ 0 \\ 3\cos t \end{pmatrix}$

9. $\mathbf{x}' = \begin{pmatrix} -1/2 & 1 & 1/2 \\ 1 & -1 & 1 \\ 1/2 & 1 & -1/2 \end{pmatrix}\mathbf{x} + \begin{pmatrix} 0 \\ -\sin t \\ 0 \end{pmatrix}$

10. Diffusion of particles on a lattice with reflecting boundaries was described in Example 3, Section 6.3. In this problem, we introduce a known particle source $x_0(t) = f(t)$ at the left end of the lattice as shown in Figure 6.6.3. Treating locations 1 and 2 as interior points and location 3 as a reflecting boundary, the system of differential equations for $\mathbf{x} = (x_1, x_2, x_3)$, the number of particles at lattice points 1, 2, and 3, respectively, is

$$\mathbf{x}' = k \begin{pmatrix} -2 & 1 & 0 \\ 1 & -2 & 1 \\ 0 & 1 & -1 \end{pmatrix}\mathbf{x} + k \begin{pmatrix} f(t) \\ 0 \\ 0 \end{pmatrix}. \qquad \text{(i)}$$

FIGURE 6.6.3   Diffusion on a lattice with a particle source at the left endpoint.

(a) Find a numerical approximation to the solution of Eq. (i) subject to the initial condition $\mathbf{x}(0) = (0, 0, 0)^T$ if $f(t) = 1$ and $k = 1$. Draw component plots of the solution and determine $\lim_{t \to \infty} \mathbf{x}(t)$.

(b) Find a numerical approximation to the solution of Eq. (i) subject to the initial condition $\mathbf{x}(0) = (0, 0, 0)^T$ if $k = 1$ and $f(t) = 1 - \cos \omega t$ for each of the cases $\omega = 1$ and $\omega = 4$. Draw component plots of the solutions in each of the two cases.

11. Find numerical approximations to the initial value problems posed in Problem 10 under the assumption that the right boundary is an absorbing boundary (see Problem 14, Section 6.3).

12. The equations presented in Section 6.1 for modeling lead uptake, subsequent exchange between tissue compartments, and removal from the human body are

$$
\begin{aligned}
x_1' &= (L + k_{12}x_2 + k_{13}x_3) - (k_{21} + k_{31} + k_{01})x_1 \\
x_2' &= k_{21}x_1 - (k_{02} + k_{12})x_2 \\
x_3' &= k_{31}x_1 - k_{13}x_3.
\end{aligned}
$$

Employing the methods of Section 6.6, find the solution of this system subject to the initial conditions $x_1(0) = 0$, $x_2(0) = 0$, and $x_3(0) = 0$ using the parameter values $k_{21} = 0.011$, $k_{12} = 0.012$, $k_{31} = 0.0039$, $k_{13} = 0.000035$, $k_{01} = 0.021$, $k_{02} = 0.016$, and input function

$$
L(t) = \begin{cases} 35, & 0 \le t \le 365, \\ 0, & t > 365. \end{cases}
$$

Solve the problem in two stages, the first over the time interval $0 \le t \le 365$ and the second over the time interval $t \ge 365$. Evaluate the solution for the first stage at $t = 365$ to provide initial conditions for the second stage of the problem. You may wish to use a computer to assist your calculations. Check your results by comparing graphs of the solutions with the graphs in Figure 6.1.4 or by comparing with an approximation obtained directly by using an initial value problem solver.

13. When viscous damping forces are included and the external force $F_2(t)$ is set to zero, the system of equations describing the motion of the coupled mass-spring system shown in Figure 6.1.1 of Section 6.1 is

$$
\mathbf{x}' = \begin{pmatrix} 0 & 0 & 1 & 0 \\ 0 & 0 & 0 & 1 \\ -(k_1 + k_2)/m_1 & k_2/m_1 & -\gamma/m_1 & 0 \\ k_2/m_2 & -(k_2 + k_3)/m_2 & 0 & -\gamma/m_2 \end{pmatrix} \mathbf{x}
$$

$$
+ \begin{pmatrix} 0 \\ 0 \\ F_1(t)/m_1 \\ 0 \end{pmatrix},
$$

where γ is the damping coefficient for both m_1 and m_2. Assume a harmonic input $F_1(t)/m_1 = e^{i\omega t}$ and use the parameter

values $m_1 = 2$, $m_2 = \frac{9}{4}$, $k_1 = 1$, $k_2 = 3$, $k_3 = \frac{15}{4}$ of Example 2, Section 6.4, to compute and plot the graphs of the gain functions $|G_1(i\omega)|$ and $|G_2(i\omega)|$ for the displacements x_1 and x_2, respectively. Compare the graphs of the gain functions with the natural frequencies of the undamped system for each of the damping coefficients $\gamma = 1$, 0.5, and 0.1.

Undetermined Coefficients. For each of the nonhomogeneous terms specified in Problems 14 through 16, use the method of undetermined coefficients to find a particular solution of

$$
\mathbf{x}' = \begin{pmatrix} -2 & 1 \\ 1 & -2 \end{pmatrix} \mathbf{x} + \mathbf{g}(t) = \mathbf{A}\mathbf{x} + \mathbf{g}(t), \tag{ii}
$$

given that the general solution of the corresponding homogeneous system $\mathbf{x}' = \mathbf{A}\mathbf{x}$ is

$$
\mathbf{x} = c_1 e^{-3t} \begin{pmatrix} 1 \\ -1 \end{pmatrix} + c_2 e^{-t} \begin{pmatrix} 1 \\ 1 \end{pmatrix}.
$$

14. $\mathbf{g}(t) = \begin{pmatrix} 0 \\ 6t \end{pmatrix} = \begin{pmatrix} 0 \\ 0 \end{pmatrix} + t \begin{pmatrix} 0 \\ 6 \end{pmatrix}$.

Since the entries of $\mathbf{g}(t)$ are linear functions of t, substitute an expression of the form

$$
\mathbf{x}_p(t) = t\mathbf{a} + \mathbf{b} = t \begin{pmatrix} a_1 \\ a_2 \end{pmatrix} + \begin{pmatrix} b_1 \\ b_2 \end{pmatrix}
$$

into Eq. (ii) and match the coefficients of the vector polynomial on both sides of the resulting equation to obtain the two systems

$$
\mathbf{A}\mathbf{a} = -\begin{pmatrix} 0 \\ 3 \end{pmatrix}, \qquad \mathbf{A}\mathbf{b} = \mathbf{a}.
$$

Solve the first equation for $\mathbf{a}$, substitute the result into the second equation, and solve for $\mathbf{b}$.

15. $\mathbf{g}(t) = \begin{pmatrix} \sin t \\ 0 \end{pmatrix} = \sin t \begin{pmatrix} 1 \\ 0 \end{pmatrix}$.

Since an entry of $\mathbf{g}(t)$ contains a sine function, substitute an expression of the form

$$
\mathbf{x}_p(t) = (\cos t)\mathbf{a} + (\sin t)\mathbf{b}
$$

$$
= \cos t \begin{pmatrix} a_1 \\ a_2 \end{pmatrix} + \sin t \begin{pmatrix} b_1 \\ b_2 \end{pmatrix}
$$

into Eq. (ii) and match the coefficients of the sine function and the cosine function on both sides of the resulting equation to obtain the coupled systems

$$
\mathbf{A}\mathbf{a} = \mathbf{b}, \qquad \mathbf{A}\mathbf{b} = -\mathbf{a} - \begin{pmatrix} 1 \\ 0 \end{pmatrix}.
$$

Show that $(\mathbf{A}^2 + \mathbf{I}_2)\mathbf{a} = -\begin{pmatrix} 1 \\ 0 \end{pmatrix}$ and solve for $\mathbf{a}$. Then substitute this result into the second equation above and solve for $\mathbf{b}$.

16. $\mathbf{g}(t) = \begin{pmatrix} 6e^{-t} \\ 0 \end{pmatrix} = e^{-t} \begin{pmatrix} 6 \\ 0 \end{pmatrix}$.

Since the homogeneous equation has a solution of the form $\mathbf{x}_p(t) = e^{-t}\mathbf{a}$, it is necessary to include both $te^{-t}\mathbf{a}$ and $e^{-t}\mathbf{b}$ in the assumed form for the particular solution,

$$\mathbf{x}_p(t) = e^{-t}\mathbf{b} + te^{-t}\mathbf{a}.$$

Substitute this expression into Eq. (ii) using the given $\mathbf{g}$ and show that $\mathbf{a}$ and $\mathbf{b}$ must be selected to satisfy

$$(\mathbf{A} + \mathbf{I}_2)\mathbf{a} = \mathbf{0}, \qquad (\mathbf{A} + \mathbf{I}_2)\mathbf{b} = \mathbf{a} - \begin{pmatrix} 2 \\ 0 \end{pmatrix}.$$

Then find $\mathbf{a}$ and $\mathbf{b}$ that satisfy both of these equations.

6.7 Defective Matrices

We conclude the chapter on systems of first order linear equations with a discussion of the case in which the real, constant matrix $\mathbf{A}$ is defective, that is, $\mathbf{A}$ has one or more defective eigenvalues. Recall that an eigenvalue is defective if its geometric multiplicity g is less than its algebraic multiplicity m. In other words, there are fewer than m linearly independent eigenvectors associated with this eigenvalue. We include this section for completeness but it may be regarded as optional for the following reasons:

1. In the majority of applications, $\mathbf{A}$ has a complete set of linearly independent eigenvectors. Thus the theory and methods presented in the preceding sections are adequate to handle these problems.

2. The repeated eigenvalue problem for linear constant coefficient systems of dimension 2 was treated in Section 3.5.

3. For linear systems of dimension larger than, say, 4, the payoff utility of finding analytic representations of fundamental sets of solutions is questionable since the complexity of the representations often requires that graphs of the components be constructed in order to help understand their behavior. Thus most scientists and engineers resort directly to numerical approximations in order to obtain the same information.

4. If it is necessary to find analytical representations of fundamental sets of solutions for systems of dimension greater than 2 in the case that $\mathbf{A}$ is defective, then a computer algebra system or the method of Laplace transforms discussed in Section 6.5 together with a computer algebra system may be used to find an analytic representation of $e^{\mathbf{A}t}$ in terms of elementary functions.

▶ Fundamental Sets for Defective Matrices. Let λ be an eigenvalue of $\mathbf{A}$ with algebraic multiplicity m. The following theorem shows how to find m linearly independent solutions of $\mathbf{x}' = \mathbf{A}\mathbf{x}$ associated with λ.

THEOREM 6.7.1

Suppose $\mathbf{A}$ is a real $n \times n$ matrix and λ is an eigenvalue of $\mathbf{A}$ with algebraic multiplicity m. Let $\mathbf{v}_1, \dots, \mathbf{v}_m$ be linearly independent solutions of $(\mathbf{A} - \lambda\mathbf{I}_n)^m\mathbf{v} = \mathbf{0}$. Then

$$\mathbf{x}_k = e^{\lambda t}\left[\mathbf{v}_k + \frac{t}{1!}(\mathbf{A} - \lambda\mathbf{I}_n)\mathbf{v}_k + \cdots + \frac{t^{m-1}}{(m-1)!}(\mathbf{A} - \lambda\mathbf{I}_n)^{m-1}\mathbf{v}_k \right], \quad k = 1, \dots, m, \qquad (1)$$

are linearly independent solutions of $\mathbf{x}' = \mathbf{A}\mathbf{x}$.

Proof

We rewrite $\mathbf{x}' = \mathbf{A}\mathbf{x}$ in the form

$$\mathbf{x}' = (\mathbf{A} - \lambda\mathbf{I}_n)\mathbf{x} + \lambda\mathbf{x} \tag{2}$$

and show that $\mathbf{x}_k$ defined by Eq. (1) satisfies Eq. (2). Differentiating the expression (1) yields

$$\frac{d\mathbf{x}_k}{dt} = e^{\lambda t}\left[(\mathbf{A} - \lambda\mathbf{I}_n)\mathbf{v}_k + \cdots + \frac{t^{m-2}}{(m-2)!}(\mathbf{A} - \lambda\mathbf{I}_n)^{m-1}\mathbf{v}_k\right] + \lambda\mathbf{x}_k. \tag{3}$$

On the other hand, multiplying $\mathbf{x}_k$ by $(\mathbf{A} - \lambda\mathbf{I}_n)$ gives

$$(\mathbf{A} - \lambda\mathbf{I}_n)\mathbf{x}_k = e^{\lambda t}\left[(\mathbf{A} - \lambda\mathbf{I}_n)\mathbf{v}_k + \cdots + \frac{t^{m-1}}{(m-1)!}(\mathbf{A} - \lambda\mathbf{I}_n)^m\mathbf{v}_k\right]. \tag{4}$$

Since $(\mathbf{A} - \lambda\mathbf{I}_n)^m\mathbf{v}_k = \mathbf{0}$, the last term within the brackets on the right-hand side of Eq. (4) equals $\mathbf{0}$. Therefore Eq. (4) reduces to

$$(\mathbf{A} - \lambda\mathbf{I}_n)\mathbf{x}_k = e^{\lambda t}\left[(\mathbf{A} - \lambda\mathbf{I}_n)\mathbf{v}_k + \cdots + \frac{t^{m-2}}{(m-2)!}(\mathbf{A} - \lambda\mathbf{I}_n)^{m-1}\mathbf{v}_k\right]. \tag{5}$$

From Eqs. (3) and (5), we see that $\dfrac{d\mathbf{x}_k}{dt} = (\mathbf{A} - \lambda\mathbf{I}_n)\mathbf{x}_k + \lambda\mathbf{x}_k$.

The set of solutions $\{\mathbf{x}_1, \ldots, \mathbf{x}_m\}$ is linearly independent on $I = (-\infty, \infty)$ since setting $t = 0$ in the statement

$$c_1\mathbf{x}_1(t) + \cdots + c_m\mathbf{x}_m(t) = \mathbf{0} \quad \text{for every } t \in I$$

yields the equation

$$c_1\mathbf{v}_1 + \cdots + c_m\mathbf{v}_m = \mathbf{0}.$$

Since $\mathbf{v}_1, \ldots, \mathbf{v}_m$ are linearly independent, the constants $c_1, \ldots, c_m$ are necessarily zero.

Remark

1. Theorem 6.7.1 applies to both defective and nondefective eigenvalues.
2. For $\mathbf{A}$, as stated in the theorem, the existence of m linearly independent solutions to $(\mathbf{A} - \lambda\mathbf{I}_n)^m\mathbf{v} = \mathbf{0}$ is guaranteed by Theorem A.4.2 in Appendix A.4.
3. Assuming that $\mathbf{A}$ has r eigenvalues $\lambda_1, \ldots, \lambda_r$ with corresponding algebraic multiplicities $m_1, \ldots, m_r$ such that $m_1 + \cdots + m_r = n$, it can be shown that the union of the solution sets of $(\mathbf{A} - \lambda_j\mathbf{I}_n)^{m_j}\mathbf{v} = \mathbf{0}$ for $j = 1, \ldots, r$, say, $\left\{\mathbf{v}_1^{(1)}, \ldots, \mathbf{v}_{m_1}^{(1)}, \ldots, \mathbf{v}_1^{(r)}, \ldots, \mathbf{v}_{m_r}^{(r)}\right\}$, is linearly independent. Thus the union of the sets of solutions obtained by applying Theorem 6.7.1 to each eigenvalue, $\left\{\mathbf{x}_1^{(1)}(t), \ldots, \mathbf{x}_{m_1}^{(1)}(t), \ldots, \mathbf{x}_1^{(r)}(t), \ldots, \mathbf{x}_{m_r}^{(r)}(t)\right\}$, is a fundamental set of solutions of $\mathbf{x}' = \mathbf{A}\mathbf{x}$.
4. That there exist solutions to $\mathbf{x}' = \mathbf{A}\mathbf{x}$ of the form (1) is made plausible by noting that $e^{\mathbf{A}t}\mathbf{v}$ is a solution of $\mathbf{x}' = \mathbf{A}\mathbf{x}$ for any vector $\mathbf{v}$. Therefore $e^{\mathbf{A}t}\mathbf{v}_k$ for $\mathbf{v}_k$ as in Theorem 6.7.1 is a solution of $\mathbf{x}' = \mathbf{A}\mathbf{x}$ and can be expressed in the form $e^{\lambda t}e^{(\mathbf{A}-\lambda\mathbf{I}_n)t}\mathbf{v}_k$. Theorem 6.7.1 shows that a series expansion of $e^{(\mathbf{A}-\lambda\mathbf{I}_n)t}\mathbf{v}_k$ based on Eq. (17) in Section 6.5 (with $\mathbf{A}$ replaced by $\mathbf{A} - \lambda\mathbf{I}_n$) is guaranteed to consist of no more than m nonzero terms.
5. If the λ in Theorem 6.7.1 is complex, $\lambda = \mu + i\nu$, $\nu \neq 0$, then $2m$ linearly independent, real-valued solutions of $\mathbf{x}' = \mathbf{A}\mathbf{x}$ associated with λ are $\{\text{Re }\mathbf{x}_1, \ldots, \text{Re }\mathbf{x}_m\}$ and $\{\text{Im }\mathbf{x}_1, \ldots, \text{Im }\mathbf{x}_m\}$.

EXAMPLE 1

Find a fundamental set of solutions for

$$\mathbf{x}' = \begin{pmatrix} 1 & -1 \\ 1 & 3 \end{pmatrix} = \mathbf{Ax}. \tag{6}$$

The characteristic equation of $\mathbf{A}$ is $p(\lambda) = \lambda^2 - 4\lambda + 4 = (\lambda - 2)^2$, so the only eigenvalue of $\mathbf{A}$ is $\lambda_1 = 2$ with algebraic multiplicity 2. Thus

$$(\mathbf{A} - \lambda_1 \mathbf{I}_2) = \begin{pmatrix} -1 & -1 \\ 1 & 1 \end{pmatrix}, \qquad (\mathbf{A} - \lambda_1 \mathbf{I}_2)^2 = \begin{pmatrix} 0 & 0 \\ 0 & 0 \end{pmatrix}.$$

Two linearly independent solutions of $(\mathbf{A} - \lambda_1 \mathbf{I}_2)^2 \mathbf{v} = \mathbf{0}$ are $\mathbf{v}_1 = (1, 0)^T$ and $\mathbf{v}_2 = (0, 1)^T$. Following Theorem 6.7.1, we find two linearly independent solutions

$$\hat{\mathbf{x}}_1(t) = e^{\lambda_1 t}[\mathbf{v}_1 + t(\mathbf{A} - \lambda_1 \mathbf{I}_2)\mathbf{v}_1]$$

$$= e^{2t}\left[\begin{pmatrix} 1 \\ 0 \end{pmatrix} + t\begin{pmatrix} -1 \\ 1 \end{pmatrix}\right] = e^{2t}\begin{pmatrix} 1 - t \\ t \end{pmatrix}$$

and

$$\hat{\mathbf{x}}_2(t) = e^{\lambda_1 t}[\mathbf{v}_2 + t(\mathbf{A} - \lambda_1 \mathbf{I}_2)\mathbf{v}_2]$$

$$= e^{2t}\left[\begin{pmatrix} 0 \\ 1 \end{pmatrix} + t\begin{pmatrix} -1 \\ 1 \end{pmatrix}\right] = e^{2t}\begin{pmatrix} -t \\ 1 + t \end{pmatrix}.$$

Comparing with the functions in the fundamental set for Eq. (6) found in Example 4, Section 6.5,

$$\mathbf{x}_1(t) = e^{2t}\begin{pmatrix} 1 \\ -1 \end{pmatrix}, \qquad \mathbf{x}_2(t) = e^{2t}\begin{pmatrix} t \\ -1 - t \end{pmatrix},$$

we see that $\hat{\mathbf{x}}_1(t) = \mathbf{x}_1(t) - \mathbf{x}_2(t)$ and $\hat{\mathbf{x}}_2(t) = -\mathbf{x}_2(t)$. Since the fundamental matrix $\hat{\mathbf{X}}(t) = [\hat{\mathbf{x}}_1(t), \hat{\mathbf{x}}_2(t)]$ satisfies $\hat{\mathbf{X}}(0) = \mathbf{I}_2$, it follows that $\hat{\mathbf{X}}(t) = e^{\mathbf{A}t}$.

EXAMPLE 2

Find a fundamental set of solutions for

$$\mathbf{x}' = \begin{pmatrix} 4 & 6 & 6 \\ 1 & 3 & 2 \\ -1 & -5 & -2 \end{pmatrix} \mathbf{x} = \mathbf{Ax}. \tag{7}$$

The characteristic polynomial of $\mathbf{A}$ is $p(\lambda) = -(\lambda^3 - 5\lambda^2 + 8\lambda - 4) = -(\lambda - 1)(\lambda - 2)^2$, so the eigenvalues of $\mathbf{A}$ are the simple eigenvalue $\lambda_1 = 1$ and $\lambda_2 = 2$ with algebraic multiplicity 2. Since the eigenvector associated with λ_1 is $\mathbf{v}_1 = (4, 1, -3)^T$, one solution of Eq. (7) is

$$\mathbf{x}_1(t) = e^t\begin{pmatrix} 4 \\ 1 \\ -3 \end{pmatrix}. \tag{8}$$

Next we compute

$$\mathbf{A} - \lambda_2 \mathbf{I}_3 = \begin{pmatrix} 2 & 6 & 6 \\ 1 & 1 & 2 \\ -1 & -5 & -4 \end{pmatrix}$$

and

$$(\mathbf{A} - \lambda_2 \mathbf{I}_3)^2 = \begin{pmatrix} 4 & -12 & 0 \\ 1 & -3 & 0 \\ -3 & 9 & 0 \end{pmatrix}.$$

Reducing $(\mathbf{A} - \lambda_2 \mathbf{I}_3)^2$ to echelon form

$$\begin{pmatrix} 1 & -3 & 0 \\ 0 & 0 & 0 \\ 0 & 0 & 0 \end{pmatrix}$$

using elementary row operations confirms that $\text{rank}[(\mathbf{A} - \lambda_2 \mathbf{I}_3)^2] = 1$. Linearly independent solutions of $(\mathbf{A} - \lambda_2 \mathbf{I}_3)^2 \mathbf{v} = \mathbf{0}$ are $\mathbf{v}_1^{(2)} = (3, 1, 0)^T$ and $\mathbf{v}_2^{(2)} = (0, 0, 1)^T$. Therefore two linearly independent solutions of Eq. (7) associated with λ_2 are

$$\mathbf{x}_2(t) = e^{\lambda_2 t} \left[\mathbf{v}_1^{(2)} + t(\mathbf{A} - \lambda_2 \mathbf{I}_3) \mathbf{v}_1^{(2)} \right]$$

$$= e^{2t} \left[\begin{pmatrix} 3 \\ 1 \\ 0 \end{pmatrix} + t \begin{pmatrix} 12 \\ 4 \\ -8 \end{pmatrix} \right] = e^{2t} \begin{pmatrix} 3 + 12t \\ 1 + 4t \\ -8t \end{pmatrix}, \qquad (9)$$

and

$$\mathbf{x}_3(t) = e^{\lambda_2 t} \left[\mathbf{v}_2^{(2)} + t(\mathbf{A} - \lambda_2 \mathbf{I}_3) \mathbf{v}_2^{(2)} \right]$$

$$= e^{2t} \left[\begin{pmatrix} 0 \\ 0 \\ 1 \end{pmatrix} + t \begin{pmatrix} 6 \\ 2 \\ -4 \end{pmatrix} \right] = e^{2t} \begin{pmatrix} 6t \\ 2t \\ 1 - 4t \end{pmatrix}. \qquad (10)$$

A fundamental set for Eq. (7) contains $\mathbf{x}_1(t)$, $\mathbf{x}_2(t)$, and $\mathbf{x}_3(t)$ given by Eqs. (8), (9), and (10), respectively.

EXAMPLE 3

Find the general solution of

$$\mathbf{x}' = \begin{pmatrix} -3 & -1 & -6 \\ -2 & -1 & -4 \\ 1 & 0 & 1 \end{pmatrix} \mathbf{x} = \mathbf{A}\mathbf{x}. \qquad (11)$$

The characteristic polynomial of $\mathbf{A}$ is $p(\lambda) = -(\lambda^3 + 3\lambda^2 + 3\lambda + 1) = -(\lambda + 1)^3$, so $\lambda_1 = -1$ is the only eigenvalue. The algebraic multiplicity of this eigenvalue is $m_1 = 3$.

Since

$$(\mathbf{A} - \lambda_1 \mathbf{I}_3)^3 = \begin{pmatrix} 0 & 0 & 0 \\ 0 & 0 & 0 \\ 0 & 0 & 0 \end{pmatrix},$$

three linearly independent solutions of $(\mathbf{A} - \lambda_1 \mathbf{I}_3)^3 \mathbf{v} = \mathbf{0}$ are

$$\mathbf{v}_1 = \begin{pmatrix} 1 \\ 0 \\ 0 \end{pmatrix}, \qquad \mathbf{v}_2 = \begin{pmatrix} 0 \\ 1 \\ 0 \end{pmatrix}, \qquad \mathbf{v}_3 = \begin{pmatrix} 0 \\ 0 \\ 1 \end{pmatrix}.$$

Using

$$\mathbf{A} - \lambda_1 \mathbf{I}_3 = \begin{pmatrix} -2 & -1 & -6 \\ -2 & 0 & -4 \\ 1 & 0 & 2 \end{pmatrix}, \qquad (\mathbf{A} - \lambda_1 \mathbf{I}_3)^2 = \begin{pmatrix} 0 & 2 & 4 \\ 0 & 2 & 4 \\ 0 & -1 & -2 \end{pmatrix},$$

we find that a fundamental set of solutions of Eq. (11) contains

$$\mathbf{x}_1(t) = e^{-t} \left[\mathbf{v}_1 + t(\mathbf{A} - \lambda_1 \mathbf{I}_3)\mathbf{v}_1 + \frac{t^2}{2!}(\mathbf{A} - \lambda_1 \mathbf{I}_3)^2 \mathbf{v}_1 \right]$$

$$= e^{-t} \left[\begin{pmatrix} 1 \\ 0 \\ 0 \end{pmatrix} + t \begin{pmatrix} -2 \\ -2 \\ 1 \end{pmatrix} \right] = e^{-t} \begin{pmatrix} 1 - 2t \\ -2t \\ t \end{pmatrix},$$

$$\mathbf{x}_2(t) = e^{-t} \left[\mathbf{v}_2 + t(\mathbf{A} - \lambda_1 \mathbf{I}_3)\mathbf{v}_2 + \frac{t^2}{2!}(\mathbf{A} - \lambda_1 \mathbf{I}_3)^2 \mathbf{v}_2 \right]$$

$$= e^{-t} \left[\begin{pmatrix} 0 \\ 1 \\ 0 \end{pmatrix} + t \begin{pmatrix} -1 \\ 0 \\ 0 \end{pmatrix} + \frac{t^2}{2!} \begin{pmatrix} 2 \\ 2 \\ -1 \end{pmatrix} \right] = e^{-t} \begin{pmatrix} t^2 - t \\ 1 + t^2 \\ -t^2/2 \end{pmatrix},$$

and

$$\mathbf{x}_3(t) = e^{-t} \left[\mathbf{v}_3 + t(\mathbf{A} - \lambda_1 \mathbf{I}_3)\mathbf{v}_3 + \frac{t^2}{2!}(\mathbf{A} - \lambda_1 \mathbf{I}_3)^2 \mathbf{v}_3 \right]$$

$$= e^{-t} \left[\begin{pmatrix} 0 \\ 0 \\ 1 \end{pmatrix} + t \begin{pmatrix} -6 \\ -4 \\ 2 \end{pmatrix} + \frac{t^2}{2!} \begin{pmatrix} 4 \\ 4 \\ -2 \end{pmatrix} \right] = e^{-t} \begin{pmatrix} 2t^2 - 6t \\ 2t^2 - 4t \\ 1 + 2t - t^2 \end{pmatrix}.$$

Note that the fundamental matrix $\mathbf{X}(t) = [\mathbf{x}_1(t), \mathbf{x}_2(t), \mathbf{x}_3(t)]$ is, in fact, equal to $e^{\mathbf{A}t}$ since $\mathbf{X}(0) = \mathbf{I}_3$. Thus the general solution of Eq. (11) is given by

$$\mathbf{x} = \mathbf{X}(t)\mathbf{c} = e^{\mathbf{A}t}\mathbf{c}.$$

EXAMPLE 4

Use the eigenvalue method to find a fundamental set of solutions of the fourth order equation

$$\frac{d^4y}{dt^4} + 2\frac{d^2y}{dt^2} + y = 0. \tag{12}$$

If we define $\mathbf{x} = (y, y', y'', y''')^T$, the dynamical system equivalent to Eq. (12) is

$$\mathbf{x}' = \begin{pmatrix} 0 & 1 & 0 & 0 \\ 0 & 0 & 1 & 0 \\ 0 & 0 & 0 & 1 \\ -1 & 0 & -2 & 0 \end{pmatrix} \mathbf{x} = \mathbf{Ax}. \tag{13}$$

Since the characteristic polynomial of $\mathbf{A}$ is $p(\lambda) = \lambda^4 + 2\lambda^2 + 1 = (\lambda - i)^2(\lambda + i)^2$, the only eigenvalues of $\mathbf{A}$ are $\lambda_1 = i$ and $\lambda_2 = -i$ with corresponding algebraic multiplicities $m_1 = 2$ and $m_2 = 2$. Using Gaussian elimination, we find two linearly independent solutions of

$$(\mathbf{A} - \lambda_1 \mathbf{I}_4)^2 \mathbf{v} = \begin{pmatrix} -1 & -2i & 1 & 0 \\ 0 & -1 & -2i & 1 \\ -1 & 0 & -3 & -2i \\ 2i & -1 & 4i & -3 \end{pmatrix} \mathbf{v} = \mathbf{0},$$

$$\mathbf{v}_1 = \begin{pmatrix} -3 \\ -2i \\ 1 \\ 0 \end{pmatrix} \quad \text{and} \quad \mathbf{v}_2 = \begin{pmatrix} -2i \\ 1 \\ 0 \\ 1 \end{pmatrix}.$$

Two complex-valued, linearly independent solutions of Eq. (13) are therefore

$$\hat{\mathbf{x}}_1(t) = e^{\lambda_1 t}[\mathbf{v}_1 + t(\mathbf{A} - \lambda_1 \mathbf{I}_4)\mathbf{v}_1] = e^{it}\left[\begin{pmatrix} -3 \\ -2i \\ 1 \\ 0 \end{pmatrix} + t \begin{pmatrix} i \\ -1 \\ -i \\ 1 \end{pmatrix} \right]$$

and

$$\hat{\mathbf{x}}_2(t) = e^{\lambda_1 t}[\mathbf{v}_2 + t(\mathbf{A} - \lambda_1 \mathbf{I}_4)\mathbf{v}_2] = e^{it}\left[\begin{pmatrix} -2i \\ 1 \\ 0 \\ 1 \end{pmatrix} + t \begin{pmatrix} -1 \\ -i \\ 1 \\ i \end{pmatrix} \right].$$

Using complex arithmetic and Euler's formula, we form a fundamental set with the following of four real-valued solutions of Eq. (13):

$$\mathbf{x}_1(t) = \text{Re}\,\hat{\mathbf{x}}_1(t) = \begin{pmatrix} -3\cos t - t\sin t \\ -t\cos t + 2\sin t \\ \cos t + t\sin t \\ t\cos t \end{pmatrix},$$

$$\mathbf{x}_2(t) = \text{Im } \hat{\mathbf{x}}_1(t) = \begin{pmatrix} t\cos t - 3\sin t \\ -2\cos t - t\sin t \\ -t\cos t + \sin t \\ t\sin t \end{pmatrix},$$

$$\mathbf{x}_3(t) = \text{Re } \hat{\mathbf{x}}_2(t) = \begin{pmatrix} -t\cos t + 2\sin t \\ \cos t + t\sin t \\ t\cos t \\ \cos t - t\sin t \end{pmatrix},$$

and

$$\mathbf{x}_4(t) = \text{Im } \hat{\mathbf{x}}_2(t) = \begin{pmatrix} -t\sin t - 2\cos t \\ \sin t - t\cos t \\ t\sin t \\ \sin t + t\cos t \end{pmatrix}.$$

The first components of $\mathbf{x}_1$, $\mathbf{x}_2$, $\mathbf{x}_3$, and $\mathbf{x}_4$ provide us with a fundamental set for Eq. (12), $y_1(t) = -3\cos t - t\sin t$, $y_2(t) = t\cos t - 3\sin t$, $y_3(t) = -t\cos t + 2\sin t$, and $y_4(t) = -t\sin t - 2\cos t$. We note that each of these solutions is a linear combination of a simpler fundamental set for Eq. (12) consisting of $\cos t$, $\sin t$, $t\cos t$, and $t\sin t$.

PROBLEMS

In each of Problems 1 through 8, find a fundamental matrix for the given system:

1. $\mathbf{x}' = \begin{pmatrix} 4 & -9 \\ 1 & -2 \end{pmatrix} \mathbf{x}$

2. $\mathbf{x}' = \begin{pmatrix} 3 & -9 \\ 1 & -3 \end{pmatrix} \mathbf{x}$

3. $\mathbf{x}' = \begin{pmatrix} 1 & 1 & 1 \\ 2 & 1 & -1 \\ -3 & 2 & 4 \end{pmatrix} \mathbf{x}$

4. $\mathbf{x}' = \begin{pmatrix} 5 & -3 & -2 \\ 8 & -5 & -4 \\ -4 & 3 & 3 \end{pmatrix} \mathbf{x}$

5. $\mathbf{x}' = \begin{pmatrix} -7 & 9 & -6 \\ -8 & 11 & -7 \\ -2 & 3 & -1 \end{pmatrix} \mathbf{x}$

6. $\mathbf{x}' = \begin{pmatrix} 5 & 6 & 2 \\ -2 & -2 & -1 \\ -2 & -3 & 0 \end{pmatrix} \mathbf{x}$

7. $\mathbf{x}' = \begin{pmatrix} -8 & -16 & -16 & -17 \\ -2 & -10 & -8 & -7 \\ -2 & 0 & -2 & -3 \\ 6 & 14 & 14 & 14 \end{pmatrix} \mathbf{x}$

8. $\mathbf{x}' = \begin{pmatrix} 1 & -1 & -2 & 3 \\ 2 & -3/2 & -1 & 7/2 \\ -1 & 1/2 & 0 & -3/2 \\ -2 & 3/2 & 3 & -7/2 \end{pmatrix} \mathbf{x}$

In each of Problems 9 and 10, find the solution of the given initial value problem. Draw the corresponding trajectory in $x_1 x_2$-space and also draw the graph of x_1 versus t.

9. $\mathbf{x}' = \begin{pmatrix} 1 & -4 \\ 4 & -7 \end{pmatrix} \mathbf{x}, \qquad \mathbf{x}(0) = \begin{pmatrix} 7 \\ 1 \end{pmatrix}$

10. $\mathbf{x}' = \begin{pmatrix} 3 & -4 \\ 1 & -1 \end{pmatrix} \mathbf{x}, \qquad \mathbf{x}(0) = \begin{pmatrix} -5 \\ 7 \end{pmatrix}$

In each of Problems 11 and 12, find the solution of the given initial value problem. Draw the corresponding trajectory in $x_1 x_2 x_3$-space and also draw the graph of x_1 versus t.

11. $\mathbf{x}' = \begin{pmatrix} 4 & 1 & 3 \\ 6 & 4 & 6 \\ -5 & -2 & -4 \end{pmatrix} \mathbf{x}, \qquad \mathbf{x}(0) = \begin{pmatrix} 1 \\ -2 \\ 5 \end{pmatrix}$
12. $\mathbf{x}' = \begin{pmatrix} 1 & 1 & 0 \\ -14 & -5 & 1 \\ 15 & 5 & -2 \end{pmatrix} \mathbf{x}, \qquad \mathbf{x}(0) = \begin{pmatrix} 5 \\ 5 \\ -4 \end{pmatrix}$

CHAPTER SUMMARY

Section 6.1 Many science and engineering problems are modeled by systems of differential equations of dimension $n > 2$: vibrating systems with two or more degrees of freedom; compartment models arising in biology, ecology, pharmacokinetics, transport theory, and chemical reactor systems; linear control systems; and electrical networks.

Section 6.2

▶ If $\mathbf{P}(t)$ and $\mathbf{g}(t)$ are continuous on I, a unique solution to the initial value problem

$$\mathbf{x}' = \mathbf{P}(t)\mathbf{x} + \mathbf{g}(t), \qquad \mathbf{x}(t_0) = \mathbf{x}_0, \quad t_0 \in I$$

exists throughout I.

▶ A set of n solutions $\mathbf{x}_1, \ldots, \mathbf{x}_n$ to the homogeneous equation

$$\mathbf{x}' = \mathbf{P}(t)\mathbf{x}, \quad \mathbf{P} \text{ continuous on } I$$

form a **fundamental set** on I if their **Wronskian** $W[\mathbf{x}_1, \ldots, \mathbf{x}_n](t)$ is nonzero for one (and hence all) $t \in I$. If $\mathbf{x}_1, \ldots, \mathbf{x}_n$ form a fundamental set of solutions to the homogeneous equation, then a **general solution** is

$$\mathbf{x} = c_1 \mathbf{x}_1(t) + \cdots + c_n \mathbf{x}_n(t),$$

where $c_1, \ldots, c_n$ are arbitrary constants. The n solutions are **linearly independent** on I if and only if their Wronskian is nonzero on I.

▶ The theory of nth order linear equations,

$$\frac{d^n y}{dt^n} + p_1(t) \frac{d^{n-1} y}{dt^{n-1}} + \cdots + p_{n-1}(t) \frac{dy}{dt} + p_n(t)y = g(t),$$

follows from the theory of first order linear systems of dimension n.

Section 6.3

▶ If $\mathbf{v}_1, \ldots, \mathbf{v}_n$ are linearly independent eigenvectors of the real, constant $n \times n$ matrix $\mathbf{A}$ and the corresponding eigenvalues $\lambda_1, \ldots, \lambda_n$ are real, then a real general solution of $\mathbf{x}' = \mathbf{A}\mathbf{x}$ is

$$\mathbf{x} = c_1 e^{\lambda_1 t} \mathbf{v}_1 + \cdots + c_n e^{\lambda_n t} \mathbf{v}_n,$$

where $c_1, \ldots, c_n$ are arbitrary real constants.

▶ Two classes of $n \times n$ matrices that have n linearly independent eigenvectors are (i) matrices with n distinct eigenvalues and (ii) **symmetric matrices**, that is, matrices satisfying $\mathbf{A}^T = \mathbf{A}$.

Section 6.4 If $\mathbf{A}$ is real, constant, and nondefective, each pair of complex conjugate eigenvalues $\mu \pm i\nu$ with corresponding eigenvectors $\mathbf{v} = \mathbf{a} \pm i\mathbf{b}$ yields two linearly independent, real vector solutions $\text{Re}\{\exp[(\mu + i\nu)t][\mathbf{a} + i\mathbf{b}]\} = \exp(\mu t)(\cos \nu t\mathbf{a} - \sin \nu t\mathbf{b})$ and $\text{Im}\{\exp[(\mu + i\nu)t][\mathbf{a} + i\mathbf{b}]\} = \exp(\mu t)(\sin \nu t\mathbf{a} + \cos \nu t\mathbf{b})$.

Section 6.5

▶ If $\mathbf{x}_1, \ldots, \mathbf{x}_n$ form a fundamental set for $\mathbf{x}' = \mathbf{P}(t)\mathbf{x}$, the $n \times n$ matrix $\mathbf{X}(t) = [\mathbf{x}_1, \ldots, \mathbf{x}_n]$ is called a **fundamental matrix** and satisfies $\mathbf{X}' = \mathbf{P}(t)\mathbf{X}$. In addition, the fundamental matrix $\boldsymbol{\Phi}(t) = \mathbf{X}(t)\mathbf{X}^{-1}(t_0)$ satisfies $\boldsymbol{\Phi}(t_0) = \mathbf{I}_n$.

▶ If $\mathbf{A}$ is a real, constant $n \times n$ matrix, the matrix exponential function

$$e^{\mathbf{A}t} = \mathbf{I}_n + \mathbf{A}t + \frac{1}{2!}\mathbf{A}^2 t^2 + \cdots + \frac{1}{n!}\mathbf{A}^n t^n + \cdots$$

$$= \mathbf{I}_n + \sum_{k=1}^{\infty} \mathbf{A}^k \frac{t^k}{k!}$$

is a fundamental matrix satisfying $e^{\mathbf{A}0} = \mathbf{I}_n$ and $(e^{\mathbf{A}t})^{-1} = e^{-\mathbf{A}t}$.

▶ Methods for computing $e^{\mathbf{A}t}$:

   ▶ If $\mathbf{X}(t)$ is any fundamental matrix for $\mathbf{x}' = \mathbf{A}\mathbf{x}$, $e^{\mathbf{A}t} = \mathbf{X}(t)\mathbf{X}^{-1}(0)$.
   ▶ If $\mathbf{A}$ is **diagonalizable**, that is, $\mathbf{T}^{-1}\mathbf{A}\mathbf{T} = \mathbf{D}$, then $e^{\mathbf{A}t} = \mathbf{T}e^{\mathbf{D}t}\mathbf{T}^{-1}$.
   ▶ Use Laplace transforms to find the solution $\boldsymbol{\Phi}(t)$ to $\mathbf{X}' = \mathbf{A}\mathbf{X}$, $\mathbf{X}(0) = \mathbf{I}_n$. Then $e^{\mathbf{A}t} = \boldsymbol{\Phi}(t)$.

Section 6.6
If $\mathbf{X}(t)$ is a fundamental matrix for the homogeneous system, the general solution of $\mathbf{x}' = \mathbf{P}(t)\mathbf{x} + \mathbf{g}(t)$, $\mathbf{x}(t_0) = \mathbf{x}_0$ is $\mathbf{x} = \mathbf{X}(t)\mathbf{X}^{-1}(t_0)\mathbf{c} + \mathbf{x}_p(t)$, where a particular solution $\mathbf{x}_p(t)$ is given by the **variation of parameters formula**

$$\mathbf{x}_p(t) = \mathbf{X}(t) \int \mathbf{X}^{-1}(t)\mathbf{g}(t)\, dt.$$

Section 6.7

▶ If $\mathbf{A}$ is a real, constant $n \times n$ matrix and λ is a real eigenvalue of $\mathbf{A}$ with algebraic multiplicity m, then m linearly independent solutions of $\mathbf{x}' = \mathbf{A}\mathbf{x}$ associated with λ are

$$\mathbf{x}_k = e^{\lambda t}\left[\mathbf{v}_k + \frac{t}{1!}(\mathbf{A} - \lambda\mathbf{I}_n)\mathbf{v}_k + \cdots + \frac{t^{m-1}}{(m-1)!}(\mathbf{A} - \lambda\mathbf{I}_n)^{m-1}\mathbf{v}_k\right],$$

$k = 1, \ldots, m$, where $\mathbf{v}_1, \ldots, \mathbf{v}_m$ are linearly independent solutions of $(\mathbf{A} - \lambda\mathbf{I}_n)^m\mathbf{v} = \mathbf{0}$. If λ is complex, $\lambda = \mu + i\nu$, $\nu \neq 0$, then $2m$ linearly independent, real-valued solutions of $\mathbf{x}' = \mathbf{A}\mathbf{x}$ associated with λ are $\{\text{Re } \mathbf{x}_1, \ldots, \text{Re } \mathbf{x}_m\}$ and $\{\text{Im } \mathbf{x}_1, \ldots, \text{Im } \mathbf{x}_m\}$.

▶ The union of the sets of solutions generated by this method form a fundamental set for $\mathbf{x}' = \mathbf{A}\mathbf{x}$.

PROJECTS

Project 1 Earthquakes and Tall Buildings

Simplistic differential equation models are often used to introduce concepts and principles that are important for understanding the dynamic behavior of complex physical systems. In this project, we employ such a model to study the response of a tall building due to horizontal seismic motion at the foundation generated by an earthquake.

Figure 6.P.1 is an illustration of a building idealized as a collection of n floors, each of mass m, connected together by vertical walls. If we neglect gravitation and restrict motion to the horizontal direction, the displacements of the floors, relative to a fixed frame of reference, are denoted by $x_1, x_2, \ldots, x_n$. At equilibrium, all of the displacements and velocities are zero and the floors are in perfect vertical alignment.

© InterNetwork Media / Photodisc / Getty Images

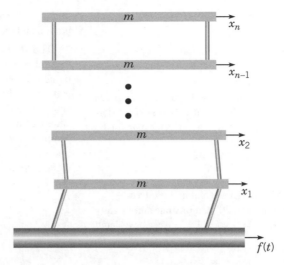

FIGURE 6.P.1 A building consisting of floors of mass m connected by stiff but flexible vertical walls.

When adjacent floors are not in alignment, we assume that the walls exert a flexural restoring force proportional to the difference in displacements between the floors with proportionality constant k. Thus the equation of motion for the jth floor is

$$mx_j'' = -k(x_j - x_{j-1}) - k(x_j - x_{j+1}), \qquad j = 2, \ldots, n-1, \tag{1}$$

whereas the equations for the first floor and nth, or top, floor are

$$mx_1'' = -k(x_1 - f(t)) - k(x_1 - x_2), \tag{2}$$

and

$$mx_n'' = -k(x_n - x_{n-1}), \tag{3}$$

respectively. The horizontal motion of the foundation generated by the earthquake is described by the input function $f(t)$.

Project 1 PROBLEMS

1. **The Undamped Building.**

(a) Show that Eqs. (1) through (3) can be expressed in matrix notation as

$$\mathbf{x}'' + \omega_0^2 \mathbf{K} \mathbf{x} = \omega_0^2 f(t) \mathbf{z}, \tag{i}$$

where

$$\omega_0^2 = \frac{k}{m}, \qquad \mathbf{x} = (x_1, x_2, \ldots, x_n)^T,$$

$$\mathbf{z} = (1, 0, \ldots, 0)^T,$$

and

$$\mathbf{K} = \begin{pmatrix} 2 & -1 & 0 & 0 & \cdots & 0 \\ -1 & 2 & -1 & 0 & \cdots & 0 \\ 0 & -1 & 2 & -1 & \cdots & 0 \\ \vdots & & \ddots & \ddots & \ddots & \vdots \\ 0 & \cdots & & -1 & 2 & -1 \\ 0 & \cdots & & 0 & -1 & 1 \end{pmatrix}. \tag{ii}$$

(b) A real $n \times n$ matrix $\mathbf{A}$ is said to be **positive definite** if $\mathbf{x}^T \mathbf{A} \mathbf{x} > 0$ for every real n-vector $\mathbf{x} \neq \mathbf{0}$. Show that $\mathbf{K}$

in (ii) satisfies

$$\mathbf{x}^T\mathbf{K}\mathbf{x} = x_1^2 + \sum_{j=1}^{n-1}(x_j - x_{j+1})^2,$$

and is therefore positive definite.

(c) Eigenvalues and eigenvectors of real symmetric matrices are real (see Appendix A.4). Show that if $\mathbf{K}$ is positive definite and λ and $\mathbf{u}$ are an eigenvalue–eigenvector pair for $\mathbf{K}$, then

$$\lambda = \frac{\mathbf{u}^T\mathbf{K}\mathbf{u}}{\mathbf{u}^T\mathbf{u}} > 0.$$

Thus all eigenvalues of $\mathbf{K}$ in (ii) are real and positive.

(d) For the cases $n = 5, 10$, and 20, demonstrate numerically that the eigenvalues of $\mathbf{K}$, $\lambda_j = \omega_j^2, j = 1, \ldots, n$ can be ordered as follows:

$$0 < \omega_1^2 < \omega_2^2 < \cdots < \omega_n^2.$$

(e) Since K is real and symmetric, it possesses a set of n orthogonal eigenvectors, $\{\mathbf{u}_1, \mathbf{u}_2, \ldots, \mathbf{u}_n\}$, that is, $\mathbf{u}_i^T\mathbf{u}_j = 0$ if $i \neq j$ (see Appendix A.4). These eigenvectors can be used to construct a **normal mode representation**,

$$\mathbf{x} = a_1(t)\mathbf{u}_1 + \cdots + a_n(t)\mathbf{u}_n, \qquad \text{(iii)}$$

of the solution of

$$\mathbf{x}'' + \omega_0^2\mathbf{K}\mathbf{x} = \omega_0^2 f(t)\mathbf{z},$$
$$\mathbf{x}(0) = \mathbf{x}_0, \qquad \text{(iv)}$$
$$\mathbf{x}'(0) = \mathbf{v}_0.$$

Substitute the representation (iii) into the differential equation and initial conditions in Eqs. (iv) and use the fact that $\mathbf{K}\mathbf{u}_j = \omega_j^2\mathbf{u}_j, j = 1, \ldots, n$ and the orthogonality of $\mathbf{u}_1, \mathbf{u}_2, \ldots, \mathbf{u}_n$ to show that for each $i = 1, \ldots, n$, the **mode amplitude** $a_i(t)$ satisfies the initial value problem

$$a_i'' + \omega_i^2\omega_0^2 a_i = \omega_0^2 f(t)z_i, \qquad a_i(0) = \alpha_i, \quad a_i'(0) = \beta_i,$$

where

$$z_i = \frac{\mathbf{u}_i^T\mathbf{z}}{\mathbf{u}_i^T\mathbf{u}_i}, \qquad \alpha_i = \frac{\mathbf{u}_i^T\mathbf{x}_0}{\mathbf{u}_i^T\mathbf{u}_i}, \qquad \beta_i = \frac{\mathbf{u}_i^T\mathbf{v}_0}{\mathbf{u}_i^T\mathbf{u}_i}.$$

(f) An unforced pure mode of vibration, say, the jth mode, can be realized by solving the initial value problem (iv) subject to the initial conditions $\mathbf{x}(0) = A_j\mathbf{u}_j$, where A_j is a mode amplitude factor, and $\mathbf{x}'(0) = \mathbf{0}$ with zero input, $f(t) = 0$. Show that, in this case, the normal mode solution of the initial value problem (iv) consists of a single term,

$$\mathbf{x}^{(j)}(t) = A_j \cos(\omega_0\omega_j t)\,\mathbf{u}_j.$$

Thus the natural frequency of the jth mode of vibration is $\omega_0\omega_j$ and the corresponding period is $2\pi/(\omega_0\omega_j)$. Assuming that $A_1 = 1$, $\omega_0 = 41$, and $n = 20$, plot a graph of the components (floor displacement versus floor number) of the first mode $\mathbf{x}^{(1)}(t)$ for several values of t over an entire cycle. Then generate analogous graphs for the second and third pure

modes of vibration. Describe and compare the modes of vibration and their relative frequencies.

2. The Building with Damping Devices. In addition to flexural restoring forces, all buildings possess intrinsic internal friction, due to structural and nonstructural elements, that causes the amplitude of vibrations to decay as these elements absorb vibrational energy. One of several techniques employed in modern earthquake-resistant design uses added damping devices such as shock absorbers between adjacent floors to artificially increase the internal damping of a building and improve its resistance to earthquake-induced motion (see Figure 6.P.2).

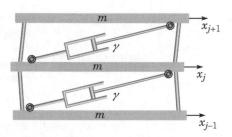

FIGURE 6.P.2 Shock absorbers are used to dampen the amplitude of relative motion between adjacent floors.

The effects of intrinsic and artificial damping are accounted for by including damping forces of the form

$$F_d^{(j)} = -\gamma(x_j' - x_{j-1}') - \gamma(x_j' - x_{j+1}'), \qquad j = 2, \ldots, n-1,$$

$$F_d^{(1)} = -\gamma(x_1' - f') - \gamma(x_1' - x_2'),$$

and

$$F_d^{(n)} = -\gamma(x_n' - x_{n-1}')$$

on the right-hand sides of Eqs. (1), (2), and (3), respectively.
(a) Show that including damping devices between each pair of adjacent floors changes the system (i) in Problem 1 to

$$\mathbf{x}'' + 2\delta\mathbf{K}\mathbf{x}' + \omega_0^2\mathbf{K}\mathbf{x} = \left[\omega_0^2 f(t) + 2\delta f'(t)\right]\mathbf{z}, \qquad \text{(i)}$$

where $\delta = \dfrac{\gamma}{2m}$.

(b) Assume an input to Eq. (i) in part (a) of the form $f(t) = e^{i\omega t}$ and show that the n-floor frequency response vector $\mathbf{G}(i\omega)$ (see Sections 4.6 and 6.6) satisfies the equation

$$\left[(\omega_0^2 + i2\delta\omega)\mathbf{K} - \omega^2\mathbf{I}_n\right]\mathbf{G}(i\omega) = (\omega_0^2 + i2\delta\omega)\mathbf{z}. \qquad \text{(ii)}$$

Thus the gain function for the frequency response of the jth floor is the absolute value of the jth component of $\mathbf{G}(i\omega)$,

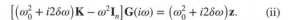

$|G_j(i\omega)|$. On the same set of coordinate axes, plot the graphs of $|G_j(i\omega)|$ versus $\omega \in [0, 6\pi]$, $j = 5, 10, 20$ using the parameter values $n = 20$, $\omega_0 = 41$, and $\delta = 10$. Repeat using $\delta = 50$ and interpret the results of your graphs.

(c) Numerically approximate the solution of Eq. (i) in part (a) subject to the initial conditions $\mathbf{x}(0) = \mathbf{0}$, $\mathbf{x}'(0) = \mathbf{0}$ using the parameter values $n = 20$, $\omega_0 = 41$, and $\delta = 10$. For the input function, use $f(t) = A \sin \omega_r t$, where $A = 2$ in and ω_r is the lowest resonant frequency of the system, that is, the value of ω at which $|G_{20}(i\omega)|$ in part (b) attains its maximum value. Plot the graphs of $x_j(t)$ versus t, $j = 5, 10, 20$ on the same set of coordinate axes. What is the amplitude of oscillatory displacement of each of these floors once the steady-state re-

sponse is attained? Repeat the simulation using $\delta = 50$ and compare your results with the case $\delta = 10$.

3. A majority of the buildings that collapsed during the Mexico City earthquake of September 19, 1985, were around 20 stories tall. The predominant period of the earthquake ground motion, around 2.0 s, was close to the natural period of vibration of these buildings. Other buildings, of different heights and with different vibrational characteristics, were often found undamaged even though they were located right next to the damaged 20-story buildings. Explain whether the model described in this project provides a possible explanation for these observations. Your argument should be supported by computer simulations and graphs.

Project 2 Controlling a Spring-Mass System to Equilibrium

Consider a system of three masses coupled together by springs or elastic bands, as depicted in Figure 6.P.3. We will assume that the magnitudes of the three masses are all equal to m and the spring constants are all equal to k. If $x_j(t)$, $j = 1, 2, 3$, represent the displacements of the masses from their equilibrium positions at time t, then the differential equations describing the motion of the masses are

$$m\frac{d^2x_1}{dt^2} = -2kx_1 + kx_2 + \hat{u}_1(t),$$

$$m\frac{d^2x_2}{dt^2} = kx_1 - 2kx_2 + kx_3 + \hat{u}_2(t), \qquad (1)$$

$$m\frac{d^2x_3}{dt^2} = kx_2 - 2kx_3 + \hat{u}_3(t).$$

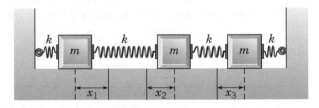

FIGURE 6.P.3 A three-mass, four-spring system.

The functions $\hat{u}_j(t)$, $j = 1, 2, 3$ represent externally applied forces acting on each of the masses. Dividing both sides of the set of equations (1) by m and using matrix notation, the system can be expressed as

$$\frac{d^2\mathbf{x}}{dt^2} = \mathbf{G}\mathbf{x} + \mathbf{u}(t), \qquad (2)$$

where

$$\mathbf{G} = \omega_0^2 \begin{pmatrix} -2 & 1 & 0 \\ 1 & -2 & 1 \\ 0 & 1 & -2 \end{pmatrix}, \qquad \mathbf{u}(t) = \begin{pmatrix} u_1(t) \\ u_2(t) \\ u_3(t) \end{pmatrix} = \frac{1}{m} \begin{pmatrix} \hat{u}_1(t) \\ \hat{u}_2(t) \\ \hat{u}_3(t) \end{pmatrix}, \qquad (3)$$

and $\omega_0^2 = k/m$. In addition to Eq. (2), we need to specify initial positions for the masses,

$$\mathbf{x}(0) = \mathbf{x}_0, \tag{4}$$

and initial velocities,

$$\frac{d\mathbf{x}}{dt}(0) = \mathbf{v}_0. \tag{5}$$

Given the system (2) and initial conditions (4) and (5), the control problem of interest is to choose the external forcing function $\mathbf{u}(t)$ in such a way as to steer or drive the system to the equilibrium state at some prescribed time $T > 0$. We are particularly interested in whether this can be done by applying external forces to only one of the masses. For example, we could require $u_2(t) \equiv 0$ and $u_3(t) \equiv 0$ and try to control the system to equilibrium by selecting an appropriate control function $u_1(t), 0 \leq t \leq T$.

Normal Mode Solutions.

Prior to addressing the control problem, we will discuss normal mode solutions of the initial value problem (2), (4), and (5). Normal mode analysis is a method for solving equations where an orthogonal basis is generated by a matrix or operator that appears in the equation. Let us consider Eqs. (2), (4), and (5) in the n-dimensional case, that is, $\mathbf{x}(t)$, $\mathbf{u}(t)$, $\mathbf{x}_0$, and $\mathbf{v}_0$ are n-vectors, while $\mathbf{G}$ is an $n \times n$ symmetric matrix. Because $\mathbf{G}$ is symmetric, it has n real eigenvalues, $\lambda_1, \ldots, \lambda_n$, with a corresponding set of orthogonal eigenvectors $S = \{\mathbf{v}_1, \ldots, \mathbf{v}_n\}$. Thus S provides us with an orthogonal basis for $\mathbf{R}^n$. Since $\mathbf{x}(t) \in \mathbf{R}^n$, it is natural to assume the following normal mode representation for $\mathbf{x}(t)$:

$$\mathbf{x}(t) = a_1(t)\mathbf{v}_1 + \cdots + a_n(t)\mathbf{v}_n. \tag{6}$$

Each vector $\mathbf{v}_j$ in the sum (6) is called a normal mode, whereas the coefficient $a_j(t)$ is the corresponding mode amplitude function. Equation (6) expresses the solution of the initial value problem (2), (4), and (5) at any time t as a superposition of the normal modes. In order to determine how the amplitude functions vary with time, we substitute the representation (6) into each of the equations (2), (4), and (5), and then use the orthogonality property of the normal modes to deduce differential equations and initial conditions for the amplitude functions. For example, substituting the sum (6) into Eq. (2) yields the equation

$$\sum_{j=1}^{n} \frac{d^2 a_j}{dt^2} \mathbf{v}_j = \sum_{j=1}^{n} a_j \mathbf{G} \mathbf{v}_j + \mathbf{u}(t). \tag{7}$$

Using the fact that $\mathbf{G}\mathbf{v}_j = \lambda_j \mathbf{v}_j$, $j = 1, \ldots, n$ permits Eq. (7) to be written as

$$\sum_{j=1}^{n} \left[\frac{d^2 a_j}{dt^2} - \lambda_j a_j \right] \mathbf{v}_j = \mathbf{u}(t). \tag{8}$$

Equation (8) may be viewed as an expansion of $\mathbf{u}(t)$ in terms of the orthogonal system S. Thus the coefficients $\dfrac{d^2 a_j}{dt^2} - \lambda_j a_j$ in the sum on the left-hand side of Eq. (8) must satisfy

$$\frac{d^2 a_i}{dt^2} - \lambda_i a_i = \frac{\langle \mathbf{v}_i, \mathbf{u}(t) \rangle}{\|\mathbf{v}_i\|^2}, \qquad i = 1, \ldots, n, \tag{9}$$

where the inner product $\langle , \rangle$ and norm $\| \cdot \|$ in Eqs. (9) refer to the Euclidean inner product and norm on $\mathbf{R}^n$, respectively. Equations (9) are easily derived by taking the inner product of both sides of Eq. (8) with $\mathbf{v}_i$ and using the fact that

$$\langle \mathbf{v}_i, \mathbf{v}_j \rangle = \begin{cases} \|\mathbf{v}_i\|^2, & j = i \\ 0, & j \neq i. \end{cases} \tag{10}$$

In a similar manner, substituting the sum (6) into Eqs. (4) and (5) yields initial conditions for each of the differential equations (9),

$$a_i(0) = \frac{\langle \mathbf{v}_i, \mathbf{x}_0 \rangle}{||\mathbf{v}_i||^2}, \qquad i = 1, \ldots, n, \tag{11}$$

and

$$\frac{da_i}{dt}(0) = \frac{\langle \mathbf{v}_i, \mathbf{v}_0 \rangle}{||\mathbf{v}_i||^2}, \qquad i = 1, \ldots, n. \tag{12}$$

Thus the mode amplitudes can be obtained by solving the initial value problems defined by Eqs. (9), (11), and (12). For our particular problem of interest, it can be shown that the eigenvalues of $\mathbf{G}$ are strictly negative (Problem 8), a fact that can be made explicit by setting $\lambda_j = -\omega_j^2, j = 1, \ldots, n$. Thus Eqs. (9) may be written as

$$\frac{d^2 a_i}{dt^2} + \omega_i^2 a_i = \frac{\langle \mathbf{v}_i, \mathbf{u}(t) \rangle}{||\mathbf{v}_i||^2}, \qquad i = 1, \ldots, n. \tag{13}$$

It is of particular interest to observe solutions of the initial value problems (13), (11), and (12) for the special case where the initial position is taken to be the jth eigenvector, $\mathbf{x}_0 = \mathbf{v}_j$, the initial velocity is taken to be zero, $\mathbf{v}_0 = \mathbf{0}$, and the forcing function is set to zero, $\mathbf{u}(t) \equiv \mathbf{0}$. Then the solutions of the initial value problems (13), (11), and (12) are given by

$$a_i(t) = \begin{cases} 0, & i \neq j \\ \cos \omega_j t, & i = j \end{cases}, \tag{14}$$

and therefore the motion consists of a single pure mode with time-varying amplitude of angular frequency ω_j,

$$\mathbf{x}(t) = \cos \omega_j t \, \mathbf{v}_j. \tag{15}$$

This observation has the following important implication for the control problem. If the jth mode has a zero component (which we may call a nodal point), for example, the kth component is zero, then a single control function applied to the kth mass will not be able to force vibrations of the jth mode to the equilibrium state. This can be shown mathematically by setting $\mathbf{u}(t) = u_k(t)\mathbf{e}_k$ and noting that $\langle \mathbf{v}_j, \mathbf{u}(t) \rangle = u_k(t)\langle \mathbf{v}_j, \mathbf{e}_k \rangle = 0$. Here, $\mathbf{e}_k$ is the n-vector with 1 in the kth entry and zeros elsewhere. Consequently, $\mathbf{u}(t) = u_k(t)\mathbf{e}_k$ cannot influence the amplitude of the jth mode.

Linear Constant Coefficient Control Problems.

We wish to recast the second order system (2) as a system of first order differential equations in order to take advantage of existing classical theory of linear control. This can be accomplished in many ways. We choose to define

$$\mathbf{y} = \begin{pmatrix} \mathbf{x} \\ \mathbf{x}' \end{pmatrix}.$$

Thus the first three components of $\mathbf{y}$ refer to the positions of the three masses, whereas the last three components refer to the velocities of the masses. Then the system (2) can be written as

$$\frac{d\mathbf{y}}{dt} = \mathbf{Ay} + \mathbf{Bu}(t), \tag{16}$$

where

$$\mathbf{A} = \begin{pmatrix} \mathbf{0} & \mathbf{I}_3 \\ \mathbf{G} & \mathbf{0} \end{pmatrix}.$$

The input matrix $\mathbf{B}$ depends on the number of controls and to what components of the system they will be applied. In the case of a scalar control $u(t)$ applied to the first mass, system (16) will be

$$\frac{d\mathbf{y}}{dt} = \mathbf{Ay} + \begin{pmatrix} 0 \\ 0 \\ 0 \\ 1 \\ 0 \\ 0 \end{pmatrix} u(t), \tag{17}$$

that is,

$$\mathbf{B} = \begin{pmatrix} 0 \\ 0 \\ 0 \\ 1 \\ 0 \\ 0 \end{pmatrix}.$$

In the case of a scalar control applied to the second mass,

$$\mathbf{B} = \begin{pmatrix} 0 \\ 0 \\ 0 \\ 0 \\ 1 \\ 0 \end{pmatrix}.$$

The initial condition for Eq. (16) obtained from Eqs. (4) and (5) is

$$\mathbf{y}(0) = \mathbf{y}_0 = \begin{pmatrix} \mathbf{x}_0 \\ \mathbf{v}_0 \end{pmatrix}. \tag{18}$$

More generally, a linear constant coefficient control system is expressed as

$$\frac{d\mathbf{y}}{dt} = \mathbf{Ay} + \mathbf{Bu}(t), \tag{19}$$

where $\mathbf{y}$ is $n \times 1$, $\mathbf{A}$ is $n \times n$, $\mathbf{B}$ is $n \times m$, and the control vector $\mathbf{u}(t)$ is $m \times 1$.

**DEFINITION
6.P.1**

The system (19) is said to be **completely controllable** at time $T > 0$ if there exists a control $\mathbf{u}(t)$ such that any arbitrary state $\mathbf{y}(0) = \mathbf{y}_0$ can be driven to the zero state $\mathbf{y}(T) = 0$.

In order to arrive at a test for controllability, we consider the following representation for the solution of the initial value problem (16) and (18):

$$\mathbf{y} = e^{\mathbf{A}t}\mathbf{y}_0 + \int_0^t e^{\mathbf{A}(t-s)}\mathbf{B}\mathbf{u}(s)\,ds. \tag{20}$$

Setting $\mathbf{y}(T) = \mathbf{0}$ yields an integral equation for $\mathbf{u}(t)$,

$$\int_0^T e^{\mathbf{A}(T-s)}\mathbf{B}\mathbf{u}(s)\,ds = -e^{\mathbf{A}T}\mathbf{y}_0. \tag{21}$$

The Cayley–Hamilton theorem from linear algebra states that if

$$p(\lambda) = \det(\lambda\mathbf{I}_n - \mathbf{A}) = \lambda^n + p_{n-1}\lambda^{n-1} + \cdots + p_1\lambda + p_0 \tag{22}$$

is the characteristic polynomial of $\mathbf{A}$, then

$$p(\mathbf{A}) = \mathbf{A}^n + p_{n-1}\mathbf{A}^{n-1} + \cdots + p_1\mathbf{A} + p_0\mathbf{I}_n = \mathbf{0}. \tag{23}$$

Remark. The characteristic polynomial of $\mathbf{A}$ defined in Appendix A.4, $\det(\mathbf{A} - \lambda\mathbf{I}_n)$, differs from the definition given in Eq. (22) by the largely irrelevant factor $(-1)^n$ since $\det(\mathbf{A} - \lambda\mathbf{I}_n) = (-1)^n \det(\lambda\mathbf{I}_n - \mathbf{A})$.

If the Cayley–Hamilton theorem is applied to the matrix exponential function $\exp[\mathbf{A}(T - s)]$, then the left-hand side of Eq. (21) may be expressed in the form

$$\mathbf{B}\int_0^T q_0(T-s)\mathbf{u}(s)\,ds + \mathbf{A}\mathbf{B}\int_0^T q_1(T-s)\mathbf{u}(s)\,ds + \cdots +$$

$$\mathbf{A}^{n-1}\mathbf{B}\int_0^T q_{n-1}(T-s)\mathbf{u}(s)\,ds = -e^{\mathbf{A}T}\mathbf{y}_0. \tag{24}$$

If the system (19) is to be controllable at time T, according to Eq. (24), we must be able to express any vector in $\mathbf{R}^n$ [the right-hand side of Eq. (24)] as a linear combination of the columns of the matrix $[\mathbf{B}, \mathbf{A}\mathbf{B}, \ldots, \mathbf{A}^{n-1}\mathbf{B}]$. This leads to the *rank test for controllability*,

THEOREM 6.P.2	The system (19) is *completely controllable at time $T > 0$ if and only if the matrix* $$\mathbf{\Psi} = [\mathbf{B}, \mathbf{A}\mathbf{B}, \ldots, \mathbf{A}^{n-1}\mathbf{B}] \tag{25}$$ *has rank n.*

The matrix $\mathbf{\Psi}$ is called the *controllability matrix* for the system (19).

Construction of the Control Function. Given that the system (19) is completely controllable, we now address the problem of determining a control that will do the job. We will restrict attention to the case where $\mathbf{B}$ is $n \times 1$ and $\mathbf{u}(t)$ is a scalar control, but the method easily extends to the case where $\mathbf{B}$ is $n \times m$ and $\mathbf{u}$ is $m \times 1$. We choose a set of functions $E = \{\phi_1(t), \phi_2(t), \ldots\}$ that are linearly independent over the time interval $[0, T]$. For example, we might choose $\phi_j(t) = \sin j\pi t/T, j = 1, 2, 3, \ldots$. We will represent the scalar control function $u(t) = u_N(t)$ as a superposition of a finite number of functions from E,

$$u_N(t) = \sum_{j=1}^N c_j\phi_j(t), \tag{26}$$

where the coefficients c_j are to be determined. The general idea is to permit construction of a sufficiently diverse class of inputs so at least one will do the job of controlling the system to equilibrium. In order to generate the vector on the right-hand side of Eq. (21), we numerically solve the initial value problem

$$\frac{d\mathbf{y}}{dt} = \mathbf{A}\mathbf{y}, \qquad \mathbf{y}(0) = \mathbf{y}_0 \tag{27}$$

over the time interval $[0, T]$. The right-hand side of Eq. (21) is the solution of the initial value problem (27) evaluated at time T and multiplied by -1,

$$-\exp{[\mathbf{A}T]}\mathbf{y}_0 = -\mathbf{y}(T).$$

Next, solve each of the N initial value problems

$$\frac{d\mathbf{y}}{dt} = \mathbf{A}\mathbf{y} + \mathbf{B}\phi_j(t), \qquad \mathbf{y}(0) = \mathbf{0}, \quad j = 1, \dots, N \tag{28}$$

over the time interval $[0, T]$. The solutions of the initial value problems (28) will be denoted by $\mathbf{z}_j(t)$. By the *principle of superposition* of solutions for linear nonhomogeneous differential equations, the solution $\mathbf{y}_N(t)$ of

$$\frac{d\mathbf{y}}{dt} = \mathbf{A}\mathbf{y} + \mathbf{B}u_N(t), \qquad \mathbf{y}(0) = \mathbf{0} \tag{29}$$

is simply

$$\mathbf{y}_N(t) = \int_0^t e^{\mathbf{A}(t-s)}\mathbf{B}u_N(s)\,ds = \sum_{j=1}^{N} c_j\mathbf{z}_j(t). \tag{30}$$

Setting $\mathbf{y}_N(T) = -\mathbf{y}(T)$ yields a linear algebraic system to be solved for $\mathbf{c} = (c_1, \dots, c_N)^T$. If $N = 6$, the system to be solved is

$$\overbrace{[\mathbf{z}_1(T), \dots, \mathbf{z}_6(T)]}^{\mathbf{Z}(T)} \overbrace{\begin{bmatrix} c_1 \\ c_2 \\ c_3 \\ c_4 \\ c_5 \\ c_6 \end{bmatrix}}^{\mathbf{c}} = -\mathbf{y}(T),$$

or, using matrix notation,

$$\mathbf{Z}(T)\mathbf{c} = -\mathbf{y}(T). \tag{31}$$

How large should N be? Since the right-hand side of Eq. (21) may be any vector in $\mathbf{R}^n$, the matrix $\mathbf{Z} = [\mathbf{z}_1(T), \mathbf{z}_2(T), \dots, \mathbf{z}_N(T)]$ must be of full rank, that is, rank $(\mathbf{Z}) = n$. Therefore we conclude that the number of inputs N must be greater than or equal to n. It is conceivable that even though the inputs $\phi_j(t)$ are linearly independent functions, the outputs $\mathbf{z}_j(T)$, $j = 1, \dots, N = n$, due to errors in numerical approximation or due to the effects of the system on the inputs, may not be sufficiently diverse so that rank $(\mathbf{Z}) < n$. Thus we take a commonsense approach. Construct $\mathbf{Z}$ with $N = n$ columns and then check to see if rank $(\mathbf{Z}) = n$. If so, then solve Eq. (31). If not, include more functions in the sum (26) (i.e., increase the value of N) until rank $(\mathbf{Z}) = n$. If $N > n$, then Eq. (31) may have many solutions, any one of which may be used to realize $u_N(t)$ in Eq. (26). The apparent freedom in the choice of the control function is exploited in the field of optimal control theory.

Project 2 PROBLEMS

1. Derive the system of equations (1) by applying Newton's second law, $ma = F$, to each of the masses. Assume that the springs follow Hooke's law: *the force exerted by a spring on a mass is proportional to the length of its departure from its equilibrium length.*

2. Find the eigenvalues and eigenvectors of the matrix **G** in (3). Assume that $\omega_0^2 = 1$. Plot the components of the eigenvectors. Verbally, and with the aid of sketches, describe the normal modes of vibration in terms of the motion of the masses relative to one another.

3. From the normal mode representation of the solution of the initial value problem (2), (4), and (5), explain why the system cannot be completely controllable by applying a control function only to the center mass.

4. Repeat Problem 2 for a system of four masses connected by springs. Give a physical interpretation of the normal modes of vibration. Does the normal mode representation rule out complete controllability for the case of a scalar control applied to any single mass? Explain.

5. Find the rank of the controllability matrix for the three-mass system (i) in the case that a single control is applied

to the first mass and (ii) in the case that a single control is applied to the second mass.

6. Find the rank of the controllability matrix for the four-mass system in all of the cases that a single control is applied to each of the four masses and compare your results with the conclusions obtained in Problem 4. (Note that by symmetry considerations, it is only necessary to consider one of the masses on the end and one of the interior masses.)

7. Prove the Cayley–Hamilton theorem for the special case that **A** is diagonalizable.

8. A symmetric matrix G is said to be *negative definite* if $\mathbf{x}^T \mathbf{G} \mathbf{x} < 0$ for every $\mathbf{x} \neq 0$. Prove that a symmetric matrix **G** is negative definite if and only if its eigenvalues are all negative. Use this result to show that **G** in (3) has negative eigenvalues.

9. For the three-mass system, find a scalar control $u(t)$ that, when applied to the first mass, will drive the system (16) from the initial state $\mathbf{y}(0) = [-1, 1, 2, 0, 0, 0]^T$ to equilibrium at time $T = 10$. Plot the graphs of $u(t)$ and $y_j(t)$, $j = 1, \ldots, 6$ to confirm that the system is actually driven to equilibrium.

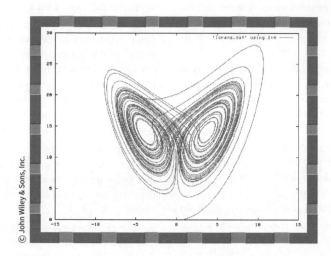

CHAPTER SEVEN

© John Wiley & Sons, Inc.

Nonlinear Differential Equations and Stability

I n this chapter we take up the investigation of nonlinear systems of differential equations. Nonlinear systems arise in many real-life situations. Examples encountered in this chapter include models for an oscillating pendulums, competing species, and predator–prey interactions. Such systems can be solved by analytical methods only in rare instances. Numerical approximation methods provide one means of dealing with nonlinear systems; these will be discussed in Chapter 8. Another approach, presented in this chapter, is geometrical in character and leads to a qualitative understanding of the behavior of solutions rather than to detailed quantitative information. A combination of methods is often needed to achieve optimal results.

7.1 Autonomous Systems and Stability

We first introduced two-dimensional systems of the form

$$\frac{dx}{dt} = F(x, y), \qquad \frac{dy}{dt} = G(x, y) \tag{1}$$

in Section 3.6. Recall that the system (1) is called **autonomous** because the functions F and G do not depend on the independent variable t. In Chapter 3 we were mainly concerned with showing how to find the solutions of homogeneous linear systems, and we presented only a few examples of nonlinear systems. Now we want to focus on the analysis of two-dimensional nonlinear systems of the form (1). Unfortunately, it is only in exceptional cases that solutions can be found by analytical methods. One alternative is to use numerical methods to approximate solutions. Software packages often include one or more algorithms, such as the Runge–Kutta method discussed in Section 8.3, for this purpose. Detailed quantitative information about solutions usually requires the use of such methods and we employed them in producing many of the figures in this chapter. Another alternative is to consider what qualitative information can be obtained about solutions without actually solving the equations. The main purpose of this chapter is to show that a good deal of information about the qualitative behavior of solutions can often be obtained with relatively little effort. The questions that we consider in this chapter are associated with the idea of stability of a solution, and the methods that we employ are basically geometrical. The numerical and geometrical approaches complement each other rather well: the numerical methods provide detailed information about a single solution,[1] whereas the geometrical methods yield qualitative information about all solutions simultaneously.

Both the concept of stability and the use of geometrical methods were introduced in Chapter 1 and have been used repeatedly in later chapters as well: in Section 2.5 for first order autonomous equations

$$\frac{dy}{dt} = f(y), \tag{2}$$

and in Chapters 3 and 6 for linear systems with constant coefficients. In this chapter we refine the ideas and extend the discussion to nonlinear autonomous systems of equations.

▶ **Autonomous Systems.** Autonomous systems occur frequently in applications. Physically, an autonomous system is one whose configuration, including physical parameters and external forces or effects, is independent of time. The response of the system to given initial conditions is then independent of the time at which the conditions are imposed.

We are concerned with systems of two simultaneous differential equations of the form (1). We assume that the functions F and G are continuous and have continuous partial derivatives in some domain D of the xy-plane. If (x_0, y_0) is a point in this domain, then by Theorem 3.6.1 there exists a unique solution $x = \phi(t)$, $y = \psi(t)$ of the system (1) satisfying the initial conditions

$$x(t_0) = x_0, \qquad y(t_0) = y_0. \tag{3}$$

The solution is defined in some time interval I that contains the point t_0.

Frequently, we will write the initial value problem (1), (3) in the vector form

$$\frac{d\mathbf{x}}{dt} = \mathbf{f}(\mathbf{x}), \qquad \mathbf{x}(t_0) = \mathbf{x}_0, \tag{4}$$

where $\mathbf{x} = x\mathbf{i} + y\mathbf{j}$, $\mathbf{f}(\mathbf{x}) = F(x, y)\mathbf{i} + G(x, y)\mathbf{j}$, and $\mathbf{x}_0 = x_0\mathbf{i} + y_0\mathbf{j}$. In this case, the solution is expressed as $\mathbf{x} = \boldsymbol{\phi}(t)$, where $\boldsymbol{\phi}(t) = \phi(t)\mathbf{i} + \psi(t)\mathbf{j}$. As usual, we interpret a solution $\mathbf{x} = \boldsymbol{\phi}(t)$ as a curve traced by a moving point in the xy-plane, the phase plane.

The autonomous system (1) has an associated direction field that is independent of time. Consequently, there is only one trajectory passing through each point (x_0, y_0) in the phase plane. In other words, all solutions that satisfy an initial condition of the form (3) lie on the same trajectory, regardless of the time t_0 at which they pass through (x_0, y_0). Thus, just as for the constant coefficient linear system

$$\frac{d\mathbf{x}}{dt} = \mathbf{A}\mathbf{x}, \tag{5}$$

a single phase portrait simultaneously displays important qualitative information about all solutions of the system (1). We will see this fact confirmed repeatedly in this chapter.

---

[1]Of course, they can be used repeatedly with different initial conditions to approximate more than one solution.

▶ Stability and Instability. The concepts of stability, asymptotic stability, and instability have already been mentioned several times in this book. It is now time to give a precise mathematical definition of these concepts, at least for autonomous systems of the form

$$\mathbf{x}' = \mathbf{f}(\mathbf{x}). \tag{6}$$

In the following definitions, and elsewhere, we use the notation $||\mathbf{x}||$ to designate the length, or magnitude, of the vector $\mathbf{x}$. If $\mathbf{x} = x\mathbf{i} + y\mathbf{j}$, then $||\mathbf{x}|| = \sqrt{x^2 + y^2}$.

The points, if any, where $\mathbf{f}(\mathbf{x}) = \mathbf{0}$ are called **critical points** of the autonomous system (6). At such points $\mathbf{x}' = \mathbf{0}$ also, so critical points correspond to constant, or equilibrium, solutions of the system of differential equations.

Roughly speaking, we have seen that a critical point is stable if all trajectories that start close to the critical point remain close to it for all future times. A critical point is asymptotically stable if all nearby trajectories not only remain nearby but also actually approach the critical point as $t \to \infty$. A critical point is unstable if at least one nearby trajectory does not remain close to the critical point as t increases.

More precisely, a critical point $\mathbf{x}_0$ of the system (6) is said to be **stable** if, given any $\epsilon > 0$, there is a $\delta > 0$ such that every solution $\mathbf{x} = \boldsymbol{\phi}(t)$ of the system (1), which at $t = 0$ satisfies

$$||\boldsymbol{\phi}(0) - \mathbf{x}_0|| < \delta, \tag{7}$$

exists for all positive t and satisfies

$$||\boldsymbol{\phi}(t) - \mathbf{x}_0|| < \epsilon \tag{8}$$

for all $t \geq 0$. A critical point that is not stable is said to be **unstable**.

Stability of a critical point is illustrated geometrically in Figures 7.1.1a and 7.1.1b. The mathematical statements (7) and (8) say that all solutions that start "sufficiently close" (i.e., within the distance δ) to $\mathbf{x}_0$ stay "close" (within the distance ϵ) to $\mathbf{x}_0$. Note that in Figure 7.1.1a the trajectory is within the circle $||\mathbf{x} - \mathbf{x}_0|| = \delta$ at $t = 0$ and, although it soon passes outside of this circle, it remains within the circle $||\mathbf{x} - \mathbf{x}_0|| = \epsilon$ for all $t \geq 0$. In fact, it eventually approaches $\mathbf{x}_0$. However, this limiting behavior is not necessary for stability, as illustrated in Figure 7.1.1b.

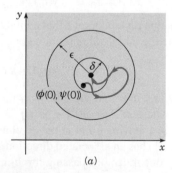 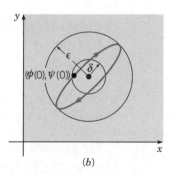

(a)	(b)

FIGURE 7.1.1 (a) Asymptotic stability. (b) Stability.

A critical point $\mathbf{x}_0$ is said to be **asymptotically stable** if it is stable and if there exists a $\delta_0 > 0$ such that, if a solution $\mathbf{x} = \boldsymbol{\phi}(t)$ satisfies

$$||\boldsymbol{\phi}(0) - \mathbf{x}_0|| < \delta_0, \tag{9}$$

then

$$\lim_{t \to \infty} \boldsymbol{\phi}(t) = \mathbf{x}_0. \tag{10}$$

Thus trajectories that start "sufficiently close" to $\mathbf{x}_0$ not only must stay "close" but also must eventually approach $\mathbf{x}_0$ as $t \to \infty$. This is the case for the trajectory in Figure 7.1.1a

but not for the one in Figure 7.1.1*b*. Note that asymptotic stability is a stronger property than stability, since a critical point must be stable before we can even talk about whether it might be asymptotically stable.

On the other hand, the limit condition (10), which is an essential feature of asymptotic stability, does not by itself imply even ordinary stability. Indeed, examples can be constructed in which all the trajectories approach $\mathbf{x}_0$ as $t \to \infty$, but for which $\mathbf{x}_0$ is not a stable critical point. Geometrically, all that is needed is a family of trajectories having members that start arbitrarily close to $\mathbf{x}_0$, and then depart an arbitrarily large distance before eventually approaching $\mathbf{x}_0$ as $t \to \infty$.

In this chapter we concentrate on two-dimensional systems, but the definitions just given are independent of the dimension of the system. If you interpret the vectors in Eqs. (6) through (10) as *n*-dimensional, then the definitions of stability, asymptotic stability, and instability apply also to *n*-dimensional systems. These definitions can be made more concrete by interpreting them in terms of a specific physical problem.

▶ The Oscillating Pendulum. The concepts of asymptotic stability, stability, and instability can be easily visualized in terms of an oscillating pendulum. Consider the configuration shown in Figure 7.1.2, in which a mass m is attached to one end of a rigid, but weightless, rod of length L. The other end of the rod is supported at the origin O, and the rod is free to rotate in the plane of the paper. The position of the pendulum is described by the angle θ between the rod and the downward vertical direction, with the counterclockwise direction taken as positive. The gravitational force mg acts downward, while the damping force $c|d\theta/dt|$, where c is positive, is always opposite to the direction of motion. We assume that θ and $d\theta/dt$ are both positive.

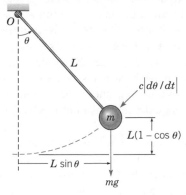

FIGURE 7.1.2 An oscillating pendulum.

The equation of motion can be quickly derived from the principle of angular momentum,[2] which states that the time rate of change of angular momentum about any point is equal to the moment of the resultant force about that point. The angular momentum about the origin, $mL^2(d\theta/dt)$, is the product of the mass m, the moment arm L, and the velocity $L\,d\theta/dt$. Thus the equation of motion is

$$mL^2\frac{d^2\theta}{dt^2} = -cL\frac{d\theta}{dt} - mgL\sin\theta. \tag{11}$$

The factors L and $L\sin\theta$ on the right side of Eq. (11) are the moment arms of the resistive force and of the gravitational force, respectively; the minus signs are due to the fact that the two forces tend to make the pendulum rotate in the clockwise (negative) direction. You

[2]Recall that in Section 4.1 we derived the pendulum equation by using Newton's second law of motion.

should verify, as an exercise, that the same equation is obtained for the other three possible sign combinations of θ and $d\theta/dt$.

By straightforward algebraic operations, we can write Eq. (11) in the standard form

$$\frac{d^2\theta}{dt^2} + \frac{c}{mL}\frac{d\theta}{dt} + \frac{g}{L}\sin\theta = 0, \tag{12}$$

or

$$\frac{d^2\theta}{dt^2} + \gamma\frac{d\theta}{dt} + \omega^2\sin\theta = 0, \tag{13}$$

where $\gamma = c/mL$ and $\omega^2 = g/L$. To convert Eq. (13) to a system of two first order equations, we let $x = \theta$ and $y = d\theta/dt$; then

$$\frac{dx}{dt} = y, \qquad \frac{dy}{dt} = -\omega^2\sin x - \gamma y. \tag{14}$$

Since γ and ω^2 are constants, the system (14) is an autonomous system of the form (1).

The critical points of Eqs. (14) are found by solving the equations

$$y = 0, \qquad -\omega^2\sin x - \gamma y = 0. \tag{15}$$

We obtain $y = 0$ and $x = \pm n\pi$, where n is an integer. These points correspond to two physical equilibrium positions: one with the mass directly below the point of support ($\theta = 0$) and the other with the mass directly above the point of support ($\theta = \pi$). Our intuition suggests that the first is stable and the second is unstable.

More precisely, if the mass is slightly displaced from the lower equilibrium position, it will oscillate back and forth with gradually decreasing amplitude, eventually approaching the equilibrium position as the initial potential energy is dissipated by the damping force. This type of motion illustrates **asymptotic stability**.

On the other hand, if the mass is slightly displaced from the upper equilibrium position, it will rapidly fall, under the influence of gravity, and will ultimately approach the lower equilibrium position in this case also. This type of motion illustrates **instability**. In practice, it is impossible to maintain the pendulum in its upward equilibrium position for very long without an external constraint mechanism, since the slightest perturbation will cause the mass to fall.

Finally, consider the ideal situation in which the damping coefficient c (or γ) is zero. In this case, if the mass is displaced slightly from its lower equilibrium position, it will oscillate indefinitely with constant amplitude about the equilibrium position. Since there is no dissipation in the system, the mass will remain near the equilibrium position but will not approach it asymptotically. This type of motion is *stable* but not asymptotically stable. In

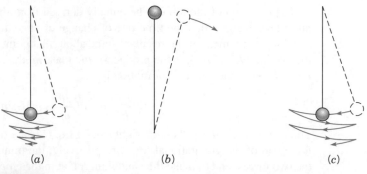

(a)   (b)   (c)

FIGURE 7.1.3 Qualitative motion of a pendulum. (*a*) With air resistance. (*b*) With or without air resistance. (*c*) Without air resistance.

general, this motion is impossible to achieve experimentally, because the slightest degree of air resistance or friction at the point of support will eventually cause the pendulum to approach its rest position.

These three types of motion are illustrated schematically in Figure 7.1.3. Solutions of the pendulum equations are discussed in more detail in the next section.

▶ The Importance of Critical Points. Critical points correspond to equilibrium solutions, that is, solutions in which $x(t)$ and $y(t)$ are constant. For such a solution, the system described by x and y is not changing; it remains in its initial state forever. It might seem reasonable to conclude that such points are not very interesting. However, recall that in Section 2.5 and later in Chapter 3, we found that the behavior of solutions in the *neighborhood* of a critical point has important implications for the behavior of solutions farther away. We will find that the same is true for nonlinear systems of the form (1). Consider the following examples.

EXAMPLE 1

Undamped Pendulum

An oscillating pendulum without damping is described by the equations

$$\frac{dx}{dt} = y, \qquad \frac{dy}{dt} = -\omega^2 \sin x, \qquad (16)$$

obtained by setting γ equal to zero in Eq. (14). The critical points for the system (16) are $(\pm n\pi, 0)$; even values of n, including zero, correspond to the downward equilbrium position and odd values of n to the upward equilibrium position. Let $\omega = 2$ and draw a phase portrait for this system. From the phase portrait, describe the behavior of solutions near each critical point and relate this behavior to the overall motion of the pendulum.

A direction field and phase portrait for the system (16) with $\omega = 2$ are shown in Figure 7.1.4. Looking first at a fairly small rectangle centered at the origin, we see that the trajectories there are closed curves resembling ellipses. These correspond to periodic solutions, or to a periodic oscillation of the pendulum about its stable downward equilibrium position. The same behavior occurs near other critical points corresponding to even values of n.

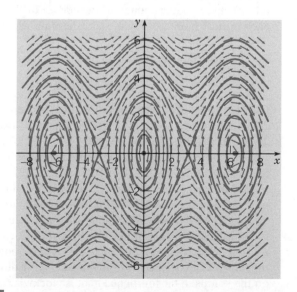

FIGURE 7.1.4 Direction field and phase portrait for the system (16) with $\omega = 2$.

The situation is different near the critical points $(\pm\pi, 0)$. In a small rectangle centered on each of these points, the trajectories display the pattern associated with saddle points. Observe that one trajectory appears to connect the points $(\pi, 0)$ and $(-\pi, 0)$, passing from the former to the latter in the lower half plane $(y < 0)$. More precisely, a solution on this trajectory approaches $(\pi, 0)$ as $t \to -\infty$ and approaches $(-\pi, 0)$ as $t \to \infty$. Similarly, there is a trajectory on which $y > 0$ that leaves $(-\pi, 0)$ as $t \to -\infty$ and approaches $(\pi, 0)$ as $t \to \infty$. These two trajectories enclose a region about the origin within which periodic motions occur. As indicated in Figure 7.1.4, a similar situation exists between each successive pair of saddle points.

Outside of these regions where periodic motions take place, the pendulum experiences a different kind of motion. If the angular velocity y is sufficiently large in magnitude, then the pendulum exhibits a whirling motion in which the angular position x steadily increases or decreases, depending on whether y is positive or negative. In the phase portrait, these are shown by the wavy curves in the upper and lower parts of the figure. The curves that "connect" pairs of saddle points are called **separatrices** because they separate the regions of periodic motions from the regions of whirling motions in the phase plane.

Finally, note that if you choose any initial point close to the origin, or close to one of the other stable equilibria, then a small oscillation will ensue. However, if you choose an initial point near one of the saddle points, there are two possibilities. If you are within the separatrices associated with that point, the motion will be a large oscillation; the pendulum will swing from near its vertical position almost all the way around, and then back and forth. However, if you start at a point just outside the separatrices, then a whirling motion will result.

EXAMPLE
2

Consider the system

$$\frac{dx}{dt} = -(2 + y)(x + y), \qquad \frac{dy}{dt} = -y(1 - x). \qquad (17)$$

Find all of the critical points for this system. Then draw a direction field and phase portrait on a rectangle large enough to include all of the critical points. From your plot, classify each critical point as to type and determine whether it is asymptotically stable, stable, or unstable.

The critical points are found by solving the equations

$$-(2 + y)(x + y) = 0, \qquad -y(1 - x) = 0. \qquad (18)$$

One way to satisfy the first equation is to choose $y = -2$. Then the second equation becomes $2(1 - x) = 0$, so $x = 1$. The first of Eqs. (18) can also be satisfied by choosing $x = -y$. Then the second equation becomes $y(1 + y) = 0$, so either $y = 0$ or else $y = -1$. Since $x = -y$, this leads to the two additional critical points $(0, 0)$ and $(1, -1)$. Thus the system (17) has three critical points: $(1, -2)$, $(0, 0)$, and $(1, -1)$. Alternatively, you can start with the second of Eqs. (18), which can be satisfied by choosing $y = 0$ or $x = 1$. By substituting each of these values in the first of Eqs. (18), you obtain the same three critical points.

Figure 7.1.5 shows a direction field and phase portrait for the system (17). If we look at the immediate neighborhood of each of the critical points, it should be clear that $(1, -2)$ is a spiral point, $(0, 0)$ is a node, and $(1, -1)$ is a saddle point. The direction of motion on the trajectories can be inferred from the underlying direction field. Thus the direction of motion on the spirals in the fourth quadrant is clockwise. The trajectories approach $(1, -2)$, albeit quite slowly, so this point is asymptotically stable. Similarly, the direction field shows that

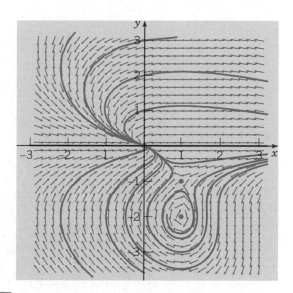

FIGURE 7.1.5 Direction field and phase portrait for the system (17).

trajectories near the node $(0, 0)$ are approaching this point; thus it is also asymptotically stable. Finally, the saddle point at $(1, -1)$ is unstable, since this is true of all saddle points. Two trajectories enter the saddle point with slopes of approximately 0.6. All other trajectories eventually leave the neighborhood of this point.

Observe that the line $y = 0$ is a trajectory; on this line $y' = 0$ and x' is positive or negative depending on whether x is less than or greater than 0. Thus no other trajectories can cross this line.

Suppose that an autonomous system has at least one asymptotically stable critical point, as in Example 2. Then it is often of interest to determine where in the phase plane the trajectories lie that ultimately approach this critical point. If a point P in the xy-plane has the property that a trajectory that starts at P approaches the critical point as $t \rightarrow \infty$, then this trajectory is said to be attracted by the critical point. The set of all such points P is called the **basin of attraction** or the **region of asymptotic stability** of the critical point. A trajectory that bounds a basin of attraction is called a **separatrix** because it separates trajectories that approach a particular critical point from other trajectories that do not do so. Determination of basins of attraction is important in understanding the large-scale behavior of the solutions of a given autonomous system.

EXAMPLE 3

For the system (17) in Example 2, describe the basins of attraction for the node and spiral point.

In Figure 7.1.6, we look more closely at the region of the phase plane containing the saddle and spiral points. The two pairs of nearby trajectories bracket the two trajectories that enter the saddle point as $t \rightarrow \infty$. In the region between these two trajectories, every trajectory approaches the spiral point. Thus this is the basin of attraction for the spiral point. It consists of an oval region surrounding the point $(1, -2)$ together with a thin tail extending to the right just below the x-axis. The two bounding trajectories, or separatrices separate the trajectories that approach the spiral point from those that approach the node.

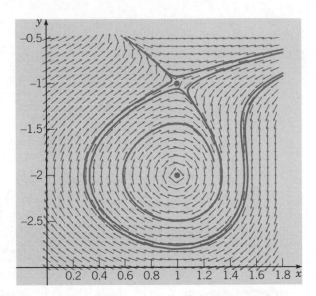

FIGURE 7.1.6   A direction field and phase portrait for the system (17) near the saddle and spiral points.

It is often the case that separatrices are trajectories that enter a saddle point; consequently, these trajectories may be of greater than average significance. From Figures 7.1.5 and 7.1.6, it appears that all trajectories that do not approach the spiral point or the saddle point ultimately approach the asymptotically stable node at the origin.

In the two preceding examples, the pattern of the trajectories near each critical point is essentially the same as one of the patterns that occur for homogeneous linear systems with constant coefficients. This situation is typical of a very large class of two-dimensional nonlinear autonomous systems. Further the behavior of solutions of an autonomous system near its critical points is one of the major factors influencing the overall behavior of the solutions throughout the xy-plane. In the next section we take up the question of finding a linear system that approximates a nonlinear system near a critical point.

PROBLEMS

For each of the systems in Problems 1 through 18:
(a) Find all the critical points (equilibrium solutions).
(b) Use a computer to draw a direction field and phase portrait for the system.
(c) From the plot(s) in part (b), determine whether each critical point is asymptotically stable, stable, or unstable, and classify it as to type.
(d) Describe the basin of attraction for each asymptotically stable critical point.

1. $dx/dt = -2y + xy, \quad dy/dt = x + 4xy$
2. $dx/dt = 1 + 5y, \quad dy/dt = 1 - 6x^2$
3. $dx/dt = 2x - x^2 - xy, \quad dy/dt = 3y - 2y^2 - 3xy$
4. $dx/dt = -(x - y)(4 - x - y), \quad dy/dt = -x(2 + y)$

5. $dx/dt = x(6 - x - y), \quad dy/dt = -x + 7y - 2xy$
6. $dx/dt = (2 - x)(y - x), \quad dy/dt = y(2 - x - x^2)$
7. $dx/dt = (2 - y)(x - y), \quad dy/dt = (1 + x)(x + y)$
8. $dx/dt = x(2 - x - y), \quad dy/dt = (1 - y)(2 + x)$
9. $dx/dt = (3 + x)(2y - x), \quad dy/dt = (2 - x)(y - x)$
10. $dx/dt = x - 12y - x^2, \quad dy/dt = (1 + y)(1 - x)$
11. $dx/dt = -y, \quad dy/dt = -3y - x(x - 1)(x - 2)$
12. $dx/dt = (2 + x)(1 - x + y),$
    $dy/dt = (y - 1)(1 + x + 2y)$
13. $dx/dt = x(8 - x - 3y), \quad dy/dt = y(3 - x)(2 + x)$
14. $dx/dt = (x - 2y + 1)(y + 1), \quad dy/dt = -y(3 - 2x - y)$
15. $dx/dt = (y - 1)(x - 1)(x + 2), \quad dy/dt = x(2 - y)$

16. $dx/dt = 25 - y^2, \quad dy/dt = (1-x)(y+x)$

17. $dx/dt = y, \quad dy/dt = x - \frac{1}{6}x^3$

18. $dx/dt = -x + y + x^2, \quad dy/dt = y - 2xy$

19. (a) Consider the equations of motion of an undamped pendulum,

$$x' = y, \qquad y' = -\omega^2 \sin x; \tag{i}$$

see Example 1. Convert this system into a single equation for dy/dx, and then show that the trajectories of the undamped pendulum satisfy the equation

$$\tfrac{1}{2}y^2 + \omega^2(1 - \cos x) = c, \tag{ii}$$

where c is a constant of integration.

(b) Multiply Eq. (ii) by mL^2 and recall that $\omega^2 = g/L$. Then, by expressing Eq. (ii) in terms of θ, obtain

$$\frac{1}{2}mL^2\left(\frac{d\theta}{dt}\right)^2 + mgL(1 - \cos\theta) = E, \tag{iii}$$

where $E = mL^2c$.

(c) Show that the first term in Eq. (iii) is the kinetic energy of the pendulum and that the second term is the potential energy due to gravity. Thus the total energy E of the pendulum is constant along any trajectory; in other words, the undamped pendulum satisfies the principle of conservation of energy. The value of E is determined by the initial conditions.

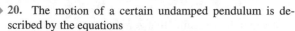

20. The motion of a certain undamped pendulum is described by the equations

$$\frac{dx}{dt} = y, \qquad \frac{dy}{dt} = -4\sin x.$$

If the pendulum is set in motion with an angular displacement A and no initial velocity, then the initial conditions are $x(0) = A, y(0) = 0$.

(a) Let $A = 0.25$ and plot x versus t. From the graph, estimate the amplitude R and period T of the resulting motion of the pendulum.

(b) Repeat part (a) for $A = 0.5, 1.0, 1.5,$ and 2.0.

(c) How do the amplitude and period of the pendulum's motion depend on the initial position A? Draw a graph to show each of these relationships. Can you say anything about the limiting value of the period as $A \to 0$?

(d) Let $A = 4$ and plot x versus t. Explain why this graph differs from those in parts (a) and (b). For what value of A does the transition take place?

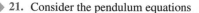

21. Consider the pendulum equations

$$\frac{dx}{dt} = y, \qquad \frac{dy}{dt} = -6\sin x.$$

If the pendulum is set in motion from its downward equilibrium position with angular velocity v, then the initial conditions are $x(0) = 0, y(0) = v$.

(a) Plot x versus t for $v = 3$ and also for $v = 6$. Explain the differing motions of the pendulum that these two graphs represent.

(b) There is a critical value of v, which we denote by v_c, such that one type of motion occurs for $v < v_c$ and a qualitatively different type of motion occurs for $v > v_c$. Determine the value of v_c to two decimal places.

22. In this problem, we derive a formula for the natural period of an undamped nonlinear pendulum. The equation of motion is

$$mL^2\frac{d^2\theta}{dt^2} + mgL\sin\theta = 0$$

obtained by setting $c = 0$ in Eq. (12). Suppose that the bob is pulled through a positive angle α and then released with zero velocity.

(a) We usually think of θ and $d\theta/dt$ as functions of t. However, if we reverse the roles of t and θ, we can regard t as a function of θ and, consequently, we can also think of $d\theta/dt$ as a function of θ. Then derive the following sequence of equations:

$$\frac{1}{2}mL^2\frac{d}{d\theta}\left[\left(\frac{d\theta}{dt}\right)^2\right] = -mgL\sin\theta,$$

$$\frac{1}{2}m\left(L\frac{d\theta}{dt}\right)^2 = mgL(\cos\theta - \cos\alpha),$$

$$dt = -\sqrt{\frac{L}{2g}}\frac{d\theta}{\sqrt{\cos\theta - \cos\alpha}}.$$

Why was the negative square root chosen in the last equation?

(b) If T is the natural period of oscillation, derive the formula

$$\frac{T}{4} = -\sqrt{\frac{L}{2g}}\int_\alpha^0 \frac{d\theta}{\sqrt{\cos\theta - \cos\alpha}}.$$

(c) By using the identities $\cos\theta = 1 - 2\sin^2(\theta/2)$ and $\cos\alpha = 1 - 2\sin^2(\alpha/2)$, followed by the change of variable $\sin(\theta/2) = k\sin\phi$ with $k = \sin(\alpha/2)$, show that

$$T = 4\sqrt{\frac{L}{g}}\int_0^{\pi/2}\frac{d\phi}{\sqrt{1 - k^2\sin^2\phi}}.$$

The integral is called the **elliptic integral** of the first kind. Note that the period depends on the ratio L/g and also on the initial displacement α through $k = \sin(\alpha/2)$.

(d) By evaluating the integral in the expression for T, obtain values for T that you can compare with the graphical estimates you obtained in Problem 20.

23. Given that $x = \phi(t), y = \psi(t)$ is a solution of the autonomous system

$$\frac{dx}{dt} = F(x, y), \qquad \frac{dy}{dt} = G(x, y)$$

for $\alpha < t < \beta$, show that $x = \Phi(t) = \phi(t - s), y = \Psi(t) = \psi(t - s)$ is a solution for $\alpha + s < t < \beta + s$ for any real number s.

24. Prove that, for the system

$$\frac{dx}{dt} = F(x, y), \qquad \frac{dy}{dt} = G(x, y),$$

there is at most one trajectory passing through a given point (x_0, y_0).

Hint: Let C_0 be the trajectory generated by the solution $x = \phi_0(t)$, $y = \psi_0(t)$, with $\phi_0(t_0) = x_0$, $\psi_0(t_0) = y_0$, and let C_1 be the trajectory generated by the solution $x = \phi_1(t)$, $y = \psi_1(t)$, with $\phi_1(t_1) = x_0$, $\psi_1(t_1) = y_0$. Use the fact that the system is autonomous, and also the existence and uniqueness theorem, to show that C_0 and C_1 are the same.

25. Prove that if a trajectory starts at a noncritical point of the system

$$\frac{dx}{dt} = F(x, y), \qquad \frac{dy}{dt} = G(x, y),$$

then it cannot reach a critical point (x_0, y_0) in a finite length of time.

Hint: Assume the contrary, that is, assume that the solution $x = \phi(t)$, $y = \psi(t)$ satisfies $\phi(a) = x_0$, $\psi(a) = y_0$. Then use the fact that $x = x_0$, $y = y_0$ is a solution of the given system satisfying the initial condition $x = x_0$, $y = y_0$ at $t = a$.

26. Assuming that the trajectory corresponding to a solution $x = \phi(t)$, $y = \psi(t)$, $-\infty < t < \infty$, of an autonomous system is closed, show that the solution is periodic.

Hint: Since the trajectory is closed, there exists at least one point (x_0, y_0) such that $\phi(t_0) = x_0$, $\psi(t_0) = y_0$, and a number $T > 0$ such that $\phi(t_0 + T) = x_0$, $\psi(t_0 + T) = y_0$. Show that $x = \Phi(t) = \phi(t + T)$ and $y = \Psi(t) = \psi(t + T)$ is a solution, and then use the existence and uniqueness theorem to show that $\Phi(t) = \phi(t)$ and $\Psi(t) = \psi(t)$ for all t.

7.2 Almost Linear Systems

In Chapter 3 we investigated solutions of the two-dimensional linear homogeneous system with constant coefficients,

$$\mathbf{x}' = \mathbf{A}\mathbf{x}. \tag{1}$$

If we assume that $\det(\mathbf{A}) \neq 0$, which is equivalent to assuming that zero is not an eigenvalue of $\mathbf{A}$, then $\mathbf{x} = \mathbf{0}$ is the only critical point (equilibrium solution) of the system (1). The stability properties of this critical point depend on the eigenvalues of $\mathbf{A}$. The results were summarized in Table 3.5.2, which is repeated here as Table 7.2.1 for convenience. We can also restate these results in the following theorem.

THEOREM 7.2.1

The critical point $\mathbf{x} = \mathbf{0}$ of the linear system (1) is asymptotically stable if the eigenvalues λ_1, λ_2 are real and negative or are complex with negative real part; stable, but not asymptotically stable, if λ_1 and λ_2 are pure imaginary; unstable if λ_1 and λ_2 are real and either is positive, or if they are complex with positive real part.

▶ **Effect of Small Perturbations.** It is apparent from this theorem or from Table 7.2.1 that the eigenvalues λ_1, λ_2 of the coefficient matrix $\mathbf{A}$ determine the type of critical point at $\mathbf{x} = \mathbf{0}$ and its stability characteristics. In turn, the values of λ_1 and λ_2 depend on the coefficients in the system (1). When such a system arises in some applied field, the values of the coefficients usually result from measurements of certain physical quantities. Such measurements are often subject to small random errors, so it is of interest to investigate whether small changes (perturbations) in the coefficients can affect the stability or instability of a critical point and/or significantly alter the pattern of trajectories.

The eigenvalues λ_1, λ_2 are the roots of the polynomial equation

$$\det(\mathbf{A} - \lambda\mathbf{I}) = 0. \tag{2}$$

| TABLE 7.2.1 | Stability properties of linear systems $\mathbf{x}' = \mathbf{Ax}$ with $\det(\mathbf{A}-\lambda\mathbf{I}) = 0$ and $\det \mathbf{A} \neq 0$. |

Eigenvalues	Type of Critical Point	Stability
$\lambda_1 > \lambda_2 > 0$	Node	Unstable
$\lambda_1 < \lambda_2 < 0$	Node	Asymptotically stable
$\lambda_2 < 0 < \lambda_1$	Saddle point	Unstable
$\lambda_1 = \lambda_2 > 0$	Proper or improper node	Unstable
$\lambda_1 = \lambda_2 < 0$	Proper or improper node	Asymptotically stable
$\lambda_1, \lambda_2 = \mu \pm i\nu$		
$\quad \mu > 0$	Spiral point	Unstable
$\quad \mu < 0$	Spiral point	Asymptotically stable
$\quad \mu = 0$	Center	Stable

It is possible to show that *small* perturbations in some or all of the coefficients are reflected in *small* perturbations in the eigenvalues. The most sensitive situation occurs when $\lambda_1 = i\nu$ and $\lambda_2 = -i\nu$, that is, when the critical point is a center and the trajectories are closed curves surrounding it. If a slight change is made in the coefficients, then the eigenvalues λ_1 and λ_2 will take on new values $\lambda_1' = \mu' + i\nu'$ and $\lambda_2' = \mu' - i\nu'$, where μ' is small in magnitude and $\nu' \cong \nu$ (see Figure 7.2.1). It is possible that $\mu' = 0$, in which case the critical point remains a center. However, in most cases $\mu' \neq 0$, and then the trajectories of the perturbed system are spirals rather than closed curves. The system is asymptotically stable if $\mu' < 0$ but unstable if $\mu' > 0$; see Problem 25.

Another slightly less sensitive case occurs if the eigenvalues λ_1 and λ_2 are equal; in this case, the critical point is a node. Small perturbations in the coefficients will normally cause the two equal roots to separate (bifurcate). If the separated roots are real, then the critical point of the perturbed system remains a node, but if the separated roots are complex conjugates, then the critical point becomes a spiral point. These two possibilities are shown schematically in Figure 7.2.2. In this case, the stability or instability of the system is not affected by small perturbations in the coefficients, but the trajectories may be altered considerably (see Problem 26).

In all other cases, the stability or instability of the system is not changed, nor is the type of critical point altered, by sufficiently small perturbations in the coefficients of the system. For example, if λ_1 and λ_2 are real, negative, and unequal, then a *small* change in the coefficients will neither change the sign of λ_1 and λ_2 nor allow them to coalesce. Thus the critical point remains an asymptotically stable node.

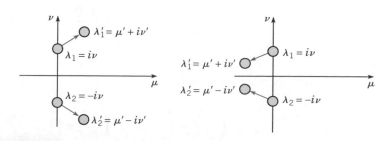

FIGURE 7.2.1 Schematic perturbation of $\lambda_1 = i\nu$, $\lambda_2 = -i\nu$.

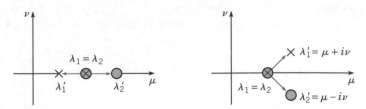

FIGURE 7.2.2 Schematic perturbation of $\lambda_1 = \lambda_2$.

▶ Linear Approximations to Nonlinear Systems. Now let us consider a nonlinear two-dimensional autonomous system

$$\mathbf{x}' = \mathbf{f}(\mathbf{x}). \tag{3}$$

Our main object is to investigate the behavior of trajectories of the system (3) near a critical point $\mathbf{x}_0$. We will seek to do this by approximating the nonlinear system (3) by an appropriate linear system, whose trajectories are easy to describe. The crucial question is whether the trajectories of the linear system are good approximations to those of the nonlinear system. Of course, we also need to know how to find the approximating linear system.

It is convenient to choose the critical point to be the origin. This involves no loss of generality, since if $\mathbf{x}_0 \neq \mathbf{0}$, it is always possible to make the substitution $\mathbf{u} = \mathbf{x} - \mathbf{x}_0$ in Eq. (3). Then $\mathbf{u}$ will satisfy an autonomous system with a critical point at the origin.

First, let us consider what it means for a nonlinear system (3) to be "close" to a linear system (1). Accordingly, suppose that

$$\mathbf{x}' = \mathbf{A}\mathbf{x} + \mathbf{g}(\mathbf{x}) \tag{4}$$

and that $\mathbf{x} = \mathbf{0}$ is an **isolated** critical point of the system (4). This means that there is some circle about the origin within which there are no other critical points. In addition, we assume that $\det \mathbf{A} \neq 0$, so $\mathbf{x} = \mathbf{0}$ is also an isolated critical point of the linear system $\mathbf{x}' = \mathbf{A}\mathbf{x}$. For the nonlinear system (4) to be close to the linear system $\mathbf{x}' = \mathbf{A}\mathbf{x}$, we must assume that $\mathbf{g}(\mathbf{x})$ is small. More precisely, we assume that the components of $\mathbf{g}$ have continuous first partial derivatives and satisfy the limit condition

$$\frac{\|\mathbf{g}(\mathbf{x})\|}{\|\mathbf{x}\|} \to 0 \qquad \text{as} \quad \mathbf{x} \to \mathbf{0}, \tag{5}$$

that is, $\|\mathbf{g}\|$ is small in comparison to $\|\mathbf{x}\|$ itself near the origin. Such a system is called an **almost linear system** in the neighborhood of the critical point $\mathbf{x} = \mathbf{0}$.

It may be helpful to express the condition (5) in scalar form. If we let $\mathbf{x}^T = (x, y)$, then $\|\mathbf{x}\| = (x^2 + y^2)^{1/2} = r$. Similarly, if $\mathbf{g}^T(\mathbf{x}) = (g_1(x, y), g_2(x, y))$, then $\|\mathbf{g}(\mathbf{x})\| = [g_1^2(x, y) + g_2^2(x, y)]^{1/2}$. Then it follows that condition (5) is satisfied if and only if

$$\frac{g_1(x, y)}{r} \to 0, \qquad \frac{g_2(x, y)}{r} \to 0 \qquad \text{as} \quad r \to 0. \tag{6}$$

EXAMPLE 1

Determine whether the system

$$\begin{pmatrix} x \\ y \end{pmatrix}' = \begin{pmatrix} 1 & 0 \\ 0 & 0.5 \end{pmatrix} \begin{pmatrix} x \\ y \end{pmatrix} + \begin{pmatrix} -x^2 - xy \\ -0.75xy - 0.25y^2 \end{pmatrix} \tag{7}$$

is almost linear in the neighborhood of the origin.

Observe that the system (7) is of the form (4), that (0, 0) is a critical point, and that det $\mathbf{A} \neq 0$. It is not hard to show that the other critical points of Eqs. (7) are (0, 2), (1, 0), and (0.5, 0.5); consequently, the origin is an isolated critical point. In checking the condition (6), it is convenient to introduce polar coordinates by letting $x = r \cos \theta$, $y = r \sin \theta$. Then

$$\frac{g_1(x, y)}{r} = \frac{-x^2 - xy}{r} = \frac{-r^2 \cos^2 \theta - r^2 \sin \theta \cos \theta}{r}$$

$$= -r(\cos^2 \theta + \sin \theta \cos \theta) \to 0$$

as $r \to 0$. In a similar way you can show that $g_2(x, y)/r \to 0$ as $r \to 0$. Hence the system (7) is almost linear near the origin.

EXAMPLE
2

The motion of a pendulum is described by the system [see Eq. (14) of Section 7.1]

$$\frac{dx}{dt} = y, \qquad \frac{dy}{dt} = -\omega^2 \sin x - \gamma y. \qquad (8)$$

The critical points are (0, 0), $(\pm \pi, 0)$, $(\pm 2\pi, 0)$, ..., so the origin is an isolated critical point of this system. Show that the system is almost linear near the origin.

To compare Eqs. (8) with Eq. (4), we must rewrite the former so that the linear and nonlinear terms are clearly identified. If we write $\sin x = x + (\sin x - x)$ and substitute this expression in the second of Eqs. (8), we obtain the equivalent system

$$\begin{pmatrix} x \\ y \end{pmatrix}' = \begin{pmatrix} 0 & 1 \\ -\omega^2 & -\gamma \end{pmatrix} \begin{pmatrix} x \\ y \end{pmatrix} - \omega^2 \begin{pmatrix} 0 \\ \sin x - x \end{pmatrix}. \qquad (9)$$

Comparing Eqs. (9) and (4), we see that $g_1(x, y) = 0$ and $g_2(x, y) = -\omega^2 (\sin x - x)$. The Taylor series for $\sin x$ implies that $\sin x - x$ behaves like $-x^3/3! = -(r^3 \cos^3 \theta)/3!$ when x is small. Consequently, $[(\sin x - x)/r] \to 0$ as $r \to 0$. Thus the conditions (6) are satisfied and the system (9) is almost linear near the origin.

Let us now return to the general nonlinear system (3), which we write in the scalar form

$$x' = F(x, y), \qquad y' = G(x, y). \qquad (10)$$

We assume that (x_0, y_0) is an isolated critical point of this system. The system (10) is almost linear in the neighborhood of (x_0, y_0) whenever the functions F and G have continuous partial derivatives up to order 2. To show this, we use Taylor expansions about the point (x_0, y_0) to write $F(x, y)$ and $G(x, y)$ in the form

$$F(x, y) = F(x_0, y_0) + F_x(x_0, y_0)(x - x_0) + F_y(x_0, y_0)(y - y_0) + \eta_1(x, y),$$

$$G(x, y) = G(x_0, y_0) + G_x(x_0, y_0)(x - x_0) + G_y(x_0, y_0)(y - y_0) + \eta_2(x, y),$$

where $\{\eta_1(x, y)/[(x - x_0)^2 + (y - y_0)^2]^{1/2}\} \to 0$ as $(x, y) \to (x_0, y_0)$, and similarly for η_2. Note that $F(x_0, y_0) = G(x_0, y_0) = 0$; also $dx/dt = d(x - x_0)/dt$ and $dy/dt = d(y - y_0)/dt$. Then the system (10) reduces to

$$\frac{d}{dt} \begin{pmatrix} x - x_0 \\ y - y_0 \end{pmatrix} = \begin{pmatrix} F_x(x_0, y_0) & F_y(x_0, y_0) \\ G_x(x_0, y_0) & G_y(x_0, y_0) \end{pmatrix} \begin{pmatrix} x - x_0 \\ y - y_0 \end{pmatrix} + \begin{pmatrix} \eta_1(x, y) \\ \eta_2(x, y) \end{pmatrix}, \qquad (11)$$

or, in vector notation,

$$\frac{d\mathbf{u}}{dt} = \frac{d\mathbf{f}}{d\mathbf{x}}(\mathbf{x}_0)\mathbf{u} + \boldsymbol{\eta}(\mathbf{x}), \tag{12}$$

where $\mathbf{u} = (x - x_0, y - y_0)^T$ and $\boldsymbol{\eta} = (\eta_1, \eta_2)^T$.

The significance of this result is twofold. First, if the functions F and G are twice differentiable, then the system (10) is almost linear, and it is unnecessary to resort to the limiting process used in Examples 1 and 2. Second, the linear system that corresponds to the nonlinear system (10) near (x_0, y_0) is given by the linear part of Eqs. (11) or (12):

$$\frac{d}{dt}\begin{pmatrix} u_1 \\ u_2 \end{pmatrix} = \begin{pmatrix} F_x(x_0, y_0) & F_y(x_0, y_0) \\ G_x(x_0, y_0) & G_y(x_0, y_0) \end{pmatrix}\begin{pmatrix} u_1 \\ u_2 \end{pmatrix}, \tag{13}$$

where $u_1 = x - x_0$ and $u_2 = y - y_0$. Equation (13) provides a simple and general method for finding the linear system corresponding to an almost linear system near a given critical point.

The matrix

$$\mathbf{J} = \begin{pmatrix} F_x & F_y \\ G_x & G_y \end{pmatrix}, \tag{14}$$

which appears as the coefficient matrix in Eq. (13), is called the **Jacobian matrix** of the functions F and G with respect to the variables x and y. We need to assume that $\det(\mathbf{J})$ is not zero at (x_0, y_0) so that this point is also an isolated critical point of the linear system (13).

EXAMPLE 3

Use Eq. (13) to find the linear system corresponding to the pendulum equations (8) near the origin; near the critical point $(\pi, 0)$.

In this case, we have, from Eq. (8),

$$F(x, y) = y, \qquad G(x, y) = -\omega^2 \sin x - \gamma y. \tag{15}$$

Since these functions are differentiable as many times as necessary, the system (8) is almost linear near each critical point. The Jacobian matrix for the system (8) is

$$\begin{pmatrix} F_x & F_y \\ G_x & G_y \end{pmatrix} = \begin{pmatrix} 0 & 1 \\ -\omega^2 \cos x & -\gamma \end{pmatrix}. \tag{16}$$

Thus, at the origin, the corresponding linear system is

$$\frac{d}{dt}\begin{pmatrix} x \\ y \end{pmatrix} = \begin{pmatrix} 0 & 1 \\ -\omega^2 & -\gamma \end{pmatrix}\begin{pmatrix} x \\ y \end{pmatrix}, \tag{17}$$

which agrees with Eq. (9).

Similarly, evaluating the Jacobian matrix at $(\pi, 0)$, we obtain

$$\frac{d}{dt}\begin{pmatrix} u \\ w \end{pmatrix} = \begin{pmatrix} 0 & 1 \\ \omega^2 & -\gamma \end{pmatrix}\begin{pmatrix} u \\ w \end{pmatrix}, \tag{18}$$

where $u = x - \pi$, $w = y$. This is the linear system corresponding to Eqs. (8) near the point $(\pi, 0)$.

We now return to the almost linear system (4). Since the nonlinear term $\mathbf{g}(\mathbf{x})$ is small compared to the linear term $\mathbf{A}\mathbf{x}$ when $\mathbf{x}$ is small, it is reasonable to hope that the trajectories of the linear system (1) are good approximations to those of the nonlinear system (4), at least near the origin. This turns out to be true in many (but not all) cases, as the following theorem states.

THEOREM 7.2.2

Let λ_1 and λ_2 be the eigenvalues of the linear system (1),

$$\mathbf{x}' = \mathbf{A}\mathbf{x},$$

corresponding to the almost linear system (4),

$$\mathbf{x}' = \mathbf{A}\mathbf{x} + \mathbf{g}(\mathbf{x}).$$

Assume that $\mathbf{x} = \mathbf{0}$ is an isolated critical point of both of these systems. Then the type and stability of $\mathbf{x} = \mathbf{0}$ for the linear system (1) and for the almost linear system (4) are as shown in Table 7.2.2.

TABLE 7.2.2 Stability and instability properties of linear and almost linear systems.

	Linear System		Almost Linear System	
λ_1, λ_2	Type	Stability	Type	Stability
$\lambda_1 > \lambda_2 > 0$	N	Unstable	N	Unstable
$\lambda_1 < \lambda_2 < 0$	N	Asymptotically stable	N	Asymptotically stable
$\lambda_2 < 0 < \lambda_1$	SP	Unstable	SP	Unstable
$\lambda_1 = \lambda_2 > 0$	PN or IN	Unstable	N or SpP	Unstable
$\lambda_1 = \lambda_2 < 0$	PN or IN	Asymptotically stable	N or SpP	Asymptotically stable
$\lambda_1, \lambda_2 = \mu \pm i\nu$				
$\quad \mu > 0$	SpP	Unstable	SpP	Unstable
$\quad \mu < 0$	SpP	Asymptotically stable	SpP	Asymptotically stable
$\quad \mu = 0$	C	Stable	C or SpP	Indeterminate

Note: N, node; IN, improper node; PN, proper node; SP, saddle point; SpP, spiral point; C, center.

The proof of Theorem 7.2.2 is beyond the scope of this book, so we will use the results without proof. Essentially, Theorem 7.2.2 says that, for small $\mathbf{x}$ (or $\mathbf{x} - \mathbf{x}_0$), the nonlinear terms are also small and do not affect the stability and type of critical point as determined by the linear terms except in the two sensitive cases discussed previously: λ_1 and λ_2 pure imaginary, and λ_1 and λ_2 real and equal. Recall that, earlier in this section, we stated that small perturbations in the coefficients of the linear system (1), and hence in the eigenvalues λ_1 and λ_2, can alter the type and stability of the critical point only in these two sensitive cases. It is reasonable to expect that the small nonlinear term in Eq. (4) might have a similar effect, at least in these two sensitive cases. This is so, but the main significance of Theorem 7.2.2 is that, in *all other cases*, the small nonlinear term does not alter the type or stability of the critical point. Thus, except in the two sensitive cases, the type and stability of the critical point of the nonlinear system (4) can be determined from a study of the much simpler linear system (1).

In a small neighborhood of the critical point, the trajectories of the nonlinear system are also similar to those of the approximating linear system. In particular, the slopes at which trajectories "enter" or "leave" the critical point are given correctly by the linear equations. Farther away, the nonlinear terms become dominant and the trajectories of the nonlinear system usually bear no resemblance to those of the linear system.

▶ Damped Pendulum. We continue the discussion of the damped pendulum begun in Examples 2 and 3. Near the origin the nonlinear equations (8) are approximated by the linear system (17). The characteristic equation for this system is

$$\lambda^2 + \gamma\lambda + \omega^2 = 0, \tag{19}$$

so the eigenvalues are

$$\lambda_1, \lambda_2 = \frac{-\gamma \pm \sqrt{\gamma^2 - 4\omega^2}}{2}. \tag{20}$$

The nature of the solutions of Eqs. (8) and (17) depends on the sign of $\gamma^2 - 4\omega^2$ as follows:

1. If $\gamma^2 - 4\omega^2 > 0$, then the eigenvalues are real, unequal, and negative. The critical point $(0, 0)$ is an asymptotically stable node of the linear system (17) and of the almost linear system (8).

2. If $\gamma^2 - 4\omega^2 = 0$, then the eigenvalues are real, equal, and negative. The critical point $(0, 0)$ is an asymptotically stable (proper or improper) node of the linear system (17). It may be either an asymptotically stable node or spiral point of the almost linear system (8).

3. If $\gamma^2 - 4\omega^2 < 0$, then the eigenvalues are complex with a negative real part. The critical point $(0, 0)$ is an asymptotically stable spiral point of the linear system (17) and of the almost linear system (8).

Thus the critical point $(0, 0)$ is a spiral point of the system (8) if the damping is small and a node if the damping is large enough. In either case, the origin is asymptotically stable.

Let us now consider the case $\gamma^2 - 4\omega^2 < 0$, corresponding to small damping, in more detail. The direction of motion on the spirals near $(0, 0)$ can be obtained directly from Eqs. (8). Consider the point at which a spiral intersects the positive y-axis ($x = 0$ and $y > 0$). At such a point it follows from Eqs. (8) that $dx/dt > 0$. Thus the point (x, y) on the trajectory is moving to the right, so the direction of motion on the spirals is clockwise.

The behavior of the pendulum near the critical points $(\pm n\pi, 0)$, with n even, is the same as its behavior near the origin. We expect this on physical grounds since all these critical points correspond to the downward equilibrium position of the pendulum. The conclusion can be confirmed by repeating the analysis carried out above for the origin. Figure 7.2.3 shows the clockwise spirals at a few of these critical points.

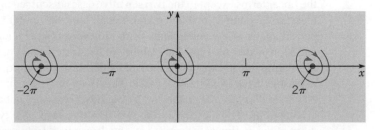

FIGURE 7.2.3 Asymptotically stable spiral points for the damped pendulum.

Now let us consider the critical point $(\pi, 0)$. Here, the nonlinear equations (8) are approximated by the linear system (18), whose eigenvalues and eigenvectors are determined from

$$\begin{pmatrix} -\lambda & 1 \\ \omega^2 & -\gamma - \lambda \end{pmatrix} \begin{pmatrix} v_1 \\ v_2 \end{pmatrix} = \begin{pmatrix} 0 \\ 0 \end{pmatrix}. \tag{21}$$

The characteristic equation of this system is

$$\lambda^2 + \gamma\lambda - \omega^2 = 0, \tag{22}$$

so the eigenvalues are

$$\lambda_1, \lambda_2 = \frac{-\gamma \pm \sqrt{\gamma^2 + 4\omega^2}}{2}. \tag{23}$$

One eigenvalue (λ_1) is positive and the other (λ_2) is negative. Therefore, regardless of the amount of damping, the critical point $x = \pi$, $y = 0$ is an (unstable) saddle point both of the linear system (18) and of the almost linear system (8).

To examine the behavior of trajectories near the saddle point $(\pi, 0)$ in more detail, we also need the eigenvectors. From the first row of Eq. (21), we have

$$-\lambda v_1 + v_2 = 0.$$

Therefore the eigenvectors are $(1, \lambda_1)$ and $(1, \lambda_2)$, corresponding to the eigenvalues λ_1 and λ_2, respectively. Consequently, the general solution of Eqs. (18) is

$$\begin{pmatrix} u \\ w \end{pmatrix} = C_1 e^{\lambda_1 t} \begin{pmatrix} 1 \\ \lambda_1 \end{pmatrix} + C_2 e^{\lambda_2 t} \begin{pmatrix} 1 \\ \lambda_2 \end{pmatrix}, \tag{24}$$

where C_1 and C_2 are arbitrary constants. Since $\lambda_1 > 0$ and $\lambda_2 < 0$, the first solution becomes unbounded and the second tends to zero as $t \to \infty$. Hence the trajectories that "enter" the saddle point are obtained by setting $C_1 = 0$. As they approach the saddle point, the entering trajectories are tangent to the line having slope $\lambda_2 < 0$. Thus one lies in the second quadrant $(C_2 < 0)$, and the other lies in the fourth quadrant $(C_2 > 0)$. For $C_2 = 0$, we obtain the pair of trajectories "exiting" from the saddle point. These trajectories have slope $\lambda_1 > 0$; one lies in the first quadrant $(C_1 > 0)$, and the other lies in the third quadrant $(C_1 < 0)$.

The situation is the same at other critical points $(n\pi, 0)$ with n odd. These all correspond to the upward equilibrium position of the pendulum, so we expect them to be unstable. The analysis at $(\pi, 0)$ can be repeated to show that they are saddle points oriented in the same way as the one at $(\pi, 0)$. Diagrams of the trajectories in the neighborhood of two saddle points are shown in Figure 7.2.4.

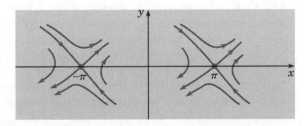

FIGURE 7.2.4 Unstable saddle points for the damped pendulum.

EXAMPLE
4

The equations of motion of a certain pendulum are

$$\frac{dx}{dt} = y, \qquad \frac{dy}{dt} = -9 \sin x - 0.2y, \tag{25}$$

where $x = \theta$ and $y = d\theta/dt$. Draw a phase portrait for this system and explain how it shows the possible motions of the pendulum.

By plotting the trajectories starting at various initial points in the phase plane, we obtain the phase portrait shown in Figure 7.2.5. As we have seen, the critical points (equilibrium solutions) are the points $(n\pi, 0)$, where $n = 0, \pm 1, \pm 2, \ldots$. Even values of n, including zero, correspond to the downward position of the pendulum, while odd values of n correspond to the upward position. Near each of the asymptotically stable critical points, the trajectories are clockwise spirals that represent a decaying oscillation about the downward equilibrium position. The wavy horizontal portions of the trajectories that occur for larger values of $|y|$ represent whirling motions of the pendulum. Note that a whirling motion cannot continue indefinitely, no matter how large $|y|$ is. Eventually, the angular velocity is so much reduced by the damping term that the pendulum can no longer go over the top, and instead begins to oscillate about its downward position.

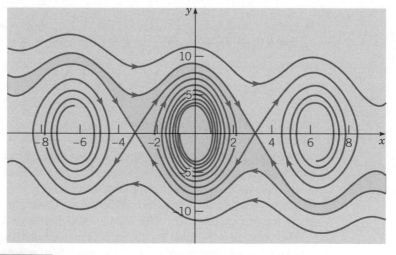

FIGURE 7.2.5   Phase portrait for the damped pendulum of Example 4.

Each of the asymptotically stable critical points has its own basin of attraction, that is, those points from which trajectories ultimately approach the given critical point. The basin of attraction for the origin is shown in color in Figure 7.2.5 and consists of a roughly oval region surrounding the origin together with two thin tails that extend infinitely far to either side. It is bounded by the trajectories (separatrices) that enter the saddle points at $(-\pi, 0)$ and $(\pi, 0)$. The basins of attraction for the other asymptotically stable critical points have the same shape as for the origin; they are simply shifted right or left by multiples of 2π. Note that it is mathematically possible (but physically unrealizable) to choose initial conditions on a separatrix so that the resulting motion leads to a balanced pendulum in a vertically upward position of unstable equilibrium.

An important difference between linear and nonlinear autonomous systems is illustrated by the pendulum equations. Recall that the linear system (1) has only the single critical point $\mathbf{x} = \mathbf{0}$ if $\det \mathbf{A} \neq 0$. Thus, if the origin is asymptotically stable, then not only do trajectories that start close to the origin approach it, but, in fact, every trajectory approaches the origin. In this case, the critical point $\mathbf{x} = \mathbf{0}$ is said to be **globally asymptotically stable**. This property of linear systems is not, in general, true for nonlinear systems. For nonlinear systems, it is important to determine (or to estimate) the basin of attraction for each asymptotically stable critical point.

PROBLEMS

In each of Problems 1 through 20:
(a) Determine all critical points of the given system of equations.
(b) Find the corresponding linear system near each critical point.
(c) Find the eigenvalues of each linear system. What conclusions can you then draw about the nonlinear system?
(d) Draw a phase portrait of the nonlinear system to confirm your conclusions, or to extend them in those cases where the linear system does not provide definite information about the nonlinear system.
(e) Draw a sketch of, or describe in words, the basin of attraction of each asymptotically stable critical point.

1. $dx/dt = -2x + y$, $dy/dt = x^2 - y$
2. $dx/dt = x - y$, $dy/dt = x - 3y + xy - 3$
3. $dx/dt = x + y^2$, $dy/dt = x + 2y$
4. $dx/dt = x - y^2$, $dy/dt = x - 2y + x^2$
5. $dx/dt = (4 + x)(y - x)$, $dy/dt = (10 - x)(y + x)$
6. $dx/dt = x - x^2 - xy$, $dy/dt = 3y - xy - 2y^2$
7. $dx/dt = 1 - y$, $dy/dt = x^2 - y^2$
8. $dx/dt = x - x^2 - 2xy$, $dy/dt = -y(x + 1)$
9. $dx/dt = -(x - y)(6 - x - y)$, $dy/dt = x(4 + y)$
10. $dx/dt = x + x^2 + y^2$, $dy/dt = y - xy$
11. $dx/dt = 2x + y + xy^3$, $dy/dt = x - 2y - xy$
12. $dx/dt = (2 + x)\sin y$, $dy/dt = 1 - x - \cos y$
13. $dx/dt = x - y^2$, $dy/dt = y - x^2$
14. $dx/dt = 3 - xy$, $dy/dt = x - 3y^3$
15. $dx/dt = -2x - y - x(x^2 + y^2)$,
    $dy/dt = x - y + y(x^2 + y^2)$
16. $dx/dt = y + x(1 - x^2 - y^2)$,
    $dy/dt = -x + y(1 - x^2 - y^2)$
17. $dx/dt = (2 + y)(y - 0.5x)$, $dy/dt = (2 - x)(y + 0.5x)$
18. $dx/dt = 4 - y^2$, $dy/dt = (1.5 + x)(y - x)$
19. $dx/dt = (1 - y)(2x - y)$, $dy/dt = (2 + x)(x - 2y)$
20. $dx/dt = 2x^2y - 3x^2 - 4y$, $dy/dt = -2xy^2 + 6xy$

21. Consider the autonomous system
$$\frac{dx}{dt} = y, \qquad \frac{dy}{dt} = x + 5x^3.$$
(a) Show that the origin is the only critical point.
(b) Find the approximating linear system near the origin. Write down the corresponding equation for dy/dx and solve it. Sketch the trajectories of the linear system and observe that the origin is a saddle point. From the parametric form of the solution, show that the trajectories that enter and leave the origin lie on the lines $y = -x$ and $y = x$, respectively.
(c) Now consider the original nonlinear equation. Write down the corresponding equation for dy/dx and solve it. Sketch the trajectories for the nonlinear system that correspond to $y = -x$ and $y = x$ for the linear system.

22. Consider the autonomous system
$$\frac{dx}{dt} = x, \qquad \frac{dy}{dt} = -5y + x^3.$$
(a) Show that the origin is the only critical point.
(b) Find the approximating linear system near the origin. Write down the corresponding equation for dy/dx and solve it. Sketch the trajectories of the linear system and observe that the origin is a saddle point. From the parametric form of the solution, show that the trajectories that enter and leave the origin lie on the y-axis and the x-axis, respectively.
(c) Now consider the original nonlinear equation. Write down the corresponding equation for dy/dx and solve it. Sketch the trajectories for the nonlinear system that correspond to $x = 0$ and $y = 0$ for the linear system.

23. The equations of motion of a certain nonlinear damped pendulum are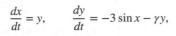
$$\frac{dx}{dt} = y, \qquad \frac{dy}{dt} = -3\sin x - \gamma y,$$
where γ is the damping coefficient. Suppose that the initial conditions are $x(0) = 0$, $y(0) = v$.
(a) For $\gamma = \frac{1}{4}$, plot x versus t for $v = 3$ and for $v = 6$. Explain these plots in terms of the motions of the pendulum that they represent. Also explain how they relate to the corresponding graphs in Problem 21 of Section 7.1.

(b) Let v_c be the critical value of the initial velocity where the transition from one type of motion to the other occurs. Determine v_c to two decimal places.

(c) Repeat part (b) for other values of γ and determine how v_c depends on γ.

24. Theorem 7.2.2 provides no information about the stability of a critical point of an almost linear system if that point is a center of the corresponding linear system. The systems

$$
\begin{aligned}
\frac{dx}{dt} &= y + x(x^2 + y^2), \\
\frac{dy}{dt} &= -x + y(x^2 + y^2)
\end{aligned}
\tag{i}
$$

and

$$
\begin{aligned}
\frac{dx}{dt} &= y - x(x^2 + y^2), \\
\frac{dy}{dt} &= -x - y(x^2 + y^2)
\end{aligned}
\tag{ii}
$$

show that this must be so.

(a) Show that $(0, 0)$ is a critical point of each system and, furthermore, is a center of the corresponding linear system.

(b) Show that each system is almost linear.

(c) Let $r^2 = x^2 + y^2$, and note that $x\, dx/dt + y\, dy/dt = r\, dr/dt$. For system (ii), show that $dr/dt < 0$ and that $r \to 0$ as $t \to \infty$; hence the critical point is asymptotically stable. For system (i), show that the solution of the initial value problem for r with $r = r_0$ at $t = 0$ becomes unbounded as $t \to \frac{1}{2}r_0^2$; hence the critical point is unstable.

25. In this problem, we show how small changes in the coefficients of a system of linear equations can affect a critical point that is a center. Consider the system

$$
\mathbf{x}' = \begin{pmatrix} 0 & 1 \\ -1 & 0 \end{pmatrix} \mathbf{x}.
$$

Show that the eigenvalues are $\pm i$ so that $(0, 0)$ is a center. Now consider the system

$$
\mathbf{x}' = \begin{pmatrix} \epsilon & 1 \\ -1 & \epsilon \end{pmatrix} \mathbf{x},
$$

where $|\epsilon|$ is arbitrarily small. Show that the eigenvalues are $\epsilon \pm i$. Thus no matter how small $|\epsilon| \neq 0$ is, the center

becomes a spiral point. If $\epsilon < 0$, the spiral point is asymptotically stable; if $\epsilon > 0$, the spiral point is unstable.

26. In this problem, we show how small changes in the coefficients of a system of linear equations can affect the nature of a critical point when the eigenvalues are equal. Consider the system

$$
\mathbf{x}' = \begin{pmatrix} -1 & 1 \\ 0 & -1 \end{pmatrix} \mathbf{x}.
$$

Show that the eigenvalues are $\lambda_1 = -1$, $\lambda_2 = -1$ so that the critical point $(0, 0)$ is an asymptotically stable node. Now consider the system

$$
\mathbf{x}' = \begin{pmatrix} -1 & 1 \\ -\epsilon & -1 \end{pmatrix} \mathbf{x},
$$

where $|\epsilon|$ is arbitrarily small. Show that if $\epsilon > 0$, then the eigenvalues are $-1 \pm i\sqrt{\epsilon}$, so that the asymptotically stable node becomes an asymptotically stable spiral point. If $\epsilon < 0$, then the eigenvalues are $-1 \pm \sqrt{|\epsilon|}$, and the critical point remains an asymptotically stable node.

27. A generalization of the damped pendulum equation discussed in the text, or a damped spring-mass system, is the Liénard equation

$$
\frac{d^2 x}{dt^2} + c(x)\frac{dx}{dt} + g(x) = 0.
$$

If $c(x)$ is a constant and $g(x) = kx$, then this equation has the form of the linear pendulum equation [replace $\sin\theta$ with θ in Eq. (13) of Section 7.1]; otherwise, the damping force $c(x)\, dx/dt$ and the restoring force $g(x)$ are nonlinear. Assume that c is continuously differentiable, g is twice continuously differentiable, and $g(0) = 0$.

(a) Write the Liénard equation as a system of two first order equations by introducing the variable $y = dx/dt$.

(b) Show that $(0, 0)$ is a critical point and that the system is almost linear in the neighborhood of $(0, 0)$.

(c) Show that if $c(0) > 0$ and $g'(0) > 0$, then the critical point is asymptotically stable, and that if $c(0) < 0$ or $g'(0) < 0$, then the critical point is unstable.

Hint: Use Taylor series to approximate c and g in the neighborhood of $x = 0$.

7.3 Competing Species

In this section we use phase plane methods to investigate some problems involving competition for scarce resources. We will express the equations in terms of two species competing for the same food supply. However the same or similar models have also been used to study other competitive situations, for example, businesses competing in the same economic markets.

The problems that we consider here involve two interacting populations and are extensions of those discussed in Section 2.5, which dealt with a single population. Although the models described here are extremely simple compared to the very complex relationships that exist in nature, it is still possible to acquire some insight into ecological principles from a study of these idealized problems.

Suppose that in some closed environment there are two similar species competing for a limited food supply—for example, two species of fish in a pond that do not prey on each other but do compete for the available food. Let x and y be the populations of the two species at time t. As discussed in Section 2.5, we assume that the population of each of the species, in the absence of the other, is governed by a logistic equation. Thus

$$\frac{dx}{dt} = x(\epsilon_1 - \sigma_1 x), \tag{1a}$$

$$\frac{dy}{dt} = y(\epsilon_2 - \sigma_2 y), \tag{1b}$$

respectively, where ϵ_1 and ϵ_2 are the growth rates of the two populations, and ϵ_1/σ_1 and ϵ_2/σ_2 are their saturation levels. In addition, when both species are present, each will impinge on the available food supply for the other. In effect, they reduce each other's growth rates and saturation populations. The simplest way to reduce the growth rate of species x due to the presence of species y is to replace the growth rate factor $\epsilon_1 - \sigma_1 x$ in Eq. (1a) by $\epsilon_1 - \sigma_1 x - \alpha_1 y$, where α_1 is a measure of the degree to which species y interferes with species x. Similarly, in Eq. (1b) we replace $\epsilon_2 - \sigma_2 y$ by $\epsilon_2 - \sigma_2 y - \alpha_2 x$, where α_2 is a measure of the degree to which species x interferes with species y. Thus we have the system of equations

$$\begin{aligned}\frac{dx}{dt} &= x(\epsilon_1 - \sigma_1 x - \alpha_1 y), \\ \frac{dy}{dt} &= y(\epsilon_2 - \sigma_2 y - \alpha_2 x).\end{aligned} \tag{2}$$

The values of the positive constants ϵ_1, σ_1, α_1, ϵ_2, σ_2, and α_2 depend on the particular species under consideration and, in general, must be determined from observations. We are interested in solutions of Eqs. (2) for which x and y are nonnegative. In the following two examples, we discuss two typical problems in some detail. At the end of the section, we return to the general equations (2).

EXAMPLE 1

Discuss the qualitative behavior of solutions of the system

$$\begin{aligned}\frac{dx}{dt} &= x(1 - x - y), \\ \frac{dy}{dt} &= y(0.75 - y - 0.5x).\end{aligned} \tag{3}$$

We find the critical points by solving the system of algebraic equations

$$x(1 - x - y) = 0, \qquad y(0.75 - y - 0.5x) = 0. \tag{4}$$

The first equation can be satisfied by choosing $x = 0$; then the second equation requires that $y = 0$ or $y = 0.75$. Similarly, the second equation can be satisfied by choosing $y = 0$, and then the first equation requires that $x = 0$ or $x = 1$. Thus we have found three critical points, namely, $(0, 0)$, $(0, 0.75)$, and $(1, 0)$. Equations (4) are also satisfied by solutions of the system

$$1 - x - y = 0, \qquad 0.75 - y - 0.5x = 0,$$

which leads to a fourth critical point $(0.5, 0.5)$. The four critical points correspond to equilibrium solutions of the system (3). The first three of these points involve the extinction of one or both species; only the last corresponds to the long-term survival of both species. Other solutions are represented as curves or trajectories in the xy-plane that describe the evolution of the populations in time. To begin to discover their qualitative behavior, we can proceed in the following way.

First, observe that the coordinate axes are themselves trajectories. This follows directly from Eqs. (3) since $dx/dt = 0$ on the y-axis (where $x = 0$) and, similarly, $dy/dt = 0$ on the x-axis. Thus no other trajectories can cross the coordinate axes. For a population problem, only nonnegative values of x and y are significant, and we conclude that any trajectory that starts in the first quadrant must remain there for all time.

A direction field for the system (3) in the positive quadrant is shown in Figure 7.3.1. The heavy dots in this figure are the critical points or equilibrium solutions. Based on the direction field, it appears that the point (0.5, 0.5) attracts other solutions and is therefore asymptotically stable, while the other three critical points are unstable. To confirm these conclusions, we can look at the linear approximations near each critical point.

The system (3) is almost linear in the neighborhood of each critical point. There are two ways to obtain the linear system near a critical point (X, Y). First, we can use the substitution $x = X + u$, $y = Y + w$ in Eqs. (3), retaining only the terms that are linear in u and w. Alternatively, we can evaluate the Jacobian matrix

$$\mathbf{J} = \begin{pmatrix} F_x(X, Y) & F_y(X, Y) \\ G_x(X, Y) & G_y(X, Y) \end{pmatrix} \tag{5}$$

to find the coefficient matrix of the approximating linear system. When several critical points are to be investigated, it is usually more efficient to use the Jacobian matrix. For the system (3),

$$F(x, y) = x(1 - x - y), \qquad G(x, y) = y(0.75 - y - 0.5x), \tag{6}$$

so by evaluating $\mathbf{J}$, we obtain the linear system

$$\frac{d}{dt}\begin{pmatrix} u \\ w \end{pmatrix} = \begin{pmatrix} 1 - 2X - Y & -X \\ -0.5Y & 0.75 - 2Y - 0.5X \end{pmatrix}\begin{pmatrix} u \\ w \end{pmatrix}, \tag{7}$$

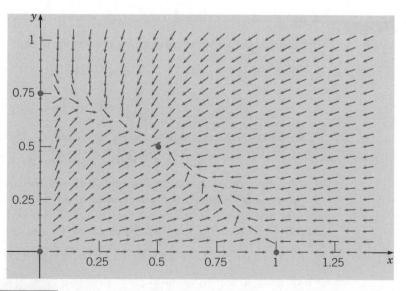

FIGURE 7.3.1 Critical points and direction field for the system (3).

where $u = x - X$ and $w = y - Y$. The system (7) is the approximate linear system near the critical point (X, Y). We will now examine each of the four critical points in turn.

(i) x = 0, y = 0. This critical point corresponds to a state in which both species die as a result of their competition. Separating the linear and nonlinear terms in the system (3) leads to

$$\frac{d}{dt}\begin{pmatrix} x \\ y \end{pmatrix} = \begin{pmatrix} 1 & 0 \\ 0 & 0.75 \end{pmatrix}\begin{pmatrix} x \\ y \end{pmatrix} - \begin{pmatrix} x^2 + xy \\ 0.5xy + y^2 \end{pmatrix}. \tag{8}$$

Alternatively, we can set $X = Y = 0$ in Eq. (7), Either way, we find that near the origin the corresponding linear system is

$$\frac{d}{dt}\begin{pmatrix} x \\ y \end{pmatrix} = \begin{pmatrix} 1 & 0 \\ 0 & 0.75 \end{pmatrix}\begin{pmatrix} x \\ y \end{pmatrix}. \tag{9}$$

The eigenvalues of the matrix in system (9) are $\lambda_1 = 1$ and $\lambda_2 = 0.75$. Thus the origin is an unstable node of the linear system (9) and of the nonlinear system (8) or (3).

Moreover, $\mathbf{v}_1 = (1, 0)^T$ and $\mathbf{v}_2 = (0, 1)^T$ are eigenvectors corresponding to λ_1 and λ_2, respectively. In the neighborhood of the origin, all trajectories are tangent to the y-axis except for one trajectory that lies along the x-axis.

The general solution of system (3) is

$$\begin{pmatrix} x \\ y \end{pmatrix} = c_1 e^t \begin{pmatrix} 1 \\ 0 \end{pmatrix} + c_2 e^{0.75t} \begin{pmatrix} 0 \\ 1 \end{pmatrix}. \tag{10}$$

(ii) x = 1, y = 0. This corresponds to a state in which species x survives the competition but species y does not. By evaluating the Jacobian matrix in system (7) for $X = 1$, $Y = 0$, we find that the linear system corresponding to the system (3) near the critical point $(1, 0)$ is

$$\frac{d}{dt}\begin{pmatrix} u \\ w \end{pmatrix} = \begin{pmatrix} -1 & -1 \\ 0 & 0.25 \end{pmatrix}\begin{pmatrix} u \\ w \end{pmatrix}. \tag{11}$$

The Jacobian matrix has eigenvalues $\lambda_1 = -1$ and $\lambda_2 = 0.25$. Since the eigenvalues have opposite signs, the point $(1, 0)$ is a saddle point, and hence is an unstable equilibrium point of the linear system (11) and of the nonlinear system (3).

The corresponding eigenvectors are $\mathbf{v}_1 = (1, 0)^T$ and $\mathbf{v}_2 = (4, -5)^T$, and the general solution of the system (11) is

$$\begin{pmatrix} u \\ w \end{pmatrix} = c_1 e^{-t} \begin{pmatrix} 1 \\ 0 \end{pmatrix} + c_2 e^{0.25t} \begin{pmatrix} 4 \\ -5 \end{pmatrix}. \tag{12}$$

The behavior of the trajectories near $(1, 0)$ can be seen from Eq. (12). If $c_2 = 0$, then there is one pair of trajectories that approaches the critical point along the x-axis. If $c_1 = 0$, then another pair of trajectories leaves the critical point tangent to the line of slope $-\frac{5}{4}$. All other trajectories also depart from the neighborhood of $(1, 0)$.

(iii) x = 0, y = 0.75. In this case, species y survives but x does not. The analysis is similar to that for the point $(1, 0)$. The corresponding linear system is

$$\frac{d}{dt}\begin{pmatrix} u \\ w \end{pmatrix} = \begin{pmatrix} 0.25 & 0 \\ -0.375 & -0.75 \end{pmatrix}\begin{pmatrix} u \\ w \end{pmatrix}. \tag{13}$$

The eigenvalues and eigenvectors are

$$\lambda_1 = 0.25, \quad \mathbf{v}_1 = \begin{pmatrix} 8 \\ -3 \end{pmatrix}; \quad \lambda_2 = -0.75, \quad \mathbf{v}_2 = \begin{pmatrix} 0 \\ 1 \end{pmatrix}, \tag{14}$$

so the general solution of Eq. (13) is

$$\begin{pmatrix} u \\ w \end{pmatrix} = c_1 e^{0.25t} \begin{pmatrix} 8 \\ -3 \end{pmatrix} + c_2 e^{-0.75t} \begin{pmatrix} 0 \\ 1 \end{pmatrix}. \tag{15}$$

From the observation that λ_1 and λ_2 have opposite signs, the point $(0, 0.75)$ is also a saddle point. The additional information provided by the eigenvectors allow us to conclude that one pair of trajectories approaches the critical point along the y-axis and another departs tangent to the line with slope $-\frac{3}{8}$. All other trajectories also leave the neighborhood of the critical point.

(iv) x = 0.5, y = 0.5. This critical point corresponds to an equilibrium state in which both species coexist (because $x > 0$ and $y > 0$). The eigenvalues and eigenvectors of the corresponding linear system

$$\frac{d}{dt} \begin{pmatrix} u \\ w \end{pmatrix} = \begin{pmatrix} -0.5 & -0.5 \\ -0.25 & -0.5 \end{pmatrix} \begin{pmatrix} u \\ w \end{pmatrix} \tag{16}$$

are

$$\lambda_1 = \frac{-2 + \sqrt{2}}{4} \cong -0.146, \quad \mathbf{v}_1 = \begin{pmatrix} \sqrt{2} \\ -1 \end{pmatrix};$$

$$\lambda_2 = \frac{-2 - \sqrt{2}}{4} \cong -0.854, \quad \mathbf{v}_2 = \begin{pmatrix} \sqrt{2} \\ 1 \end{pmatrix}. \tag{17}$$

Therefore the general solution of Eq. (16) is

$$\begin{pmatrix} u \\ w \end{pmatrix} = c_1 e^{-0.146t} \begin{pmatrix} \sqrt{2} \\ -1 \end{pmatrix} + c_2 e^{-0.854t} \begin{pmatrix} \sqrt{2} \\ 1 \end{pmatrix}. \tag{18}$$

Since both eigenvalues are negative, the critical point $(0.5, 0.5)$ is an asymptotically stable node of the linear system (16) and of the nonlinear system (3). All trajectories approach the critical point as $t \to \infty$. One pair of trajectories approaches the critical point along the line with slope $\sqrt{2}/2$ determined from the eigenvector $\mathbf{v}_2$. All other trajectories approach the critical point tangent to the line with slope $-\sqrt{2}/2$ determined from the eigenvector $\mathbf{v}_1$.

A phase portrait for the system (3) is shown in Figure 7.3.2. By looking closely at the trajectories near each critical point, you can see that they behave in the manner predicted by the linear system near that point. In addition, note that the quadratic terms on the right side of Eqs. (3) are all negative. Since for x and y large and positive, these terms are the dominant ones, it follows that far from the origin in the first quadrant both x' and y' are negative, that is, the trajectories are directed inward. Thus all trajectories that start at a point (x_0, y_0) with $x_0 > 0$ and $y_0 > 0$ eventually approach the point $(0.5, 0.5)$.

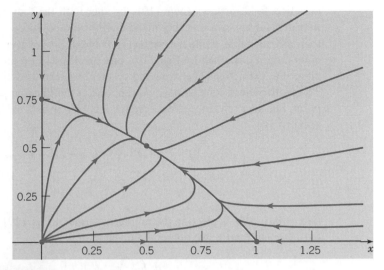

FIGURE 7.3.2 A phase portrait of the system (3).

EXAMPLE
2

Discuss the qualitative behavior of the solutions of the system

$$\frac{dx}{dt} = x(1 - x - y),$$
$$\frac{dy}{dt} = y(0.8 - 0.6y - x),$$
(19)

when x and y are nonnegative. Observe that this system is also a special case of Eq. (2) for two competing species.

Once again, there are four critical points, namely, $(0, 0)$, $(1, 0)$, $(0, \frac{4}{3})$, and $(0.5, 0.5)$, corresponding to the four equilibrium solutions of the system (19). Figure 7.3.3 shows a

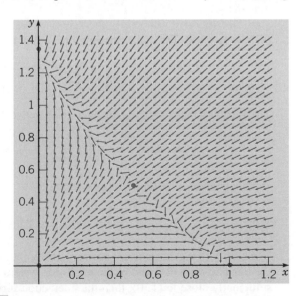

FIGURE 7.3.3 Critical points and direction field for the system (19).

direction field for the system (19), together with the four critical points. From the direction field, it appears that the mixed equilibrium solution (0.5, 0.5) is a saddle point, and therefore unstable, while the points (1, 0) and (0, $\frac{4}{3}$) are asymptotically stable. Thus, for the competition described by Eqs. (19), one species will eventually overwhelm the other and drive it to extinction. The surviving species is determined by the initial state of the system. To confirm these conclusions, we can look at the linear approximations near each critical point. For later reference, we record the Jacobian matrix for Eqs. (19) evaluated at a critical point (X, Y):

$$\begin{pmatrix} F_x(X, Y) & F_y(X, Y) \\ G_x(X, Y) & G_y(X, Y) \end{pmatrix} = \begin{pmatrix} 1 - 2X - Y & -X \\ -Y & 0.8 - 1.2Y - X \end{pmatrix}. \tag{20}$$

(i) x = 0, y = 0. Neglecting the nonlinear terms in Eqs. (19), or setting $X = 0$ and $Y = 0$ in the Jacobian matrix, we obtain the linear system

$$\frac{d}{dt} \begin{pmatrix} x \\ y \end{pmatrix} = \begin{pmatrix} 1 & 0 \\ 0 & 0.8 \end{pmatrix} \begin{pmatrix} x \\ y \end{pmatrix}, \tag{21}$$

which is valid near the origin. The eigenvalues and eigenvectors of the matrix in system (21) are

$$\lambda_1 = 1, \qquad \mathbf{v}_1 = \begin{pmatrix} 1 \\ 0 \end{pmatrix}; \qquad \lambda_2 = 0.8, \qquad \mathbf{v}_2 = \begin{pmatrix} 0 \\ 1 \end{pmatrix}, \tag{22}$$

so the general solution of linear system (21) is

$$\begin{pmatrix} x \\ y \end{pmatrix} = c_1 e^t \begin{pmatrix} 1 \\ 0 \end{pmatrix} + c_2 e^{0.8t} \begin{pmatrix} 0 \\ 1 \end{pmatrix}. \tag{23}$$

Since this Jacobian matrix has two positive eigenvalues, the origin is an unstable node of the linear system (21) and also of the nonlinear system (19). The eigenvectors tell us that all trajectories leave the origin tangent to the y-axis except for one trajectory that lies along the x-axis.

(ii) x = 1, y = 0. The corresponding linear system is

$$\frac{d}{dt} \begin{pmatrix} u \\ w \end{pmatrix} = \begin{pmatrix} -1 & -1 \\ 0 & -0.2 \end{pmatrix} \begin{pmatrix} u \\ w \end{pmatrix}. \tag{24}$$

The Jacobian matrix has eigenvalues and eigenvectors

$$\lambda_1 = -1, \qquad \mathbf{v}_1 = \begin{pmatrix} 1 \\ 0 \end{pmatrix}; \qquad \lambda_2 = -0.2, \qquad \mathbf{v}_2 = \begin{pmatrix} 5 \\ -4 \end{pmatrix}. \tag{25}$$

and the general solution of linear system (24) is

$$\begin{pmatrix} u \\ w \end{pmatrix} = c_1 e^{-t} \begin{pmatrix} 1 \\ 0 \end{pmatrix} + c_2 e^{-0.2t} \begin{pmatrix} 5 \\ -4 \end{pmatrix}. \tag{26}$$

Since both eigenvalues are negative, the point (1, 0) is an asymptotically stable node of the linear system (24) and of the nonlinear system (19). If the initial values of x and y are sufficiently close to (1, 0), then the interaction process leads ultimately to that state, that is, to the survival of species x and to the extinction of species y. The pair of trajectories with $c_2 = 0$ approaches the critical point along the x-axis. All other trajectories approach $(1, 0)$ tangent to the line with slope $-\frac{4}{5}$ that is determined by the eigenvector $\mathbf{v}_2$.

(iii) $x = 0$, $y = \frac{4}{3}$. The analysis in this case is similar to that for the point $(1, 0)$. The appropriate linear system is

$$\frac{d}{dt}\begin{pmatrix} u \\ w \end{pmatrix} = \begin{pmatrix} -1/3 & 0 \\ -4/3 & -4/5 \end{pmatrix}\begin{pmatrix} u \\ w \end{pmatrix}. \tag{27}$$

The eigenvalues and eigenvectors of this system are

$$\lambda_1 = -\frac{1}{3}, \qquad \mathbf{v}_1 = \begin{pmatrix} 1 \\ -20/7 \end{pmatrix}; \qquad \lambda_2 = -\frac{4}{5}, \qquad \mathbf{v}_2 = \begin{pmatrix} 0 \\ 1 \end{pmatrix}, \tag{28}$$

and its general solution is

$$\begin{pmatrix} u \\ w \end{pmatrix} = c_1 e^{-t/3}\begin{pmatrix} 1 \\ -20/7 \end{pmatrix} + c_2 e^{-4t/5}\begin{pmatrix} 0 \\ 1 \end{pmatrix}. \tag{29}$$

Thus the critical point $(0, 4/3)$ is an asymptotically stable node of both the linear system (27) and the nonlinear system (19). All nearby trajectories approach the critical point tangent to the line with slope $-20/7$ except for one pair of trajectories that lies along the y-axis.

(iv) $x = 0.5$, $y = 0.5$. The corresponding linear system is

$$\frac{d}{dt}\begin{pmatrix} u \\ w \end{pmatrix} = \begin{pmatrix} -0.5 & -0.5 \\ -0.5 & -0.3 \end{pmatrix}\begin{pmatrix} u \\ w \end{pmatrix}. \tag{30}$$

The eigenvalues and eigenvectors are

$$\lambda_1 = \frac{-4 + \sqrt{26}}{10} \cong 0.1099, \qquad \mathbf{v}_1 = \begin{pmatrix} 5 \\ -1 - \sqrt{26} \end{pmatrix} \cong \begin{pmatrix} 5 \\ -6.0990 \end{pmatrix},$$

$$\lambda_2 = \frac{-4 - \sqrt{26}}{10} \cong -0.9099, \qquad \mathbf{v}_2 = \begin{pmatrix} 5 \\ -1 + \sqrt{26} \end{pmatrix} \cong \begin{pmatrix} 5 \\ 4.0990 \end{pmatrix}, \tag{31}$$

so the general solution is

$$\begin{pmatrix} u \\ w \end{pmatrix} = c_1 e^{0.1099t}\begin{pmatrix} 5 \\ -6.0990 \end{pmatrix} + c_2 e^{-0.9099t}\begin{pmatrix} 5 \\ 4.0990 \end{pmatrix}. \tag{32}$$

Since the eigenvalues are of opposite sign, the critical point $(0.5, 0.5)$ is a saddle point and therefore is unstable, as we had surmised earlier. One pair of trajectories approaches the critical point as $t \to \infty$; all others depart from it. The trajectories that approach the critical point are tangent to the line with slope $(\sqrt{26} - 1)/5 \cong 0.8198$ determined from the eigenvector $\mathbf{v}_2$. Likewise, eigenvector $\mathbf{v}_1$ informs us the trajectories departing from $(0.5, 0.5)$ are tangent to the line with slope $(-\sqrt{26} - 1)/5 \cong 1.2198$.

A phase portrait for the system (19) is shown in Figure 7.3.4. Near each of the critical points the trajectories of the nonlinear system behave as predicted by the corresponding linear approximation. Of particular interest is the pair of trajectories that enter the saddle point. These trajectories form a separatrix that divides the first quadrant into two basins of attraction. Trajectories starting above the separatrix ultimately approach the node at $(0, 4/3)$, while trajectories starting below the separatrix approach the node at $(1, 0)$. If the initial state lies precisely on the separatrix, then the solution (x, y) will approach the saddle point as $t \to \infty$. However the slightest perturbation as one follows this trajectory will dislodge the point (x, y) from the separatrix and cause it to approach one of the nodes instead. Thus, in practice, one species will survive the competition and the other will not.

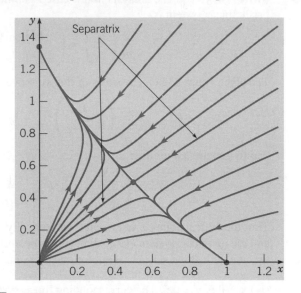

FIGURE 7.3.4 A phase portrait of the system (19).

Examples 1 and 2 show that in some cases the competition between two species leads to an equilibrium state of coexistence, whereas in other cases the competition results in the eventual extinction of one of the species. To understand more clearly how and why this happens, and to learn how to predict which situation will occur, it is useful to look again at the general system (2). There are four cases to be considered, depending on the relative orientation of the lines

$$\epsilon_1 - \sigma_1 x - \alpha_1 y = 0 \qquad \text{and} \qquad \epsilon_2 - \sigma_2 y - \alpha_2 x = 0, \tag{33}$$

as shown in Figure 7.3.5. These lines are called the x and y **nullclines**, respectively, because x' is zero on the first and y' is zero on the second. In each part of Figure 7.3.5, the x-nullcline is the solid line and the y-nullcline is the dashed line.

Let (X, Y) denote any critical point in any one of the four cases. As in Examples 1 and 2, the system (2) is almost linear in the neighborhood of this point because the right side of each differential equation is a quadratic polynomial. To study the system (2) in the neighborhood of this critical point, we can look at the corresponding linear system

$$\frac{d}{dt}\begin{pmatrix} u \\ w \end{pmatrix} = \begin{pmatrix} \epsilon_1 - 2\sigma_1 X - \alpha_1 Y & -\alpha_1 X \\ -\alpha_2 Y & \epsilon_2 - 2\sigma_2 Y - \alpha_2 X \end{pmatrix}\begin{pmatrix} u \\ w \end{pmatrix}. \tag{34}$$

We now use Eq. (34) to determine the conditions under which the model described by Eqs. (2) permits the coexistence of the two species x and y. Of the four possible cases shown in Figure 7.3.5, coexistence is possible only in cases (c) and (d). In these cases, the nonzero values of X and Y are readily obtained by solving the algebraic equations (33); the result is

$$X = \frac{\epsilon_1\sigma_2 - \epsilon_2\alpha_1}{\sigma_1\sigma_2 - \alpha_1\alpha_2}, \qquad Y = \frac{\epsilon_2\sigma_1 - \epsilon_1\alpha_2}{\sigma_1\sigma_2 - \alpha_1\alpha_2}. \tag{35}$$

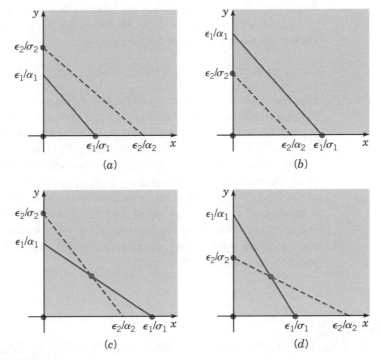

FIGURE 7.3.5 The various cases for the competing species system (2).

Further, since $\epsilon_1 - \sigma_1 X - \alpha_1 Y = 0$ and $\epsilon_2 - \sigma_2 Y - \alpha_2 X = 0$, Eq. (34) immediately reduces to

$$\frac{d}{dt}\begin{pmatrix} u \\ w \end{pmatrix} = \begin{pmatrix} -\sigma_1 X & -\alpha_1 X \\ -\alpha_2 Y & -\sigma_2 Y \end{pmatrix}\begin{pmatrix} u \\ w \end{pmatrix}. \tag{36}$$

The eigenvalues of the system (36) are found from the equation

$$\lambda^2 + (\sigma_1 X + \sigma_2 Y)\lambda + (\sigma_1 \sigma_2 - \alpha_1 \alpha_2)XY = 0. \tag{37}$$

Thus

$$\lambda_{1,2} = \frac{-(\sigma_1 X + \sigma_2 Y) \pm \sqrt{(\sigma_1 X + \sigma_2 Y)^2 - 4(\sigma_1 \sigma_2 - \alpha_1 \alpha_2)XY}}{2}. \tag{38}$$

If $\sigma_1 \sigma_2 - \alpha_1 \alpha_2 < 0$, then the radicand of Eq. (38) is positive and greater than $(\sigma_1 X + \sigma_2 Y)^2$. Thus the eigenvalues are real and of opposite sign. Consequently, the critical point (X, Y) is an unstable saddle point, and coexistence is not possible. This is the case in Example 2, where $\sigma_1 = 1$, $\alpha_1 = 1$, $\sigma_2 = 0.6$, $\alpha_2 = 1$, and $\sigma_1 \sigma_2 - \alpha_1 \alpha_2 = -0.4$.

On the other hand, if $\sigma_1 \sigma_2 - \alpha_1 \alpha_2 > 0$, then the radicand of Eq. (38) is less than $(\sigma_1 X + \sigma_2 Y)^2$. Thus the eigenvalues are real, negative, and unequal, or complex with a negative real part. A straightforward analysis of the radicand of Eq. (38) shows that the eigenvalues cannot be complex (see Problem 7). Thus the critical point is an asymptotically stable node, and sustained coexistence is possible. This is illustrated by Example 1, where $\sigma_1 = 1$, $\alpha_1 = 1$, $\sigma_2 = 1$, $\alpha_2 = 0.5$, and $\sigma_1 \sigma_2 - \alpha_1 \alpha_2 = 0.5$.

Let us relate this result to Figures 7.3.5c and 7.3.5d. In Figure 7.3.5c we have

$$\frac{\epsilon_1}{\sigma_1} > \frac{\epsilon_2}{\alpha_2} \quad \text{or} \quad \epsilon_1 \alpha_2 > \epsilon_2 \sigma_1 \qquad \text{and} \qquad \frac{\epsilon_2}{\sigma_2} > \frac{\epsilon_1}{\alpha_1} \quad \text{or} \quad \epsilon_2 \alpha_1 > \epsilon_1 \sigma_2. \tag{39}$$

These inequalities, coupled with the condition that X and Y given by Eqs. (35) be positive, yield the inequality $\sigma_1\sigma_2 < \alpha_1\alpha_2$. Hence, in this case, the critical point is a saddle point. On the other hand, in Figure 7.3.5d we have

$$\frac{\epsilon_1}{\sigma_1} < \frac{\epsilon_2}{\alpha_2} \quad \text{or} \quad \epsilon_1\alpha_2 < \epsilon_2\sigma_1 \quad \text{and} \quad \frac{\epsilon_2}{\sigma_2} < \frac{\epsilon_1}{\alpha_1} \quad \text{or} \quad \epsilon_2\alpha_1 < \epsilon_1\sigma_2. \tag{40}$$

Now the condition that X and Y be positive yields $\sigma_1\sigma_2 > \alpha_1\alpha_2$. Hence the critical point is asymptotically stable. For this case, we can also show that the other critical points $(0, 0)$, $(\epsilon_1/\sigma_1, 0)$, and $(0, \epsilon_2/\sigma_2)$ are unstable. Thus for any positive initial values of x and y, the two populations approach the equilibrium state of coexistence given by Eqs. (35).

Equations (2) provide the biological interpretation of the result that coexistence occurs or not depending on whether $\sigma_1\sigma_2 - \alpha_1\alpha_2$ is positive or negative. The σ's are a measure of the inhibitory effect that the growth of each population has on itself, while the α's are a measure of the inhibiting effect that the growth of each population has on the other species. Thus, when $\sigma_1\sigma_2 > \alpha_1\alpha_2$, interaction (competition) is "weak" and the species can coexist; when $\sigma_1\sigma_2 < \alpha_1\alpha_2$, interaction (competition) is "strong" and the species cannot coexist—one must die out.

PROBLEMS

Each of Problems 1 through 6 can be interpreted as describing the interaction of two species with populations x and y. In each of these problems, carry out the following steps.
(a) Draw a direction field and describe how solutions seem to behave.
(b) Find the critical points.
(c) For each critical point, find the corresponding linear system. Find the eigenvalues and eigenvectors of the linear system, classify each critical point as to type, and determine whether it is asymptotically stable, stable, or unstable.
(d) Sketch the trajectories in the neighborhood of each critical point.
(e) Compute and plot enough trajectories of the given system to show clearly the behavior of the solutions.
(f) Determine the limiting behavior of x and y as $t \to \infty$, and interpret the results in terms of the populations of the two species.

1. $dx/dt = x(1.5 - x - 0.5y)$, $dy/dt = y(2 - y - 0.75x)$
2. $dx/dt = x(1.5 - x - 0.5y)$, $dy/dt = y(2 - 0.5y - 1.5x)$
3. $dx/dt = x(1.5 - 0.5x - y)$, $dy/dt = y(2 - y - 1.125x)$
4. $dx/dt = x(1.5 - 0.5x - y)$, $dy/dt = y(0.75 - y - 0.125x)$
5. $dx/dt = x(1 - x - y)$, $dy/dt = y(1.5 - y - x)$
6. $dx/dt = x(1 - x + 0.5y)$, $dy/dt = y(2.5 - 1.5y + 0.25x)$
7. Show that

$$(\sigma_1 X + \sigma_2 Y)^2 - 4(\sigma_1\sigma_2 - \alpha_1\alpha_2)XY$$
$$= (\sigma_1 X - \sigma_2 Y)^2 + 4\alpha_1\alpha_2 XY.$$

Hence conclude that the eigenvalues given by Eq. (38) can never be complex.

8. Consider the system (2) in the text, and assume that $\sigma_1\sigma_2 - \alpha_1\alpha_2 = 0$.
(a) Find all the critical points of the system. Observe that the result depends on whether $\sigma_1\epsilon_2 - \alpha_2\epsilon_1$ is zero.
(b) If $\sigma_1\epsilon_2 - \alpha_2\epsilon_1 > 0$, classify each critical point and determine whether it is asymptotically stable, stable, or unstable. Note that Problem 5 is of this type. Then do the same if $\sigma_1\epsilon_2 - \alpha_2\epsilon_1 < 0$.
(c) Analyze the qualitative behavior of the trajectories when $\sigma_1\epsilon_2 - \alpha_2\epsilon_1 = 0$.

9. Consider the system (3) in Example 1 of the text. Recall that this system has an asymptotically stable critical point at $(0.5, 0.5)$, corresponding to the stable coexistence of the two population species. Now suppose that immigration or emigration occurs at the constant rates of δa and δb for the species x and y, respectively. In this case, Eqs. (3) are replaced by

$$\frac{dx}{dt} = x(1 - x - y) + \delta a,$$
$$\frac{dy}{dt} = y(0.75 - y - 0.5x) + \delta b. \tag{i}$$

The question is what effect this has on the location of the stable equilibrium point.
(a) To find the new critical point, we must solve the equations

$$x(1 - x - y) + \delta a = 0,$$
$$y(0.75 - y - 0.5x) + \delta b = 0. \tag{ii}$$

One way to proceed is to assume that x and y are given by power series in the parameter δ; thus

$$x = x_0 + x_1\delta + \cdots, \qquad y = y_0 + y_1\delta + \cdots. \qquad \text{(iii)}$$

Substitute Eqs. (iii) into Eqs. (ii) and collect terms according to powers of δ.

(b) From the constant terms (the terms not involving δ), show that $x_0 = 0.5$ and $y_0 = 0.5$, thus confirming that, in the absence of immigration or emigration, the critical point is $(0.5, 0.5)$.

(c) From the terms that are linear in δ, show that

$$x_1 = 4a - 4b, \qquad y_1 = -2a + 4b. \qquad \text{(iv)}$$

(d) Suppose that $a > 0$ and $b > 0$ so that immigration occurs for both species. Show that the resulting equilibrium solution may represent an increase in both populations, or an increase in one but a decrease in the other. Explain intuitively why this is a reasonable result.

10. The system

$$x' = -y, \qquad y' = -\gamma y - x(x - 0.15)(x - 3)$$

results from an approximation to the Hodgkin–Huxley equations, which model the transmission of neural impulses along an axon.

(a) Find the critical points and classify them by investigating the approximate linear system near each one.

(b) Draw phase portraits for $\gamma = 0.8$ and for $\gamma = 1.5$.

(c) Consider the trajectory that leaves the critical point $(2, 0)$. Find the value of γ for which this trajectory ultimately approaches the origin as $t \to \infty$. Draw a phase portrait for this value of γ.

Bifurcation Points. Consider the system

$$x' = F(x, y, \alpha), \qquad y' = G(x, y, \alpha), \qquad \text{(i)}$$

where α is a parameter. The equations

$$F(x, y, \alpha) = 0, \qquad G(x, y, \alpha) = 0 \qquad \text{(ii)}$$

determine the x and y nullclines, respectively. Any point where an x nullcline and a y nullcline intersect is a critical point. As α varies and the configuration of the nullclines changes, it may well happen that, at a certain value of α, two critical points coalesce into one. For further variations in α, the critical point may disappear altogether, or two critical points may reappear, often with stability characteristics different from before they coalesced. Of course, the process may occur in the reverse order. For a certain value of α, two formerly nonintersecting nullclines may come together, creating a critical point, which, for further changes in α, may split into two. A value of α at which critical points coalesce, and possibly are lost or gained, is a bifurcation point. Since a phase portrait of a system is very dependent on the location and nature of the critical points, an understanding of bifurcations is essential to an understanding of the global behavior of the system's solutions. Problems 11 through 17 illustrate some possibilities that involve bifurcations.

In each of Problems 11 through 14:
(a) Sketch the nullclines and describe how the critical points move as α increases.
(b) Find the critical points.
(c) Let $\alpha = 2$. Classify each critical point by investigating the corresponding approximate linear system. Draw a phase portrait in a rectangle containing the critical points.
(d) Find the bifurcation point α_0 at which the critical points coincide. Locate this critical point and find the eigenvalues of the approximate linear system. Draw a phase portrait.
(e) For $\alpha > \alpha_0$, there are no critical points. Choose such a value of α and draw a phase portrait.

11. $x' = -6x + y + x^2, \qquad y' = \frac{3}{2}\alpha - y$

12. $x' = \frac{3}{2}\alpha - y, \qquad y' = -x + y + x^2$

13. $x' = -3x + y + x^2, \qquad y' = -\alpha - x + y$

14. $x' = -\alpha - 2x + y, \qquad y' = -4x + y + x^2$

In each of Problems 15 and 16:
(a) Find the critical points.
(b) Determine the value of α, denoted by α_0, where two critical points coincide.
(c) By finding the approximating linear systems and their eigenvalues, determine how the stability properties of these two critical points change as α passes through the bifurcation point α_0.
(d) Draw phase portraits for values of α near α_0 to show how the transition through the bifurcation point occurs.

15. $x' = (3 + x)(1 - x + y),$
$\qquad y' = (y - 1)(1 + x + \alpha y); \qquad \alpha > 0$

16. $x' = y(\alpha - 2x + 3y), \qquad y' = (4 - x)(3 + y)$

17. Suppose that a certain pair of competing species are described by the system

$$\frac{dx}{dt} = x(4 - x - y),$$

$$\frac{dy}{dt} = y(2 + 2\alpha - y - \alpha x),$$

where $\alpha > 0$ is a parameter.
(a) Find the critical points. Note that $(2, 2)$ is a critical point for all values of α.
(b) Determine the nature of the critical point $(2, 2)$ for $\alpha = 0.75$ and for $\alpha = 1.25$. There is a value of α between 0.75 and 1.25 where the nature of the critical point changes abruptly. Denote this value by α_0; it is also called a bifurcation point.
(c) Find the approximate linear system near the point $(2, 2)$ in terms of α.
(d) Find the eigenvalues of the linear system in part (c) as functions of α. Then determine α_0.
(e) Draw phase portraits near $(2, 2)$ for $\alpha = \alpha_0$ and for values of α slightly less than, and slightly greater than, α_0. Explain how the transition in the phase portrait takes place as α passes through α_0.

7.4 Predator–Prey Equations

In the preceding section we discussed a model of two species that interact by competing for a common food supply or other natural resource. In this section we investigate the situation in which one species (the predator) preys on the other species (the prey), while the prey lives on a different source of food. For example, consider foxes and rabbits in a closed forest. The foxes prey on the rabbits, and the rabbits live on the vegetation in the forest. Other examples are bass in a lake as predators and redear as prey, or ladybugs as predators and aphids as prey. We emphasize again that a model involving only two species cannot fully describe the complex relationships among species that actually occur in nature. Nevertheless the study of simple models is the first step toward an understanding of more complicated phenomena.

We denote by x and y the populations of the prey and predator, respectively, at time t. In constructing a model of the interaction of the two species, we make the following assumptions:

1. In the absence of the predator, the prey grows at a rate proportional to the current population; thus $dx/dt = ax$, $a > 0$, when $y = 0$.
2. In the absence of the prey, the predator dies out; thus $dy/dt = -cy$, $c > 0$, when $x = 0$.
3. Each encounter between predator and prey tends both to promote the growth of the predator and to inhibit the growth of the prey. The number of encounters between predator and prey is proportional to the product of their populations. Thus the growth rate of the predator includes a term of the form γxy, whereas the growth rate of the prey includes a term of the form $-\alpha xy$, where γ and α are positive constants.

As a consequence of these assumptions, we are led to the equations

$$\frac{dx}{dt} = ax - \alpha xy = x(a - \alpha y),$$
$$\frac{dy}{dt} = -cy + \gamma xy = y(-c + \gamma x). \tag{1}$$

The constants a, c, α, and γ are all positive; a and c are the growth rate of the prey and the death rate of the predator, respectively, and α and γ are measures of the effect of the interaction between the two species. In specific situations values for a, α, c, and γ must be determined by observation.

The pair of nonlinear equations (1) are known as the **Lotka–Volterra equations**. Although these are rather simple equations, they have been used as a preliminary model for many predator–prey relationships. Ways of making them more realistic are discussed at the end of this section and in the problems. Our goal here is to determine the qualitative behavior of the solutions (trajectories) of the system (1) for arbitrary positive initial values of x and y. We do this first for a specific example and then return to the general equations (1) at the end of the section.

EXAMPLE 1

Discuss the solutions of the system

$$\frac{dx}{dt} = x(1 - 0.5y) = x - 0.5xy = F(x, y),$$
$$\frac{dy}{dt} = y(-0.75 + 0.25x) = -0.75y + 0.25xy = G(x, y) \tag{2}$$

for x and y positive.

The critical points of this system are the solutions of the algebraic equations

$$x(1 - 0.5y) = 0, \qquad y(-0.75 + 0.25x) = 0. \tag{3}$$

Thus the critical points are $(0, 0)$ and $(3, 2)$. Figure 7.4.1 shows the critical points and a direction field for the system (2). From this figure, it appears that trajectories in the first quadrant encircle the critical point $(3, 2)$. Whether the trajectories are actually closed curves, or whether they spiral slowly in or out cannot be definitely determined from the direction field. The origin appears to be a saddle point. Just as for the competition equations in Section 7.3, the coordinate axes are trajectories of Eqs. (1) or (2). Consequently, no other trajectory can cross a coordinate axis, which means that every solution starting in the first quadrant remains there for all time.

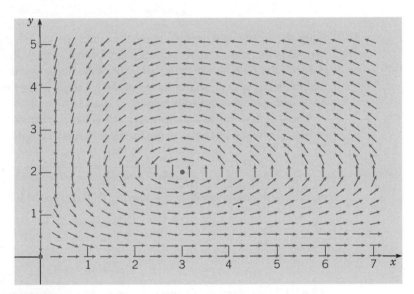

FIGURE 7.4.1   Critical points and direction field for the predator–prey system (2).

Next we examine the local behavior of solutions near each critical point. Near the origin we can neglect the nonlinear terms in Eqs. (2) to obtain the corresponding linear system

$$\frac{d}{dt}\begin{pmatrix} x \\ y \end{pmatrix} = \begin{pmatrix} 1 & 0 \\ 0 & -0.75 \end{pmatrix}\begin{pmatrix} x \\ y \end{pmatrix}. \tag{4}$$

The eigenvalues and eigenvectors of Eq. (4) are

$$\lambda_1 = 1, \qquad \mathbf{v}_1 = \begin{pmatrix} 1 \\ 0 \end{pmatrix}; \qquad \lambda_2 = -0.75, \qquad \mathbf{v}_2 = \begin{pmatrix} 0 \\ 1 \end{pmatrix}, \tag{5}$$

so its general solution is

$$\begin{pmatrix} x \\ y \end{pmatrix} = c_1 e^t \begin{pmatrix} 1 \\ 0 \end{pmatrix} + c_2 e^{-0.75t} \begin{pmatrix} 0 \\ 1 \end{pmatrix}. \tag{6}$$

Thus the origin is a saddle point both of the linear system (4) and of the nonlinear system (2), and therefore is unstable. One pair of trajectories enters the origin along the y-axis and another leaves the origin along the x-axis. All other trajectories also depart from the neighborhood of the origin.

To examine the critical point $(3, 2)$, let us first find the Jacobian matrix

$$\begin{pmatrix} F_x & F_y \\ G_x & G_y \end{pmatrix} = \begin{pmatrix} 1 - 0.5y & -0.5x \\ 0.25y & -0.75 + 0.25x \end{pmatrix}. \tag{7}$$

Evaluating this matrix at the point $(3, 2)$, we obtain the linear system

$$\frac{d}{dt}\begin{pmatrix} u \\ w \end{pmatrix} = \begin{pmatrix} 0 & -1.5 \\ 0.5 & 0 \end{pmatrix}\begin{pmatrix} u \\ w \end{pmatrix}, \tag{8}$$

where $u = x - 3$ and $w = y - 2$. The eigenvalues of this system are

$$\lambda_1 = \frac{\sqrt{3}\,i}{2} \quad \text{and} \quad \lambda_2 = -\frac{\sqrt{3}\,i}{2}. \tag{9}$$

Since the eigenvalues are imaginary, the critical point $(3, 2)$ is a center of the linear system (8) and is therefore a stable critical point for that system. Recall from Section 7.2 that this is one of the cases in which the behavior of the linear system may or may not carry over to the nonlinear system, so the nature of the point $(3, 2)$ for the nonlinear system (2) cannot be determined from this information.

The simplest way to find the trajectories of the linear system (8) is to divide the second of Eqs. (8) by the first so as to obtain the differential equation

$$\frac{dw}{du} = \frac{dw/dt}{du/dt} = \frac{0.5u}{-1.5w} = -\frac{u}{3w},$$

or, after separating variables and moving all terms to the left-hand side,

$$u\,du + 3w\,dw = 0. \tag{10}$$

Consequently,

$$u^2 + 3w^2 = k, \tag{11}$$

where k is an arbitrary nonnegative constant of integration. Thus the trajectories of the linear system (8) lie on ellipses centered at the critical point and elongated somewhat in the horizontal direction.

Now let us return to the nonlinear system (2). Dividing the second of Eqs. (2) by the first, we obtain

$$\frac{dy}{dx} = \frac{y(-0.75 + 0.25x)}{x(1 - 0.5y)}. \tag{12}$$

Equation (12) is a separable equation and can be put in the form

$$\frac{1 - 0.5y}{y}\,dy = \frac{-0.75 + 0.25x}{x}\,dx,$$

from which it follows that

$$0.75\ln x + \ln y - 0.5y - 0.25x = c, \tag{13}$$

where c is a constant of integration. Although by using only elementary functions we cannot solve Eq. (13) explicitly for either variable in terms of the other, it is possible to show that the graph of the equation for a fixed value of c is a closed curve surrounding the critical point $(3, 2)$. Thus the critical point is also a center of the nonlinear system (2), and the predator and prey populations exhibit a cyclic variation.

Figure 7.4.2 shows a phase portrait of the system (2). For some initial conditions, the trajectory represents small variations in x and y about the critical point, and is almost elliptical

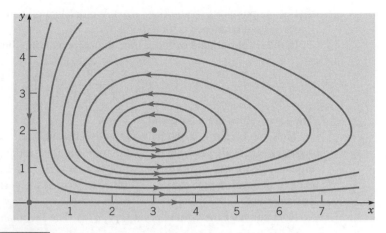

FIGURE 7.4.2 A phase portrait of the system (2).

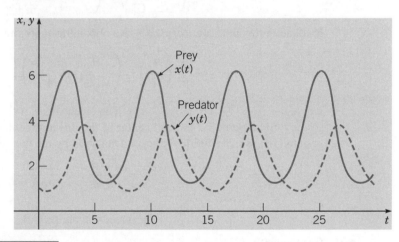

FIGURE 7.4.3 Variations of the prey (blue, solid) and predator (grey, dashed) populations with time for the system (2).

in shape, as the linear analysis suggests. For other initial conditions, the oscillations in x and y are more pronounced, and the shape of the trajectory is significantly different from an ellipse. Observe that the trajectories are traversed in the counterclockwise direction. The dependence of x and y on t for a typical set of initial conditions is shown in Figure 7.4.3. Note that x and y are periodic functions of t, as they must be since the trajectories are closed curves. Further the oscillation of the predator population lags behind that of the prey. Starting from a state in which both predator and prey populations are relatively small, the prey first increase because there is little predation. Then the predators, with abundant food, increase in population also. This causes heavier predation, and the prey tend to decrease. Finally, with a diminished food supply, the predator population also decreases, and the system returns to the original state.

The general system (1) can be analyzed in exactly the same way as in the example. The critical points of the system (1) are the solutions of

$$x(a - \alpha y) = 0, \qquad y(-c + \gamma x) = 0,$$

that is, the points $(0, 0)$ and $(c/\gamma, a/\alpha)$. We first examine the solutions of the corresponding linear system near each critical point.

In the neighborhood of the origin, the corresponding linear system is

$$\frac{d}{dt}\begin{pmatrix} x \\ y \end{pmatrix} = \begin{pmatrix} a & 0 \\ 0 & -c \end{pmatrix}\begin{pmatrix} x \\ y \end{pmatrix}. \tag{14}$$

The eigenvalues are $\lambda_1 = a$ and $\lambda_2 = -c$.

Thus the origin is a saddle point and hence unstable. Because an eigenvector for the negative eigenvalue is $\mathbf{v}_2 = (0, 1)^T$, entrance to the saddle point is along the y-axis. As an eigenvector for the positive eigenvalue is $\mathbf{v}_1 = (1, 0)^T$, there is one trajectory that exits the critical point along the x-axis. All other trajectories depart from the neighborhood of the critical point.

Next consider the critical point $(c/\gamma, a/\alpha)$. The Jacobian matix is

$$\begin{pmatrix} F_x & F_y \\ G_x & G_y \end{pmatrix} = \begin{pmatrix} a - \alpha y & -\alpha x \\ \gamma y & -c + \gamma x \end{pmatrix},$$

Evaluating this matrix at $(c/\gamma, a/\alpha)$, we obtain the approximate linear system

$$\frac{d}{dt}\begin{pmatrix} u \\ w \end{pmatrix} = \begin{pmatrix} 0 & -\alpha c/\gamma \\ \gamma a/\alpha & 0 \end{pmatrix}\begin{pmatrix} u \\ w \end{pmatrix}, \tag{15}$$

where $u = x - c/\gamma$ and $w = y - a/\alpha$. The eigenvalues of the system (15) are $\lambda = \pm i\sqrt{ac}$, so the critical point is a (stable) center of the linear system. To find the trajectories of the system (15), we can divide the second equation by the first to obtain

$$\frac{dw}{du} = \frac{dw/dt}{du/dt} = -\frac{(\gamma a/\alpha)u}{(\alpha c/\gamma)w}, \tag{16}$$

or

$$\gamma^2 au\, du + \alpha^2 cw\, dw = 0. \tag{17}$$

Consequently,

$$\gamma^2 au^2 + \alpha^2 cw^2 = k, \tag{18}$$

where k is a nonnegative constant of integration. Thus the trajectories of the linear system (15) are ellipses, just as in the example.

Returning briefly to the nonlinear system (1), observe that it can be reduced to the single equation

$$\frac{dy}{dx} = \frac{dy/dt}{dx/dt} = \frac{y(-c + \gamma x)}{x(a - \alpha y)}. \tag{19}$$

Equation (21) is separable and has the solution

$$a \ln y - \alpha y + c \ln x - \gamma x = C, \tag{20}$$

where C is a constant of integration. Again it is possible to show that, for fixed C, the graph of Eq. (20) is a closed curve surrounding the critical point $(c/\gamma, a/\alpha)$. Thus this critical point is also a center for the general nonlinear system (1).

The cyclic variation of the predator and prey populations can be analyzed in more detail when the deviations from the point $(c/\gamma, a/\alpha)$ are small and the linear system (15) can be used. The solution of the system (15) can be written in the form

$$u = \frac{c}{\gamma}K \cos(\sqrt{ac}\, t + \phi), \qquad w = \frac{a}{\alpha}\sqrt{\frac{c}{a}}K \sin(\sqrt{ac}\, t + \phi), \tag{21}$$

where the constants K and ϕ are determined by the initial conditions. Thus

$$x = \frac{c}{\gamma} + \frac{c}{\gamma}K\cos(\sqrt{ac}\,t + \phi),$$

$$y = \frac{a}{\alpha} + \frac{a}{\alpha}\sqrt{\frac{c}{a}}K\sin(\sqrt{ac}\,t + \phi). \tag{22}$$

These equations are good approximations for the nearly elliptical trajectories close to the critical point $(c/\gamma, a/\alpha)$. We can use them to draw several conclusions about the cyclic variation of the predator and prey on such trajectories.

1. The sizes of the predator and prey populations vary sinusoidally with period $2\pi/\sqrt{ac}$. This period of oscillation is independent of the initial conditions.

2. The predator and prey populations are out of phase by one-quarter of a cycle. The prey leads and the predator lags, as explained in the example.

3. The amplitudes of the oscillations are Kc/γ for the prey and $a\sqrt{c}K/\alpha\sqrt{a}$ for the predator and hence depend on the initial conditions as well as on the parameters of the problem.

4. The average populations of prey and predator over one complete cycle are c/γ and a/α, respectively. These are the same as the equilibrium populations (see Problem 10).

Cyclic variations of predator and prey as predicted by Eqs. (1) have been observed in nature. One striking example is described by Odum (pp. 191–192). Based on the records of the Hudson Bay Company of Canada, the abundance of lynx and snowshoe hare, as indicated by the number of pelts turned in over the period 1845–1935, shows a distinct periodic variation with a period of 9 to 10 years. The peaks of abundance are followed by very rapid declines, and the peaks of abundance of the lynx and hare are out of phase, with that of the hare preceding that of the lynx by a year or more.

Since the critical point $(c/\gamma, a/\alpha)$ is a center, we expect that small perturbations of the Lotka–Volterra equations may well lead to significant changes in the solutions. Put another way, unless the Lotka–Volterra equations exactly describe a given predator–prey relationship, the actual fluctuations of the populations may differ substantially from those predicted by the Lotka–Volterra equations, due to small inaccuracies in the model equations. This situation has led to many attempts[3] to replace the Lotka–Volterra equations by other systems that are less susceptible to the effects of small perturbations. Problem 13 introduces one such alternative model.

Another criticism of the Lotka–Volterra equations is that in the absence of the predator, the prey will grow without bound. This can be corrected by allowing for the natural inhibiting effect that an increasing population has on the growth rate of the population. For example, the first of Eqs. (1) can be modified so that when $y = 0$, it reduces to a logistic equation for x. The effects of this modification are explored in Problems 11 and 12. Problems 14 through 16 deal with harvesting in a predator–prey relationship. The results may seem rather counterintuitive.

Finally, we repeat a warning stated earlier: relationships among species in the natural world are often complex and subtle. You should not expect too much of a simple system of two differential equations in describing such relationships. Even if you are convinced that the general form of the equations is sound, the determination of numerical values for the coefficients may present serious difficulties.

---

[3]See the book by Brauer and Castillo-Chávez, listed in the references, for an extensive discussion of alternative models for predator–prey relationships.

PROBLEMS

Each of Problems 1 through 5 can be interpreted as describing the interaction of two species with population densities x and y. In each of these problems, carry out the following steps:

(a) Draw a direction field and describe how solutions seem to behave.

(b) Find the critical points.

(c) For each critical point, find the corresponding linear system. Find the eigenvalues and eigenvectors of the linear system. Classify each critical point as to type, and determine whether it is asymptotically stable, stable, or unstable.

(d) Sketch the trajectories in the neighborhood of each critical point.

(e) Draw a phase portrait for the system.

(f) Determine the limiting behavior of x and y as $t \to \infty$ and interpret the results in terms of the populations of the two species.

1. $dx/dt = x(1.5 - 0.5y), \quad dy/dt = y(-0.5 + x)$

2. $dx/dt = x(1 - 0.5y), \quad dy/dt = y(-0.25 + 0.5x)$

3. $dx/dt = x(1 - 0.5x - 0.5y), \quad dy/dt = y(-0.25 + 0.5x)$

4. $dx/dt = x(1.125 - x - 0.5y), \quad dy/dt = y(-1 + x)$

5. $dx/dt = x(-1 + 2.5x - 0.3y - x^2),$
   $dy/dt = y(-1.5 + x)$

6. In this problem, we examine the phase difference between the cyclic variations of the predator and prey populations as given by Eqs. (22) of this section. Suppose we assume that $K > 0$ and that t is measured from the time that the prey population (x) is a maximum; then $\phi = 0$. Show that the predator population (y) is a maximum at $t = \pi/(2\sqrt{ac}) = T/4$, where T is the period of the oscillation. When is the prey population increasing most rapidly? decreasing most rapidly? a minimum? Answer the same questions for the predator population. Draw a typical elliptic trajectory enclosing the point $(c/\gamma, a/\alpha)$, and mark these points on it.

7. (a) Find the ratio of the amplitudes of the oscillations of the prey and predator populations about the critical point $(c/\gamma, a/\alpha)$, using the approximation (22), which is valid for small oscillations. Observe that the ratio is independent of the initial conditions.

(b) Evaluate the ratio found in part (a) for the system (2).

(c) Estimate the amplitude ratio for the solution of the nonlinear system (2) shown in Figure 7.4.3. Does the result agree with that obtained from the linear approximation?

(d) Determine the prey–predator amplitude ratio for other solutions of the system (2), that is, for solutions satisfying other initial conditions. Is the ratio independent of the initial conditions?

8. (a) Find the period of the oscillations of the prey and predator populations, using the approximation (22), which is

valid for small oscillations. Note that the period is independent of the amplitude of the oscillations.

(b) For the solution of the nonlinear system (2) shown in Figure 7.4.3, estimate the period as well as possible. Is the result the same as for the linear approximation?

(c) Calculate other solutions of the system (2), that is, solutions satisfying other initial conditions, and determine their periods. Is the period the same for all initial conditions?

9. Consider the system

$$\frac{dx}{dt} = ax\left(1 - \frac{y}{2}\right), \qquad \frac{dy}{dt} = by\left(-1 + \frac{x}{3}\right),$$

where a and b are positive constants. Observe that this system is the same as in the example in the text if $a = 1$ and $b = 0.75$. Suppose the initial conditions are $x(0) = 5$ and $y(0) = 2$.

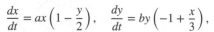

(a) Let $a = 1$ and $b = 1$. Plot the trajectory in the phase plane and determine (or estimate) the period of the oscillation.

(b) Repeat part (a) for $a = 3$ and $a = \frac{1}{3}$, with $b = 1$.

(c) Repeat part (a) for $b = 3$ and $b = \frac{1}{3}$, with $a = 1$.

(d) Describe how the period and the shape of the trajectory depend on a and b.

10. The average sizes of the prey and predator populations are defined as

$$\bar{x} = \frac{1}{T}\int_A^{A+T} x(t)\, dt, \qquad \bar{y} = \frac{1}{T}\int_A^{A+T} y(t)\, dt,$$

respectively, where T is the period of a full cycle, and A is any nonnegative constant.

(a) Using the approximation (22), which is valid near the critical point, show that $\bar{x} = c/\gamma$ and $\bar{y} = a/\alpha$.

(b) For the solution of the nonlinear system (2) shown in Figure 7.4.3, estimate $\bar{x}$ and $\bar{y}$ as well as you can. Try to determine whether $\bar{x}$ and $\bar{y}$ are given by c/γ and a/α, respectively, in this case.

Hint: Consider how you might estimate the value of an integral even though you do not have a formula for the integrand.

(c) Calculate other solutions of the system (2), that is, solutions satisfying other initial conditions, and determine $\bar{x}$ and $\bar{y}$ for these solutions. Are the values of $\bar{x}$ and $\bar{y}$ the same for all solutions?

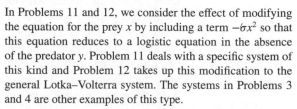

In Problems 11 and 12, we consider the effect of modifying the equation for the prey x by including a term $-\sigma x^2$ so that this equation reduces to a logistic equation in the absence of the predator y. Problem 11 deals with a specific system of this kind and Problem 12 takes up this modification to the general Lotka–Volterra system. The systems in Problems 3 and 4 are other examples of this type.

11. Consider the system

$$x' = x(1 - \sigma x - 0.5y), \qquad y' = y(-0.75 + 0.25x),$$

where $\sigma > 0$. Observe that this system is a modification of the system (2) in Example 1.

(a) Find all of the critical points. How does their location change as σ increases from zero? Observe that there is a critical point in the interior of the first quadrant only if $\sigma < \frac{1}{3}$.

(b) Determine the type and stability property of each critical point. Find the value $\sigma_1 < \frac{1}{3}$ where the nature of the critical point in the interior of the first quadrant changes. Describe the change that takes place in this critical point as σ passes through σ_1.

(c) Draw a direction field and phase portrait for a value of σ between zero and σ_1; for a value of σ between σ_1 and $\frac{1}{3}$.

(d) Describe the effect on the two populations as σ increases from zero to $\frac{1}{3}$.

12. Consider the system

$$\frac{dx}{dt} = x(a - \sigma x - \alpha y), \qquad \frac{dy}{dt} = y(-c + \gamma x),$$

where a, σ, α, c, and γ are positive constants.

(a) Find all critical points of the given system. How does their location change as σ increases from zero? Assume that $a/\sigma > c/\gamma$, or that $\sigma < a\gamma/c$. Why is this assumption necessary?

(b) Determine the nature and stability characteristics of each critical point.

(c) Show that there is a value of σ between zero and $a\gamma/c$ where the critical point in the interior of the first quadrant changes from a spiral point to a node.

(d) Describe the effect on the two populations as σ increases from zero to $a\gamma/c$.

13. In the Lotka–Volterra equations, the interaction between the two species is modeled by terms proportional to the product xy of the respective populations. If the prey population is much larger than the predator population, this may overstate the interaction; for example, a predator may hunt only when it is hungry, and ignore the prey at other times. In this problem, we consider an alternative model of a type proposed by Rosenzweig and MacArthur.[4]

(a) Consider the system

$$x' = x\left(1 - 0.2x - \frac{2y}{x+6}\right),$$

$$y' = y\left(-0.25 + \frac{x}{x+6}\right).$$

Find all of the critical points of this system.

(b) Determine the type and stability characteristics of each critical point.

(c) Draw a direction field and phase portrait for this system.

Harvesting in a Predator–Prey Relationship. In a predator–prey situation it may happen that one or perhaps both species are valuable sources of food (for example). Or,

the prey may be regarded as a pest, leading to efforts to reduce their number. In a constant-effort model of harvesting, we introduce a term $-E_1 x$ in the prey equation and a term $-E_2 y$ in the predator equation. A constant-yield model of harvesting is obtained by including the term $-H_1$ in the prey equation and the term $-H_2$ in the predator equation. The constants E_1, E_2, H_1, and H_2 are always nonnegative. Problems 14 and 15 deal with constant-effort harvesting, and Problem 16 with constant-yield harvesting.

14. Applying a constant-effort model of harvesting to the Lotka–Volterra equations (1), we obtain the system

$$x' = x(a - \alpha y - E_1), \qquad y' = y(-c + \gamma x - E_2).$$

When there is no harvesting, the equilibrium solution is $(c/\gamma, a/\alpha)$.

(a) Before doing any mathematical analysis, think about the situation intuitively. How do you think the populations will change if the prey alone is harvested? if the predator alone is harvested? if both are harvested?

(b) How does the equilibrium solution change if the prey is harvested, but not the predator ($E_1 > 0$, $E_2 = 0$)?

(c) How does the equilibrium solution change if the predator is harvested, but not the prey ($E_1 = 0$, $E_2 > 0$)?

(d) How does the equilibrium solution change if both are harvested ($E_1 > 0$, $E_2 > 0$)?

15. If we modify the Lotka–Volterra equations by including a self-limiting term $-\sigma x^2$ in the prey equation, and then assume constant-effort harvesting, we obtain the equations

$$x' = x(a - \sigma x - \alpha y - E_1),$$

$$y' = y(-c + \gamma x - E_2).$$

In the absence of harvesting, an equilibrium solution is $x = c/\gamma$, $y = (a/\alpha) - (\sigma c)/(\alpha \gamma)$.

(a) How does this equilibrium solution change if the prey is harvested, but not the predator ($E_1 > 0$, $E_2 = 0$)?

(b) How does this equilibrium solution change if the predator is harvested, but not the prey ($E_1 = 0$, $E_2 > 0$)?

(c) How does this equilibrium solution change if both are harvested ($E_1 > 0$, $E_2 > 0$)?

16. In this problem, we apply a constant-yield model of harvesting to the situation in Example 1. Consider the system

$$x' = x(1 - 0.5y) - H_1,$$

$$y' = y(-0.75 + 0.25x) - H_2,$$

where H_1 and H_2 are nonnegative constants. Recall that if $H_1 = H_2 = 0$, then (3, 2) is an equilibrium solution for this system.

(a) Before doing any mathematical analysis, think about the situation intuitively. How do you think the populations will change if the prey alone is harvested? if the predator alone is harvested? if both are harvested?

---

[4] See the book by Brauer and Castillo-Chávez for further details.

(b) How does the equilibrium solution change if the prey is harvested, but not the predator ($H_1 > 0, H_2 = 0$)?

(c) How does the equilibrium solution change if the predator is harvested, but not the prey ($H_1 = 0, H_2 > 0$)?

(d) How does the equilibrium solution change if both are harvested ($H_1 > 0, H_2 > 0$)?

7.5 Periodic Solutions and Limit Cycles

In this section we discuss further the possible existence of periodic solutions of two-dimensional autonomous systems

$$\mathbf{x}' = \mathbf{f}(\mathbf{x}). \tag{1}$$

Such solutions satisfy the relation

$$\mathbf{x}(t + T) = \mathbf{x}(t) \tag{2}$$

for all t and for some positive constant T called the period. The corresponding trajectories are *closed curves* in the phase plane. Periodic solutions often play an important role in physical problems because they represent phenomena that occur repeatedly. In many situations, a periodic solution represents a "final state" that is approached by all "neighboring" solutions as the transients due to the initial conditions die out.

A special case of a periodic solution is a constant solution $\mathbf{x} = \mathbf{x}_0$, which corresponds to a critical point of the autonomous system. As a constant solution satisfies Eq. (2) for any $T > 0$, such a solution is periodic. In this section when we speak of a periodic solution, we mean a *nonconstant* periodic solution. In this case, the period T is usually chosen as the smallest positive number for which Eq. (2) is valid.

Recall that the solutions of the linear autonomous system

$$\mathbf{x}' = \mathbf{A}\mathbf{x} \tag{3}$$

are periodic if and only if the eigenvalues of $\mathbf{A}$ are pure imaginary. In this case the critical point at the origin is a center. We emphasize that if the eigenvalues of $\mathbf{A}$ are pure imaginary, then every solution of the linear system (3) is periodic, whereas if the eigenvalues are not pure imaginary, then there are no (nonconstant) periodic solutions. The predator–prey equations discussed in Section 7.4, although nonlinear, behave similarly; all solutions in the first quadrant are periodic. The following example illustrates a different way in which periodic solutions of nonlinear autonomous systems can occur.

EXAMPLE 1

Discuss the solutions of the system

$$\begin{pmatrix} x \\ y \end{pmatrix}' = \begin{pmatrix} x + y - x(x^2 + y^2) \\ -x + y - y(x^2 + y^2) \end{pmatrix}. \tag{4}$$

It is not difficult to show that $(0, 0)$ is the only critical point of the system (4), and the system is almost linear in the neighborhood of the origin. The corresponding linear system

$$\begin{pmatrix} x \\ y \end{pmatrix}' = \begin{pmatrix} 1 & 1 \\ -1 & 1 \end{pmatrix} \begin{pmatrix} x \\ y \end{pmatrix} \tag{5}$$

has eigenvalues $1 \pm i$. Therefore the origin is an unstable spiral point for both the linear system (5) and the nonlinear system (4). Thus any solution that starts near the origin in the phase plane will spiral away from the origin. Since there are no other critical points, we might think that all solutions of Eq. (4) correspond to trajectories that spiral out to infinity. However we now show that this is incorrect, because far away from the origin the trajectories are directed inward.

It is convenient to introduce polar coordinates r and θ, where

$$x = r\cos\theta, \qquad y = r\sin\theta, \tag{6}$$

and $r \geq 0$. If we multiply the first of Eq. (4) by x, the second by y, and add, we then obtain

$$x\frac{dx}{dt} + y\frac{dy}{dt} = (x^2 + y^2) - (x^2 + y^2)^2. \tag{7}$$

Since $r^2 = x^2 + y^2$ and $r\,(dr/dt) = x\,(dx/dt) + y\,(dy/dt)$, it follows from Eq. (7) that

$$r\frac{dr}{dt} = r^2(1 - r^2). \tag{8}$$

This is similar to the equations discussed in Section 2.5. The critical points (for $r \geq 0$) are the origin and the point $r = 1$, which corresponds to the unit circle in the phase plane. From Eq. (8), it follows that $dr/dt > 0$ if $r < 1$ and $dr/dt < 0$ if $r > 1$. Thus inside the unit circle the trajectories are directed outward, while outside the unit circle they are directed inward. Apparently, the circle $r = 1$ is a limiting trajectory for this system.

To determine an equation for θ, we multiply the first of Eq. (4) by y, the second by x, and subtract, obtaining

$$y\frac{dx}{dt} - x\frac{dy}{dt} = x^2 + y^2. \tag{9}$$

Upon calculating dx/dt and dy/dt from Eqs. (6), we find that the left side of Eq. (9) is $-r^2(d\theta/dt)$, so Eq. (9) reduces to

$$\frac{d\theta}{dt} = -1. \tag{10}$$

The system of equations (8), (10) for r and θ is equivalent to the original system (4). One solution of the system (8), (10) is

$$r = 1, \qquad \theta = -t + t_0, \tag{11}$$

where t_0 is an arbitrary constant. As t increases, a point satisfying Eqs. (11) moves clockwise around the unit circle. Thus the autonomous system (4) has a periodic solution. Other solutions can be obtained by solving Eq. (8) by separation of variables; if $r \neq 0$ and $r \neq 1$, then

$$\frac{dr}{r(1 - r^2)} = dt. \tag{12}$$

Equation (12) can be solved by using partial fractions to rewrite the left side and then integrating. By performing these calculations, we find that the solution of Eqs. (10) and (12) is

$$r = \frac{1}{\sqrt{1 + c_0 e^{-2t}}}, \qquad \theta = -t + t_0, \tag{13}$$

where c_0 and t_0 are arbitrary constants. The solution (13) also contains the solution (11), which is obtained by setting $c_0 = 0$ in the first of Eqs. (13).

The solution satisfying the initial conditions $r = \rho$, $\theta = \alpha$ at $t = 0$ is given by

$$r = \frac{1}{\sqrt{1 + [(1/\rho^2) - 1]e^{-2t}}}, \qquad \theta = -(t - \alpha). \tag{14}$$

If $\rho < 1$, then $r \to 1$ from the inside as $t \to \infty$; if $\rho > 1$, then $r \to 1$ from the outside as $t \to \infty$. Thus, in all cases, the trajectories spiral toward the circle $r = 1$ as $t \to \infty$. Several trajectories are shown in Figure 7.5.1.

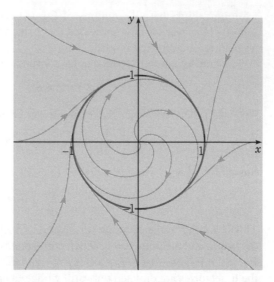

FIGURE 7.5.1 Trajectories of the system (4); the unit circle is a limit cycle.

In this example, the circle $r = 1$ not only corresponds to periodic solutions of the system (4), but it also attracts other nonclosed trajectories that spiral toward it as $t \to \infty$. In general, a closed trajectory in the phase plane such that other nonclosed trajectories spiral toward it, either from the inside or the outside, as $t \to \infty$, is called a **limit cycle**. Thus the circle $r = 1$ is a limit cycle for the system (4). If all trajectories that start near a closed trajectory (both inside and outside) spiral toward the closed trajectory as $t \to \infty$, then the limit cycle is **asymptotically stable**. Since the limiting trajectory is itself a periodic orbit rather than an equilibrium point, this type of stability is often called **orbital stability**. If the trajectories on one side spiral toward the closed trajectory, while those on the other side spiral away as $t \to \infty$, then the limit cycle is said to be **semistable**. If the trajectories on both sides of the closed trajectory spiral away as $t \to \infty$, then the closed trajectory is **unstable**. It is also possible to have closed trajectories that other trajectories neither approach nor depart from—for example, the periodic solutions of the predator–prey equations in Section 7.4. In this case, the closed trajectory is **stable**.

In Example 1, the existence of an asymptotically stable limit cycle was established by solving the equations explicitly. Unfortunately, this is usually not possible, so it is worthwhile to know general theorems concerning the existence or nonexistence of limit cycles of nonlinear autonomous systems. In discussing these theorems, it is convenient to rewrite the system (1) in the scalar form

$$\frac{dx}{dt} = F(x, y), \qquad \frac{dy}{dt} = G(x, y). \tag{15}$$

THEOREM 7.5.1

Let the functions F and G have continuous first partial derivatives in a domain D of the xy-plane. A closed trajectory of the system (15) must necessarily enclose at least one critical (equilibrium) point. If it encloses only one critical point, the critical point cannot be a saddle point.

Although we omit the proof of this theorem, it is easy to show examples of it. One is given by Example 1 and Figure 7.5.1 in which the closed trajectory encloses the critical point $(0, 0)$, an (unstable) spiral point. Another example is the system of predator–prey equations in Section 7.4; see Figure 7.4.2. Each closed trajectory surrounds the critical point $(3, 2)$; in this case, the critical point is a center.

Theorem 7.5.1 is also useful in a negative sense. If a given region contains no critical points, then there can be no closed trajectory lying entirely in the region. The same conclusion is true if the region contains only one critical point, and this point is a saddle point. For instance, in Example 2 of Section 7.3, an example of competing species, the only critical point in the interior of the first quadrant is the saddle point $(0.5, 0.5)$. Therefore this system has no closed trajectory lying in the first quadrant.

A second result about the nonexistence of closed trajectories is given in the following theorem.

THEOREM 7.5.2	Let the functions F and G have continuous first partial derivatives in a simply connected domain D of the xy-plane. If $F_x + G_y$ has the same sign throughout D, then there is no closed trajectory of the system (15) lying entirely in D.

A simply connected two-dimensional domain is one with no holes. Theorem 7.5.2 is a straightforward consequence of Green's theorem in the plane; see Problem 13. Note that if $F_x + G_y$ changes sign in the domain, then no conclusion can be drawn; there may or may not be closed trajectories in D.

To illustrate Theorem 7.5.2, consider the system (4). A routine calculation shows that

$$F_x(x, y) + G_y(x, y) = 2 - 4(x^2 + y^2) = 2(1 - 2r^2), \tag{16}$$

where, as usual, $r^2 = x^2 + y^2$. Hence $F_x + G_y$ is positive for $0 \le r < 1/\sqrt{2}$, so there is no closed trajectory in this circular disk. Of course, we showed in Example 1 that there is no closed trajectory in the larger region $r < 1$. This illustrates that the information given by Theorem 7.5.2 may not be the best possible result. Again referring to Eq. (16), note that $F_x + G_y < 0$ for $r > 1\sqrt{2}$. However the theorem is not applicable in this case because this annular region is not simply connected. Indeed, as shown in Example 1, it does contain a limit cycle.

The following theorem gives conditions that guarantee the existence of a closed trajectory.

THEOREM 7.5.3	(**Poincaré–Bendixson Theorem**) Let the functions F and G have continuous first partial derivatives in a domain D of the xy-plane. Let D_1 be a bounded subdomain in D, and let R be the region that consists of D_1 plus its boundary (all points of R are in D). Suppose that R contains no critical point of the system (15). If there exists a constant t_0 such that $x = \phi(t)$, $y = \psi(t)$ is a solution of the system (15) that exists and stays in R for all $t \ge t_0$, then either $x = \phi(t)$, $y = \psi(t)$ is a periodic solution (closed trajectory), or $x = \phi(t)$, $y = \psi(t)$ spirals toward a closed trajectory as $t \to \infty$. In either case, the system (15) has a periodic solution in R.

Note that if R does contain a closed trajectory, then necessarily, by Theorem 7.5.1, this trajectory must enclose a critical point. However this critical point cannot be in R. Thus R cannot be simply connected; it must have a hole.

As an application of the Poincaré–Bendixson theorem, consider again the system (4). Since the origin is a critical point, it must be excluded. For instance, we can consider the region R defined by $0.5 \le r \le 2$. Next, we must show that there is a solution whose trajectory stays in R for all t greater than or equal to some t_0. This follows immediately from Eq. (8). For $r = 0.5$, $dr/dt > 0$, so r increases, while for $r = 2$, $dr/dt < 0$, so r decreases. Thus any trajectory that crosses the boundary of R is entering R. Consequently, any solution of Eq. (4) that starts in R at $t = t_0$ cannot leave but must stay in R for $t > t_0$. Of course, other numbers could be used instead of 0.5 and 2; all that is important is that $r = 1$ is included.

One should not infer from this discussion of the preceding theorems that it is easy to determine whether a given nonlinear autonomous system has periodic solutions; often it is not a simple matter at all. Theorems 7.5.1 and 7.5.2 are frequently inconclusive, and for Theorem 7.5.3 it is often difficult to determine a region R and a solution that always remains within it.

We close this section with another example of a nonlinear system that has a limit cycle.

EXAMPLE 2

The van der Pol equation

$$u'' - \mu(1 - u^2)u' + u = 0, \tag{17}$$

where μ is a nonnegative constant, describes the current u in a triode oscillator. Discuss the solutions of this equation.

If $\mu = 0$, Eq. (17) reduces to $u'' + u = 0$, whose solutions are sine or cosine waves of period 2π. For $\mu > 0$, the second term on the left side of Eq. (17) must also be considered. This is the resistance term, proportional to u', with a coefficient $-\mu(1 - u^2)$ that depends on u. For large u, this term is positive and acts as usual to reduce the amplitude of the response. However, for small u, the resistance term is negative and so causes the response to grow. This suggests that perhaps there is a solution of intermediate size that other solutions approach as t increases.

To analyze Eq. (17) more carefully, we write it as a two-dimensional system by introducing the variables $x = u$, $y = u'$. Then it follows that

$$x' = y, \qquad y' = -x + \mu(1 - x^2)y. \tag{18}$$

The only critical point of the system (18) is the origin. Near the origin the corresponding linear system is

$$\begin{pmatrix} x \\ y \end{pmatrix}' = \begin{pmatrix} 0 & 1 \\ -1 & \mu \end{pmatrix} \begin{pmatrix} x \\ y \end{pmatrix}, \tag{19}$$

whose eigenvalues are $(\mu \pm \sqrt{\mu^2 - 4})/2$. Thus the origin is an unstable spiral point for $0 < \mu < 2$ and an unstable node for $\mu \ge 2$. In all cases, a solution that starts near the origin grows as t increases.

With regard to periodic solutions, Theorems 7.5.1 and 7.5.2 provide only partial information. From Theorem 7.5.1, we conclude that if there are closed trajectories, they must enclose the origin. Next we calculate $F_x(x, y) + G_y(x, y)$, with the result that

$$F_x(x, y) + G_y(x, y) = \mu(1 - x^2). \tag{20}$$

Then it follows from Theorem 7.5.2 that closed trajectories, if there are any, are not contained in the strip $|x| < 1$, where $F_x + G_y > 0$.

The application of the Poincaré–Bendixson theorem to this problem is not nearly as simple as for Example 1. If we introduce polar coordinates, we find that the equation for the radial variable r is

$$r' = \mu(1 - r^2 \cos^2 \theta)r \sin^2 \theta. \tag{21}$$

Again, consider an annular region R given by $r_1 \leq r \leq r_2$, where r_1 is small and r_2 is large. When $r = r_1$, the linear term on the right side of Eq. (21) dominates, and $r' > 0$ except on the x-axis, where $\sin\theta = 0$ and consequently $r' = 0$ also. Thus trajectories are entering R at every point on the circle $r = r_1$, except possibly for those on the x-axis, where the trajectories are tangent to the circle. When $r = r_2$, the cubic term on the right side of Eq. (21) is the dominant one. Thus $r' < 0$, except for points on the x-axis where $r' = 0$ and for points near the y-axis where $r^2 \cos^2\theta < 1$ and the linear term makes $r' > 0$. Thus, no matter how large a circle is chosen, there will be points on it (namely, the points on or near the y-axis) where trajectories are leaving R. Therefore the Poincaré–Bendixson theorem is not applicable unless we consider more complicated regions.

It is possible to show, by a more intricate analysis, that the van der Pol equation does have a unique limit cycle. However, we will not follow this line of argument further. We turn instead to a different approach in which we plot numerically computed solutions. Experimental observations indicate that the van der Pol equation has an asymptotically stable periodic solution whose period and amplitude depend on the parameter μ. By looking at graphs of trajectories in the phase plane and of u versus t, we can gain some understanding of this periodic behavior.

Figure 7.5.2 shows two trajectories of the van der Pol equation in the phase plane for $\mu = 0.2$. The trajectory starting near the origin spirals outward in the clockwise direction. This is consistent with the behavior of the linear approximation near the origin. The other trajectory passes through $(-3, 2)$ and spirals inward, again in the clockwise direction. Both trajectories approach a closed curve that corresponds to an asymptotically stable periodic solution. In Figure 7.5.3, we show the plots of u versus t for the solutions corresponding to the trajectories in Figure 7.5.2. The solution that is initially smaller gradually increases in amplitude, whereas the larger solution gradually decays. Both solutions approach a stable periodic motion that corresponds to the limit cycle. Figure 7.5.3 also shows that there is a phase difference between the two solutions as they approach the limit cycle. The plots of u versus t are nearly sinusoidal in shape, consistent with the nearly circular limit cycle in this case.

Figures 7.5.4 and 7.5.5 show similar plots for the case $\mu = 1$. Trajectories again move clockwise in the phase plane, but the limit cycle is considerably different from a circle. The plots of u versus t tend more rapidly to the limiting oscillation, and again show a phase difference. The oscillations are somewhat less symmetric in this case, rising somewhat more steeply than they fall.

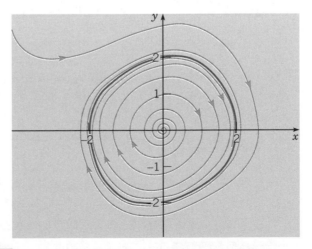

FIGURE 7.5.2   Trajectories of the van der Pol equation (17) for $\mu = 0.2$.

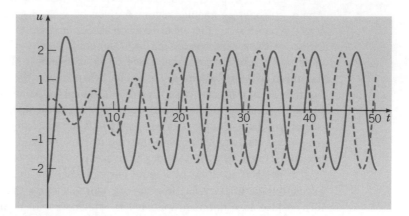

FIGURE 7.5.3 Plots of u versus t for the trajectories in Figure 7.5.2.

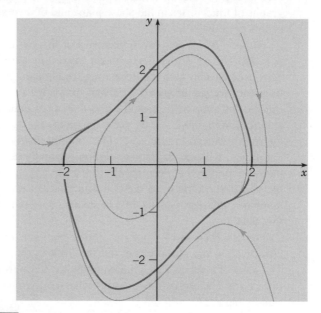

FIGURE 7.5.4 Trajectories of the van der Pol equation (17) for $\mu = 1$.

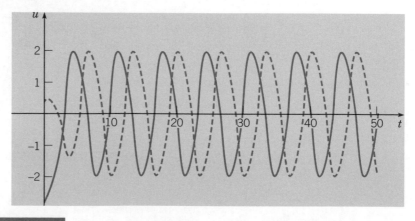

FIGURE 7.5.5 Plots of u versus t for the trajectories in Figure 7.5.4.

Figure 7.5.6 shows the phase plane for $\mu = 5$. The motion remains clockwise, and the limit cycle is even more elongated, especially in the y direction. A plot of u versus t is shown in Figure 7.5.7. Although the solution starts far from the limit cycle, the limiting oscillation is virtually reached in a fraction of a period. Starting from one of its extreme values on the x-axis in the phase plane, the solution moves toward the other extreme position slowly at first, but once a certain point on the trajectory is reached, the remainder of the transition is completed very swiftly. The process is then repeated in the opposite direction. The waveform of the limit cycle, as shown in Figure 7.5.7, is quite different from a sine wave.

These graphs clearly show that, in the absence of external excitation, the van der Pol oscillator has a certain characteristic mode of vibration for each value of μ. The graphs of u versus t show that the amplitude of this oscillation changes very little with μ, but the period

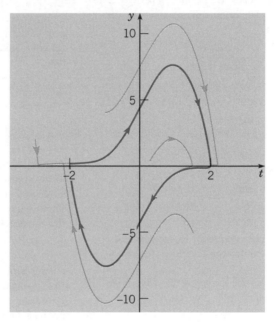

FIGURE 7.5.6 Trajectories of the van der Pol equation (17) for $\mu = 5$.

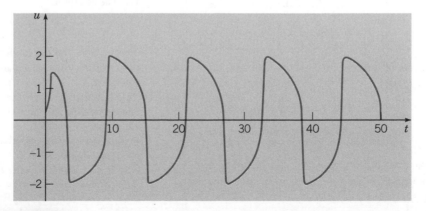

FIGURE 7.5.7 Plot of u versus t for the outward spiralling trajectory in Figure 7.5.6.

increases as μ increases. At the same time, the waveform changes from one that is very nearly sinusoidal with gentle transitions between intervals where the solution is increasing and decreasing, to waveforms where these transitions are more abrupt. (See Figure 7.5.7.)

The presence of a single periodic motion that attracts all (nearby) solutions, that is, an asymptotically stable limit cycle, is one of the characteristic phenomena associated with nonlinear differential equations.

PROBLEMS

In each of Problems 1 through 6, an autonomous system is expressed in polar coordinates. Determine all periodic solutions, all limit cycles, and determine their stability characteristics.

1. $dr/dt = 3r^2(1 - r^2)$, $\qquad d\theta/dt = 1$
2. $dr/dt = r(5 - r)^2$, $\qquad d\theta/dt = -1$
3. $dr/dt = r(r - 2)(r - 5)$, $\qquad d\theta/dt = 1$
4. $dr/dt = r(1 - r)(r - 2)$, $\qquad d\theta/dt = -1$
5. $dr/dt = \sin 3\pi r$, $\qquad d\theta/dt = 1$
6. $dr/dt = r|r - 3|(r - 7)$, $\qquad d\theta/dt = -1$

7. If $x = r \cos\theta$, $y = r \sin\theta$, show that
$$y(dx/dt) - x(dy/dt) = -r^2(d\theta/dt).$$

8. (a) Show that the system
$$\frac{dx}{dt} = y + \frac{xf(r)}{r}, \qquad \frac{dy}{dt} = -x + \frac{yf(r)}{r}$$

has periodic solutions corresponding to the zeros of $f(r)$. What is the direction of motion on the closed trajectories in the phase plane?

(b) Let $f(r) = r(r - 6)^2(r^2 - 6r + 5)$. Determine all periodic solutions and determine their stability characteristics.

9. Determine the periodic solutions, if any, of the system
$$\frac{dx}{dt} = -y + \frac{x}{\sqrt{x^2 + y^2}}(x^2 + y^2 - 5),$$
$$\frac{dy}{dt} = x + \frac{y}{\sqrt{x^2 + y^2}}(x^2 + y^2 - 5).$$

10. Using Theorem 7.5.2, show that the linear autonomous system
$$\frac{dx}{dt} = a_{11}x + a_{12}y, \qquad \frac{dy}{dt} = a_{21}x + a_{22}y$$

does not have a periodic solution (other than $x = 0$, $y = 0$) if $a_{11} + a_{22} \neq 0$.

In each of Problems 11 and 12, show that the given system has no periodic solutions other than constant solutions:

11. $dx/dt = 4x + y + 3x^3 - y^2$,
    $dy/dt = -x + 5y + x^2y + y^3/3$

12. $dx/dt = -2x - 3y - xy^2$, $\qquad dy/dt = y + x^3 - x^2y$

13. Prove Theorem 7.5.2 by completing the following argument. According to Green's theorem in the plane, if C is a sufficiently smooth simple closed curve, and if F and G are continuous and have continuous first partial derivatives, then
$$\int_C [F(x, y)\, dy - G(x, y)\, dx]$$
$$= \iint_R [F_x(x, y) + G_y(x, y)]\, dA,$$

where C is traversed counterclockwise and R is the region enclosed by C. Assume that $x = \phi(t)$, $y = \psi(t)$ is a solution of the system (15) that is periodic with period T. Let C be the closed curve given by $x = \phi(t)$, $y = \psi(t)$ for $0 \leq t \leq T$. Show that the line integral is zero for this curve. Then show that the conclusion of Theorem 7.5.2 must follow.

14. (a) By examining the graphs of u vs. t in Figures 7.5.3, 7.5.5, and 7.5.7, estimate the period T of the van der Pol oscillator in these cases.

(b) Calculate and plot the graphs of solutions of the van der Pol equation for other values of the parameter μ. Estimate the period T in these cases also.

(c) Plot the estimated values of T versus μ. Describe how T depends on μ.

15. The equation

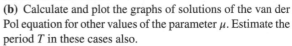

$$u'' - \mu\left(1 - \tfrac{1}{3}u'^2\right)u' + u = 0$$

is often called the **Rayleigh equation.**

(a) Write the Rayleigh equation as a system of two first order equations.

(b) Show that the origin is the only critical point of this system. Determine its type and whether it is asymptotically stable, stable, or unstable.

(c) Let $\mu = 1$. Choose initial conditions and compute the corresponding solution of the system on an interval such as $0 \leq t \leq 20$ or longer. Plot u versus t and also plot the trajectory in the phase plane. Observe that the trajectory approaches a closed curve (limit cycle). Estimate the amplitude A and the period T of the limit cycle.

(d) Repeat part (c) for other values of μ, such as $\mu = 0.2$, 0.5, 2, and 5. In each case, estimate the amplitude A and the period T.

(e) Describe how the limit cycle changes as μ increases. For example, make a table of values and/or plot A and T as functions of μ.

16. Consider the system of equations

$$x' = \mu x + y - x(x^2 + y^2),$$
$$y' = -x + \mu y - y(x^2 + y^2),$$ (i)

where μ is a parameter of unspecified sign. Observe that this system is the same as the one in Example 1, except for the introduction of μ.

(a) Show that the origin is the only critical point.

(b) Find the linear system that approximates Eqs. (i) near the origin and find its eigenvalues. Determine the type and stability of the critical point at the origin. How does this classification depend on μ?

(c) Referring to Example 1 if necessary, rewrite Eqs. (i) in polar coordinates.

(d) Show that when $\mu > 0$, there is a periodic solution $r = \sqrt{\mu}$. By solving the system found in part (c), or by plotting numerically computed solutions, conclude that this periodic solution attracts all other nonzero solutions.

Note: As the parameter μ increases through the value zero, the previously asymptotically stable critical point at the origin loses its stability, and simultaneously a new asymptotically stable solution (the limit cycle) emerges. Thus the point $\mu = 0$ is a bifurcation point; this type of bifurcation is called a **Hopf bifurcation**.

 17. Consider the van der Pol system

$$x' = y, \qquad y' = -x + \mu(1 - x^2)y,$$

where we now allow the parameter μ to be any real number.

(a) Show that the origin is the only critical point. Determine its type and stability property, and how these depend on μ.

(b) Let $\mu = -1$; draw a phase portrait and conclude that there is a periodic solution that surrounds the origin. Observe that this periodic solution is unstable. Compare your plot with Figure 7.5.4.

(c) Draw a phase portrait for a few other negative values of μ. Describe how the shape of the periodic solution changes with μ.

(d) Consider small positive or negative values of μ. By drawing phase portraits, determine how the periodic solution changes as $\mu \to 0$. Compare the behavior of the van der Pol system as μ increases through zero with the behavior of the system in Problem 16.

Problems 18 and 19 extend the consideration of the Rosenzweig–MacArthur predator–prey model introduced in Problem 13 of Section 7.4.

18. Consider the system

$$x' = x\left(2.4 - 0.2x - \frac{2y}{x+6}\right),$$
$$y' = y\left(-0.25 + \frac{x}{x+6}\right).$$

Observe that this system differs from that in Problem 13 of Section 7.4 only in the growth rate for the prey.

(a) Find all of the critical points.

(b) Determine the type and stability of each critical point.

(c) Draw a phase portrait in the first quadrant and conclude that there is an asymptotically stable limit cycle. Thus this model predicts a stable long-term oscillation of the prey and predator populations.

19. Consider the system

$$x' = x\left(a - 0.2x - \frac{2y}{x+6}\right),$$
$$y' = y\left(-0.25 + \frac{x}{x+6}\right),$$

where a is a positive parameter. Observe that this system includes the one in Problem 18 above and also the one in Problem 13 in Section 7.4.

(a) Find all of the critical points.

(b) Consider the critical point in the interior of the first quadrant. Find the eigenvalues of the approximate linear system. Determine the value a_0 where this critical point changes from asymptotically stable to unstable.

(c) Draw a phase portrait for a value of a slightly greater than a_0. Observe that a limit cycle has appeared. How does the limit cycle change as a increases further?

20. There are certain chemical reactions in which the constituent concentrations oscillate periodically over time. The system

$$x' = 1 - (b+1)x + \frac{x^2 y}{4}, \qquad y' = bx - \frac{x^2 y}{4}$$

is a special case of a model, known as the Brusselator, of this kind of reaction. Assume that b is a positive parameter, and consider solutions in the first quadrant of the xy-plane.

(a) Show that the only critical point is $(1, 4b)$.

(b) Find the eigenvalues of the approximate linear system at the critical point.

(c) Classify the critical point as to type and stability. How does the classification depend on b?

(d) As b increases through a certain value b_0, the critical point changes from asymptotically stable to unstable. What is that value b_0?

(e) Plot trajectories in the phase plane for values of b slightly less than and slightly greater than b_0. Observe the limit cycle when $b > b_0$; the Brusselator has a Hopf bifurcation point at b_0.

(f) Plot trajectories for several values of $b > b_0$ and observe how the limit cycle deforms as b increases.

21. The system

$$x' = 3\left(x + y - \tfrac{1}{3}x^3 - k\right),$$
$$y' = -\tfrac{1}{3}(x + 0.8y - 0.7)$$

is a special case of the Fitzhugh–Nagumo equations, which model the transmission of neural impulses along an axon. The parameter k is the external stimulus.

(a) Show that the system has one critical point regardless of the value of k.

(b) Find the critical point for $k = 0$ and show that it is an asymptotically stable spiral point. Repeat the analysis for $k = 0.5$ and show that the critical point is now an unstable spiral point. Draw a phase portrait for the system in each case.

(c) Find the value k_0 where the critical point changes from asymptotically stable to unstable. Find the critical point and draw a phase portrait for the system for $k = k_0$.

(d) For $k > k_0$, the system exhibits an asymptotically stable limit cycle; the system has a Hopf bifurcation point at k_0. Draw a phase portrait for $k = 0.4, 0.5$, and 0.6. Observe that the limit cycle is not small when k is near k_0. Also plot x versus t and estimate the period T in each case.

(e) As k increases further, there is a value k_1 at which the critical point again becomes asymptotically stable and the limit cycle vanishes. Find k_1.

7.6 Chaos and Strange Attractors: The Lorenz Equations

In principle, the methods described in this chapter for two-dimensional autonomous systems can be applied to higher dimensional systems as well. In practice, several difficulties arise when we try to do this. One problem is that there is simply a greater number of possible cases that can occur, and the number increases with the dimension of the system and its phase space. Another problem is the difficulty of graphing trajectories accurately in a phase space of more than two dimensions. Even in three dimensions, it may not be easy to construct a clear and understandable plot of the trajectories, and it becomes more difficult as the number of variables increases. Finally, and this has been clearly realized only in the last 30 years or so, there are different and very complex phenomena that can occur, and frequently do occur, in systems of three or more dimensions that are not present in two-dimensional systems. Our goal in this section is to provide a brief introduction to some of these phenomena by discussing one particular three-dimensional autonomous system that has been intensively studied.

An important problem in meteorology, and in other applications of fluid dynamics, concerns the motion of a layer of fluid, such as the earth's atmosphere, that is warmer at the bottom than at the top; see Figure 7.6.1. If the vertical temperature difference ΔT is small, then there is a linear variation of temperature with altitude, but no significant motion of the fluid layer. However, if ΔT is large enough, then the warmer air rises, displacing the cooler air above it, and a steady convective motion results. If the temperature difference increases further, then eventually the steady convective flow breaks up and a more complex and turbulent motion ensues.

FIGURE 7.6.1 A layer of fluid heated from below.

While investigating this phenomenon, Edward N. Lorenz was led (by a process too involved to describe here) to the nonlinear autonomous three-dimensional system

$$\frac{dx}{dt} = \sigma(-x + y), \qquad \frac{dy}{dt} = rx - y - xz, \qquad \frac{dz}{dt} = -bz + xy. \tag{1}$$

Equations (1) are now commonly referred to as the Lorenz equations.[5] Observe that the second and third equations involve quadratic nonlinearities. However, except for being a three-dimensional system, superficially the Lorenz equations appear no more complicated than the competing species or predator–prey equations discussed in Sections 7.3 and 7.4. The variable x in Eqs. (1) is related to the intensity of the fluid motion, while the variables y and z are related to the temperature variations in the horizontal and vertical directions. The Lorenz equations also involve three parameters σ, r, and b, all of which are real and positive. The parameters σ and b depend on the material and geometrical properties of the fluid layer. For the earth's atmosphere, reasonable values of these parameters are $\sigma = 10$ and $b = \frac{8}{3}$; they will be assigned these values in much of what follows in this section. The parameter r, on the other hand, is proportional to the temperature difference ΔT, and our purpose is to investigate how the nature of the solutions of Eqs. (1) changes with r.

The first step in analyzing the Lorenz equations is to locate the critical points by solving the algebraic system

$$\begin{aligned} \sigma x - \sigma y &= 0, \\ rx - y - xz &= 0, \\ -bz + xy &= 0. \end{aligned} \tag{2}$$

From the first equation, we have $y = x$. Then, eliminating y from the second and third equations, we obtain

$$x(r - 1 - z) = 0, \tag{3}$$

$$-bz + x^2 = 0. \tag{4}$$

One way to satisfy Eq. (3) is to choose $x = 0$. Then it follows that $y = 0$ and, from Eq. (4), $z = 0$. Alternatively, we can satisfy Eq. (3) by choosing $z = r - 1$. Then Eq. (4) requires that $x = \pm\sqrt{b(r - 1)}$ and then $y = \pm\sqrt{b(r - 1)}$ also. Observe that these expressions for x and y are real only when $r \geq 1$. Thus $(0, 0, 0)$, which we will denote by P_1, is a critical point for all values of r, and it is the only critical point for $r < 1$. However, when $r > 1$, there are also two other critical points, namely, $[\sqrt{b(r - 1)}, \sqrt{b(r - 1)}, r - 1]$, and $[-\sqrt{b(r - 1)}, -\sqrt{b(r - 1)}, r - 1]$. We will denote the latter two points by P_2 and P_3, respectively. Note that all three critical points coincide when $r = 1$. As r increases through the value 1, the critical point P_1 at the origin *bifurcates*, and the critical points P_2 and P_3 come into existence.

Next we will determine the local behavior of solutions in the neighborhood of each critical point. Although much of the following analysis can be carried out for arbitrary values of σ and b, we will simplify our work by using the values $\sigma = 10$ and $b = \frac{8}{3}$. Near the origin (the critical point P_1) the approximating linear system is

$$\begin{pmatrix} x \\ y \\ z \end{pmatrix}' = \begin{pmatrix} -10 & 10 & 0 \\ r & -1 & 0 \\ 0 & 0 & -8/3 \end{pmatrix} \begin{pmatrix} x \\ y \\ z \end{pmatrix}. \tag{5}$$

---

[5]A very thorough treatment of the Lorenz equations appears in the book by Sparrow listed in the references.

The eigenvalues are determined from the equation

$$\begin{vmatrix} -10-\lambda & 10 & 0 \\ r & -1-\lambda & 0 \\ 0 & 0 & -8/3-\lambda \end{vmatrix} = -\left(\frac{8}{3}+\lambda\right)[\lambda^2+11\lambda-10(r-1)] = 0. \qquad (6)$$

Therefore

$$\lambda_1 = -\frac{8}{3}, \qquad \lambda_2 = \frac{-11-\sqrt{81+40r}}{2}, \qquad \lambda_3 = \frac{-11+\sqrt{81+40r}}{2}. \qquad (7)$$

Note that all three eigenvalues are negative for $r < 1$. For example, when $r = \frac{1}{2}$, the eigenvalues are $\lambda_1 = -\frac{8}{3}$, $\lambda_2 = -10.52494$, $\lambda_3 = -0.47506$. Hence the origin is asymptotically stable for this range of r, both for the linear approximation (5) and for the original system (1). However λ_3 changes sign when $r = 1$ and is positive for $r > 1$. The value $r = 1$ corresponds to the initiation of convective flow in the physical problem described earlier. The origin is unstable for $r > 1$. All solutions starting near the origin tend to grow, except for those lying precisely in the plane determined by the eigenvectors associated with λ_1 and λ_2 (or, for the nonlinear system (1), in a certain surface tangent to this plane at the origin).

Next, consider the neighborhood of the critical point $P_2 = [\sqrt{8(r-1)/3}, \sqrt{8(r-1)/3}, r-1]$ for $r > 1$. If u, v, and w are the perturbations from the critical point in the x, y, and z directions, respectively, then the approximating linear system is

$$\begin{pmatrix} u \\ v \\ w \end{pmatrix}' = \begin{pmatrix} -10 & 10 & 0 \\ 1 & -1 & -\sqrt{8(r-1)/3} \\ \sqrt{8(r-1)/3} & \sqrt{8(r-1)/3} & -8/3 \end{pmatrix} \begin{pmatrix} u \\ v \\ w \end{pmatrix}. \qquad (8)$$

The eigenvalues of the coefficient matrix of Eq. (8) are determined from the equation

$$3\lambda^3 + 41\lambda^2 + 8(r+10)\lambda + 160(r-1) = 0, \qquad (9)$$

which is obtained by straightforward algebraic steps that are omitted here. The solutions of Eq. (9) depend on r in the following way:

1. For $1 < r < r_1 \cong 1.3456$, there are three negative real eigenvalues.
2. For $r_1 < r < r_2 \cong 24.737$, there is one negative real eigenvalue and two complex eigenvalues with negative real part.
3. For $r_2 < r$, there is one negative real eigenvalue and two complex eigenvalues with positive real part.

The same results are obtained for the critical point P_3. Thus there are several different situations.

1. For $0 < r < 1$, the only critical point is P_1 and it is asymptotically stable. All solutions approach this point (the origin) as $t \to \infty$.
2. For $1 < r < r_1$, the critical points P_2 and P_3 are asymptotically stable and P_1 is unstable. All nearby solutions approach one or the other of the points P_2 and P_3 exponentially.
3. For $r_1 < r < r_2$, the critical points P_2 and P_3 are asymptotically stable and P_1 is unstable. All nearby solutions approach one or the other of the points P_2 and P_3; most of them spiral inward to the critical point.
4. For $r_2 < r$, all three critical points are unstable. Most solutions near P_2 or P_3 spiral away from the critical point.

However this is by no means the end of the story. Let us consider solutions for r somewhat greater than r_2. In this case, P_1 has one positive eigenvalue and each of P_2 and P_3 has a pair of complex eigenvalues with a positive real part. A trajectory can approach any one of the critical points only on certain highly restricted paths. The slightest deviation from these paths causes the trajectory to depart from the critical point. Since none of the critical points is asymptotically stable, one might expect that most trajectories would approach infinity for large t. However it can be shown that all solutions remain bounded as $t \to \infty$; see Problem 4. In fact, it can be shown that all solutions ultimately approach a certain limiting set of points that has zero volume. Indeed, this is true, not only for $r > r_2$, but also for all positive values of r.

A plot of computed values of x versus t for a typical solution with $r > r_2$ is shown in Figure 7.6.2. Note that the solution oscillates back and forth between positive and negative values in a rather erratic manner. Indeed, the graph of x versus t resembles a random vibration, although the Lorenz equations are entirely deterministic and the solution is completely determined by the initial conditions. Nevertheless the solution also exhibits a certain *regularity* in that the frequency and amplitude of the oscillations are essentially constant in time.

The solutions of the Lorenz equations are also extremely sensitive to perturbations in the initial conditions. Figure 7.6.3 shows the graphs of computed values of x versus t for

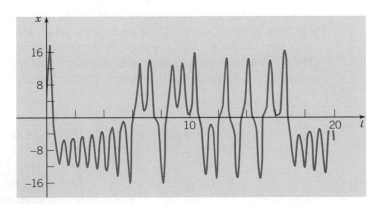

FIGURE 7.6.2   A plot of x versus t for the Lorenz equations (1) with $r = 28$; the initial point is $(5, 5, 5)$.

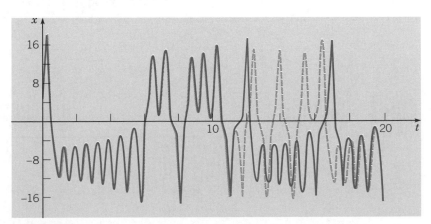

FIGURE 7.6.3   Plots of x versus t for two initially nearby solutions of Lorenz equations with $r = 28$; the initial point is $(5, 5, 5)$ for the dashed curve and is $(5.01, 5, 5)$ for the solid curve.

the two solutions whose initial points are (5, 5, 5) and (5.01, 5, 5). The dashed graph is the same as the one in Figure 7.6.2, while the solid graph starts at a nearby point. The two solutions remain close until t is near 10, after which they are quite different and, indeed, seem to have no relation to each other. It was this property that particularly attracted the attention of Lorenz in his original study of these equations, and caused him to conclude that detailed long-range weather predictions are probably not possible.

The attracting set in this case, although of zero volume, has a rather complicated structure and is called a **strange attractor**. The term **chaotic** has come into general use to describe solutions such as those shown in Figures 7.6.2 and 7.6.3.

To determine how and when the strange attractor is created, it is illuminating to investigate solutions for smaller values of r. For $r = 21$, solutions starting at three different initial points are shown in Figure 7.6.4. For the initial point (3, 8, 0), the solution begins to converge to the point P_3 almost at once; see Figure 7.6.4a. For the second initial point (5, 5, 5), there is a fairly short interval of transient behavior, after which the solution converges to P_2; see Figure 7.6.4b. However, as shown in Figure 7.6.4c, for the third initial point (5, 5, 10), there is a much longer interval of transient chaotic behavior before the solution eventually converges to P_2. As r increases, the duration of the chaotic transient behavior also increases. When $r = r_3 \cong 24.06$, the chaotic transients appear to continue indefinitely, and the strange attractor comes into being.

One can also show the trajectories of the Lorenz equations in the three-dimensional phase space, or at least projections of them in various planes. Figures 7.6.5 and 7.6.6 show projections in the xy- and xz-planes, respectively, of the trajectory starting at (5, 5, 5). Observe that the graphs in these figures appear to cross over themselves repeatedly, but this cannot be true for the actual trajectories in three-dimensional space because of the general

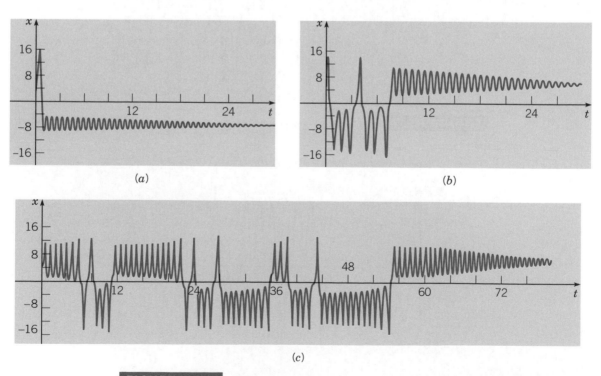

(a)

(b)

(c)

FIGURE 7.6.4   Plots of x versus t for three solutions of Lorenz equations with $r = 21$. (a) Initial point is (3, 8, 0). (b) Initial point is (5, 5, 5). (c) Initial point is (5, 5, 10).

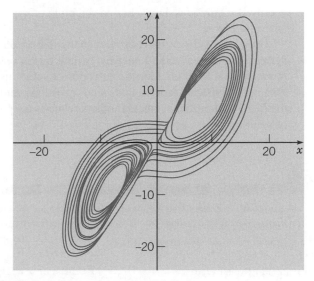

FIGURE 7.6.5 Projections of a trajectory of the Lorenz equations (with $r = 28$) in the xy-plane.

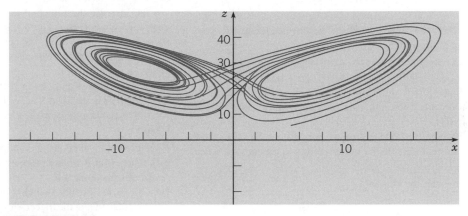

FIGURE 7.6.6 Projections of a trajectory of the Lorenz equations (with $r = 28$) in the xz-plane.

uniqueness theorem. The apparent crossings are due wholly to the two-dimensional character of the figures.

The sensitivity of solutions to perturbations of the initial data also has implications for numerical computations, such as those reported here. Different step sizes, different numerical algorithms, or even the execution of the same algorithm on different machines will introduce small differences in the computed solution, which eventually lead to large deviations. For example, the exact sequence of positive and negative loops in the calculated solution depends strongly on the precise numerical algorithm and its implementation, as well as on the initial conditions. However the general appearance of the solution and the structure of the attracting set are independent of all these factors.

Solutions of the Lorenz equations for other parameter ranges exhibit other interesting types of behavior. For example, for certain values of r greater than r_2, intermittent bursts of chaotic behavior separate long intervals of apparently steady periodic oscillation. For other ranges of r, solutions show a period-doubling property. Some of these features are taken up in the problems.

Since about 1975, the Lorenz equations and other higher dimensional autonomous systems have been studied intensively, and this is one of the most active areas of current mathematical research. Chaotic behavior of solutions appears to be much more common than was suspected at first, and many questions remain unanswered. Some of these are mathematical in nature, whereas others relate to the physical applications or interpretations of solutions.

PROBLEMS

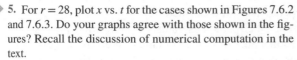

Problems 1 through 3 ask you to fill in some of the details of the analysis of the Lorenz equations in this section:

1. (a) Show that the eigenvalues of the linear system (5), valid near the origin, are given by Eq. (7).
(b) Determine the corresponding eigenvectors.
(c) Determine the eigenvalues and eigenvectors of the system (5) in the case where $r = 28$.

2. (a) Show that the linear approximation valid near the critical point P_2 is given by Eq. (8).
(b) Show that the eigenvalues of system (8) satisfy Eq. (9).
(c) For $r = 28$, solve Eq. (9) and thereby determine the eigenvalues of the system (8).

3. (a) By solving Eq. (9) numerically, show that the real part of the complex roots changes sign when $r \cong 24.737$.
(b) Show that a cubic polynomial $x^3 + Ax^2 + Bx + C$ has one real zero and two pure imaginary zeros only if $AB = C$.
(c) By applying the result of part (b) to Eq. (9), show that the real part of the complex roots changes sign when $r = \frac{470}{19}$.

4. Consider the ellipsoid

$$V(x, y, z) = 2rx^2 + \sigma y^2 + \sigma(z - 3r)^2 = c > 0.$$

(a) Calculate

$$\frac{dV}{dt} = \frac{\partial V}{\partial x}\frac{dx}{dt} + \frac{\partial V}{\partial y}\frac{dy}{dt} + \frac{\partial V}{\partial z}\frac{dz}{dt}$$

along trajectories of the Lorenz equations (1).
(b) Determine a sufficient condition on c so that every trajectory crossing $V(x, y, z) = c$ is directed inward.
(c) Evaluate the condition found in part (b) for the case $\sigma = 10$, $b = \frac{8}{3}$, $r = 28$.

In each of Problems 5 through 7, carry out the indicated investigations of the Lorenz equations.

5. For $r = 28$, plot x vs. t for the cases shown in Figures 7.6.2 and 7.6.3. Do your graphs agree with those shown in the figures? Recall the discussion of numerical computation in the text.

6. For $r = 28$, plot the projections in the xy- and xz-planes, respectively, of the trajectory starting at the point $(5, 5, 5)$. Do the graphs agree with those in Figures 7.6.5 and 7.6.6?

7. (a) For $r = 21$, plot x versus t for the solutions starting at the initial points $(3, 8, 0)$, $(5, 5, 5)$, and $(5, 5, 10)$. Use a t interval of at least $0 \le t \le 30$. Compare your graphs with those in Figure 7.6.4.
(b) Repeat the calculation in part (a) for $r = 22$, $r = 23$, and $r = 24$. Increase the t interval as necessary so that you can determine when each solution begins to converge to one of the critical points. Record the approximate duration of the chaotic transient in each case. Describe how this quantity depends on the value of r.
(c) Repeat the calculations in parts (a) and (b) for values of r slightly greater than 24. Try to estimate the value of r for which the duration of the chaotic transient approaches infinity.

8. For certain r intervals, or windows, the Lorenz equations exhibit a period-doubling property. Careful calculations may reveal this phenomenon.
(a) One period-doubling window contains the value $r = 100$. Let $r = 100$ and plot the trajectory starting at $(5, 5, 5)$ or some other initial point of your choice. Does the solution appear to be periodic? What is the period?
(b) Repeat the calculation in part (a) for slightly smaller values of r. When $r \cong 99.98$, you may be able to observe that the period of the solution doubles. Try to observe this result by performing calculations with nearby values of r.
(c) As r decreases further, the period of the solution doubles repeatedly. The next period doubling occurs at about $r = 99.629$. Try to observe this by plotting trajectories for nearby values of r.

9. Now consider values of r slightly larger than those in Problem 8.
(a) Plot trajectories of the Lorenz equations for values of r between 100 and 100.78. You should observe a steady periodic solution for this range of r values.
(b) Plot trajectories for values of r between 100.78 and 100.8. Determine, as best you can, how and when the periodic trajectory breaks up.

CHAPTER SUMMARY

Nonlinear two-dimensional autonomous systems have the form

$$\frac{dx}{dt} = F(x, y), \qquad \frac{dy}{dt} = G(x, y),$$

or, in vector notation,

$$\frac{d\mathbf{x}}{dt} = \mathbf{f}(\mathbf{x}).$$

The first four sections in this chapter deal mainly with approximating a nonlinear system by a linear one. The last two sections introduce phenomena that occur only in nonlinear systems.

Section 7.1

▶ **Critical points**, $\mathbf{x}_0$, of the system $\mathbf{x}' = \mathbf{f}(\mathbf{x})$ satisfy $\mathbf{f}(\mathbf{x}_0) = \mathbf{0}$. In this case, the constant solution $\mathbf{x}(t) = \mathbf{x}_0$ is an **equilibrium solution**.

▶ Formal definitions of **stability**, **asymptotic stability**, and **instability** of critical points are given. Stability and asymptotic stability are illustrated by an undamped and a damped simple pendulum, respectively, about its downward equilibrium position. Instability is illustrated by a pendulum, damped or undamped, about its upward equilibrium position.

▶ Examples illustrate **basins of attraction** and their boundaries, called **separatrices**.

Section 7.2

▶ If F and G are twice differentiable, then the nonlinear autonomous system $\mathbf{x}' = \mathbf{f}(\mathbf{x})$ can be approximated near a critical point $\mathbf{x}_0$ by a linear system $\mathbf{u}' = \mathbf{A}\mathbf{u}$, where $\mathbf{u} = \mathbf{x} - \mathbf{x}_0$. The coefficient matrix $\mathbf{A}$ is the Jacobian matrix $\mathbf{J}$ evaluated at $\mathbf{x}_0$. Thus

$$\mathbf{A} = \mathbf{J}\left(\mathbf{x}_0\right) = \begin{pmatrix} F_x\left(x_0, y_0\right) & F_y\left(x_0, y_0\right) \\ G_x\left(x_0, y_0\right) & G_y\left(x_0, y_0\right) \end{pmatrix}.$$

▶ Classification of a critical point $\mathbf{x}_0$ as a node, improper node, proper node, saddle point, spiral point, or center depends upon the eigenvalues of the Jacobian matrix $\mathbf{J}(\mathbf{x}_0)$.

TABLE 7.2.2 — Stability and instability properties of linear and almost linear systems.

λ_1, λ_2	Linear System Type	Linear System Stability	Almost Linear System Type	Almost Linear System Stability
$\lambda_1 > \lambda_2 > 0$	N	Unstable	N	Unstable
$\lambda_1 < \lambda_2 < 0$	N	Asymptotically stable	N	Asymptotically stable
$\lambda_2 < 0 < \lambda_1$	SP	Unstable	SP	Unstable
$\lambda_1 = \lambda_2 > 0$	PN or IN	Unstable	N or SpP	Unstable
$\lambda_1 = \lambda_2 < 0$	PN or IN	Asymptotically stable	N or SpP	Asymptotically stable
$\lambda_1, \lambda_2 = \mu \pm i\nu$				
$\mu > 0$	SpP	Unstable	SpP	Unstable
$\mu < 0$	SpP	Asymptotically stable	SpP	Asymptotically stable
$\mu = 0$	C	Stable	C or SpP	Indeterminate

Note: N, node; IN, improper node; PN, proper node; SP, saddle point; SpP, spiral point; C, center.

▶ Theorem 7.2.2 states that the trajectories of the nonlinear system locally resemble those of the linear approximation, except possibly in the cases where the eigenvalues of the linear system are either pure imaginary or real and equal. Thus, in most cases, the linear system is a good local approximation to the nonlinear system.

Section 7.3 Application: Competing Species

▶ The equations

$$\frac{dx}{dt} = x(\epsilon_1 - \sigma_1 x - \alpha_1 y), \qquad \frac{dy}{dt} = y(\epsilon_2 - \sigma_2 y - \alpha_2 x)$$

are often used as a model of competition, such as between two species in nature or perhaps between two businesses.

▶ Examples show that sometimes the two competitors can coexist in a stable manner, but sometimes one will overwhelm the other and drive it to extinction. The analysis in this section explains why this happens and enables you to predict which outcome will occur for a given system.

Section 7.4 Application: Predator–Prey

▶ The predator–prey, or **Lotka–Volterra**, equations

$$\frac{dx}{dt} = x(a - \alpha y), \qquad \frac{dy}{dt} = y(-c + \gamma x)$$

are a starting point for the study of the relation between a prey x and its predator y.

▶ The solutions of this system exhibit a cyclic variation about a critical point (a center) in the first quadrant. This type of behavior has sometimes been observed in nature.

Section 7.5 Nonlinear systems, unlike linear systems, sometimes have periodic solutions, or **limit cycles**, that attract other nearby solutions.

▶ Several theorems specify conditions under which limit cycles do, or do not, exist.

▶ The **van der Pol equation** (written in system form)

$$x' = y, \qquad y' = -x + \mu\left(1 - x^2\right)y$$

is an important equation that illustrates the occurrence of a limit cycle.

Section 7.6 In three or more dimensions, there is the possibility that solutions may be **chaotic**. In addition to critical points and limit cycles, solutions may converge to sets of points known as **strange attractors**.

▶ The **Lorenz equations**, arising in a study of the atmosphere,

$$\frac{dx}{dt} = \sigma(-x + y), \qquad \frac{dy}{dt} = rx - y - xz, \qquad \frac{dz}{dt} = -bz + xy$$

provide an example of the occurence of chaos in a relatively simple three-dimensional nonlinear system.

PROJECTS | Project 1 Modeling of Epidemics

© John Wiley & Sons, Inc.

Infectious disease is disease caused by a biological agent (virus, bacterium, or parasite) that can be spread directly or indirectly from one organism to another. A sudden outbreak of infectious disease that spreads rapidly and affects a large number of people, animals, or plants in a particular area for a limited period of time is referred to as an *epidemic*. Mathematical models are used to help understand the dynamics of an epidemic, to design treatment and control strategies (such as a vaccination program or quarantine policy), and to help forecast whether an epidemic will occur. In this project, we consider two simple models that highlight some important principles of epidemics.

The SIR Model. Most mathematical models of disease assume that the population is subdivided into a set of distinct compartments, or classes. The class in which an individual resides at time t depends on that individual's experience with respect to the disease. The simplest of these models classifies individuals as either susceptible, infectious, or removed from the population following the infectious period (see Figure 7.P.1).

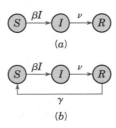

(a)

(b)

FIGURE 7.P.1   (a) The SIR epidemic model and (b) the SIRS epidemic model.

Accordingly, we define the state variables

$S(t)$ = number of susceptible individuals at time t,

$I(t)$ = number of infected individuals at time t,

$R(t)$ = number of post-infective individuals removed from the population at time t (due to immunity, quarantine, or death).

Susceptible individuals are able to catch the disease, after which they move into the infectious class. Infectious individuals spread the disease to susceptibles, and remain in the infectious class for a period of time (the infectious period) before moving into the removed class. Individuals in the removed class consist of those who can no longer acquire or spread the disease. The mathematical model (referred to as the SIR model) describing the temporal evolution of the sizes of the classes is based on the following assumptions:

1. The rate at which susceptibles become infected is proportional to the number of encounters between susceptible and infected individuals, which in turn is proportional to the product of the two populations, βSI. Larger values of β correspond to higher contact rates between infecteds and susceptibles.

2. The rate of transition from class I to class R is proportional to I, that is, vI. The biological meaning of v is that $1/v$ is the average length of the infectious period.

3. During the time period over which the disease evolves there is no immigration, emigration, births, or deaths except possibly from the disease.

With these assumptions, the differential equations that describe the number of individuals in the three classes are

$$
\begin{aligned}
S' &= -\beta IS, \\
I' &= \beta IS - vI, \\
R' &= vI.
\end{aligned}
\tag{1}
$$

It is convenient to restrict analysis to the first two equations in Eq. (1) since they are independent of R,

$$
\begin{aligned}
S' &= -\beta IS, \\
I' &= \beta IS - vI.
\end{aligned}
\tag{2}
$$

The SIRS Model. A slight variation in the SIR model results by assuming that individuals in the R class are temporarily immune, say, for an average length of time $1/\gamma$, after which they rejoin the class of susceptibles. The governing equations in this scenario, referred to as the SIRS model, are

$$S' = -\beta IS + \gamma R,$$
$$I' = \beta IS - \nu I, \tag{3}$$
$$R' = \nu I - \gamma R.$$

Project 1 PROBLEMS

1. Assume that $S(0) + I(0) + R(0) = N$, that is, the total size of the population at time $t = 0$ is N. Show that $S(t) + I(t) + R(t) = N$ for all $t > 0$ for both the SIR and SIRS models.

2. The triangular region $\Gamma = \{(S, I) : 0 \le S + I \le N\}$ in the SI-plane is depicted in Figure 7.P.2. Use an analysis based strictly on direction fields to show that no solution of the system (2) can leave the set Γ. More precisely, show that each point on the boundary of Γ is either a critical point of the system (2), or else the direction field vectors point toward the interior of Γ or are parallel to the boundary of Γ.

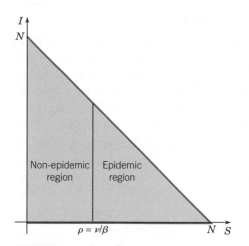

FIGURE 7.P.2   The state variables S and I for the SIR and SIRS models must lie in the region $\Gamma = \{(S, I) : 0 \le S + I \le N\}$.

3. If epidemics are identified with solution trajectories in which the number of infected individuals initially increases, reaches a maximum, and then decreases, use a nullcline analysis to show that an epidemic occurs if and only if $S(0) > \rho = \nu/\beta$. Assume that $\nu/\beta < N$. Thus $\rho = \nu/\beta$ is, in effect, a threshold value of susceptibles separating Γ into an epidemic region and a nonepidemic region. Explain how the size of the nonepidemic region depends on contact rate and length of infection period.

4. Find an equation of the form $H(S, I) = c$ satisfied by the solutions of Eq. (2). Then construct a phase portrait within Γ for the system (2) consisting of trajectories emanating from points along the upper boundary $S + I = N$ of Γ corresponding to initial states in which $R(0) = 0$.

5. In the SIR system (1), describe qualitatively the asymptotic behavior of S, I, and R as $t \to \infty$. In particular, answer the question "Does everyone get infected?" Then explain the statement "The epidemic does not die out due to the lack of susceptibles, but rather due to a lack of infectives."

6. Vaccinated individuals are protected from acquiring the disease and are, in effect, removed from participating in the transmission of the disease. Explain how an epidemic can be avoided by vaccinating a sufficiently large fraction p of the population, but it is not necessary to vaccinate the entire population.

7. Use the equation $S + I + R = N$ to reduce the SIRS model (3) to a system of dimension 2. Then use the qualitative methods of Chapter 7 and numerical simulations to discover as much as you can about the properties of solutions of the system (3). Compare and contrast your findings with the properties of solutions of the SIR model.

Project 2 Harvesting in a Competitive Environment

Consider again the system [Eq. (2) of Section 7.3]

$$\frac{dx}{dt} = x(\epsilon_1 - \sigma_1 x - \alpha_1 y), \qquad \frac{dy}{dt} = y(\epsilon_2 - \sigma_2 y - \alpha_2 x), \tag{1}$$

which models competition between two species. To be specific, suppose that x and y are the populations of two species of fish in a pond, lake, or ocean. Suppose further that species x is a good source of nourishment, so it is desirable to harvest members of x for food. Intuitively, it may seem reasonable to believe that if x is harvested too aggressively, then its numbers may be reduced to the point where it is no longer able to survive the competition with y

and will decline to possible extinction. So the policy issue is how to determine a harvest rate that will provide useful food without threatening the long-term survival of the species. There are two simple models that have been used to investigate harvesting in a competitive situation: a constant-effort model and a constant-yield model. The first of these is described in Problems 1 through 3, and the second in Problem 4.

Project 2 PROBLEMS

1. Consider again the system

$$\frac{dx}{dt} = x(1 - x - y),$$
$$\frac{dy}{dt} = y(0.75 - y - 0.5x),$$ (i)

which appeared in Example 1 of Section 7.3. A constant-effort model, applied to the species x alone, assumes that the rate of growth of x is altered by including the term $-Ex$, where E is a positive constant measuring the effort invested in harvesting members of species x. This assumption means that for a given effort E, the rate of catch is proportional to the population x, and that for a given population x, the rate of catch is proportional to the effort E. Based on this assumption, Eqs. (i) are replaced by

$$\frac{dx}{dt} = x(1 - x - y) - Ex = x(1 - E - x - y),$$
$$\frac{dy}{dt} = y(0.75 - y - 0.5x).$$ (ii)

(a) For $E = 0$, the critical points of Eqs. (ii) are as in Example 1 of Section 7.3. As E increases, some critical points move, while others remain fixed. Which ones move and how?
(b) For a certain value of E, denoted by E_0, the asymptotically stable node originally at (0.5, 0.5) coincides with the saddle point (0, 0.75). Find the value of E_0.
(c) Draw a direction field and/or a phase portrait for $E = E_0$ and for values of E slightly less than and slightly greater than E_0.
(d) How does the nature of the critical point (0, 0.75) change as E passes through E_0?
(e) What happens to the species x for $E > E_0$?

2. Consider the system

$$\frac{dx}{dt} = x(1 - x - y),$$
$$\frac{dy}{dt} = y(0.8 - 0.6y - x),$$ (iii)

which appeared in Example 2 of Section 7.3. If constant-effort harvesting is applied to species x, then the modified equations are

$$\frac{dx}{dt} = x(1 - x - y) - Ex = x(1 - E - x - y),$$
$$\frac{dy}{dt} = y(0.8 - 0.6y - x).$$ (iv)

(a) For $E = 0$, the critical points of Eqs. (iv) are as in Example 2 of Section 7.3. As E increases, some critical points move, while others remain fixed. Which ones move and how?
(b) For a certain value of E, denoted by E_0, the saddle point originally at (0.5, 0.5) coincides with the asymptotically stable node originally at (1, 0). Find the value of E_0.
(c) Draw a direction field and/or a phase portrait for $E = E_0$ and for values of E slightly less than, and slightly greater than, E_0. Estimate the basin of attraction for each asymptotically stable critical point.
(d) Consider the asymptotically stable node originally at (1, 0). How does the nature of this critical point change as E passes through E_0?
(e) What happens to the species x for $E > E_0$?

3. Consider the system (i) in Problem 1, and assume now that both x and y are harvested, with efforts E_1 and E_2, respectively. Then the modified equations are

$$\frac{dx}{dt} = x(1 - E_1 - x - y),$$
$$\frac{dy}{dt} = y(0.75 - E_2 - y - 0.5x).$$ (v)

(a) When $E_1 = E_2 = 0$, there is an asymptotically stable node at (0.5, 0.5). Find conditions on E_1 and E_2 that permit the continued long-term survival of both species.
(b) Use the conditions found in part (a) to sketch the region in the E_1E_2-plane that corresponds to the long-term survival of both species. Also identify regions where one species survives but not the other, and a region where both decline to extinction.

4. A constant-yield model, applied to species x, assumes that dx/dt is reduced by a positive constant H, the yield rate. For the situation described by Eqs. (i), the modified equations are

$$\frac{dx}{dt} = x(1 - x - y) - H,$$
$$\frac{dy}{dt} = y(0.75 - y - 0.5x).$$ (vi)

(a) For $H = 0$, the x-nullclines are the lines $x = 0$ and $x + y = 1$. For $H > 0$, show that the x-nullcline is a hyperbola whose asymptotes are $x = 0$ and $x + y = 1$.
(b) How do the critical points move as H increases from zero?
(c) For a certain value of H, denoted by H_c, the asymptotically stable node originally at (0.5, 0.5) coincides with the

saddle point originally at (0, 0.75). Determine the value of H_c. Also determine the values of x and y where the two critical points coincide.

(d) Where are the critical points for $H > H_c$? Classify them as to type.

(e) What happens to species x for $H > H_c$? What happens to species y?

(f) Draw a direction field and/or phase portrait for $H = H_c$ and for values of H slightly less than, and slightly greater than, H_c.

Project 3 The Rössler System

The system

$$x' = -y - z, \qquad y' = x + ay, \qquad z' = b + z(x - c), \tag{1}$$

where a, b, and c are positive parameters, is known as the Rössler[6] system. It is a relatively simple system, consisting of two linear equations and a third equation with a single quadratic nonlinearity. In the following problems, we ask you to carry out some numerical investigations of this system, with the goal of exploring its period-doubling property. To simplify matters, set $a = 0.25$, $b = 0.5$, and let $c > 0$ remain arbitrary.

Project 3 PROBLEMS

1. **(a)** Show that there are no critical points when $c < \sqrt{0.5}$, one critical point for $c = \sqrt{0.5}$, and two critical points when $c > \sqrt{0.5}$.

(b) Find the critical point(s) and determine the eigenvalues of the associated Jacobian matrix when $c = \sqrt{0.5}$ and when $c = 1$.

(c) How do you think trajectories of the system will behave for $c = 1$? Plot the trajectory starting at the origin. Does it behave the way that you expected?

(d) Choose one or two other initial points and plot the corresponding trajectories. Do these plots agree with your expectations?

2. **(a)** Let $c = 1.3$. Find the critical points and the corresponding eigenvalues. What conclusions, if any, can you draw from this information?

(b) Plot the trajectory starting at the origin. What is the limiting behavior of this trajectory? To see the limiting behavior clearly, you may wish to choose a t-interval for your plot so that the initial transients are eliminated.

(c) Choose one or two other initial points and plot the corresponding trajectories. Are the limiting behavior(s) the same as in part (b)?

(d) Observe that there is a limit cycle whose basin of attraction is fairly large (although not all of xyz-space). Draw a plot of x, y, or z versus t and estimate the period of motion around the limit cycle.

3. The limit cycle found in Problem 2 comes into existence as a result of a Hopf bifurcation at a value c_1 of c between 1 and 1.3. Determine, or at least estimate more precisely, the

value of c_1. There are several ways in which you might do this.

(a) Draw plots of trajectories for different values of c.

(b) Calculate eigenvalues at critical points for different values of c.

(c) Use the result of Problem 3(b) in Section 7.6.

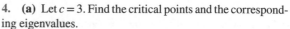

4. **(a)** Let $c = 3$. Find the critical points and the corresponding eigenvalues.

(b) Plot the trajectory starting at the point (1, 0, −2). Observe that the limit cycle now consists of two loops before it closes; it is often called a 2-cycle.

(c) Plot x, y, or z versus t and show that the period of motion on the 2-cycle is very nearly double the period of the simple limit cycle in Problem 2. There has been a period-doubling bifurcation of cycles for a certain value of c between 1.3 and 3.

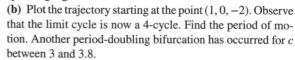

5. **(a)** Let $c = 3.8$. Find the critical points and the corresponding eigenvalues.

(b) Plot the trajectory starting at the point (1, 0, −2). Observe that the limit cycle is now a 4-cycle. Find the period of motion. Another period-doubling bifurcation has occurred for c between 3 and 3.8.

(c) For $c = 3.85$, show that the limit cycle is an 8-cycle. Verify that its period is very close to eight times the period of the simple limit cycle in Problem 2.

Note: As c increases further, there is an accelerating cascade of period-doubling bifurcations. The bifurcation values of c converge to a limit, which marks the onset of chaos.

---

[6]See the book by Strogatz for a more extensive discussion and further references.

© Cristian Andrei Matei / Getty Images, Inc.

Numerical Methods

A differential equation of the form $\frac{dy}{dt} = f(t, y)$, or a system of such equations, quite often cannot be solved explicitly using the methods established thus far in this text. In such cases, having an alternative approach that enables us to approximate the solution is desirable. There are various ways of approximating certain integrals that, in turn, lead to numerical methods used to construct such approximate solutions. Some well-known methods are developed and analyzed in this chapter.

8.1 Numerical Approximations: Euler's Method

We have seen in Section 1.2 that by drawing a direction field, we can visualize qualitatively the behavior of solutions of a differential equation. In fact, if we use a fairly fine grid, then we obtain a direction field such as the one in Figure 8.1.1 which corresponds to the differential equation

$$\frac{dy}{dt} + \frac{1}{2}y = \frac{3}{2} - t. \tag{1}$$

Many tangent line segments at successive values of t almost touch each other in this figure. It takes only a bit of imagination to consider starting at a point on the y-axis and linking line segments for consecutive t-values in the grid, thereby producing a piecewise linear graph. Such a graph would apparently be an approximation to a solution of the differential equation. Of course, this raises some questions, including the following:

1. Can we carry out this linking of tangent lines in a simple and systematic manner?

2. If so, does the resulting piecewise linear function provide an approximation to an actual solution of the differential equation?

3. If so, can we say anything about the accuracy of the approximation?

It turns out that the answer to each question is affirmative. We will take up the first question here, and return to the other two in Section 8.2.

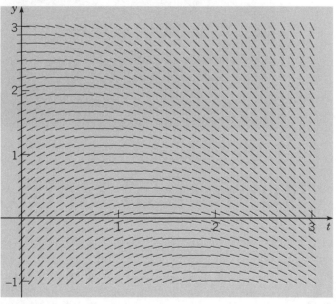

FIGURE 8.1.1   A direction field for $y' + \frac{1}{2}y = \frac{3}{2} - t$.

Suppose that we have a differential equation

$$\frac{dy}{dt} = f(t, y) \tag{2}$$

and a starting point, given by the initial condition

$$y(t_0) = y_0. \tag{3}$$

Suppose also that we have chosen a sequence of points $t_0, t_1, t_2, \ldots, t_n, \ldots$. Let the solution of the initial value problem (2), (3) be denoted by $y = \phi(t)$. Then the line tangent to the graph of $\phi(t)$ at the initial point (t_0, y_0) has the slope $f(t_0, y_0)$, and the equation of this tangent line is

$$y = y_0 + f(t_0, y_0)(t - t_0). \tag{4}$$

We can use the tangent line (4) to approximate the solution $\phi(t)$ in the interval $t_0 \le t \le t_1$. In particular, if we evaluate Eq. (4) at $t = t_1$, we obtain the value

$$y_1 = y_0 + f(t_0, y_0)(t_1 - t_0), \tag{5}$$

which is an approximation to the solution value $\phi(t_1)$; see Figure 8.1.2.

To proceed further, we can try to repeat the process. Unfortunately, we do not know the value $\phi(t_1)$ of the solution at t_1. The best we can do is to use the approximate value y_1

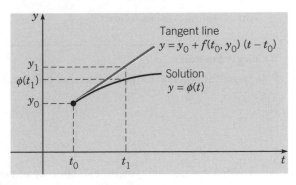

FIGURE 8.1.2 A tangent line approximation.

instead. Thus we construct the line through (t_1, y_1) with the slope $f(t_1, y_1)$:

$$y = y_1 + f(t_1, y_1)(t - t_1). \tag{6}$$

To approximate the value of $\phi(t)$ at t_2, we use Eq. (6), obtaining

$$y_2 = y_1 + f(t_1, y_1)(t_2 - t_1). \tag{7}$$

Continuing in this manner, we use the value of y calculated at each step to determine the slope of the approximation for the next step. The general expression for y_{n+1} in terms of t_n, t_{n+1}, and y_n is

$$y_{n+1} = y_n + f(t_n, y_n)(t_{n+1} - t_n), \qquad n = 0, 1, 2, \dots. \tag{8}$$

Equation (8) is known as Euler's formula. If there is a uniform step size h between the points $t_0, t_1, t_2, \dots$, then Euler's method is summarized by the equations

$$t_{n+1} = t_n + h, \qquad y_{n+1} = y_n + hf(t_n, y_n), \qquad n = 0, 1, 2, \dots. \tag{9}$$

By introducing the notation $f_n = f(t_n, y_n)$, we can rewrite Eqs. (9) as

$$t_{n+1} = t_n + h, \qquad y_{n+1} = y_n + hf_n, \qquad n = 0, 1, 2, \dots. \tag{10}$$

To use Euler's method, you simply evaluate Eqs. (9) repeatedly [or use Eq. (8) if the step size is not constant]. The result at each step is used to execute the next step. In this way you generate a sequence of values $y_1, y_2, y_3, \dots$ that approximate the values of the solution $\phi(t)$ at the points $t_1, t_2, t_3, \dots$. If, instead of a sequence of points, you need a function to approximate the solution $\phi(t)$, then you can use the piecewise linear function constructed from the collection of tangent line segments. That is, let y be given by Eq. (4) in $[t_0, t_1]$, by Eq. (6) in $[t_1, t_2]$, and in general by

$$y = y_n + f(t_n, y_n)(t - t_n) \tag{11}$$

in $[t_n, t_{n+1}]$.

This is summarized as follows:

Euler's Formula

Suppose the solution of the initial value problem

$$\begin{cases} \dfrac{dy}{dt} = f(t, y) \\[2mm] y(t_0) = y_0 \end{cases} \tag{12}$$

is denoted $y = \phi(t)$ and you have a sequence of points $t_0 < t_1 < t_2 < \cdots < t_n < \cdots$. For $n = 0, 1, 2, \ldots$, we have the following:

Approximation of $y = \phi(t)$ at $t = t_{n+1}$:

$$y_{n+1} = y_n + f(t_n, y_n)(t_{n+1} - t_n). \tag{13}$$

Linear approximation of $\phi(t)$ on the interval $[t_n, t_{n+1}]$:

$$y(t) = y_n + f(t_n, y_n)(t - t_n). \tag{14}$$

Special case: If a uniform step size h is used, then $t_{n+1} - t_n = h$, for all n, and so Eq. (13) simplifies to

$$y_{n+1} = y_n + h f(t_n, y_n).$$

EXAMPLE 1

Use Euler's method with a step size $h = 0.2$ to approximate the solution of the initial value problem

$$\frac{dy}{dt} + \frac{1}{2}y = \frac{3}{2} - t, \qquad y(0) = 1 \tag{15}$$

on the interval $0 \leq t \leq 1$.

For later comparison, the exact solution of the initial value problem (15) is

$$y = 7 - 2t - 6e^{-t/2}. \tag{16}$$

This solution can be found by using the method of integrating factors discussed in Section 2.2.

To use Euler's method, note that $f(t, y) = -\frac{1}{2}y + \frac{3}{2} - t$, so using the initial values $t_0 = 0$, $y_0 = 1$, we have

$$f_0 = f(t_0, y_0) = f(0, 1) = -0.5 + 1.5 - 0 = 1.0.$$

Thus, from Eq. (4), the tangent line approximation for t in [0, 0.2] is

$$y = 1 + 1.0(t - 0) = 1 + t. \tag{17}$$

Setting $t = 0.2$ in Eq. (17), we find the approximate value y_1 of the solution at $t = 0.2$, namely, $y_1 = 1.2$.

At the next step we have

$$f_1 = f(t_1, y_1) = f(0.2, 1.2) = -0.6 + 1.5 - 0.2 = 0.7,$$

and from Eq. (6),

$$y = 1.2 + 0.7(t - 0.2) = 1.06 + 0.7t \tag{18}$$

for t in [0.2, 0.4]. Evaluating the expression in Eq. (18) at $t = 0.4$, we obtain $y_2 = 1.34$.

Continuing in this manner, we obtain the results shown in Table 8.1.1. The first column contains the t-values separated by the step size $h = 0.2$. The third column shows the corresponding y-values computed from Euler's formula (10). The fourth column displays the tangent line approximations found from Eq. (11). The second column contains values of the solution (16) of the initial value problem (15), correct to five decimal places. The solution (16) and the tangent line approximation are also plotted in Figure 8.1.3.

TABLE 8.1.1	Results of Euler's method with $h = 0.2$ for $\frac{dy}{dt} + \frac{1}{2}y = \frac{3}{2} - t, \quad y(0) = 1$.		
t	Exact	Euler with $h = 0.2$	Tangent line
0.0	1.00000	1.00000	$y = 1 + t$
0.2	1.17098	1.20000	$y = 1.06 + 0.7t$
0.4	1.28762	1.34000	$y = 1.168 + 0.43t$
0.6	1.35509	1.42600	$y = 1.3138 + 0.187t$
0.8	1.37808	1.46340	$y = 1.48876 - 0.0317t$
1.0	1.36082	1.45706	

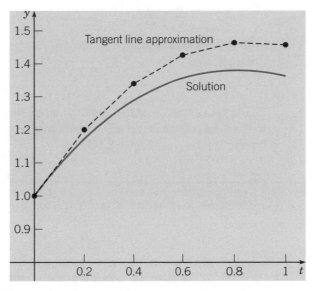

FIGURE 8.1.3 Plots of the solution and a tangent line approximation for the initial value problem (15).

As you can see from Table 8.1.1 and Figure 8.1.3, the approximations given by Euler's method for this problem are greater than the corresponding values of the exact solution. This is because the graph of the solution is concave down and therefore the tangent line approximations lie above the graph. It is also clear that the step size $h = 0.2$ is too large to produce a good approximation to the solution (16) on the interval [0, 1]. Better results can be obtained by using a smaller step size, with a corresponding increase in the number of computational steps, as we will see in Example 2.

Euler's method dates from about 1768 and is the oldest numerical method for approximating the solution of a differential equation. It is simple in concept and easy to execute.

The simplicity of Euler's method makes it a good way to begin to explore the numerical approximation of solutions of relatively simple differential equations.

The purpose of Example 1 is to show you the details of implementing a few steps of Euler's method so that it will be clear exactly what computations are being executed. Of course, computations such as these are usually done on a computer. Some software packages include code for the Euler method, whereas others do not. In any case, it is easy to write a computer program to carry out the calculations required to produce results such as those in Table 8.1.1. The outline of such a program is given below; the specific instructions can be written in any high-level programming language.

Pseudo-Code for the Euler Method

Step 1. define $f(t, y)$

Step 2. input initial values $t0$ and $y0$

Step 3. input step size h and number of steps n

Step 4. output $t0$ and $y0$

Step 5. for j from 1 to n **do**

Step 6. $k1 = f(t, y)$

         $y = y + h * k1$

         $t = t + h$

Step 7. output t and y

Step 8. end

The output of this algorithm can be numbers listed on the screen or printed on a printer, as in the third column of Table 8.1.1. Alternatively, the calculated results can be displayed in graphical form, as in Figure 8.1.3.

EXAMPLE 2

Consider again the initial value problem (15),

$$\frac{dy}{dt} + \frac{1}{2}y = \frac{3}{2} - t, \qquad y(0) = 1.$$

Use Euler's method with various step sizes to calculate approximate values of the solution for $0 \le t \le 5$. Compare the calculated results with the corresponding values of the exact solution (16),

$$y = \phi(t) = 7 - 2t - 6e^{-t/2}.$$

We used step sizes $h = 0.1, 0.05, 0.025,$ and 0.01, corresponding respectively, to 50, 100, 200, and 500 steps, to go from $t = 0$ to $t = 5$. Some of the results of these calculations, along with the values of the exact solution, are presented in Table 8.1.2. All computed entries are rounded to five decimal places, although more digits were retained in the intermediate calculations.

In Figure 8.1.4, we have plotted the absolute value of the error (i.e., the difference between the exact solution and its approximations) for each value of h and for each value of t, as recorded in Table 8.1.2. The lines in this graph do not necessarily represent the error accurately in between the data points, but are included to make the plot more visually understandable.

What conclusions can we draw from the data in Table 8.1.2 and from Figure 8.1.4. In the first place, for a fixed value of t, the computed approximate values become more accurate

TABLE 8.1.2	A comparison of exact solution with Euler's method for several step sizes h for $\frac{dy}{dt} + \frac{1}{2}y = \frac{3}{2} - t$, $y(0) = 1$.				
t	Exact	$h = 0.1$	$h = 0.05$	$h = 0.025$	$h = 0.01$
0.0	1.00000	1.00000	1.00000	1.00000	1.00000
1.0	1.36082	1.40758	1.38387	1.37227	1.36538
2.0	0.79272	0.84908	0.82061	0.80659	0.79825
3.0	−0.33878	−0.28783	−0.31349	−0.32618	−0.33375
4.0	−1.81201	−1.77107	−1.79163	−1.80184	−1.80795
5.0	−3.49251	−3.46167	−3.47710	−3.48481	−3.48943

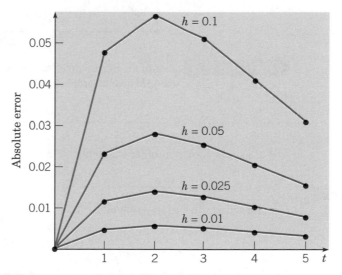

FIGURE 8.1.4 Plot of the absolute value of the error in using Euler's method for the initial value problem (15) for $h = 0.1$, 0.05, 0.025, and 0.01.

as the step size h decreases. This is what we would expect, of course, but it is encouraging that the data confirm our expectations. For example, for $t = 1$, the approximate value with $h = 0.1$ is too large by about 3.43%, whereas the value with $h = 0.01$ is too large by only 0.34%. In this case, reducing the step size by a factor of 10 (and performing 10 times as many computations) also reduces the error by a factor of about 10. A second observation is that, for a fixed step size h, the approximations become more accurate as t increases. For instance, for $h = 0.1$, the error for $t = 5$ is only about 0.88%, compared with 3.43% for $t = 1$. For the data we have recorded, the maximum error occurs at $t = 2$ in each case. An examination of data at intermediate points not recorded in Table 8.1.2 would reveal where the maximum error occurs for a given step size and how large it is.

A plot of the maximum recorded error (i.e., the error at $t = 2$) versus the step size h is shown in Figure 8.1.5. Each data point lies very close to a straight line through the origin, which means that the maximum error is very nearly proportional to h. From Figure 8.1.5 or from the data in Table 8.1.2, you can conclude that the value of the proportionality constant is about 0.56.

All in all, Euler's method seems to work rather well for this problem. Reasonably good results are obtained even for a moderately large step size $h = 0.1$, and the approximation can be improved by decreasing h.

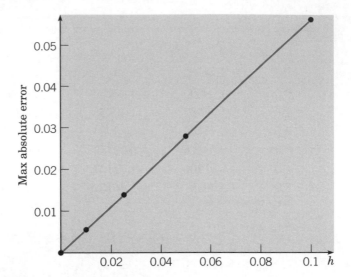

FIGURE 8.1.5 Plot of the maximum error versus step size h in using Euler's method for the initial value problem (15).

Let us now look at another example.

EXAMPLE 3

Consider the initial value problem

$$\frac{dy}{dt} - 2y = 4 - t, \qquad y(0) = 1. \tag{19}$$

Use Euler's method with several step sizes to find approximate values of the solution on the interval $0 \le t \le 5$. Compare the results with the corresponding values of the exact solution.

The solution of the initial value problem (19) is

$$y = -\frac{7}{4} + \frac{1}{2}t + \frac{11}{4}e^{2t}. \tag{20}$$

It can be easily found by using the method discussed in Section 2.2.

Using the same range of step sizes as in Example 2, we obtain the results presented in Table 8.1.3.

TABLE 8.1.3 A comparison of exact solution with Euler's method for several step sizes h for $\frac{dy}{dt} - 2y = 4 - t$, $y(0) = 1$.

t	Exact	$h = 0.1$	$h = 0.05$	$h = 0.025$	$h = 0.01$
0.0	1.000000	1.000000	1.000000	1.000000	1.000000
1.0	19.06990	15.77728	17.25062	18.10997	18.67278
2.0	149.3949	104.6784	123.7130	135.5440	143.5835
3.0	1109.179	652.5349	837.0745	959.2580	1045.395
4.0	8197.884	4042.122	5633.351	6755.175	7575.577
5.0	60573.53	25026.95	37897.43	47555.35	54881.32

The data in Table 8.1.3 again confirm our expectation that, for a given value of t, accuracy improves as the step size h is reduced. For example, for $t = 1$, the percentage error diminishes from 17.3% when $h = 0.1$ to 2.1% when $h = 0.01$. However the error increases fairly rapidly as t increases for a fixed h. Even for $h = 0.01$, the error at $t = 5$ is 9.4%, and it is much greater for larger step sizes. This is shown in Figure 8.1.6, which shows the absolute error versus t for each value of h. The maximum error always occurs at $t = 5$ and is plotted against h in Figure 8.1.7. Again, the data points lie approximately on a straight line through

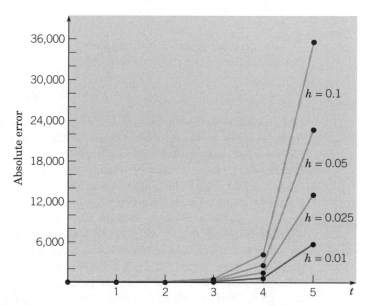

FIGURE 8.1.6 Plot of the absolute value of the error in using Euler's method for the initial value problem (19) for $h = 0.1$, 0.05, 0.025, and 0.01.

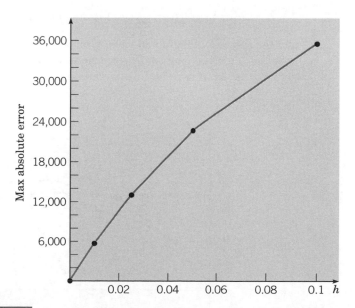

FIGURE 8.1.7 Plot of the maximum error versus step size h in using Euler's method for the initial value problem (19).

the origin, so the maximum error is nearly proportional to the step size, as in Example 2. Now, however, the proportionality constant is greater than 50,000, or about 100,000 times greater than in Example 2.

Of course, the accuracy that is needed depends on the purpose for which the results are intended, but the errors in Table 8.1.3 are too large for most scientific or engineering applications. To improve the situation, one might either try even smaller step sizes or else restrict the computations to a rather short interval away from the initial point. Nevertheless it is clear that Euler's method is much less effective in this example than in Example 2.

To understand better what is happening in these examples, let us look again at Euler's method for the general initial value problem

$$\frac{dy}{dt} = f(t, y), \qquad y(t_0) = y_0, \tag{21}$$

whose solution we denote by $\phi(t)$. Recall that a first order differential equation has an infinite family of solutions, indexed by an arbitrary constant c, and that the initial condition picks out one member of this infinite family by determining the value of c. Thus $\phi(t)$ is the member of the infinite family of solutions that satisfies the initial condition $\phi(t_0) = y_0$.

At the first step, Euler's method uses the tangent line approximation to the graph of $y = \phi(t)$ passing through the initial point (t_0, y_0), and this produces the approximate value y_1 at t_1. Usually, $y_1 \neq \phi(t_1)$, so at the second step, Euler's method uses the tangent line approximation not to $y = \phi(t)$, but to a nearby solution $y = \phi_1(t)$ that passes through the point (t_1, y_1). So it is at each following step. Euler's method uses a succession of tangent line approximations to a sequence of different solutions $\phi(t)$, $\phi_1(t)$, $\phi_2(t)$, ... of the differential equation. At each step the tangent line is constructed to the solution passing through the point determined by the result of the preceding step, as shown in Figure 8.1.8. The quality of the approximation after many steps depends strongly on the behavior of the set of solutions that pass through the points (t_n, y_n) for $n = 1, 2, 3, \ldots$.

In Examples 1 and 2, the general solution of the differential equation is

$$y = 7 - 2t + ce^{-t/2} \tag{22}$$

and the solution of the initial value problem (15) corresponds to $c = -6$. This family of solutions is a converging family since the term involving the arbitrary constant c approaches zero as $t \to \infty$. It does not matter very much which solutions we are approximating by

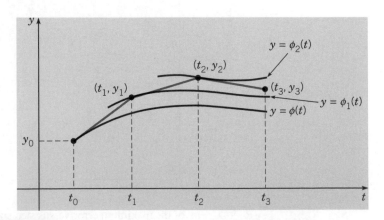

FIGURE 8.1.8   The Euler method.

tangent lines in the implementation of Euler's method, since all the solutions are getting closer and closer to each other as t increases.

On the other hand, in Example 3 the general solution of the differential equation is

$$y = -\frac{7}{4} + \frac{1}{2}t + ce^{2t}, \tag{23}$$

and this is a diverging family. Note that solutions corresponding to two nearby values of c separate arbitrarily far as t increases. In Example 3, we are trying to follow the solution for $c = \frac{11}{4}$, but, in the use of Euler's method, we are actually at each step following another solution that separates from the desired one faster and faster as t increases. This explains why the errors in Example 3 are so much larger than those in Example 2.

In using a numerical procedure such as the Euler method, one must always keep in mind the question of whether the results are accurate enough to be useful. In the preceding examples, the accuracy of the numerical results could be ascertained directly by a comparison with the solution obtained analytically. Of course, usually the analytical solution is not available if a numerical procedure is to be employed, so what is needed are bounds for, or at least estimates of, the error that do not require a knowledge of the exact solution. In Sections 8.2 and 8.3, we present some information on the analysis of errors and also discuss other algorithms that are computationally much more efficient than the Euler method. However, the best that we can expect, or hope for, from a numerical approximation is that it reflects the behavior of the actual solution. Thus a member of a diverging family of solutions will always be harder to approximate than a member of a converging family. Finally, remember that drawing a direction field is often a helpful first step in understanding the behavior of differential equations and their solutions.

PROBLEMS

Many of the problems in this section call for fairly extensive numerical computations. The amount of computing that is reasonable for you to do depends strongly on the type of computing equipment that you have. A few steps of the requested calculations can be carried out on almost any pocket calculator—or even by hand if necessary. To do more, you will need a computer or at least a programmable calculator.

Remember also that numerical results may vary somewhat depending on how your program is constructed and on how your computer executes arithmetic steps, rounds off, and so forth. Minor variations in the last decimal place may be due to such causes and do not necessarily indicate that something is amiss. Answers in the back of the book are recorded to six digits in most cases, although more digits were retained in the intermediate calculations.

In each of Problems 1 through 4:
(a) Find approximate values of the solution of the given initial value problem at $t = 0.1, 0.2, 0.3,$ and 0.4 using the Euler method with $h = 0.1$.
(b) Repeat part (a) with $h = 0.05$. Compare the results with those found in (a).
(c) Repeat part (a) with $h = 0.025$. Compare the results with those found in (a) and (b).
(d) Find the solution $y = \phi(t)$ of the given problem and evaluate $\phi(t)$ at $t = 0.1, 0.2, 0.3,$ and 0.4. Compare these values with the results of (a), (b), and (c).

1. $y' = 3 + t - y$, $\qquad y(0) = 1$
2. $y' = 2y - 1$, $\qquad y(0) = 1$
3. $y' = 0.5 - t + 2y$, $\qquad y(0) = 1$
4. $y' = 3 \cos t - 2y$, $\qquad y(0) = 0$

In each of Problems 5 through 10, draw a direction field for the given differential equation and state whether you think that the solutions are converging or diverging.

5. $y' = 5 - 3\sqrt{y}$
6. $y' = y(3 - ty)$
7. $y' = (4 - ty)/(1 + y^2)$
8. $y' = -ty + 0.1y^3$
9. $y' = t^2 + y^2$
10. $y' = (y^2 + 2ty)/(3 + t^2)$

In each of Problems 11 through 14, use Euler's method to find approximate values of the solution of the given initial value problem at $t = 0.5, 1, 1.5, 2, 2.5,$ and 3:
(a) With $h = 0.1$
(b) With $h = 0.05$
(c) With $h = 0.025$
(d) With $h = 0.01$

11. $y' = 5 - 3\sqrt{y}$, $\qquad y(0) = 2$
12. $y' = y(3 - ty)$, $\qquad y(0) = 0.5$

13. $y' = (4 - ty)/(1 + y^2)$, $\quad y(0) = -2$

14. $y' = -ty + 0.1y^3$, $\quad y(0) = 1$

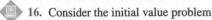 15. Consider the initial value problem

$$y' = 3t^2/(3y^2 - 4), \qquad y(1) = 0.$$

(a) Use Euler's method with $h = 0.1$ to obtain approximate values of the solution at $t = 1.2, 1.4, 1.6$, and 1.8.

(b) Repeat part (a) with $h = 0.05$.

(c) Compare the results of parts (a) and (b). Note that they are reasonably close for $t = 1.2, 1.4$, and 1.6 but are quite different for $t = 1.8$. Also note (from the differential equation) that the line tangent to the solution is parallel to the y-axis when $y = \pm 2/\sqrt{3} \cong \pm 1.155$. Explain how this might cause such a difference in the calculated values.

 16. Consider the initial value problem

$$y' = t^2 + y^2, \qquad y(0) = 1.$$

Use Euler's method with $h = 0.1, 0.05, 0.025$, and 0.01 to explore the solution of this problem for $0 \le t \le 1$. What is your best estimate of the value of the solution at $t = 0.8$? At $t = 1$? Are your results consistent with the direction field in Problem 9?

17. Consider the initial value problem

$$y' = (y^2 + 2ty)/(3 + t^2), \qquad y(1) = 2.$$

Use Euler's method with $h = 0.1, 0.05, 0.025$, and 0.01 to explore the solution of this problem for $1 \le t \le 3$. What is your best estimate of the value of the solution at $t = 2.5$? At $t = 3$? Are your results consistent with the direction field in Problem 10?

18. Consider the initial value problem

$$y' = -ty + 0.1y^3, \qquad y(0) = \alpha,$$

where α is a given number.

(a) Draw a direction field for the differential equation (or re-examine the one from Problem 8). Observe that there is a critical value of α in the interval $2 \le \alpha \le 3$ that separates converging solutions from diverging ones. Call this critical value α_0.

(b) Use Euler's method with $h = 0.01$ to estimate α_0. Do this by restricting α_0 to an interval $[a, b]$, where $b - a = 0.01$.

19. Consider the initial value problem

$$y' = y^2 - t^2, \qquad y(0) = \alpha,$$

where α is a given number.

(a) Draw a direction field for the differential equation. Observe that there is a critical value of α in the interval $0 \le \alpha \le 1$ that separates converging solutions from diverging ones. Call this critical value α_0.

(b) Use Euler's method with $h = 0.01$ to estimate α_0. Do this by restricting α_0 to an interval $[a, b]$, where $b - a = 0.01$.

8.2 Accuracy of Numerical Methods

In Section 8.1, we introduced the Euler, or tangent line, method for approximating the solution of an initial value problem

$$\frac{dy}{dt} = f(t, y), \tag{1}$$

$$y(t_0) = y_0. \tag{2}$$

This method involves the repeated evaluation of the expressions

$$t_{n+1} = t_n + h, \tag{3}$$

$$y_{n+1} = y_n + hf(t_n, y_n) \tag{4}$$

for $n = 0, 1, 2, \ldots$. The result is a set of approximate values $y_1, y_2, \ldots$ at the mesh points $t_1, t_2, \ldots$. We assume, for simplicity, that the step size h is constant, although this is not necessary. In this section we will begin to investigate the errors that may occur in this numerical approximation process.

Some examples of Euler's method appear in Section 8.1. As another example, consider the initial value problem

$$\frac{dy}{dt} = 1 - t + 4y, \tag{5}$$

$$y(0) = 1. \tag{6}$$

Equation (5) is a first order linear equation, and it is easily verified that the solution satisfying the initial condition (6) is

$$y = \phi(t) = \tfrac{1}{4}t - \tfrac{3}{16} + \tfrac{19}{16}e^{4t}. \tag{7}$$

Since the exact solution is known, we do not need numerical methods to approximate the solution of the initial value problem (5), (6). On the other hand, the availability of the exact solution makes it easy to determine the accuracy of any numerical procedure that we use on this problem. We will use this problem in this section and the next to illustrate and compare different numerical methods. The solutions of Eq. (5) diverge rather rapidly from each other, so we should expect that it will be fairly difficult to approximate the solution (7) well over any considerable interval. Indeed, this is the reason for choosing this particular problem; it will be relatively easy to observe the benefits of using more accurate methods.

EXAMPLE 1

Using the Euler formula (4) and step sizes $h = 0.05, 0.025, 0.01$, and 0.001, determine approximate values of the solution $y = \phi(t)$ of the problem (5), (6) on the interval $0 \leq t \leq 2$.

The indicated calculations were carried out on a computer, and some of the results are shown in Table 8.2.1. Their accuracy is not particularly impressive. For $h = 0.01$, the percentage error is 3.85% at $t = 0.5$, 7.49% at $t = 1.0$, and 14.4% at $t = 2.0$. The corresponding percentage errors for $h = 0.001$ are 0.40%, 0.79%, and 1.58%, respectively. Observe that if $h = 0.001$, then it requires 2000 steps to traverse the interval from $t = 0$ to $t = 2$. Thus considerable computation is needed to obtain even reasonably good accuracy for this problem using the Euler method. When we discuss other numerical approximation methods in Section 8.3, we will find that it is possible to obtain comparable or better accuracy with much larger step sizes and many fewer computational steps.

TABLE 8.2.1

A comparison of results for the numerical solution of $\frac{dy}{dt} = 1 - t + 4y$, $y(0) = 1$ using the Euler method for different step sizes h.

t	$h = 0.05$	$h = 0.025$	$h = 0.01$	$h = 0.001$	Exact
0.0	1.0000000	1.0000000	1.0000000	1.0000000	1.0000000
0.1	1.5475000	1.5761188	1.5952901	1.6076289	1.6090418
0.2	2.3249000	2.4080117	2.4644587	2.5011159	2.5053299
0.3	3.4333560	3.6143837	3.7390345	3.8207130	3.8301388
0.4	5.0185326	5.3690304	5.6137120	5.7754844	5.7942260
0.5	7.2901870	7.9264062	8.3766865	8.6770692	8.7120041
1.0	45.588400	53.807866	60.037126	64.382558	64.897803
1.5	282.07187	361.75945	426.40818	473.55979	479.25919
2.0	1745.6662	2432.7878	3029.3279	3484.1608	3540.2001

To begin to investigate the errors in using numerical approximations, and also to suggest ways to construct more accurate algorithms, it is helpful to mention some alternative ways to look at the Euler method.

First, let us write the differential equation (1) at the point $t = t_n$ in the form

$$\frac{d\phi}{dt}(t_n) = f[t_n, \phi(t_n)]. \tag{8}$$

Then we approximate the derivative in Eq. (8) by the corresponding (forward) difference quotient, obtaining

$$\frac{\phi(t_{n+1}) - \phi(t_n)}{t_{n+1} - t_n} \cong f[t_n, \phi(t_n)]. \tag{9}$$

Finally, if we replace $\phi(t_{n+1})$ and $\phi(t_n)$ by their approximate values y_{n+1} and y_n, respectively, and solve for y_{n+1}, we obtain the Euler formula (4).

Another way to proceed is to write the problem as an integral equation. Since $y = \phi(t)$ is a solution of the initial value problem (1), (2), by integrating from t_n to t_{n+1}, we obtain

$$\int_{t_n}^{t_{n+1}} \phi'(t) \, dt = \int_{t_n}^{t_{n+1}} f[t, \phi(t)] \, dt,$$

or

$$\phi(t_{n+1}) = \phi(t_n) + \int_{t_n}^{t_{n+1}} f[t, \phi(t)] \, dt. \tag{10}$$

The integral in Eq. (10) is represented geometrically as the area under the curve in Figure 8.2.1 between $t = t_n$ and $t = t_{n+1}$. If we approximate the integral by replacing $f[t, \phi(t)]$ by its value $f[t_n, \phi(t_n)]$ at $t = t_n$, then we are approximating the actual area by the area of the shaded rectangle. In this way we obtain

$$\phi(t_{n+1}) \cong \phi(t_n) + f[t_n, \phi(t_n)](t_{n+1} - t_n)$$
$$= \phi(t_n) + hf[t_n, \phi(t_n)]. \tag{11}$$

Finally, to obtain an approximation y_{n+1} for $\phi(t_{n+1})$, we make a second approximation by replacing $\phi(t_n)$ by its approximate value y_n in Eq. (11). This gives the Euler formula $y_{n+1} = y_n + hf(t_n, y_n)$. A more accurate algorithm can be obtained by approximating the integral more accurately. This is discussed in Section 8.3.

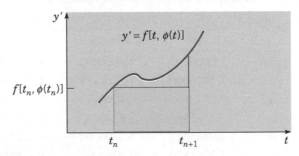

FIGURE 8.2.1 Integral derivation of the Euler method.

A third approach is to assume that the solution $y = \phi(t)$ has a Taylor series about the point t_n. Then

$$\phi(t_n + h) = \phi(t_n) + \phi'(t_n)h + \phi''(t_n)\frac{h^2}{2!} + \cdots,$$

or

$$\phi(t_{n+1}) = \phi(t_n) + f[t_n, \phi(t_n)]h + \phi''(t_n)\frac{h^2}{2!} + \cdots. \tag{12}$$

If the series is terminated after the first two terms, and $\phi(t_{n+1})$ and $\phi(t_n)$ are replaced by their approximate values y_{n+1} and y_n, we again obtain the Euler formula (4). If more terms in the series are retained, a more accurate formula is obtained. Further, by using a Taylor series with a remainder, it is possible to estimate the magnitude of the error in the formula.

▶ **Errors in Numerical Approximations.** The use of a numerical procedure, such as the Euler method, to approximate the solution of an initial value problem raises a number of questions that must be answered before the numerical approximation can be accepted as satisfactory. One of these is the question of **convergence**. That is, as the step size h tends to zero, do the values of the numerical approximation $y_1, y_2, \ldots, y_n, \ldots$ approach the corresponding values of the actual solution? If we assume that the answer is affirmative, there remains the important practical question of how rapidly the numerical approximation converges to the solution. In other words, how small a step size is needed in order to guarantee a given level of accuracy? We want to use a step size that is small enough to ensure the required accuracy, but not too small. An unnecessarily small step size slows down the calculations, makes them more expensive, and may even cause a loss of accuracy in some cases.

There are two fundamental sources of error in approximating the solution of an initial value problem numerically. Let us first assume that our computer is such that we can carry out all computations with complete accuracy; that is, we can retain infinitely many decimal places. The difference E_n between the solution $y = \phi(t)$ of the initial value problem (1), (2) and its numerical approximation y_n at t_n is given by

$$E_n = \phi(t_n) - y_n \tag{13}$$

and is known as the **global truncation error**. It arises from two causes: One, at each step we use an approximate formula to determine y_{n+1}; two, the input data at each step are only approximately correct. For example, in calculating y_{n+1}, we use y_n rather than (the unknown) $\phi(t_n)$ and, in general, $\phi(t_n)$ is not equal to y_n. If we assume that $y_n = \phi(t_n)$, then the only error in going one further step is due to the use of an approximate formula. This error is known as the **local truncation error** e_n.

The second fundamental source of error is that we carry out the computations in arithmetic with only a finite number of digits. This leads to a **round-off error** R_n defined by

$$R_n = y_n - Y_n, \tag{14}$$

where Y_n is the value *actually computed* from the given numerical method.

The absolute value of the total error in computing $\phi(t_n)$ is given by

$$|\phi(t_n) - Y_n| = |\phi(t_n) - y_n + y_n - Y_n|. \tag{15}$$

Making use of the triangle inequality, $|a + b| \leq |a| + |b|$, we obtain, from Eq. (15),

$$|\phi(t_n) - Y_n| \leq |\phi(t_n) - y_n| + |y_n - Y_n|$$
$$\leq |E_n| + |R_n|. \tag{16}$$

Thus the total error is bounded by the sum of the absolute values of the global truncation and round-off errors. For the numerical procedures discussed in this book, it is possible to obtain useful estimates of the global truncation error. However we limit our discussion primarily to the local truncation error, which is somewhat simpler. The round-off error is clearly more random in nature. It depends on the type of computer used, the sequence in which the computations are carried out, the method of rounding off, and so forth. An analysis of round-off error is beyond the scope of this book, but it is possible to say more about it than one might at first expect.[1] Some of the dangers from round-off error are discussed in Problems 23 through 25.

▶ Local Truncation Error for the Euler Method. Let us assume that the solution $y = \phi(t)$ of the initial value problem (1), (2) has a continuous second derivative in the interval of interest. To ensure this, we can assume that f, f_t, and f_y are continuous. Observe that if f has these properties and if ϕ is a solution of the initial value problem (1), (2), then

$$\phi'(t) = f[t, \phi(t)],$$

and, by the chain rule,

$$\phi''(t) = f_t[t, \phi(t)] + f_y[t, \phi(t)]\phi'(t)$$

$$= f_t[t, \phi(t)] + f_y[t, \phi(t)]f[t, \phi(t)]. \tag{17}$$

Since the right side of this equation is continuous, ϕ'' is also continuous.

Then, making use of a Taylor polynomial with a remainder to expand ϕ about t_n, we obtain

$$\phi(t_n + h) = \phi(t_n) + \phi'(t_n)h + \tfrac{1}{2}\phi''(\bar{t}_n)h^2, \tag{18}$$

where $\bar{t}_n$ is some point in the interval $t_n < \bar{t}_n < t_n + h$. Subtracting Eq. (4) from Eq. (18), and noting that $\phi(t_n + h) = \phi(t_{n+1})$ and $\phi'(t_n) = f[t_n, \phi(t_n)]$, we find that

$$\phi(t_{n+1}) - y_{n+1} = [\phi(t_n) - y_n] + h\{f[t_n, \phi(t_n)] - f(t_n, y_n)\} + \tfrac{1}{2}\phi''(\bar{t}_n)h^2. \tag{19}$$

To compute the local truncation error, we apply Eq. (19) to the true solution $\phi(t)$; that is, we take y_n to be $\phi(t_n)$. Then we immediately see from Eq. (19) that the local truncation error e_{n+1} is

$$e_{n+1} = \phi(t_{n+1}) - y_{n+1} = \tfrac{1}{2}\phi''(\bar{t}_n)h^2. \tag{20}$$

Thus the local truncation error for the Euler method is proportional to the square of the step size h, and the proportionality factor depends on the second derivative of the solution ϕ. The expression given by Eq. (20) depends on n and, in general, is different for each step. A uniform bound, valid on an interval $[a, b]$, is given by

$$|e_n| \le Mh^2/2, \tag{21}$$

where M is the maximum of $|\phi''(t)|$ on the interval $[a, b]$. Since Eq. (21) is based on a consideration of the worst possible case—that is, the largest possible value of $|\phi''(t)|$—it may well be a considerable overestimate of the actual local truncation error in some parts of the interval $[a, b]$. The primary difficulty in using Eq. (20) or (21) lies in estimating $|\phi''(t)|$ or M. However, the central fact expressed by these equations is that the local truncation error is proportional to h^2. Thus, if h is multiplied by $\frac{1}{2}$, then the error is multiplied by $\frac{1}{4}$, and so on. Further, the proportionality factor depends on the second derivative of the solution, so Euler's method works best on problems whose solutions have relatively small second derivatives.

One use of Eq. (21) is to choose a step size that will result in a local truncation error no greater than some given tolerance level. For example, if the local truncation error must be no greater than ϵ, then from Eq. (21) we have

$$h \le \sqrt{2\epsilon/M}. \tag{22}$$

---

[1]See, for example, the book by Henrici listed in the references.

More important than the local truncation error is the global truncation error E_n. The analysis for estimating E_n is more difficult than that for e_n. However, knowing the local truncation error, we can make an *intuitive* estimate of the global truncation error at a fixed $T > t_0$ as follows. Suppose that we take n steps in going from t_0 to $T = t_0 + nh$. In each step, the error is at most $Mh^2/2$; thus the error in n steps is at most $nMh^2/2$. Noting that $n = (T - t_0)/h$, we find that the global truncation error for the Euler method in going from t_0 to T is bounded by

$$n\frac{Mh^2}{2} = (T - t_0)\frac{Mh}{2}. \tag{23}$$

This argument is not complete since it does not take into account the effect that an error at one step will have in succeeding steps. Nevertheless it can be shown that the global truncation error in using the Euler method on a finite interval is no greater than a constant times h. The Euler method is called a first order method because its global truncation error is proportional to the first power of the step size.

Because it is more accessible, we will hereafter use the local truncation error as our principal measure of the accuracy of a numerical method and for comparing different methods. If we have a priori information about the solution of the given initial value problem, we can use the result (20) to obtain more precise information about how the local truncation error varies with t. As an example, consider the illustrative problem

$$y' = 1 - t + 4y, \qquad y(0) = 1 \tag{24}$$

on the interval $0 \le t \le 2$. Let $y = \phi(t)$ be the solution of the initial value problem (24). Then, as noted previously,

$$\phi(t) = (4t - 3 + 19e^{4t})/16$$

and therefore

$$\phi''(t) = 19e^{4t}.$$

Equation (20) then states that

$$e_{n+1} = \frac{19e^{4\bar{t}_n}h^2}{2}, \qquad t_n < \bar{t}_n < t_n + h. \tag{25}$$

The appearance of the factor 19 and the rapid growth of e^{4t} explain why the results in Table 8.2.1 are not very accurate.

For instance, using $h = 0.05$, the error in the first step is

$$e_1 = \phi(t_1) - y_1 = \frac{19e^{4\bar{t}_0}(0.0025)}{2}, \qquad 0 < \bar{t}_0 < 0.05.$$

It is clear that e_1 is positive, and since $e^{4\bar{t}_0} < e^{0.2}$, we have

$$e_1 \le \frac{19e^{0.2}(0.0025)}{2} \cong 0.02901. \tag{26}$$

Note also that $e^{4\bar{t}_0} > 1$; hence $e_1 > 19(0.0025)/2 = 0.02375$. The actual error is 0.02542. It follows from Eq. (25) that the error becomes progressively worse with increasing t. This is also clearly shown by the results in Table 8.2.1. Similar computations for bounds for the local truncation error give

$$1.0617 \cong \frac{19e^{3.8}(0.0025)}{2} \le e_{20} \le \frac{19e^4(0.0025)}{2} \cong 1.2967 \tag{27}$$

in going from 0.95 to 1.0 and

$$57.96 \cong \frac{19e^{7.8}(0.0025)}{2} \leq e_{40} \leq \frac{19e^{8}(0.0025)}{2} \cong 70.80 \tag{28}$$

in going from 1.95 to 2.0.

These results indicate that, for this problem, the local truncation error is about 2500 times larger near $t = 2$ than near $t = 0$. Thus, to reduce the local truncation error to an acceptable level throughout $0 \leq t \leq 2$, one must choose a step size h based on an analysis near $t = 2$. Of course, this step size will be much smaller than necessary near $t = 0$. For example, to achieve a local truncation error of 0.01 for this problem, we need a step size of about 0.00059 near $t = 2$ and a step size of about 0.032 near $t = 0$. The use of a uniform step size that is smaller than necessary over much of the interval results in more calculations than necessary, more time consumed, and possibly more danger of unacceptable round-off errors.

Another approach is to keep the local truncation error approximately constant throughout the interval by gradually reducing the step size as t increases. In the example problem, we would need to reduce h by a factor of about 50 in going from $t = 0$ to $t = 2$. A method that provides for variations in the step size is called **adaptive**. All modern computer codes for solving differential equations have the capability of adjusting the step size as needed. We will return to this idea in the next section.

PROBLEMS

In each of Problems 1 through 6, find approximate values of the solution of the given initial value problem at $t = 0.1, 0.2, 0.3,$ and 0.4.
(a) Use the Euler method with $h = 0.05$.
(b) Use the Euler method with $h = 0.025$.

1. $y' = 3 + t - y, \qquad y(0) = 1$
2. $y' = 5t - 3\sqrt{y}, \qquad y(0) = 2$
3. $y' = 2y - 3t, \qquad y(0) = 1$
4. $y' = 2t + e^{-ty}, \qquad y(0) = 1$
5. $y' = (y^2 + 2ty)/(3 + t^2), \qquad y(0) = 0.5$
6. $y' = (t^2 - y^2)\sin y, \qquad y(0) = -1$

In each of Problems 7 through 12, find approximate values of the solution of the given initial value problem at $t = 0.5,$ 1.0, 1.5, and 2.0.
(a) Use the Euler method with $h = 0.025$.
(b) Use the Euler method with $h = 0.0125$.

7. $y' = 0.5 - t + 2y, \qquad y(0) = 1$
8. $y' = 5t - 3\sqrt{y}, \qquad y(0) = 2$
9. $y' = \sqrt{t + y}, \qquad y(0) = 3$
10. $y' = 2t + e^{-ty}, \qquad y(0) = 1$
11. $y' = (4 - ty)/(1 + y^2), \qquad y(0) = -2$
12. $y' = (y^2 + 2ty)/(3 + t^2), \qquad y(0) = 0.5$

13. Complete the calculations leading to the entries in columns three and four of Table 8.2.1.

14. Using three terms in the Taylor series given in Eq. (12) and taking $h = 0.1$, determine approximate values of the solution of the illustrative example $y' = 1 - t + 4y, y(0) = 1$ at $t = 0.1$ and 0.2. Compare the results with those using the Euler method and with the exact values.
Hint: If $y' = f(t, y)$, what is y''?

In each of Problems 15 and 16, estimate the local truncation error for the Euler method in terms of the solution $y = \phi(t)$. Obtain a bound for e_{n+1} in terms of t and $\phi(t)$ that is valid on the interval $0 \leq t \leq 1$. By using a formula for the solution, obtain a more accurate error bound for e_{n+1}. For $h = 0.1$, compute a bound for e_1 and compare it with the actual error at $t = 0.1$. Also compute a bound for the error e_4 in the fourth step.

15. $y' = 2y - 1, \qquad y(0) = 1$
16. $y' = 0.5 - t + 2y, \qquad y(0) = 1$

In each of Problems 17 through 20, obtain a formula for the local truncation error for the Euler method in terms of t and the solution ϕ.

17. $y' = t^2 + y^2, \qquad y(0) = 1$
18. $y' = 5t - 3\sqrt{y}, \qquad y(0) = 2$
19. $y' = \sqrt{t + y}, \qquad y(1) = 3$
20. $y' = 2t + e^{-ty}, \qquad y(0) = 1$

21. Consider the initial value problem

$$y' = \cos 5\pi t, \qquad y(0) = 1.$$

(a) Determine the solution $y = \phi(t)$ and draw a graph of $y = \phi(t)$ for $0 \le t \le 1$.

(b) Determine approximate values of $\phi(t)$ at $t = 0.2, 0.4$, and 0.6 using the Euler method with $h = 0.2$. Draw a broken-line graph for the approximate solution and compare it with the graph of the exact solution.

(c) Repeat the computation of part (b) for $0 \le t \le 0.4$, but take $h = 0.1$.

(d) Show by computing the local truncation error that neither of these step sizes is sufficiently small. Determine a value of h to ensure that the local truncation error is less than 0.05 throughout the interval $0 \le t \le 1$. That such a small value of h is required results from the fact that $\max |\phi''(t)|$ is large.

 22. Using a step size $h = 0.05$ and the Euler method, but retaining only three digits throughout the computations, determine approximate values of the solution at $t = 0.1, 0.2, 0.3$, and 0.4 for each of the following initial value problems.

(a) $y' = 1 - t + 4y$, $\quad y(0) = 1$
(b) $y' = 3 + t - y$, $\quad y(0) = 1$
(c) $y' = 2y - 3t$, $\quad y(0) = 1$

Compare the results with those obtained in Example 1 and in Problems 1 and 3. The small differences between some of those results rounded to three digits and the present results are due to round-off error. The round-off error would become important if the computation required many steps.

23. The following problem illustrates a danger that occurs because of round-off error when nearly equal numbers are subtracted and the difference is then multiplied by a large number. Evaluate the quantity

$$1000 \cdot \begin{vmatrix} 6.010 & 18.04 \\ 2.004 & 6.000 \end{vmatrix}$$

in the following ways:

(a) First round each entry in the determinant to two digits.
(b) First round each entry in the determinant to three digits.
(c) Retain all four digits. Compare this value with the results in parts (a) and (b).

24. The distributive law $a(b - c) = ab - ac$ does not hold, in general, if the products are rounded off to a smaller number of digits. To show this in a specific case, take $a = 0.22$, $b = 3.19$, and $c = 2.17$. After each multiplication, round off the last digit.

 25. In this section we stated that the global truncation error for the Euler method applied to an initial value problem over a fixed interval is no more than a constant times the step size h. In this problem, we show you how to obtain some experimental evidence in support of this statement. Consider the initial value problem in Example 1 for which some numerical approximations are given in Table 8.2.1. Observe that, for each step size, the maximum error E occurs at the endpoint $t = 2$. Now let us assume that $E = Ch^p$, where the constants C and p are to be determined. By taking the logarithm of each side of this equation, we obtain

$$\ln E = \ln C + p \ln h,$$

which is the equation of a straight line in the $(\ln h)$ $(\ln E)$-plane. The slope of this line is the value of the exponent p and the intercept on the $(\ln E)$-axis determines the value of C.

(a) Using the data in Table 8.2.1, calculate the maximum error E for each of the given values of h.

(b) Plot $\ln E$ versus $\ln h$ for the four data points that you obtained in part (a).

(c) Do the points in part (b) lie approximately on a single straight line? If so, then this is evidence that the assumed expression for E is correct.

(d) Estimate the slope of the line in part (c). If the statement in the text about the magnitude of the global truncation error is correct, then the slope should be no greater than 1.

Note: Your estimate of the slope p depends on how you choose the straight line. If you have a curve-fitting routine in your software, you can use it to determine the straight line that best fits the data. Otherwise, you may wish to resort to less precise methods. For example, you could calculate the slopes of the line segments joining (one or more) pairs of data points, and then average your results.

8.3 Improved Euler and Runge–Kutta Methods

Since for many problems the Euler method requires a very small step size to produce sufficiently accurate results, much effort has been devoted to the development of more efficient methods. In this section we will discuss two of these methods. Consider the initial value problem

$$\frac{dy}{dt} = f(t, y), \qquad y(t_0) = y_0 \tag{1}$$

and let $y = \phi(t)$ denote its solution. Recall from Eq. (10) of Section 8.2 that, by integrating the given differential equation from t_n to t_{n+1}, we obtain

$$\phi(t_{n+1}) = \phi(t_n) + \int_{t_n}^{t_{n+1}} f[t, \phi(t)]\, dt. \tag{2}$$

The Euler formula

$$y_{n+1} = y_n + hf(t_n, y_n) \tag{3}$$

is obtained by replacing $f[t, \phi(t)]$ in Eq. (2) by its approximate value $f(t_n, y_n)$ at the left endpoint of the interval of integration.

▶ Improved Euler Formula. A better approximate formula can be obtained if the integrand in Eq. (2) is approximated more accurately. One way to do this is to replace the integrand by the average of its values at the two endpoints, namely, $\{f[t_n, \phi(t_n)] + f[t_{n+1}, \phi(t_{n+1})]\}/2$. This is equivalent to approximating the area under the curve in Figure 8.3.1 between $t = t_n$ and $t = t_{n+1}$ by the area of the shaded trapezoid. Further we replace $\phi(t_n)$ and $\phi(t_{n+1})$ by their respective approximate values y_n and y_{n+1}. In this way we obtain, from Eq. (2),

$$y_{n+1} = y_n + \frac{f(t_n, y_n) + f(t_{n+1}, y_{n+1})}{2} h. \tag{4}$$

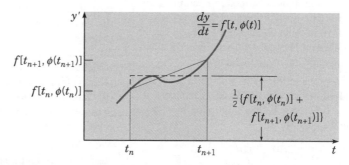

FIGURE 8.3.1   Derivation of the improved Euler method.

Since the unknown y_{n+1} appears as one of the arguments of f on the right side of Eq. (4), this equation defines y_{n+1} implicitly rather than explicitly. Depending on the nature of the function f, it may be fairly difficult to solve Eq. (4) for y_{n+1}. This difficulty can be overcome by replacing y_{n+1} on the right side of Eq. (4) by the value obtained using the Euler formula (3). Thus

$$y_{n+1} = y_n + \frac{f(t_n, y_n) + f[t_n + h, y_n + hf(t_n, y_n)]}{2} h$$

$$= y_n + \frac{f_n + f(t_n + h, y_n + hf_n)}{2} h, \tag{5}$$

where t_{n+1} has been replaced by $t_n + h$.

Equation (5) gives an explicit formula for computing y_{n+1}, the approximate value of $\phi(t_{n+1})$, in terms of the data at t_n. This formula is known as the **improved Euler formula** or the **Heun formula**. The improved Euler formula is an example of a two-stage method; that is, we first calculate $y_n + hf_n$ from the Euler formula and then use this result to calculate y_{n+1} from Eq. (5). The improved Euler formula (5) does represent an improvement over the Euler formula (3) because the local truncation error in using Eq. (5) is proportional to h^3, whereas for the Euler method it is proportional to h^2. This error estimate for the improved

Euler formula is established in Problem 17. It can also be shown that for a finite interval the global truncation error for the improved Euler formula is bounded by a constant times h^2, so this method is a second order method. Note that this greater accuracy is achieved at the expense of more computational work, since it is now necessary to evaluate $f(t, y)$ twice in order to go from t_n to t_{n+1}.

If $f(t, y)$ depends only on t and not on y, then solving the differential equation $y' = f(t, y)$ reduces to integrating $f(t)$. In this case, the improved Euler formula (5) becomes

$$y_{n+1} - y_n = (h/2)[f(t_n) + f(t_n + h)], \tag{6}$$

which is just the trapezoid rule for numerical integration.

This is summarized as follows:

Improved Euler's Formula

Suppose the solution of the initial value problem

$$\begin{cases} \dfrac{dy}{dt} = f(t, y) \\ y(t_0) = y_0 \end{cases} \tag{7}$$

is denoted $y = \phi(t)$ and you have a sequence of points $t_0 < t_1 < t_2 < \cdots < t_n < \cdots$. For $n = 0, 1, 2, \ldots$, we have the following:

Approximation of $y = \phi(t)$ at $t = t_{n+1}$:

$$y_{n+1} = y_n + \frac{f(t_n, y_n) + f[t_n + h, \ y_n + hf(t_n, y_n)]}{2} h \tag{8}$$

where $t_{n+1} - t_n = h$.

Linear approximation of $\phi(t)$ on the interval $[t_n, t_{n+1}]$:

$$y(t) = y_n + f(t_n, y_n)(t - t_n). \tag{9}$$

Special case: If $f(t, y) = f(t)$, then Eq. (8) simplifies to

$$y_{n+1} = y_n + \frac{h}{2}[f(t_n) + f(t_n + h)]. \tag{10}$$

EXAMPLE 1

Use the improved Euler formula (5) to calculate approximate values of the solution of the initial value problem

$$y' = 1 - t + 4y, \qquad y(0) = 1. \tag{11}$$

To make clear exactly what computations are required, we show a couple of steps in detail. For this problem, $f(t, y) = 1 - t + 4y$; hence

$$f_n = 1 - t_n + 4y_n$$

and

$$f(t_n + h, y_n + hf_n) = 1 - (t_n + h) + 4(y_n + hf_n).$$

Further $t_0 = 0$, $y_0 = 1$, and $f_0 = 1 - t_0 + 4y_0 = 5$. If $h = 0.025$, then

$$f(t_0 + h, y_0 + hf_0) = 1 - 0.025 + 4[1 + (0.025)(5)] = 5.475.$$

Then, from Eq. (5),

$$y_1 = 1 + (0.5)(5 + 5.475)(0.025) = 1.1309375. \tag{12}$$

At the second step we must calculate

$$f_1 = 1 - 0.025 + 4(1.1309375) = 5.49875,$$

$$y_1 + hf_1 = 1.1309375 + (0.025)(5.49875) = 1.26840625,$$

and

$$f(t_2, y_1 + hf_1) = 1 - 0.05 + 4(1.26840625) = 6.023625.$$

Then, from Eq. (5),

$$y_2 = 1.1309375 + (0.5)(5.49875 + 6.023625)(0.025) = 1.2749671875. \tag{13}$$

Further results for $0 \leq t \leq 2$ obtained by using the improved Euler method with $h = 0.025$ and $h = 0.01$ are given in Table 8.3.1. To compare the results of the improved Euler method with those of the Euler method, note that the improved Euler method requires two evaluations of f at each step, whereas the Euler method requires only one. This is significant because typically most of the computing time in each step is spent in evaluating f, so counting these evaluations is a reasonable way to estimate the total computing effort. Thus, for a given step size h, the improved Euler method requires twice as many evaluations of f as the Euler method. Alternatively, the improved Euler method for step size h requires the same number of evaluations of f as the Euler method with step size $h/2$.

TABLE 8.3.1 A comparison of results using the Euler and improved Euler methods for the initial value problem $y' = 1 - t + 4y$, $y(0) = 1$.

	Euler		Improved Euler		
t	$h = 0.01$	$h = 0.001$	$h = 0.025$	$h = 0.01$	Exact
0	1.0000000	1.0000000	1.0000000	1.0000000	1.0000000
0.1	1.5952901	1.6076289	1.6079462	1.6088585	1.6090418
0.2	2.4644587	2.5011159	2.5020618	2.5047827	2.5053299
0.3	3.7390345	3.8207130	3.8228282	3.8289146	3.8301388
0.4	5.6137120	5.7754844	5.7796888	5.7917911	5.7942260
0.5	8.3766865	8.6770692	8.6849039	8.7074637	8.7120041
1.0	60.037126	64.382558	64.497931	64.830722	64.897803
1.5	426.40818	473.55979	474.83402	478.51588	479.25919
2.0	3029.3279	3484.1608	3496.6702	3532.8789	3540.2001

By referring to Table 8.3.1, you can see that the improved Euler method with $h = 0.025$ gives much better results than the Euler method with $h = 0.01$. Note that to reach $t = 2$ with these step sizes, the improved Euler method requires 160 evaluations of f, while the Euler method requires 200. More noteworthy is that the improved Euler method with $h = 0.025$ is also slightly more accurate than the Euler method with $h = 0.001$ (2000 evaluations of f). In other words, with something like one-twelfth of the computing effort, the improved Euler method yields results for this problem that are comparable to, or a bit better than, those generated by the Euler method. This illustrates that, compared to the Euler method, the improved Euler method is clearly more efficient, yielding substantially better results or requiring much less total computing effort, or both.

The percentage errors at $t = 2$ for the improved Euler method are 1.23% for $h = 0.025$ and 0.21% for $h = 0.01$.

▶ Variation of Step Size. In Section 8.2, we mentioned the possibility of adjusting the step size as a calculation proceeds so as to maintain the local truncation error at a more or less constant level. The goal is to use no more steps than necessary and, at the same time, to keep some control over the accuracy of the approximation. Here, we will describe how this can be done. Suppose that after n steps we have reached the point (t_n, y_n). We choose a step size h and calculate y_{n+1}. Next we need to estimate the error we have made in calculating y_{n+1}. Not knowing the actual solution, the best that we can do is to use a more accurate method and repeat the calculation starting from (t_n, y_n). For example, if we used the Euler method for the original calculation, we might repeat it with the improved Euler method. Then the difference between the two calculated values is an estimate e_{n+1}^{est} of the error in using the original method. If the estimated error is different from the error tolerance ϵ, then we adjust the step size and repeat the calculation. To make this adjustment efficiently, it is crucial to know how the local truncation error e_{n+1} depends on the step size h. For the Euler method, the local truncation error is proportional to h^2, so to bring the estimated error down (or up) to the tolerance level ϵ, we must multiply the original step size by the factor $\sqrt{\epsilon / e_{n+1}^{est}}$.

To illustrate this procedure, consider the example problem (11):

$$y' = 1 - t + 4y, \qquad y(0) = 1.$$

You can verify that after one step with $h = 0.1$ we obtain the values 1.5 and 1.595 from the Euler method and the improved Euler method, respectively. Thus the estimated error in using the Euler method is 0.095. If we have chosen an error tolerance of 0.05, for instance, then we need to adjust the step size downward by the factor $\sqrt{0.05/0.095} \cong 0.73$. Rounding downward to be conservative, let us choose the adjusted step size $h = 0.07$. Then, from the Euler formula, we obtain

$$y_1 = 1 + (0.07)f(0, 1) = 1.35 \cong \phi(0.07).$$

Using the improved Euler method, we obtain $y_1 = 1.39655$, so the estimated error in using the Euler formula is 0.04655, which is slightly less than the specified tolerance. The actual error, based on a comparison with the solution itself, is somewhat greater, namely, 0.05122.

We can follow the same procedure at each step of the calculation, thereby keeping the local truncation error approximately constant throughout the entire numerical process. Modern adaptive codes for solving differential equations adjust the step size as they proceed in very much this way, although they use more accurate formulas than the Euler and improved Euler formulas. Consequently, they are able to achieve both efficiency and accuracy by using very small steps only where they are really needed.

▶ Runge–Kutta Method. The Euler and improved Euler methods belong to what is now called the Runge–Kutta class of numerical approximation methods. Here, we discuss the method originally developed by Runge and Kutta. This method is now called the classic fourth order, four-stage Runge–Kutta method, but it is often referred to simply as *the* Runge–Kutta method, and we will follow this practice for brevity. This method has a local truncation error that is proportional to h^5. Thus it is two orders of magnitude more accurate than the improved Euler method and three orders of magnitude better than the Euler method. It is relatively simple to use and is sufficiently accurate to handle many problems efficiently. This is especially true of adaptive Runge–Kutta methods, in which provision is made to vary the step size as needed. We return to this issue at the end of the section.

The Runge–Kutta method involves a weighted average of values of $f(t, y)$ at different points in the interval $t_n \leq t \leq t_{n+1}$. It is given by

$$t_{n+1} = t_n + h, \qquad y_{n+1} = y_n + h\left(\frac{k_{n1} + 2k_{n2} + 2k_{n3} + k_{n4}}{6}\right), \tag{14}$$

where

$$\begin{aligned}
k_{n1} &= f(t_n, y_n), \\
k_{n2} &= f\left(t_n + \tfrac{1}{2}h, y_n + \tfrac{1}{2}hk_{n1}\right), \\
k_{n3} &= f\left(t_n + \tfrac{1}{2}h, y_n + \tfrac{1}{2}hk_{n2}\right), \\
k_{n4} &= f\left(t_n + h, y_n + hk_{n3}\right).
\end{aligned} \tag{15}$$

The sum $(k_{n1} + 2k_{n2} + 2k_{n3} + k_{n4})/6$ can be interpreted as an average slope. Note that k_{n1} is the slope at the left end of the interval, k_{n2} is the slope at the midpoint using the Euler formula to go from t_n to $t_n + h/2$, k_{n3} is a second approximation to the slope at the midpoint, and k_{n4} is the slope at $t_n + h$ using the Euler formula and the slope k_{n3} to go from t_n to $t_n + h$.

Although in principle it is not difficult to show that Eq. (14) differs from the Taylor expansion of the solution ϕ by terms that are proportional to h^5, the algebra is rather lengthy.[2] Thus we will simply accept the fact that the local truncation error in using Eq. (14) is proportional to h^5 and that for a finite interval the global truncation error is at most a constant times h^4. The earlier description of this method as a fourth order, four-stage method reflects the facts that the global truncation error is of fourth order in the step size h and that there are four intermediate stages in the calculation (the calculation of $k_{n1}, \ldots, k_{n4}$).

Clearly, the Runge–Kutta formula, Eqs. (14) and (15), is more complicated than the formulas discussed previously. This is of relatively little significance, however, since it is not hard to write a computer program to implement this method. Such a program has the same structure as the algorithm for the Euler method outlined in Section 8.1.

Note that if f does not depend on y, then

$$k_{n1} = f(t_n), \qquad k_{n2} = k_{n3} = f(t_n + h/2), \qquad k_{n4} = f(t_n + h), \tag{16}$$

and Eq. (14) reduces to

$$y_{n+1} - y_n = (h/6)\left[f(t_n) + 4f(t_n + h/2) + f(t_n + h)\right]. \tag{17}$$

Equation (17) can be identified as Simpson's rule for the approximate evaluation of the integral of $y' = f(t)$. The fact that Simpson's rule has an error proportional to h^5 is consistent with the local truncation error in the Runge–Kutta formula.

---

[2]See, for example, Chapter 3 of the book by Henrici listed in the references.

The Runge-Kutta method is summarized as follows:

Runge–Kutta Method

Suppose the solution of the initial value problem

$$\begin{cases} \dfrac{dy}{dt} = f(t, y) \\ y(t_0) = y_0 \end{cases} \tag{18}$$

is denoted $y = \phi(t)$ and you have a sequence of points $t_0 < t_1 < t_2 < \cdots < t_n < \cdots$. For $n = 0, 1, 2, \ldots$, we have the following:

Approximation of $y = \phi(t)$ at $t = t_{n+1}$:

$$y_{n+1} = y_n + h\left(\frac{k_{n1} + 2k_{n2} + 2k_{n3} + k_{n4}}{6} \right), \tag{19}$$

where

$$\begin{cases} k_{n1} = f(t_n, y_n) \\ k_{n2} = f\left(t_n + \tfrac{1}{2}h, y_n + \tfrac{1}{2}hk_{n1} \right) \\ k_{n3} = f\left(t_n + \tfrac{1}{2}h, y_n + \tfrac{1}{2}hk_{n2} \right) \\ k_{n4} = f(t_n + h, y_n + hk_{n3}). \end{cases} \tag{20}$$

Linear approximation of $\phi(t)$ on the interval $[t_n, t_{n+1}]$:

$$y(t) = y_n + f(t_n, y_n)(t - t_n). \tag{21}$$

Special case: If $f(t, y) = f(t)$, then Eq. (19) simplifies to

$$y_{n+1} = y_n + \frac{h}{6}\left[f(t_n) + 4f\left(t_n + \frac{h}{2} \right) + f(t_n + h) \right]. \tag{22}$$

EXAMPLE 2

Use the Runge–Kutta method to calculate approximate values of the solution $y = \phi(t)$ of the initial value problem

$$y' = 1 - t + 4y, \qquad y(0) = 1. \tag{23}$$

Taking $h = 0.2$, we have

$$\begin{array}{ll} k_{01} = f(0, 1) = 5; & hk_{01} = 1.0, \\ k_{02} = f(0 + 0.1, 1 + 0.5) = 6.9; & hk_{02} = 1.38, \\ k_{03} = f(0 + 0.1, 1 + 0.69) = 7.66; & hk_{03} = 1.532, \\ k_{04} = f(0 + 0.2, 1 + 1.532) = 10.928. & \end{array}$$

Thus

$$\begin{aligned} y_1 &= 1 + (0.2/6)[5 + 2(6.9) + 2(7.66) + 10.928] \\ &= 1 + 1.5016 = 2.5016. \end{aligned}$$

Further results using the Runge–Kutta method with $h = 0.2$, $h = 0.1$, and $h = 0.05$ are given in Table 8.3.2. Note that the Runge–Kutta method yields a value at $t = 2$ that differs from the exact solution by only 0.122% if the step size is $h = 0.1$, and by only 0.00903% if $h = 0.05$. In the latter case, the error is less than 1 part in 10,000, and the calculated value at $t = 2$ is correct to four digits.

For comparison, note that both the Runge–Kutta method with $h = 0.05$ and the improved Euler method with $h = 0.025$ require 160 evaluations of f to reach $t = 2$. The improved Euler method yields a result at $t = 2$ that is in error by 1.23%. Although this error may be acceptable for some purposes, it is more than 135 times the error yielded by the Runge–Kutta method with comparable computing effort. Note also that the Runge–Kutta method with $h = 0.2$, or 40 evaluations of f, produces a value at $t = 2$ with an error of 1.40%, which is only slightly greater than the error in the improved Euler method with $h = 0.025$, or 160 evaluations of f. Thus we see again that a more accurate algorithm is more efficient; it produces better results with similar effort, or similar results with less effort.

TABLE 8.3.2

A comparison of results using the improved Euler and Runge–Kutta methods for the initial value problem $\frac{dy}{dt} = 1 - t + 4y$, $y(0) = 1$.

	Improved Euler	Runge–Kutta			Exact
t	$h = 0.025$	$h = 0.2$	$h = 0.1$	$h = 0.05$	
0	1.0000000	1.0000000	1.0000000	1.0000000	1.0000000
0.1	1.6079462		1.6089333	1.6090338	1.6090418
0.2	2.5020618	2.5016000	2.5050062	2.5053060	2.5053299
0.3	3.8228282		3.8294145	3.8300854	3.8301388
0.4	5.7796888	5.7776358	5.7927853	5.7941197	5.7942260
0.5	8.6849039		8.7093175	8.7118060	8.7120041
1.0	64.497931	64.441579	64.858107	64.894875	64.897803
1.5	474.83402		478.81928	479.22674	479.25919
2.0	3496.6702	3490.5574	3535.8667	3539.8804	3540.2001

The classic Runge–Kutta method suffers from the same shortcoming as other methods with a fixed step size for problems in which the local truncation error varies widely over the interval of interest. That is, a step size that is small enough to achieve satisfactory accuracy in some parts of the interval may be much smaller than necessary in other parts of the interval. This has stimulated the development of adaptive Runge–Kutta methods that provide for modifying the step size automatically as the computation proceeds, so as to maintain the local truncation error near or below a specified tolerance level. As explained earlier in this section, this requires the estimation of the local truncation error at each step. One way to do this is to repeat the computation with a fifth order method—which has a local truncation error proportional to h^6—and then to use the difference between the two results as an estimate of the error. If this is done in a straightforward (unsophisticated) manner, then the use of the fifth order method requires at least five more evaluations of f at each step, in addition to those required originally by the fourth order method. However, if we make an appropriate choice of the intermediate points and the weighting coefficients in the expressions for $k_{n1}, \ldots$ in a certain fourth order Runge–Kutta method, then these expressions can be used again, together with one additional stage, in a corresponding fifth order method. This results in a substantial gain in efficiency. It turns out that this can be done in more than one way.[3] The resulting adaptive Runge–Kutta methods are very powerful and efficient means of numerically approximating the solutions of an enormous class of initial value problems. Specific implementations of one or more of them are widely available in commercial software packages.

PROBLEMS

In each of Problems 1 through 6, find approximate values of the solution of the given initial value problem at $t = 0.1, 0.2, 0.3$, and 0.4. Compare the results with those obtained by the Euler method in Section 8.2 and with the exact solution (if available).

(a) Use the improved Euler method with $h = 0.05$.
(b) Use the improved Euler method with $h = 0.025$.
(c) Use the improved Euler method with $h = 0.0125$.
(d) Use the Runge–Kutta method with $h = 0.1$.
(e) Use the Runge–Kutta method with $h = 0.05$.

1. $y' = 3 + t - y$, $\quad y(0) = 1$
2. $y' = 5t - 3\sqrt{y}$, $\quad y(0) = 2$
3. $y' = 2y - 3t$, $\quad y(0) = 1$
4. $y' = 2t + e^{-ty}$, $\quad y(0) = 1$
5. $y' = (y^2 + 2ty)/(3 + t^2)$, $\quad y(0) = 0.5$
6. $y' = (t^2 - y^2)\sin y$, $\quad y(0) = -1$

In each of Problems 7 through 12, find approximate values of the solution of the given initial value problem at $t = 0.5$, 1.0, 1.5, and 2.0.

(a) Use the improved Euler method with $h = 0.025$.
(b) Use the improved Euler method with $h = 0.0125$.
(c) Use the Runge–Kutta method with $h = 0.1$.
(d) Use the Runge–Kutta method with $h = 0.05$.

7. $y' = 0.5 - t + 2y$, $\quad y(0) = 1$
8. $y' = 5t - 3\sqrt{y}$, $\quad y(0) = 2$
9. $y' = \sqrt{t + y}$, $\quad y(0) = 3$
10. $y' = 2t + e^{-ty}$, $\quad y(0) = 1$
11. $y' = (4 - ty)/(1 + y^2)$, $\quad y(0) = -2$
12. $y' = (y^2 + 2ty)/(3 + t^2)$, $\quad y(0) = 0.5$

13. Complete the calculations leading to the entries in columns four and five of Table 8.3.1.

14. Confirm the results in Table 8.3.2 by executing the indicated computations.

15. Consider the initial value problem
$$y' = t^2 + y^2, \qquad y(0) = 1.$$
(a) Draw a direction field for this equation.

(b) Use the Runge–Kutta method to find approximate values of the solution at $t = 0.8, 0.9$, and 0.95. Choose a small enough step size so that you believe your results are accurate to at least four digits.

(c) Try to extend the calculations in part (b) to obtain an accurate approximation to the solution at $t = 1$. If you encounter difficulties in doing this, explain why you think this happens. The direction field in part (a) may be helpful.

16. Consider the initial value problem
$$y' = 3t^2/(3y^2 - 4), \qquad y(0) = 0.$$
(a) Draw a direction field for this equation.

(b) Estimate how far the solution can be extended to the right. Let t_M be the right endpoint of the interval of existence of this solution. What happens at t_M to prevent the solution from continuing farther?

(c) Use the Runge–Kutta method with various step sizes to determine an approximate value of t_M.

(d) If you continue the computation beyond t_M, you can continue to generate values of y. What significance, if any, do these values have?

(e) Suppose that the initial condition is changed to $y(0) = 1$. Repeat parts (b) and (c) for this problem.

17. In this problem, we establish that the local truncation error for the improved Euler formula is proportional to h^3. If we assume that the solution ϕ of the initial value problem $y' = f(t, y)$, $y(t_0) = y_0$ has derivatives that are continuous through the third order (f has continuous second partial derivatives), then it follows that
$$\phi(t_n + h) = \phi(t_n) + \phi'(t_n)h + \frac{\phi''(t_n)}{2!}h^2 + \frac{\phi'''(\bar{t}_n)}{3!}h^3,$$
where $t_n < \bar{t}_n < t_n + h$. Assume that $y_n = \phi(t_n)$.

(a) Show that, for y_{n+1} as given by Eq. (5),
$$e_{n+1} = \phi(t_{n+1}) - y_{n+1}$$
$$= \frac{\phi''(t_n)h - \{f[t_n + h, y_n + hf(t_n, y_n)] - f(t_n, y_n)\}}{2!}h$$
$$+ \frac{\phi'''(\bar{t}_n)h^3}{3!}. \tag{i}$$

---

[3] The first widely used fourth and fifth order Runge–Kutta pair was developed by Erwin Fehlberg in the late 1960s. Its popularity was considerably enhanced by the appearance in 1977 of its Fortran implementation RKF45 by Lawrence F. Shampine and H. A. Watts.

(b) Making use of the facts that $\phi''(t) = f_t[t, \phi(t)] + f_y[t, \phi(t)]\phi'(t)$ and that the Taylor approximation with a remainder for a function $F(t, y)$ of two variables is

$$F(a + h, b + k) = F(a, b) + F_t(a, b)h + F_y(a, b)k$$
$$+ \frac{1}{2!}(h^2 F_{tt} + 2hk F_{ty} + k^2 F_{yy})\Big|_{x=\xi, y=\eta}$$

where ξ lies between a and $a + h$ and η lies between b and $b + k$, show that the first term on the right side of Eq. (i) is proportional to h^3 plus higher order terms. This is the desired result.

(c) Show that if $f(t, y)$ is linear in t and y, then $e_{n+1} = \phi'''(\bar{t}_n)h^3/6$, where $t_n < \bar{t}_n < t_{n+1}$.

Hint: What are f_{tt}, f_{ty}, and f_{yy}?

18. Consider the improved Euler method for solving the illustrative initial value problem $y' = 1 - t + 4y$, $y(0) = 1$. Using the result of Problem 17(c) and the exact solution of the initial value problem, determine e_{n+1} and a bound for the error at any step on $0 \le t \le 2$. Compare this error with the one obtained in Eq. (25) of Section 8.2 using the Euler method.

Also obtain a bound for e_1 for $h = 0.05$, and compare it with Eq. (26) of Section 8.2.

In each of Problems 19 and 20, use the actual solution $\phi(t)$ to determine e_{n+1} and a bound for e_{n+1} at any step on $0 \le t \le 1$ for the improved Euler method for the given initial value problem. Also obtain a bound for e_1 for $h = 0.1$, and compare it with the similar estimate for the Euler method and with the actual error using the improved Euler method.

19. $y' = 2y - 1$, $\qquad y(0) = 1$

20. $y' = 0.5 - t + 2y$, $\qquad y(0) = 1$

In each of Problems 21 through 24, carry out one step of the Euler method and of the improved Euler method, using the step size $h = 0.1$. Suppose that a local truncation error no greater than 0.0025 is required. Estimate the step size that is needed for the Euler method to satisfy this requirement at the first step.

21. $y' = 0.5 - t + 2y$, $\qquad y(0) = 1$

22. $y' = 5t - 3\sqrt{y}$, $\qquad y(0) = 2$

23. $y' = \sqrt{t + y}$, $\qquad y(0) = 3$

24. $y' = (y^2 + 2ty)/(3 + t^2)$, $\qquad y(0) = 0.5$

8.4 Numerical Methods for Systems of First Order Equations

In Sections 8.1–8.3 we have discussed numerical methods for approximating the solutions of initial value problems for a single first order differential equation. These methods can also be applied to a system of first order equations. The algorithms are the same for nonlinear and for linear equations, so we will not restrict ourselves to linear equations in this section. We consider a system of two first order equations

$$x' = f(t, x, y), \qquad y' = g(t, x, y), \tag{1}$$

with the initial conditions

$$x(t_0) = x_0, \qquad y(t_0) = y_0. \tag{2}$$

The functions f and g are assumed to satisfy the conditions of Theorem 3.6.1 so that the initial value problem (1), (2) has a unique solution in some interval of the t-axis containing the point t_0. We wish to determine approximate values $x_1, x_2, \ldots, x_n, \ldots$ and $y_1, y_2, \ldots, y_n, \ldots$ of the solution $x = \phi(t)$, $y = \psi(t)$ at the points $t_n = t_0 + nh$ with $n = 1, 2, \ldots$.

In vector notation, the initial value problem (1), (2) can be written as

$$\mathbf{x}' = \mathbf{f}(t, \mathbf{x}), \qquad \mathbf{x}(t_0) = \mathbf{x}_0, \tag{3}$$

where $\mathbf{x}$ is the vector with components x and y, $\mathbf{f}$ is the vector function with components f and g, and $\mathbf{x}_0$ is the vector with components x_0 and y_0. The Euler, improved Euler, and Runge–Kutta methods can be readily generalized to handle systems of two (or more) equations. All that is needed (formally) is to replace the scalar variable x by the vector $\mathbf{x}$ and the scalar function f by the vector function $\mathbf{f}$ in the appropriate equations. For example, the scalar Euler formula

$$t_{n+1} = t_n + h, \qquad x_{n+1} = x_n + hf_n \tag{4}$$

is replaced by

$$t_{n+1} = t_n + h, \qquad \mathbf{x}_{n+1} = \mathbf{x}_n + h\mathbf{f}_n, \tag{5}$$

where $\mathbf{f}_n = \mathbf{f}(t_n, \mathbf{x}_n)$. In component form we have

$$\begin{pmatrix} x_{n+1} \\ y_{n+1} \end{pmatrix} = \begin{pmatrix} x_n \\ y_n \end{pmatrix} + h \begin{pmatrix} f(t_n, x_n, y_n) \\ g(t_n, x_n, y_n) \end{pmatrix}. \tag{6}$$

The initial conditions are used to determine $\mathbf{f}_0$, which is the vector tangent to the trajectory of the solution $\mathbf{x} = \begin{bmatrix} \phi(t) \\ \psi(t) \end{bmatrix}$ at the initial point in the xy-plane. We move in the direction of this tangent vector for a time step h in order to find the next point $\mathbf{x}_1$. Then we calculate a new tangent vector $\mathbf{f}_1$, move along it for a time step h to find $\mathbf{x}_2$, and so forth.

In a similar way, the Runge–Kutta method (Section 8.3) can be extended to a system. For the step from t_n to t_{n+1} we have

$$t_{n+1} = t_n + h, \qquad \mathbf{x}_{n+1} = \mathbf{x}_n + (h/6)(\mathbf{k}_{n1} + 2\mathbf{k}_{n2} + 2\mathbf{k}_{n3} + \mathbf{k}_{n4}), \tag{7}$$

where

$$\begin{aligned} \mathbf{k}_{n1} &= \mathbf{f}(t_n, \mathbf{x}_n), \\ \mathbf{k}_{n2} &= \mathbf{f}[t_n + (h/2), \mathbf{x}_n + (h/2)\mathbf{k}_{n1}], \\ \mathbf{k}_{n3} &= \mathbf{f}[t_n + (h/2), \mathbf{x}_n + (h/2)\mathbf{k}_{n2}], \\ \mathbf{k}_{n4} &= \mathbf{f}(t_n + h, \mathbf{x}_n + h\mathbf{k}_{n3}). \end{aligned} \tag{8}$$

The vector equations (3), (5), (7), and (8) are, in fact, valid in any number of dimensions. All that is needed is to interpret the vectors as having n components rather than two.

EXAMPLE 1

Determine approximate values of the solution $x = \phi(t)$, $y = \psi(t)$ of the initial value problem

$$x' = -x + 4y, \qquad y' = x - y, \tag{9}$$

$$x(0) = 2, \qquad y(0) = -0.5, \tag{10}$$

at the point $t = 0.2$. Use the Euler method with $h = 0.1$ and the Runge–Kutta method with $h = 0.2$. Compare the results with the values of the exact solution:

$$\phi(t) = \frac{e^t + 3e^{-3t}}{2}, \qquad \psi(t) = \frac{e^t - 3e^{-3t}}{4}. \tag{11}$$

Note that the differential equations (9) form a linear homogeneous system with constant coefficients. Consequently, the solution is easily found by using the methods described in Chapter 3. In particular, the eigenvalues of the coefficient matrix are 1 and -3, so the origin is a saddle point for this system.

To approximate the solution numerically, let us first use the Euler method. For this problem, $f_n = -x_n + 4y_n$ and $g_n = x_n - y_n$; hence

$$f_0 = -2 + (4)(-0.5) = -4, \qquad g_0 = 2 - (-0.5) = 2.5.$$

Then, from the Euler formulas (5) and (6), we obtain

$$x_1 = 2 + (0.1)(-4) = 1.6, \qquad y_1 = -0.5 + (0.1)(2.5) = -0.25.$$

At the next step

$$f_1 = -1.6 + (4)(-0.25) = -2.6, \qquad g_1 = 1.6 - (-0.25) = 1.85.$$

Consequently,

$$x_2 = 1.6 + (0.1)(-2.6) = 1.34, \qquad y_2 = -0.25 + (0.1)(1.85) = -0.065.$$

The values of the exact solution, correct to six decimal places, are $\phi(0.2) = 1.433919$ and $\psi(0.2) = -0.106258$. Thus the values calculated from the Euler method are in error by about 0.0939 and 0.0413, respectively, corresponding to percentage errors of about 6.5% and 38.8%.

Now let us use the Runge–Kutta method to approximate $\phi(0.2)$ and $\psi(0.2)$. With $h = 0.2$, we obtain the following values from Eqs. (8):

$$\mathbf{k}_{01} = \begin{pmatrix} f(2, -0.5) \\ g(2, -0.5) \end{pmatrix} = \begin{pmatrix} -4 \\ 2.5 \end{pmatrix};$$

$$\mathbf{k}_{02} = \begin{pmatrix} f(1.6, -0.25) \\ g(1.6, -0.25) \end{pmatrix} = \begin{pmatrix} -2.6 \\ 1.85 \end{pmatrix};$$

$$\mathbf{k}_{03} = \begin{pmatrix} f(1.74, -0.315) \\ g(1.74, -0.315) \end{pmatrix} = \begin{pmatrix} -3.00 \\ 2.055 \end{pmatrix};$$

$$\mathbf{k}_{04} = \begin{pmatrix} f(1.4, -0.089) \\ g(1.4, -0.089) \end{pmatrix} = \begin{pmatrix} -1.756 \\ 1.489 \end{pmatrix}.$$

Then, substituting these values in Eq. (7), we obtain

$$\mathbf{x}_1 = \begin{pmatrix} 2 \\ -0.5 \end{pmatrix} + \frac{0.2}{6} \begin{pmatrix} -16.956 \\ 11.799 \end{pmatrix} = \begin{pmatrix} 1.4348 \\ -0.1067 \end{pmatrix}.$$

These values of x_1 and y_1 are in error by about 0.000881 and 0.000442, respectively, corresponding to percentage errors of about 0.0614% and 0.416%.

This example again illustrates the increase in accuracy and sharper approximations that can be obtained by using a more accurate approximation method, such as the Runge–Kutta method. In the calculations we have just outlined, the Runge–Kutta method requires only twice as many function evaluations as the Euler method, but the error in the Runge–Kutta method is about 100 times less than in the Euler method.

Of course, these computations can be continued and Figure 8.4.1 shows the results of extending them as far as $t = 1$. The Euler method with $h = 0.1$ gives values that are qualitatively fairly accurate, but are quantitatively in error by several percent. The Runge–Kutta approximations with $h = 0.2$ differ from the true values of the solution only in the third or fourth decimal place, and are indistinguishable from them in the plot.

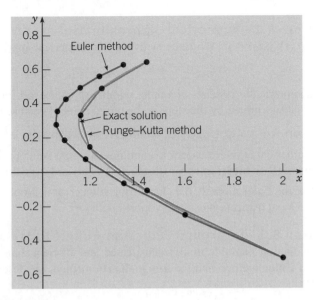

FIGURE 8.4.1 Approximations to the solution of the initial value problem (9), (10) using the Euler method ($h = 0.1$) and the Runge–Kutta method ($h = 0.2$).

PROBLEMS

 In each of Problems 1 through 6, determine approximate values of the solution $x = \phi(t)$, $y = \psi(t)$ of the given initial value problem at $t = 0.2, 0.4, 0.6, 0.8$, and 1.0. Compare the results obtained by different methods and different step sizes.

(a) Use the Euler method with $h = 0.1$.

(b) Use the Runge–Kutta method with $h = 0.2$.

(c) Use the Runge–Kutta method with $h = 0.1$.

1. $x' = x + y + t$, $y' = 4x - 2y$;

$x(0) = 1$, $y(0) = 0$

2. $x' = 2x + ty$, $y' = xy$;

$x(0) = 1$, $y(0) = 1$

3. $x' = -tx - y - 1$, $y' = x$;

$x(0) = 1$, $y(0) = 1$

4. $x' = x - y + xy$, $y' = 3x - 2y - xy$;

$x(0) = 0$, $y(0) = 1$

5. $x' = x(1 - 0.5x - 0.5y)$, $y' = y(-0.25 + 0.5x)$;

$x(0) = 4$, $y(0) = 1$

6. $x' = \exp(-x + y) - \cos x$, $y' = \sin(x - 3y)$;

$x(0) = 1$, $y(0) = 2$

7. Consider the example problem $x' = -x + 4y$, $y' = x - y$ with the initial conditions $x(0) = 2$ and $y(0) = -0.5$. Use the Runge–Kutta method to find approximate values of the solution of this problem on the interval $0 \leq t \leq 1$. Start with $h = 0.2$ and then repeat the calculation with step sizes $h = 0.1, 0.05, \ldots$, each half as long as in the preceding case. Continue the process until the first six digits of the solution at $t = 1$ are unchanged for successive step sizes. Determine whether these digits are accurate by comparing them with the exact solution given in Eqs. (11) in the text.

8. Consider the initial value problem

$$x'' + t^2 x' + 3x = t; \qquad x(0) = 1, \quad x'(0) = 2.$$

Convert this problem to a system of two first order equations, and determine approximate values of the solution at $t = 0.5$ and $t = 1.0$ using the Runge–Kutta method with $h = 0.1$.

CHAPTER SUMMARY

Differential equations are used to model systems that change continuously in time. Numerical methods can be used to gain insight on the behavior of a solution of a differential equation.

Section 8.1 Numerical Approximations For the initial value problem $y' = f(t, y)$, $y(t_0) = y_0$, the Euler method is the numerical approximation algorithm

$$t_{n+1} = t_n + h, \qquad y_{n+1} = y_n + hf(t_n, y_n), \qquad n = 0, 1, 2, \ldots$$

that geometrically consists of a finite number of connected line segments, each of whose slopes is determined by the slope at the initial point of each segment.

Section 8.2 Accuracy of Numerical Methods

▶ Numerical approximations to solutions of initial value problems involve two types of error: (i) **truncation error** (local and global) and (ii) **round-off error**.

▶ For the Euler method, the global truncation error is proportional to the step size h and the local truncation error is proportional to h^2.

Section 8.3 Improved Euler and Runge–Kutta Methods Two numerical approximation methods more sophisticated and efficient than the Euler method, the **improved Euler** method and the **Runge–Kutta** method, are described and illustrated by examples.

Section 8.4 Numerical Methods for Systems of First Order Equations
The Euler and Runge–Kutta methods described in Sections 8.1–8.3 are extended to systems of first order equations, and are illustrated for a typical two-dimensional system.

PROJECTS

© Cristian Andrei Matei /
Getty Images, Inc.

Project 1 Designing a Drip Dispenser for a Hydrology Experiment

In order to make laboratory measurements of water filtration and saturation rates in various types of soils under the condition of steady rainfall, a hydrologist wishes to design drip-dispensing containers in such a way that the water drips out at a nearly constant rate. The containers are supported above glass cylinders that contain the soil samples (Figure 8.P.1). The hydrologist elects to use the following differential equation, based on Torricelli's principle (see Problem 6, Section 2.3), to help solve the design problem:

$$A(h)\frac{dh}{dt} = -\alpha a \sqrt{2gh}. \tag{1}$$

In Eq. (1), $h(t)$ is the height of the liquid surface above the dispenser outlet at time t, $A(h)$ is the cross-sectional area of the dispenser at height h, a is the area of the outlet, and α is a measured contraction coefficient that accounts for the observed fact that the cross section of the (smooth) outflow stream is smaller than a. Note that the hydrologist is using a laminar flow model as a guide in designing the shape of the container. Forces due to surface tension at the tiny outlet are ignored in the design problem. Once the shape of the container has been determined, the outlet aperture is adjusted to a desired drip rate that will remain nearly constant for an extended period of time. Of course, since surface tension forces are not accounted for in Eq. (1), the equation is not a valid model for the output flow rate when the aperture is so small that the water drips out. Nevertheless, once the hydrologist sees and interprets the results based on her design strategy, she feels justified in using Eq. (1).

FIGURE 8.P.1 Water dripping into a soil sample.

Project 1 PROBLEMS

1. Assume that the shape of the dispensers are surfaces of revolution so that $A(h) = \pi[r(h)]^2$, where $r(h)$ is the radius of the container at height h. For each of the h-dependent cross-sectional radii prescribed below in (i)–(v):

(a) Create a surface plot of the surface of revolution, and
(b) Find numerical approximations of solutions of Eq. (1) for $0 \le t \le 60$:

   i. $r(h) = r_1, \qquad 0 \le h \le H$

   ii. $r(h) = r_0 + (r_1 - r_0)h/H, \qquad 0 \le h \le H$

   iii. $r(h) = r_0 + (r_1 - r_0)\sqrt{h/H}, \qquad 0 \le h \le H$

   iv. $r(h) = r_0 \exp\left[(h/H)\ln(r_1/r_0)\right], \qquad 0 \le h \le H$

   v. $r(h) = \dfrac{r_0 r_1 H}{r_1 H - (r_1 - r_0)h}, \qquad 0 \le h \le H$

Use the parameter values specified as follows:

$$r_0 = 0.1 \text{ ft}$$
$$r_1 = 1 \text{ ft}$$
$$\alpha = 0.6$$
$$a = 0.1 \text{ ft}^2$$

In addition, use the initial condition $h(0) = H$, where the initial height H of water in each of the containers is determined by requiring that the initial volume of water satisfies

$$V(0) = \int_0^H \pi r^2(h)\, dh = 1 \text{ ft}^3.$$

Determine the qualitative shape of the container such that the output flow rate given by the right-hand side of Eq. (1), $F_R(t) = \alpha a\sqrt{2gh(t)}$, varies slowly during the early stages of the experiment. As a design criterion, consider plotting the ratio $R = F_R(t)/F_R(0)$ for $0 \le t \le 60$, where values of R near 1 are most desirable. Based on the results of your computer experiments, sketch the shape of what a suitable container should look like.

2. After viewing the results of her computer experiments, it slowly dawns on the hydrologist that the "optimal shape" of the container is consistent with what would be expected based on the conceptualization that the water in the ideal container would consist of a collection of small parcels of water of mass m, all possessing the same amount of potential energy. If laboratory spatial constraints were not an issue, what would be the ideal "shape" of each container? Perform a computer experiment that supports your conclusions based on potential energy considerations.

Project 2 Monte Carlo Option Pricing: Pricing Financial Options by Flipping a Coin

A discrete model for change in price of a stock over a time interval $[0, T]$ is

$$S_{n+1} = S_n + \mu S_n \Delta t + \sigma S_n \varepsilon_{n+1}\sqrt{\Delta t}, \quad S_0 = s, \tag{1}$$

where $S_n = S(t_n)$ is the stock price at time $t_n = n\Delta t$, $n = 0, \dots, N-1$, $\Delta t = T/N$, μ is the annual growth rate of the stock, and σ is a measure of the stock's annual price volatility or tendency to fluctuate. Highly volatile stocks have large values for σ, for example, values ranging from 0.2 to 0.4. Each term in the sequence $\varepsilon_1, \varepsilon_2, \dots$ takes on the value 1 or -1

depending on whether the outcome of a coin tossing experiment is heads or tails, respectively. Thus, for each $n = 1, 2, \ldots,$

$$\varepsilon_n = \begin{cases} 1 & \text{with probability} = \frac{1}{2} \\ -1 & \text{with probability} = \frac{1}{2}. \end{cases} \tag{2}$$

A sequence of such numbers can easily be created by using one of the random number generators available in most mathematical computer software applications. Given such a sequence, the difference equation (1) can then be used to simulate a **sample path** or **trajectory** of stock prices, $\{s, S_1, S_2, \ldots, S_N\}$. The "random" terms $\sigma S_n \varepsilon_{n+1} \sqrt{\Delta t}$ on the right-hand side of (1) can be thought of as "shocks" or "disturbances" that model fluctuations in the stock price. By repeatedly simulating stock price trajectories and computing appropriate averages, it is possible to obtain estimates of the price of a **European call option**, a type of financial derivative. A statistical simulation algorithm of this type is called a **Monte Carlo method**.

A European call option is a contract between two parties, a holder and a writer, whereby, for a premium paid to the writer, the holder acquires the right (but not the obligation) to purchase the stock at a future date T (the **expiration date**) at a price K (the **strike price**) agreed upon in the contract. If the buyer elects to exercise the option on the expiration date, the writer is obligated to sell the underlying stock to the buyer at the price K. Thus the option has, associated with it, a **payoff function**

$$f(S) = \max(S - K, 0), \tag{3}$$

where $S = S(T)$ is the price of the underlying stock at the time T when the option expires (see Figure 8.P.2).

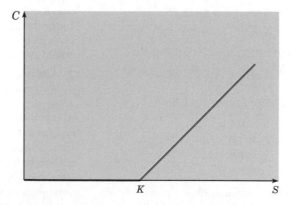

FIGURE 8.P.2   The value of a call option at expiration is $C = \max(S - K, 0)$, where K is the strike price of the option and $S = S(T)$ is the stock price at expiration.

Equation (3) is the value of the option at time T since, if $S(T) > K$, the holder can purchase, at price K, stock with market value $S(T)$ and thereby make a profit equal to $S(T) - K$, not counting the option premium. If $S(T) < K$, the holder will simply let the option expire since it would be irrational to purchase stock at a price that exceeds the market value. The option valuation problem is to determine the correct and fair price of the option at the time that the holder and writer enter into the contract.[4]

---

[4]The 1997 Nobel Prize in Economics was awarded to Robert C. Merton and Myron S. Scholes for their work, along with Fischer Black, in developing the Black–Scholes options pricing model.

To estimate the price of a call option using a Monte Carlo method, an ensemble

$$\left\{ S_N^{(k)} = S^{(k)}(T), \quad k = 1, \ldots, M \right\}$$

of M stock prices at expiration is generated using the difference equation

$$S_{n+1}^{(k)} = S_n^{(k)} + rS_n^{(k)}\Delta t + \sigma S_n^{(k)}\varepsilon_{n+1}^{(k)}\sqrt{\Delta t}, \quad S_0^{(k)} = s. \tag{4}$$

For each $k = 1, \ldots, M$, the difference equation (4) is identical to Eq. (1), except that the growth rate μ is replaced by the annual rate of interest r that it costs the writer to borrow money. Option pricing theory requires that the average value of the payoffs $\{f(S_N^{(k)}), k = 1, \ldots, M\}$ be equal to the compounded total return obtained by investing the option premium, $\hat{C}(s)$, at rate r over the life of the option,

$$\frac{1}{M}\sum_{k=1}^{M} f(S_N^{(k)}) = (1 + r\Delta t)^N \hat{C}(s). \tag{5}$$

Solving (5) for $\hat{C}(s)$ yields the Monte Carlo estimate

$$\hat{C}(s) = (1 + r\Delta t)^{-N}\left\{ \frac{1}{M}\sum_{k=1}^{M} f(S_N^{(k)}) \right\} \tag{6}$$

for the option price. Thus the Monte Carlo estimate $\hat{C}(s)$ is the present value of the average of the payoffs computed using the rules of compound interest.

Project 2 PROBLEMS

1. Show that Euler's method applied to the differential equation

$$\frac{dS}{dt} = \mu S \tag{i}$$

yields Eq. (1) in the absence of random disturbances, that is, when $\sigma = 0$.

2. Simulate five sample trajectories of Eq. (1) for the following parameter values and plot the trajectories on the same set of coordinate axes: $\mu = 0.12$, $\sigma = 0.1$, $T = 1$, $s = \$40$, $N = 254$. Then repeat the experiment using the value $\sigma = 0.25$ for the volatility. Do the sample trajectories generated in the latter case appear to exhibit a greater degree of variability in their behavior?

Hint: For the ε_n's, it is permissible to use a random number generator that creates normally distributed random numbers with mean zero and variance 1.

3. Use the difference equation (4) to generate an ensemble of stock prices $S_N^{(k)} = S^{(k)}(N\Delta t), k = 1, \ldots, M$ (where $T = N\Delta t$) and then use formula (6) to compute a Monte Carlo estimate of the value of a five-month call option ($T = \frac{5}{12}$ years)

for the following parameter values: $r = 0.06$, $\sigma = 0.2$, and $K = \$50$. Find estimates corresponding to current stock prices of $S(0) = s = \$45$, $\$50$, and $\$55$. Use $N = 200$ time steps for each trajectory and $M \cong 10,000$ sample trajectories for each Monte Carlo estimate.[5] Check the accuracy of your results by comparing the Monte Carlo approximation with the value computed from the exact Black–Scholes formula

$$C(s) = \frac{s}{2}\,\mathrm{erfc}\left(-\frac{d_1}{\sqrt{2}}\right) - \frac{K}{2}\,e^{-rT}\,\mathrm{erfc}\left(-\frac{d_2}{\sqrt{2}}\right), \tag{ii}$$

where

$$d_1 = \frac{1}{\sigma\sqrt{T}}\left[\ln\left(\frac{s}{k}\right) + \left(r + \frac{\sigma^2}{2}\right)T\right],$$

$$d_2 = d_1 - \sigma\sqrt{T}$$

and $\mathrm{erfc}(x)$ is the complementary error function,

$$\mathrm{erfc}(x) = \frac{2}{\sqrt{\pi}}\int_x^\infty e^{-t^2}\,dt.$$

4. **Variance Reduction by Antithetic Variates.** A simple and widely used technique for increasing the efficiency

---

[5] As a rule of thumb, you may assume that the sampling error in these Monte Carlo estimates is proportional to $1/\sqrt{M}$. Using software packages such as MATLAB that allow vector operations where all M trajectories can be simulated simultaneously greatly speeds up the calculations.

and accuracy of Monte Carlo simulations in certain situations with little additional increase in computational complexity is the method of antithetic variates. For each $k = 1, \ldots, M$, use the sequence $\left\{ \varepsilon_1^{(k)}, \ldots, \varepsilon_{N-1}^{(k)} \right\}$ in Eq. (4) to simulate a payoff $f(S_N^{(k+)})$ and also use the sequence $\left\{ -\varepsilon_1^{(k)}, \ldots, -\varepsilon_{N-1}^{(k)} \right\}$ in Eq. (4) to simulate an associated payoff $f(S_N^{(k-)})$. Thus, the payoffs are simulated in pairs $\left\{ f(S_N^{(k+)}), f(S_N^{(k-)}) \right\}$. A modified Monte Carlo estimate is then computed by replacing each payoff $f(S_N^{(k)})$ in Eq. (6) by the average $[f(S_N^{(k+)}) + f(S_N^{(k-)})]/2$,

$$\hat{C}_{AV}(s) = \frac{\dfrac{1}{M} \displaystyle\sum_{k=1}^{M} \dfrac{f(S_N^{(k+)}) + f(S_N^{(k-)})}{2}}{(1 + r\Delta t)^N}. \tag{iii}$$

Use the parameters specified in Problem 3 to compute several (say, 20 or so) option price estimates using Eq. (6) and an equivalent number of option price estimates using (iii). For each of the two methods, plot a histogram of the estimates and compute the mean and standard deviation of the estimates. Comment on the accuracies of the two methods.

Matrices and Linear Algebra

lementary matrix methods were introduced in Chapter 3 to facilitate the study of first order linear systems of dimension 2. In Appendix A, we present several additional concepts and tools from matrix theory necessary for the study of first order linear systems of dimension $n \geq 2$ taken up in Chapter 6. Readers with an adequate background in matrix methods and linear algebra may wish to proceed directly to Chapter 6, referring back to this appendix for reference or review if necessary.

A.1 Matrices

The idea of the matrix as a mathematical object is most easily conveyed by considering a type of problem that frequently occurs in science and engineering, the linear algebraic system of equations. For example, consider the system

$$
\begin{array}{rcrcrcrcr}
x_1 & + & 2x_2 & - & x_3 & + & 2x_4 & = & 0, \\
2x_1 & + & x_2 & + & 2x_3 & - & 3x_4 & = & 0, \\
-x_1 & + & 3x_2 & + & x_3 & + & x_4 & = & 0, \\
x_1 & + & 9x_2 & + & 3x_3 & + & x_4 & = & 0.
\end{array}
\tag{1}
$$

Solving the system (1) means explicitly characterizing the set of all values of x_1, x_2, x_3, and x_4 that simultaneously satisfy all four equations. The solution may simply consist of numerical values for the unknowns, or it may require a mathematical description of a more complex solution set. All information required to solve the problem resides in the coefficients of the unknowns, whereas the names for the unknowns are largely irrelevant. Not only is it particularly convenient to store the coefficients in a rectangular array called a matrix, but efficient algorithms also exist for systematically manipulating the entries in the array to arrive at a mathematical characterization of all solutions. The matrix of coefficients for the system (1), displayed in an array identical to the pattern in which they appear, is

$$
\begin{pmatrix}
1 & 2 & -1 & 2 \\
2 & 1 & 2 & -3 \\
-1 & 3 & 1 & 1 \\
1 & 9 & 3 & 1
\end{pmatrix}.
\tag{2}
$$

Solution algorithms are applied to arrays of this type.

Matrices also play an important role in differential equations. Recall from Chapter 3 that the solution for a linear, constant coefficient system of dimension 2,

$$\mathbf{x}' = \begin{pmatrix} a_{11} & a_{12} \\ a_{21} & a_{22} \end{pmatrix} \mathbf{x} = \mathbf{A}\mathbf{x}, \tag{3}$$

is based on the eigenvalue method, that is, finding the eigenvalues and eigenvectors of the 2×2 matrix $\mathbf{A}$. In Chapter 6, the eigenvalue method is also used to find solutions of linear, constant coefficient systems of dimension n. In this appendix, we develop those matrix concepts and methods pursuant to that goal.

**DEFINITION
A.1.1**

A matrix is an array of mathematical objects, or elements, arranged in m rows and n columns.

The elements may be real or complex numbers such as the matrix (2). The elements may also be functions or contain symbols such as

$$\begin{pmatrix} e^{-t} & e^{2t} \\ e^{-t} & 2e^{2t} \end{pmatrix}$$

and

$$\begin{pmatrix} 3-\lambda & -1 & 2 \\ 0 & -1-\lambda & -4 \\ 1 & -2 & 1-\lambda \end{pmatrix}.$$

We will designate matrices by boldfaced capitals $\mathbf{A}, \mathbf{B}, \mathbf{C}, \dots$ such as

$$\mathbf{A} = \begin{pmatrix} a_{11} & a_{12} & \cdots & a_{1n} \\ a_{21} & a_{22} & \cdots & a_{2n} \\ \vdots & \vdots & & \vdots \\ a_{m1} & a_{m2} & \cdots & a_{mn} \end{pmatrix}. \tag{4}$$

The rows of $\mathbf{A}$ are the m horizontal n-tuples

$$(a_{11}, a_{12}, \dots, a_{1n}), \quad (a_{21}, a_{22}, \dots, a_{2n}), \quad \dots, \quad (a_{m1}, a_{m2}, \dots, a_{mn}).$$

The columns of $\mathbf{A}$ are the n vertical m-tuples

$$\begin{pmatrix} a_{11} \\ a_{21} \\ \vdots \\ a_{m1} \end{pmatrix}, \begin{pmatrix} a_{12} \\ a_{22} \\ \vdots \\ a_{m2} \end{pmatrix}, \dots, \begin{pmatrix} a_{1n} \\ a_{2n} \\ \vdots \\ a_{mn} \end{pmatrix}.$$

The element lying in the ith row and jth column is denoted by a_{ij}, the first subscript identifying its row and the second its column. Frequently, the condensed notation $\mathbf{A} = (a_{ij})$ is used, indicating that $\mathbf{A}$ is the matrix whose generic element is a_{ij}. In this case, the **dimensions** of $\mathbf{A}$ (number of rows and columns), if not explicitly stated, are understood by the context in which the matrix is used.

Two matrices of special importance are square matrices and vectors:

1. Matrices that have the same number of rows and columns—that is, $m = n$—are called **square matrices**. The matrix (2) is an example of a square matrix of dimension 4.

2. Matrices that have only one column—that is, the dimensions are $n \times 1$—are called **column vectors** or simply **vectors**. A vector with n entries is called an n-vector. Examples of column vectors are the following 2-vector, 3-vector, and 4-vector, respectively,

$$\begin{pmatrix} 1 \\ 2 \end{pmatrix}, \quad \begin{pmatrix} -1 \\ 0 \\ 3 \end{pmatrix}, \quad \begin{pmatrix} 1 \\ 2 \\ -4 \\ 3 \end{pmatrix}.$$

We denote (column) vectors by boldfaced lowercase letters, $\mathbf{x}, \mathbf{y}, \mathbf{z}, \ldots$. The general form of an n-vector is

$$\mathbf{x} = \begin{pmatrix} x_1 \\ x_2 \\ \vdots \\ x_n \end{pmatrix}.$$

The number x_j is commonly called the jth **component** of the column vector.

▶ Matrix Algebra.

DEFINITION A.1.2	**Equality.** Two $m \times n$ matrices $\mathbf{A} = (a_{ij})$ and $\mathbf{B} = (b_{ij})$ are *equal* if all corresponding elements are equal—that is, if $a_{ij} = b_{ij}$ for each i and j.

DEFINITION A.1.3	**Zero Matrix.** The **zero matrix**, denoted by $\mathbf{0}$, is the matrix (or vector) whose elements are zero. Thus $\mathbf{0} = (z_{ij})$, where $z_{ij} = 0$ for each i and j.

Some of the usual arithmetic operations for numbers can be extended to matrices in a natural way.

DEFINITION A.1.4	**Matrix Addition.** The sum of two $m \times n$ matrices $\mathbf{A}$ and $\mathbf{B}$ is defined to be the matrix obtained by adding corresponding elements: $$\mathbf{A} + \mathbf{B} = (a_{ij}) + (b_{ij}) = (a_{ij} + b_{ij}). \qquad (5)$$

Thus, if $\mathbf{C} = \mathbf{A} + \mathbf{B}$, the (i, j)th element of $\mathbf{C}$ is the sum of the (i, j)th element of $\mathbf{A}$ and the (i, j)th element of $\mathbf{B}$. For example,

$$\begin{pmatrix} -1 & 3 & 0 \\ 2 & 1 & 4 \end{pmatrix} + \begin{pmatrix} 2 & 5 & -7 \\ -3 & 4 & 1 \end{pmatrix} = \begin{pmatrix} 1 & 8 & -7 \\ -1 & 5 & 5 \end{pmatrix}.$$

If the dimensions of two matrices are not identical, the matrices cannot be added.

Note that $\mathbf{A} + \mathbf{0} = \mathbf{A}$. For this reason, the zero matrix $\mathbf{0}$ is often called the **additive identity**.

In matrix algebra, it is traditional to use the term **scalar** to refer to a real or complex number. The product of a matrix and a scalar is defined in a natural way.

**DEFINITION
A.1.5**

Product of a Matrix and a Scalar. The product of a matrix $\mathbf{A}$ and a scalar α is defined as follows:

$$\alpha\mathbf{A} = \alpha(a_{ij}) = (\alpha a_{ij}). \tag{6}$$

Therefore the (i, j)th entry of $\alpha\mathbf{A}$ is simply the (i, j)th entry of $\mathbf{A}$ multiplied by α. For example,

$$-3\begin{pmatrix} -2 & 1 & 0 \\ 2 & -1 & 4 \\ 1 & 0 & 5 \end{pmatrix} = \begin{pmatrix} 6 & -3 & 0 \\ -6 & 3 & -12 \\ -3 & 0 & -15 \end{pmatrix}.$$

The negative of a matrix $\mathbf{A} = (a_{ij})$, denoted by $-\mathbf{A}$, is the matrix whose (i, j)th entry is $-a_{ij}$. We then define the **difference** of two matrices, $\mathbf{A} - \mathbf{B}$, by

$$\mathbf{A} - \mathbf{B} = \mathbf{A} + (-\mathbf{B}).$$

Thus

$$\mathbf{A} - \mathbf{B} = (a_{ij}) + (-b_{ij}) = (a_{ij} - b_{ij}).$$

The following theorem gives the algebraic properties of matrices for operations restricted to matrix addition and the multiplication of matrices by scalars. The properties follow readily from the properties of real or complex numbers and the above definitions for matrix addition and multiplication of matrices by scalars.

**THEOREM
A.1.6**

If $\mathbf{A}$, $\mathbf{B}$, and $\mathbf{C}$ are $m \times n$ matrices and α and β are scalars, then

  a. $\mathbf{A} + \mathbf{B} = \mathbf{B} + \mathbf{A}$

  b. $\mathbf{A} + (\mathbf{B} + \mathbf{C}) = (\mathbf{A} + \mathbf{B}) + \mathbf{C}$

  c. $\alpha(\beta\mathbf{A}) = (\alpha\beta)\mathbf{A}$

  d. $(\alpha + \beta)\mathbf{A} = \alpha\mathbf{A} + \beta\mathbf{A}$

  e. $\alpha(\mathbf{A} + \mathbf{B}) = \alpha\mathbf{A} + \alpha\mathbf{B}$

  f. $(-1)\mathbf{A} = -\mathbf{A}$

  g. $-(-\mathbf{A}) = \mathbf{A}$

  h. $0\mathbf{A} = \mathbf{0}$

  i. $\alpha\mathbf{0} = \mathbf{0}$

Note that the properties stated in Theorem A.1.6 simply extend the properties of equivalent operations for real or complex numbers. However the operation of matrix multiplication does not have the same properties as multiplication of real or complex numbers.

We first define the matrix product of a $1 \times n$ matrix $\mathbf{x}$ with an $n \times 1$ matrix $\mathbf{y}$ by

$$\mathbf{xy} = (x_1, \ldots, x_n) \begin{pmatrix} y_1 \\ \vdots \\ y_n \end{pmatrix} = x_1 y_1 + \cdots + x_n y_n.$$

This is simply the familiar dot or inner product from calculus extended to vectors with n components. The product of an $m \times n$ matrix and an $n \times p$ matrix is defined next.

DEFINITION A.1.7

Matrix Multiplication. If $\mathbf{A}$ is an $m \times n$ matrix and $\mathbf{B}$ is an $n \times p$ matrix, then $\mathbf{AB}$ is defined to be the $m \times p$ matrix $\mathbf{C} = (c_{ij})$ satisfying

$$c_{ij} = \sum_{k=1}^{n} a_{ik} b_{kj}, \qquad 1 \leq i \leq m, \quad 1 \leq j \leq p. \tag{7}$$

Note that the matrix product $\mathbf{AB}$ is defined only if the number of columns in the first factor $\mathbf{A}$ is equal to the number of rows in the second factor $\mathbf{B}$. It is convenient to think of c_{ij} defined in Eq. (7) as the dot product of the ith row of $\mathbf{A}$ with the jth column of $\mathbf{B}$,

$$c_{ij} = (a_{i1}, \ldots, a_{in}) \begin{pmatrix} b_{1j} \\ \vdots \\ b_{nj} \end{pmatrix} = a_{i1} b_{1j} + \cdots + a_{in} b_{nj}.$$

A simple mnemonic device for determining if the matrix product is defined, as well as the dimensions of the resultant product, consists of writing the dimensions of each factor adjacent to one another in the same order that the matrices are to be multiplied,

$$\underbrace{\mathbf{A}}_{m \times n} \quad \underbrace{\mathbf{B}}_{n \times p} \quad = \quad \underbrace{\mathbf{C}}_{m \times p}. \tag{8}$$

The product is defined if the interior dimensions n match as indicated in Eq. (8).

EXAMPLE 1

Let

$$\mathbf{A} = \begin{pmatrix} 2 & -1 \\ 3 & 1 \end{pmatrix} \qquad \text{and} \qquad \mathbf{B} = \begin{pmatrix} 1 & 0 & 2 \\ 2 & 1 & -1 \end{pmatrix}.$$

Since $\mathbf{A}$ is 2×2 and $\mathbf{B}$ is 2×3, $\mathbf{AB}$ is defined and is 2×3. From Eq. (7), we have

$$\mathbf{AB} = \begin{pmatrix} 2-2 & 0-1 & 4+1 \\ 3+2 & 0+1 & 6-1 \end{pmatrix} = \begin{pmatrix} 0 & -1 & 5 \\ 5 & 1 & 5 \end{pmatrix}.$$

The product $\mathbf{BA}$ is not defined because the number of columns of $\mathbf{B}$ (three) is not equal to the number of rows of $\mathbf{A}$ (two).

Example 1 shows that matrix multiplication does not share the property of commutativity possessed by multiplication of numbers since $\mathbf{BA}$ is not defined. Even in the case that both products $\mathbf{AB}$ and $\mathbf{BA}$ exist, the two products are usually not equal.

EXAMPLE
2

Consider the matrices

$$\mathbf{A} = \begin{pmatrix} 1 & -2 & 1 \\ 0 & 2 & -1 \\ 2 & 1 & 1 \end{pmatrix} \quad \text{and} \quad \mathbf{B} = \begin{pmatrix} 2 & 1 & -1 \\ 1 & -1 & 0 \\ 2 & -1 & 1 \end{pmatrix}.$$

Again, from the definition of multiplication given in Eq. (7), we have

$$\mathbf{AB} = \begin{pmatrix} 2-2+2 & 1+2-1 & -1+0+1 \\ 0+2-2 & 0-2+1 & 0+0-1 \\ 4+1+2 & 2-1-1 & -2+0+1 \end{pmatrix}$$

$$= \begin{pmatrix} 2 & 2 & 0 \\ 0 & -1 & -1 \\ 7 & 0 & -1 \end{pmatrix}.$$

Similarly, we find that

$$\mathbf{BA} = \begin{pmatrix} 0 & -3 & 0 \\ 1 & -4 & 2 \\ 4 & -5 & 4 \end{pmatrix}.$$

Clearly, $\mathbf{AB} \neq \mathbf{BA}$.

From these examples we conclude that, in general,

$$\mathbf{AB} \neq \mathbf{BA}.$$

The following theorem states properties of matrix multiplication that are shared in common with multiplication of numbers.

THEOREM
A.1.8

Suppose that $\mathbf{A}$, $\mathbf{B}$, and $\mathbf{C}$ are matrices for which the following products are defined and let α be a scalar. Then

  a. $(\mathbf{AB})\mathbf{C} = \mathbf{A}(\mathbf{BC})$
  b. $\mathbf{A}(\alpha\mathbf{B}) = (\alpha\mathbf{A})\mathbf{B} = \alpha(\mathbf{AB})$
  c. $\mathbf{A}(\mathbf{B} + \mathbf{C}) = \mathbf{AB} + \mathbf{AC}$
  d. $(\mathbf{A} + \mathbf{B})\mathbf{C} = \mathbf{AC} + \mathbf{BC}$

▶ **Special Matrices.**   **The identity matrix.** Denote the Kronecker delta by δ_{ij},

$$\delta_{ij} = \begin{cases} 1, & \text{if } i = j \\ 0, & \text{if } i \neq j \end{cases}.$$

The $n \times n$ matrix

$$\mathbf{I}_n = (\delta_{ij}) = \begin{pmatrix} 1 & 0 & \cdots & 0 \\ 0 & 1 & \cdots & 0 \\ \vdots & \vdots & \ddots & \vdots \\ 0 & 0 & \cdots & 1 \end{pmatrix} \tag{9}$$

is called the **multiplicative identity**, or simply the **identity matrix**. If A is an $n \times n$ matrix, then from the definition of matrix multiplication we have

$$\mathbf{I}_n\mathbf{A} = \mathbf{A}\mathbf{I}_n = \mathbf{A}.$$

To see that $\mathbf{I}_n\mathbf{A} = \mathbf{A}$, we note that the (i, j)th entry of $\mathbf{I}_n\mathbf{A}$ is

$$\sum_{k=1}^{n} \delta_{ik}a_{kj} = a_{ij}$$

since $\delta_{ik} = 0$ for all $k \neq i$. The equality $\mathbf{A}\mathbf{I}_n = \mathbf{A}$ is established in the same way. In particular, for any n-vector $\mathbf{x}$,

$$\mathbf{I}_n\mathbf{x} = \mathbf{x}.$$

Diagonal and Triangular Matrices. If A is an $n \times n$ matrix, the set of elements a_{jj}, $j = 1, \ldots, n$ is referred to as the **principal diagonal**, or simply, the **diagonal**, of the matrix. If all elements off the diagonal are zero, that is, $a_{ij} = 0$ whenever $i \neq j$, then the matrix is called a **diagonal matrix**. For example,

$$\mathbf{A} = \begin{pmatrix} a_{11} & 0 & 0 & 0 \\ 0 & a_{22} & 0 & 0 \\ 0 & 0 & a_{33} & 0 \\ 0 & 0 & 0 & a_{44} \end{pmatrix}$$

is a diagonal matrix, which we can write in abbreviated form as $\mathbf{A} = \text{diag}(a_{11}, a_{22}, a_{33}, a_{44})$.

If all the elements of a square matrix below (or above) the principal diagonal are zero, then the matrix is called an **upper** (**lower**) **triangular matrix**. Examples of upper and lower triangular matrices are

$$\begin{pmatrix} a_{11} & a_{12} & a_{13} \\ 0 & a_{22} & a_{23} \\ 0 & 0 & a_{33} \end{pmatrix} \quad \text{and} \quad \begin{pmatrix} a_{11} & 0 & 0 \\ a_{21} & a_{22} & 0 \\ a_{31} & a_{32} & a_{33} \end{pmatrix},$$

respectively.

The Transpose of a Matrix. If A is an $m \times n$ matrix, then the $n \times m$ matrix formed by interchanging the rows and columns of $\mathbf{A}$ is called the **transpose** of $\mathbf{A}$ and is denoted by $\mathbf{A}^T$. Thus if $\mathbf{A} = (a_{ij})$, $\mathbf{A}^T = (a_{ji})$. For example, if

$$\mathbf{A} = \begin{pmatrix} -1 & 4 & -3 \\ 2 & 5 & -7 \end{pmatrix},$$

then

$$\mathbf{A}^T = \begin{pmatrix} -1 & 2 \\ 4 & 5 \\ -3 & -7 \end{pmatrix}.$$

THEOREM A.1.9

Properties of Transposes.

a. $\mathbf{I}_n^T = \mathbf{I}_n$

b. For any matrix $\mathbf{A}$, $(\mathbf{A}^T)^T = \mathbf{A}$.

c. If $\mathbf{AB}$ is defined, then $(\mathbf{AB})^T = \mathbf{B}^T\mathbf{A}^T$.

Parts (a) and (b) are obvious. To prove part (c), assume that $\mathbf{A}$ is $m \times n$ and that $\mathbf{B}$ is $n \times p$. Note that the ijth entry of $(\mathbf{AB})^T$ is the jith entry of $\mathbf{AB}$, $\sum_{k=1}^{n} a_{jk}b_{ki}$. On the other hand, the ijth entry of $\mathbf{B}^T\mathbf{A}^T$ is the dot product of the ith row of $\mathbf{B}^T$ with the jth column of $\mathbf{A}^T$, or equivalently, the dot product of the ith column of $\mathbf{B}$ with the jth row of $\mathbf{A}$, $\sum_{k=1}^{n} b_{ki}a_{jk} = \sum_{k=1}^{n} a_{jk}b_{ki}$. Thus, for each i and j, the (i, j)th entries of $(\mathbf{AB})^T$ and $\mathbf{B}^T\mathbf{A}^T$ are equal.

Symmetric Matrices. An $n \times n$ matrix $\mathbf{A}$ is said to be symmetric if $\mathbf{A} = \mathbf{A}^T$. In this case, $a_{ij} = a_{ji}$ for all i and j with the principal diagonal serving as the axis of symmetry. An example of a symmetric matrix is

$$A = \begin{pmatrix} 2 & -3 & 1 & -2 \\ -3 & 0 & 4 & -5 \\ 1 & 4 & -3 & 8 \\ -2 & -5 & 8 & 1 \end{pmatrix}.$$

Matrix Inverse. The square matrix $\mathbf{A}$ is said to be **nonsingular** or **invertible** if there is another matrix $\mathbf{B}$ such that

$$\mathbf{AB} = \mathbf{BA} = \mathbf{I}_n.$$

If there is such a $\mathbf{B}$, there is only one. To see this, suppose that there is also a matrix $\mathbf{C}$ such that

$$\mathbf{AC} = \mathbf{CA} = \mathbf{I}_n.$$

Then

$$\mathbf{B} = \mathbf{BI}_n = \mathbf{B}(\mathbf{AC}) = (\mathbf{BA})\mathbf{C} = \mathbf{I}_n\mathbf{C} = \mathbf{C}$$

where the third equality follows from part (a) of Theorem A.1.8. The unique inverse of $\mathbf{A}$, if it exists, is called the **multiplicative inverse**, or simply the **inverse**, of $\mathbf{A}$, and we write $\mathbf{B} = \mathbf{A}^{-1}$. Then

$$\mathbf{AA}^{-1} = \mathbf{A}^{-1}\mathbf{A} = \mathbf{I}_n. \tag{10}$$

Whereas all nonzero scalars possess inverses, there are lots of nonzero matrices that do not have inverses. Matrices that do not have an inverse are called **singular** or **noninvertible**.

EXAMPLE 3

Demonstrate that

$$\mathbf{A} = \begin{pmatrix} 1 & 2 \\ 2 & 2 \end{pmatrix} \quad \text{and} \quad \mathbf{B} = \begin{pmatrix} -1 & 1 \\ 1 & -\frac{1}{2} \end{pmatrix}$$

are both nonsingular by showing that $\mathbf{AB} = \mathbf{BA} = \mathbf{I}_2$.

We have

$$\begin{pmatrix} 1 & 2 \\ 2 & 2 \end{pmatrix} \begin{pmatrix} -1 & 1 \\ 1 & -\frac{1}{2} \end{pmatrix} = \begin{pmatrix} 1 & 0 \\ 0 & 1 \end{pmatrix} = \begin{pmatrix} -1 & 1 \\ 1 & -\frac{1}{2} \end{pmatrix} \begin{pmatrix} 1 & 2 \\ 2 & 2 \end{pmatrix}.$$

Thus both $\mathbf{A}$ and $\mathbf{B}$ are nonsingular, $\mathbf{A}^{-1} = \mathbf{B}$, and $\mathbf{B}^{-1} = \mathbf{A}$.

THEOREM
A.1.10

Properties of Matrix Inverses.

a. I_n is nonsingular and $I_n^{-1} = I_n$.

b. If **A** and **B** are nonsingular $n \times n$ matrices, then **AB** is nonsingular and

$$(\mathbf{AB})^{-1} = \mathbf{B}^{-1}\mathbf{A}^{-1}.$$

c. If **A** is nonsingular, so is $\mathbf{A}^{-1}$, and $(\mathbf{A}^{-1})^{-1} = \mathbf{A}$.

d. If **A** is nonsingular, so is $\mathbf{A}^T$, and $(\mathbf{A}^T)^{-1} = (\mathbf{A}^{-1})^T$.

e. If **A** and **B** are $n \times n$ matrices, either of which is singular, then **AB** and **BA** are both singular.

Proof

Part (a) follows from the fact that $\mathbf{I}_n\mathbf{I}_n = \mathbf{I}_n$.
To prove part (b), we use the fact that $\mathbf{A}^{-1}$ and $\mathbf{B}^{-1}$ exist to write

$$(\mathbf{AB})(\mathbf{B}^{-1}\mathbf{A}^{-1}) = \mathbf{A}(\mathbf{B}^{-1}\mathbf{B})\mathbf{A}^{-1} = \mathbf{AI}_n\mathbf{A}^{-1} = \mathbf{AA}^{-1} = \mathbf{I}_n.$$

Similarly, $(\mathbf{B}^{-1}\mathbf{A}^{-1})(\mathbf{AB}) = \mathbf{I}_n$, so $\mathbf{B}^{-1}\mathbf{A}^{-1}$ is the unique inverse of **AB**.
Part (c) follows from $\mathbf{AA}^{-1} = \mathbf{A}^{-1}\mathbf{A} = \mathbf{I}_n$.
Part (d) follows from the two statements

$$\mathbf{I}_n = \mathbf{I}_n^T = (\mathbf{AA}^{-1})^T = (\mathbf{A}^{-1})^T\mathbf{A}^T,$$

and

$$\mathbf{I}_n = \mathbf{I}_n^T = (\mathbf{A}^{-1}\mathbf{A})^T = \mathbf{A}^T(\mathbf{A}^{-1})^T,$$

where we have used part (c) of Theorem A.1.9. Thus the unique inverse of $\mathbf{A}^T$ is $(\mathbf{A}^{-1})^T$.
A proof of (e) will be possible once we have discussed determinants.

PROBLEMS

1. Given the matrices

$$\mathbf{A} = \begin{pmatrix} 2 & 2 \\ -1 & 3 \end{pmatrix}, \quad \mathbf{B} = \begin{pmatrix} -1 & 1 & -2 \\ 1 & 3 & 4 \end{pmatrix},$$

$$\mathbf{x} = \begin{pmatrix} 2 \\ 3 \end{pmatrix}, \quad \mathbf{y} = (1, 2, -1),$$

compute each of the following for which the indicated operations are defined:

(a) 4**A**   (b) $-2\mathbf{x}$ (c) **A** + **B**

(d) **Ax**   (e) **Bx**   (f) **AB**

(g) **BA**   (h) $\mathbf{B}^T\mathbf{A}$ (i) $\mathbf{yB}^T$

(j) $(\mathbf{AB})^T$ (k) $\mathbf{x}^T\mathbf{Ax}$ (l) $\mathbf{xBy}^T$

2. If $\mathbf{A} = \begin{pmatrix} 1 & -2 & 0 \\ 3 & -2 & -1 \\ -2 & 1 & 3 \end{pmatrix}$ and if $\mathbf{B} = \begin{pmatrix} 1 & 0 & 2 \\ 0 & 1 & -1 \\ 2 & -1 & 3 \end{pmatrix}$, find

(a) 2**A** + **B**   (b) **A** − 4**B**

(c) **AB**   (d) **BA**

(e) $(\mathbf{A} + \mathbf{B})^T$ (f) $(\mathbf{B} + \mathbf{A})\mathbf{A}^T$

(g) $\mathbf{A}^2 = \mathbf{AA}$ (h) **AB** − **BA**

(i) $\mathbf{B}^3 - 5\mathbf{B}^2 + 2\mathbf{B} + 2\mathbf{I}_3$

3. Demonstrate that

$$\mathbf{A} = \begin{pmatrix} 2 & 2 & 3 \\ 1 & 0 & 1 \\ 1 & 1 & 1 \end{pmatrix} \quad \text{and} \quad \mathbf{B} = \begin{pmatrix} -1 & 1 & 2 \\ 0 & -1 & 1 \\ 1 & 0 & -2 \end{pmatrix}$$

are both nonsingular by showing that $\mathbf{AB} = \mathbf{BA} = \mathbf{I}_3$.

4. Prove each of the following laws of matrix algebra:

(a) $\mathbf{A} + \mathbf{B} = \mathbf{B} + \mathbf{A}$

(b) $\mathbf{A} + (\mathbf{B} + \mathbf{C}) = (\mathbf{A} + \mathbf{B}) + \mathbf{C}$

(c) $\alpha(\mathbf{A} + \mathbf{B}) = \alpha\mathbf{A} + \alpha\mathbf{B}$

(d) $(\alpha + \beta)\mathbf{A} = \alpha\mathbf{A} + \beta\mathbf{A}$

(e) $\mathbf{A}(\mathbf{BC}) = (\mathbf{AB})\mathbf{C}$

(f) $\mathbf{A}(\mathbf{B} + \mathbf{C}) = \mathbf{AB} + \mathbf{BC}$

5. If $\mathbf{A} = \text{diag}(a_{11}, \ldots, a_{nn})$, under what conditions is $\mathbf{A}$ nonsingular? What is $\mathbf{A}^{-1}$ in this case?

6. Prove that sums and products of upper (lower) triangular matrices are upper (lower) triangular.

7. Let $\mathbf{A} = \text{diag}(a_{11}, \ldots, a_{nn})$ be a diagonal matrix. Prove that $\mathbf{A}^k = \text{diag}(a_{11}^k, \ldots, a_{nn}^k)$, where

$$\mathbf{A}^2 = \mathbf{AA}, \mathbf{A}^3 = \mathbf{AAA}, \ldots.$$

8. Prove that if $\mathbf{A}$ is symmetric and nonsingular, then $\mathbf{A}^{-1}$ is symmetric.

9. Two square matrices $\mathbf{A}$ and $\mathbf{B}$ are said to commute if $\mathbf{AB} = \mathbf{BA}$. If $\mathbf{A}$ and $\mathbf{B}$ are symmetric, prove that $\mathbf{AB}$ is symmetric if and only if $\mathbf{A}$ and $\mathbf{B}$ commute.

10. If $\mathbf{A}$ is any square matrix, show each of the following:

(a) $\mathbf{AA}^T$ and $\mathbf{A}^T\mathbf{A}$ are both symmetric.

(b) $\mathbf{A} + \mathbf{A}^T$ is symmetric.

A.2 Systems of Linear Algebraic Equations, Linear Independence, and Rank

In this section, we describe an efficient algorithm for finding solutions, if they exist, to a set of m simultaneous linear algebraic equations in n variables,

$$
\begin{aligned}
a_{11}x_1 + a_{12}x_2 + \cdots + a_{1n}x_n &= b_1, \\
a_{21}x_1 + a_{22}x_2 + \cdots + a_{2n}x_n &= b_2, \\
\vdots \qquad \vdots \qquad\qquad \vdots \quad\ &\ \ \vdots \\
a_{m1}x_1 + a_{m2}x_2 + \cdots + a_{mn}x_n &= b_m,
\end{aligned}
\tag{1}
$$

or, in matrix notation,

$$\mathbf{Ax} = \mathbf{b}. \tag{2}$$

We begin with an example to help motivate an extremely useful algorithm for solving linear algebraic systems of equations.

EXAMPLE 1

Use the method of elimination to solve the system of equations

$$
\begin{aligned}
x_1 - x_2 + 2x_3 + x_4 &= 3, \\
2x_1 - 3x_2 + 6x_3 + 5x_4 &= 4, \\
-2x_1 + 4x_2 - 8x_3 - 8x_4 &= -2.
\end{aligned}
\tag{3}
$$

If we subtract two times the first row from the second row, we obtain

$$
\begin{aligned}
x_1 - x_2 + 2x_3 + x_4 &= 3, \\
0x_1 - x_2 + 2x_3 + 3x_4 &= -2, \\
-2x_1 + 4x_2 - 8x_3 - 8x_4 &= -2.
\end{aligned}
$$

Adding two times the first row to the third row gives

$$
\begin{array}{rrrrrl}
x_1 & - & x_2 & + & 2x_3 & + & x_4 & = & 3, \\
0x_1 & - & x_2 & + & 2x_3 & + & 3x_4 & = & -2, \\
0x_1 & + & 2x_2 & - & 4x_3 & - & 6x_4 & = & 4.
\end{array}
$$

Multiplying the second row by -1 yields

$$
\begin{array}{rrrrrl}
x_1 & - & x_2 & + & 2x_3 & + & x_4 & = & 3, \\
0x_1 & + & x_2 & - & 2x_3 & - & 3x_4 & = & 2, \\
0x_1 & + & 2x_2 & - & 4x_3 & - & 6x_4 & = & 4.
\end{array}
$$

Next, subtract two times row two from the third row to obtain

$$
\begin{array}{rrrrrl}
x_1 & - & x_2 & + & 2x_3 & + & x_4 & = & 3, \\
0x_1 & + & x_2 & - & 2x_3 & - & 3x_4 & = & 2, \\
0x_1 & + & 0x_2 & + & 0x_3 & + & 0x_4 & = & 0.
\end{array}
$$

Thus, the nonzero equations satisfied by the unknowns are

$$
\begin{array}{rrrrrl}
x_1 & - & x_2 & + & 2x_3 & + & x_4 & = & 3, \\
& & x_2 & - & 2x_3 & - & 3x_4 & = & 2.
\end{array}
\tag{4}
$$

To solve the system (4), we can choose two of the unknowns arbitrarily and then use the equations to solve for the remaining two unknowns. If we let $x_3 = \alpha$ and $x_4 = \beta$, where α and β are arbitrary, it then follows that

$$
\begin{aligned}
x_2 &= 2\alpha + 3\beta + 2, \\
x_1 &= 2\alpha + 3\beta + 2 - 2\alpha - \beta + 3 = 2\beta + 5.
\end{aligned}
$$

If we write the solution in vector notation, we have

$$
\mathbf{x} =
\begin{pmatrix}
2\beta + 5 \\
2\alpha + 3\beta + 2 \\
\alpha \\
\beta
\end{pmatrix}
= \alpha \underbrace{\begin{pmatrix} 0 \\ 2 \\ 1 \\ 0 \end{pmatrix}}_{\mathbf{v}_1}
+ \beta \underbrace{\begin{pmatrix} 2 \\ 3 \\ 0 \\ 1 \end{pmatrix}}_{\mathbf{v}_2}
+ \underbrace{\begin{pmatrix} 5 \\ 2 \\ 0 \\ 0 \end{pmatrix}}_{\mathbf{v}_p}.
\tag{5}
$$

The solution set of Eq. (3) is therefore

$$
\left\{ \mathbf{x} : \mathbf{x} = \alpha \mathbf{v}_1 + \beta \mathbf{v}_2 + \mathbf{v}_p, \ -\infty < \alpha, \beta < \infty \right\},
$$

where the vectors $\mathbf{v}_1$, $\mathbf{v}_2$, and $\mathbf{v}_p$ are as indicated in Eq. (5). This example illustrates both a systematic solution algorithm and the general structure of solution sets (provided any solutions actually exist) for linear systems of equations. If we denote the matrix of coefficients in Eq. (3) by

$$\mathbf{A} = \begin{pmatrix} 1 & -1 & 2 & 1 \\ 2 & -3 & 6 & 5 \\ -2 & 4 & -8 & -8 \end{pmatrix}$$

and the vector on the right-hand side of Eq. (3) by

$$\mathbf{b} = \begin{pmatrix} 3 \\ 4 \\ -2 \end{pmatrix},$$

it is easy to verify that $\mathbf{v}_1$ and $\mathbf{v}_2$ are both solutions of the homogeneous equation $\mathbf{Ax} = \mathbf{0}$, while $\mathbf{v}_p$ is a solution of the nonhomogeneous equation $\mathbf{Ax} = \mathbf{b}$. It is common to refer to the solution representation (5) as the **general solution** of Eq. (3). It is a two-parameter family of solutions (the parameters are α and β) representing all possible solutions of $\mathbf{Ax} = \mathbf{b}$.

Motivated by the above example, we now describe, in more detail, a simple but powerful algorithm for solving linear systems called **Gaussian elimination** or **row reduction**. It is a systematic formalization of the operations applied to the system in the example above. We first note that all of the work is performed on the coefficients and the right-hand sides of the equations. The computations are made more efficient by ignoring the names of the unknowns and dealing only with the **augmented matrix** obtained by appending the column m-vector $\mathbf{b}$ of right-hand sides to the $m \times n$ coefficient matrix $\mathbf{A}$. We will denote the augmented matrix by

$$(\mathbf{A}|\mathbf{b}). \tag{6}$$

For example, the augmented matrix for the system (3) is

$$\begin{pmatrix} 1 & -1 & 2 & 1 & \bigm| & 3 \\ 2 & -3 & 6 & 5 & \bigm| & 4 \\ -2 & 4 & -8 & -8 & \bigm| & -2 \end{pmatrix}.$$

In the augmented matrix, a vertical line is normally used to separate the right-hand side of the system from the matrix of coefficients $\mathbf{A}$. The augmented matrix is a rectangular array of numbers, that is, a matrix, to which the elementary row operations described below may be applied. Always keep in mind that $(\mathbf{A}|\mathbf{b})$ is really equivalent to $\mathbf{Ax} = \mathbf{b}$, which is, in turn, matrix notation for the set of equations (1). In the case of a homogeneous system of equations, that is, when $\mathbf{b} = \mathbf{0}$, it is not necessary to augment $\mathbf{A}$ with a column of zeros but sometimes it may be done for the sake of clarity. The permissible row operations used in Gaussian elimination are formally described by the following definition.

DEFINITION A.2.1

The **elementary row operations** used in Gaussian elimination are:

E1. Any row of Eq. (6) may be multiplied by a scalar and the result added to another row of Eq. (6).

E2. Any two rows of Eq. (6) may be interchanged.

E3. Any row of Eq. (6) may be multiplied by a nonzero scalar.

DEFINITION A.2.2

The **solution set** of the system $\mathbf{Ax} = \mathbf{b}$ is the set of all n-vectors $\mathbf{x}$ that satisfy $\mathbf{Ax} = \mathbf{b}$.

Part of the utility of Gaussian elimination is explained by the following theorem.

THEOREM A.2.3

If an elementary row operation is applied to Eq. (6), the solution set of the resulting system of equations is identical to the solution set of Eq. (6). That is, elementary row operations do not alter the solution set of a linear system of equations.

Proof

Suppose $\hat{\mathbf{x}} = (\hat{x}_1, \ldots, \hat{x}_n)^T$ is a solution of $\mathbf{Ax} = \mathbf{b}$ and α times row i is added to row j so that row j is replaced by

$$(a_{j1} + \alpha a_{i1})x_1 + \cdots + (a_{jn} + \alpha a_{in})x_n = b_j + \alpha b_i. \tag{7}$$

Since

$$a_{j1}\hat{x}_1 + \cdots + a_{jn}\hat{x}_n = b_j \tag{8}$$

and

$$a_{i1}\hat{x}_1 + \cdots + a_{in}\hat{x}_n = b_i, \tag{9}$$

it is obvious that

$$(a_{j1} + \alpha a_{i1})\hat{x}_1 + \cdots + (a_{jn} + \alpha a_{in})\hat{x}_n = b_j + \alpha b_i. \tag{10}$$

On the other hand, if $\hat{x}$ satisfies Eq. (7) and satisfies

$$a_{i1}x_1 + \cdots + a_{in}x_n = b_i,$$

then $\hat{x}$ also satisfies

$$a_{j1}x_1 + \cdots + a_{jn}x_n = b_j$$

so no new solutions are introduced by a row operation of type E1. We leave it to you to show that the operations of interchanging two rows or multiplying a row by a nonzero scalar will also not alter the solution set.

DEFINITION A.2.4

If a matrix $\mathbf{B}$ can be obtained from a matrix $\mathbf{A}$ by a finite sequence of elementary row operations, the matrices $\mathbf{A}$ and $\mathbf{B}$ are said to be **row equivalent**.

In view of Theorem A.2.3, we see that if $(\mathbf{A}|\mathbf{b})$ and $(\mathbf{B}|\mathbf{c})$ are row equivalent, then $\mathbf{Ax} = \mathbf{b}$ and $\mathbf{Bx} = \mathbf{c}$ have identical solution sets.

▶ **The Row Echelon Form of a Matrix.** Certain types of structures present in linear systems of equations may be usefully exploited in the solution process. For example, suppose the augmented matrix of a system of equations has the form

$$\begin{pmatrix} 1 & -2 & 3 & 7 \\ 0 & 1 & -1 & -2 \\ 0 & 0 & 1 & 1 \end{pmatrix}. \tag{11}$$

The system of equations represented by the augmented matrix (11) has an upper triangular structure,

$$\begin{aligned} x_1 \; - \; 2x_2 \; + \; 3x_3 \; &= \; 7, \\ x_2 \; - \; x_3 \; &= \; -2, \\ x_3 \; &= \; 1. \end{aligned} \tag{12}$$

In solving for the unknowns, it is most sensible to begin with the third equation that gives us $x_3 = 1$ directly. Substituting this value into the second equation, we obtain

$$x_2 = -2 + x_3 = -2 + 1 = -1.$$

Finally, substituting the values obtained for x_2 and x_3 into the first equation gives

$$x_1 = 7 + 2x_2 - 3x_3 = 7 - 2 - 3 = 2.$$

Thus, we obtain the solution of the system (12)

$$x = \begin{pmatrix} 2 \\ -1 \\ 1 \end{pmatrix}.$$

The technique used to solve for the unknowns in this example is known as the **method of backsubstitution**. It can be applied to systems of equations that have an upper triangular structure such as Eqs. (12) or, more generally, a row echelon structure that will be described below. This example suggests the following strategy for solving a system of equations that is not necessarily in the desired form:

1. Use elementary row operations, which do not alter the solution set, to reduce the system to upper triangular form or, more generally, row echelon form.
2. Solve the resulting system using backsubstitution.

We illustrate the algorithm on the following system of equations:

$$\begin{aligned} x_1 \; - \; 2x_2 \; + \; 3x_3 \; &= \; 7, \\ -x_1 \; + \; x_2 \; - \; 2x_3 \; &= \; -5, \\ 2x_1 \; - \; x_2 \; - \; x_3 \; &= \; 4. \end{aligned} \tag{13}$$

The augmented matrix for the system (13) is

$$\left(\begin{array}{ccc|c} 1 & -2 & 3 & 7 \\ -1 & 1 & -2 & -5 \\ 2 & -1 & -1 & 4 \end{array} \right). \tag{14}$$

We now perform row operations on the matrix (14) with the goal of introducing zeros in the lower left part of the matrix. Each step is described and the result recorded below.

a. Add the first row to the second row and add (-2) times the first row to the third row:

$$\left(\begin{array}{ccc|c} 1 & -2 & 3 & 7 \\ 0 & -1 & 1 & 2 \\ 0 & 3 & -7 & -10 \end{array} \right)$$

b. Multiply the second row by -1:

$$\begin{pmatrix} 1 & -2 & 3 & 7 \\ 0 & 1 & -1 & -2 \\ 0 & 3 & -7 & -10 \end{pmatrix}$$

c. Add (-3) times the second row to the third row:

$$\begin{pmatrix} 1 & -2 & 3 & 7 \\ 0 & 1 & -1 & -2 \\ 0 & 0 & -4 & -4 \end{pmatrix}$$

d. Divide the third row by -4:

$$\begin{pmatrix} 1 & -2 & 3 & 7 \\ 0 & 1 & -1 & -2 \\ 0 & 0 & 1 & 1 \end{pmatrix} \tag{15}$$

Thus we have arrived at the augmented matrix (11) for the upper triangular system (12) on which we illustrated the method of backsubstitution. This demonstrates the feasibility of the algorithm on a particular example.

We now describe the desired structure of the target matrix of the solution algorithm based on elementary row operations in more detail. Let $\mathbf{A}$ be an $m \times n$ matrix. Whether the matrix is augmented or not is irrelevant. We will say that a row of $\mathbf{A}$ is a **zero row** if each of its elements is zero. If a row has at least one nonzero entry, then that row is a **nonzero row**. The **leading entry** of a nonzero row is the first nonzero entry in the row, reading from left to right. A zero row has no leading (nonzero) entry. As an example, consider the matrix

$$\begin{pmatrix} 0 & 0 & 0 & 1 \\ 0 & -1 & 0 & 1 \\ 0 & 0 & 0 & 0 \\ 2 & 0 & 0 & 0 \end{pmatrix}.$$

Rows 1, 2, and 4 are nonzero rows, whereas row 3 is a zero row. The leading entries in rows 1, 2, and 4 are 1, -1, and 2, respectively.

**DEFINITION
A.2.5**

An $m \times n$ matrix $\mathbf{A}$ is said to be in **row echelon form** (or just echelon form) if it has the following properties:

    i. For some integer r, $0 \le r \le m$, the first r rows of $\mathbf{A}$ are nonzero.

    ii. Each leading entry lies to the right of the leading entry of the previous row.

    iii. Below each leading entry is a column of zeros.

    iv. After the first r rows, each row consists entirely of zeros, that is, zero rows are placed at the bottom of the matrix.

For example, the following matrices are all in row echelon form:

$$\begin{pmatrix} 2 & -2 \\ 0 & 3 \end{pmatrix} \quad \begin{pmatrix} 0 & 3 & 0 & 2 \\ 0 & 0 & 2 & -3 \\ 0 & 0 & 0 & 7 \end{pmatrix} \quad \begin{pmatrix} 1 & -2 & 4 \\ 0 & 0 & 5 \\ 0 & 0 & 0 \\ 0 & 0 & 0 \end{pmatrix}.$$

However, the matrix

$$\begin{pmatrix} 2 & 3 & -2 \\ 0 & 0 & 0 \\ 1 & 5 & -9 \end{pmatrix}$$

is not in row echelon form, since a nonzero row appears beneath a zero row and the leading entry in the third row does not lie to the right of the leading entry in the first row.

The transformation of a given matrix into a matrix that is in row echelon form by a sequence of elementary row operations is referred to as **row reduction** or **Gaussian elimination**.

EXAMPLE
2

Use Gaussian elimination to find the general solution of the linear system

$$\begin{pmatrix} 1 & -2 & 1 & 0 \\ 2 & 1 & -1 & 3 \\ 1 & 3 & -2 & 3 \end{pmatrix} \mathbf{x} = \begin{pmatrix} 1 \\ -1 \\ -2 \end{pmatrix} \tag{16}$$

by reducing the augmented matrix

$$\left(\begin{array}{cccc|c} 1 & -2 & 1 & 0 & 1 \\ 2 & 1 & -1 & 3 & -1 \\ 1 & 3 & -2 & 3 & -2 \end{array} \right) \tag{17}$$

to row echelon form.

Row echelon form is attained by the following row operations.

a. Subtract 2 times row one from row two:

$$\left(\begin{array}{cccc|c} 1 & -2 & 1 & 0 & 1 \\ 0 & 5 & -3 & 3 & -3 \\ 1 & 3 & -2 & 3 & -2 \end{array} \right)$$

b. Subtract row one from row three:

$$\left(\begin{array}{cccc|c} 1 & -2 & 1 & 0 & 1 \\ 0 & 5 & -3 & 3 & -3 \\ 0 & 5 & -3 & 3 & -3 \end{array} \right)$$

c. Subtract row two from row three:

$$\left(\begin{array}{cccc|c} 1 & -2 & 1 & 0 & 1 \\ 0 & 5 & -3 & 3 & -3 \\ 0 & 0 & 0 & 0 & 0 \end{array} \right) \tag{18}$$

System (18) is in row echelon form and represents the system

$$x_1 - 2x_2 + x_3 \qquad = 1,$$
$$5x_2 - 3x_3 + 3x_4 = -3. \tag{19}$$

Equation (19) consists of two equations for four unknowns. We are free to choose any two of the unknowns arbitrarily. We set $x_3 = \alpha$ and $x_4 = \beta$, where α and β are arbitrary scalars. The second equation in (19) yields $x_2 = -\frac{3}{5} + \frac{3}{5}\alpha - \frac{3}{5}\beta$. Backsubstituting this expression for x_2 into the first equation in (19) then yields $x_1 = -\frac{1}{5} + \frac{1}{5}\alpha - \frac{6}{5}\beta$. The solution $\mathbf{x}$ may be expressed in the form

$$\mathbf{x} = \alpha \underbrace{\begin{pmatrix} \frac{1}{5} \\ \frac{3}{5} \\ 1 \\ 0 \end{pmatrix}}_{\mathbf{v}_1} + \beta \underbrace{\begin{pmatrix} -\frac{6}{5} \\ -\frac{3}{5} \\ 0 \\ 1 \end{pmatrix}}_{\mathbf{v}_2} + \underbrace{\begin{pmatrix} -\frac{1}{5} \\ -\frac{3}{5} \\ 0 \\ 0 \end{pmatrix}}_{\mathbf{v}_p}. \tag{20}$$

If $\mathbf{A}$ is the matrix of coefficients and $\mathbf{b}$ is the vector of right-hand sides in Eq. (16), then $\mathbf{Av}_1 = \mathbf{0}$, $\mathbf{Av}_2 = \mathbf{0}$, and $\mathbf{Av}_p = \mathbf{b}$.

▶ Linearly Dependent and Linearly Independent Sets of Vectors.

DEFINITION A.2.6

A **linear combination** of vectors $\mathbf{v}_1, \ldots, \mathbf{v}_k$ is an expression of the form

$$c_1 \mathbf{v}_1 + \cdots + c_k \mathbf{v}_k,$$

where $c_1, \ldots, c_k$ are any scalars.

DEFINITION A.2.7

A set of k vectors $\mathbf{v}_1, \ldots, \mathbf{v}_k$ is said to be **linearly dependent** if there exists a set of (real or complex) numbers $c_1, \ldots, c_k$, at least one of which is nonzero, such that

$$c_1 \mathbf{v}_1 + \cdots + c_k \mathbf{v}_k = \mathbf{0}. \tag{21}$$

On the other hand, if the only set $c_1, \ldots, c_k$ for which Eq. (21) is satisfied is $c_1 = c_2 = \cdots = c_k = 0$, then the set $\mathbf{v}_1, \ldots, \mathbf{v}_k$ is said to be **linearly independent**.

Assuming that each $\mathbf{v}_j$, $j = 1, \ldots, k$ is an m-vector, using matrix notation, Eq. (21) can be written as

$$\begin{pmatrix} v_{11}c_1 & + & \cdots & + & v_{1k}c_k \\ \vdots & & & & \vdots \\ v_{m1}c_1 & + & \cdots & + & v_{mk}c_k \end{pmatrix} = \mathbf{Vc} = \mathbf{0}. \tag{22}$$

Therefore, to test whether the set $\mathbf{v}_1, \ldots, \mathbf{v}_k$ is linearly dependent or linearly independent, we find the general solution of $\mathbf{Vc} = \mathbf{0}$. If the only solution is the zero vector, $\mathbf{c} = \mathbf{0}$, then the set is linearly independent. If there are nonzero solutions, then the set is linearly dependent.

EXAMPLE 3

Determine whether the set of vectors

$$
\mathbf{v}_1 = \begin{pmatrix} 1 \\ 2 \\ -1 \end{pmatrix}, \qquad
\mathbf{v}_2 = \begin{pmatrix} 2 \\ 1 \\ 3 \end{pmatrix}, \qquad
\mathbf{v}_3 = \begin{pmatrix} -4 \\ 1 \\ -11 \end{pmatrix}
\tag{23}
$$

is linearly independent or linearly dependent.

To determine whether the set is linearly dependent, we seek constants c_1, c_2, c_3 such that $c_1\mathbf{v}_1 + c_2\mathbf{v}_2 + c_3\mathbf{v}_3 = \mathbf{Vc} = \mathbf{0}$. Equivalently, we look for the general solution of

$$
\begin{pmatrix} 1 & 2 & -4 \\ 2 & 1 & 1 \\ -1 & 3 & -11 \end{pmatrix}
\begin{pmatrix} c_1 \\ c_2 \\ c_3 \end{pmatrix} =
\begin{pmatrix} 0 \\ 0 \\ 0 \end{pmatrix}.
\tag{24}
$$

We use elementary row operations to reduce the corresponding augmented matrix

$$
\left(\begin{array}{ccc|c} 1 & 2 & -4 & 0 \\ 2 & 1 & 1 & 0 \\ -1 & 3 & -11 & 0 \end{array} \right)
$$

to row echelon form.

a. Add (-2) times the first row to the second row, and add the first row to the third row:

$$
\left(\begin{array}{ccc|c} 1 & 2 & -4 & 0 \\ 0 & -3 & 9 & 0 \\ 0 & 5 & -15 & 0 \end{array} \right)
$$

b. Divide the second row by (-3), then add (-5) times the second row to the third row:

$$
\left(\begin{array}{ccc|c} 1 & 2 & -4 & 0 \\ 0 & 1 & -3 & 0 \\ 0 & 0 & 0 & 0 \end{array} \right)
$$

The last matrix is equivalent to the system

$$
\begin{aligned}
c_1 + 2c_2 - 4c_3 &= 0 \\
c_2 - 3c_3 &= 0.
\end{aligned}
\tag{25}
$$

There are two equations and three unknowns so we are able to set $c_3 = \alpha$, where α is an arbitrary number. From the second equation in the system (25), we have $c_2 = 3c_3 = 3\alpha$, and from the first equation, we obtain $c_1 = 4c_3 - 2c_2 = -2\alpha$. The general solution of Eq. (24) is $\mathbf{c} = \alpha(-2, 3, 1)^T$. Choosing α to be any nonzero scalar, say, $\alpha = 1$, exhibits a nonzero solution $\mathbf{c} = (-2, 3, 1)^T$ for Eq. (24). Thus $-2\mathbf{v}_1 + 3\mathbf{v}_2 + \mathbf{v}_3 = \mathbf{0}$, so the given set of vectors is linearly dependent.

▶ The Rank of a Matrix. We now introduce a property of a matrix that is very useful in the study of linear algebraic systems.

**DEFINITION
A.2.8**

Let **B** be any matrix that is row equivalent to **A** and is in row echelon form. Then the rank of the matrix **A**, denoted by rank(**A**), is the number of nonzero rows in the matrix **B**.

For the rank of an $m \times n$ matrix **A** to be well defined, it is necessary that all matrices row equivalent to **A** have the same rank, and therefore rank refers to a property of the set of all matrices row equivalent to **A**. The property is indeed well defined, although we do not prove it here.

**EXAMPLE
4**

Find the rank of the matrix

$$\mathbf{A} = \begin{pmatrix} 2 & 0 & 1 & -1 \\ 1 & -3 & 4 & 2 \\ 3 & 3 & -2 & -4 \\ 4 & 0 & 2 & -2 \end{pmatrix}.$$

Using elementary row operations, **A** is reduced to the following matrix in row echelon form:

$$\begin{pmatrix} 1 & -3 & 4 & 2 \\ 0 & 6 & -7 & -5 \\ 0 & 0 & 0 & 0 \\ 0 & 0 & 0 & 0 \end{pmatrix} \tag{26}$$

and therefore rank(**A**) = 2.

**THEOREM
A.2.9**

If **B** is an $n \times n$ matrix in row echelon form, then the columns of **B** are linearly independent if and only if rank(**B**) = n.

Proof

Suppose that the columns of **B** are linearly independent. Then the only solution of $\mathbf{Bc} = \mathbf{0}$ is $\mathbf{c} = \mathbf{0}$. If rank(**B**) = $r < n$, then **B** must have the form

$$\mathbf{B} = \begin{pmatrix} * & * & & & & & * \\ 0 & * & * & & & & * \\ \vdots & & \ddots & & & & * \\ 0 & \cdots & 0 & * & * & \cdots & * \\ 0 & \cdots & 0 & 0 & 0 & \cdots & 0 \\ \vdots & & \vdots & \vdots & \vdots & & \vdots \\ 0 & \cdots & 0 & 0 & 0 & \cdots & 0 \end{pmatrix} \tag{27}$$

with one or more zero rows. Nonzero rows in the matrix (27) are those containing asterisks, although not all elements represented by an $*$ are necessarily nonzero. Therefore it is possible to choose $n - r \geq 1$ components of $\mathbf{c}$ arbitrarily with the remaining r components determined by the r nonzero equations using backsubstitution. Any nonzero choice for these coefficients shows that the columns of $\mathbf{B}$ are linearly dependent, contradicting the original assumption. Thus rank$(\mathbf{B}) = n$. Conversely, suppose that rank$(\mathbf{B}) = n$. Then the form of $\mathbf{B}$ must be

$$\mathbf{B} = \begin{pmatrix} \times & * & * & \cdots & * \\ 0 & \times & * & \cdots & * \\ \vdots & & & & \vdots \\ 0 & \cdots & & & \times \end{pmatrix},$$

where each entry denoted by $\times$ is nonzero, whereas entries denoted by $*$ may be zero or nonzero. Using backsubstitution, the upper triangular structure of $\mathbf{Bc} = \mathbf{0}$ implies that $\mathbf{c} = \mathbf{0}$ is the only solution, and therefore the columns of $\mathbf{B}$ are linearly independent.

THEOREM A.2.10

Let $\mathbf{A}$ be an $n \times n$ matrix. Then rank$(\mathbf{A}) = n$ if and only if the columns of $\mathbf{A}$ are linearly independent.

Proof

Assume that rank$(\mathbf{A}) = n$ and consider the equation $\mathbf{Ac} = \mathbf{0}$. Let $\mathbf{B}$ be a matrix that is row equivalent to $\mathbf{A}$ and in row echelon form. By Theorem A.2.9, rank$(\mathbf{B}) = n$ so the only solution of $\mathbf{Bc} = \mathbf{0}$ is $\mathbf{c} = \mathbf{0}$. It follows that the only solution of $\mathbf{Ac} = \mathbf{0}$ is $\mathbf{c} = \mathbf{0}$ since the solution sets of $\mathbf{Ac} = \mathbf{0}$ and $\mathbf{Bc} = \mathbf{0}$ are identical.

On the other hand, if the columns of $\mathbf{A}$ are linearly independent, then the only solution of $\mathbf{Ac} = \mathbf{0}$ is $\mathbf{c} = \mathbf{0}$. Therefore the only solution of $\mathbf{Bc} = \mathbf{0}$ is $\mathbf{c} = \mathbf{0}$ so rank$(\mathbf{B}) = n$ by Theorem A.2.9. It follows that rank$(\mathbf{A}) = $ rank$(\mathbf{B}) = n$ by the definition of rank.

▶ Solution Sets of $Ax = 0$.

THEOREM A.2.11

Assume that $\mathbf{A}$ is an $n \times n$ matrix.

(i) If rank$(\mathbf{A}) = n$, then the unique solution of $\mathbf{Ax} = \mathbf{0}$ is $\mathbf{x} = \mathbf{0}$.

(ii) If rank$(\mathbf{A}) = r < n$, then the general solution of $\mathbf{Ax} = \mathbf{0}$ has the form

$$\mathbf{x} = \alpha_1 \mathbf{v}_1 + \cdots + \alpha_{n-r} \mathbf{v}_{n-r},$$

where $\mathbf{v}_1, \ldots, \mathbf{v}_{n-r}$ is a set of linearly independent solutions of $\mathbf{Ax} = \mathbf{0}$.

Proof

If rank$(\mathbf{A}) = n$, the columns of $\mathbf{A}$ are linearly independent by Theorem A.2.10. Thus $\mathbf{x} = \mathbf{0}$ is the only solution of $\mathbf{Ax} = \mathbf{0}$.

If rank($\mathbf{A}$) = $r < n$, any matrix $\mathbf{B}$ that is row equivalent to $\mathbf{A}$ and in row echelon form will have r nonzero rows and $n - r \geq 1$ rows of zeros,

$$\mathbf{B} = \begin{pmatrix} * & * & & & & & & * \\ 0 & * & * & & & & & * \\ \vdots & & \ddots & & & & & \vdots \\ 0 & \cdots & 0 & * & * & \cdots & & * \\ 0 & \cdots & 0 & 0 & 0 & \cdots & & 0 \\ \vdots & & & \vdots & \vdots & \vdots & & \vdots \\ 0 & \cdots & 0 & 0 & 0 & \cdots & & 0 \end{pmatrix},$$

where the rows containing the symbols $*$ are nonzero but not all elements represented by $*$ are necessarily nonzero. Since the system $\mathbf{Bx} = \mathbf{0}$ consists of n unknowns and r equations, we are able to choose $n - r$ components of $\mathbf{x}$ to be arbitrary constants, say, $\alpha_1, \ldots, \alpha_{n-r}$. The remaining r components of $\mathbf{x}$ are obtained by backsubstitution using the r nonzero equations and each of these components is a linear combination of $\alpha_1, \ldots, \alpha_{n-r}$. Solutions of $\mathbf{Bx} = \mathbf{0}$ may then be represented in the form

$$\mathbf{x} = \alpha_1 \mathbf{v}_1 + \cdots + \alpha_{n-r} \mathbf{v}_{n-r}. \tag{28}$$

Since $\alpha_1, \ldots, \alpha_{n-r}$ are arbitrary, for each $j = 1, \ldots, n - r$, we set $\alpha_j = 1$ and $\alpha_k = 0$ if $k \neq j$ to show that $\mathbf{Av}_j = \mathbf{0}$ for each $j = 1, \ldots, n - r$. Furthermore the set $\mathbf{v}_1, \ldots, \mathbf{v}_{n-r}$ is linearly independent. If not, then at least one vector in the set, say, $\mathbf{v}_{n-r}$, may be expressed as a linear combination of the other vectors,

$$\mathbf{v}_{n-r} = c_1 \mathbf{v}_1 + \cdots + c_{n-r-1} \mathbf{v}_{n-r-1}. \tag{29}$$

Substituting the right-hand side of Eq. (29) for $\mathbf{v}_{n-r}$ in Eq. (28) shows that the general solution of $\mathbf{Bx} = \mathbf{0}$ would then contain only $n - r - 1$ arbitrary parameters

$$\mathbf{x} = (\alpha_1 + \alpha_{n-r} c_1)\mathbf{v}_1 + \cdots + (\alpha_{n-r-1} + \alpha_{n-r} c_{n-r-1})\mathbf{v}_{n-r-1} = \beta_1 \mathbf{v}_1 + \cdots + \beta_{n-r-1} \mathbf{v}_{n-r-1},$$

implying that rank($\mathbf{A}$) = $r + 1$ instead of r, as we had originally assumed. Since the solution sets of $\mathbf{Ax} = \mathbf{0}$ and $\mathbf{Bx} = \mathbf{0}$ are identical, the general solution of $\mathbf{Ax} = \mathbf{0}$ is also given by Eq. (28).

EXAMPLE 5

Find the general solution of $\mathbf{Ax} = \mathbf{0}$, where $\mathbf{A}$ is the matrix in Example 4.

The matrix (26) that is row equivalent to $\mathbf{A}$ and in row echelon form represents the set of equations

$$x_1 - 3x_2 + 4x_3 + 2x_4 = 0,$$
$$6x_2 - 7x_3 - 5x_4 = 0.$$

Since $n = 4$ and $r = $ rank($\mathbf{A}$) = 2, we can choose $n - r = 2$ components of $\mathbf{x}$ arbitrarily, say, $x_3 = \alpha_1$ and $x_4 = \alpha_2$. It follows from the second equation that $x_2 = \frac{7}{6}\alpha_1 + \frac{5}{6}\alpha_2$. From the first equation, we find that $x_1 = -\frac{1}{2}\alpha_1 + \frac{1}{2}\alpha_2$. Thus the general solution of $\mathbf{Ax} = \mathbf{0}$ is

$$\mathbf{x} = \begin{pmatrix} -\alpha_1/2 + \alpha_2/2 \\ 7\alpha_1/6 + 5\alpha_2/6 \\ \alpha_1 \\ \alpha_2 \end{pmatrix} = \alpha_1 \underbrace{\begin{pmatrix} -\frac{1}{2} \\ \frac{7}{6} \\ 1 \\ 0 \end{pmatrix}}_{\mathbf{v}_1} + \alpha_2 \underbrace{\begin{pmatrix} \frac{1}{2} \\ \frac{5}{6} \\ 0 \\ 1 \end{pmatrix}}_{\mathbf{v}_2}.$$

It is easy to check that $\mathbf{v}_1$ and $\mathbf{v}_2$ are linearly independent and satisfy $\mathbf{Ax} = \mathbf{0}$.

▶ Solution Sets of $\mathsf{Ax} = \mathsf{b}$. We introduce some convenient terminology for discussing the problem of solving the nonhomogeneous system $\mathbf{Ax} = \mathbf{b}$.

**DEFINITION
A.2.12**

The **span** of a set of vectors $\{\mathbf{v}_1, \dots, \mathbf{v}_k\}$ is the set of all possible linear combinations of the given vectors. The span is denoted by span$\{\mathbf{v}_1, \dots, \mathbf{v}_k\}$ and a vector $\mathbf{b}$ is in this span if

$$\mathbf{b} = x_1\mathbf{v}_1 + x_2\mathbf{v}_2 + \cdots + x_k\mathbf{v}_k$$

for some scalars $x_1, \dots, x_k$.

Let the columns of the $n \times n$ matrix $\mathbf{A}$ be denoted by the set of n-vectors $\mathbf{a}_1, \dots, \mathbf{a}_n$. We will denote the set of all possible linear combinations of the columns of $\mathbf{A}$ by $S_{\mathrm{col}}(\mathbf{A})$,

$$S_{\mathrm{col}}(\mathbf{A}) = \text{span}\,\{\mathbf{a}_1, \dots, \mathbf{a}_n\}.$$

In terms of the columns of $\mathbf{A}$, the equation $\mathbf{Ax} = \mathbf{b}$ may be expressed in the form

$$x_1\mathbf{a}_1 + x_2\mathbf{a}_2 + \cdots + x_n\mathbf{a}_n = \mathbf{b}.$$

Thus $\mathbf{Ax} = \mathbf{b}$ has a solution if and only if $\mathbf{b} \in S_{\mathrm{col}}(\mathbf{A})$.

Since the solution set of $\mathbf{Ax} = \mathbf{b}$ is unaltered by elementary row operations, we examine the solvability problem under the assumption that the augmented matrix $(\mathbf{A}|\mathbf{b})$ has been brought into row echelon form $(\mathbf{A}^*|\mathbf{b}^*)$, where $\mathbf{A}^* = (a_{ij}^*)$ and $\mathbf{b}^* = (b_i^*)$. If $(\mathbf{A}^*|\mathbf{b}^*)$ is in row echelon form, it is easy to ascertain whether $\mathbf{b}^* \in S_{\mathrm{col}}(\mathbf{A}^*)$ and therefore whether $\mathbf{Ax} = \mathbf{b}$ has any solutions. A few examples using 3×3 matrices help make this clear.

If rank$(\mathbf{A}) = 3$, then $(\mathbf{A}^*|\mathbf{b}^*)$ must have the form

$$\begin{pmatrix} a_{11}^* & a_{12}^* & a_{13}^* & \Big| & b_1^* \\ 0 & a_{22}^* & a_{23}^* & \Big| & b_2^* \\ 0 & 0 & a_{33}^* & \Big| & b_3^* \end{pmatrix}, \tag{30}$$

where each of the diagonal entries a_{11}^*, a_{22}^*, and a_{33}^* is nonzero. It is clear that, for any 3-vector $\mathbf{b}^*$, the equations represented by the augmented matrix (30),

$$a_{11}^*x_1 + a_{12}^*x_2 + a_{13}^*x_3 = b_1^*,$$
$$a_{22}^*x_2 + a_{23}^*x_3 = b_2^*,$$
$$a_{33}^*x_3 = b_3^*,$$

have a solution $\hat{\mathbf{x}} = (\hat{x}_1, \hat{x}_2, \hat{x}_3)$ that can be found by backsubstitution. Thus for any 3-vector $\mathbf{b}^*$, there exist scalars $\hat{x}_1$, $\hat{x}_2$, and $\hat{x}_3$ such that

$$\mathbf{A}^*\hat{\mathbf{x}} = \hat{x}_1 \begin{pmatrix} a_{11}^* \\ 0 \\ 0 \end{pmatrix} + \hat{x}_2 \begin{pmatrix} a_{12}^* \\ a_{22}^* \\ 0 \end{pmatrix} + \hat{x}_3 \begin{pmatrix} a_{13}^* \\ a_{23}^* \\ a_{33}^* \end{pmatrix} = \begin{pmatrix} b_1^* \\ b_2^* \\ b_3^* \end{pmatrix} = \mathbf{b}^*;$$

equivalently, $\mathbf{b}^* \in S_{\text{col}}(\mathbf{A}^*)$. Since $\mathbf{A}\mathbf{x} = \mathbf{b}$ is row equivalent to $\mathbf{A}^*\mathbf{x} = \mathbf{b}^*$, it follows that $\mathbf{A}\hat{\mathbf{x}} = \mathbf{b}$ and $\mathbf{b} \in S_{\text{col}}(\mathbf{A})$.

Next suppose that $\text{rank}(\mathbf{A}) = 2$, $\text{rank}(\mathbf{A}|\mathbf{b}) = 2$, and $(\mathbf{A}^*|\mathbf{b}^*)$ has the form

$$\begin{pmatrix} a_{11}^* & a_{12}^* & a_{13}^* & \bigg| & b_1^* \\ 0 & a_{22}^* & a_{23}^* & \bigg| & b_2^* \\ 0 & 0 & 0 & \bigg| & 0 \end{pmatrix}, \tag{31}$$

where $a_{11}^* \neq 0$ and at least one of a_{22}^* and a_{23}^* is nonzero. The equations represented by the augmented matrix (31) are

$$a_{11}^* x_1 + a_{12}^* x_2 + a_{13}^* x_3 = b_1^*,$$
$$a_{22}^* x_2 + a_{23}^* x_3 = b_2^*.$$

Thus if $\mathbf{b}^* = (b_1^*, b_2^*, 0)^T$, there exist scalars $\hat{x}_1$, $\hat{x}_2$, and $\hat{x}_3$ such that

$$\mathbf{A}^*\hat{\mathbf{x}} = \hat{x}_1 \begin{pmatrix} a_{11}^* \\ 0 \\ 0 \end{pmatrix} + \hat{x}_2 \begin{pmatrix} a_{12}^* \\ a_{22}^* \\ 0 \end{pmatrix} + \hat{x}_3 \begin{pmatrix} a_{13}^* \\ a_{23}^* \\ 0 \end{pmatrix} = \begin{pmatrix} b_1^* \\ b_2^* \\ 0 \end{pmatrix} = \mathbf{b}^*;$$

equivalently, $\mathbf{b}^* \in S_{\text{col}}(\mathbf{A}^*)$. By row equivalence of $\mathbf{A}^*\mathbf{x} = \mathbf{b}^*$ and $\mathbf{A}\mathbf{x} = \mathbf{b}$, we have $\mathbf{A}\hat{\mathbf{x}} = \mathbf{b}$. The same conclusion holds if $\text{rank}(\mathbf{A}) = \text{rank}(\mathbf{A}|\mathbf{b}) = 2$ and $(\mathbf{A}^*|\mathbf{b}^*)$ has the form

$$\begin{pmatrix} 0 & a_{12}^* & a_{13}^* & \bigg| & b_1^* \\ 0 & 0 & a_{23}^* & \bigg| & b_2^* \\ 0 & 0 & 0 & \bigg| & 0 \end{pmatrix}.$$

Now suppose $\text{rank}(\mathbf{A}) = 2$ but $\text{rank}(\mathbf{A}|\mathbf{b}) = 3$ and $(\mathbf{A}^*|\mathbf{b}^*)$ has the form

$$\begin{pmatrix} a_{11}^* & a_{12}^* & a_{13}^* & \bigg| & b_1^* \\ 0 & a_{22}^* & a_{23}^* & \bigg| & b_2^* \\ 0 & 0 & 0 & \bigg| & b_3^* \end{pmatrix},$$

which represents the equations

$$x_1 \begin{pmatrix} a_{11}^* \\ 0 \\ 0 \end{pmatrix} + x_2 \begin{pmatrix} a_{12}^* \\ a_{22}^* \\ 0 \end{pmatrix} + x_3 \begin{pmatrix} a_{13}^* \\ a_{23}^* \\ 0 \end{pmatrix} = \begin{pmatrix} b_1^* \\ b_2^* \\ b_3^* \end{pmatrix}. \tag{32}$$

Since the third component of each column of $\mathbf{A}^*$ is equal to 0, while the third component of $\mathbf{b}^*$ is nonzero, it is clear that there is no choice of x_1, x_2, and x_3 that will make statement (32) true. The same conclusion holds if $(\mathbf{A}^*|\mathbf{b}^*)$ has the form

$$\begin{pmatrix} 0 & a_{12}^* & a_{13}^* & \bigg| & b_1^* \\ 0 & 0 & a_{23}^* & \bigg| & b_2^* \\ 0 & 0 & 0 & \bigg| & b_3^* \end{pmatrix}.$$

Thus, whenever rank($\mathbf{A}$) = 2 and rank($\mathbf{A}|\mathbf{b}$) = 3, then $\mathbf{b}^* \notin S_{col}(\mathbf{A}^*)$ and therefore $\mathbf{Ax} = \mathbf{b}$ has no solution. We proceed to the general case.

THEOREM A.2.13

If $\mathbf{A}$ is an $n \times n$ matrix such that rank($\mathbf{A}$) = n and $\mathbf{b}$ is an n-vector, then $\mathbf{Ax} = \mathbf{b}$ has a unique solution.

Proof

If rank($\mathbf{A}$) = n, elementary row operations reduce ($\mathbf{A}|\mathbf{b}$) to row echelon form

$$\begin{pmatrix} \times & * & * & \cdots & * & \bigm| & * \\ 0 & \times & * & \cdots & * & \bigm| & * \\ \vdots & & & & \vdots & \bigm| & \vdots \\ 0 & \cdots & & & \times & \bigm| & * \end{pmatrix},$$

where each of the entries denoted by $\times$ is nonzero, otherwise rank($\mathbf{A}$) < n. It follows that a solution of this system, say, $\hat{\mathbf{x}}_1$, can be found by backsubstitution. This solution is also a solution of the row equivalent system $\mathbf{Ax} = \mathbf{b}$. If there were another solution of $\mathbf{Ax} = \mathbf{b}$, say, $\hat{\mathbf{x}}_2$, then $\mathbf{A}(\hat{\mathbf{x}}_1 - \hat{\mathbf{x}}_2) = \mathbf{A}\hat{\mathbf{x}}_1 - \mathbf{A}\hat{\mathbf{x}}_2 = \mathbf{b} - \mathbf{b} = \mathbf{0}$. Since rank($\mathbf{A}$) = n, the columns of $\mathbf{A}$ are linearly independent so it must be the case that $\hat{\mathbf{x}}_1 - \hat{\mathbf{x}}_2 = \mathbf{0}$.

THEOREM A.2.14

Let $\mathbf{A}$ be an $n \times n$ matrix and let $\mathbf{b}$ be an n-vector.

(i) If rank($\mathbf{A}$) = rank($\mathbf{A}|\mathbf{b}$) = $r < n$, then $\mathbf{Ax} = \mathbf{b}$ has a general solution of the form

$$\mathbf{x} = \alpha_1 \mathbf{v}_1 + \cdots + \alpha_{n-r} \mathbf{v}_{n-r} + \mathbf{v}_p,$$

where the n-vectors $\mathbf{v}_1, \ldots, \mathbf{v}_{n-r}$ are linearly independent solutions of $\mathbf{Ax} = \mathbf{0}$, and $\mathbf{v}_p$ satisfies $\mathbf{Av}_p = \mathbf{b}$.

(ii) If rank($\mathbf{A}$) = $r < n$ but rank($\mathbf{A}|\mathbf{b}$) = $r + 1$, then $\mathbf{Ax} = \mathbf{b}$ has no solution.

Proof

Suppose that rank($\mathbf{A}$) = $r < n$. If the augmented matrix ($\mathbf{A}|\mathbf{b}$) is reduced to row echelon form, there are only two possibilities. Either rank($\mathbf{A}|\mathbf{b}$) = r,

$$\begin{pmatrix} \times & \times & & & & \times & \bigm| & * \\ 0 & \times & \times & & & \times & \bigm| & * \\ \vdots & & \ddots & & & \times & \bigm| & \vdots \\ 0 & \cdots & 0 & \times & \times & \cdots & \times & \bigm| & * \\ 0 & \cdots & 0 & 0 & 0 & \cdots & 0 & \bigm| & 0 \\ \vdots & & \vdots & \vdots & \vdots & & \vdots & \bigm| & \vdots \\ 0 & \cdots & 0 & 0 & 0 & \cdots & 0 & \bigm| & 0 \end{pmatrix},$$

or rank$(\mathbf{A}|\mathbf{b}) = r + 1$,

$$\begin{pmatrix} \times & \times & & & & & \times & \bigg| & * \\ 0 & \times & \times & & & & \times & \bigg| & * \\ \vdots & & \ddots & & & & \times & \bigg| & \vdots \\ 0 & \cdots & 0 & \times & \times & \cdots & \times & \bigg| & * \\ 0 & \cdots & 0 & 0 & 0 & \cdots & 0 & \bigg| & \times \\ \vdots & & \vdots & \vdots & \vdots & & \vdots & \bigg| & \vdots \\ 0 & \cdots & 0 & 0 & 0 & \cdots & 0 & \bigg| & 0 \end{pmatrix}, \tag{33}$$

where in each row at least one of the elements denoted by the symbol $\times$ is nonzero. In the former case, it is possible by backsubstitution to find a $\mathbf{v}_p$ such that $\mathbf{A}\mathbf{v}_p = \mathbf{b}$. Now let $\hat{\mathbf{x}}$ be any vector that satisfies $\mathbf{A}\mathbf{x} = \mathbf{b}$. Then $\mathbf{A}(\hat{\mathbf{x}} - \mathbf{v}_p) = \mathbf{A}\hat{\mathbf{x}} - \mathbf{A}\mathbf{v}_p = \mathbf{b} - \mathbf{b} = \mathbf{0}$. From statement (ii) of Theorem A.2.11, it follows that

$$\hat{\mathbf{x}} - \mathbf{v}_p = \alpha_1\mathbf{v}_1 + \cdots + \alpha_{n-r}\mathbf{v}_{n-r},$$

where $\mathbf{v}_1, \ldots, \mathbf{v}_{n-r}$ are linearly independent and $\mathbf{A}\mathbf{v}_j = \mathbf{0}, j = 1, \ldots, n - r$. It follows that any solution $\hat{\mathbf{x}}$ of $\mathbf{A}\mathbf{x} = \mathbf{b}$ must be of the form

$$\hat{\mathbf{x}} = \alpha_1\mathbf{v}_1 + \cdots + \alpha_{n-r}\mathbf{v}_{n-r} + \mathbf{v}_p,$$

where set $\mathbf{v}_1, \ldots, \mathbf{v}_{n-r}$ are linearly independent solutions of $\mathbf{A}\mathbf{x} = \mathbf{0}$ and $\mathbf{v}_p$ is a particular solution of $\mathbf{A}\mathbf{x} = \mathbf{b}$, that is, $\mathbf{A}\mathbf{v}_p = \mathbf{b}$.

If the row echelon form of $(\mathbf{A}|\mathbf{b})$ is given by the augmented matrix (33), the last column cannot be represented as a linear combination of the columns to the left of the last column. Thus there exists no solution to $\mathbf{A}\mathbf{x} = \mathbf{b}$.

EXAMPLE 6

Consider the matrix $\mathbf{A}$ and the vectors $\mathbf{b}_1$ and $\mathbf{b}_2$ given by

$$\mathbf{A} = \begin{pmatrix} 1 & 2 & 1 \\ 0 & 1 & -1 \\ -1 & 1 & -4 \end{pmatrix}, \qquad \mathbf{b}_1 = \begin{pmatrix} 4 \\ 0 \\ -4 \end{pmatrix}, \qquad \mathbf{b}_2 = \begin{pmatrix} 1 \\ -1 \\ 1 \end{pmatrix}.$$

Find the general solution of $\mathbf{A}\mathbf{x} = \mathbf{b}_j, j = 1, 2$ or else determine that there is no solution.

We augment the matrix $\mathbf{A}$ with both $\mathbf{b}_1$ and $\mathbf{b}_2$,

$$\begin{pmatrix} 1 & 2 & 1 & \bigg| & 4 & 1 \\ 0 & 1 & -1 & \bigg| & 0 & -1 \\ -1 & 1 & -4 & \bigg| & -4 & 1 \end{pmatrix}.$$

Using elementary row operations, we reduce this matrix to one in row echelon form,

$$\begin{pmatrix} 1 & 2 & 1 & \bigg| & 4 & 1 \\ 0 & 1 & -1 & \bigg| & 0 & -1 \\ 0 & 0 & 0 & \bigg| & 0 & 5 \end{pmatrix}. \tag{34}$$

From Eq. (34), we see that $\text{rank}(\mathbf{A}|\mathbf{b}_1) = \text{rank}(\mathbf{A}) = 2$, whereas $\text{rank}(\mathbf{A}|\mathbf{b}_2) = 3 = \text{rank}(\mathbf{A}) + 1$. Thus $\mathbf{b}_2 \notin S_{\text{col}}(\mathbf{A})$ and there is no solution to $\mathbf{A}\mathbf{x} = \mathbf{b}_2$. However the reduced system of equations row equivalent to $\mathbf{A}\mathbf{x} = \mathbf{b}_1$ consists of $x_1 + 2x_2 + x_3 = 4$ and $x_2 - x_3 = 0$ that has the general solution

$$\mathbf{x} = \alpha \begin{pmatrix} -3 \\ 1 \\ 1 \end{pmatrix} + \begin{pmatrix} 4 \\ 0 \\ 0 \end{pmatrix}.$$

PROBLEMS

1. In each case, reduce $\mathbf{A}$ to row echelon form and determine $\text{rank}(\mathbf{A})$.

(a) $\mathbf{A} = \begin{pmatrix} 1 & -3 & 4 \\ 2 & -6 & 8 \end{pmatrix}$

(b) $\mathbf{A} = \begin{pmatrix} 0 & 0 & 1 \\ 0 & 1 & 0 \\ 1 & 0 & 0 \end{pmatrix}$

(c) $\mathbf{A} = \begin{pmatrix} 1 & 2 & 1 \\ 2 & -1 & 3 \\ 1 & -5 & 3 \end{pmatrix}$

(d) $\mathbf{A} = \begin{pmatrix} 1 & 0 & -2 & -1 \\ -1 & 1 & 1 & 0 \\ 1 & 1 & -3 & -2 \\ 2 & -3 & -1 & 1 \end{pmatrix}$

In each of Problems 2 through 5, if there exist solutions of the homogeneous system of linear equations other than $\mathbf{x} = \mathbf{0}$, express the general solution as a linear combination of linearly independent column vectors.

2.
$$\begin{aligned} x_1 \quad\quad - x_3 &= 0 \\ 3x_1 + x_2 + x_3 &= 0 \\ -x_1 + x_2 + 2x_3 &= 0 \end{aligned}$$

3.
$$\begin{aligned} x_1 + 2x_2 - x_3 &= 0 \\ 2x_1 + x_2 + x_3 &= 0 \\ x_1 - x_2 + 2x_3 &= 0 \end{aligned}$$

4.
$$\begin{aligned} 3x_1 \quad\quad - x_3 &= 0 \\ 2x_1 - x_2 + 2x_3 &= 0 \\ x_1 + x_2 - 3x_3 &= 0 \end{aligned}$$

5.
$$\begin{aligned} x_1 - 2x_2 \quad\quad + x_4 &= 0 \\ 2x_1 + x_2 + x_3 - x_4 &= 0 \\ x_1 + 2x_2 + x_3 - 2x_4 &= 0 \\ 3x_1 + 3x_2 + 2x_3 - 3x_4 &= 0 \end{aligned}$$

In each of Problems 6 through 9, find the general solution of the given set of equations, or else show that there is no solution.

6.
$$\begin{aligned} 2x_1 + x_2 + x_3 &= 2 \\ -x_1 \quad\quad + x_3 &= 1 \\ x_1 + x_2 + 2x_3 &= 3 \end{aligned}$$

7.
$$\begin{aligned} 2x_1 + x_2 + x_3 &= 0 \\ -x_1 \quad\quad + x_3 &= -1 \\ x_1 + x_2 + 2x_3 &= 1 \end{aligned}$$

8.
$$\begin{aligned} -2x_1 \quad\quad + x_3 &= 1 \\ 3x_2 - x_3 &= 2 \\ -x_1 + x_2 + 2x_3 &= 3 \end{aligned}$$

9.
$$\begin{aligned} x_1 - x_2 + x_3 + x_4 &= -1 \\ x_2 + x_3 + 3x_4 &= 2 \\ x_1 \quad\quad + 2x_3 + 4x_4 &= 1 \\ x_2 + x_3 + 3x_4 &= 2 \end{aligned}$$

In each of Problems 10 through 14, determine whether the members of the given set of vectors are linearly independent. If they are linearly dependent, find a linear relation among them.

10. $\mathbf{v}_1 = (1, 1, 0)^T$, $\quad \mathbf{v}_2 = (0, 1, 1)^T$, $\quad \mathbf{v}_3 = (1, 0, 1)^T$

11. $\mathbf{v}_1 = (2, 1, 0)^T$, $\quad \mathbf{v}_2 = (0, 1, 0)^T$, $\quad \mathbf{v}_3 = (-1, 2, 0)^T$

12. $\mathbf{v}_1 = (1, 2, 2, 3)^T$, $\quad \mathbf{v}_2 = (-1, 0, 3, 1)^T$, $\mathbf{v}_3 = (-2, -1, 1, 0)^T$, $\quad \mathbf{v}_4 = (-3, 0, -1, 3)^T$

13. $\mathbf{v}_1 = (1, 2, -1, 0)^T$, $\mathbf{v}_2 = (2, 3, 1, -1)^T$,
$\mathbf{v}_3 = (-1, 0, 2, 2)^T$, $\mathbf{v}_4 = (3, -1, 1, 3)^T$

14. $\mathbf{v}_1 = (1, 2, -2)^T$, $\mathbf{v}_2 = (3, 1, 0)^T$,
$\mathbf{v}_3 = (2, -1, 1)^T$, $\mathbf{v}_4 = (4, 3, -2)^T$

In each of Problems 15 through 17, determine whether $\mathbf{b}_j \in S_{\text{col}}(\mathbf{A}), j = 1, 2$. If so, express $\mathbf{b}_j$ as a linear combination of the columns of $\mathbf{A}$.

15. $\mathbf{A} = \begin{pmatrix} 2 & 2 & -2 \\ -1 & -3 & -1 \\ 1 & -1 & -2 \end{pmatrix}$, $\mathbf{b}_1 = \begin{pmatrix} 2 \\ 5 \\ 3 \end{pmatrix}$,

$\mathbf{b}_2 = \begin{pmatrix} 1 \\ 1 \\ 1 \end{pmatrix}$

16. $\mathbf{A} = \begin{pmatrix} 1 & -1 & -3 & 3 \\ 2 & 0 & -4 & 2 \\ -1 & 2 & 4 & -5 \end{pmatrix}$, $\mathbf{b}_1 = \begin{pmatrix} -1 \\ 1 \\ 0 \end{pmatrix}$,

$\mathbf{b}_2 = \begin{pmatrix} 2 \\ 0 \\ -4 \end{pmatrix}$

17. $\mathbf{A} = \begin{pmatrix} 1 & 2 & -1 & 2 \\ -1 & 1 & 2 & 2 \\ 0 & 1 & 3 & 4 \\ 2 & 0 & -1 & 1 \end{pmatrix}$, $\mathbf{b}_1 = \begin{pmatrix} 4 \\ -5 \\ -3 \\ 7 \end{pmatrix}$,

$\mathbf{b}_2 = \begin{pmatrix} 0 \\ 1 \\ 1 \\ 2 \end{pmatrix}$

A.3 Determinants and Inverses

With each square matrix $\mathbf{A}$, we associate a number called its **determinant**, denoted by det $\mathbf{A}$. When $\mathbf{A}$ is represented in array form

$$\mathbf{A} = \begin{pmatrix} a_{11} & a_{12} & \cdots & a_{1n} \\ a_{21} & a_{22} & \cdots & a_{2n} \\ \vdots & \vdots & & \vdots \\ a_{n1} & a_{n2} & \cdots & a_{nn} \end{pmatrix},$$

we will denote its determinant by enclosing the array between vertical bars,

$$\det \mathbf{A} = |\mathbf{A}| = \begin{vmatrix} a_{11} & a_{12} & \cdots & a_{1n} \\ a_{21} & a_{22} & \cdots & a_{2n} \\ \vdots & \vdots & & \vdots \\ a_{n1} & a_{n2} & \cdots & a_{nn} \end{vmatrix}.$$

Definitions of determinants of 1×1 and 2×2 matrices are

$$\det(a_{11}) = a_{11}, \qquad \begin{vmatrix} a_{11} & a_{12} \\ a_{21} & a_{22} \end{vmatrix} = a_{11}a_{22} - a_{12}a_{21},$$

respectively. If $\mathbf{A}$ is 3×3, then det $\mathbf{A}$ is defined by

$$|\mathbf{A}| = a_{11}a_{22}a_{33} - a_{11}a_{23}a_{32} - a_{12}a_{21}a_{33} \\ + a_{12}a_{23}a_{31} + a_{13}a_{21}a_{32} - a_{13}a_{22}a_{31}. \tag{1}$$

It is easy to check that the sum on the right-hand side of Eq. (1) is equal to the sum

$$|\mathbf{A}| = (-1)^{(1+1)} a_{11} \underbrace{\begin{vmatrix} a_{22} & a_{23} \\ a_{32} & a_{33} \end{vmatrix}}_{M_{11}} + (-1)^{(1+2)} a_{12} \underbrace{\begin{vmatrix} a_{21} & a_{23} \\ a_{31} & a_{33} \end{vmatrix}}_{M_{12}} + (-1)^{(1+3)} a_{13} \underbrace{\begin{vmatrix} a_{21} & a_{22} \\ a_{31} & a_{32} \end{vmatrix}}_{M_{13}},$$

which can, in turn, be written as

$$|\mathbf{A}| = \sum_{j=1}^{3} (-1)^{1+j} a_{1j} M_{1j}, \tag{2}$$

where M_{1j} is the determinant of the 2×2 submatrix obtained by deleting the 1st row and jth column of $\mathbf{A}$, that is, the row and column in which a_{1j} resides. The same value for the determinant is obtained by other expansions similar to the pattern exhibited in Eq. (2). For example, the sum

$$\sum_{i=1}^{3} (-1)^{i+2} a_{i2} M_{i2}$$

$$= (-1)^{(1+2)} a_{12} \underbrace{\begin{vmatrix} a_{21} & a_{23} \\ a_{31} & a_{33} \end{vmatrix}}_{M_{12}} + (-1)^{(2+2)} a_{22} \underbrace{\begin{vmatrix} a_{11} & a_{13} \\ a_{31} & a_{33} \end{vmatrix}}_{M_{22}} + (-1)^{(3+2)} a_{32} \underbrace{\begin{vmatrix} a_{11} & a_{13} \\ a_{21} & a_{23} \end{vmatrix}}_{M_{32}},$$

where M_{i2} is the determinant of the 2×2 submatrix obtained by deleting the ith row and 2nd column of $\mathbf{A}$, is also easily shown to yield the sum on the right-hand side of Eq. (1).

Determinants of square matrices of higher order are defined recursively by following the pattern illustrated above for 3×3 matrices. If $\mathbf{A}$ is $n \times n$, denote by M_{ij} the determinant of the $(n-1) \times (n-1)$ submatrix of $\mathbf{A}$ obtained by deleting the ith row and jth column from $\mathbf{A}$. M_{ij} is called the **minor** of a_{ij}. The **cofactor** of a_{ij} is the quantity $(-1)^{i+j} M_{ij}$. Then the **cofactor expansion** of $|\mathbf{A}|$ along the ith row is defined by

$$|\mathbf{A}| = \sum_{j=1}^{n} (-1)^{i+j} a_{ij} M_{ij}. \tag{3}$$

Similarly, the **cofactor expansion** of $|\mathbf{A}|$ along the jth column is defined by

$$|\mathbf{A}| = \sum_{i=1}^{n} (-1)^{i+j} a_{ij} M_{ij}. \tag{4}$$

The determinant of each $(n-1) \times (n-1)$ minor in Eqs. (3) and (4) is, in turn, defined by a cofactor expansion along either a row or column. For each minor, this will require a sum of $(n-1)$ minors of dimension $(n-2) \times (n-2)$. This reduction process continues until one gets down to a sum of $n!/2$ determinants of 2×2 matrices. Though we do not prove it here, the value of the determinant obtained by a cofactor expansion is independent of the row or column along which the expansions are performed.

EXAMPLE 1

Evaluate $|\mathbf{A}|$ if

$$\mathbf{A} = \begin{pmatrix} 1 & -1 & 2 & 4 \\ -1 & 3 & -2 & 1 \\ 0 & 2 & 1 & 0 \\ -3 & 1 & 1 & -1 \end{pmatrix}.$$

To evaluate $|\mathbf{A}|$, we can expand by cofactors along any row or column. It is prudent to expand along the third row because this row has two zero elements, thereby reducing the labor. Thus

$$|\mathbf{A}| = \sum_{j=1}^{4} a_{3j} M_{3j}$$

$$= -2 \begin{vmatrix} 1 & 2 & 4 \\ -1 & -2 & 1 \\ -3 & 1 & -1 \end{vmatrix} + 1 \begin{vmatrix} 1 & -1 & 4 \\ -1 & 3 & 1 \\ -3 & 1 & -1 \end{vmatrix}.$$

These 3×3 determinants can be evaluated directly from the definition of a 3×3 determinant, or we can expand each (say, along the first row) to get

$$|\mathbf{A}| = -2 \left\{ \begin{vmatrix} -2 & 1 \\ 1 & -1 \end{vmatrix} - 2 \begin{vmatrix} -1 & 1 \\ -3 & -1 \end{vmatrix} + 4 \begin{vmatrix} -1 & -2 \\ -3 & 1 \end{vmatrix} \right\}$$

$$+ 1 \left\{ \begin{vmatrix} 3 & 1 \\ 1 & -1 \end{vmatrix} + \begin{vmatrix} -1 & 1 \\ -3 & -1 \end{vmatrix} + 4 \begin{vmatrix} -1 & 3 \\ -3 & 1 \end{vmatrix} \right\}$$

$$= -2 \{1 - 8 - 28\} + \{-4 + 4 + 32\} = 102.$$

Given a square matrix, most commonly used software packages and many calculators provide functions for evaluating determinants. The following theorem lists a number of important properties of determinants that can be used to simplify the computations required for their evaluation.

THEOREM A.3.1

Properties of Determinants. Let $\mathbf{A}$ and $\mathbf{B}$ be $n \times n$ matrices. Then

1. If $\mathbf{B}$ is obtained from $\mathbf{A}$ by adding a constant multiple of one row (or column) to another row (or column), then $|\mathbf{B}| = |\mathbf{A}|$.
2. If $\mathbf{B}$ is obtained from $\mathbf{A}$ by interchanging two rows or two columns, then $|\mathbf{B}| = -|\mathbf{A}|$.
3. If $\mathbf{B}$ is obtained from $\mathbf{A}$ by multiplying any row or any column by a scalar α, then $|\mathbf{B}| = \alpha|\mathbf{A}|$.
4. If $\mathbf{A}$ has a zero row or zero column, then $|\mathbf{A}| = 0$.
5. If $\mathbf{A}$ has two identical rows (or two identical columns), then $|\mathbf{A}| = 0$.
6. If one row (or column) of $\mathbf{A}$ is a constant multiple of another row (or column), then $|\mathbf{A}| = 0$.

In the previous section, we found that triangular matrices were very important. The determinant of a triangular matrix is also very easy to evaluate.

THEOREM
A.3.2

The determinant of a triangular matrix is the product of its diagonal elements.

Proof

Suppose that $\mathbf{A}$ is upper triangular. By repeatedly computing cofactor expansions along the first column, we find that

$$\begin{vmatrix} a_{11} & a_{12} & \cdots & a_{1n} \\ 0 & a_{22} & \cdots & a_{2n} \\ 0 & 0 & \cdots & a_{3n} \\ \vdots & \vdots & & \vdots \\ 0 & 0 & \cdots & a_{nn} \end{vmatrix} = a_{11} \begin{vmatrix} a_{22} & \cdots & a_{2n} \\ \vdots & \ddots & \vdots \\ 0 & \cdots & a_{nn} \end{vmatrix}$$

$$= a_{11}a_{22} \begin{vmatrix} a_{33} & \cdots & a_{3n} \\ \vdots & \ddots & \vdots \\ 0 & \cdots & a_{nn} \end{vmatrix}$$

$$= \cdots = a_{11}a_{22} \cdots a_{nn}.$$

In a similar way, it can be shown that the determinant of a lower triangular matrix is also the product of its diagonal elements.

A matrix can often be reduced to upper triangular form by a sequence of elementary row operations of type E1, that is, adding scalar multiples of one row to another row. By property 1 of Theorem A.3.1, the determinant of the resultant triangular matrix will be equal to the determinant of the original matrix. Thus, using elementary row operations of type E1 combined with property 1 of Theorem A.3.1, provides a computationally efficient method for computing the determinant of a matrix.

EXAMPLE
2

In the matrix $\mathbf{A}$ of Example 1, if we add row one to row two and then add 3 times row one to row four, we get

$$\det \mathbf{A} = \begin{vmatrix} 1 & -1 & 2 & 4 \\ -1 & 3 & -2 & 1 \\ 0 & 2 & 1 & 0 \\ -3 & 1 & 1 & -1 \end{vmatrix} = \begin{vmatrix} 1 & -1 & 2 & 4 \\ 0 & 2 & 0 & 5 \\ 0 & 2 & 1 & 0 \\ 0 & -2 & 7 & 11 \end{vmatrix},$$

where we have used property 1 of Theorem A.3.1. Then, subtracting row two from row three followed by adding row two to row four gives us

$$\det \mathbf{A} = \begin{vmatrix} 1 & -1 & 2 & 4 \\ 0 & 2 & 0 & 5 \\ 0 & 0 & 1 & -5 \\ 0 & 0 & 7 & 16 \end{vmatrix},$$

where we have again used property 1 of Theorem A.3.1. By subtracting 7 times row three from row four in the last result, we reduce the problem to evaluating the determinant of an upper triangular matrix,

$$\det \mathbf{A} = \begin{vmatrix} 1 & -1 & 2 & 4 \\ 0 & 2 & 0 & 5 \\ 0 & 0 & 1 & -5 \\ 0 & 0 & 0 & 51 \end{vmatrix} = 1 \cdot 2 \cdot 1 \cdot 51 = 102.$$

THEOREM A.3.3

$|\mathbf{A}^T| = |\mathbf{A}|$.

Proof

The cofactor expansion along the ith column of $\mathbf{A}^T$ is equal to the cofactor expansion along the ith row of $\mathbf{A}$.

We state without proof the following important and frequently used result.

THEOREM A.3.4

$|\mathbf{AB}| = |\mathbf{A}||\mathbf{B}|$.

The next theorem shows that determinants of two row equivalent $n \times n$ matrices are either both zero or both nonzero.

THEOREM A.3.5

Let $\mathbf{A}$ be an $n \times n$ matrix and let $\mathbf{B}$ be row equivalent to $\mathbf{A}$. Then $|\mathbf{A}| = 0$ if and only if $|\mathbf{B}| = 0$. In particular, if $\mathbf{A}_E$ is a matrix in row echelon form that is row equivalent to $\mathbf{A}$, then $|\mathbf{A}| = 0$ if and only if $|\mathbf{A}_E| = 0$.

Proof

We examine the effect of each type of elementary row operation on the value of a determinant.

If $\mathbf{B}$ is obtained from $\mathbf{A}$ by adding a scalar multiple of one row to another row, then $|\mathbf{B}| = |\mathbf{A}|$ by property 1 of Theorem A.3.1.

If $\mathbf{B}$ is obtained from $\mathbf{A}$ by interchanging two rows of $\mathbf{A}$, then $|\mathbf{B}| = -|\mathbf{A}|$ by property 2 of Theorem A.3.1.

If $\mathbf{B}$ is obtained from $\mathbf{A}$ by multiplying a row of $\mathbf{A}$ by a nonzero scalar α, then $|\mathbf{B}| = \alpha|\mathbf{A}|$ by property 3 of Theorem A.3.1.

In general, if $\mathbf{B}$ is obtained from $\mathbf{A}$ by a finite sequence of elementary row operations, then

$$|\mathbf{B}| = \alpha_k \alpha_{k-1} \cdots \alpha_1 |\mathbf{A}|.$$

where each α_j, $j = 1, \ldots, k$ is a nonzero constant. Thus, if $|\mathbf{B}| \neq 0$, then $|\mathbf{A}| \neq 0$. On the other hand, if $|\mathbf{B}| = 0$, then $|\mathbf{A}| = 0$. As a special case, this result holds if $\mathbf{B} = \mathbf{A}_E$.

The determinant can be used to test whether the set of column vectors of a square matrix is linearly independent or linearly dependent.

**THEOREM
A.3.6**

The columns of $\mathbf{A}$ are linearly independent if and only if $|\mathbf{A}| \neq 0$.

Proof

Suppose that the columns of $\mathbf{A}$ are linearly independent. Then the only solution of $\mathbf{Ac} = \mathbf{0}$ is $\mathbf{c} = \mathbf{0}$. Let $\mathbf{B}$ be row equivalent to $\mathbf{A}$ and in row echelon form. Since the solution sets of $\mathbf{Bc} = \mathbf{0}$ and $\mathbf{Ac} = \mathbf{0}$ are identical, the only solution of $\mathbf{Bc} = \mathbf{0}$ is $\mathbf{c} = \mathbf{0}$. This means that the columns of $\mathbf{B}$ are linearly independent. By Theorem A.2.9, rank$(\mathbf{B}) = n$. Since $\mathbf{B}$ is triangular, the diagonal entries must all be nonzero. Consequently, $|\mathbf{B}| \neq 0$ and the row equivalence of $\mathbf{A}$ and $\mathbf{B}$ implies that $|\mathbf{A}| \neq 0$ by Theorem A.3.5.

Conversely, if $|\mathbf{A}| \neq 0$ and $\mathbf{B}$ is row equivalent to $\mathbf{A}$ and in row echelon form, then $|\mathbf{B}| \neq 0$ by Theorem A.3.5. Thus the diagonal entries of $\mathbf{B}$ must all be nonzero, which implies that rank$(\mathbf{B}) = n$. By Theorem A.2.9, the columns $\mathbf{B}$ are linearly independent and the only solution of $\mathbf{Bc} = \mathbf{0}$ is $\mathbf{c} = \mathbf{0}$. Thus the only solution of $\mathbf{Ac} = \mathbf{0}$ is $\mathbf{c} = \mathbf{0}$. Consequently, the columns of $\mathbf{A}$ are linearly independent.

**EXAMPLE
3**

Use the determinant to show that the three vectors in Example 3 of Section A.2,

$$\mathbf{v}_1 = \begin{pmatrix} 1 \\ 2 \\ -1 \end{pmatrix}, \qquad \mathbf{v}_2 = \begin{pmatrix} 2 \\ 1 \\ 3 \end{pmatrix}, \qquad \mathbf{v}_3 = \begin{pmatrix} -4 \\ 1 \\ 11 \end{pmatrix},$$

are linearly dependent.

We juxtapose the vectors to form the 3×3 matrix $\mathbf{A} = [\mathbf{v}_1, \mathbf{v}_2, \mathbf{v}_3]$. Calculation of the determinant is then simplified by adding appropriate scalar multiples of the first row to the second and third rows followed by a cofactor expansion along the first column,

$$|\mathbf{A}| = \begin{vmatrix} 1 & 2 & -4 \\ 2 & 1 & 1 \\ -1 & 3 & -11 \end{vmatrix} = \begin{vmatrix} 1 & 2 & -4 \\ 0 & -3 & 9 \\ 0 & 5 & -15 \end{vmatrix} = \begin{vmatrix} -3 & 9 \\ 5 & -15 \end{vmatrix} = 45 - 45 = 0.$$

Since $|\mathbf{A}| = 0$, the set $\{\mathbf{v}_1, \mathbf{v}_2, \mathbf{v}_3\}$ is linearly dependent.

▶ **The Determinant Test for Solvability of Ax = b.**

**THEOREM
A.3.7**

Let $\mathbf{A}$ be an $n \times n$ matrix and let $\mathbf{b}$ be an n-vector.

(i) If $|\mathbf{A}| \neq 0$, then rank$(\mathbf{A}) = n$ and $\mathbf{Ax} = \mathbf{b}$ has a unique solution. In particular, the unique solution of $\mathbf{Ax} = \mathbf{0}$ is $\mathbf{x} = \mathbf{0}$.

(ii) If $|\mathbf{A}| = 0$, then $r = $ rank$(\mathbf{A}) < n$. In this case, either

   (a) rank$(\mathbf{A}|\mathbf{b}) = r + 1$ and $\mathbf{Ax} = \mathbf{b}$ has no solution, or

   (b) rank$(\mathbf{A}|\mathbf{b}) = r$ and the general solution of $\mathbf{Ax} = \mathbf{b}$ is of the form

$$\mathbf{x} = \alpha_1 \mathbf{v}_1 + \cdots + \alpha_{n-r} \mathbf{v}_{n-r} + \mathbf{v}_p, \tag{5}$$

where $\mathbf{v}_1, \ldots, \mathbf{v}_{n-r}$ are linearly independent solutions of $\mathbf{Ax} = \mathbf{0}$, and $\mathbf{Av}_p = \mathbf{b}$.

In particular, if $|\mathbf{A}| = 0$, then the general solution of the homogeneous equation $\mathbf{Ax} = \mathbf{0}$ is of the form

$$\mathbf{x} = \alpha_1 \mathbf{v}_1 + \cdots + \alpha_{n-r} \mathbf{v}_{n-r},$$

where $\mathbf{v}_1, \ldots, \mathbf{v}_{n-r}$ are linearly independent solutions of $\mathbf{Ax} = \mathbf{0}$.

Proof

Assume that $|\mathbf{A}| \neq 0$. If $\mathbf{A}_E$ is row equivalent to $\mathbf{A}$ and in row echelon form, then $|\mathbf{A}_E| \neq 0$ by Theorem A.3.5. Since $\mathbf{A}_E$ is upper triangular and $|\mathbf{A}_E| \neq 0$, all of the diagonal entries are nonzero. Thus $\mathrm{rank}(\mathbf{A}_E) = \mathrm{rank}(\mathbf{A}) = n$. It follows from Theorem A.2.13 that $\mathbf{Ax} = \mathbf{b}$ has a unique solution. If $\mathbf{b} = \mathbf{0}$, that unique solution is obviously $\mathbf{x} = \mathbf{0}$. Thus statement (i) of the theorem is proved.

Now suppose that $|\mathbf{A}| = 0$. Then $|\mathbf{A}_E| = 0$ by Theorem A.3.5. Since $\mathbf{A}_E$ is upper triangular and $|\mathbf{A}_E| = 0$, at least one of the diagonal entries is equal to zero. Consequently, $r = \mathrm{rank}(\mathbf{A}) < n$. If rank $(\mathbf{A}|\mathbf{b}) = r + 1$, then $\mathbf{Ax} = \mathbf{b}$ has no solution by part (ii) of Theorem A.2.14. On the other hand, if rank $(\mathbf{A}|\mathbf{b}) = r$, then by part (i) of Theorem A.2.14 the general solution of $\mathbf{Ax} = \mathbf{b}$ has the form

$$\mathbf{x} = \alpha_1 \mathbf{v}_1 + \cdots + \alpha_{n-r} \mathbf{v}_{n-r} + \mathbf{v}_p, \tag{6}$$

where $\mathbf{v}_1, \ldots, \mathbf{v}_{n-r}$ are linearly independent solutions of $\mathbf{Ax} = \mathbf{0}$, and $\mathbf{Av}_p = \mathbf{b}$. Thus statement (ii) of the theorem is proved.

In the special case of the homogeneous equation $\mathbf{Ax} = \mathbf{0}$, if $|\mathbf{A}| = 0$, then by part (ii) of Theorem A.2.11 the general solution of $\mathbf{Ax} = \mathbf{0}$ has the form

$$\mathbf{x} = \alpha_1 \mathbf{v}_1 + \cdots + \alpha_{n-r} \mathbf{v}_{n-r}, \tag{7}$$

where $\mathbf{v}_1, \ldots, \mathbf{v}_{n-r}$ are linearly independent solutions of $\mathbf{Ax} = \mathbf{0}$.

▶ Matrix Inverses.

THEOREM A.3.8

An $n \times n$ matrix is nonsingular if and only if $|\mathbf{A}| \neq 0$.

Proof

Assume that $|\mathbf{A}| \neq 0$. We seek a matrix $\mathbf{B} = [\mathbf{b}_1, \ldots, \mathbf{b}_n]$ such that

$$\mathbf{AB} = [\mathbf{Ab}_1, \ldots, \mathbf{Ab}_n] = \mathbf{I}_n = [\mathbf{e}_1, \ldots, \mathbf{e}_n],$$

where

$$\mathbf{e}_1 = (1, 0, 0, \ldots, 0)^T, \quad \mathbf{e}_2 = (0, 1, 0, \ldots, 0)^T, \quad \ldots, \quad \mathbf{e}_n = (0, \ldots, 0, 1)^T.$$

Since $|\mathbf{A}| \neq 0$, Theorem A.3.7 implies that there is a unique solution $\mathbf{b}_k$ to each of the systems $\mathbf{Ab}_k = \mathbf{e}_k, k = 1, \ldots, n$. Thus there is a unique matrix $\mathbf{B}$ such that $\mathbf{AB} = \mathbf{I}_n$. Since $|\mathbf{A}^T| = |\mathbf{A}| \neq 0$, the same argument produces a unique matrix $\mathbf{C}^T$ such that $\mathbf{A}^T\mathbf{C}^T = \mathbf{I}_n$, or equivalently, $\mathbf{CA} = \mathbf{I}$. It follows that $\mathbf{C} = \mathbf{CI}_n = \mathbf{C}(\mathbf{AB}) = (\mathbf{CA})\mathbf{B} = \mathbf{I}_n\mathbf{B} = \mathbf{B}$.

Conversely, if $\mathbf{A}$ is nonsingular, there is a matrix $\mathbf{B}$ such that $\mathbf{AB} = \mathbf{I}_n$. By Theorem A.3.4, $|\mathbf{AB}| = |\mathbf{A}||\mathbf{B}| = 1$, so $|\mathbf{A}| \neq 0$.

Thus, whenever $|\mathbf{A}| \neq 0$, the inverse of $\mathbf{A}$ can be found by reducing the augmented matrix $(\mathbf{A}|\mathbf{I}_n)$ to $(\mathbf{I}_n|\mathbf{B})$ using elementary row operations. Since the jth column of $\mathbf{B}$ is the solution of $\mathbf{Ab}_j = \mathbf{e}_j$, $j = 1, \ldots, n$, it follows that $\mathbf{A}^{-1} = \mathbf{B}$.

**EXAMPLE
4**

Find the inverse of

$$\mathbf{A} = \begin{pmatrix} 1 & -1 & -1 \\ 3 & -1 & 2 \\ 2 & 2 & 3 \end{pmatrix}.$$

We begin by forming the augmented matrix $(\mathbf{A}|\mathbf{I}_3)$:

$$(\mathbf{A}|\mathbf{I}_3) = \begin{pmatrix} 1 & -1 & -1 & | & 1 & 0 & 0 \\ 3 & -1 & 2 & | & 0 & 1 & 0 \\ 2 & 2 & 3 & | & 0 & 0 & 1 \end{pmatrix}.$$

The matrix $\mathbf{A}$ can be transformed into $\mathbf{I}_3$ by the following sequence of elementary row operations, and at the same time, $\mathbf{I}_3$ is transformed into $\mathbf{A}^{-1}$.

a. Obtain zeros in the off-diagonal positions in the first column by adding (-3) times the first row to the second row and adding (-2) times the first row to the third row:

$$\begin{pmatrix} 1 & -1 & -1 & | & 1 & 0 & 0 \\ 0 & 2 & 5 & | & -3 & 1 & 0 \\ 0 & 4 & 5 & | & -2 & 0 & 1 \end{pmatrix}.$$

b. Obtain a 1 in the diagonal position in the second column by multiplying the second row by $\frac{1}{2}$:

$$\begin{pmatrix} 1 & -1 & -1 & | & 1 & 0 & 0 \\ 0 & 1 & \frac{5}{2} & | & -\frac{3}{2} & \frac{1}{2} & 0 \\ 0 & 4 & 5 & | & -2 & 0 & 1 \end{pmatrix}.$$

c. Obtain zeros in the off-diagonal positions in the second column by adding the second row to the first row and adding (-4) times the second row to the third row:

$$\begin{pmatrix} 1 & 0 & \frac{3}{2} & | & -\frac{1}{2} & \frac{1}{2} & 0 \\ 0 & 1 & \frac{5}{2} & | & -\frac{3}{2} & \frac{1}{2} & 0 \\ 0 & 0 & -5 & | & 4 & -2 & 1 \end{pmatrix}.$$

d. Obtain a 1 in the diagonal position of the third column by multiplying the third row by $-\frac{1}{5}$:

$$\begin{pmatrix} 1 & 0 & \frac{3}{2} & | & -\frac{1}{2} & \frac{1}{2} & 0 \\ 0 & 1 & \frac{5}{2} & | & -\frac{3}{2} & \frac{1}{2} & 0 \\ 0 & 0 & 1 & | & -\frac{4}{5} & \frac{2}{5} & -\frac{1}{5} \end{pmatrix}.$$

e. Obtain zeros in the off-diagonal positions in the third column by adding $(-\frac{3}{2})$ times the third row to the first row and adding $(-\frac{5}{2})$ times the third row to the second row:

$$\left(\begin{array}{ccc|ccc} 1 & 0 & 0 & \frac{7}{10} & -\frac{1}{10} & \frac{3}{10} \\ 0 & 1 & 0 & \frac{1}{2} & -\frac{1}{2} & \frac{1}{2} \\ 0 & 0 & 1 & -\frac{4}{5} & \frac{2}{5} & -\frac{1}{5} \end{array}\right).$$

The last of these matrices is $(\mathbf{I}|\mathbf{A}^{-1})$, a fact that can be verified directly by multiplying the matrices $\mathbf{A}^{-1}$ and $\mathbf{A}$. This example was made slightly simpler by the fact that the original matrix $\mathbf{A}$ had a 1 in the upper left corner $(a_{11} = 1)$. If this is not the case, then the first step is to produce a 1 there by multiplying the first row by $1/a_{11}$, as long as $a_{11} \neq 0$. If $a_{11} = 0$, then the first row must be interchanged with some other row to bring a nonzero element into the upper left position before proceeding.

We finally note that if $\mathbf{A}$ or $\mathbf{B}$ is singular, then $|\mathbf{AB}| = |\mathbf{A}|\,|\mathbf{B}| = |\mathbf{BA}| = 0$ and therefore Theorem A.3.8 implies that both $\mathbf{AB}$ and $\mathbf{BA}$ are singular. Thus part (e) of Theorem A.1.10 is proved.

PROBLEMS

In each of Problems 1 through 10, use elementary row and column operations to simplify the task of evaluating the determinant by cofactor expansions. If $\mathbf{A}$ is nonsingular, find $\mathbf{A}^{-1}$.

1. $\mathbf{A} = \begin{pmatrix} 1 & 4 \\ -2 & 3 \end{pmatrix}$

2. $\mathbf{A} = \begin{pmatrix} 3 & -1 \\ 6 & 2 \end{pmatrix}$

3. $\mathbf{A} = \begin{pmatrix} 1 & 2 & 3 \\ 2 & 4 & 5 \\ 3 & 5 & 6 \end{pmatrix}$

4. $\mathbf{A} = \begin{pmatrix} 1 & 1 & -1 \\ 2 & -1 & 1 \\ 1 & 1 & 2 \end{pmatrix}$

5. $\mathbf{A} = \begin{pmatrix} 1 & 2 & 1 \\ -2 & 1 & 8 \\ 1 & -2 & -7 \end{pmatrix}$

6. $\mathbf{A} = \begin{pmatrix} 2 & 1 & 0 \\ 0 & 2 & 1 \\ 0 & 0 & 2 \end{pmatrix}$

7. $\mathbf{A} = \begin{pmatrix} 1 & 1 & -1 \\ 2 & 1 & 0 \\ 3 & -2 & 1 \end{pmatrix}$

8. $\mathbf{A} = \begin{pmatrix} 2 & 3 & 1 \\ -1 & 2 & 1 \\ 4 & -1 & -1 \end{pmatrix}$

9. $\mathbf{A} = \begin{pmatrix} 1 & 0 & 0 & -1 \\ 0 & -1 & 1 & 0 \\ -1 & 0 & 1 & 0 \\ 0 & 1 & -1 & 1 \end{pmatrix}$

10. $\mathbf{A} = \begin{pmatrix} 1 & -1 & 2 & 0 \\ -1 & 2 & -4 & 2 \\ 1 & 0 & 1 & 3 \\ -2 & 2 & 0 & -1 \end{pmatrix}$

11. Let

$$\mathbf{A} = \begin{pmatrix} 1 & 3 & 0 \\ -1 & 2 & 4 \\ 1 & -2 & 0 \end{pmatrix} \quad \text{and} \quad \mathbf{B} = \begin{pmatrix} 2 & 0 & 1 \\ -1 & 3 & 2 \\ 4 & 0 & 0 \end{pmatrix}.$$

Verify that $|\mathbf{AB}| = |\mathbf{A}|\,|\mathbf{B}|$.

12. If $\mathbf{A}$ is nonsingular, show that $|\mathbf{A}^{-1}| = 1/|\mathbf{A}|$.

In each of Problems 13 through 16, find all values of λ such that the given matrix is singular.

13. $\begin{pmatrix} -\lambda & -1 & -3 \\ 2 & 3-\lambda & 3 \\ -2 & 1 & 1-\lambda \end{pmatrix}$

14. $\begin{pmatrix} 4-\lambda & 6 & 6 \\ 1 & 3-\lambda & 2 \\ -1 & -4 & -3-\lambda \end{pmatrix}$

15. $\begin{pmatrix} 1-\lambda & 1 & 0 & -1 \\ 0 & -1-\lambda & 3 & 4 \\ 0 & 0 & -2-\lambda & -3 \\ 0 & 0 & 0 & 1-\lambda \end{pmatrix}$

A.4 The Eigenvalue Problem

Recall that in Chapter 3 we used the eigenvalue method to find solutions of homogeneous linear first order systems of dimension two,

$$\mathbf{x}' = \begin{pmatrix} a_{11} & a_{12} \\ a_{21} & a_{22} \end{pmatrix} \mathbf{x} = \mathbf{A}\mathbf{x}. \tag{1}$$

Nonzero solutions of Eq. (1) of the form $\mathbf{x} = e^{\lambda t}\mathbf{v}$, where $\mathbf{v}$ is a nonzero 2-vector, exist if and only if the scalar λ and the vector $\mathbf{v}$ satisfy the equation

$$\mathbf{A}\mathbf{v} = \lambda\mathbf{v},$$

or equivalently,

$$(\mathbf{A} - \lambda\mathbf{I}_2)\mathbf{v} = \mathbf{0}. \tag{2}$$

The eigenvalue method generalizes, in a natural way, to homogeneous linear constant coefficient systems of dimension n

$$\mathbf{x}' = \begin{pmatrix} a_{11} & \cdots & a_{1n} \\ \vdots & & \vdots \\ a_{n1} & \cdots & a_{nn} \end{pmatrix} \mathbf{x}. \tag{3}$$

As in the case for systems of dimension 2, nonzero solutions of Eq. (3) of the form $\mathbf{x} = e^{\lambda t}\mathbf{v}$, where $\mathbf{v}$ is a nonzero n-vector, exist if and only if λ and $\mathbf{v}$ satisfy

$$(\mathbf{A} - \lambda\mathbf{I}_n)\mathbf{v} = \mathbf{0}, \tag{4}$$

where $\mathbf{A}$ is the $n \times n$ matrix of coefficients on the right-hand side of Eq. (3),

$$\mathbf{A} = \begin{pmatrix} a_{11} & \cdots & a_{1n} \\ \vdots & & \vdots \\ a_{n1} & \cdots & a_{nn} \end{pmatrix}.$$

From Theorem A.3.7, nonzero solutions of Eq. (4) exist if and only if

$$\det(\mathbf{A} - \lambda\mathbf{I}_n) = 0. \tag{5}$$

Just as in Chapter 3, values of λ that satisfy Eq. (5) are called **eigenvalues** of the matrix $\mathbf{A}$, and the nonzero solutions of Eq. (4) that are obtained by using such a value of λ are called the **eigenvectors** corresponding to that eigenvalue.

In the general $n \times n$ case, Eq. (5) is a polynomial equation $p(\lambda) = 0$, where

$$p(\lambda) = (-1)^n(\lambda^n + p_{n-1}\lambda^{n-1} + \cdots + p_1\lambda + p_0) \tag{6}$$

is called the **characteristic polynomial** of $\mathbf{A}$. The equation $p(\lambda) = 0$ is called the **characteristic equation** of $\mathbf{A}$. By the fundamental theorem of algebra, we know that $p(\lambda)$ has a factorization over the complex numbers of the form

$$p(\lambda) = (-1)^n(\lambda - \lambda_1)^{m_1} \cdots (\lambda - \lambda_k)^{m_k}, \tag{7}$$

where $\lambda_i \neq \lambda_j$ if $i \neq j$. Thus the eigenvalues of $\mathbf{A}$ are λ_j, $j = 1, \ldots, k$. For each $j = 1, \ldots, k$, the power m_j of the factor $\lambda - \lambda_j$ in Eq. (7) is called the **algebraic multiplicity** of the eigenvalue λ_j. Since $p(\lambda)$ is a polynomial of degree n, it must be the case that

$$m_1 + \cdots + m_k = n.$$

Having found the eigenvalues $\lambda_1, \ldots, \lambda_k$ and their algebraic multiplicities by completely factoring $p(\lambda)$, the next step is to find the maximum number of linearly independent eigenvectors that belong to each eigenvalue. Thus, for each $j = 1, \ldots, k$, we find the general solution of

$$(\mathbf{A} - \lambda_j \mathbf{I}_n)\mathbf{v} = \mathbf{0}. \tag{8}$$

If we denote the rank of $(\mathbf{A} - \lambda_j \mathbf{I}_n)$ by r_j, then from Theorem A.2.11 we know that the general solution of Eq. (8) will have the form

$$\mathbf{v} = \alpha_1 \mathbf{v}_1^{(j)} + \cdots + \alpha_{n-r_j} \mathbf{v}_{n-r_j}^{(j)}, \tag{9}$$

where $\mathbf{v}_1^{(j)}, \ldots, \mathbf{v}_{n-r_j}^{(j)}$ are linearly independent solutions of $(\mathbf{A} - \lambda_j \mathbf{I}_n)\mathbf{v} = \mathbf{0}$. The vectors $\mathbf{v}_1^{(j)}, \ldots, \mathbf{v}_{n-r_j}^{(j)}$ are the maximal set of linearly independent eigenvectors that belong to the eigenvalue λ_j. The maximal number of linearly independent eigenvectors, $g_j = n - r_j$, that belong to the eigenvalue λ_j is called the **geometric multiplicity** of the eigenvalue λ_j. If the geometric multiplicity of eigenvalue λ_j is equal to the algebraic multiplicity, that is, $g_j = m_j$, the eigenvalue λ_j is said to be **nondefective**. Otherwise λ_j is said to be **defective**. If $g_j = m_j$ for each $j = 1, \ldots, k$, then $\mathbf{A}$ is said to be a **nondefective matrix**. If $g_j < m_j$ for some j, then the matrix $\mathbf{A}$ is said to be **defective**.

EXAMPLE 1

Find the eigenvalues and eigenvectors of the matrix

$$\mathbf{A} = \begin{pmatrix} 2 & -3 & -1 \\ 0 & -1 & 0 \\ -1 & 1 & 2 \end{pmatrix}. \tag{10}$$

The eigenvalues λ and the eigenvectors $\mathbf{v}$ satisfy the equation $(\mathbf{A} - \lambda \mathbf{I}_3)\mathbf{v} = \mathbf{0}$, or

$$\begin{pmatrix} 2 - \lambda & -3 & -1 \\ 0 & -1 - \lambda & 0 \\ -1 & 1 & 2 - \lambda \end{pmatrix} \begin{pmatrix} v_1 \\ v_2 \\ v_3 \end{pmatrix} = \begin{pmatrix} 0 \\ 0 \\ 0 \end{pmatrix}. \tag{11}$$

The eigenvalues are the roots of the characteristic equation

$$p(\lambda) = \det(\mathbf{A} - \lambda \mathbf{I}_3) = \begin{vmatrix} 2 - \lambda & -3 & -1 \\ 0 & -1 - \lambda & 0 \\ -1 & 1 & 2 - \lambda \end{vmatrix} = -\lambda^3 + 3\lambda^2 + \lambda - 3. \tag{12}$$

The roots of Eq. (12) are $\lambda_1 = -1$, $\lambda_2 = 1$, and $\lambda_3 = 3$. Equivalently, the characteristic polynomial of $\mathbf{A}$, $p(\lambda)$, has the factorization $p(\lambda) = -(\lambda + 1)(\lambda - 1)(\lambda - 3)$. The eigenvalues are said to be **simple** since the algebraic multiplicity of each one is equal to 1. Since each eigenvalue has at least one eigenvector, the geometric multiplicity of each eigenvalue will also be 1. Hence the matrix $\mathbf{A}$ is nondefective and possesses three linearly independent eigenvectors.

To find the eigenvector corresponding to the eigenvalue λ_1, we substitute $\lambda = -1$ in Eq. (11). This gives the system

$$\begin{pmatrix} 3 & -3 & -1 \\ 0 & 0 & 0 \\ -1 & 1 & 3 \end{pmatrix} \begin{pmatrix} v_1 \\ v_2 \\ v_3 \end{pmatrix} = \begin{pmatrix} 0 \\ 0 \\ 0 \end{pmatrix}.$$

We can reduce this to the equivalent system

$$\begin{pmatrix} 1 & -1 & -3 \\ 0 & 0 & 1 \\ 0 & 0 & 0 \end{pmatrix} \begin{pmatrix} v_1 \\ v_2 \\ v_3 \end{pmatrix} = \begin{pmatrix} 0 \\ 0 \\ 0 \end{pmatrix}$$

by elementary row operations. Thus the components of $\mathbf{v}$ must satisfy the equations $v_1 - v_2 - 3v_3 = 0$ and $v_3 = 0$. The general solution is given by $\mathbf{v} = \alpha(1, 1, 0)^T$ where α is arbitrary. Setting $\alpha = 1$ yields the eigenvector

$$\mathbf{v}_1 = \begin{pmatrix} 1 \\ 1 \\ 0 \end{pmatrix}$$

corresponding to the eigenvalue $\lambda = -1$.

For $\lambda = 1$, Eqs. (11) reduce to the pair of equations $v_1 - v_3 = 0$ and $v_2 = 0$ with general solution $\mathbf{v} = \alpha(1, 0, 1)^T$. Choosing $\alpha = 1$ gives the eigenvector

$$\mathbf{v}_2 = \begin{pmatrix} 1 \\ 0 \\ 1 \end{pmatrix}$$

that belongs to the eigenvalue $\lambda = 1$.

Finally, substituting $\lambda = 3$ into Eq. (11) and reducing the system to row echelon form by elementary row operations yields $v_1 + v_3 = 0$ and $v_2 = 0$ with general solution $\mathbf{v} = \alpha(1, 0, -1)^T$. Setting $\alpha = 1$ gives the eigenvector

$$\mathbf{v}_3 = \begin{pmatrix} 1 \\ 0 \\ -1 \end{pmatrix}$$

associated with the eigenvalue $\lambda = 3$.

Since $g_j = m_j = 1, j = 1, \ldots 3$, all of the eigenvalues of $\mathbf{A}$ are nondefective and therefore $\mathbf{A}$ is a nondefective matrix.

EXAMPLE
2

Find the eigenvalues and eigenvectors of the matrix

$$\mathbf{A} = \begin{pmatrix} 4 & 6 & 6 \\ 1 & 3 & 2 \\ -1 & -5 & -2 \end{pmatrix}. \tag{13}$$

The characteristic polynomial $p(\lambda) = \det(\mathbf{A} - \lambda\mathbf{I}_3)$ of $\mathbf{A}$ is

$$p(\lambda) = -\lambda^3 + 5\lambda^2 - 8\lambda + 4 = -(\lambda - 1)(\lambda - 2)^2$$

so the eigenvalues are $\lambda_1 = 1$ and $\lambda_2 = 2$ with algebraic multiplicities 1 and 2, respectively. The system $(\mathbf{A} - \lambda_1\mathbf{I}_3)\mathbf{v} = \mathbf{0}$ is reduced by elementary row operations to $v_1 + 2v_2 + 2v_3 = 0$ and $3v_2 + v_3 = 0$ with general solution $\mathbf{v} = \alpha(4, 1, -3)^T$. Setting $\alpha = 1$ gives us the eigenvector

$$\mathbf{v}_1 = \begin{pmatrix} 4 \\ 1 \\ -3 \end{pmatrix}$$

belonging to λ_1.

To find the eigenvector(s) belonging to λ_2, we find the general solution of $(\mathbf{A} - \lambda_2 \mathbf{I}_3)\mathbf{v} = \mathbf{0}$. Row reduction leads to the system $v_1 + v_2 + 2v_3 = 0$ and $2v_2 + v_3 = 0$. Since rank $(\mathbf{A} - \lambda_2 \mathbf{I}_3) = 2$, we realize that $\lambda = 2$ can have only one eigenvector, that is, λ_2 has a geometric multiplicity equal to 1 and is therefore a defective eigenvalue. The general solution of $(\mathbf{A} - \lambda_2 \mathbf{I}_3)\mathbf{v} = \mathbf{0}$ is $\mathbf{v} = \alpha(3, 1, -2)^T$. Setting $\alpha = 1$ yields the eigenvector

$$\mathbf{v}_2 = \begin{pmatrix} 3 \\ 1 \\ -2 \end{pmatrix}$$

for the eigenvalue λ_2. Since $g_2 = 1$ and $m_2 = 2$, the eigenvalue λ_2 is defective. Therefore, the matrix $\mathbf{A}$ in this example is a defective matrix.

▶ Complex Eigenvalues. If the entries of the matrix $\mathbf{A}$ are real, then the coefficients of the characteristic polynomial $p(\lambda) = \det(\mathbf{A} - \lambda \mathbf{I}_n)$ of $\mathbf{A}$ are also real. Thus if $\mu + iv$ is a complex eigenvalue of $\mathbf{A}$ and $v \neq 0$, then $\mu - iv$ must also be an eigenvalue of $\mathbf{A}$. To see this, suppose $\lambda = \mu + iv$ with $v \neq 0$ is an eigenvalue of $\mathbf{A}$ with corresponding eigenvector $\mathbf{v}$ so that

$$\mathbf{A}\mathbf{v} = (\mu + iv)\mathbf{v}.$$

If we take the complex conjugate on both sides of this equation we get

$$\mathbf{A}\bar{\mathbf{v}} = (\mu - iv)\bar{\mathbf{v}}, \tag{14}$$

where we have used the fact that $\bar{\mathbf{A}} = \mathbf{A}$ since $\mathbf{A}$ is real along with the property that, if z_1 and z_2 are complex numbers, then $\overline{z_1 z_2} = \bar{z}_1 \bar{z}_2$. Eq. (14) shows that not only is $\bar{\lambda} = \mu - iv$ an eigenvalue of $\mathbf{A}$, its eigenvector $\bar{\mathbf{v}}$ is the conjugate of the eigenvector belonging to $\lambda = \mu + iv$.

EXAMPLE 3

Find the eigenvalues and eigenvectors of the matrix

$$\mathbf{A} = \begin{pmatrix} -\frac{1}{2} & \frac{7}{2} & -3 \\ -\frac{1}{2} & -\frac{1}{2} & -1 \\ \frac{1}{2} & \frac{3}{2} & -2 \end{pmatrix}.$$

The characteristic polynomial of $\mathbf{A}$ is

$$p(\lambda) = -\lambda^3 - 3\lambda^2 - 7\lambda - 5 = -(\lambda + 1)[\lambda - (-1 + 2i)][\lambda - (-1 - 2i)]$$

so the eigenvalues of $\mathbf{A}$ are $\lambda_1 = -1, \lambda_2 = -1 + 2i$, and $\lambda_3 = \bar{\lambda}_2 = -1 - 2i$. The eigenvector belonging to λ_1 is found to be

$$\mathbf{v}_1 = \begin{pmatrix} 1 \\ -1 \\ -1 \end{pmatrix}.$$

The equations represented by $(\mathbf{A} - \lambda_2 \mathbf{I}_3)\mathbf{v} = \mathbf{0}$ are reduced, via elementary row operations and following the rules of complex arithmetic, to $v_1 - (2 + i)v_3 = 0$ and $v_2 - iv_3 = 0$. The general solution of this system of equations is $\mathbf{v} = \alpha(2 + i, i, 1)^T$. Choosing $\alpha = 1$ yields the complex eigenvector

$$\mathbf{v}_2 = \begin{pmatrix} 2+i \\ i \\ 1 \end{pmatrix}.$$

The eigenvector belonging to λ_3 is therefore

$$\mathbf{v}_3 = \bar{\mathbf{v}}_2 = \begin{pmatrix} 2-i \\ -i \\ 1 \end{pmatrix}.$$

▶ Symmetric Matrices. Recall that a symmetric matrix $\mathbf{A}$ satisfies $\mathbf{A}^T = \mathbf{A}$, that is, $a_{ji} = a_{ij}$. The properties of eigenvectors of symmetric matrices require that we introduce the notion of angle between vectors. We define the **inner** or **dot product** of two n-vectors $\mathbf{u}$ and $\mathbf{v}$ by

$$\langle \mathbf{u}, \mathbf{v} \rangle = u_1 v_1 + u_2 v_2 + \cdots + u_n v_n.$$

Note that if $\mathbf{v} \neq \mathbf{0}$, then $\langle \mathbf{v}, \mathbf{v} \rangle = v_1^2 + \cdots + v_n^2 > 0$.

**DEFINITION
A.4.1**

Two n-vectors $\mathbf{u}$ and $\mathbf{v}$ are said to be **orthogonal** if $\langle \mathbf{u}, \mathbf{v} \rangle = 0$.

If a set of nonzero n-vectors $\{\mathbf{v}_1, \ldots, \mathbf{v}_k\}$ are pairwise orthogonal, that is, $\langle \mathbf{v}_i, \mathbf{v}_j \rangle = 0$ whenever $i \neq j$, then the set is also linearly independent. To see this, we take the inner product of both sides of the equation

$$c_1 \mathbf{v}_1 + \cdots + c_k \mathbf{v}_k = \mathbf{0}$$

with $\mathbf{v}_j$ for each $j = 1, \ldots, k$,

$$c_1 \langle \mathbf{v}_1, \mathbf{v}_j \rangle + \cdots + c_j \langle \mathbf{v}_j, \mathbf{v}_j \rangle + \cdots + c_k \langle \mathbf{v}_k, \mathbf{v}_j \rangle = \langle \mathbf{0}, \mathbf{v}_j \rangle,$$

or since $\langle \mathbf{v}_i, \mathbf{v}_j \rangle = 0$ for all $i \neq j$, the equation reduces to

$$c_j \langle \mathbf{v}_j, \mathbf{v}_j \rangle = 0.$$

Since $\langle \mathbf{v}_j, \mathbf{v}_j \rangle > 0$, $c_j = 0$ for each $j = 1, \ldots, k$.

Eigenvalues and eigenvectors of symmetric matrices have the following useful properties:

1. All eigenvalues are real.
2. There always exists a full, or complete, set of n linearly independent eigenvectors, regardless of the algebraic multiplicities of the eigenvalues. Thus symmetric matrices are always nondefective.
3. If $\mathbf{v}_1$ and $\mathbf{v}_2$ are eigenvectors that correspond to different eigenvalues, then $\langle \mathbf{v}_1, \mathbf{v}_2 \rangle = 0$. Thus if all eigenvalues are simple, then the associated eigenvectors form an orthogonal set of vectors.
4. Corresponding to an eigenvalue of algebraic multiplicity m, it is possible to choose m eigenvectors that are mutually orthogonal. Thus the complete set of n eigenvectors of a symmetric matrix can always be chosen to be orthogonal.

EXAMPLE
4

Find the eigenvalues and eigenvectors of the matrix

$$\mathbf{A} = \begin{pmatrix} 0 & 1 & 1 \\ 1 & 0 & 1 \\ 1 & 1 & 0 \end{pmatrix}. \tag{15}$$

The characteristic polynomial of $\mathbf{A}$ is

$$p(\lambda) = \det(\mathbf{A} - \lambda \mathbf{I}_3) = -\lambda^3 + 3\lambda + 2 = -(\lambda - 2)(\lambda + 1)^2.$$

Thus the eigenvalues of $\mathbf{A}$ are $\lambda_1 = 2$ and $\lambda_2 = -1$ with algebraic multiplicities 1 and 2, respectively. To find the eigenvector corresponding to λ_1, we use elementary row operations to reduce the system $(\mathbf{A} - \lambda_1 \mathbf{I}_3)\mathbf{v} = \mathbf{0}$ to the system $2v_1 - v_2 - v_3 = 0$ and $v_2 - v_3 = 0$ whose general solution then yields the eigenvector

$$\mathbf{v}_1 = \begin{pmatrix} 1 \\ 1 \\ 1 \end{pmatrix}.$$

Elementary row operations reduce $(\mathbf{A} - \lambda_2 \mathbf{I}_3)\mathbf{v} = \mathbf{0}$ to the single nonzero equation $v_1 + v_2 + v_3 = 0$, which has the general solution

$$\mathbf{v} = \alpha \begin{pmatrix} 1 \\ 0 \\ -1 \end{pmatrix} + \beta \begin{pmatrix} 0 \\ 1 \\ -1 \end{pmatrix},$$

where α and β are arbitrary scalars. First choosing $\alpha = 1$ and $\beta = 0$ and then $\alpha = 0$ and $\beta = 1$ yields a pair of linearly independent eigenvectors

$$\mathbf{v}_2 = \begin{pmatrix} 1 \\ 0 \\ -1 \end{pmatrix}, \qquad \hat{\mathbf{v}}_3 = \begin{pmatrix} 0 \\ 1 \\ -1 \end{pmatrix}$$

associated with λ_2. Since $\mathbf{v}_1$ belongs to a different eigenvalue, $\langle \mathbf{v}_1, \mathbf{v}_2 \rangle = 0$ and $\langle \mathbf{v}_1, \hat{\mathbf{v}}_3 \rangle = 0$ in accordance with property 3 of eigenvectors of symmetric matrices. However, $\mathbf{v}_2$ and $\hat{\mathbf{v}}_3$ are not orthogonal since $\langle \mathbf{v}_2, \hat{\mathbf{v}}_3 \rangle = 1$. For a second eigenvector belonging to λ_2, we could assume that $\mathbf{v}_3$ is the linear combination

$$\mathbf{v}_3 = \alpha \begin{pmatrix} 1 \\ 0 \\ -1 \end{pmatrix} + \beta \begin{pmatrix} 0 \\ 1 \\ -1 \end{pmatrix}$$

instead, and choose α and β to satisfy the orthogonality condition

$$\langle \mathbf{v}_2, \mathbf{v}_3 \rangle = 2\alpha + \beta = 0.$$

Then $\mathbf{v}_2$ and $\mathbf{v}_3$ will be orthogonal eigenvectors of λ_2, provided that $\beta = -2\alpha \neq 0$. For example, choosing $\alpha = 1$ requires that $\beta = -2$ and gives

$$\mathbf{v}_3 = \begin{pmatrix} 1 \\ -2 \\ 1 \end{pmatrix}.$$

Then, in agreement with property 4 above, $\{\mathbf{v}_1, \mathbf{v}_2, \mathbf{v}_3\}$ is a complete orthogonal set of eigenvectors for the matrix $\mathbf{A}$.

▶ Diagonalizable Matrices. If an $n \times n$ matrix $\mathbf{A}$ has n linearly independent eigenvectors, $\mathbf{v}_1, \ldots, \mathbf{v}_n$, and we form a matrix $\mathbf{T}$ using these vectors as columns, that is,

$$\mathbf{T} = [\mathbf{v}_1, \ldots, \mathbf{v}_n] = \begin{pmatrix} v_{11} & \cdots & v_{1n} \\ \vdots & & \vdots \\ v_{n1} & \cdots & v_{nn} \end{pmatrix}, \tag{16}$$

then

$$\mathbf{T}^{-1}\mathbf{A}\mathbf{T} = \mathbf{D} = \begin{pmatrix} \lambda_1 & 0 & \cdots & 0 \\ 0 & \lambda_2 & \cdots & 0 \\ \vdots & & & \vdots \\ 0 & 0 & \cdots & \lambda_n \end{pmatrix}. \tag{17}$$

To see why Eq. (17) is true, we first note that $\mathbf{T}^{-1}\mathbf{v}_j = \mathbf{e}_j$ for each $j = 1, \ldots, n$, where $\mathbf{e}_1, \ldots, \mathbf{e}_n$ are the column vectors of $\mathbf{I}_n$. This is evident by comparing, column by column, the right-hand sides of

$$\mathbf{T}^{-1}\mathbf{T} = \mathbf{T}^{-1}[\mathbf{v}_1, \ldots, \mathbf{v}_n] = [\mathbf{T}^{-1}\mathbf{v}_1, \ldots, \mathbf{T}^{-1}\mathbf{v}_n]$$

and

$$\mathbf{T}^{-1}\mathbf{T} = \mathbf{I}_n = [\mathbf{e}_1, \ldots, \mathbf{e}_n].$$

We now write the product $\mathbf{A}\mathbf{T}$ in the form

$$\mathbf{A}\mathbf{T} = \mathbf{A}[\mathbf{v}_1, \ldots, \mathbf{v}_n] = [\mathbf{A}\mathbf{v}_1, \ldots, \mathbf{A}\mathbf{v}_n] = [\lambda_1\mathbf{v}_1, \ldots, \lambda_n\mathbf{v}_n]$$

by using the relations $\mathbf{A}\mathbf{v}_j = \lambda_j\mathbf{v}_j, j = 1, \ldots, n$. Premultiplying $\mathbf{A}\mathbf{T}$ by $\mathbf{T}^{-1}$ then gives

$$\mathbf{T}^{-1}\mathbf{A}\mathbf{T} = \mathbf{T}^{-1}[\lambda_1\mathbf{v}_1, \ldots, \lambda_n\mathbf{v}_n] = [\lambda_1\mathbf{T}^{-1}\mathbf{v}_1, \ldots, \lambda_n\mathbf{T}^{-1}\mathbf{v}_n]$$
$$= [\lambda_1\mathbf{e}_1, \ldots, \lambda_n\mathbf{e}_n] = \mathbf{I}_n\mathbf{D} = \mathbf{D}.$$

Thus, if $\mathbf{A}$ has n linearly independent eigenvectors, $\mathbf{A}$ can be transformed into a diagonal matrix by the process shown in Eq. (17). In this case, we say that the matrix $\mathbf{A}$ is **diagonalizable**. In linear algebra, two $n \times n$ matrices $\mathbf{A}$ and $\mathbf{B}$ are said to be **similar** if there exists an invertible $n \times n$ matrix $\mathbf{T}$ such that $\mathbf{T}^{-1}\mathbf{A}\mathbf{T} = \mathbf{B}$, and the transformation is known as a **similarity transformation**. Thus Eq. (17) states that nondefective matrices are similar to diagonal matrices.

▶ Jordan Forms. As discussed above, an $n \times n$ matrix $\mathbf{A}$ can be diagonalized if it has a full complement of n linearly independent eigenvectors. If there is a shortage of eigenvectors (because one or more eigenvalues are defective), then $\mathbf{A}$ can always be transformed, via a similarity transformation, into a nearly diagonal matrix called its Jordan form, which has the eigenvalues of $\mathbf{A}$ on the main diagonal, ones in certain positions on the diagonal above the main diagonal, and zeros elsewhere. For example, we consider the matrix

$$\mathbf{A} = \begin{pmatrix} -3 & 0 & -1 \\ 1 & -2 & 1 \\ 1 & 0 & -1 \end{pmatrix} \tag{18}$$

with characteristic polynomial $p(\lambda) = -(\lambda + 2)^3$. Thus $\lambda_1 = -2$ is an eigenvalue with algebraic multiplicity $m_1 = 3$. Applying row reduction to $(\mathbf{A} - \lambda_1\mathbf{I}_3)\mathbf{v} = \mathbf{0}$ leads to the system $v_1 + v_3 = 0$ so the general solution is

$$\mathbf{v} = \alpha \begin{pmatrix} -1 \\ 0 \\ 1 \end{pmatrix} + \beta \begin{pmatrix} 0 \\ 1 \\ 0 \end{pmatrix}. \tag{19}$$

Choosing $\alpha = 1$ and $\beta = 1$ gives the eigenvector

$$\mathbf{v}_1 = \begin{pmatrix} -1 \\ 1 \\ 1 \end{pmatrix},$$

whereas choosing $\alpha = 1$ and $\beta = 0$ gives a second linearly independent eigenvector

$$\mathbf{v}_2 = \begin{pmatrix} -1 \\ 0 \\ 1 \end{pmatrix}.$$

Thus the geometric multiplicity of $\lambda_1 = -2$ is $g_1 = 2$ and $\mathbf{A}$ is not diagonalizable. However, if

$$\mathbf{w} = \begin{pmatrix} 0 \\ 0 \\ 1 \end{pmatrix}$$

is then selected as a solution of

$$(\mathbf{A} - \lambda_1 \mathbf{I}_3)\mathbf{w} = \mathbf{v}_1 \tag{20}$$

and the matrix

$$\mathbf{T} = [\mathbf{v}_1, \mathbf{w}, \mathbf{v}_2] = \begin{pmatrix} -1 & 0 & -1 \\ 1 & 0 & 0 \\ 1 & 1 & 1 \end{pmatrix}$$

is formed from $\mathbf{v}_1$, $\mathbf{w}$, and $\mathbf{v}_2$, then

$$\mathbf{T}^{-1}\mathbf{A}\mathbf{T} = \mathbf{J} = \begin{pmatrix} -2 & 1 & 0 \\ 0 & -2 & 0 \\ 0 & 0 & -2 \end{pmatrix}.$$

Thus $\mathbf{A}$ is similar to a matrix with the eigenvalue $\lambda_1 = -2$ along the diagonal, a one in the second column of the first row, and zeros elsewhere. The vector $\mathbf{w}$ satisfying Eq. (20) is called a **generalized eigenvector** belonging to λ_1.

However, instead of pursuing a discussion of generalized eigenvectors here, it is knowledge of the existence and structure of the Jordan form that is most relevant to our approach to finding fundamental sets of solutions of $\mathbf{x}' = \mathbf{A}\mathbf{x}$ for the case of defective $\mathbf{A}$ in Section 6.7. If the distinct eigenvalues of $\mathbf{A}$ are denoted by $\lambda_1, \ldots, \lambda_r$ with corresponding algebraic multiplicities $m_1, \ldots, m_r$ (such that $m_1 + \cdots + m_r = n$), then there is a nonsingular $n \times n$ matrix $\mathbf{T}$ such that

$$\mathbf{T}^{-1}\mathbf{A}\mathbf{T} = \mathbf{J}, \tag{21}$$

where $\mathbf{J}$ is block diagonal,

$$\mathbf{J} = \begin{pmatrix} \mathbf{J}_1 & \mathbf{0} & \mathbf{0} & \cdots & \mathbf{0} \\ \mathbf{0} & \mathbf{J}_2 & \mathbf{0} & \cdots & \mathbf{0} \\ \mathbf{0} & \mathbf{0} & \mathbf{J}_3 & \cdots & \mathbf{0} \\ \vdots & \vdots & \vdots & \ddots & \vdots \\ \mathbf{0} & \mathbf{0} & \mathbf{0} & \cdots & \mathbf{J}_r \end{pmatrix}. \tag{22}$$

For each $k = 1, \ldots, r$, the Jordan block $\mathbf{J}_k$ is an $m_k \times m_k$ matrix of the form

$$\mathbf{J}_k = \begin{pmatrix} \lambda_k & * & 0 & 0 & \cdots & 0 \\ 0 & \lambda_k & * & 0 & \cdots & 0 \\ 0 & 0 & \ddots & \ddots & & 0 \\ \vdots & \vdots & & & & \vdots \\ & & & & \lambda_k & * \\ 0 & 0 & & \cdots & & \lambda_k \end{pmatrix}, \tag{23}$$

where the asterisks represent ones or zeros. Note that the zeros that appear in Eq. (22) are block matrices of appropriate dimension that consist entirely of zero entries. This structural result underlies the following theorem that is used in Section 6.7 to help construct fundamental sets of solutions for $\mathbf{x}' = \mathbf{A}\mathbf{x}$ in the case that $\mathbf{A}$ is defective.

THEOREM
A.4.2

If λ is an eigenvalue of $\mathbf{A}$ and the algebraic multiplicity of λ equals m, then $\text{rank}(\mathbf{A} - \lambda \mathbf{I}_n)^m = n - m$. Thus the general solution of

$$(\mathbf{A} - \lambda \mathbf{I}_n)^m \mathbf{v} = \mathbf{0} \tag{24}$$

can be expressed as
$$\mathbf{v} = \alpha_1 \mathbf{v}_1 + \cdots + \alpha_m \mathbf{v}_m,$$

where $\{\mathbf{v}_1, \ldots, \mathbf{v}_m\}$ is a linearly independent set of solutions to Eq. (24).

Rather than prove this theorem, we consider the following example that shows why it is true and how it depends on being able to relate $\mathbf{A}$ to its Jordan form. Suppose that

$$\mathbf{T}^{-1}\mathbf{A}\mathbf{T} = \mathbf{J}, \tag{25}$$

where

$$\mathbf{J} = \begin{pmatrix} \mathbf{J}_1 & \mathbf{0} \\ \mathbf{0} & \mathbf{J}_2 \end{pmatrix}, \qquad \mathbf{J}_1 = \begin{pmatrix} \lambda_1 & * & 0 \\ 0 & \lambda_1 & * \\ 0 & 0 & \lambda_1 \end{pmatrix}, \qquad \text{and} \qquad \mathbf{J}_2 = \begin{pmatrix} \lambda_2 & * \\ 0 & \lambda_2 \end{pmatrix},$$

and the asterisks represent either ones or zeros in the Jordan blocks $\mathbf{J}_1$ and $\mathbf{J}_2$. Thus the algebraic multiplicity of λ_1 is $m_1 = 3$ and the algebraic multiplicity of λ_2 is $m_2 = 2$. From Eq. (25), we may write

$$\mathbf{T}^{-1}\left(\mathbf{A} - \lambda_1 \mathbf{I}_5\right)\mathbf{T} = \mathbf{J} - \lambda_1 \mathbf{I}_5$$

and consequently,

$$\mathbf{T}^{-1}\left(\mathbf{A} - \lambda_1 \mathbf{I}_5\right)^3 \mathbf{T} = \left(\mathbf{J}\lambda_1 - \mathbf{I}_5\right)^3.$$

We state, without proof, that, since $\mathbf{T}$ is nonsingular, the rank of $(\mathbf{A} - \lambda_1 \mathbf{I}_5)^3$ is equal to the rank of $(\mathbf{J} - \lambda_1 \mathbf{I}_5)^3$. Thus it is only necessary to examine the rank of $(\mathbf{J} - \lambda_1 \mathbf{I}_5)^3$. The block diagonal structure of $\mathbf{J} - \lambda_1 \mathbf{I}_5$ allows us to write

$$(\mathbf{J} - \lambda_1 \mathbf{I}_5)^3 = \begin{pmatrix} (\mathbf{J}_1 - \lambda_1 \mathbf{I}_3)^3 & \mathbf{0} \\ \mathbf{0} & (\mathbf{J}_2 - \lambda_1 \mathbf{I}_2)^3 \end{pmatrix}.$$

Since

$$\mathbf{J}_1 - \lambda_1 \mathbf{I}_3 = \begin{pmatrix} 0 & * & 0 \\ 0 & 0 & * \\ 0 & 0 & 0 \end{pmatrix},$$

it follows that

$$(\mathbf{J}_1 - \lambda_1 \mathbf{I}_3)^2 = \begin{pmatrix} 0 & 0 & * \\ 0 & 0 & 0 \\ 0 & 0 & 0 \end{pmatrix}$$

and

$$(\mathbf{J}_1 - \lambda_1 \mathbf{I}_3)^3 = \begin{pmatrix} 0 & 0 & 0 \\ 0 & 0 & 0 \\ 0 & 0 & 0 \end{pmatrix}.$$

On the other hand,

$$(\mathbf{J}_2 - \lambda_1 \mathbf{I}_2)^3 = \begin{pmatrix} (\lambda_2 - \lambda_1)^3 & \times \\ 0 & (\lambda_2 - \lambda_1)^3 \end{pmatrix},$$

where the value of the entry represented by the symbol $\times$ is irrelevant. Since λ_1 and λ_2 are distinct, $\lambda_2 - \lambda_1 \neq 0$. Thus $\text{rank}(\mathbf{A} - \lambda_1 \mathbf{I}_5)^3 = \text{rank}(\mathbf{J} - \lambda_1 \mathbf{I}_5)^3 = 5 - 3 = 2$. Note that we are comparing the number of unknowns with the number of independent nonzero equations when we attempt to find the general solution of $(\mathbf{A} - \lambda_1 \mathbf{I}_5)^3 \mathbf{v} = \mathbf{0}$. Since there are only two independent nonzero rows in the reduced form of $(\mathbf{A} - \lambda_1 \mathbf{I}_5)^3 \mathbf{v} = \mathbf{0}$ and there are five unknowns, we are able to choose three components of $\mathbf{v}$ arbitrarily. The nonzero rows in the reduced form of $(\mathbf{A} - \lambda_1 \mathbf{I}_5)^3 \mathbf{v} = \mathbf{0}$ will determine the remaining two unknown components. Thus the general solution of $(\mathbf{A} - \lambda_1 \mathbf{I}_5)^3 \mathbf{v} = \mathbf{0}$ will be of the form $\mathbf{v} = \alpha_1 \mathbf{v}_1 + \alpha_2 \mathbf{v}_2 + \alpha_3 \mathbf{v}_3$, where $\mathbf{v}_1$, $\mathbf{v}_2$, and $\mathbf{v}_3$ are linearly independent.

PROBLEMS

In each of Problems 1 through 10, find all eigenvalues and eigenvectors of the given matrix.

1. $\begin{pmatrix} 5 & -1 \\ 3 & 1 \end{pmatrix}$

2. $\begin{pmatrix} 3 & -2 \\ 4 & -1 \end{pmatrix}$

3. $\begin{pmatrix} 1 & 0 & 0 \\ 2 & 1 & -2 \\ 3 & 2 & 1 \end{pmatrix}$

4. $\begin{pmatrix} 3 & 2 & 2 \\ 1 & 4 & 1 \\ -2 & -4 & -1 \end{pmatrix}$

5. $\begin{pmatrix} -3 & -7 & -5 \\ 2 & 4 & 3 \\ 1 & 2 & 2 \end{pmatrix}$

6. $\begin{pmatrix} 7 & -2 & -4 \\ 3 & 0 & -2 \\ 6 & -2 & -3 \end{pmatrix}$

7. $\begin{pmatrix} -2 & 2 & -3 \\ 2 & 1 & -6 \\ -1 & -2 & 0 \end{pmatrix}$

8. $\begin{pmatrix} 4 & 2 & 3 \\ 2 & 1 & 0 \\ 1 & -2 & 0 \end{pmatrix}$

9. $\begin{pmatrix} 11 & -4 & -7 \\ 7 & -2 & -5 \\ 10 & -4 & -6 \end{pmatrix}$

10. $\begin{pmatrix} 1 & 0 & 1 \\ 0 & 2 & 0 \\ 1 & 0 & -1 \end{pmatrix}$

18. $\begin{pmatrix} -3 & 2 & 1 & -2 \\ 0 & -1 & 0 & 0 \\ -2 & 2 & 0 & -2 \\ 0 & 0 & 0 & -1 \end{pmatrix}$

In each of Problems 11 through 16, find the eigenvalues and a complete orthogonal set of eigenvectors for the given symmetric matrix.

11. $\begin{pmatrix} 3 & 0 & 0 \\ 0 & 3 & 4 \\ 0 & 4 & 3 \end{pmatrix}$

12. $\begin{pmatrix} 2 & 4 & -6 \\ 4 & 2 & -6 \\ -6 & -6 & -15 \end{pmatrix}$

19. $\begin{pmatrix} 5 & -5 & -3 & 4 \\ 0 & -1 & 0 & 0 \\ 6 & -6 & -4 & 6 \\ 0 & -1 & 0 & 1 \end{pmatrix}$

13. $\begin{pmatrix} 0 & 2 & 2 \\ 2 & 0 & 2 \\ 2 & 2 & 0 \end{pmatrix}$

14. $\begin{pmatrix} 1 & 1 & 1 \\ 1 & 1 & 1 \\ 1 & 1 & 1 \end{pmatrix}$

15. $\begin{pmatrix} 1 & 0 & 0 \\ 0 & 1 & 0 \\ 0 & 0 & 1 \end{pmatrix}$

16. $\begin{pmatrix} 3 & 2 & 4 \\ 2 & 0 & 2 \\ 4 & 2 & 3 \end{pmatrix}$

20. $\begin{pmatrix} -2 & \frac{1}{3} & \frac{2}{3} & -\frac{2}{3} \\ 2 & -3 & -2 & 0 \\ -2 & \frac{4}{3} & \frac{2}{3} & -\frac{2}{3} \\ -1 & \frac{1}{3} & \frac{2}{3} & -\frac{5}{3} \end{pmatrix}$

In each of Problems 17 through 20, use a computer to find the eigenvalues and eigenvectors for the given matrix.

17. $\begin{pmatrix} -1 & 0 & -1 & 0 \\ 0 & 1 & 1 & 0 \\ -1 & 1 & 2 & 1 \\ 0 & 0 & 1 & -1 \end{pmatrix}$

A N S W E R S

CHAPTER 1   INTRODUCTION

Section 1.1   Mathematical Models and Solutions   *page 10*

1. (a) $u' = -k(u - 70)$, $u(0) = 200$ (b) $\ln(13/10)/\ln(13/12) \approx 3.28$ minutes

3. 4:08pm

5. (a) $2\ln 18 \approx 5.78$ months (b) $2\ln(900/(900 - p_0))$ months (c) $p_0 = 900(1 - e^{-6}) \approx 898$

7. (a) $r = -\ln(82.04/100)/7 \approx 0.02828$ day^{-1} (b) $Q(t) = 100e^{-0.02828t}$, t is measured in days (c) $-\ln(1/2)/0.02828 \approx 24.5$ days

9. (a) $mv' = -mg + kv^2$ (b) $-\sqrt{mg/k}$ (c) 0.0002 kg/m

11. $Q' = 1/2 + (1/4)\sin t - Q/50$, $Q(0) = 50$

13. (a) $C(t) = C_0 e^{-kt}$ (b) $C_2 = C_0 + C_0 e^{-kT}$ (c) $C_n = C_0(1 + e^{-kT} + \ldots + e^{-k(n-1)T})$, $\lim_{n\to\infty} C_n = C_0/(1 - e^{-kT})$

17. 10.78 years; $25,872

19. $dV/dt = -kV^{2/3}$, $k > 0$

Section 1.2   Qualitative Methods: Phase Lines and Direction Fields   *page 25*

1. $y = 1$ asymptotically stable, $y = 0$, $y = 2$ unstable

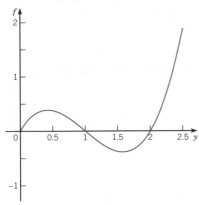

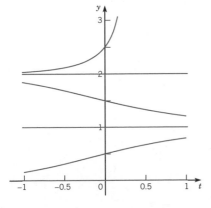

3. $y = 0$ asymptotically stable

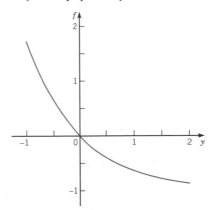

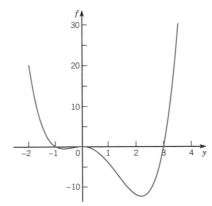

5. $y = -1$ asymptotically stable, $y = 0$ semistable, $y = 3$ unstable

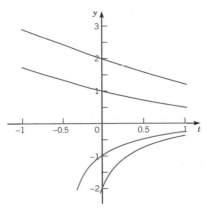

601

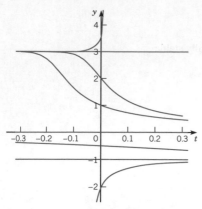

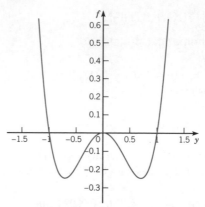

9. $y = -1$ asymptotically stable, $y = 0$ semistable, $y = 1$ unstable

7. $y = -a/b$ asymptotically stable, $y = 0$ unstable, graph depicts $a = 1, b = 2$

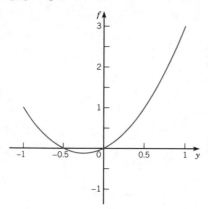

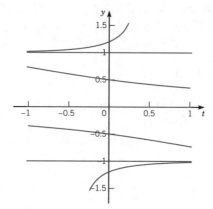

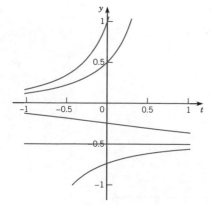

11. $y = 0$ asymptotically stable, $y = b^2/a^2$ unstable, graph depicts $a = 1, b = 2$

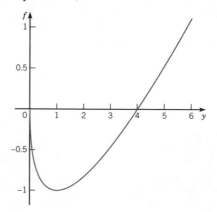

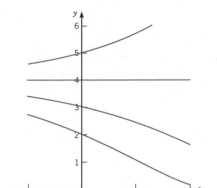

13. $y = 0$, $y = 1$ semistable

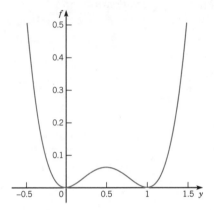

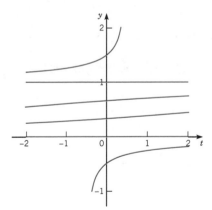

15. $y \to \infty$ when $y_0 > 3/2$, $y \to -\infty$ when $y_0 < 3/2$, $y \to 3/2$ when $y_0 = 3/2$

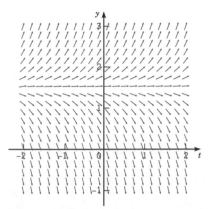

17. $y \to -1/2$

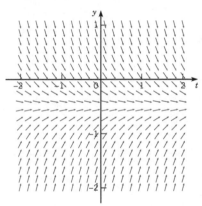

19. $y \to \infty$ when $y_0 > -2$, $y \to -\infty$ when $y_0 < -2$, $y \to -2$ when $y_0 = -2$

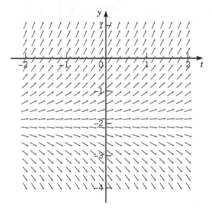

21. $y \to \infty$ when $y_0 > 5$, $y \to 5$ when $y_0 = 5$, $y \to 0$ when $y_0 < 5$

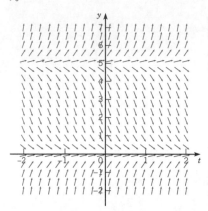

23. $y \to \infty$ when $y_0 > 2$, $y \to 2$ when $0 < y_0 \le 2$, $y \to 0$ when $y_0 = 0$, $y \to -\infty$ when $y_0 < 0$

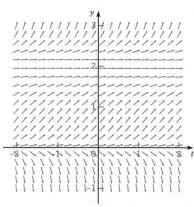

25. (c)　　　　**27.** (b)　　　　**29.** (e)

33. (a) $y = 0$ unstable, $y = K$ asymptotically stable, graph depicts $K = 4$

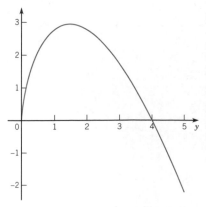

35. (b) $C = 1/(1 + \alpha)$, asymptotically stable

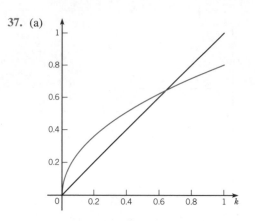

37. (a)

Section 1.3　Definitions, Classification, and Terminology　*page 35*

1. second order, linear

3. fourth order, linear

5. second order, nonlinear

7. nonhomogeneous

9. nonhomogeneous

11. nonhomogeneous

21. $r = -2$

23. $r = -3, r = 2$

25. $r = -2, r = -1$

27. $C = 1$

29. $C = 1/2 - \sin 1$

31. $C = 1$

33. (a) $c_1 = 3$, $c_2 = -2$ (b) $c_1 = 1$, $c_2 = -5$

CHAPTER 2　FIRST ORDER DIFFERENTIAL EQUATIONS

Section 2.1　Separable Equations　*page 44*

1. $5y^2 - 2x^5 = c$, $y \ne 0$.

3. $y^{-2} + 2\cos x = c$, if $y \ne 0$; $y = 0$.

5. $8\tan y + \sin 4x - 4x = c$, if $\cos y \ne 0$; $y = \pm(2n + 1)\pi/2$.

7. $y^2 + 1 = ce^{e^{x^2}}$

9. $3y + y^3 - x^3 = c$

11. $y = (4x^{3/2}/3 + c)^2$; $y = 0$

13. (a) $y = 1/(6x^2 - x - 8)$

(b)

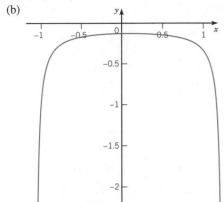

(c) $(1 - \sqrt{193})/12 < x < (1 + \sqrt{193})/12$

15. (a) $y = \sqrt{2(1-x)e^x - 1}$

(b)

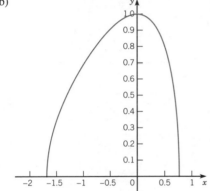

(c) $-1.68 < x < 0.77$, approximately

17. (a) $y = -\sqrt{3\ln(1 + x^2) + 49}$

(b)

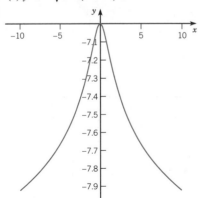

(c) $-\infty < x < \infty$

19. (a) $y = 2/(-2x^4 - 2x^2 + 3)$

(b)

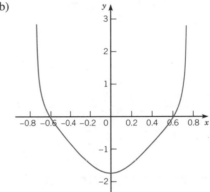

(c) $\sqrt{(-1 + \sqrt{7})/2} < x < \infty$

21. (a) $y = -\tan(\ln(\cos(2x))/2 + \pi/3)$

(b)

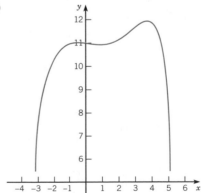

(c) $-\pi/4 < x < \pi/4$

23. (a) $y = 11/2 + \sqrt{x^3 - e^x + 125/4}$

(b)

(c) $-3.14 < x < 5.10$, approximately

25. (a) $y = -3/4 + (1/4)\sqrt{65 - 8e^x - 8e^{-x}}$

(b)

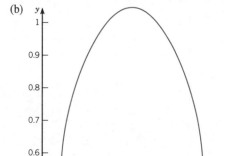

(c) $-\ln 8 < x < \ln 8$

27. (a) $y = (\pi - \arcsin(3\cos^2 x))/3$

(b)

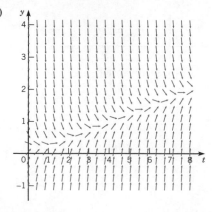

(c) $|x - \pi/2| < 0.62$ approximately

29. $4y^3 - 6y^2 - x^3 - x - 8 = 0; -2 < x < \infty$

31. $y = -2/(x^2 + 4x - 2); x = -2$

33. $y = -5 + \sqrt{16 + \sin 2x}; x = \pi/4 + n\pi$

35. (a) $y \to 4$ if $y_0 > 0$; $y = 0$ if $y_0 = 0$; $y \to -\infty$ if $y_0 < 0$
(b) $T \approx 3.29527$

37. $x = (c/a)y + (ad - bc)/a^2 \ln|ay + b| + k; a \neq 0,$
$ay + b \neq 0$

Section 2.2 Linear Equations: Method of Integrating Factors   *page 54*

1. (a)

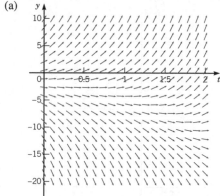

(b) all solutions converge to an increasing function
(c) $y = ce^{-4t} + t/4 + e^{-2t}/2 - 1/16$

3. (a)

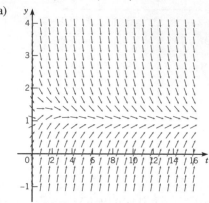

(b) all solutions converge to $y = 1$ (c) $y = ce^{-t} + t^2 e^{-t}/2 + 1$

5. (a)

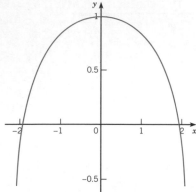

(b) some of the solutions increase without bound, some decrease without bound (c) $y = ce^{2t} - 3e^t$

7. (a)

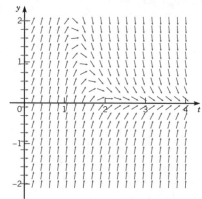

(b) all solutions converge to 0 (c) $y = ce^{-t^2} + 8t^2 e^{-t^2}$

9. (a)

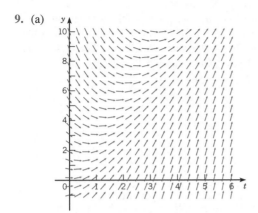

(b) all solutions increase without bound (c) $y = ce^{-t/2} + 3t - 6$

11. (a)

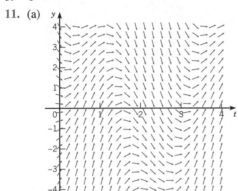

(b) all solutions converge to an oscillatory function
(c) $y = ce^{-t} - 2\cos 2t + \sin 2t$

13. $y = 3e^t - 2e^{2t} + 2te^{2t}$

15. $y = 1/4 + t^2/6 - t/5 + t^{-4}/30$

17. $y = 2e^{2t} + te^{2t}$

19. $y = -e^{-t}/t^4 - e^{-t}/t^3$

21. (a)

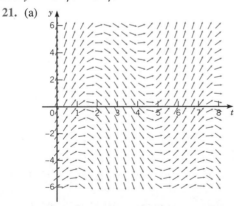

solutions increase or decrease without bound, $a_0 \approx -1$
(b) $y = (a + 9/10)e^{t/3} - (9/10)\cos t + (27/10)\sin t$
(c) $a_0 = -9/10$

23. (a)

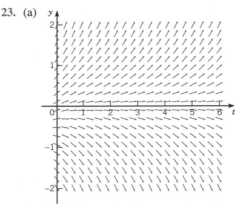

solutions increase or decrease without bound, $a_0 \approx 0$
(b) $y = [(2 + a(3\pi + 4))e^{2t/3} - 2e^{-\pi t/2}]/(3\pi + 4)$ (c) $a_0 = -2/(3\pi + 4)$

25. (a)

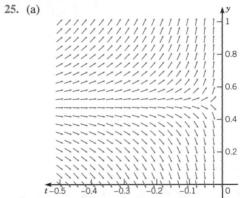

solutions increase or decrease without bound, $a_0 \approx 0.5$
(b) $y = t^{-2}(a\pi^2/4 - \cos t)$ (c) $a_0 = 4/\pi^2$

27. $(1.3643, 0.8201)$ approximately

29. (a) $y = ce^{-t/4} + 12 + (8\cos 2t + 64\sin 2t)/65$, solution converges toward an oscillating function
(b) approximately 10.066

31. $y_0 = -16/3$, solution decreases without bound

35. $y' + y = 3 - t$

37. $y' + y = 2 - 2t - t^2$

41. $y = ct^{-1} + (3/4)t^{-1}\cos 2t + (3/2)\sin 2t$

43. $y = ce^{-t/2} + 3t^2 - 12t + 24$

Section 2.3   Modeling with First Order Equations   *page 65*

1. $t = 50\ln 50 \approx 195.6$ min

3. $Q = 40(e^{-3/10} - e^{-1/2}) \approx 5.37$ lb

5. (a) $Q(t) = 25 - (625/2501)\cos t + (25/5002)\sin t + (63150/2501)e^{-t/50}$

(b)

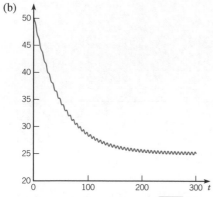

(c) level = 25, amplitude = $25\sqrt{2501}/5002 \approx 0.24995$

7. (a) $V = 2000G$, graph depicts $G = 5$

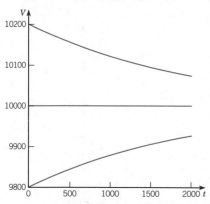

(b) $V(t) = 2000G + 20Ge^{-t/2000}$ (c) $G = 6$ gal

9. (a) $Q(t) = 2500(1 - e^{-t/1000})$ m
(b) $t = 1000\ln(1000/999) \approx 1$ min (c) 1 min

11. (a) $S(t) = k(e^{rt} - 1)/r$ (b) $k \approx \$6,060.99$ (c) 6.92%

13. (a) $t \approx 135.36$ months (b) $152,698.56

15. (a) $1.2097 \cdot 10^{-4}$ year^{-1} (b) $Q(t) = Q_0 e^{-1.2097t/10^4}$
(c) 5,730 years

17. (a) $\tau \approx 2.9632$; no (b) $\tau = 10\ln 2 \approx 6.9315$
(c) $\tau \approx 6.3804$

(d)

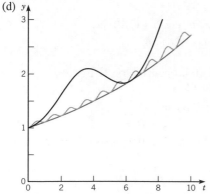

19. $t = \ln(13/8)/\ln(13/12) \approx 6.07$ min

21. (a) $c(t) = k + (P/r) + (c_0 - k - (P/r))e^{-rt/V}$;
$\lim_{t\to\infty} c(t) = k + (P/r)$ (b) $T = (V\ln 2)/r$, $T = (V\ln 10)/r$
(c) Superior: $T = 430.9$ years, Michigan: $T = 71.4$ years,
Erie: $T = 6.1$ years, Ontario: $T = 17.6$ years

23. (a) 50.22 m (b) 5.56 s

(c)

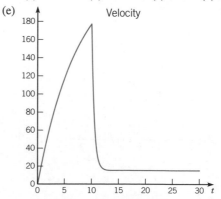

25. (a) 176.7 ft/s (b) 1074.5 ft (c) 15 ft/s (d) 256.6 s

(e)

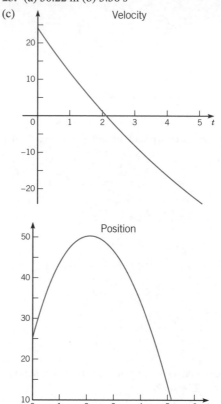

27. (a) $x_m = mv_0/k - (m^2g/k^2)\ln(1 + kv_0/mg)$; $t_m = (m/k)$
$\ln(1 + kv_0/mg)$

29. (a) $v_L = 2a^2g(\rho' - \rho)/9\mu$ (b) $e = 4\pi a^3 g(\rho' - \rho)/3E$

31. (a) $v = R\sqrt{2g/(R + x)}$ (b) 50.6 hours

33. (a) $v = (u\cos A)e^{-rt}$, $w = -g/r + (u\sin A + g/r)e^{-rt}$
(b) $x = u\cos A(1 - e^{-rt})/r$, $y = -gt/r + (u\sin A + g/r)$
$(1 - e^{-rt})/r + h$

(c)
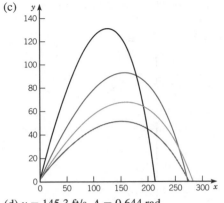

(d) $u = 145.3$ ft/s, $A = 0.644$ rad

Section 2.4 Differences Between Linear and Nonlinear Equations *page 79*

1. $0 < t < 3$

3. $\pi/2 < t < 3\pi/2$

5. $-2 < t < 2$

7. $2t + 5y \neq 0$

9. $t^2 - y^2 \neq 1$, $y \neq 0$, $t \neq 0$

11. $y \neq 0$, $y \neq 3$

13. (a) no (b) $[(2/3)(t - (1/2))]^{3/2}$ (c) $|y(2)| \leq (4/3)^{3/2}$

15. $y^2 = -4t^2 + y_0^2$, defined for $t^2 \leq y_0^2/4$

17. $y = y_0/\sqrt{2ty_0^2 + 1}$, defined for $t > -1/(2y_0^2)$ for $y_0 \neq 0$; for $-\infty < t < \infty$ for $y_0 = 0$

19.
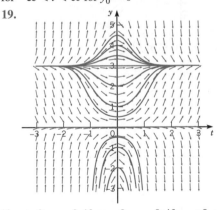

if $y_0 > 0$, $y \to 3$; if $y_0 = 0$, $y \to 0$; if $y_0 < 0$, $y \to -\infty$

21.
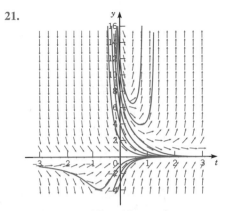

if $y_0 > 9$, $y \to \infty$; if $y_0 \leq 9$, $y \to 0$

27. $y = (1 - e^{-2t})/2$ for $0 \leq t \leq 1$; $y = (e^2 - 1)e^{-2t}/2$ for $t > 1$

Section 2.5 Autonomous Equations and Population Dynamics *page 91*

1. (a) $\tau = (1/r)\ln 4$; 55.452 years (b) $T = (1/r)\ln[\beta(1 - \alpha)/(1 - \beta)\alpha]$; 175.78 years

3. (a) $y = Ke^{\ln(y_0/K)e^{-rt}}$ (b) $y(2) \approx 0.7153K \approx 57.6 \cdot 10^6$ kg (c) $\tau \approx 2.215$ years

5. (b) $k^2/2g(\alpha a)^2$

7. (a) $y = y_0 e^{-\beta t}$ (b) $x = x_0 e^{-\alpha y_0(1 - e^{-\beta t})/\beta}$ (c) $x_0 e^{-\alpha y_0/\beta}$

9. (a) $\lim_{t \to \infty} x(t) = \min(p, q)$; $x(t) = pq(e^{\alpha(q-p)t} - 1)/(qe^{\alpha(q-p)t} - p)$ (b) $\lim_{t \to \infty} x(t) = p$; $x(t) = p^2\alpha t/(p\alpha t + 1)$

11. (a) $a \leq 0$: $y = 0$ asymptotically stable; $a > 0$: $y = 0$ unstable, $y = \pm\sqrt{a}$ asymptotically stable

(b) $a = 1$

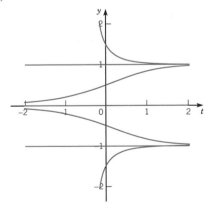

$a = 0$

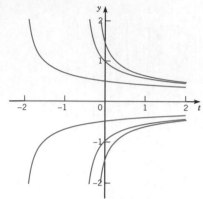

$a = -1$

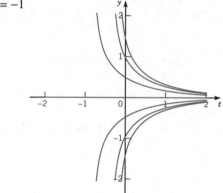

(c)

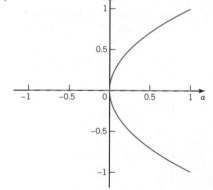

Section 2.6 Exact Equations and Integrating Factors *page 100*

1. (a) exact (b) $x^2 + 3x + y^2 - 2y = c$
(c)

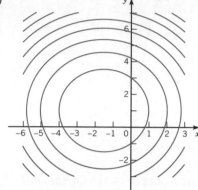

3. (a) exact (b) $x^3 - x^2y + 2x + 2y^3 + 3y = c$
(c)

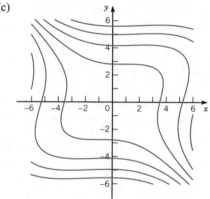

5. (a) exact (b) $2x^2 + 2xy + 3y^2/2 = c$
(c)

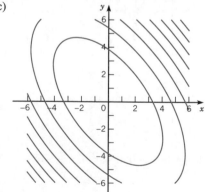

7. (a) exact (b) $e^x \sin y + 2y \cos x = c$
(c)

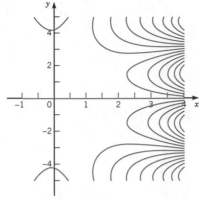

9. (a) exact (b) $e^{xy} \cos 2x + x^2 - 3y = c$
(c)

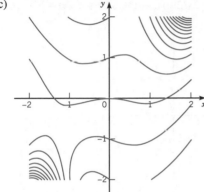

11. (a) not exact
13. $y = (x + \sqrt{28 - 3x^2})/2, |x| < \sqrt{28/3}$
15. $b = 3, x^2y^2 + 2x^3y = c$
19. (b) $x^2 - 1/y^2 + 2\ln|y| = c, y = 0$
(c)

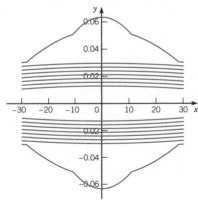

21. (b) $xy^2 - e^y(y^2 - 2y + 2) = c$
(c)

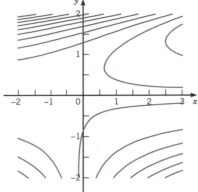

25. (a) $\mu(x) = e^{3x}, (3x^2y + y^3)e^{3x} = c$
(b)

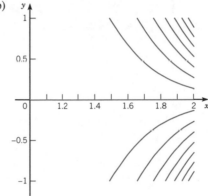

27. (a) $\mu(y) = y, xy - \sin y + y \cos y = c$
(b)

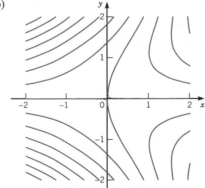

29. (a) $\mu(y) = \sin y$, $e^x \sin y + y^2 = c$

(b)

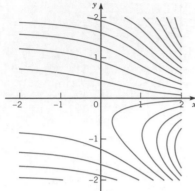

31. (a) $\mu(xy) = xy$, $x^3 y + 3x^2 + y^3 = c$

(b)

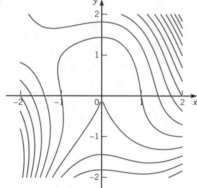

Section 2.7 Substitution Methods *page 108*

1. (a) not homogeneous

3. (a) homogeneous (b) $3x^2/y^2 + \ln(y^2 x^2) = C$

(c)

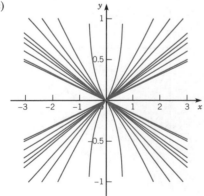

5. (a) homogeneous (b) $y = x \sin(\ln |x| + C)$

(c)

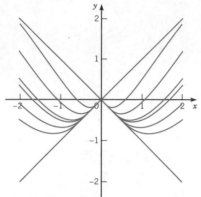

7. (a) homogeneous (b) $(1/2 - 9/(2\sqrt{29})) \ln(1 + \sqrt{29} - 2y/x) + (1/2 + 9/(2\sqrt{29})) \ln(-1 + \sqrt{29} + 2y/x) + \ln x = C$

(c)

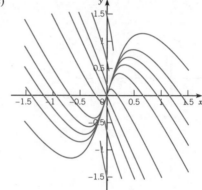

9. (a) homogeneous (b) $\ln |1 - x^2/y^2| + 2x/y + \ln |x| = C$

(c)

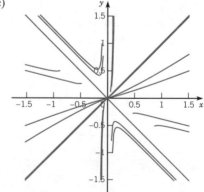

11. $y = x\sqrt{1/4 + \ln(x^2/4)}$; $(2e^{-1/8}, \infty)$

13. (a) $y' + (1/t)y = ty^2$ (b) $y = 1/(Ct - t^2)$; $y = 0$
(c)

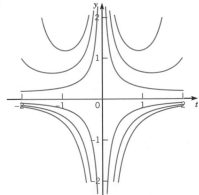

15. (a) $y' + (3/t)y = ty^2$ (b) $y = 1/(Ct^3 - t^3 \ln t)$; $y = 0$
(c)

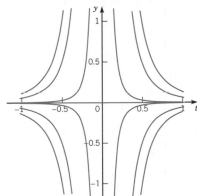

17. (a) $y' + 4t/(5(1 + t^2))y = 4t/(5(1 + t^2))y^4$ (b) $y^{-3} = 1 + C(1 + t^2)^{6/5}$; $y = 0$
(c)

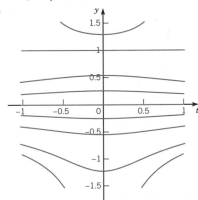

19. (a) $y' - y = y^{1/2}$ (b) $y = (Ce^{t/2} - 1)^2$; $y = 0$
(c)

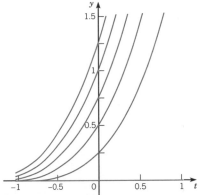

21. (a) $y' - \epsilon y = -\sigma y^3$ (b) $y = ((\sigma/\epsilon) + Ce^{-2\epsilon t})^{-1/2}$; $y = 0$

23. (b) $y = 4t - (e^{-5t^2/2} \int_0^t e^{5s^2/2}\, ds)^{-1}$

25. (a) linear (b) $x = e^y + Ce^{-2y}$

27. (a) linear (b) $y = 1 - 1/x + Ce^{-x}/x$

29. (a) separable, linear (b) $y = Ce^{2\sqrt{x}}$

31. (a) exact, Bernoulli (b) $x \ln x + xy^2 = C$

33. (a) separable (b) $y = C - \ln(\ln x)$

35. (a) Bernoulli, homogeneous (b) $y^2 = Cx^{5/2} - 8x^2$

CHAPTER 3   SYSTEMS OF TWO FIRST ORDER EQUATIONS

Section 3.1   Systems of Two Linear Algebraic Equations   *page 128*

1. (a) $x_1 = 2, x_2 = 1$ (b) intersecting

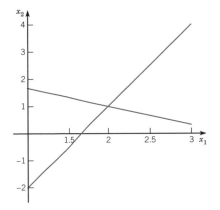

3. (a) $x_1 = 0, x_2 = 0$ (b) intersecting

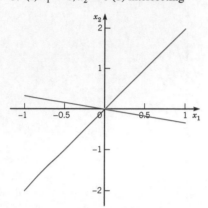

5. (a) $x_1 = -1, x_2 = -2$ (b) intersecting

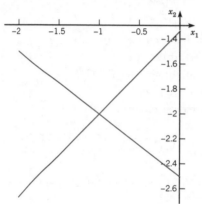

7. (a) $x_1 = c, x_2 = 2c/3 - 2$; c arbitrary (b) coincident

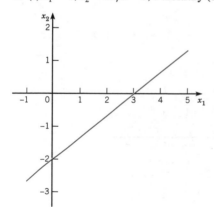

9. (a) $x_1 = 2, x_2 = 2$ (b) intersecting

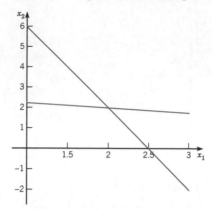

11. (a) $x_1 = 0, x_2 = 0$ (b) intersecting

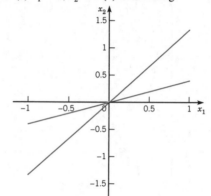

13. $\lambda_1 = 2, \mathbf{v}_1 = \begin{pmatrix} 2 \\ 1 \end{pmatrix}$; $\lambda_2 = -1, \mathbf{v}_2 = \begin{pmatrix} 1 \\ 2 \end{pmatrix}$

15. $\lambda_1 = \lambda_2 = 1, \mathbf{v}_1 = \begin{pmatrix} 2 \\ 1 \end{pmatrix}$

17. $\lambda_1 = -1 + 2i, \mathbf{v}_1 = \begin{pmatrix} 2i \\ 1 \end{pmatrix}$; $\lambda_2 = -1 - 2i, \mathbf{v}_2 = \begin{pmatrix} -2i \\ 1 \end{pmatrix}$

19. $\lambda_1 = \lambda_2 = -1, \mathbf{v}_1 = \begin{pmatrix} 2 \\ 1 \end{pmatrix}$

21. $\lambda_1 = i, \mathbf{v}_1 = \begin{pmatrix} 2 + i \\ 1 \end{pmatrix}$; $\lambda_2 = -i, \mathbf{v}_2 = \begin{pmatrix} 2 - i \\ 1 \end{pmatrix}$

23. $\lambda_1 = -3, \mathbf{v}_1 = \begin{pmatrix} -1 \\ 4 \end{pmatrix}$; $\lambda_2 = 2, \mathbf{v}_2 = \begin{pmatrix} 1 \\ 1 \end{pmatrix}$

25. $\lambda_1 = \lambda_2 = -1/2, \mathbf{v}_1 = \begin{pmatrix} 1 \\ 1 \end{pmatrix}$

27. $\lambda_1 = -2, \mathbf{v}_1 = \begin{pmatrix} -4 \\ 9 \end{pmatrix}$; $\lambda_2 = 0, \mathbf{v}_2 = \begin{pmatrix} -4 \\ 3 \end{pmatrix}$

29. $\lambda_1 = 3i, \mathbf{v}_1 = \begin{pmatrix} -1 - 3i \\ 5 \end{pmatrix}$; $\lambda_2 = -3i, \mathbf{v}_2 = \begin{pmatrix} -1 + 3i \\ 5 \end{pmatrix}$

31. $\lambda_1 = 2, \mathbf{v}_1 = \begin{pmatrix} 1 \\ 1 \end{pmatrix}; \lambda_2 = 1/2, \mathbf{v}_2 = \begin{pmatrix} -1 \\ 1 \end{pmatrix}$

33. (a) $\lambda = (-1 \pm \sqrt{25 + 4\alpha})/2$ (b) $\alpha > -25/4$ two distinct real eigenvalues; $\alpha = -25/4$ one repeated real eigenvalue; $\alpha < -25/4$ two distinct complex eigenvalues

35. (a) $\lambda = (1 + \alpha \pm \sqrt{25 - 2\alpha + \alpha^2})/2$ (b) two distinct real eigenvalues for any α

Section 3.2 Systems of Two First Order Linear Differential Equations *page 142*

1. autonomous, nonhomogeneous;

$$\mathbf{x}' = \begin{pmatrix} 0 & 1 \\ 1 & 0 \end{pmatrix} \mathbf{x} + \begin{pmatrix} 0 \\ 4 \end{pmatrix}$$

3. nonautonomous, homogeneus; $\mathbf{x}' = \begin{pmatrix} -2t & 1 \\ 3 & -1 \end{pmatrix} \mathbf{x}$

5. autonomous, homogeneus; $\mathbf{x}' = \begin{pmatrix} 3 & -1 \\ 1 & 2 \end{pmatrix} \mathbf{x}$

7. nonautonomous, nonhomogeneous;

$$\mathbf{x}' = \begin{pmatrix} 1 & 1 \\ -2 & \sin t \end{pmatrix} \mathbf{x} + \begin{pmatrix} 4 \\ 0 \end{pmatrix}$$

9. (b)

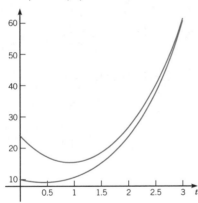

11. (b)

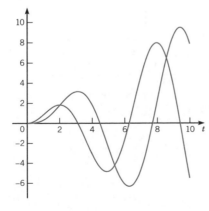

13. (T_a, T_a)

15. (a) $(2, 1)$

(b)

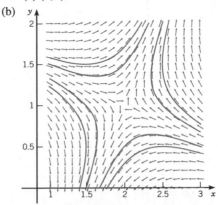

(c) almost all solutions depart from the critical point

17. (a) $(5, 9)$

(b)

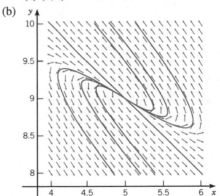

(c) other solutions depart from the critical point

19. (a) $(2, 1)$

(b)

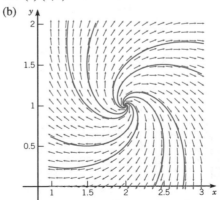

(c) other solutions spiral away from the critical point

21. $x_1' = x_2, x_2' = -2x_1 - 0.5x_2$

23. $x_1' = x_2, x_2' = (0.25 - t^2)/t^2 x_1 - (1/t)x_2$

25. $x_1' = x_2,$ $x_2' = -4x_1 - 0.25x_2 + 2\cos 3t,$ $x_1(0) = 1,$ $x_2(0) = -2;$ $\mathbf{x}' = \begin{pmatrix} 0 & 1 \\ -4 & -0.25 \end{pmatrix}\mathbf{x} + \begin{pmatrix} 0 \\ 2\cos 3t \end{pmatrix},$

$\mathbf{x}(0) = \begin{pmatrix} 1 \\ -2 \end{pmatrix}$

31. (a) $Q_1' = 3q_1 - \frac{1}{15}Q_1 + \frac{1}{100}Q_2,$ $Q_1(0) = Q_1^0,$ $Q_2' = q_2 + \frac{1}{30}Q_1 - \frac{3}{100}Q_2,$ $Q_2(0) = Q_2^0$ (b) $Q_1^E = 6(9q_1 + q_2),$ $Q_2^E = 20(3q_1 + 2q_2)$ (c) no (d) $10/9 \le Q_2^E/Q_1^E \le 20/3$

Section 3.3   Homogeneous Linear Systems with Constant Coefficients   *page 165*

1. $\mathbf{x} = c_1 e^{-t}\begin{pmatrix} 1 \\ 2 \end{pmatrix} + c_2 e^{2t}\begin{pmatrix} 2 \\ 1 \end{pmatrix}$

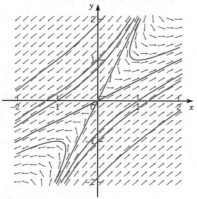

3. $\mathbf{x} = c_1 e^{-t}\begin{pmatrix} 1 \\ 3 \end{pmatrix} + c_2 e^{t}\begin{pmatrix} 1 \\ 1 \end{pmatrix}$

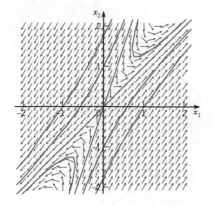

5. $\mathbf{x} = c_1 e^{-2t}\begin{pmatrix} 1 \\ 2 \end{pmatrix} + c_2\begin{pmatrix} 3 \\ 4 \end{pmatrix}$

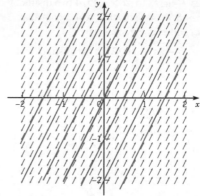

7. $\mathbf{x} = c_1 e^{2t}\begin{pmatrix} 1 \\ 1 \end{pmatrix} + c_2 e^{t/2}\begin{pmatrix} -1 \\ 1 \end{pmatrix}$

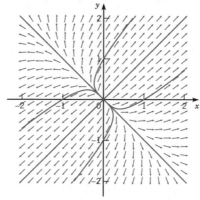

9. $\mathbf{x} = c_1 e^{t/2}\begin{pmatrix} -1 \\ 1 \end{pmatrix} + c_2 e^{t/4}\begin{pmatrix} -3 \\ 2 \end{pmatrix}$

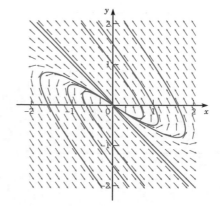

11. $\mathbf{x} = c_1 e^{-t} \begin{pmatrix} 1 \\ 1 \end{pmatrix} + c_2 e^{3t} \begin{pmatrix} 1 \\ 5 \end{pmatrix}$

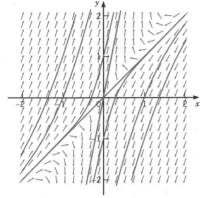

15. $\mathbf{x} = -\dfrac{3}{2} e^{2t} \begin{pmatrix} 1 \\ 3 \end{pmatrix} + \dfrac{7}{2} e^{4t} \begin{pmatrix} 1 \\ 1 \end{pmatrix}$

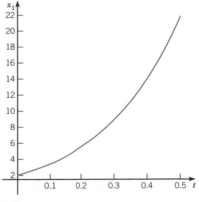

13. $\mathbf{x} = 7e^{-t} \begin{pmatrix} 1 \\ 1 \end{pmatrix} - 2e^{-2t} \begin{pmatrix} 2 \\ 3 \end{pmatrix}$

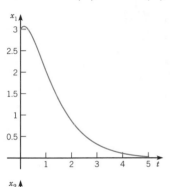

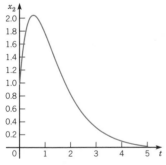

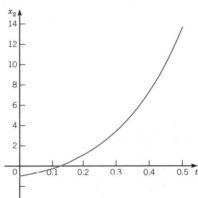

17. (a)

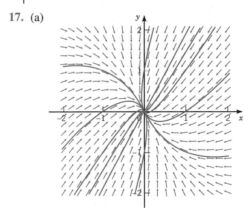

(b)

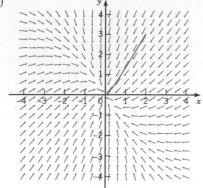

(c)

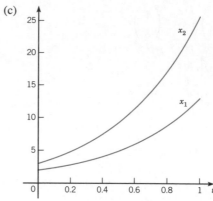

(c)

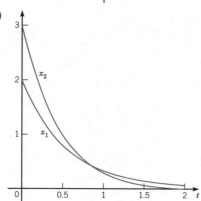

21. (a)

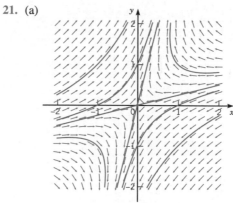

19. (a)

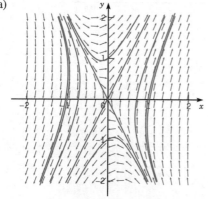

(b)

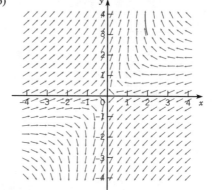

(b)

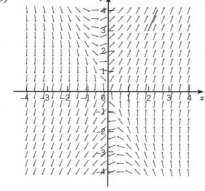

(c)

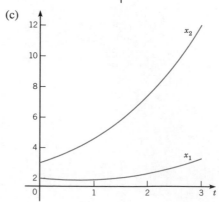

23. (a)

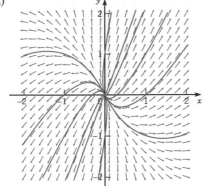

(b)

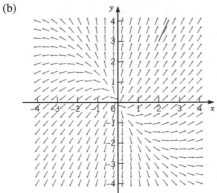

(c)

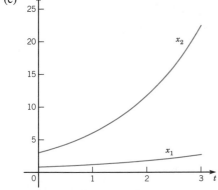

25. (a) $\mathbf{x}' = \begin{pmatrix} 0 & 1 \\ -10 & -7 \end{pmatrix}\mathbf{x}$ (b) $\mathbf{x} = c_1 e^{-2t}\begin{pmatrix} -1 \\ 2 \end{pmatrix} +$

$c_2 e^{-5t}\begin{pmatrix} -1 \\ 5 \end{pmatrix}$ (c) $y = c_1 e^{-2t} + c_2 e^{-5t}$

(d)

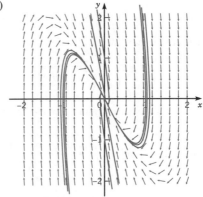

(e) $(0, 0)$ nodal sink

27. (a) $\mathbf{x}' = \begin{pmatrix} 0 & 1 \\ -12 & 7 \end{pmatrix}\mathbf{x}$

(b) $\mathbf{x} = c_1 e^{3t}\begin{pmatrix} 1 \\ 3 \end{pmatrix} + c_2 e^{4t}\begin{pmatrix} 1 \\ 4 \end{pmatrix}$ (c) $y = c_1 e^{3t} + c_2 e^{4t}$

(d)

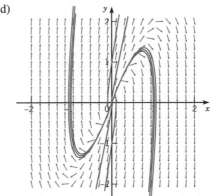

(e) $(0, 0)$ nodal source

29. (a) $\mathbf{x}' = \begin{pmatrix} 0 & 1 \\ -1/3 & -7/6 \end{pmatrix}\mathbf{x}$ (b) $\mathbf{x} = c_1 e^{-2t/3}\begin{pmatrix} -3 \\ 2 \end{pmatrix} +$

$c_2 e^{-t/2}\begin{pmatrix} -2 \\ 1 \end{pmatrix}$ (c) $y = c_1 e^{-2t/3} + c_2 e^{-t/2}$

(d)

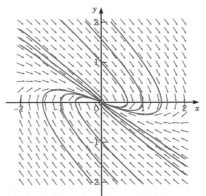

(e) $(0, 0)$ nodal sink

31. (a) If $\mathbf{u} = \mathbf{x} + \hat{\mathbf{u}}$, where $\hat{\mathbf{u}} = \begin{pmatrix} T_a \\ T_a \end{pmatrix}$, then $\mathbf{x}' = \mathbf{Kx}$.

The eigenvalues of $\mathbf{K}$ are the roots of

$\det(\mathbf{K} - \lambda\mathbf{I}) = \lambda^2 + (k_1 + k_2 + \epsilon k_2)\lambda + \epsilon k_1 k_2 = 0$,

$$\lambda_1, \lambda_2 = \frac{1}{2}\left(-(k_1 + k_2) - \epsilon k_2 \pm (k_1 + k_2)\sqrt{1 + \frac{2\epsilon k_2(k_2 - k_1)}{(k_1 + k_2)^2}}\right).$$

Then $\mathbf{x} = c_1 e^{\lambda_1 t}\mathbf{v}_1 + c_2 e^{\lambda_2 t}\mathbf{v}_2$, where $\mathbf{v}_1$ and $\mathbf{v}_2$ are the corresponding eigenvectors.

(b) For small ϵ, use the approximation $\sqrt{1 + \frac{2\epsilon k_2(k_2 - k_1)}{(k_1 + k_2)^2}} \approx$
$1 + \frac{\epsilon k_2(k_2 - k_1)}{(k_1 + k_2)^2}$.

(d) $\quad \mathbf{u} \approx c_1 e^{-(k_1 + k_2)t}\begin{pmatrix} 1 \\ -\epsilon k_2/(k_1 + k_2) \end{pmatrix} + c_2 e^{-\epsilon k_1 k_2 t/(k_1 + k_2)}$
$\begin{pmatrix} k_2/(k_1 + k_2) \\ 1 \end{pmatrix} + \begin{pmatrix} T_a \\ T_a \end{pmatrix}$

(e) For small ϵ and large t, $\mathbf{u} \approx c_2 e^{-\epsilon k_1 k_2 t/(k_1 + k_2)}\begin{pmatrix} k_2/(k_1 + k_2) \\ 1 \end{pmatrix}$

$+ \begin{pmatrix} T_a \\ T_a \end{pmatrix}$, so $u_1 \approx \frac{k_1}{k_1 + k_2}T_a + \frac{k_2}{k_1 + k_2}u_2$; u_1 is a weighted average of u_2 and T_a. For small ϵ, $c_2 \approx u_{20} - T_a$.

33. (a) $\left(\frac{1}{CR_2} - \frac{R_1}{L}\right)^2 - \frac{4}{CL} \geq 0$

Section 3.4 Complex Eigenvalues *page 177*

1. $\mathbf{x} = c_1 e^t \begin{pmatrix} \cos 2t \\ \cos 2t + \sin 2t \end{pmatrix} + c_2 e^t \begin{pmatrix} \sin 2t \\ \sin 2t - \cos 2t \end{pmatrix}$

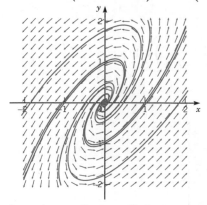

increasing amplitude oscillations

3. $\mathbf{x} = c_1 \begin{pmatrix} 5\cos t \\ 2\cos t + \sin t \end{pmatrix} + c_2 \begin{pmatrix} 5\sin t \\ 2\sin t - \cos t \end{pmatrix}$

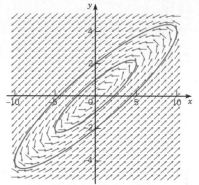

periodic solutions

5. $\mathbf{x} = c_1 e^{-t}\begin{pmatrix} \cos t \\ 2\cos t + \sin t \end{pmatrix} + c_2 e^{-t}\begin{pmatrix} \sin t \\ 2\sin t - \cos t \end{pmatrix}$

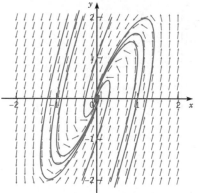

decreasing amplitude oscillations

7. $\mathbf{x} = e^{-t}\begin{pmatrix} 4\cos 2t + 6\sin 2t \\ -3\cos 2t + 2\sin 2t \end{pmatrix}$

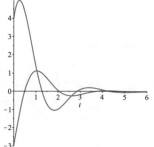

decreasing amplitude oscillations

9. $\mathbf{x} = e^{-t} \begin{pmatrix} \cos t - 3 \sin t \\ \cos t - \sin t \end{pmatrix}$

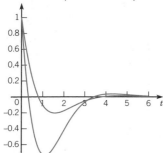

decreasing amplitude oscillations

11. (a) $\lambda = -1/4 \pm i$

(b)

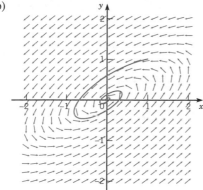

(c)

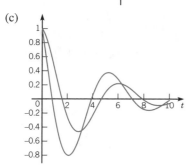

13. (a) $\lambda = \alpha \pm i$ (b) $\alpha = 0$

(c)

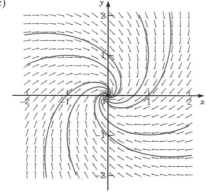

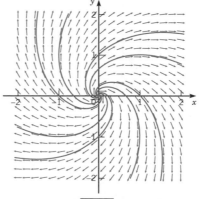

15. (a) $\lambda = \pm\sqrt{4 - 5\alpha}$ (b) $\alpha = \frac{4}{5}$

(c)

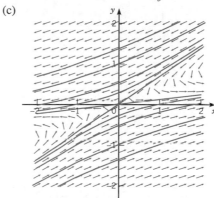

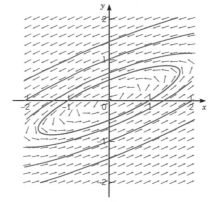

17. (a) $\lambda = -1 \pm \sqrt{-\alpha}$ (b) $\alpha = -1, 0$

(c)

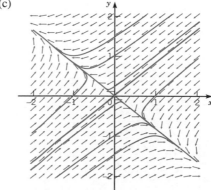

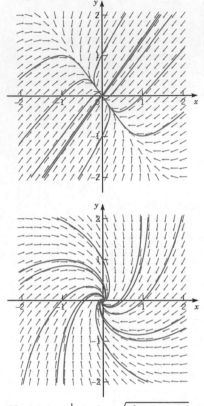

19. (a) $\lambda = \frac{1}{2}\alpha - 2 \pm \sqrt{\alpha^2 + 8\alpha - 24}$ (b) $\alpha = -4 - 2\sqrt{10}$, $-4 + 2\sqrt{10}$, $\frac{5}{2}$

(c)

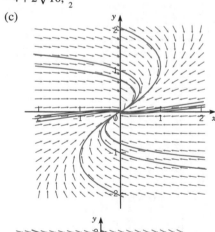

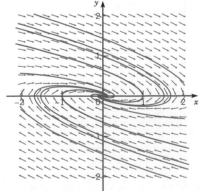

21. (b)
$$\binom{i}{v} = c_1 e^{-t/2} \binom{\cos(t/2)}{4\sin(t/2)} + c_2 e^{-t/2} \binom{\sin(t/2)}{-4\cos(t/2)}$$
(c) $c_1 = 2$, $c_2 = -\frac{3}{4}$ in answer to part (b) (d) $\lim_{t\to\infty} i(t) = \lim_{t\to\infty} v(t) = 0$; no

Section 3.5 Repeated Eigenvalues *page 188*

1. $\mathbf{x} = c_1 e^t \binom{2}{1} + c_2 \left[te^t \binom{2}{1} + e^t \binom{1}{0} \right]$

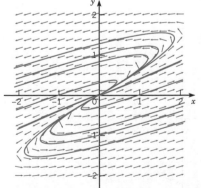

nonzero solutions grow or decrease without bound

3. $\mathbf{x} = c_1 e^{-t} \begin{pmatrix} 2 \\ 1 \end{pmatrix} + c_2 \left[t e^{-t} \begin{pmatrix} 2 \\ 1 \end{pmatrix} + e^{-t} \begin{pmatrix} 0 \\ 2 \end{pmatrix} \right]$

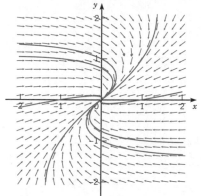

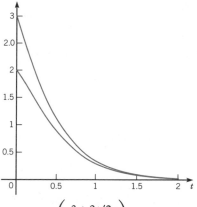

solutions converge to zero

5. $\mathbf{x} = c_1 e^{-2t} \begin{pmatrix} 1 \\ 2 \end{pmatrix} + c_2 \left[t e^{-2t} \begin{pmatrix} 1 \\ 2 \end{pmatrix} + e^{-2t} \begin{pmatrix} 0 \\ -2 \end{pmatrix} \right]$

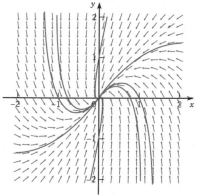

9. $\mathbf{x} = e^{t/2} \begin{pmatrix} 3 + 3t/2 \\ -2 - 3t/2 \end{pmatrix}$

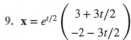

solutions converge to zero

7. $\mathbf{x} = e^{-3t} \begin{pmatrix} 3 + 4t \\ 2 + 4t \end{pmatrix}$

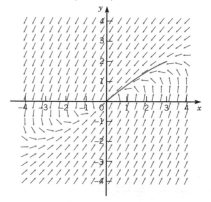

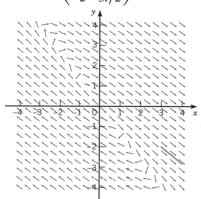

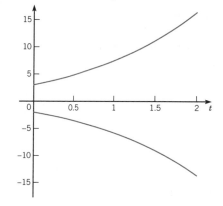

11. $\mathbf{x} = e^{-t/2}\begin{pmatrix} 3 - 5t \\ 1 - 5t \end{pmatrix}$

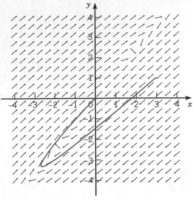

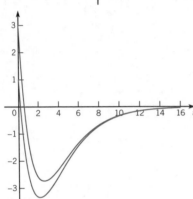

13. (b) $i(t) = e^{-t/2}(1 + t)$, $v(t) = 2e^{-t/2}(1 - t)$

Section 3.6 A Brief Introduction to Nonlinear Systems *page 195*

1. (a) $H(x, y) = y/x^2$

(b)

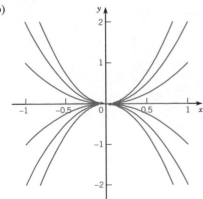

(c)

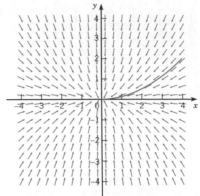

3. (a) $H(x, y) = x^2 y$

(b)

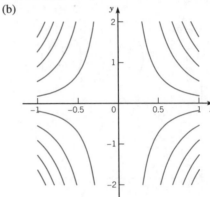

(c)

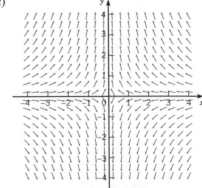

5. (a) $H(x, y) = y^2 - 4x^2$

(b)

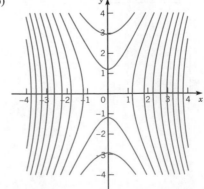

(c)

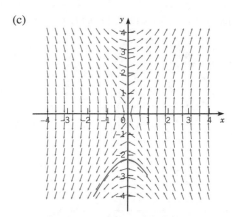

7. (a) $(0, 0)$ (b) $H(x, y) = x^2 - 4xy + y^2$

(c)

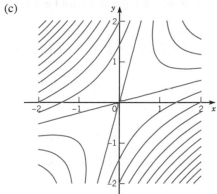

(d) $(0, 0)$ is a saddle point

9. (a) $(0, 0)$ (b) $H(x, y) = x^2 - 2xy + 2y^2$

(c)

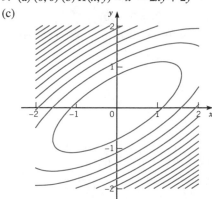

(d) $(0, 0)$ is a center

11. (a) $(0, 0)$, $(2, 3)$, $(-2, 3)$ (b) $H(x, y) = x^2 y^2 - 3x^2 y - 2y^2$

(c)

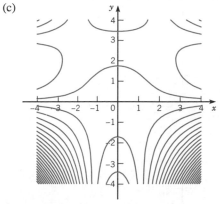

(d) $(0, 0)$, $(2, 3)$, and $(-2, 3)$ are saddle points

13. (a) $\left(-\frac{1}{2}, 1\right)$, $(0, 0)$

(b)

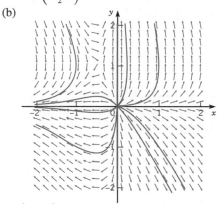

(c) $\left(-\frac{1}{2}, 1\right)$ is a saddle point; $(0, 0)$ is an unstable node

15. (a) $(0, 0)$, $(0, 2)$, $\left(\frac{1}{2}, \frac{1}{2}\right)$, $(1, 0)$

(b)

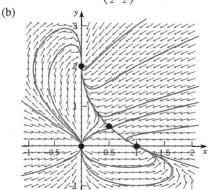

(c) $(0, 0)$ is an unstable node; $(1, 0)$ and $(0, 2)$ are asymptotically stable nodes; $\left(\frac{1}{2}, \frac{1}{2}\right)$ is a saddle point

17. (a) $(0,0)$, $(1 - \sqrt{2}, 1 + \sqrt{2})$, $(1 + \sqrt{2}, 1 - \sqrt{2})$

(b)

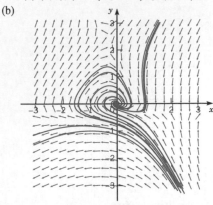

(c) $(0,0)$ is an asymptotically stable spiral point; $(1 - \sqrt{2}, 1 + \sqrt{2})$ and $(1 + \sqrt{2}, 1 - \sqrt{2})$ are saddle points

19. (a) $(0,0), (0,1), \left(\frac{1}{2}, \frac{1}{2}\right), \left(-\frac{1}{2}, \frac{1}{2}\right)$

(b)

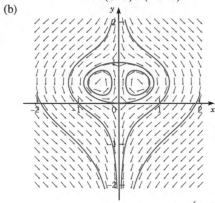

(c) $(0,0)$ and $(0,1)$ are saddle points; $\left(\frac{1}{2}, \frac{1}{2}\right)$ and $\left(-\frac{1}{2}, \frac{1}{2}\right)$ are centers

21. (a)

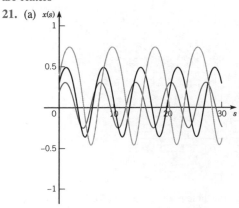

(b) the period increases as the amplitude increases

23. (a) $(0,0)$ is a saddle point; $(-1,0)$ and $(1,0)$ are asymptotically stable spiral points. (b) (i) $(1,0)$, (ii) $(-1,0)$, (iii) $(-1,0)$, (iv) $(1,0)$

25. critical points: $(0,0)$, $(0, 3/4)$, $(1/2, 1/2)$, $(1,0)$

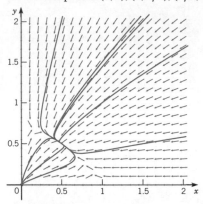

The limiting behavior of the solutions shows that the populations co-exist.

CHAPTER 4 SECOND ORDER LINEAR EQUATIONS

Section 4.1 Definitions and Examples *page 214*

1. linear; homogeneous

3. linear; homogeneous

5. nonlinear

7. linear; nonhomogeneous when $d \neq 0$, homogeneous when $d = 0$

9. $k = 140$ N/m

11. $y'' + 49y = 0$, $y(0) = 0$, $y'(0) = 0.05$

13. $Q'' + (4 \times 10^6)Q = 0$, $Q(0) = 10^{-6}$, $Q'(0) = 0$

15. $y'' + 4y' + 128y = 0$, $y(0) = 0$, $y'(0) = \frac{1}{4}$

17. $Q'' + 1500Q' + (5 \times 10^5)Q = 0$, $Q(0) = 10^{-6}$, $Q'(0) = 0$

19. (a) $mx'' + kx + \epsilon x^3 = 0$ (b) $mx'' + kx = 0$

21. $\rho l u'' + \rho_0 g u = 0$

23. (a)

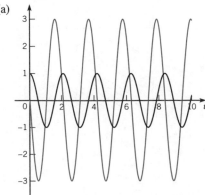

(b)

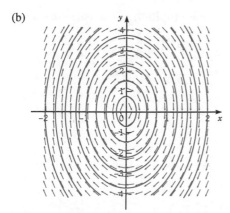

(c) $(0, 0)$ center, stable

25. (a)

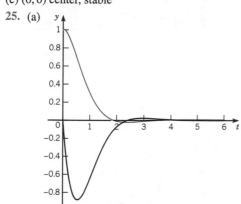

(b)

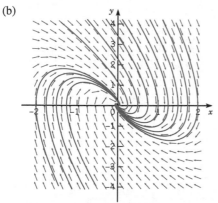

(c) $(0, 0)$ spiral, asymptotically stable

27. The frequency of the linear pendulum is higher than the frequency of the nonlinear pendulum. The difference between frequencies increases with the amplitude of oscillations.

Section 4.2 Theory of Second Order Linear Homogeneous Equations *page 226*

1. $I = (0, \infty)$
3. $I = (0, 4)$
5. $I = (0, \infty)$
7. $I = (-1, 1)$
9. $-\frac{7}{2}e^{t/2}$
11. e^{-4t}
13. $-e^{2t}$
17. no

19. $te^t + ct$
21. $-4(t \cos t - \sin t)$
23. yes
25. yes
29. $y_2 = t^3$
31. $y_2 = t^{-1} \ln t$
33. $y_2 = \cos x^2$
35. $y_2 = x^{1/4} e^{-2\sqrt{x}}$
37. -

Section 4.3 Linear Homogeneous Equations with Constant Coefficients *page 239*

1. (a) $y = c_1 e^{-3t} + c_2 e^t$ (b) $(0, 0)$ saddle point, unstable
(c) $\mathbf{x} = c_1 e^{-3t} \begin{pmatrix} 1 \\ -3 \end{pmatrix} + c_2 e^t \begin{pmatrix} 1 \\ 1 \end{pmatrix}$

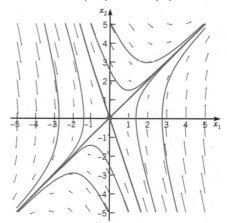

3. (a) $y = c_1 e^{2t} + c_2 t e^{2t}$ (b) $(0, 0)$ improper node, unstable
(c) $\mathbf{x} = c_1 e^{2t} \begin{pmatrix} 1 \\ 2 \end{pmatrix} + c_2 e^{2t} \begin{pmatrix} t \\ 1 + 2t \end{pmatrix}$

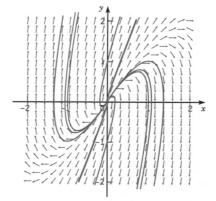

5. (a) $y = c_1 e^t \sin t + c_2 e^t \cos t$ (b) $(0,0)$ spiral point, unstable (c) $\mathbf{x} = c_1 e^t \begin{pmatrix} \sin t \\ \sin t + \cos t \end{pmatrix} + c_2 e^t \begin{pmatrix} \cos t \\ \cos t - \sin t \end{pmatrix}$

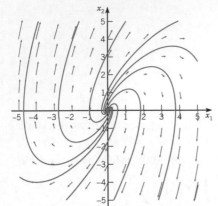

7. (a) $y = c_1 e^{t/2} + c_2 t e^{t/2}$ (b) $(0,0)$ improper node, unstable (c) $\mathbf{x} = c_1 e^{t/2} \begin{pmatrix} 1 \\ \frac{1}{2} \end{pmatrix} + c_2 e^{t/2} \begin{pmatrix} t \\ 1 + (t/2) \end{pmatrix}$

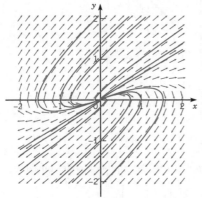

9. (a) $y = c_1 e^{t/2} + c_2 e^{-t/3}$ (b) $(0,0)$ saddle point, unstable (c) $\mathbf{x} = c_1 e^{t/2} \begin{pmatrix} 1 \\ \frac{1}{2} \end{pmatrix} + c_2 e^{-t/3} \begin{pmatrix} 1 \\ -\frac{1}{3} \end{pmatrix}$

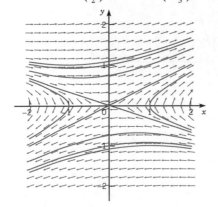

11. (a) $y = c_1 e^{-4t} + c_2 e^{2t}$ (b) $(0,0)$ saddle point, unstable (c) $\mathbf{x} = c_1 e^{-4t} \begin{pmatrix} 1 \\ -4 \end{pmatrix} + c_2 e^{2t} \begin{pmatrix} 1 \\ 2 \end{pmatrix}$

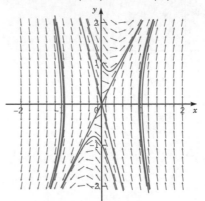

13. (a) $y = c_1 + c_2 e^{-5t}$ (b) $(x_1, 0)$ nonisolated, stable (c) $\mathbf{x} = c_1 \begin{pmatrix} 1 \\ 0 \end{pmatrix} + c_2 e^{-5t} \begin{pmatrix} 1 \\ -5 \end{pmatrix}$

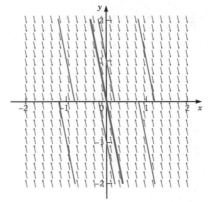

15. (a) $y = c_1 e^{2t/5} + c_2 t e^{2t/5}$ (b) $(0,0)$ improper node, unstable (c) $\mathbf{x} = c_1 e^{2t/5} \begin{pmatrix} 1 \\ \frac{2}{5} \end{pmatrix} + c_2 e^{2t/5} \begin{pmatrix} t \\ 1 + (2t/5) \end{pmatrix}$

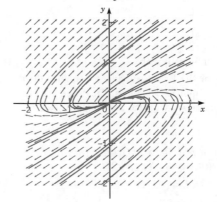

17. (a) $y = c_1 e^{-3t} \sin 2t + c_2 e^{-3t} \cos 2t$ (b) $(0,0)$ spiral point, asymptotically stable (c) $\mathbf{x} = c_1 e^{-3t} \begin{pmatrix} \sin 2t \\ 2\cos 2t - 3\sin 2t \end{pmatrix} -$

$c_2 e^{-3t} \begin{pmatrix} \cos 2t \\ 3\cos 2t + 2\sin 2t \end{pmatrix}$

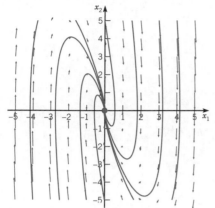

19. (a) $\lambda_1 = (9 + 3\sqrt{5})/2$, $\lambda_2 = (9 - 3\sqrt{5})/2$; $y = c_1 e^{\lambda_1 t} + c_2 e^{\lambda_2 t}$ (b) $(0,0)$ node, unstable (c) $\mathbf{x} = c_1 e^{\lambda_1 t} \begin{pmatrix} 1 \\ \lambda_1 \end{pmatrix} + c_2 e^{\lambda_2 t} \begin{pmatrix} 1 \\ \lambda_2 \end{pmatrix}$

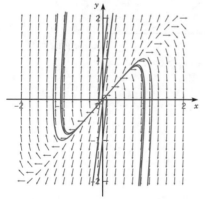

21. (a) $y = c_1 e^{-2t} + c_2 t e^{-2t}$ (b) $(0,0)$ improper node, asymptotically stable (c) $\mathbf{x} = c_1 e^{-2t} \begin{pmatrix} 1 \\ -2 \end{pmatrix} + c_2 e^{-2t} \begin{pmatrix} t \\ 1 - 2t \end{pmatrix}$

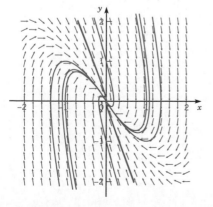

23. (a) $y = c_1 \sin(3t/2) + c_2 \cos(3t/2)$ (b) $(0,0)$ center, stable (c) $\mathbf{x} = c_1 \begin{pmatrix} \sin(3t/2) \\ \frac{3}{2}\cos(3t/2) \end{pmatrix} + c_2 \begin{pmatrix} \cos(3t/2) \\ -\frac{3}{2}\sin(3t/2) \end{pmatrix}$

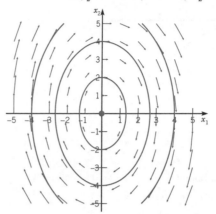

25. (a) $y = c_1 e^{-t/2} \sin t + c_2 e^{-t/2} \cos t$ (b) $(0,0)$ spiral point, asymptotically stable (c) $\mathbf{x} = c_1 e^{-t/2} \begin{pmatrix} \sin t \\ \cos t - \frac{1}{2}\sin t \end{pmatrix} -$

$c_2 e^{-t/2} \begin{pmatrix} \cos t \\ \sin t + \frac{1}{2}\cos t \end{pmatrix}$

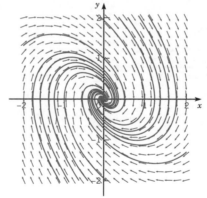

27. $y = e^t$,

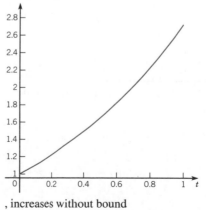

, increases without bound

29. $y = 2e^{2t/3} - \frac{7}{3}te^{2t/3}$,

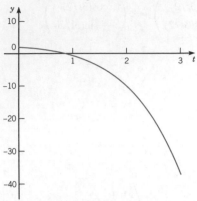

, decreases without bound

31. $y(t) = 2e^{-2t}\sin t + e^{-2t}\cos t$,

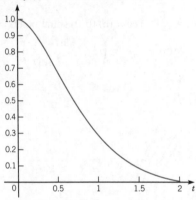

, decaying oscillation

33. $y = 2te^{-3t}$,

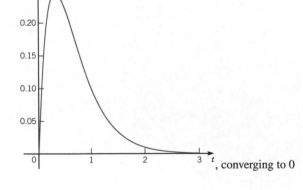

, converging to 0

35. $y = -1 - e^{-3t}$,

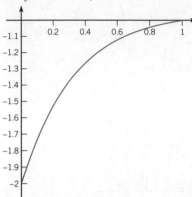

, converging to -1

37. $y = 7e^{-2(t+1)} + 5te^{-2(t+1)}$,

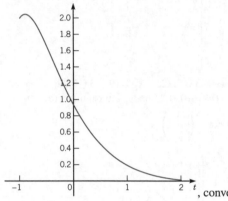

, converging to 0

39. $y = \frac{5}{2}e^{-t/2}\sin t + 3e^{-t/2}\cos t$,

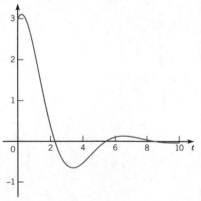

, decaying oscillation

41. $y = \frac{1}{10} e^{9-9t} + \frac{9}{10} e^{-1+t}$,

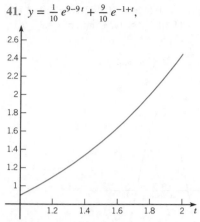

, increases without bound

43. $y = -\frac{1}{2} e^{1+(1/2)t} + \frac{3}{2} e^{-1-(1/2)t}$,

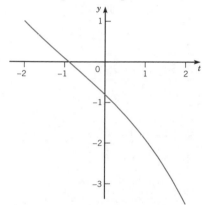

, decreases without bound

45. $y'' + 4y' + 4 = 0$

47. $y \to 0$ for $\alpha < 0$; y becomes unbounded for $\alpha > 3$

51. (a) $b > 0$ and $0 < c < b^2/4a$ (b) $c < 0$ (c) $b < 0$ and $0 < c < b^2/4a$

55. $y = c_1/x^2 + c_2/x$ **57.** $y = c_1 x^{-1} + c_2 x^6$

59. $y = c_1 x^3 + c_2 x^3 \ln x$

61. $y = c_1 x^{3/2} \sin\left(\frac{\sqrt{3}}{2} \ln x\right) + c_2 x^{3/2} \cos\left(\frac{\sqrt{3}}{2} \ln x\right)$

63. $y = -x^{-1/2} \sin(2 \ln x) + 2x^{-1/2} \cos(2 \ln x)$,

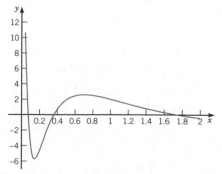

, growing amplitude oscillation

65. $y = x^{-1} \cos(2 \ln x)$,

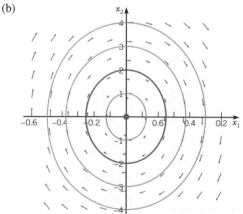

, growing amplitude oscillation

Section 4.4   Mechanical and Electrical Vibrations   *page 250*

1. $y = 3\sqrt{2} \cos(2t - \pi/4)$

3. $y = 2\sqrt{5} \cos(3t - \delta)$, $\delta = -\arctan(\frac{1}{2}) \approx -0.4636$

5. (a) $y = \frac{1}{4} \cos(8t)$ ft, t in s, $\omega = 8$ rad/s, $T = \pi/4$ s, $R = \frac{1}{4}$ ft,

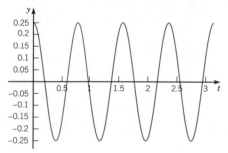

(b)

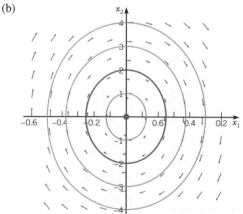

7. $y = (\sqrt{2}/8) \sin(8\sqrt{2}t) - \frac{1}{12} \cos(8\sqrt{2}t)$ ft, t in s, $\omega = 8\sqrt{2}$ rad/s, $T = \sqrt{2}\pi/8$ s, $R = \sqrt{11/288} \approx 0.1954$ ft, $\delta = \pi - \arctan(3/\sqrt{2}) \approx 2.0113$

9. (a) $y = e^{-10t}\left[2\cos(4\sqrt{6}t) + (5/\sqrt{6})\sin(4\sqrt{6}t)\right]$ cm, t ins; $v = 4\sqrt{6}$ rad/s, $T_d = \sqrt{6}\pi/12$ s, $T_d/T = \frac{7}{2}\sqrt{6} \approx 1.4289$, $\tau \approx 0.4045$

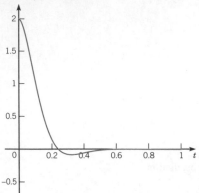

(b)

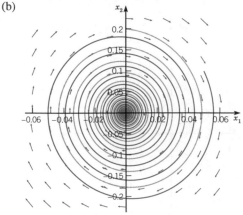

11. (a) $y \approx 0.057198e^{-0.15t}\cos(3.87008t - 0.50709)$ m, t in s; $v = 3.87008$ rad/s, $v/\omega_0 = 3.87008/\sqrt{15} \approx 0.99925$

(b)

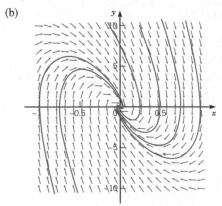

13. $\gamma = \sqrt{20/9} \approx 1.4907$

17. $\gamma = 8$ lb-s/ft

23. $\gamma = 5$ lb-s/ft

25. (a) $\tau \approx 41.715$ (b) $\gamma = 0.5$, $\tau \approx 20.402$; $\gamma = 1.0$, $\tau \approx 9.168$; $\gamma = 1.5$, $\tau \approx 7.184$ (d) $\gamma_0 \approx 1.73$, min $\tau \approx 4.87$ (e) $\tau = (2/\gamma)\ln(400/\sqrt{4 - \gamma^2})$

27. $\rho l u'' + \rho_0 g u = 0$, $T = 2\pi\sqrt{\rho l/\rho_0 g}$

29. (a) $y = \sqrt{2}\sin\sqrt{2}t$

(b)

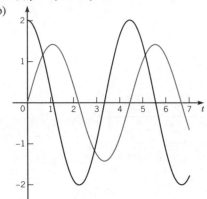

(c)

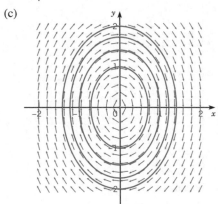

31. (b) $y = a\cos(\sqrt{k/mt}) + b\sqrt{m/k}\sin(\sqrt{k/mt})$
(c) $(ka^2 + mb^2)/2$

Section 4.5 Nonhomogeneous Equations; Method of Undetermined Coefficients *page 260*

1. $y = c_1 e^{3t} + c_2 e^{-t} - e^{2t}$

3. $y = c_1 e^{3t} + c_2 e^{-t} + \frac{3}{16}te^{-t} + \frac{3}{8}t^2 e^{-t}$

5. $y = c_1\cos 3t + c_2\sin 3t + \frac{1}{162}(9t^2 - 6t + 1)e^{3t} + \frac{2}{3}$

7. $y = c_1 e^t + c_2 e^{4t} - 2te^t/3$

9. $y = c_1 e^{-t} + c_2 te^{-t} + 3t^2 e^{-t}/2$

11. $y = c_1 e^{-t} + c_2 e^{-t/2} + t^2 - 6t + 14 - \frac{3}{10}\sin t - \frac{9}{10}\cos t$

13. $u = c_1\cos\omega_0 t + c_2\sin\omega_0 t + (\omega_0^2 - \omega^2)^{-1}\cos\omega t$

15. $y = c_1 e^{-t/2}\cos(\sqrt{15}t/2) + c_2 c_1 e^{-t/2}\sin(\sqrt{15}t/2) + \frac{1}{6}e^t - \frac{1}{4}e^{-t}$

17. $y = e^t - \frac{1}{2}e^{-2t} - t - \frac{1}{2}$

19. $y = -3e^t + 4te^t + \frac{1}{6}t^3 e^t + 4$

21. $y = -\frac{1}{8}\sin 2t + 2\cos 2t - \frac{3}{4}t\cos 2t$

23. (a) $Y(t) = t(A_0 t^4 + A_1 t^3 + A_2 t^2 + A_3 t + A_4) + t(B_0 t^2 + B_1 t + B_2)e^{-3t} + D\sin 3t + E\cos 3t$ (b) $A_0 = \frac{2}{15}$, $A_1 = -\frac{2}{9}$, $A_2 = \frac{8}{27}$, $A_3 = -\frac{8}{27}$, $A_4 = \frac{16}{81}$, $B_0 = -\frac{1}{9}$, $B_1 = -\frac{1}{9}$, $B_2 = -\frac{2}{27}$, $D = -\frac{1}{18}$, $E = -\frac{1}{18}$

25. (a) $Y(t) = e^t(A\cos 2t + B\sin 2t) + (D_0 t + D_1)e^{2t}\sin t + (E_0 t + E_1)e^{2t}\cos t$ (b) $A = -\frac{1}{20}$, $B = -\frac{3}{20}$, $D_0 = -\frac{3}{2}$, $D_1 = -5$, $E_0 = \frac{3}{2}$, $E_1 = \frac{1}{2}$

27. (a) $Y(t) = A_0 t^2 + A_1 t + A_2 + t^2(B_0 t + B_1)e^{2t} + (D_0 t + D_1)\sin 2t + (E_0 t + E_1)\cos 2t$ (b) $A_0 = \frac{1}{2}$, $A_1 = 1$, $A_2 = \frac{3}{4}$, $B_0 = \frac{2}{3}$, $B_1 = 0$, $D_0 = 0$, $D_1 = -\frac{1}{16}$, $E_0 = \frac{1}{8}$, $E_1 = \frac{1}{16}$

29. (a) $Y(t) = (A_0 t^2 + A_1 t + A_2)e^t\sin 2t + (B_0 t^2 + B_1 t + B_2)e^t\cos 2t + e^{-t}(D\cos t + E\sin t) + Fe^t$ (b) $A_0 = \frac{1}{52}$, $A_1 = \frac{10}{169}$, $A_2 = -\frac{1233}{35152}$, $B_0 = -\frac{5}{52}$, $B_1 = \frac{73}{676}$, $B_2 = -\frac{4105}{35152}$, $D = -\frac{3}{2}$, $E = \frac{3}{2}$, $F = \frac{2}{3}$

31. (a) - (b) $w = -\frac{2}{5} + c_1 e^{5t}$ (c) -

33. $y = c_1 x^{-1} + c_2 x^{-5} + \frac{1}{12}x$

35. $y = c_1\cos(2\ln x) + c_2\sin(2\ln x) + \frac{1}{3}\sin(\ln x)$

37. $y =$
$$\begin{cases} t, & 0 \le t \le \pi \\ -(1 + \pi/2)\sin t - (\pi/2)\cos t + (\pi/2)e^{\pi-t}, & t > \pi \end{cases}$$

Section 4.6   Forced Vibrations, Frequency Response, and Resonance   *page 272*

1. $-2\sin 7t\sin 4t$

3. $2\cos(9\pi t/2)\cos(5\pi t/2)$

5. $y'' + 256y = 56\cos 3t$, $y(0) = 1$, $y'(0) = 0$, y in ft, t in s

7. (a) $y = \frac{191}{247}\cos 16t + \frac{56}{247}\cos 3t$

(b)

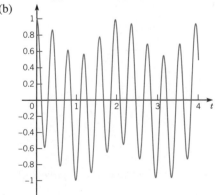

(c) $G(i\omega) = \dfrac{1}{256 - \omega^2}$, $|G(i\omega)| = \dfrac{1}{|256 - \omega^2|}$, $\phi(\omega) =$
$$\begin{cases} 0 & \omega < 16 \\ \pi/2 & \omega = 16 \\ \pi & \omega > 16 \end{cases}, \quad \omega_{\max} = 16$$

9. $y = \frac{8}{3}(\cos 7t - \cos 8t)$ ft, t in s

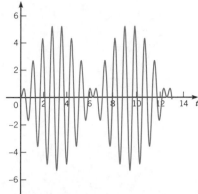

11. (a) $y = \frac{180}{901}\cos 2t + \frac{6}{901}\sin 2t$ ft, t in s (b) $m = 4$ slugs

15. $Q(t) = 10^{-6}(3e^{-4000t}/4 - 3e^{-1000t} + 9/4)$ coulombs, t in s, $Q(0.001) \approx 1.1601 \times 10^{-6}$; $Q(0.01) \approx 2.2499 \times 10^{-6}$; $Q(t) \to (9/4) \times 10^{-6}$ as $t \to \infty$

17. (a) $y = \dfrac{3(\cos t - \cos\omega t)}{\omega^2 - 1}$

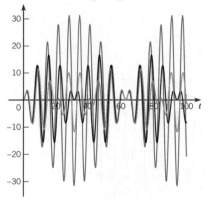

19.

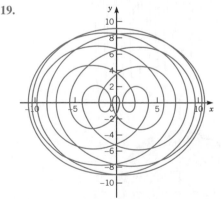

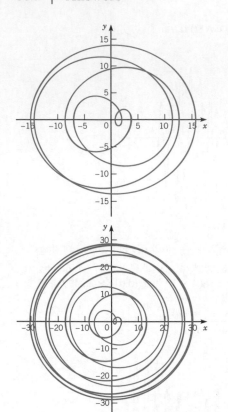

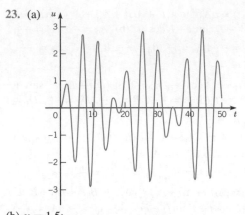

23. (a)

(b) $\omega = 1.5$:

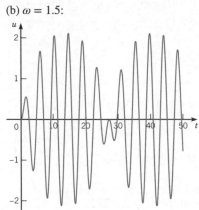

$\omega = 1.9$:

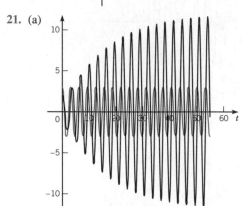

21. (a)

(b)

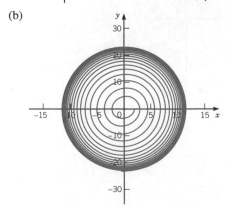

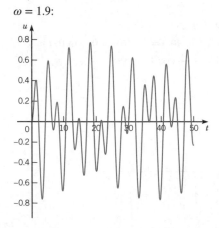

Section 4.7   Variation of Parameters   *page 280*

3. $\mathbf{x} = \begin{pmatrix} 2t + te^{-t} \\ -(1 + t) \end{pmatrix}$

5. $\mathbf{x} = \begin{pmatrix} t\cos t \\ -t\sin t \end{pmatrix}$

7. $\mathbf{x} = \begin{pmatrix} 2t - 2e^t + e^{-t}t + e^{-t} \\ -t - 1 + 2e^t \end{pmatrix}$

9. $\mathbf{x} = \begin{pmatrix} \sin t + \cos t + t \cos t \\ \cos t - \sin t - t \sin t \end{pmatrix}$

11. $y = -\frac{2}{3} t e^{-t}$

13. $y = 2t^2 e^{t/2}$

15. $y = c_1 \cos 2t + c_2 \sin 2t + (\sin 2t) \ln(\tan 2t + \sec 2t) - \frac{3}{4}$

17. $y = c_1 \cos 2t + c_2 \sin 2t + 2 \sin(t/2) + 6 \sin(3t/2) + 3 \ln(\tan(t/4)) \sin 2t$

19. $y = c_1 e^t + c_2 t e^t - \frac{1}{2} e^t \ln(1 + t^2) + t e^t \arctan t$

21. $y = c_1 \cos 2t + c_2 \sin 2t + \frac{1}{2} \int [\sin 2(t - s)] g(s) \, ds$

23. $Y(t) = \frac{1}{2}(t - 1) e^{2t}$

25. $Y(x) = -\frac{3}{2} x^{1/2} \cos x$

27. $Y(x) = x^{-1/2} \int t^{-3/2} \sin(x - t) g(t) \, dt$

29. $y = c_1 x^2 + c_2 x^2 \ln x + \frac{1}{6} x^2 (\ln x)^3$

31. $y = c_1 t^{-1} + c_2 t^{-5} + \frac{1}{2} t$

35. $y = (b - a)^{-1} \int_{t_0}^t [e^{b(t-s)} - e^{a(t-s)}] g(s) \, ds$

37. $y = \int_{t_0}^t (t - s) e^{a(t-s)} g(s) \, ds$

39. -

41. $y = c_1 e^t + c_2 t - \frac{1}{2}(2t - 1) e^{-t}$

CHAPTER 5   THE LAPLACE TRANSFORM

Section 5.1   Definition of the Laplace Transform   *page 302*

1. piecewise continuous

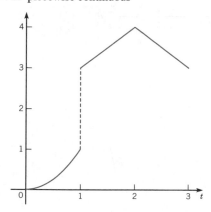

3. piecewise continuous

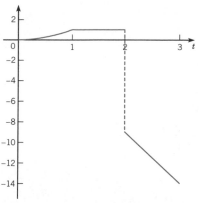

5. yes; $M = 0$, $K = 2$, $a = 3$

7. yes; $M = 0$, $K = 1$, $a = 5$

9. no

11. yes; $M = 0$, $K = 1$, $a = 0$

13. (a) $1/s^2$ (b) $2/s^3$ (c) $n!/s^{n+1}$

15. $\dfrac{e^{-s}}{s} - \dfrac{e^{-2s}}{s}$

17. $\dfrac{2}{s^3} + \left(\dfrac{3}{s} - \dfrac{3}{s^2} - \dfrac{2}{s^3} \right) e^{-s} + \left(\dfrac{3}{s} + \dfrac{1}{s^2} \right) e^{-2s}$

19. $\dfrac{b}{s^2 - b^2}$, $s > |b|$

21. $\dfrac{b}{(s - a)^2 - b^2}$, $s - a > |b|$

23. $\dfrac{b}{(s - a)^2 + b^2}$, $s > a$

25. $\dfrac{1}{(s - a)^2}$, $s > a$

27. $\dfrac{s^2 + a^2}{(s - a)^2 (s + a)^2}$, $s > |a|$

29. $\dfrac{2a(3s^2 - a^2)}{(s^2 + a^2)^3}$, $s > 0$

31. converges

33. diverges

Section 5.2   Properties of the Laplace Transform   *page 309*

1. $\dfrac{3}{(s + 2)^2 + 9}$

3. $\dfrac{720}{s^7} - \dfrac{8}{s^3} + \dfrac{5}{s}$

5. $\dfrac{720}{(s + 2)^7} + \dfrac{12}{(s + 2)^4} + \dfrac{1}{s + 2}$

7. $2 \dfrac{3 s^2 b - b^3}{\left(s^2 + b^2 \right)^3}$

9. $2\dfrac{b(s-a)}{\left[(s-a)^2 + b^2\right]^2}$

13. $\dfrac{18s + 15}{9s^2 + 12s + 4}$

15. $\dfrac{24s + 20}{6s^2 + 5s + 1}$

17. $\dfrac{s^4 - 5s^3 + 7s^2 + 2}{s^3\left(s^2 - 5s - 6\right)}$

19. $\dfrac{s^5 + 8s^3 + 16s + 2s^4 + 17s^2 + 28}{\left(s^2 + 4\right)^2\left(s^2 + 2s + 5\right)}$

21. $\dfrac{1 + 9(1 + s)^2}{(1 + s)^2(s^4 - 6)}$

23. $\dfrac{9s^2 + 2s + 1}{s(s^2 + 16)} - \dfrac{e^{-\pi s}}{s(s^2 + 16)}$

25. $\dfrac{1}{s^2(s^2 + 4)} - \dfrac{e^{-s}}{s^2(s^2 + 4)}$

27. $F_0\left[\dfrac{1}{s^2(ms^2 + \gamma s + k)} - \dfrac{e^{-sT}(1 + sT)}{s^2(ms^2 + \gamma s + k)}\right]$

29. (a) $Y' + s^2 Y = s$ (b) $s^2 Y'' + 2sY' - [s^2 + \alpha(\alpha + 1)]Y = -1$

Section 5.3 The Inverse Laplace Transform *page 319*

1. $a = 7$, $b = -4$

3. $a = 2, b = -5, c = 3$

5. $a = 2, b = 1, c = -2$

7. $a_1 = -1, a_2 = 10, a_3 = -24, b_1 = 0, b_2 = 0, b_3 = 0$

9. $6\sin 5t$

11. $-\frac{2}{5}e^{-4t} + \frac{2}{5}e^t$

13. $5e^{-5t}\cos 7t$

15. $2e^t\cos t + 3e^t\sin t$

17. $-2e^{-2t}\cos t + 5e^{-2t}\sin t$

19. $2e^{2t} - 3e^{-2t} + e^{-t}$

21. $4e^{4t}\sin 3t + 12e^{-3t}$

23. $-3t + e^{-t}\cos t$

25. $(1 - \frac{7}{3}t)e^{-t}\cos 3t - (\frac{8}{9} + \frac{4}{3}t)e^{-t}\sin 3t$

27. $\left(\frac{1}{250} + \frac{13}{25}t - \frac{1}{40}t^2\right)e^t\cos t + \left(-\frac{261}{500} + \frac{3}{200}t + \frac{1}{5}t^2\right)e^t\sin t - \left(\frac{1}{250} + \frac{1}{100}t\right)e^{-2t}$

Section 5.4 Solving Differential Equations with Laplace Transforms *page 327*

1. $6e^{-2t} + 2e^{6t}$

3. $e^{4t}\sin 3t$

5. $2e^t\cos\sqrt{3}t - \frac{2}{3}\sqrt{3}e^t\sin\sqrt{3}t$

7. $(\omega^2 - 4)^{-1}\left[(\omega^2 - 5)\cos\omega t + \cos 2t\right]$

9. $\frac{7}{5}e^t\sin t - \frac{1}{5}e^t\cos t + \frac{1}{5}e^{-t}$

11. $e^t t - t^2 e^t + \frac{2}{3}t^3 e^t$

13. $\cos\sqrt{3}t$

15. $\mathbf{y} = \begin{pmatrix} 4e^{2t} - 3e^t \\ 6e^{2t} - 6e^t \end{pmatrix}$

17. $\mathbf{y} = \begin{pmatrix} 5\cos 6t + 4\sin 6t \\ 4\cos 6t - 5\sin 6t \end{pmatrix}$

19. $\mathbf{y} = \begin{pmatrix} -64te^{-6t} \\ (1 - 8t)e^{-6t} \end{pmatrix}$

21. $\mathbf{y} = \begin{pmatrix} -\frac{4}{25}e^{-t} + \left(-\frac{82}{15}t - \frac{589}{225}\right)e^{4t} - \frac{2}{9}e^t \\ \frac{1}{25}e^{-t} + \left(\frac{641}{225} - \frac{82}{15}t\right)e^{4t} - \frac{8}{9}e^t \end{pmatrix}$

23. $\mathbf{y} = \begin{pmatrix} -\frac{1}{20}e^{-3t} + \frac{1}{4}e^{-t} - \frac{1}{5}\cos t + \frac{1}{10}\sin t \\ -\frac{3}{10}\cos t + \frac{2}{5}\sin t + \frac{1}{20}e^{-3t} + \frac{1}{4}e^{-t} \end{pmatrix}$

25. $\begin{pmatrix} x \\ y \end{pmatrix} = \begin{pmatrix} -\frac{1}{5}\sqrt{10}\sin\left(\frac{1}{2}\sqrt{10}t\right) + \sin t \\ \frac{1}{5}\sqrt{10}\sin\left(\frac{1}{2}\sqrt{10}t\right) \end{pmatrix}$

Section 5.5 Discontinuous Functions and Periodic Functions *page 336*

1.

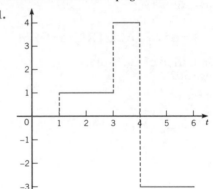

3.

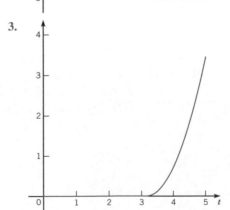

5.

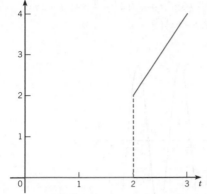

7. $F(s) = \dfrac{120e^{-9s}}{s^6}$

9. $F(s) = \dfrac{e^{-s\pi}}{s^2} - \dfrac{\pi e^{-2s\pi}}{s} - \dfrac{e^{-2s\pi}}{s^2}$

11. $F(s) = -\dfrac{e^{-3s}}{s} + \dfrac{e^{-3s}}{s^2} - \dfrac{e^{-4s}}{s} - \dfrac{e^{-4s}}{s^2}$

13. $f(t) = e^{5(t-1)}(t-1)^5 u_1(t)$

15. $f(t) = 2e^{t-2}\cos(t-2)u_2(t)$

17. $f(t) = \frac{1}{2}\left(e^{3(t-1)} + e^{t-1}\right)u_1(t)$

19. $F(s) = \dfrac{1 - e^{-s}}{s}, \; s > 0$

21. $F(s) = \dfrac{1}{s}\left[1 - e^{-s} + \cdots + e^{-2ns} - e^{-(2n+1)s}\right] = \dfrac{1 - e^{-(2n+2)s}}{s(1 + e^{-s})}$, $s > 0$

23. $\mathcal{L}\{f(t)\} = \dfrac{1 - (1+s)e^{-s}}{s^2(1 - e^{-s})}, \; s > 0$

25. (a) $\mathcal{L}\{f(t)\} = s^{-1}(1 - e^{-s}), \; s > 0$

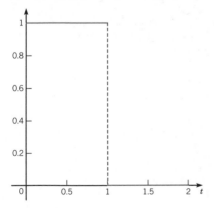

(b) $\mathcal{L}\{g(t)\} = s^{-2}(1 - e^{-s}), \; s > 0$

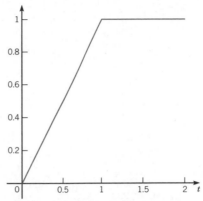

(c) $\mathcal{L}\{h(t)\} = s^{-2}(1 - e^{-s})^2, \; s > 0$

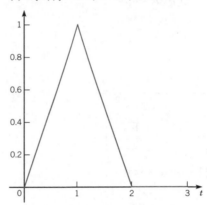

Section 5.6   Differential Equations with Discontinuous Forcing Functions *page 342*

1. $y = 1 + 4\cos t + 3\sin t - u_{\pi/2}(t)(1 - \sin t)$

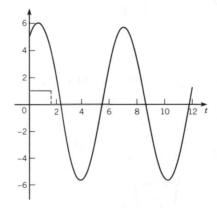

3. $y = \frac{1}{6}[1 - u_{2\pi}(t)](2\sin t - \sin 2t)$

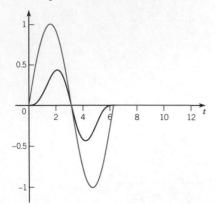

5. $y = \frac{1}{2} + \frac{1}{2}e^{-2t} - e^{-t} - u_{10}(t)[\frac{1}{2} + \frac{1}{2}e^{-2(t-10)} - e^{-(t-10)}]$

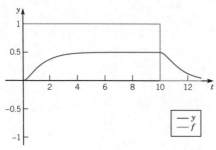

7. $y = \cos t + u_{3\pi}(t)[1 - \cos(t - 3\pi)]$

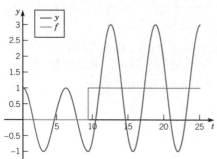

9. $y = 6\cos t + \frac{15}{2}\sin t + \frac{1}{2}t - \frac{1}{2}u_6(t)[t - 6 - \sin(t - 6)]$

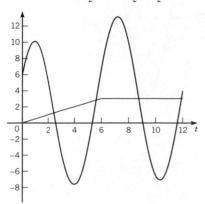

11. $y = 3\cos 2t + \frac{7}{2}\sin 2t + u_\pi(t)[\frac{1}{4} - \frac{1}{4}\cos(2t - 2\pi)] -$
$u_{3\pi}(t)[\frac{1}{4} - \frac{1}{4}\cos(2t - 6\pi)]$

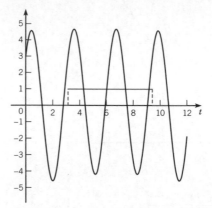

13. $y = h(t) - u_\pi(t)h(t - \pi)$, $\quad h(t) = (3 - 4\cos t + \cos 2t)/12$

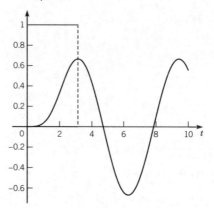

15. $g(t) = [u_{t_0}(t)(t - t_0) - 2u_{t_0+k}(t)(t - t_0 - k) + u_{t_0+2k}(t)(t - t_0 - 2k)](h/k)$

17. (a) $k = 5$

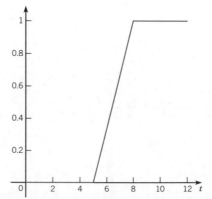

(b) $y = [u_5(t)h(t - 5) - u_{5+k}(t)h(t - 5 - k)]/k$, $\quad h(t) = \frac{1}{4}t - \frac{1}{8}\sin 2t$

(c)

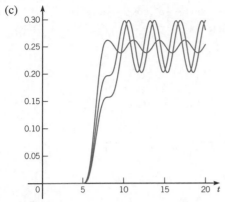

$A = |\sin k|/4k$

19. (a) $y = 1 - \cos t + 2\sum_{k=1}^{n}(-1)^k u_{k\pi}(t)[1 - \cos(t - k\pi)]$

(b)

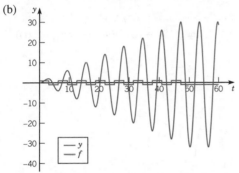

(c) As $n \to \infty$, the amplitude of the solution increases

21. (a)

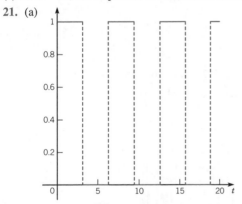

(b) $y = 1 - \cos t + \sum_{k=1}^{n}(-1)^k u_{k\pi}(t)[1 - \cos(t - k\pi)]$

(c)

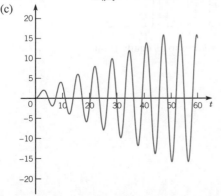

23. (a) $y = 1 - \cos t + 2\sum_{k=1}^{n}(-1)^k u_{11k/4}(t)[1 - \cos(t - \frac{11}{4}k)]$

(b)

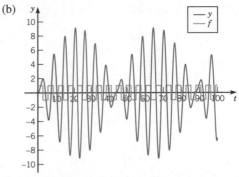

(c) slow period ≈ 88, fast period ≈ 6 (d) slow period 88.247, fast period 5.8656

Section 5.7   Impulse Functions   *page 350*

1. $y = e^{-t}\sin t + u_{\pi}(t)e^{-(t-\pi)}\sin(t - \pi)$

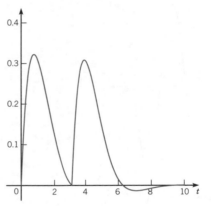

3. $y = -\frac{1}{2}e^{-2t} + \frac{1}{2}e^{-t} + u_5(t)[-e^{-2(t-5)} + e^{-(t-5)}] + u_{10}(t)$ $[\frac{1}{2} + \frac{1}{2}e^{-2(t-10)} - e^{-(t-10)}]$

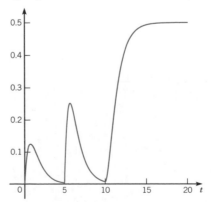

5. $y = \frac{1}{4}\sin t - \frac{1}{4}\cos t + \frac{1}{4}e^{-t}\cos\sqrt{2}t +$
$(1/\sqrt{2})u_{3\pi}(t)e^{-(t-3\pi)}\sin\sqrt{2}(t-3\pi)$

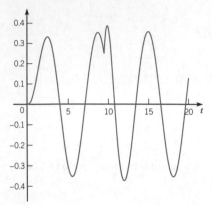

7. $y = \sin t + u_{2\pi}(t)\sin(t - 2\pi)$

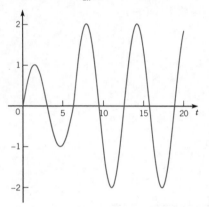

9. $y = u_{\pi/2}(t)[1 - \cos(t - \pi/2)] +$
$3u_{3\pi/2}(t)\sin(t - 3\pi/2) - u_{2\pi}(t)[1 - \cos(t - 2\pi)]$

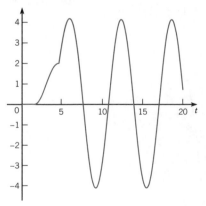

11. $y = \frac{1}{5}\cos t + \frac{2}{5}\sin t - \frac{1}{5}e^{-t}\cos t - \frac{3}{5}e^{-t}\sin t +$
$u_{\pi/2}(t)e^{-(t-\pi/2)}\sin(t - \pi/2)$

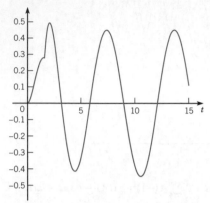

13. (a) $-e^{-T/4}\delta(t - 5 - T)$, $T = 8\pi/\sqrt{15}$
(b) $y = \frac{2}{\sqrt{15}}e^{-(t-5)/4}\sin(\sqrt{15}(t-5)/4)(u_5(t) - u_{5+8\pi/\sqrt{15}}(t))$

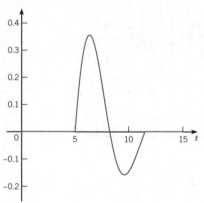

15. (a) $k_1 \approx 6.9557$ (b) $k_1 \approx 6.4746$ (c) $k_1 = 6$
17. (b) $y = \sum_{k=1}^{20} u_{k\pi}(t)\sin(t - k\pi)$

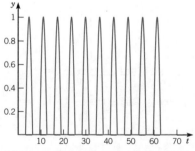

(c) the oscillator returns to its equilibrium

19. (b) $y = \sum_{k=1}^{20} u_{k\pi/2}(t) \sin(t - k\pi/2)$

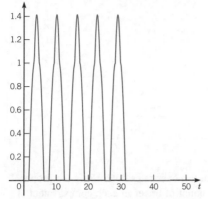

(c) the oscillator returns to its equilibrium

21. (b) $y = \sum_{k=1}^{15} u_{(2k-1)\pi}(t) \sin[t - (2k-1)\pi]$

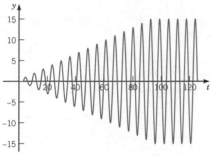

(c) the oscillator continues to oscillate at a constant amplitude

23. (b)

$$y = \frac{20}{\sqrt{399}} \sum_{k=1}^{20} (-1)^{k+1} u_{k\pi}(t) e^{-(t-k\pi)/20} \frac{\sin\sqrt{399}(t - k\pi)}{20}$$

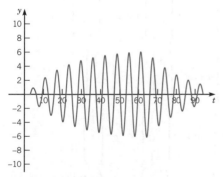

(c) the solution oscillates to zero

Section 5.8   Convolution Integrals and Their Applications   *page 359*

3. $F(s) = 24/s^4(s^2 + 36)$

5. $F(s) = 1/s^2(s - 1)$

7. $f(t) = \frac{1}{2} \int_0^t (t - \tau)^2 \sin\tau d\tau$

9. $f(t) = \frac{1}{12} \int_0^t (t - \tau)^3 e^{-3(t-\tau)} \sin 2\tau d\tau$

11. $f(t) = \int_0^t \sin(t - \tau) \sin(\tau)d\tau$

13. (c) $\displaystyle\int_0^1 u^m(1 - u)^n du = \frac{\Gamma(m + 1)\Gamma(n + 1)}{\Gamma(m + n + 2)}$

15. $y = \frac{1}{4} \int_0^t e^{-3(t-\tau)} \sin 4(t - \tau) \sin\alpha\tau d\tau$

17. $y = e^{-t/2} \cos t - \frac{1}{2} e^{-t/2} \sin t +$ $\int_0^t e^{-(t-\tau)/2} \sin(t - \tau)[1 - u_\pi(\tau)]d\tau$

19. $y = 2e^{-t} - e^{-2t} + \int_0^t [e^{-(t-\tau)} - e^{-2(t-\tau)}] \cos\alpha\tau d\tau$

21. $y = \frac{32}{15} \cos t - \frac{2}{15} \cos 4t + \frac{1}{60} \int_0^t [4 \sin(t-\tau) - \sin 4(t-\tau)] g(\tau)d\tau$

23. $\Phi(s) = \dfrac{F(s)}{1 + K(s)}$

25. (a) $\phi(t) = \cos t$ (b) $\phi''(t) + \phi(t) = 0$, $\phi(0) = 1$, $\phi'(0) = 0$ (c) $\phi(t) = \cos t$

27. (a) $\phi(t) = (1 - 2t + t^2)e^{-t}$ (b) $\phi''(t) + 2\phi'(t) + \phi(t) = 2e^{-t}$, $\phi(0) = 1$, $\phi'(0) = -3$ (c) $\phi(t) = (1 - 2t + t^2)e^{-t}$

29. (a) $\phi(t) = \cos t$ (b) $\phi^{(4)}(t) - \phi(t) = 0$, $\phi(0) = 1$, $\phi'(0) = 0$, $\phi''(0) = -1$, $\phi'''(0) = 0$ (c) $\phi(t) = \cos t$

Section 5.9   Linear Systems and Feedback Control   *page 369*

1. $H = \dfrac{H_1 + G_2}{1 + H_1 G_1}$

5. $a_0, a_1, a_2 > 0$ and $a_1 a_2 > a_0$

7. 0

9. 2

11. 2

13. $-4 < K < 20$

15. $-16 < K < 29$

CHAPTER 6   SYSTEMS OF FIRST ORDER LINEAR EQUATIONS

Section 6.1   Definitions and Examples   *page 388*

1. (a) $\begin{pmatrix} 7e^t & 5e^{-t} & 13e^{2t} \\ 2e^t & 10e^{-t} & 2e^{2t} \\ 11e^t & 0 & 2e^{2t} \end{pmatrix}$

(b)
$\begin{pmatrix} -2 + 2e^{2t} + 4e^{3t} & 1 + 6e^{-2t} - e^t & 2e^t + 4e^{3t} - e^{4t} \\ -1 + 10e^{2t} - 4e^{3t} & 5 + 3e^{-2t} + e^t & e^t + 20e^{3t} + e^{4t} \\ -3 - 2e^{2t} + 20e^{3t} & -1 + 9e^{-2t} - 5e^t & 3e^t - 4e^{3t} - 5e^{4t} \end{pmatrix}$

(c) $\begin{pmatrix} e^t & -2e^{-t} & 2e^{2t} \\ 5e^t & -e^{-t} & -2e^{2t} \\ -e^t & -3e^{-t} & 10e^{2t} \end{pmatrix}$

(d) $(e-1) \begin{pmatrix} 1 & 2e^{-1} & \frac{1}{2}(e+1) \\ 5 & e^{-1} & -\frac{1}{2}(e+1) \\ -1 & 3e^{-1} & \frac{5}{2}(e+1) \end{pmatrix}$

5. $\mathbf{x}' = \begin{pmatrix} 0 & 1 & 0 \\ 0 & 0 & 1 \\ -8/t & 0 & -\sin t/t \end{pmatrix} \mathbf{x} + \begin{pmatrix} 0 \\ 0 \\ \cos t/t \end{pmatrix}$

7. $\mathbf{x}' = \begin{pmatrix} 0 & 1 & 0 \\ 0 & 0 & 1 \\ -t^2 & -t^2 & -t \end{pmatrix} \mathbf{x} + \begin{pmatrix} 0 \\ 0 \\ \ln t \end{pmatrix}$

9. $\mathbf{x}' = \begin{pmatrix} 0 & 1 & 0 & 0 & 0 & 0 \\ 0 & 0 & 1 & 0 & 0 & 0 \\ 0 & 0 & 0 & 1 & 0 & 0 \\ 0 & 0 & 0 & 0 & 1 & 0 \\ 0 & 0 & 0 & 0 & 0 & 1 \\ -3/(x^2-2) & 0 & -x^2/(x^2-2) & 0 & 0 & 0 \end{pmatrix} \mathbf{x}$

11. $\mathbf{x}' = \begin{pmatrix} -k_{01} - k_{21} - k_{31} & k_{12} & k_{13} \\ k_{21} & -k_{02} - k_{12} & 0 \\ k_{31} & 0 & -k_{13} \end{pmatrix} \mathbf{x}$
$+ \begin{pmatrix} L \\ 0 \\ 0 \end{pmatrix}$

13. (b) $\mathbf{x} = \dfrac{\alpha}{L_{12} + L_{21}} \begin{pmatrix} L_{12} + L_{21} \exp[-(L_{12}+L_{21})t] \\ L_{21} - L_{21} \exp[-(L_{12}+L_{21})t] \end{pmatrix}$ (c)
$\bar{x}_1 = \alpha L_{12}/(L_{12}+L_{21})$, $\bar{x}_2 = \alpha L_{21}/(L_{12}+L_{21})$; increasing $L_{12} + L_{21}$ increases the rate of approach to equilibrium

(d)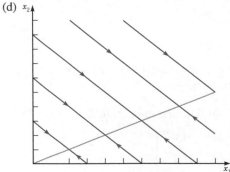

(e)

the ratio of fractional turnover rates determines the proportions of the initial amount of tracer in each compartment at equilibrium

15. $C = (0 \quad 1 \quad 0 \quad 0)$

Section 6.2   Basic Theory of First Order Linear Systems   *page 398*

1. $-\infty < t < \infty$

3. $t > 1$, or $0 < t < 1$, or $t < 0$

5. $\ldots, -3\pi/2 < x < -\pi/2, -\pi/2 < x < 1, 1 < x < \pi/2, \pi/2 < x < 3\pi/2, \ldots$

9. yes

11. $W = 1$

13. $W = -48e^{-4t}$

15. $W = 6x$

Section 6.3   Homogeneous Linear Systems with Constant Coefficients   *page 408*

1. $\mathbf{x} = c_1 e^{-3t} \begin{pmatrix} 1 \\ 1 \\ 1 \end{pmatrix} + c_2 e^{-4t} \begin{pmatrix} 1 \\ 0 \\ -1 \end{pmatrix} + c_3 e^{-6t} \begin{pmatrix} 1 \\ -2 \\ 1 \end{pmatrix}$

3. $\mathbf{x} = c_1 e^{-2t} \begin{pmatrix} 1 \\ 1 \\ 0 \end{pmatrix} + c_2 e^{-2t} \begin{pmatrix} 0 \\ 1 \\ 2 \end{pmatrix} + c_3 e^{7t} \begin{pmatrix} 2 \\ -2 \\ 1 \end{pmatrix}$

5. $\mathbf{x} = c_1 e^{5t} \begin{pmatrix} 1 \\ -2 \\ 1 \end{pmatrix} + c_2 e^{-5t} \begin{pmatrix} 1 \\ 0 \\ -1 \end{pmatrix} + c_3 e^{8t} \begin{pmatrix} 1 \\ 1 \\ 1 \end{pmatrix}$

7. $\mathbf{x} = c_1 e^{-t} \begin{pmatrix} 3 \\ -4 \\ -2 \end{pmatrix} + c_2 e^{-2t} \begin{pmatrix} 4 \\ -5 \\ -7 \end{pmatrix} + c_3 e^{2t} \begin{pmatrix} 0 \\ 1 \\ -1 \end{pmatrix}$

9. $\mathbf{x} = e^{2t}\begin{pmatrix} 11 \\ 11 \\ 0 \end{pmatrix} + e^{t}\begin{pmatrix} 0 \\ -10 \\ 5 \end{pmatrix}$

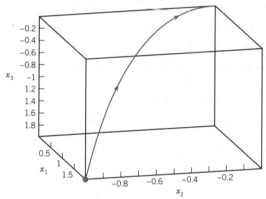

all three components increase without bound

11. $\mathbf{x} = e^{-2t}\begin{pmatrix} 0 \\ -1 \\ 0 \end{pmatrix} + 2e^{-4t}\begin{pmatrix} 1 \\ 0 \\ -1 \end{pmatrix}$

the solution converges to 0

13. $\mathbf{x}' = \begin{pmatrix} -1 & 1 & 0 & 0 \\ 1 & -2 & 1 & 0 \\ 0 & 1 & -2 & 1 \\ 0 & 0 & 1 & -1 \end{pmatrix}\mathbf{x}$

$\mathbf{x} = c_1\begin{pmatrix} 1 \\ 1 \\ 1 \\ 1 \end{pmatrix} + c_2 e^{-(2-\sqrt{2})kt}\begin{pmatrix} 1-\sqrt{2} \\ -3+2\sqrt{2} \\ 3-2\sqrt{2} \\ -1+\sqrt{2} \end{pmatrix} + c_3 e^{-2kt}\begin{pmatrix} 1 \\ -1 \\ -1 \\ 1 \end{pmatrix}$

$+ c_4 e^{-(2+\sqrt{2})kt}\frac{1}{4}\begin{pmatrix} 1+\sqrt{2} \\ -3-2\sqrt{2} \\ 3+2\sqrt{2} \\ -1-\sqrt{2} \end{pmatrix}$

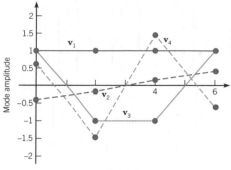

long-term decay rate is controlled by the negative eigenvalue nearest zero, $-2+\sqrt{2}$.

15. $\mathbf{u}_1 = \begin{pmatrix} 1 \\ 1 \\ 0 \end{pmatrix}, \quad \mathbf{u}_2 = \begin{pmatrix} 1 \\ 0 \\ 1 \end{pmatrix}, \quad \mathbf{u}_3 = \begin{pmatrix} -1 \\ -1 \\ 1 \end{pmatrix}$

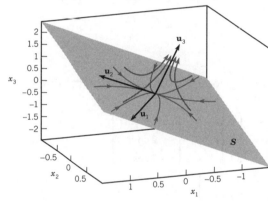

$\mathbf{x} = c_1 e^{-2t}\mathbf{u}_1 + c_2 e^{-t}\mathbf{u}_2 + c_3 e^{t}\mathbf{u}_3$. If $\mathbf{x}_0 \notin S$, $\mathbf{x}(t)$ is asymptotic to line determined by $\mathbf{u}_3$ as $t \to \infty$.

17. $R_1 \xrightarrow{k_1} R_2 \xrightarrow{k_2} R_3$

$\mathbf{m} = m_0\begin{pmatrix} 0 \\ 0 \\ 1 \end{pmatrix} + m_0 e^{-k_1 t}\begin{pmatrix} 1 \\ k_1/(k_2-k_1) \\ -k_2/(k_2-k_1) \end{pmatrix}$

$+ m_0 e^{-k_2 t}\begin{pmatrix} 0 \\ -k_1/(k_2-k_1) \\ k_1/(k_2-k_1) \end{pmatrix}, \quad k_2 \neq k_1$

$\mathbf{m} = m_0\begin{pmatrix} 0 \\ 0 \\ 1 \end{pmatrix} + m_0 e^{-k_1 t}\begin{pmatrix} 1 \\ k_1 t \\ -1-k_1 t \end{pmatrix}, \quad k_2 = k_1$

19. $e^{-3t}\begin{pmatrix} 1 \\ -2 \\ 0 \\ 1 \end{pmatrix}$, $e^{3t}\begin{pmatrix} 0 \\ 0 \\ 1 \\ 0 \end{pmatrix}$, $e^{3t}\begin{pmatrix} 1 \\ 0 \\ 0 \\ -1 \end{pmatrix}$, $e^{3t}\begin{pmatrix} 1 \\ 1 \\ 0 \\ 1 \end{pmatrix}$

21. $e^{-4t}\begin{pmatrix} 1 \\ 0 \\ 0 \\ 1 \end{pmatrix}$, $e^{-2t}\begin{pmatrix} 1 \\ 0 \\ -1 \\ 0 \end{pmatrix}$, $e^{2t}\begin{pmatrix} 1 \\ 1 \\ -1 \\ -1 \end{pmatrix}$, $e^{4t}\begin{pmatrix} 0 \\ 1 \\ 0 \\ 1 \end{pmatrix}$

23. $e^{-2t}\begin{pmatrix} 1 \\ -2 \\ 2 \\ -2 \\ 1 \end{pmatrix}$, $e^{-2t}\begin{pmatrix} 0 \\ 1 \\ 0 \\ 1 \\ 0 \end{pmatrix}$, $e^{-t}\begin{pmatrix} -1 \\ 0 \\ 0 \\ 1 \\ -1 \end{pmatrix}$, $e^{t}\begin{pmatrix} 2 \\ 0 \\ 0 \\ 0 \\ 1 \end{pmatrix}$,

$e^{2t}\begin{pmatrix} 1 \\ 1 \\ -1 \\ 1 \\ 0 \end{pmatrix}$

Section 6.4   Nondefective Matrices with Complex Eigenvalues   *page 419*

1. $\mathbf{x} = c_1 e^{-t}\begin{pmatrix} 1 \\ 0 \\ 1 \end{pmatrix} + c_2 e^{-t}\begin{pmatrix} \cos t + \sin t \\ 2\cos t \\ -2\cos t \end{pmatrix}$

$+ c_3 e^{-t}\begin{pmatrix} \sin t - \cos t \\ 2\sin t \\ -2\sin t \end{pmatrix}$

3. $\mathbf{x} = c_1 e^{-t}\begin{pmatrix} 1 \\ 1 \\ -1 \end{pmatrix} + c_2 e^{-t}\begin{pmatrix} -\sin 2t \\ \cos 2t \\ -\sin 2t \end{pmatrix} + c_3 e^{-t}\begin{pmatrix} \cos 2t \\ \sin 2t \\ \cos 2t \end{pmatrix}$

5. $\mathbf{x} = c_1 e^{-t}\begin{pmatrix} 1 \\ 0 \\ -1 \end{pmatrix} + c_2 e^{-t}\begin{pmatrix} 2\cos 3t + 2\sin 3t \\ 3\cos 3t \\ -\sin 3t \end{pmatrix}$

$+ c_3 e^{-t}\begin{pmatrix} 2\sin 3t - 2\cos 3t \\ 3\sin 3t \\ \cos 3t \end{pmatrix}$

7. $\mathbf{x} = c_1 e^{-2t}\begin{pmatrix} 4 \\ -5 \\ -7 \end{pmatrix} + c_2 e^{-t}\begin{pmatrix} 3 \\ -4 \\ -2 \end{pmatrix} + c_3 e^{2t}\begin{pmatrix} 0 \\ 1 \\ -1 \end{pmatrix}$

9. (a) $\mathbf{a} = \begin{pmatrix} 5 \\ 1 \\ 2 \end{pmatrix}$, $\mathbf{b} = \begin{pmatrix} 0 \\ -2 \\ 1 \end{pmatrix}$

(b) and (c)

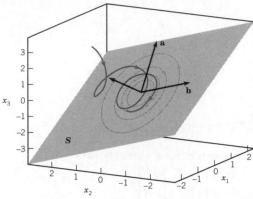

$\mathbf{x} = c_1 e^{-t}\begin{pmatrix} 1 \\ -1 \\ -1 \end{pmatrix} + c_2\begin{pmatrix} 5\cos 4t \\ \cos 4t - 2\sin 4t \\ 2\cos 4t + \sin 4t \end{pmatrix}$

$+ c_3\begin{pmatrix} 5\sin 4t \\ \sin 4t + 2\cos 4t \\ 2\sin 4t - \cos 4t \end{pmatrix}$ If $\mathbf{x}_0 \notin S$, the solution trajectory approaches a closed curve in S.

13. (a) $\lambda_1 = i$, $\mathbf{v}_1 = \begin{pmatrix} 1 \\ 1 \\ i \\ i \end{pmatrix}$; $\lambda_2 = -i$, $\mathbf{v}_2 = \begin{pmatrix} 1 \\ 1 \\ -i \\ -i \end{pmatrix}$; $\lambda_3 = \sqrt{3}i$,

$\mathbf{v}_3 = \begin{pmatrix} 1 \\ -1 \\ \sqrt{3}i \\ -\sqrt{3}i \end{pmatrix}$; $\lambda_4 = -\sqrt{3}i$, $\mathbf{v}_4 = \begin{pmatrix} 1 \\ -1 \\ -\sqrt{3}i \\ \sqrt{3}i \end{pmatrix}$

(b) $\mathbf{x} = c_1\begin{pmatrix} \cos t \\ \cos t \\ -\sin t \\ -\sin t \end{pmatrix} + c_2\begin{pmatrix} \sin t \\ \sin t \\ \cos t \\ \cos t \end{pmatrix} + c_3\begin{pmatrix} \cos\sqrt{3}t \\ -\cos\sqrt{3}t \\ -\sqrt{3}\sin\sqrt{3}t \\ \sqrt{3}\sin\sqrt{3}t \end{pmatrix}$

$+ c_4\begin{pmatrix} \sin\sqrt{3}t \\ -\sin\sqrt{3}t \\ \sqrt{3}\cos\sqrt{3}t \\ -\sqrt{3}\cos\sqrt{3}t \end{pmatrix}$

(c)

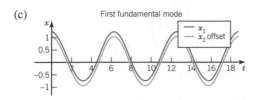

First fundamental mode

Second fundamental mode

(d)

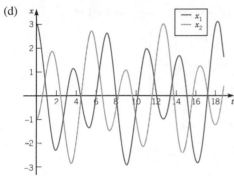

15. $e^{-t}\begin{pmatrix}1\\0\\0\\-1\end{pmatrix}$, $e^{t}\begin{pmatrix}1\\1\\0\\1\end{pmatrix}$, $e^{-t}\begin{pmatrix}\cos 4t\\-\sin 4t\\\cos 4t\\-\sin 4t\end{pmatrix}$, $e^{-t}\begin{pmatrix}\sin 4t\\\cos 4t\\\sin 4t\\\cos 4t\end{pmatrix}$

17. $\begin{pmatrix}\cos 2t - \sin 2t\\\cos 2t + \sin 2t\\-\cos 2t\\\cos 2t\end{pmatrix}$, $\begin{pmatrix}\sin 2t + \cos 2t\\\sin 2t - \cos 2t\\-\sin 2t\\\sin 2t\end{pmatrix}$,

$\begin{pmatrix}\cos 3t - \sin 3t\\-2\sin 3t\\-2\sin 3t\\0\end{pmatrix}$, $\begin{pmatrix}\sin 3t + \cos 3t\\2\cos 3t\\2\cos 3t\\0\end{pmatrix}$

Section 6.5 Fundamental Matrices and the Exponential of a Matrix *page 430*

1. $e^{At} = \begin{pmatrix} -\frac{1}{3}e^{-t} + \frac{4}{3}e^{2t} & \frac{2}{3}e^{-t} - \frac{2}{3}e^{2t} \\ -\frac{2}{3}e^{-t} + \frac{2}{3}e^{2t} & \frac{4}{3}e^{-t} - \frac{1}{3}e^{2t} \end{pmatrix}$

3. $e^{At} = \begin{pmatrix} (1+2t)e^{t} & -4te^{t} \\ te^{t} & (-2t+1)e^{t} \end{pmatrix}$

5. $e^{At} = \begin{pmatrix} \cos(t/2) + 2\sin(t/2) & -5\sin(t/2) \\ \sin(t/2) & \cos(t/2) - 2\sin(t/2) \end{pmatrix}$

7. $e^{At} = \begin{pmatrix} e^{3t}(\cosh t + 2\sinh t) & -\sinh te^{3t} \\ 3\sinh te^{3t} & e^{3t}(\cosh t - 2\sinh t) \end{pmatrix}$

9. $e^{At} = \begin{pmatrix} \cosh t + 2\sinh t & -\sinh t \\ 3\sinh t & \cosh t - 2\sinh t \end{pmatrix}$

11. $e^{At} = \begin{pmatrix} e^{-5t/2}(\cos(\sqrt{15}t/2) - \frac{1}{\sqrt{15}}\sin(\sqrt{15}t/2)) \\ -\frac{2}{\sqrt{15}}e^{-5t/2}\sin(\sqrt{15}t/2) \\ \frac{8}{\sqrt{15}}e^{-5t/2}\sin(\sqrt{15}t/2) \\ e^{-5t/2}(\cos(\sqrt{15}t/2) + \frac{1}{\sqrt{15}}\sin(\sqrt{15}t/2)) \end{pmatrix}$

13. $e^{At} = \begin{pmatrix} -2e^{-2t} + 3e^{-t} & 2e^{-3t/2}\sinh\left(\frac{1}{2}t\right) \\ \frac{3}{2}e^{2t} + \frac{5}{2}e^{-2t} - 4e^{-t} & \frac{13}{12}e^{2t} + \frac{5}{4}e^{-2t} - \frac{4}{3}e^{-t} \\ -\frac{3}{2}e^{2t} + \frac{7}{2}e^{-2t} - 2e^{-t} & \frac{7}{4}e^{-2t} - \frac{2}{3}e^{-t} - \frac{13}{12}e^{2t} \end{pmatrix}$

$\begin{pmatrix} 2e^{-3t/2}\sinh\left(\frac{1}{2}t\right) \\ \frac{5}{4}e^{-2t} - \frac{4}{3}e^{-t} + \frac{1}{12}e^{2t} \\ -\frac{1}{12}e^{2t} + \frac{7}{4}e^{-2t} - \frac{2}{3}e^{-t} \end{pmatrix}$

15. $\mathbf{x} = \begin{pmatrix} 4e^{-3t}\cos 3t - 3e^{-3t}\sin 3t \\ \frac{4}{3}e^{-3t}\sin 3t + e^{-3t}\cos 3t \end{pmatrix}$

17. $e^{At} = \begin{pmatrix} (-t+1)e^{-3t} & -te^{-3t} \\ te^{-3t} & (t+1)e^{-3t} \end{pmatrix}$

19. $e^{At} = \begin{pmatrix} e^{t}\cos t - 2e^{t}\sin t & -5e^{t}\sin t \\ e^{t}\sin t & e^{t}\cos t + 2e^{t}\sin t \end{pmatrix}$

21. (c) $\mathbf{x} = c_1 = \begin{pmatrix} u_0 \\ v_0 \end{pmatrix}\cos\omega t + \begin{pmatrix} v_0 \\ -\omega^2 u_0 \end{pmatrix}\dfrac{\sin\omega t}{\omega}$

Section 6.6 Nonhomogeneous Linear Systems *page 436*

3. $\mathbf{x} = c_1 e^{2t}\begin{pmatrix}\sqrt{3}\\1\end{pmatrix} + c_2 e^{-2t}\begin{pmatrix}1\\-\sqrt{3}\end{pmatrix} + \begin{pmatrix}-\frac{2}{3}e^{t} - e^{-t}\\-\frac{1}{\sqrt{3}}e^{t} + \frac{2}{\sqrt{3}}e^{-t}\end{pmatrix}$

5. $\mathbf{x} = c_1 e^{2t}\begin{pmatrix}1\\1\end{pmatrix} + c_2 e^{-3t}\begin{pmatrix}1\\-4\end{pmatrix} + \begin{pmatrix}\frac{1}{2}e^{t}\\-e^{-2t}\end{pmatrix}$

7. $\mathbf{x} = c_1 e^{-t} \begin{pmatrix} 0 \\ 1 \\ 1 \end{pmatrix} + c_2 e^{-2t} \begin{pmatrix} -1 \\ 2 \\ -1 \end{pmatrix} + c_3 e^{-t} \begin{pmatrix} 1 \\ 0 \\ 1 \end{pmatrix}$

$+ \begin{pmatrix} \frac{7}{8} + \frac{1}{4}t - \frac{11}{4}e^{-3t} \\ -\frac{3}{4} + \frac{11}{2}e^{-3t} + \frac{1}{2}t \\ -\frac{1}{8} - \frac{33}{4}e^{-3t} + \frac{1}{4}t \end{pmatrix}$

9. $\mathbf{x} = c_1 e^{t} \begin{pmatrix} 1 \\ 1 \\ 1 \end{pmatrix} + c_2 e^{-t} \begin{pmatrix} 1 \\ 0 \\ -1 \end{pmatrix} + c_3 e^{-2t} \begin{pmatrix} 1 \\ -2 \\ 1 \end{pmatrix}$

$+ \begin{pmatrix} \frac{1}{10}\cos t + \frac{3}{10}\sin t \\ -\frac{1}{10}\sin t + \frac{3}{10}\cos t \\ \frac{1}{10}\cos t + \frac{3}{10}\sin t \end{pmatrix}$

11.

$\lim_{t\to\infty} \mathbf{x}(t) = \left(\frac{3}{4}, \frac{1}{2}, \frac{1}{4}\right)^{T}$

13.

15. $\mathbf{a} = \begin{pmatrix} -\frac{3}{10} \\ -\frac{1}{5} \end{pmatrix}, \quad \mathbf{b} = \begin{pmatrix} \frac{2}{5} \\ \frac{1}{10} \end{pmatrix}$

Section 6.7 Defective Matrices *page 444*

1. $e^{\mathbf{A}t} = \begin{pmatrix} (3t+1)e^{t} & -9te^{t} \\ te^{t} & (-3t+1)e^{t} \end{pmatrix}$

3. $e^{\mathbf{A}t} = \begin{pmatrix} (-t+1)e^{2t} & te^{2t} \\ -\frac{1}{2}(-4t+t^2)e^{2t} & \left(\frac{1}{2}t^2 - t + 1\right)e^{2t} \\ \frac{1}{2}(t^2-6t)e^{2t} & -\frac{1}{2}(-4t+t^2)e^{2t} \end{pmatrix}$

$\begin{matrix} te^{2t} \\ \frac{1}{2}(-2t+t^2)e^{2t} \\ \left(1+2t-\frac{1}{2}t^2\right)e^{2t} \end{matrix}$

5. $e^{\mathbf{A}t} = \begin{pmatrix} \frac{7}{3}e^{-t} + \left(-2t-\frac{4}{3}\right)e^{2t} & -2e^{-t} + (3t+2)e^{2t} \\ \frac{14}{9}e^{-t} + \left(-\frac{10}{3}t-\frac{14}{9}\right)e^{2t} & -\frac{4}{3}e^{-t} + \left(\frac{7}{3}+5t\right)e^{2t} \\ -2te^{2t} & 3te^{2t} \end{pmatrix}$

$\begin{matrix} e^{-t} + (-1-3t)e^{2t} \\ \frac{2}{3}e^{-t} + \left(-\frac{2}{3}-5t\right)e^{2t} \\ (1-3t)e^{2t} \end{matrix}$

7. $e^{\mathbf{A}t} =$

$\begin{pmatrix} -5e^{-t}\cos t - e^{-t}\sin t + 6e^{-2t} & -9e^{-t}\cos t - 7e^{-t}\sin t + 9e^{-2t} \\ 3e^{-2t} - 3e^{-t}\cos t + e^{-t}\sin t & 8e^{-2t} - e^{-t}\sin t - 7e^{-t}\cos t \\ -e^{-2t} + e^{-t}\cos t - 3e^{-t}\sin t & -5e^{-2t} - 5e^{-t}\sin t + 5e^{-t}\cos t \\ 4e^{-t}\cos t + 2e^{-t}\sin t - 4e^{-2t} & 6e^{-t}\cos t + 8e^{-t}\sin t - 6e^{-2t} \end{pmatrix}$

$\begin{matrix} -9e^{-t}\cos t - 7e^{-t}\sin t + 9e^{-2t} & 12e^{-2t} - 12e^{-t}\cos t - 5e^{-t}\sin t \\ 7e^{-2t} - e^{-t}\sin t - 7e^{-t}\cos t & 8e^{-2t} + e^{-t}\sin t - 8e^{-t}\cos t \\ -4e^{-2t} - 5e^{-t}\sin t + 5e^{-t}\cos t & -4e^{-2t} - 7e^{-t}\sin t + 4e^{-t}\cos t \\ 6e^{-t}\cos t + 8e^{-t}\sin t - 6e^{-2t} & -8e^{-2t} + 9e^{-t}\cos t + 7e^{-t}\sin t \end{matrix}$

9. $\mathbf{x} = e^{-3t} \begin{pmatrix} 7 + 24t \\ 1 + 24t \end{pmatrix}$

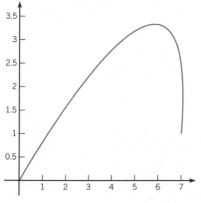

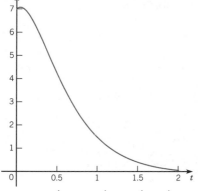

11. $\mathbf{x} = e^{t} \begin{pmatrix} 1 + 16t \\ -32 \\ 15 - 16t \end{pmatrix} + e^{2t} \begin{pmatrix} 0 \\ 30 \\ -10 \end{pmatrix}$

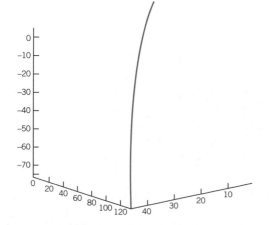

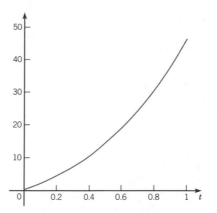

CHAPTER 7 NONLINEAR DIFFERENTIAL EQUATIONS AND STABILITY

Section 7.1 Autonomous Systems and Stability *page 464*

1. (a) $(0, 0)$, $(2, -\frac{1}{4})$

(b)

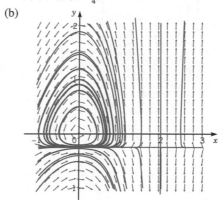

(c) $(0, 0)$ center, stable; $(2, -\frac{1}{4})$ saddle, unstable

3. (a) $(0, 0)$, $(0, \frac{3}{2})$, $(2, 0)$, $(-1, 3)$

(b)

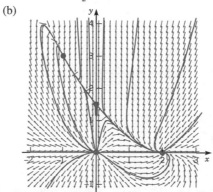

(c) $(0, 0)$, node, unstable; $(0, \frac{3}{2})$, saddle point, unstable; $(2, 0)$, node, asymptotically stable; $(-1, 3)$, node, asymptotically stable

5. (a) $(0,0)$, $(3,3)$, $(7,-1)$

(b)

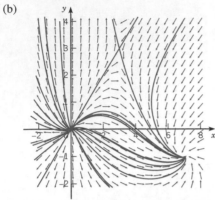

(c) $(0,0)$, node, unstable; $(3,3)$, saddle point, unstable; $(7,-1)$, spiral point, asymptotically stable

7. (a) $(0,0)$, $(-1,-1)$, $(-1,2)$, $(-2,2)$

(b)

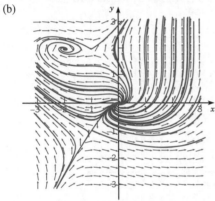

(c) $(0,0)$, spiral point, unstable; $(-1,-1)$, saddle point, unstable; $(-1,2)$, saddle point, unstable; $(-2,2)$, spiral point, asymptotically stable

9. (a) $(0,0)$, $(-3,-3)$, $(2,1)$

(b)

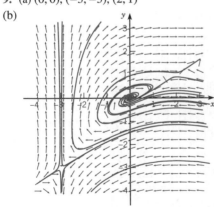

(c) $(0,0)$, spiral point, unstable; $(-3,-3)$, saddle point, unstable; $(2,1)$, saddle point, unstable

11. (a) $(0,0)$, $(1,0)$, $(2,0)$

(b)

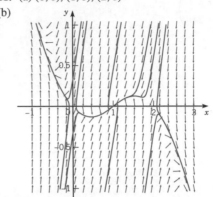

(c) $(0,0)$, saddle point, unstable; $(1,0)$, node, asymptotically stable; $(2,0)$, saddle point, unstable

13. (a) $(0,0)$, $(-2,\frac{10}{3})$, $(8,0)$, $(3,\frac{5}{3})$

(b)

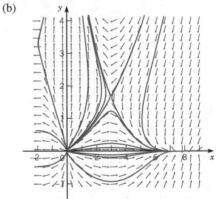

(c) $(0,0)$, node, unstable; $(-2,\frac{10}{3})$, saddle point, unstable; $(8,0)$, node, stable; $(3,\frac{5}{3})$, saddle point, unstable

15. (a) $(0,1)$, $(1,2)$, $(-2,2)$

(b)

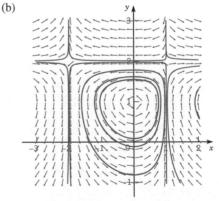

(c) $(0,1)$, center, stable; $(1,2)$, saddle point, unstable; $(-2,2)$, saddle point, unstable

17. (a) $(0,0)$, $(\sqrt{6},0)$, $(-\sqrt{6},0)$, center, stable

(b)

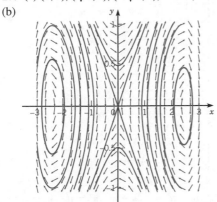

(c) $(0,0)$, saddle point, unstable; $(\sqrt{6},0)$, center, stable; $(-\sqrt{6},0)$, center, stable

21. (a)

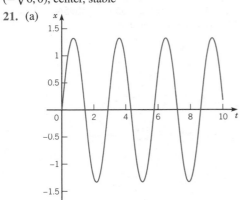

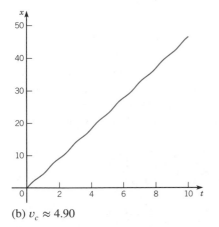

(b) $v_c \approx 4.90$

Section 7.2 Almost Linear Systems *page 475*

1. (a, b, c) $(0,0)$; $u' = -2u + w$, $w' = -w$; $\lambda = -1, -2$; node, asymptotically stable $(2,4)$; $u' = -2u + w$, $w' = 4u - w$; $\lambda = (-3 \pm \sqrt{17})/2$; saddle point, unstable

(d)

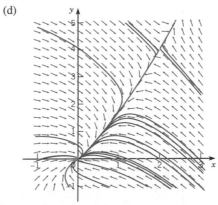

3. (a, b, c) $(0,0)$; $u' = u$, $w' = u + 2w$; $\lambda = 1, 2$; node, unstable $(-4, 2)$; $u' = u + 4w$, $w' = u + 2w$; $\lambda = (3 \pm \sqrt{33})/2$; saddle point, unstable

(d)

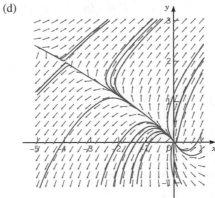

5. (a, b, c) $(0,0)$; $u' = -4u + 4w$, $w' = 10u + 10w$; $\lambda = 3 \pm \sqrt{89}$; saddle point, unstable $(-4, 4)$; $u' = 8u$, $w' = 14u + 14w$; $\lambda = 14, 8$; node, unstable $(10, 10)$; $u' = -14u + 14w$, $w' = -20u$; $\lambda = -7 \pm \sqrt{231}i$; spiral point, asymptotically stable

(d)

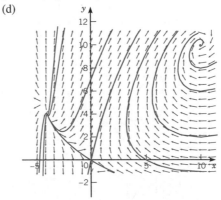

7. (a, b, c) $(1, 1)$; $u' = -w$, $w' = 2u - 2w$; $\lambda = -1 \pm i$; spiral point, asymptotically stable $(-1, 1)$; $u' = -w$, $w' = -2u - 2w$; $\lambda = -1 \pm \sqrt{3}$; saddle point, unstable

(d)

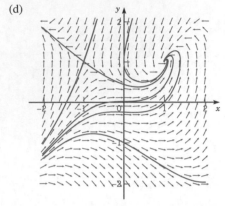

(d)

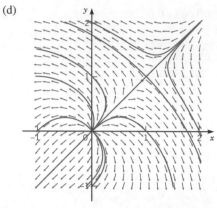

9. (a, b, c) $(0,0)$; $u' = -6u + 6w$, $w' = 4u$; $\lambda = -3 \pm \sqrt{33}$; saddle point, unstable $(0,6)$; $u' = -6u - 6w$, $w' = 10u$; $\lambda = -3 \pm \sqrt{51}\,i$; spiral point, asymptotically stable $(-4,-4)$; $u' = -14u + 14w$, $w' = -4w$; $\lambda = -14, -4$; node, asymptotically stable $(10,-4)$; $u' = 14u + 14w$, $w' = 10w$; $\lambda = 14, 10$; node, unstable

(d)

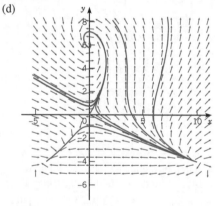

11. (a, b, c) $(0,0)$; $u' = 2u + w$, $w' = u - 2w$; $\lambda = \pm\sqrt{5}$; saddle point, unstable $(-1.1935, -1.4797)$; $u' = -1.2399u - 6.8393w$, $w' = 2.4797u - 0.80655w$; $\lambda = -1.0232 \pm 4.1125i$; spiral point, asymptotically stable

(d)

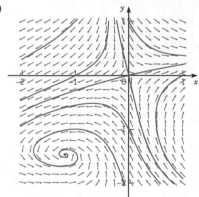

13. (a, b, c) $(0,0)$; $u' = u$, $w' = w$; $\lambda = 1, 1$; node or spiral point, unstable $(1,1)$; $u' = u - 2w$, $w' = -2u + w$; $\lambda = 3, -1$; saddle point, unstable

15. (a, b, c) $(0,0)$; $u' = -2u - w$, $w' = u - w$; $\lambda = (-3 \pm \sqrt{3}\,i)/2$; spiral point, asymptotically stable $(-0.33076, 1.0924)$ and $(0.33076, -1.0924)$; $u' = -3.5216u - 0.27735w$, $w' = 0.27735u + 2.6895w$; $\lambda = -3.5092, 2.6771$; saddle point, unstable

(d)

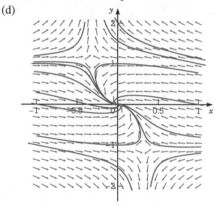

17. (a, b, c) $(0,0)$; $u' = -u + 2w$, $w' = u + 2w$; $\lambda = 2, -1$; saddle point, unstable $(2,1)$; $u' = -\frac{3}{2}u + 2w$, $w' = -2u$; $\lambda = (-6 \pm \sqrt{55}\,i)/4$; spiral point, asymptotically stable $(2,-2)$; $u' = -3w$, $w' = u$; $\lambda = \pm\sqrt{3}\,i$; center or spiral point, indeterminate $(4,-2)$; $u' = -4w$, $w' = -u - 2w$; $\lambda = -1 \pm \sqrt{5}$; saddle point, unstable

(d)

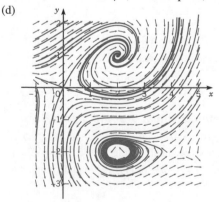

19. (a, b, c) $(0,0)$; $u' = 2u - w$, $w' = 2u - 4w$; $\lambda = -1 \pm \sqrt{7}$; saddle point, unstable $(2, 1)$; $u' = -3w$, $w' = 4u - 8w$; $\lambda = -2, -6$; node, asymptotically stable $(-2, 1)$; $u' = 5w$, $w' = -4u$; $\lambda = \pm 2\sqrt{5}\,i$; center or spiral point, indeterminate $(-2, -4)$; $u' = 10u - 5w$, $w' = 6u$; $\lambda = 5 \pm \sqrt{5}\,i$; spiral point, unstable

(d)

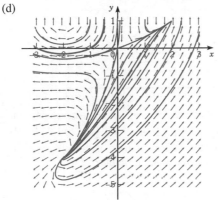

21. (b) $x' = y$, $y' = x$; $dy/dx = x/y$; $y^2 - x^2 = c$

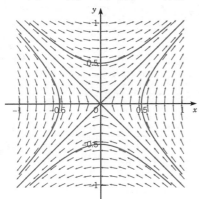

(c) $dy/dx = (x + 2x^3)/y$; $y^2 - x^2 - x^4 = c$

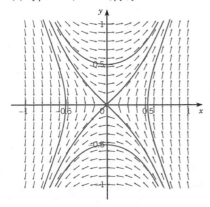

23. (a)

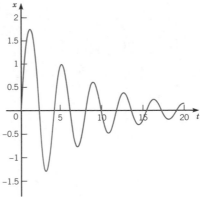

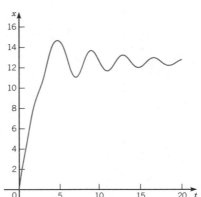

(b) $v_c \approx 3.98$

27. (a) $dx/dt = y$, $dy/dt = -g(x) - c(x)y$

Section 7.3 Competing Species *page 486*

1. (a)

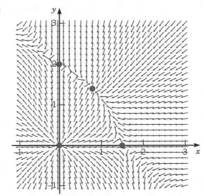

(b, c) $(0,0)$; $u' = \frac{3}{2}u$, $w' = 2w$; $\lambda = \frac{3}{2}, 2$; node, unstable $(0, 2)$; $u' = \frac{1}{2}u$, $w' = -\frac{3}{2}u - 2w$; $\lambda = \frac{1}{2}, -2$; saddle point, unstable $(\frac{3}{2}, 0)$; $u' = -\frac{3}{2}u - \frac{3}{4}w$, $w' = \frac{7}{8}w$; $\lambda = -\frac{3}{2}, \frac{7}{8}$; saddle point, unstable $(\frac{4}{5}, \frac{7}{5})$; $u' = -\frac{4}{5}u - \frac{2}{5}w$, $w' = -\frac{21}{20}u - \frac{7}{5}w$; $\lambda = (-22 \pm \sqrt{204})/20$; node, asymptotically stable

(d, e)

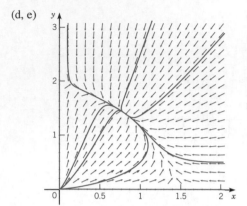

(f) $\lim_{t\to\infty}(x,y) = (\frac{4}{5}, \frac{7}{5})$; coexistence

3. (a)

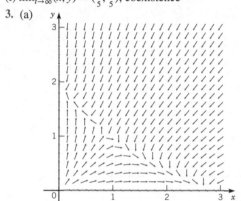

(b, c) $(0,0)$; $u' = \frac{3}{2}u$, $w' = 2w$; $\lambda = \frac{3}{2}, 2$; node, unstable $(0,2)$; $u' = -\frac{1}{2}u$, $w' = -\frac{9}{4}u - 2w$; $\lambda = -\frac{1}{2}$, -2; node, asymptotically stable $(3,0)$; $u' = -\frac{3}{2}u - 3w$, $w' = -\frac{11}{8}w$; $\lambda = -\frac{3}{2}, -\frac{11}{8}$; node, asymptotically stable $(\frac{4}{5}, \frac{11}{10})$; $u' = -\frac{2}{5}u - \frac{4}{5}w$, $w' = -\frac{99}{80}u - \frac{11}{10}w$; $\lambda = -1.80475, 0.30475$; saddle point, unstable

(d, e)

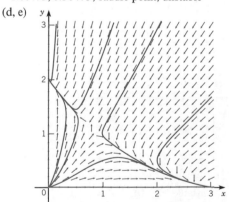

(f) limit depends on initial conditions; one species dies out

5. (a)

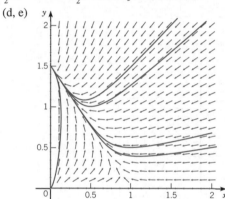

(b, c) $(0,0)$; $u' = u$, $w' = \frac{3}{2}w$; $\lambda = 1, \frac{3}{2}$; node, unstable $(0, \frac{3}{2})$; $u' = -\frac{1}{2}u$, $w' = -\frac{3}{2}u - \frac{3}{2}w$; $\lambda = -\frac{1}{2}, -\frac{3}{2}$; node, asymptotically stable $(1,0)$; $u' = -u - w$, $w' = \frac{1}{2}w$; $\lambda = -1, \frac{1}{2}$; saddle point, unstable

(d, e)

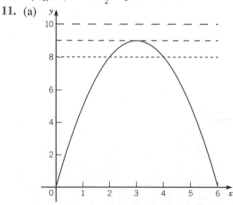

(f) $\lim_{t\to\infty}(x,y) = (0, \frac{3}{2})$; species x dies out

11. (a)

(b) $(3 - \sqrt{9 - \frac{3}{2}\alpha}, \frac{3}{2}\alpha)$, $(3 + \sqrt{9 - \frac{3}{2}\alpha}, \frac{3}{2}\alpha)$ (c) $(3 - \sqrt{6}, 3)$ is an asymptotically stable node; $(3 + \sqrt{6}, 3)$ is a saddle point

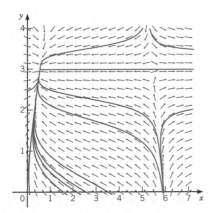

(d) $\alpha_0 = 6$; critical point is $(3, 9)$; $\lambda = 0, -1$

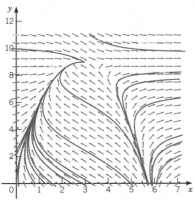

(e)

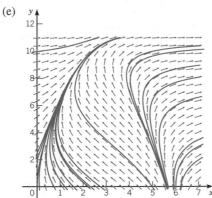

13. (a)

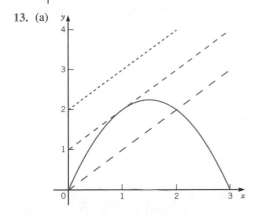

(b) $(1 - \sqrt{1 - \alpha}, 1 + \alpha - \sqrt{1 - \alpha})$, $(1 + \sqrt{1 - \alpha}, 1 + \alpha + \sqrt{1 - \alpha})$ (c) no equilibrium points

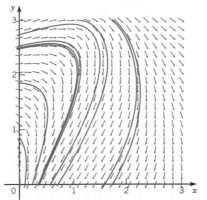

(d) $\alpha_0 = 1$; critical point is $(1, 2)$; $\lambda = 0, 0$

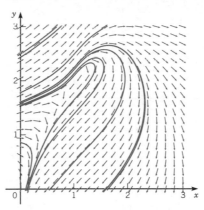

(e)

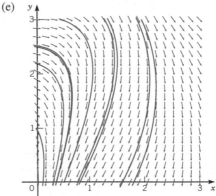

15. (a) $P_1(2, 1), P_2(-3, 1), P_3(-3, 2/\alpha), P_4((\alpha - 1)/(\alpha + 1), -2/(\alpha + 1))$ (b) P_2 and P_3 coincide when $\alpha_0 = 2$. (c) For $\alpha < 2$, P_2 is a saddle point and P_3 is an unstable node; for $\alpha > 2$, P_2 is an unstable node and P_3 is a saddle point.

(d)

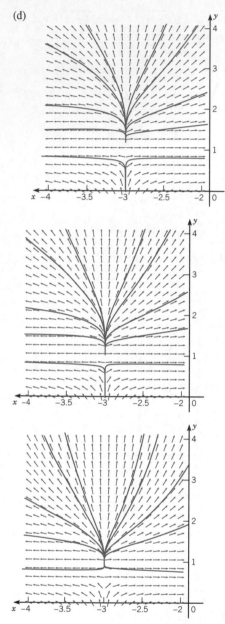

(e)

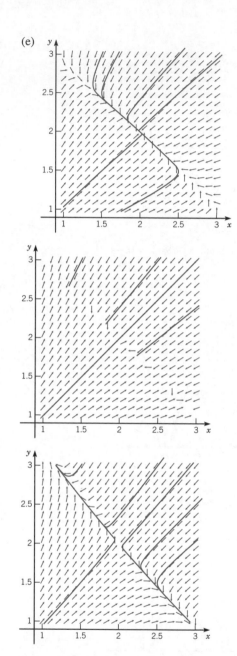

17. (a) $(0, 0)$, $(0, 2 + 2\alpha)$, $(4, 0)$, $(2, 2)$ (b) $\alpha = 0.75$, asymptotically stable node; $\alpha = 1.25$, (unstable) saddle point (c) $u' = -2u - 2w$, $w' = -2\alpha u - 2w$ (d) $\lambda = -2 \pm 2\sqrt{\alpha}$; $\alpha_0 = 1$

Section 7.4   Predator-Prey Equations   *page 494*

1. (a)

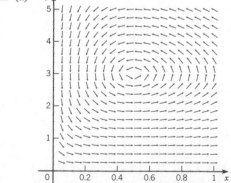

(b, c) $(0,0)$; $u' = \frac{3}{2}u$, $w' = -\frac{1}{2}w$; $\lambda = \frac{3}{2}, -\frac{1}{2}$; saddle point, unstable $(\frac{1}{2}, 3)$; $u' = -\frac{1}{4}w$, $w' = 3u$; $\lambda = \pm\sqrt{3}\,i/2$; center or spiralpoint, indeterminate

(d, e)

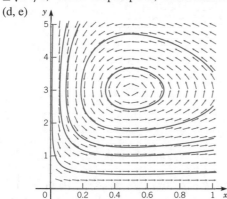

(f) solutions oscillate

3. (a)

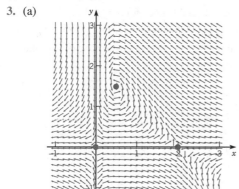

(b, c) $(0,0)$; $u' = u$, $w' = -\frac{1}{4}w$; $\lambda = 1, -\frac{1}{4}$; saddle point, unstable $(2,0)$; $u' = -u - w$, $w' = \frac{3}{4}w$; $\lambda = -1, \frac{3}{4}$; saddle point, unstable $(\frac{1}{2}, \frac{3}{2})$; $u' = -\frac{1}{4}u - \frac{1}{4}w$, $w' = \frac{3}{4}u$; $\lambda = (-1 \pm \sqrt{11}\,i)/8$; spiral point, asymptotically stable

(d, e)

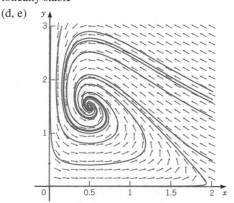

(f) $\lim_{t\to\infty}(x, y) = (\frac{1}{2}, \frac{3}{2})$; coexistence

5. (a)

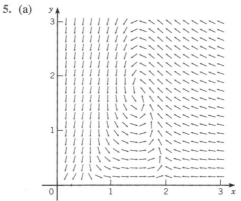

(b, c) $(0,0)$; $u' = -u$, $w' = -\frac{3}{2}w$; $\lambda = -1, -\frac{3}{2}$; node, asymptotically stable $(\frac{1}{2}, 0)$; $u' = \frac{3}{4}u - \frac{3}{20}w$, $w' = -w$; $\lambda = -1, \frac{3}{4}$; saddle point, unstable $(2,0)$; $u' = -3u - \frac{3}{5}w$, $w' = \frac{1}{2}w$; $\lambda = -3, \frac{1}{2}$; saddle point, unstable $(\frac{3}{2}, \frac{5}{3})$; $u' = -\frac{3}{4}u - \frac{9}{20}w$, $w' = \frac{5}{3}u$; $\lambda = (-3 \pm \sqrt{39}\,i)/8$;

spiral point, asymptotically stable

(d, e)

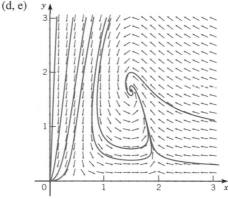

(f) depending on the initial conditions, $\lim_{t\to\infty}(x, y) = (0, 0)$ or $\lim_{t\to\infty}(x, y) = (\frac{3}{2}, \frac{5}{3})$; mutual extinction or coexistence

7. (a) $\sqrt{c}\,\alpha/\sqrt{a}\,\gamma$ (b) $\sqrt{3}$ (c) - (d) The ratio of prey amplitude to predator amplitude increases very slowly as the initial point moves away from the equilibrium point.

9. (a)

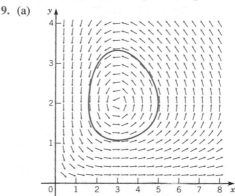

$T \approx 6.5$

(b)
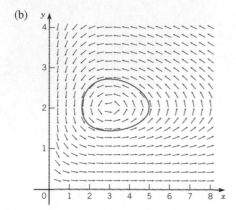

$T \approx 3.7$

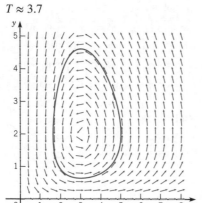

$T \approx 11.5$

(c)
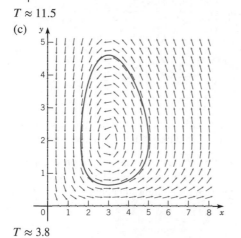

$T \approx 3.8$

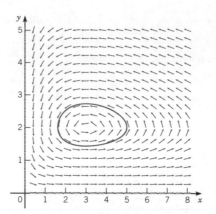

$T \approx 11.1$

11. (a) $P_1(0,0)$, $P_2(1/\sigma, 0)$, $P_3(3, 2 - 6\sigma)$; P_2 moves to the left and P_3 moves down; they coincide at $(3, 0)$ when $\sigma = \frac{1}{3}$. (b) P_1 is a saddle point. P_2 is a saddle point for $\sigma < \frac{1}{3}$ and an asymptotically stable node for $\sigma > \frac{1}{3}$. P_3 is an asymptotically stable spiral point for $\sigma < \sigma_1 = (\sqrt{7/3} - 1)/2 \approx 0.2638$, an asymptotically stable node for $\sigma_1 < \sigma < \frac{1}{3}$, and a saddle point for $\sigma > \frac{1}{3}$.

(c)
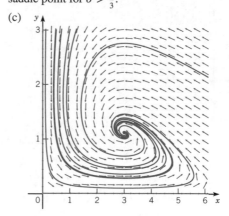

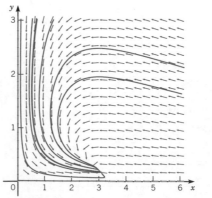

13. (a, b) $P_1(0,0)$ is a saddle point; $P_2(5,0)$ is a saddle point; $P_3(2,2.4)$ is an asymptotically stable spiral point

(c)

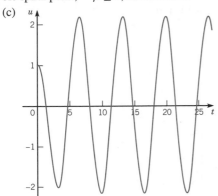

15. (a) same prey, fewer predators (b) more prey, fewer predators (c) more prey, even fewer predators

Section 7.5 Periodic Solutions and Limit Cycles *page 504*

1. $r = 1$, $\theta = t + t_0$, asymptotically stable limit cycle

3. $r = 2$, $\theta = t + t_0$, asymptotically stable limit cycle; $r = 5$, $\theta = t + t_0$, unstable periodic solution

5. $r = (2n-1)/3$, $\theta = t + t_0$, $n = 1, 2, 3, \ldots$, asymptotically stable limit cycle; $r = 2n/3$, $\theta = t + t_0$, $n = 1, 2, 3, \ldots$, unstable periodic solution

9. $r = \sqrt{5}$, $\theta = t + t_0$, unstable periodic solution

15. (a) $x' = y$, $y' = -x + \mu y - \mu y^3/3$ (b) $0 < \mu < 2$, unstable spiral point; $\mu \geq 2$, unstable node

(c)

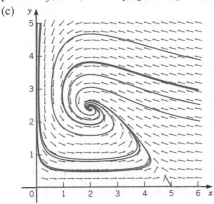

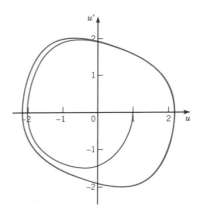

$A \approx 2.16$, $T \approx 6.65$ (d) $\mu = 0.2$, $A \approx 1.99$, $T \approx 6.31$; $\mu = 0.5$, $A \approx 2.03$, $T \approx 6.39$; $\mu = 2$, $A \approx 2.60$, $T \approx 7.65$; $\mu = 5$, $A \approx 4.36$, $T \approx 11.60$

17. (a) The origin is an asymptotically stable node for $\mu < -2$, an asymptotically stable spiral point for $-2 < \mu < 0$, an unstable spiral point for $0 < \mu < 2$, and an unstable node for $\mu > 2$.

(b)

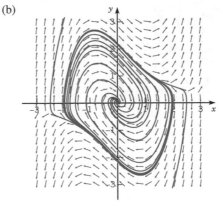

(c) $\mu = -1/2$:

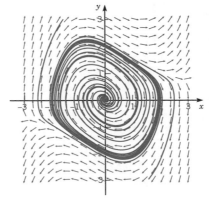

$\mu = -3/2$:

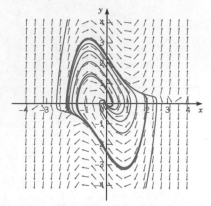

(d) $\mu = -1/10$:

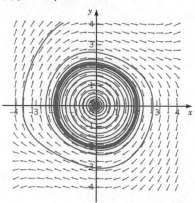

$\mu = 1/10$:

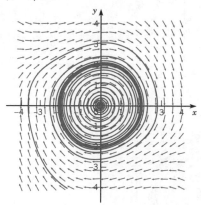

19. (a) $(0,0)$, $(5a,0)$, $(2, 4a - 1.6)$ (b) $\lambda = -0.25 + 0.125a \pm 0.25\sqrt{220 - 400a + 25a^2}$; $a_0 = 2$

(c)

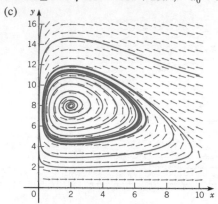

21. (b) $k = 0$, $(1.1994, -0.62426)$;

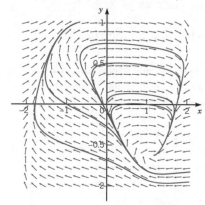

$k = 0.5$, $(0.80485, -0.13106)$

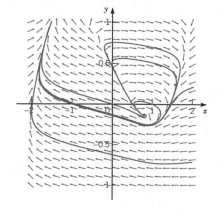

(c) $k_0 \approx 0.3465$, $(0.95450, -0.31813)$

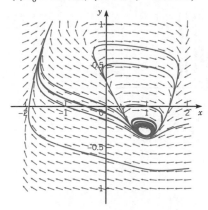

(d) $k = 0.4$, $T \approx 11.23$;

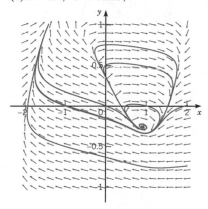

$k = 0.5$, $T \approx 10.37$;

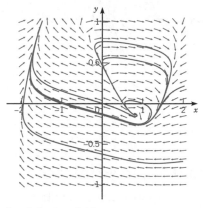

$k = 0.6$, $T \approx 9.93$

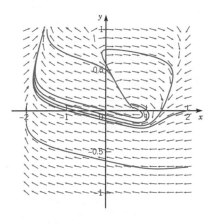

(e) $k_1 \approx 1.4035$

Section 7.6   Chaos and Strange Attractors: The Lorenz Equations   *page 512*

1. (b) $\lambda = \lambda_1$, $\mathbf{v}_1 = (0, 0, 1)^T$; $\lambda = \lambda_2$, $\mathbf{v}_2 = (20, 9 - \sqrt{81 + 40r}, 0)^T$; $\lambda = \lambda_3$, $\mathbf{v}_3 = (20, 9 + \sqrt{81 + 40r}, 0)^T$
(c) $\lambda_1 \approx -2.6667$, $\mathbf{v}_1 = (0, 0, 1)^T$; $\lambda_2 \approx -22.8277$, $\mathbf{v}_2 \approx (20, -25.6554, 0)^T$; $\lambda_3 \approx 11.8277$, $\mathbf{v}_3 \approx (20, 43.6554, 0)^T$

5.

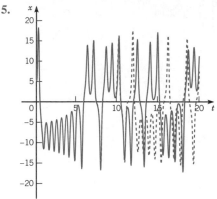

7. (a)

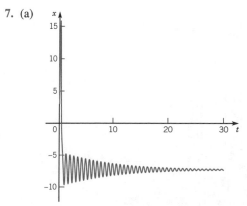

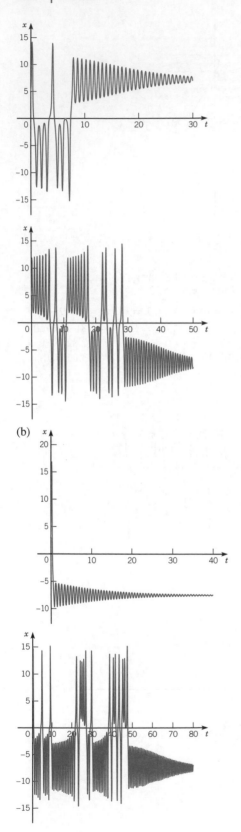

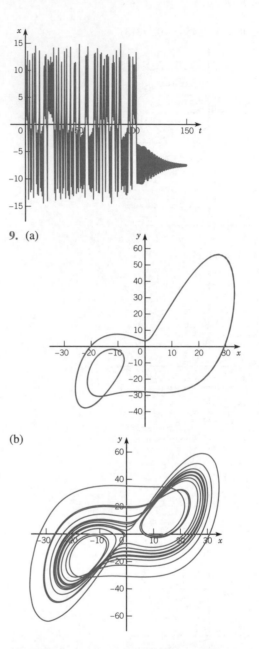

9. (a)

(b)

CHAPTER 8 NUMERICAL METHODS

Section 8.1 Numerical Approximations: Euler's Method *page 529*

1. (a) 1.2, 1.39, 1.571, 1.7439
(b) 1.1975, 1.38549, 1.56491, 1.73658
(c) 1.19631, 1.38335, 1.56200, 1.73308
(d) 1.19516, 1.38127, 1.55918, 1.72968
3. (a) 1.25, 1.54, 1.878, 2.2736
(b) 1.26, 1.5641, 1.92156, 2.34359
(c) 1.26551, 1.57746, 1.94586, 2.38287
(d) 1.2714, 1.59182, 1.97212, 2.42554

5.

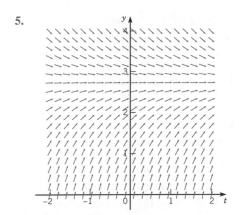

converge for $y \geq 0$; undefined for $y < 0$

7.

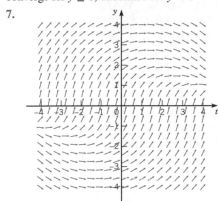

converge

9.

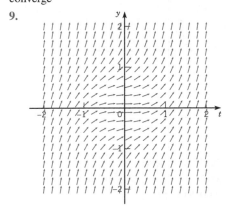

diverge

11. (a) 2.30800, 2.49006, 2.60023, 2.66773, 2.70939, 2.73521
(b) 2.30167, 2.48263, 2.59352, 2.66227, 2.70519, 2.73209
(c) 2.29864, 2.47903, 2.59024, 2.65958, 2.70310, 2.73053
(d) 2.29686, 2.47691, 2.58830, 2.65798, 2.70185, 2.72959

13. (a) −1.48849, −0.412339, 1.04687, 1.43176, 1.54438, 1.51971
(b) −1.46909, −0.287883, 1.05351, 1.42003, 1.53000, 1.50549
(c) −1.45865, −0.217545, 1.05715, 1.41486, 1.52334, 1.49879
(d) −1.45212, −0.173376, 1.05941, 1.41197, 1.51949, 1.49490

15. (a) −0.166134, −0.410872, −0.804660, 4.15867
(b) −0.174652, −0.434238, −0.889140, −3.09810

17. A reasonable estimate for y at $t = 2.5$ is between 18 and 19. No reliable estimate is possible at $t = 3$ from the specified data.

19. (a)

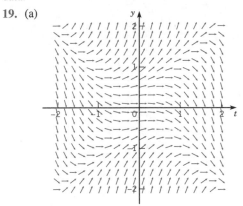

(b) $0.67 < \alpha_0 < 0.68$

Section 8.2 Accuracy of Numerical Methods *page 536*

1. (a) 1.1975, 1.38549, 1.56491, 1.73658
(b) 1.19631, 1.38335, 1.56200, 1.73308

3. (a) 1.2025, 1.41603, 1.64289, 1.88590
(b) 1.20388, 1.41936, 1.64896, 1.89572

5. (a) 0.509239, 0.522187, 0.539023, 0.559936
(b) 0.509701, 0.523155, 0.540550, 0.562089

7. (a) 2.90330, 7.53999, 19.4292, 50.5614
(b) 2.93506, 7.70957, 20.1081, 52.9779

9. (a) 3.95713, 5.09853, 6.41548, 7.90174
(b) 3.95965, 5.10371, 6.42343, 7.91255

11. (a) −1.45865, −0.217545, 1.05715, 1.41487
(b) −1.45322, −0.180813, 1.05903, 1.41244

15. $e_{n+1} = [2\phi(\bar{t}_n) - 1]h^2$, $|e_{n+1}| \leq \left[1 + 2\max_{0 \leq t \leq 1} |\phi(t)|\right] h^2$,

$e_{n+1} = e^{2\bar{t}_n} h^2$, $|e_1| \leq 0.012$, $|e_4| \leq 0.022$

17. $e_{n+1} = [\bar{t}_n + \bar{t}_n^2 \phi(\bar{t}_n) + \phi^3(\bar{t}_n)]h^2$

19. $e_{n+1} = (1 + [\bar{t}_n + \phi(\bar{t}_n)]^{-1/2})h^2/4$

21. (a) $\phi(t) = 1 + (1/5\pi)\sin 5\pi t$

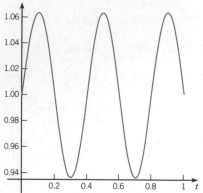

(b) 1.2, 1.0, 1.2

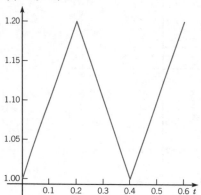

(c) 1.0, 1.1, 1.1, 1.0, 1.0 (d) $h < 1/\sqrt{50\pi} \approx 0.08$

23. (a) 0   (b) 60   (c) −92.16

25. (a) 56.0393, 510.8722, 1107.4123, 1794.5339

(b)

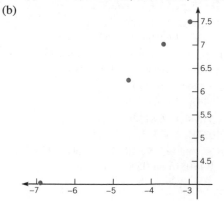

(c) yes (d) slope ≈ 0.9

Section 8.3   Improved Euler and Runge-Kutta Methods   *page 545*

1. (a) 1.19512, 1.38120, 1.55909, 1.72956
(b) 1.19515, 1.38125, 1.55916, 1.72965
(c) 1.19516, 1.38126, 1.55918, 1.72967
(d) 1.19516, 1.38127, 1.55918, 1.72968
(e) 1.19516, 1.38127, 1.55918, 1.72968

3. (a) 1.20526, 1.42273, 1.65511, 1.90570
(b) 1.20533, 1.42290, 1.65542, 1.90621
(c) 1.20534, 1.42294, 1.65550, 1.90634
(d) 1.20535, 1.42295, 1.65553, 1.90638
(e) 1.20535, 1.42296, 1.65553, 1.90638

5. (a) 0.510164, 0.524126, 0.542083, 0.564251
(b) 0.510168, 0.524135, 0.542100, 0.564277
(c) 0.510169, 0.524137, 0.542104, 0.564284
(d) 0.510170, 0.524138, 0.542105, 0.564286
(e) 0.520169, 0.524138, 0.542105, 0.564286

7. (a) 2.96719, 7.88313, 20.8114, 55.5106
(b) 2.96800, 7.88755, 20.8294, 55.5758
(c) 2.96825, 7.88889, 20.8349, 55.5957
(d) 2.96828, 7.88904, 20.8355, 55.5980

9. (a) 3.96217, 5.10887, 6.43134, 7.92332
(b) 3.96218, 5.10889, 6.43138, 7.92337
(c) 3.96219, 5.10890, 6.43139, 7.92338
(d) 3.96219, 5.10890, 6.43139, 7.92338

11. (a) −1.44768, −0.144478, 1.06004, 1.40960
(b) −1.44765, −0.143690, 1.06072, 1.40999
(c) −1.44764, −0.143543, 1.06089, 1.41008
(d) −1.44764, −0.143427, 1.06095, 1.41011

15. (a)

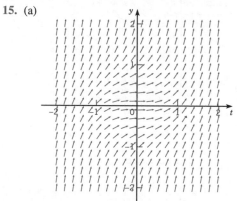

(b) 5.848616, 14.304785, 50.436365

19. $e_{n+1} = (2h^3/3)e^{2\bar{t}_n}$, $|e_{n+1}| \le 4.92604h^3$ on $0 \le t \le 1$, $|e_1| \le 0.000814269$

21. $h \approx 0.071$

23. $h \approx 0.081$

Section 8.4   Numerical Methods for Systems of First Order Equations   *page 549*

1. (a) 1.26, 0.76; 1.7714, 1.4824; 2.58991, 2.3703; 3.82374, 3.60413; 5.64246, 5.38885
(b) 1.32493, 0.758933; 1.93679, 1.57919; 2.93414, 2.66099; 4.48318, 4.22639; 6.84236, 6.56452

(c) 1.32489, 0.759516; 1.9369, 1.57999; 2.93459, 2.66201; 4.48422, 4.22784; 6.8444, 6.56684

3. (a) 0.582, 1.18; 0.117969, 1.27344; −0.336912, 1.27382; −0.730007, 1.18572; −1.02134, 1.02371
(b) 0.568451, 1.15775; 0.109776, 1.22556; −0.32208, 1.20347; −0.681296, 1.10162; −0.937852, 0.937852
(c) 0.56845, 1.15775; 0.109773, 1.22557; −0.322081, 1.20347; −0.681291, 1.10161; −0.937841, 0.93784

5. (a) 2.96225, 1.34538; 2.34119, 1.67121; 1.90236, 1.97158; 1.56602, 2.23895; 1.29768, 2.46732
(b) 3.06339, 1.34858; 2.44497, 1.68638; 1.9911, 2.00036; 1.63818, 2.27981; 1.3555, 2.5175
(c) 3.06314, 1.34899; 2.44465, 1.68699; 1.99075, 2.00107; 1.63781, 2.28057; 1.35514, 2.51827

7. For $h = 0.05$ and 0.025: $x = 1.43383$, $y = 0.642230$. These results agree with the exact solution to six digits.

CHAPTER 9 SERIES SOLUTIONS OF SECOND ORDER LINEAR EQUATIONS

Section 9.1 Review of Power Series *page 9-11*

1. $\rho = 1$

3. $\rho = \infty$

5. $\rho = \frac{1}{2}$

7. $\rho = 3$

9. $\sum_{n=0}^{\infty}(-1)^n \frac{x^{2n+1}}{(2n+1)!}$, $\rho = \infty$

11. $x = 1 + (x - 1)$, $\rho = \infty$

13. $\sum_{n=1}^{\infty}(-1)^{n+1} \frac{(x-1)^n}{n}$, $\rho = 1$

15. $\sum_{n=0}^{\infty}(-1)^n (1/9)^{n+1} x^{4n+1}$, $\rho = \sqrt{3}$

17. $y' = 1 + 4x + 9x^2 + 16x^3 + \ldots + (n+1)^2 x^n + \ldots = \sum_{n=0}^{\infty}(n+1)^2 x^n$, $y'' = 4 + 18x + 48x^2 + 100x^3 + \ldots + (n+2)^2(n+1)x^n + \ldots = \sum_{n=0}^{\infty}(n+2)^2(n+1)x^n$

21. $\sum_{n=0}^{\infty}(n+2)(n+1)a_{n+2}x^n$

23. $\sum_{n=0}^{\infty}(n+1)a_n x^n$

25. $\sum_{n=0}^{\infty}\left[(n+2)(n+1)a_{n+2} + na_n\right]x^n$

27. $\sum_{n=0}^{\infty}\left[n(n+1)a_{n+1} + a_n\right]x^n$

Section 9.2 Series Solutions Near an Ordinary Point, Part I *page 9-23*

1. $a_{n+2} = \frac{a_n}{(n+1)(n+2)}$, $y_1(x) = 1 + \frac{x^2}{2!} + \frac{x^4}{4!} + \frac{x^6}{6!} + \ldots = \sum_{k=0}^{\infty}\frac{x^{2k}}{(2k)!}$, $y_2(x) = x + \frac{x^3}{3!} + \frac{x^5}{5!} + \frac{x^7}{7!} + \ldots = \sum_{k=0}^{\infty}\frac{x^{2k+1}}{(2k+1)!}$

3. $a_{n+2} = -\frac{5a_{n+1}}{n+2} - \frac{6a_n}{(n+1)(n+2)}$, $y_1(x) = 1 - 3x^2 + 5x^3 - \frac{19}{4}x^4 + \frac{13}{4}x^5 + \ldots$, $y_2(x) = x - \frac{5}{2}x^2 + \frac{19}{6}x^3 - \frac{65}{24}x^4 + \frac{211}{120}x^5 + \ldots$

5. $a_{n+2} = -\frac{na_n}{(n+1)(n+2)}$, $y_1(x) = 1$, $y_2(x) = \sum_{k=0}^{\infty}(-1)^k \frac{(2k-1)!!x^{2k+1}}{(2k+1)!}$

7. $a_{n+2} = -\frac{a_{n-1}}{(n+1)(n+2)}$, $y_1(x) = \left[1 - \frac{x^3}{2 \cdot 3} + \frac{x^6}{2 \cdot 3 \cdot 5 \cdot 6} - \cdots + (-1)^n \frac{x^{3n}}{2 \cdot 3 \cdots (3n-1)(3n)} + \cdots\right]$, $y_2(x) = \left[x - \frac{x^4}{3 \cdot 4} + \frac{x^7}{3 \cdot 4 \cdot 6 \cdot 7} - \cdots + (-1)^n \frac{x^{3n+1}}{3 \cdot 4 \cdots (3n)(3n+1)} + \cdots\right]$

9. $a_{n+2} = \frac{a_{n+1} + a_n}{n+2}$, $y_1(x) = 1 + \frac{(x-1)^2}{2} + \frac{(x-1)^3}{6} + \frac{(x-1)^4}{6} + \ldots$, $y_2(x) = (x-1) + \frac{(x-1)^2}{2} + \frac{(x-1)^3}{2} + \frac{(x-1)^4}{4} + \ldots$

11. $a_{n+2} = \frac{na_{n+1}}{n+2} - \frac{a_n}{(n+2)(n+1)}$, $y_1(x) = 1 - \frac{x^2}{2} - \frac{x^3}{6} - \frac{x^4}{24} + \ldots$, $y_2(x) = x - \frac{x^3}{6} - \frac{x^4}{12} - \frac{x^5}{24} + \ldots$

13. $a_{n+2} = -\frac{a_n}{n+1}$, $y_1(x) = 1 - \frac{x^2}{1} + \frac{x^4}{1 \cdot 3} - \frac{x^6}{1 \cdot 3 \cdot 5} + \ldots = 1 + \sum_{n=1}^{\infty}\frac{(-1)^n x^{2n}}{1 \cdot 3 \cdot 5 \ldots (2n-1)}$, $y_2(x) = x - \frac{x^3}{2} + \frac{x^5}{2 \cdot 4} - \frac{x^7}{2 \cdot 4 \cdot 6} + \ldots = x + \sum_{n=1}^{\infty}\frac{(-1)^n x^{2n+1}}{2 \cdot 4 \cdot 6 \ldots (2n)}$

15. $a_{n+2} = -\frac{(n-2)(n-3)}{(n+1)(n+2)}a_n$, $y_1(x) = 1 - 3x^2$, $y_2(x) = x - \frac{x^3}{3}$

17. $a_{n+2} = \frac{(n+1)a_n}{3(n+2)}$, $y_1(x) = 1 + \frac{x^2}{6} + \frac{x^4}{24} + \frac{5x^6}{432} + \ldots = 1 + \sum_{n=1}^{\infty}\frac{3 \cdot 5 \ldots (2n-1)x^{2n}}{3^n \cdot 2 \cdot 4 \ldots (2n)}$, $y_2(x) = x + \frac{2x^3}{9} + \frac{8x^5}{135} + \frac{16x^7}{945} + \ldots = x + \sum_{n=1}^{\infty}\frac{2 \cdot 4 \cdot 6 \ldots (2n)x^{2n+1}}{3^n \cdot 3 \cdot 5 \ldots (2n+1)}$

REFERENCES

Differential Equations and Dynamical Systems

Arnol'd, Vladimir I., *Ordinary Differential Equations* (New York/Berlin: Springer-Verlag, 1992). Translation by Roger Cooke of the third Russian edition.

Birkhoff, G., and Rota, G.-C., *Ordinary Differential Equations* (4th ed.) (New York: Wiley, 1989).

Brauer, Fred and Nohel, John A., *The Qualitative Theory of Ordinary Differential Equations: An Introduction* (New York: W. A. Benjamin, Inc., 1969; New York: Dover, 1989).

Coddington, E. A., *An Introduction to Ordinary Differential Equations* (Englewood Cliffs, NJ: Prentice-Hall, 1961; New York: Dover, 1989).

Coddington, E. A., and Levinson, N., *Theory of Ordinary Differential Equations* (New York: McGraw-Hill, 1955).

Cole, R. H., *Theory of Ordinary Differential Equations* (New York: Irvington, 1968).

Guckenheimer, John and Holmes, Philip, *Nonlinear Oscillations, Dynamical Systems, and Bifurcations of Vector Fields* (New York/Berlin: Springer-Verlag, 1983).

Hirsch, Morris W., Smale, Stephen, and Devaney, Robert L., *Differential Equations, Dynamical Systems, and an Introduction to Chaos* (2nd ed.) (San Diego, CA/London: Elsevier Academic Press, 2004).

Hochstadt, H., *Differential Equations: A Modern Approach* (New York: Holt, 1964; New York: Dover, 1975).

Miller, R. K., and Michel, A. N., *Ordinary Differential Equations* (New York: Academic Press, 1982).

Rainville, E. D., *Intermediate Differential Equations* (2nd ed.) (New York: Macmillan, 1964).

Sparrow, Colin, *The Lorenz Equations: Bifurcations, Chaos, and Strange Attractors* (New York/Berlin: Springer-Verlag, 1982).

Strogatz, Steven H., *Nonlinear Dyamics and Chaos* (Reading, MA: Addison-Wesley, 1994).

Yosida, K., *Lectures on Differential and Integral Equations* (New York: Wiley-Interscience, 1960; New York: Dover, 1991).

Perturbation Methods, Dimensional Analysis, and Scaling

Holmes, M. H., *Introduction to Perturbation Methods* (New York: Springer-Verlag, 1995).

Lin, C. C. and Segel, L. A., *Mathematics Applied to Deterministic Problems in the Natural Sciences* (Philadelphia: SIAM, 1989).

Handbooks, Tables, and Special Functions

Abramowitz, M., and Stegun, I. A. (eds.), *Handbook of Mathematical Functions* (New York: Dover, 1965); originally published by the National Bureau of Standards, Washington, DC, 1964.

Copson, E. T., *An Introduction to the Theory of a Complex Variable* (Oxford: Oxford University, 1935).

Hochstadt, H., *Special Functions of Mathematical Physics* (New York: Holt, 1961).

Jahnke, E., and Emde, F., *Tables of Functions with Formulae and Curves* (Leipzig: Teubner, 1938; New York: Dover, 1945).

Advanced Calculus

Buck, R. C., and Buck, E. F., *Advanced Calculus* (3rd ed.) (New York: McGraw-Hill, 1978).

Courant, R., and John, F., *Introduction to Calculus and Analysis* (New York: Wiley-Interscience, 1965; reprinted by Springer-Verlag, New York, 1989).

Kaplan, W., *Advanced Calculus* (5th ed.) (Reading, MA: Addison-Wesley, 2003).

Fourier Series

Carslaw, H. S., *Introduction to the Theory of Fourier's Series and Integrals* (3rd ed.) (Cambridge: Cambridge University Press, 1930; New York: Dover, 1952).

Churchill, R. V., and Brown, J. W., *Fourier Series and Boundary Value Problems* (6th ed.) (New York: McGraw-Hill, 2000).

Partial Differential Equations

Haberman, R., *Elementary Applied Partial Differential Equations* (3rd ed.) (Englewood Cliffs, NJ: Prentice-Hall, 1998).

Pinsky, M. A., *Partial Differential Equations and Boundary Value Problems with Applications* (3rd ed.) (Boston: WCB/McGraw-Hill, 1998).

Powers, D. L., *Boundary Value Problems* (4th ed.) (San Diego: Academic Press, 1999).

Sagan, H., *Boundary and Eigenvalue Problems in Mathematical Physics* (New York: Wiley, 1961; New York: Dover, 1989).

Strauss, W. A., *Partial Differential Equations, an Introduction* (New York: Wiley, 1992).

Weinberger, H. F., *A First Course in Partial Differential Equations* (New York: Wiley, 1965; New York: Dover, 1995).

Numerical Analysis

Ascher, Uri M., and Petzold, Linda R., *Computer Methods for Ordinary Differential Equations and Differential-Algebraic Equations* (Philadelphia: Society for Industrial and Applied Mathematics, 1998).

Henrici, Peter, *Discrete Variable Methods in Ordinary Differential Equations* (New York: Wiley, 1962).

Shampine, Lawrence F., *Numerical Solution of Ordinary Differential Equations* (New York: Chapman and Hall, 1994).

Mathematical Modeling

Bailey, N. T. J., *The Mathematical Theory of Infectious Diseases and Its Applications* (2nd ed.) (New York: Hafner Press, 1975).

Brauer, Fred and Castillo-Chávez, Carlos, *Mathematical Models in Population Biology and Epidemiology* (New York/Berlin: Springer-Verlag, 2001).

Clark, Colin W., *Mathematical Bioeconomics* (2nd ed.) (New York: Wiley-Interscience, 1990).

Odum, E. P., *Fundamentals of Ecology* (3rd ed.) (Philadelphia: Saunders, 1971).

INDEX